SOIL CONDITIONS AND PLANT GROWTH

Frontispiece. A podsol developed under 48-year-old Scots pine (*Pinus sylvestris*) in a fluvio-glacial gravel derived from Old Red Sandstone and Highland Rocks. The previous crop was also Scots pine. Speymouth Forest, Morayshire.

Soil Conditions and Plant Growth

TENTH EDITION

by E. WALTER RUSSELL

C.M.G., M.A., PH.D.

Emeritus Professor, University of Reading

FIRST TO SEVENTH EDITIONS

BY SIR E. JOHN RUSSELL

D.SC., F.R.S.

LONGMAN

LONGMAN GROUP LIMITED
London

Associated companies, branches and representatives
throughout the world

First published (in Monographs in Biochemistry) by E. J. Russell	*1912*
Second Edition	*1915*
Third Edition	*1917*
Fourth Edition (in Rothamsted Monographs on Agricultural Science)	*1921*
Fifth Edition	*1927*
Sixth Edition	*1932*
Seventh Edition	*1937*
Eighth Edition rewritten and revised by E. W. Russell	*1950*
Ninth Edition	*1961*
Tenth Edition	*1973*

ISBN 0582 44048 3

TENTH EDITION © E. W. RUSSELL 1973

Printed in Great Britain by
William Clowes & Sons, Limited,
London, Beccles and Colchester

Preface to tenth edition

The importance of the inter-relations between soil and plant in determining plant growth is now so well accepted that very large resources all over the world are devoted to their study, with the consequence that rapid advances are being made in our understanding of them. These advances render the text of the previous edition of this book sufficiently out-of-date to necessitate a complete re-writing of most sections, to allow a proper discussion of the ideas and theories that are currently receiving most attention. It is also gratifying to see how rapidly the concepts of present-day physics and chemistry are being used to give an exact quantitative understanding of many factors relevant in soil–plant inter-relations, but these applications often require advanced mathematical techniques which are difficult for research workers unfamiliar with them to understand. I have attempted to give as accurate a description as I can of the physical and physico-chemical properties used in these studies, but have usually omitted any discussion of the basic mathematical techniques needed to obtain the final quantitative results.

One of the functions of this book has been to give a critical account of our present knowledge of the topics discussed, so wherever possible references are given for all the statements made. This has always been an essential feature of the book in the past, and I consider it should remain so in the future. But the amount of original work published each year is so enormous that I have made no attempt to be familiar with it all; instead I have been obliged to give references to work with which I am familiar, knowing that often the references given are neither the earliest nor the most suitable for the particular statement made or experimental result quoted. I hope any author whose work has been ignored but which is more relevant or earlier than work quoted will appreciate the reason for its omission.

I have taken the opportunity in this edition to convert all relevant experimental data into metric, and usually S.I., units, the principal exceptions being the results of experiments of purely historical interest. This means that many field experiment results given in this edition are not given in the original units, so the reader must be aware of the possibility of error creeping into the conversions.

It is again a very great pleasure to acknowledge the most generous help I have received from all of my colleagues whom I have consulted. They have supplied me with information, readily provided me with illustrations, and often read and commented on many of my drafts; and I trust they will forgive me if they find either that I have not always taken their advice or have failed to appreciate the significance of their suggestions. I cannot possibly make adequate acknowledgement to all of them, though I would like to make personal acknowledgement to Drs D. S. Jenkinson, G. E. G. Mattingly and H. L. Penman of Rothamsted, Mr B. W. Avery of the Soil Survey of England and Wales, Dr R. Scott Russell and some members of his staff, particularly Drs D. T. Clarkson and M. C. Drew, at the ARC Letcombe Laboratory, Mr P. H. Nye of the Soil Science Laboratory, Oxford University, Dr F. N. Ponnemperuma of the International Rice Research Institute, Emeritus Professor G. W. Leeper of Melbourne University, and Professor D. J. Greenland and his staff, particularly Drs P. J. Harris and C. J. M. Mott of the Department of Soil Science, Reading University. Finally I am once again greatly indebted to my wife for preparing the Author Index.

Department of Soil Science   E. W. RUSSELL
    University of Reading

Contents

Plates

1

Historical and introductory

In all ages the growth of plants has interested thoughtful men. The mystery of the change of an apparently lifeless seed to a vigorous growing plant never loses its freshness, and constitutes, indeed, no small part of the charm of gardening. The economic problems are of vital importance, and become more and more urgent as time goes on and populations increase and their needs become more complex.

There was an extensive literature on agriculture in Roman times which maintained a pre-eminent position until comparatively recently. In this we find collected many of the facts which it has subsequently been the business of agricultural experts to classify and explain. The Roman literature was collected and condensed into one volume about the year 1240 by a senator of Bologna, Petrus Crescentius, whose book[1] was one of the most popular treatises on agriculture of any time, being frequently copied, and in the early days of printing, passing through many editions—some of them very hand-some, and ultimately giving rise to the large standard European treatises of the sixteenth and seventeenth centuries. Many other agricultural books ap-peared in the fifteenth and early sixteenth centuries, notably in Italy, and later in France. In some of these are found certain ingenious speculations that have been justified by later work. Such, for instance, is Palissy's remark-able statement in 1563: 'You will admit that when you bring dung into the field it is to return to the soil something that has been taken away. . . . When a plant is burned it is reduced to a salty ash called alcaly by apothecaries and philosophers. . . . Every sort of plant without exception contains some kind of salt. Have you not seen certain labourers when sowing a field with wheat for the second year in succession, burn the unused wheat straw which had been taken from the field? In the ashes will be found the salt that the straw took out of the soil; if this is put back the soil is improved. Being burnt on the ground it serves as manure because it returns to the soil those substances that had been taken away.' But for every speculation that has been confirmed will be found many that have not, and the beginnings of agricultural chemistry

1 *Ruralium commodorum libri duodecim*, Augsburg, 1471, and many subsequent editions.

must be sought later, when men had learnt the necessity for carrying on experiments.

The search for the 'principle' of vegetation, 1630–1750[1]

It was probably very early discovered that manures, composts, dead animal bodies, and parts of animals, such as blood, all increased the fertility of the land; and this was the basis of the ancient saying that 'corruption is the mother of vegetation'. Yet the early investigators consistently ignored this ancient wisdom when they sought for the 'principle' of vegetation to account for the phenomena of soil fertility and plant growth. Thus the great Francis Bacon, Lord Verulam, believed that water formed the 'principal nourishment' of plants, the purpose of the soil being to keep them upright and protect them from excessive cold or heat, though he also considered that each plant drew a 'particular juyce' from the soil for its sustenance, thereby impoverishing the soil for that particular plant and similar ones, but not necessarily for other plants. Van Helmont (1577–1644) regarded water as the sole nutrient for plants, and his son thus records his famous Brussels experiment: 'I took an earthen vessel in which I put 200 pounds of soil dried in an oven, then I moistened with rain-water and pressed hard into it a shoot of willow weighing 5 pounds. After exactly five years the tree that had grown up weighed 169 pounds and about three ounces. But the vessel had never received anything but rain-water or distilled water to moisten the soil when this was necessary, and it remained full of soil, which was still tightly packed, and, lest any dust from outside should get into the soil, it was covered with a sheet of iron coated with tin but perforated with many holes. I did not take the weight of the leaves that fell in the autumn. In the end I dried the soil once more and got the same 200 pounds that I started with, less about two ounces. Therefore the 164 pounds of wood, bark, and root arose from the water alone.'[2]

The experiment is simple and convincing, and satisfied Robert Boyle,[3] who repeated it with 'squash, a kind of Italian pompion' and obtained similar results. Boyle further distilled the plants and concluded, quite justifiably from his premises, that the products obtained, 'salt, spirit, earth, and even oil (though that be thought of all bodies the most opposite to water), may be produced out of water'. Nevertheless, the conclusion is incorrect, because two factors had escaped van Helmont's notice—the parts played by the air and by the missing two ounces of soil. But the history of this experiment is

---

1 A more detailed account of the British contribution to the development of agricultural science, particularly as it affects crop production, is given in Sir E. John Russell: *A History of Agricultural Science in Great Britiain*, Allen and Unwin, 1966, which takes the story up to 1955.

2 *Ortus medicinae*, pp. 84–90. *Complexionum atque mistionum elementalium figmentum*, Amsterdam, 1652.

3 *The Sceptical Chymist*, Pt. II, 1661.

thoroughly typical of experiments in agricultural chemistry generally: in no other subject is it so easy to overlook a vital factor and draw from good experiments a conclusion that appears to be absolutely sound, but is in reality entirely wrong.

Some years later J. R. Glauber[1] set up the hypothesis that saltpetre is the 'principle' of vegetation. Having obtained saltpetre from the earth cleared out from cattle sheds, he argued that it must have come from the urine or droppings of the animals, and must, therefore, be contained in the animal's food, i.e. in plants. He also found that additions of saltpetre to the soil produced enormous increases in crop. He connected these two observations and supposed that saltpetre is the essential principle of vegetation. The fertility of the soil and the value of manures (he mentions dung, feathers, hair, horn, bones, cloth cuttings) are entirely due to saltpetre.

This view was supported by John Mayow's experiments.[2] He estimated the amounts of nitre in the soil at different times of the year, and showed that it occurs in greatest quantity in spring when plants are just beginning to grow, but is not to be found 'in soil on which plants grow abundantly, the reason being that all the nitre of the soil is sucked out by the plants'. J. A. Külbel,[3] on the other hand, regarded a *magma unguinosum* obtainable from humus as the 'principle' sought for.

The most accurate work in this period was published by John Woodward[4] in a remarkable paper. Setting out from the experiments of van Helmont and of Boyle, but apparently knowing nothing of the work of Glauber and of Mayow, he grew spearmint in water obtained from various sources with the following results among others:

Source of water	*Weight of plants*		*Gained in 77 days*	*Expense of water (i.e. transpiration)*	*Proportion of increase of plant to expense of water*
	When put in	*When taken out*			
	grains	grains	grains	grains	
Rain-water	$28\frac{1}{4}$	$45\frac{3}{4}$	$17\frac{1}{2}$	3004	1 to $171\frac{23}{35}$
River Thames	28	54	26	2493	1 to $95\frac{23}{26}$
Hyde Park conduit	110	249	139	13140	1 to $94\frac{74}{139}$
Hyde Park conduit plus $1\frac{1}{2}$ oz. garden mould	92	376	284	14950	1 to $52\frac{182}{284}$

Now all these plants had abundance of water, therefore all should have made equal growth had nothing more been needed. The amount of growth,

1 *Des Teutschlandts Wohlfart (Erster Theil), das dritte Capitel. De concentratione Vegetabilium, Miraculum Mundi*, Amsterdam, 1656.
2 *Tractatus quinque medico-physici*, 1674 (Alembic Club reprint, Edinburgh, 1907).
3 *Cause de la fertilité des terres*, Bordeaux, 1741.
4 *Phil. Trans. Roy. Soc.*, 1699, **21**, 382.

however, increased with the impurity of the water. 'Vegetables', he concludes, 'are not formed of water, but of a certain peculiar terrestrial matter. It has been shown that there is a considerable quantity of this matter contained in rain, spring and river water, that the greatest part of the fluid mass that ascends up into plants does not settle there but passes through their pores and exhales up into the atmosphere: that a great part of the terrestrial matter, mixed with the water, passes up into the plant along with it, and that the plant is more or less augmented in proportion as the water contains a greater or less quantity of that matter; from all of which we may reasonably infer, that earth, and not water, is the matter that constitutes vegetables.'

He discusses the use of manures and the fertility of the soil from this point of view, attributing the well-known falling off in crop yield when plants are grown for successive years on unmanured land to the circumstance that 'the vegetable matter that it at first abounded in being extracted from it by those successive crops, is most of it borne off. . . . The land may be brought to produce another series of the same vegetables, but not until it is supplied with a new fund of matter, of like sort with that it at first contained; which supply is made several ways, either by the ground's being fallow some time, until the rain has poured down a fresh stock upon it; or by the tiller's care in manuring it.' The best manures, he continues, are parts either of vegetables or of animals, which ultimately are derived from vegetables.

In his celebrated textbook of chemistry, H. Boerhaave[1] taught that plants absorb the juices of the earth and then work them up into food. The raw material, the 'prime radical juice of vegetables, is a compound from all the three kingdoms, viz. *fossil* bodics and putrified parts of *animals* and *vegetables*'. This 'we look upon as the *chyle of the plant*; being chiefly found in the first order of vessels, viz. in the roots and the body of the plant, which answers to the stomach and intestines of an animal.'

For many years no such outstanding work as that of Glauber and Woodward was published, if we except Stephen Hales's *Vegetable Staticks* in 1727, the interest of which is physiological rather than agricultural.[2] Advances were, however, being made in agricultural practice. One of the most important was the introduction of the drill and the horse hoe by Jethro Tull, an Oxford man of a strongly practical turn of mind, who insisted on the vital importance of getting the soil into a fine, crumbly state for plant growth. Tull was more than an inventor; he discussed in most picturesque language the sources of fertility in the soil.[3] In his view it was not the juices of the earth, but the very minute particles of soil loosened by the action of moisture, that constituted the 'proper pabulum' of plants. The pressure caused by the swelling of the growing roots forced these particles into the 'lacteal mouths of the roots', where they entered the circulatory system. All plants lived on

1 *A New Method of Chemistry*, London, 1727.
2 He shows, however, that air is 'wrought into the composition' of plants.
3 *Horse Hoeing Husbandry*, London, 1731.

these particles, i.e. on the same kind of food; it was incorrect to assert, as some had done, that different kinds of plants fed as differently as horses and dogs, each taking its appropriate food and no other. Plants will take in anything that comes their way, good or bad. A rotation of crops is not a necessity, but only a convenience. Conversely, any soil will nourish any plant if the temperature and water supply are properly regulated. Hoeing increased the surface of the soil or the 'pasture of the plant', and also enabled the soil better to absorb the nutritious vapours condensed from the air. Dung acted in the same way, but was more costly and less efficient.

So much were Tull's writings esteemed, Cobbett tells us, that they were 'plundered by English writers not a few and by Scotch in whole bandittis'.

The position at the end of this period cannot better be summed up than in Tull's own words: 'It is agreed that all the following materials contribute in some manner to the increase of plants, but it is disputed which of them is that very increase or food: (1) nitre, (2) water, (3) air, (4) fire, (5) earth.'

The search for plant nutrients

The phlogistic period, 1750–1800

Great interest was taken in agriculture in this country during the latter half of the eighteenth century. 'The farming tribe', writes Arthur Young during this period, 'is now made up of all ranks, from a duke to an apprentice.' Many experiments were conducted, facts were accumulated, books written, and societies formed for promoting agriculture. The Edinburgh Society, established in 1755 for the improvement of arts and manufactures, induced Francis Home[1] 'to try how far chymistry will go in settling the principles of agriculture'. The whole art of agriculture, he says, centres in one point: the nourishing of plants. Investigation of fertile soils showed that they contain oil, which is therefore a food of plants. But when a soil has been exhausted by cropping, it recovers its fertility on exposure to air,[2] which therefore supplies another food. Home made pot experiments to ascertain the effect of various substances on plant growth. 'The more they [i.e. farmers] know of the effects of different bodies on plants, the greater chance they have to discover the nourishment of plants, at least this is the only road.' Saltpetre, epsom salt, vitriolated tartar (i.e. potassium sulphate) all lead to increased plant growth, yet they are three distinct salts. Olive oil was also useful. It is thus clear that plant food is not one thing only, but several; he enumerates six: air, water, earth, salts of different kinds, oil, and fire in a fixed state. As further proof he shows that 'all vegetables and vegetable juices afford those very principles, and no other, by all the chymical experiments which have yet been made on them with or without fire'.

1 *The Principles of Agriculture and Vegetation*, Edinburgh, 1757.
2 Recorded by most early writers, e.g. Evelyn (*Terra, a philosophical discourse of earth*, 1674).

The book is a great advance on anything that had gone before it, not only because it recognises that plant nutrition depends on several factors, but because it indicates so clearly the two methods to be followed in studying the problem—pot cultures and plant analysis. Subsequent investigators, J. G. Wallerius,[1] the Earl of Dundonald[2] and R. Kirwan[3] added new details but no new principles. The problem, indeed, was carried as far as was possible until further advances were made in plant physiology and in chemistry. The writers just mentioned are, however, too important to be passed over completely. Wallerius, in 1761, professor of chemistry at Upsala, after analysing plants to discover the materials on which they live, and arguing that *Nutritio non fieri potest a rebus heterogeneis, sed homogeneis*, concludes that humus, being *homogeneous*, is the source of their food—the *nutritiva*—while the other soil constituents are *instrumentalia*, making the proper food mixture, dissolving and attenuating it, till it can enter the plant root. Thus chalk and probably salts help in dissolving the 'fatness' of the humus. Clay helps to retain the 'fatness' and prevent it being washed away by rain: sand keeps the soil open and pervious to air. The Earl of Dundonald, in 1795, adds alkaline phosphates to the list of nutritive salts, but he attaches chief importance to humus as plant food. The 'oxygenation' process going on in the soil makes the organic matter insoluble and therefore useless for the plant; lime, 'alkalis and other saline substances' dissolve it and change it to plant food; hence these substances should be used alternately with dung as manure. Manures were thus divided, as by Wallerius, into two classes: those that afford plant food, and those that have some indirect effect.

Throughout this period it was believed that plants could generate alkalis. 'Alkalis', wrote Kirwan in 1796, 'seem to be the product of the vegetable process, for either none, or scarce any, is found in the soils, or in rain water.' In like manner Lampadius thought he had proved that plants could generate silica. The theory that plants agreed in all essentials with animals was still accepted by many men of science; some interesting developments were made by Erasmus Darwin.[4]

Between 1770 and 1800 work was done on the effect of vegetation on air that was destined to revolutionise the ideas of the function of plants in the economy of nature, but its agricultural significance was not recognised until later. Joseph Priestley,[5] knowing that the atmosphere becomes vitiated by animal respiration, combustion, putrefaction, etc., and realising that some natural purification must go on, or life would no longer be possible, was led to try the effect of sprigs of living mint on vitiated air. He found that the mint

1 *Agriculturae Fundamenta Chemica: Akerbrukets Chemiska Grunder*, Upsala, 1761.
2 *A Treatise Showing the Intimate Connection that Subsists between Agriculture and Chemistry, etc.*, London, 1795.
3 *The Manures most Advantageously Applicable to the Various Sorts of Soils and the Cause of Their Beneficial Effects in each Particular Instance*, 4th edn., London, 1796.
4 *Phytologia, or the Philosophy of Agriculture and Gardening*, London, 1800.
5 *Experiments and Observations on Different Kinds of Air*, London, 1775.

made the air purer, and concludes 'that plants, instead of affecting the air in the same manner with animal respiration, reverse the effects of breathing, and tend to keep the atmosphere pure and wholesome, when it is become noxious in consequence of animals either living, or breathing, or dying, and putrefying in it'. But he had not yet discovered oxygen, and so could not give precision to his discovery: and when, later on, he did discover oxygen and learn how to estimate it, he unfortunately failed to confirm his earlier results because he overlooked a vital factor, the necessity of light. He was therefore unable to answer Scheele, who had insisted that plants, like animals, vitiate the air. It was Jan Ingen-Housz[1] who reconciled both views and showed that purification goes on in light only, whilst vitiation takes place in the darkness. Jean Senebier at Geneva had also arrived at the same result. He also studied the converse problem—the effect of air on the plant, and in 1782[2] argued that the increased weight of the tree in van Helmont's experiment (p. 2) came from the fixed air. 'Si donc l'air fixe, dissous dans l'eau de l'atmosphère, se combine dans la parenchyme avec la lumière et tous les autres éléments de la plante; si le phlogistique de cet air fixe est sûrement précipité dans les organes de la plante, si ce précipité reste, comme on le voit, puisque cet air fixe sort des plantes sous la forme d'air déphlogistiqué, il est clair que l'air fixe, combiné dans la plante avec la lumière, y laisse une matière qui n'y seroit pas, et mes expériences sur l'étoilement suffisent pour le démontrer.' Later on Senebier translated his work into the modern terms of Lavoisier's system.

The modern period, 1800–60

The foundation of plant physiology

We have seen that Home in 1757 pushed his inquiries as far as the methods in vogue would permit, and in consequence no marked advance was made for forty years. A new method was wanted before further progress could be made, or before the new idea introduced by Senebier could be developed. Fortunately, this was soon forthcoming, in 1804. To Théodore de Saussure,[3] son of the well-known de Saussure of Geneva, is due the quantitative experimental method which more than anything else has made modern agricultural chemistry possible; which formed the basis of subsequent work by Boussingault, Liebig, Lawes and Gilbert, and, indeed, still remains our safest method of investigation. Senebier tells us that the elder de Saussure was well acquainted with his work, and it is therefore not surprising that the son attacked two problems that Senebier had also studied—the effect of air on

1 *Experiments upon Vegetables, Discovering their Great Power of Purifying Common Air in the Sunshine and of Injuring in the Shade and at Night*, London, 1779.
2 *Mémoires Physico-chimiques*, 1782.
3 *Recherches chimiques sur la végétation*, Paris, 1804.

plants and the nature and origin of salts in plants. De Saussure grew plants
in air or in known mixtures of air and carbon dioxide, and measured the gas
changes by eudiometric analysis and the changes in the plant by 'carbon-
isation'. He was thus able to demonstrate the central fact of plant respiration—the absorption of oxygen and the evolution of carbon dioxide, and
further to show the decomposition of carbon dioxide and evolution of
oxygen in light. Carbon dioxide in small quantities was a vital necessity for
plants, and they perished if it was artificially removed from the air. It furnished
them not only with carbon, but also with some oxygen. Water is also decomposed and fixed by plants. On comparing the amount of dry matter
gained from these sources with the amount of material that can enter through
the roots even under the most favourable conditions, he concludes that the
soil furnished only a very small part of the plant food. Small as it is, however,
this part is indispensable: it supplies nitrogen—*une partie essentielle des
végétaux*—which, as he had shown, was not assimilated direct from the air;
and also ash constituents, *qui peuvent contribuer à former, comme dans les
animaux, leur parties solides ou osseuses.* Further, he shows that the root is
not a mere filter allowing any and every liquid to enter the plant; it has a
special action and takes in water more readily than dissolved matter, thus
effecting a concentration of the solution surrounding it; different salts, also,
are absorbed to a different extent. Passing next to the composition of the
plant ash, he shows that it is not constant, but varies with the nature of the
soil and the age of the plant; it consists mainly, however, of alkalis and
phosphates. All the constituents of the ash occur in humus. If a plant is
grown from seed in water there is no gain in ash: the amount found at the
end of the plant's growth is the same as was present in the seed excepting for
a relatively small amount falling on the plant as dust. Thus he disposes
finally of the idea that the plant *generated* potash.

After the somewhat lengthy and often wearisome works of the earlier
writers it is very refreshing to turn to de Saussure's concise and logical argu-
ments and the ample verification he gives at every stage. But for years his
teachings were not accepted, nor were his methods followed.

The two great books on agricultural chemistry then current still belonged
to the old period. A. von Thaer and Humphry Davy, while much in advance
of Wallerius, the textbook writer of 1761, nevertheless did not realise the
fundamental change introduced by de Saussure; it has always been the fate
of agricultural science to lag behind pure science. Thaer published his
Grundsätze de rationellen Landwirtschaft in 1809–12: it had a great success
on the Continent as a good, practical handbook, and was translated into
English as late as 1844 by Cuthbert Johnson. In it he adopted the prevailing
view that plants draw their carbon and other nutrients from the soil humus.
'Die Fruchtbarkeit des Bodens', he says, 'hängt eigentlich ganz vom Humus
ab. Denn ausser Wasser ist er es allein, der den Pflanzen Nahrung gibt. So
wie der Humus eine Erzeugung des Lebens ist, so ist er auch eine Bedingung

des Lebens. Er gibt den Organismen die Nahrung. Ohne ihn lässt sich kein individuelles Leben denken.' Humphry Davy's book[1] grew out of the lectures which he gave annually at the Royal Institution on agricultural chemistry between 1802 and 1812; it forms the last textbook of the older period. While no great advance was made by Davy himself he carefully sifted the facts and hypotheses of previous writers, and gives us an account, which, however defective in places, represents the best accepted knowledge of the time, set out in the new chemical language. His great name gave the subject an importance it would not otherwise have had.[2] He did not accept de Saussure's conclusion that plants obtain their carbon chiefly from the carbonic acid of the air: some plants, he says, appear to be supplied with carbon chiefly from this source, but in general he supposes the carbon to be taken in through the roots. Oils are good manures because of the carbon and hydrogen they contain; soot is valuable, because its carbon is 'in a state in which it is capable of being rendered soluble by the action of oxygen and water'. Lime is useful because it dissolves hard vegetable matter. Once the organic matter has dissolved there is no advantage in letting it decompose further: putrid urine is less useful as manure than fresh urine, while it is quite wrong to cause farmyard manure to ferment before it is applied to the land. All these ideas have been given up, and, indeed, there never was any sound experimental evidence to support them. It is even arguable that they would not have persisted so long as they did had it not been for Davy's high reputation. His insistence on the importance of the physical properties of soils—their relationship to heat and to water—was more fortunate and marks the beginning of soil physics, afterwards developed considerably by Gustav Schübler.[3] On the Continent, to an even greater extent than in England, it was held that plants drew their carbon and other nutrients from the soil humus, a view supported by the very high authority of J. J. Berzelius.[4]

The foundation of agricultural science

Hitherto experiments had been conducted either in the laboratory or in small pots: about 1834, however, J. B. Boussingault, who was already known as an adventurous traveller in South America, began a series of field experiments on his farm at Bechelbronn in Alsace. These were the first of their kind: to Boussingault, therefore, belongs the honour of having introduced the method by which the new agricultural science was to be developed. He reintroduced

---

1 *Elements of Agricultural Chemistry*, London, 1813.
2 Thus Charles Lamb, *Essays of Elia* (1820–23) in the 'Old and New Schoolmaster', writes:
   'The modern schoolmaster is required to know a little of everything because his pupil is required not to be entirely ignorant of anything. He is to know something of pneumatics, of chemistry, the quality of soils, etc. . . .'
3 *Grundsätze der Agrikulturchemie in Näherer Beziehung auf Land- und Fortswirtschaftliche Gewerbe*, Leipzig, 1838.
4 *Traité de Chimie*, Brussels, 1838.

the quantitative methods of de Saussure, weighed and analysed the manures used and the crop obtained, and at the end of the rotation drew up a balance sheet, showing how far the manures had satisfied the needs of the crop and how far other sources of supply—air, rain and soil—had been drawn upon. The results of one experiment are given in Table 1.1.[1] At the end of the period

TABLE 1.1   Statistics of a rotation

| | Weight in kg/ha of | | | | | |
	Dry matter	Carbon	Hydrogen	Oxygen	Nitrogen	Mineral matter
1 Beets	3172	1357·7	184·0	1376·7	53·9	199·8
2 Wheat	3006	1431·6	164·4	1214·9	31·3	163·8
3 Clover hay	4029	1909·7	201·5	1523·0	84·6	310·2
4 Wheat	4208	2004·2	230·0	1700·7	43·8	229·3
Turnips (catch crop)	716	307·2	39·3	302·9	12·2	54·4
5 Oats	2347	1182·3	137·3	890·9	28·4	108·0
Total during rotation	17478	8192·7	956·5	7009·0	254·2	1065·5
Added in manure	10161	3637·6	426·8	2621·5	203·2	3271·9
Difference not accounted for taken from air, rain or soil	+7317	+4555·1	+529·7	+4387·5	+51·0	−2206·4

the soil had returned to its original state of productiveness, hence the dry matter, carbon, hydrogen and oxygen not accounted for by the manure must have been supplied by the air and rain, and not by the soil. On the other hand, the manure afforded more mineral matter than the crop took off, the balance remaining in the soil. Other things being equal, he argued that the best rotation is one which yields the greatest amount of organic matter over and above what is present in the manure. No fewer than five rotations were studied, but it will suffice to set out only the nitrogen statistics (Table 1.2), which show a marked gain of nitrogen when the newer rotations are adopted, but not where wheat only is grown.

   Now the rotation has not impoverished the soil, hence he concludes that 'l'azote peut entrer directement dans l'organisme des plantes, si leur parties vertes sont aptes à le fixer'. Boussingault's work covers the whole range of agriculture and deals with the composition of crops at different stages of their growth with soils, and with problems in animal nutrition. Unfortunately the classic farm of Bechelbronn did not remain a centre of agricultural research and the experiments came to an end after the war of 1870. Some of the work

1 *Ann. Chim. Phys.* (III), 1841, **1**, 208.

TABLE 1.2   Nitrogen statistics of various rotations

Rotation	*kg/ha*		Excess in crop over that supplied in manure	
	Nitrogen in manure	*Nitrogen in crop*	*Per rotation*	*Per annum*
(1) Potatoes, (2) wheat, (3) clover, (4) wheat, turnips,* (5) oats	203·2	250·7	47·5	9·5
(1) Beets, (2) wheat, (3) clover, (4) wheat, turnips,* (5) oats	203·2	254·2	51·0	10·2
(1) Potatoes, (2) wheat, (3) clover, (4) wheat, turnips,* (5) peas, (6) rye	243·8	353·6	109·8	18·3
Jerusalem artichokes, two years	188·2	274·2	86·0	43·0†
(1) Dunged fallow, (2) wheat, (3) wheat	82·8	87·4	4·6	1·5
Lucerne, five years	224·0	1078·0	854·0	170·8

* Catch crop, i.e. taken in autumn after the wheat.
† This crop does not belong to the Leguminosae, but it is possible that the nitrogen came from the soil, and that impoverishment was going on.

was summarised by J. B. A. Dumas and Boussingault[1] in a very striking essay that has been curiously overlooked by agricultural chemists.

During this period (1830–40) Carl Sprengel was studying the ash constituents of plants, which he considered were probably essential to nutrition.[2] Schübler was working at soil physics, and a good deal of other work was quietly being done. No particularly important discoveries were being made, no controversies were going on, and no great amount of interest was taken in the subject.

But all this was changed in 1840 when Liebig's famous report to the British Association[3] upon the state of organic chemistry, published as *Chemistry in its Application to Agriculture and Physiology* in 1840, came like a thunderbolt upon the world of science. With polished invective and a fine sarcasm he holds up to scorn the plant physiologists of his day for their continued adhesion, in spite of accumulated evidence, to the view that plants derive their carbon from the soil and not from the carbonic acid of the air. 'All explanations of chemists must remain without fruit, and useless, because, even to the great leaders in physiology, carbonic acid, ammonia, acids and bases are sounds without meaning, words without sense, terms of an unknown language, which awake no thoughts and no associations.' The experi-

1 *Essai de Statique Chimique des Êtres Organisés*, Paris, 1841.
2 *Chemie für Landwirthe, Forstmänner und Cameralisten*, Göttingen, 1832.
3 There is no record of this Report ever having been presented to the Association.

ments quoted by the physiologists in support of their view are all 'valueless for the decision of any question'. 'These experiments are considered by them as convincing proofs, whilst they are fitted only to awake pity.' Liebig's ridicule did what neither de Saussure's nor Boussingault's logic had done: it finally killed the humus theory. Only the boldest would have ventured after this to assert that plants derive their carbon from any source other than carbon dioxide, although it must be admitted that we have no proof that plants really do obtain all their carbon in this way. Thirty years later, in fact, L. Grandeau[1] adduced evidence that humus may, after all, contribute something to the carbon supply, and his view found some acceptance in France;[2] for this also, however, convincing proof is lacking. But for the time carbon dioxide was considered to be the sole source of the carbon of plants. Hydrogen and oxygen came from water, and nitrogen from ammonia. Certain mineral substances were essential: alkalis were needed for neutralisation of the acids made by plants in the course of their vital processes, phosphates were necessary for seed formation, and potassium silicates for the development of grasses and cereals. The evidence lay in the composition of the ash: plants might absorb anything soluble from the soil, but they excreted from their roots whatever was non-essential. The fact of a substance being present was therefore sufficient proof of its necessity.

Plants, Liebig argued, have an inexhaustible supply of carbonic acid in the air. But time is saved in the early stages of plant growth if carbonic acid is being generated in the soil, for it enters the plant roots and affords extra nutrient over and above what the small leaves are taking in. Hence a supply of humus, which continuously yields carbonic acid, is advantageous. Further, the carbonic acid attacks and dissolves some of the alkali compounds of the soil and thus increases the mineral food supply. The true function of humus is to evolve carbonic acid.

The alkali compounds of the soil are not all equally soluble. A weathering process has to go on, which is facilitated by liming and cultivation, whereby the comparatively insoluble compounds are broken down to a more soluble state. The final solution is effected by acetic acid excreted by the plant roots, and the dissolved material now enters the plant.

The nitrogen is taken up as ammonia, which may come from the soil, from added manure, or from the air. In order that a soil may remain fertile it is necessary and sufficient to return in the form of manure the mineral constituents and the nitrogen that have been taken away. When sufficient crop analyses have been made it will be possible to draw up tables showing the farmer precisely what he must add in any particular case.

An artificial manure known as Liebig's patent manure was made up on these lines and placed on the market.

1 *Comp. Rend.,* 1872, **74**, 988; *Publication de la Station Agronomique de l'Est,* 1872.
2 See, for example, L. Cailletet (*Comp. Rend.*, 1911, **152**, 1215), Jules Lefèvre (ibid., 1905, **141**, 211), and J. Laurent (*Rev. gén. bot.*, 1904, **16**, 14).

Liebig's book was meant to attract attention to the subject, and it did; it rapidly went through several editions, and as time went on Liebig developed his thesis, and gave it a quantitative form: 'The crops on a field diminish or increase in exact proportion to the diminution or increase of the mineral substances conveyed to it in manure.' He further adds what afterwards became known as the Law of the Minimum, 'by the deficiency or absence of *one* necessary constituent, all the others being present, the soil is rendered barren for all those crops to the life of which *that one* constituent is indispensable'. These and other amplifications in the third edition, 1843, gave rise to much controversy. So much did Liebig insist, and quite rightly, on the necessity for alkalis and phosphates, and so impressed was he by the gain of nitrogen in meadow land supplied with alkalis and phosphates alone, and by the continued fertility of some of the fields of Virginia and Hungary and the meadows of Holland, that he began more and more to regard the atmosphere as the source of nitrogen for plants. Some of the passages of the first and second editions urging the necessity of ammoniacal manures were deleted from the third and later editions. 'If the soil be suitable, if it contain a sufficient quantity of alkalis, phosphates, and sulphates, nothing will be wanting. The plants will derive their ammonia from the atmosphere as they do carbonic acid', he writes in the *Farmer's Magazine*.[1] Ash analysis led him to consider the turnip as one of the plants 'which contain the least amount of phosphates and therefore require the smallest quantity for their development'. These and other practical deductions were seized upon and shown to be erroneous by J. B. Lawes and J. H. Gilbert,[2] who had for some years been conducting vegetation experiments. Lawes does not discuss the theory as such, but tests the deductions Liebig himself draws and finds them wrong. Further trouble was in store for Liebig; his patent manure when tried in practice *had failed*. This was unfortunate, and the impression in England at any rate was, in Philip Pusey's words: 'The mineral theory, too hastily adopted by Liebig, namely, that crops rise and fall in direct proportion to the quantity of mineral substances present in the soil, or to the addition or abstraction of these substances which are added in the manure, has received its death-blow from the experiments of Mr Lawes.'

And yet the failure of the patent manure was not entirely the fault of the theory, but only affords further proof of the numerous pitfalls of the subject. The manure was sound in that it contained potassium compounds and phosphates (it ought, of course, to have contained nitrogen compounds), but it was unfortunately rendered insoluble by fusion with lime and calcium phosphate so that it should not too readily wash out in the drainage water. Not till J. T. Way had shown in 1850[3] that *soil precipitates soluble salts of*

1 *Farmer's Magazine*, 1847, **16**, 511. A good summary of Liebig's position is given in his *Familiar Letters on Chemistry*, 3rd edn, 1851, 34th letter, p. 519.
2 *J. Roy. Agric. Soc. Eng.*, 1847, **8**, 226; 1851, **12**, 1; 1855, **16**, 411.
3 *J. Roy. Agric. Soc. Eng.*, 1850, **11**, 313; 1852, **13**, 123.

ammonium, potassium and phosphates was the futility of the fusion process discovered, and Liebig[1] saw the error he had made.

Meanwhile the great field experiments at Rothamsted had been started by Lawes and Gilbert in 1843. These experiments were conducted on the same general lines as those begun earlier by Boussingault, but they have the advantage that they are still going on, having been continued year after year on the same ground without alteration, except in occasional details, since 1852. The mass of data now accumulated is considerable and it is being treated by modern statistical methods. Certain conclusions are so obvious, however, that they can be drawn on mere inspection of the data. By 1855 the following points were definitely settled:[2]

1   Crops require phosphates and salts of the alkalis, but the composition of the ash does not afford reliable information as to the amounts of each constituent needed, e.g. turnips require large amounts of phosphates, although only little is present in their ash. Some of the results are:

Composition of ash, per cent (1860 crop)—		Yield of turnips, tons per acre (1843)—	
K_2O	44·8	Unmanured	4·5
P_2O_5	7·9	Superphosphate	12·8
		Superphosphate + potassic salts	11·9

2   Non-leguminous crops require a supply of some nitrogenous compounds, nitrates and ammonium salts being almost equally good. Without an adequate supply no increases of growth are obtained, even when ash constituents are added. The amount of ammonia obtainable from the atmosphere is insufficient for the needs of crops. Leguminous crops behave abnormally.
3   Soil fertility may be maintained for some years at least by means of artificial manures.
4   The beneficial effect of fallowing lies in the increase brought about in the available nitrogen compounds in the soil.

Although many of Liebig's statements were shown to be wrong, the main outline of his theory as first enunciated stands. It is no detraction that de Saussure had earlier published a somewhat similar, but less definite view of nutrition: Liebig had brought matters to a head and made men look at their cherished, but unexamined, convictions. The effect of the stimulus he gave can hardly be over-estimated, and before he had finished, the essential facts of plant nutrition were settled and the lines were laid down along which scientific manuring was to be developed. The water cultures of Knop and

1 *Familiar Letters on Chemistry*, 3rd edn, London, 1851.
2 Lawes and Gilbert's papers are collected in ten volumes of *Rothamsted Memoirs*, and the general results of their experiments are summarised by Hall in *The Book of the Rothamsted Experiments*. A detailed investigation of the early experiments of Lawes in their relation to the discovery of superphosphate has been made by Max Speter in *Superphosphate*, 1935, **8**.

other plant physiologists showed conclusively that potassium, magnesium, calcium, iron, phosphorus, along with sulphur, carbon, nitrogen, hydrogen and oxygen are all necessary for plant life. The list differs from Liebig's only in the addition of iron and the withdrawal of silica; but even silica, although not strictly essential, is advantageous for the nutrition of cereals.

In two directions, however, the controversies went on for many years. Farmers were slow to believe that 'chemical manures' could ever do more than stimulate the crop, and declared they must ultimately exhaust the ground. The Rothamsted plots falsified this prediction; manured year after year with the same substances and sown always with the same crops, they even now, after a hundred years of chemical manuring, continue to produce good crops, although secondary effects have sometimes set in. In France the great missionary was Georges Ville,[1] whose lectures were given at the experimental farm at Vincennes during 1867 and 1874–75. He went even further than Lawes and Gilbert, and maintained that artificial manures were not only more remunerative than dung, but were the only way of keeping up fertility. In recommending mixtures of salts for manure he was not guided by ash analysis but by field trials. For each crop one of the four constituents, nitrogen compounds, phosphates, lime and potassium compounds (he did not consider it necessary to add any others to his manures) was found by trial to be more wanted than the others and was therefore called the 'dominant' constituent. Thus for wheat he obtained the following results, and therefore concluded that on his soil wheat required a good supply of nitrogen, less phosphate, and still less potassium:

Constituent	Crop per acre bushels
Normal manure	43
Manure without lime	41
Manure without potash	31
Manure without phosphate	26·5
Manure without nitrogen	14
Soil without manure	12

Other experiments of the same kind showed that nitrogen was the dominant for all cereals and beetroot, potassium for potatoes and vines, phosphates for turnips and swedes. An excess of the dominant constituent was always added to the crop manure. The composition of the soil had to be taken into account, but soil analysis was no good for the purpose. Instead he drew up a simple scheme of plot trials to enable farmers to determine for themselves just what nutrient was lacking in their soil. His method was thus essentially empirical, but it still remains the best we have; his view that chemical manures

1 *On Artificial Manures, Their Chemical Selection and Scientific Application to Agriculture.* Trans. by W. Crookes, London, 1879.

are always better and cheaper than dung is, however, too narrow and has not survived.

The second controversy dealt with the source of nitrogen in plants. Priestley had stated that a plant of *Epilobium hirsutum* placed in a small vessel absorbed during the course of the month seven-eighths of the air present. De Saussure, however, denied that plants assimilated gaseous nitrogen. J. B. Boussingault's pot experiments[1] showed that peas and clover could get nitrogen from the air while wheat could not, and his rotation experiments emphasised this distinction. He himself did not make as much of this discovery as he might have done, but later[2] fully realised its importance.

Liebig, as we have seen, maintained that ammonia, but not gaseous nitrogen, was taken up by plants, a view confirmed by Lawes, Gilbert and E. Pugh[3] in the most rigid demonstration that had yet been attempted. Plants of several natural orders, including the Leguminosae, were grown in surroundings free from ammonia or any other nitrogen compound. The soil was burnt to remove all traces of nitrogen compounds, while the plants were kept throughout the experiment under glass shades, but supplied with washed and purified air and with pure water. In spite of the ample supply of mineral food the plants languished and died: the conclusion seemed irresistible that plants could not utilise gaseous nitrogen. For all non-leguminous crops this conclusion agreed with the results of field trials. But there remained the very troublesome fact that leguminous crops required no nitrogenous manure and yet they contained large quantities of nitrogen, and also enriched the soil considerably in this element. Where then had the nitrogen come from? The amount of combined nitrogen brought down by the rain was found to be far too small to account for the result. For years experiments were carried on, but the problem remained unsolved. Looking back over the papers[4] one can see how very close some of the older investigators were to the discovery of the cause of the mystery: in particular J. Lachmann[5] carefully examined the structure of the nodules, which he associated with the nutrition of the plant, and showed that they contained 'vibrionenartige' organisms. His paper, however, was published in an obscure journal and attracted little attention. W. O. Atwater in 1881 and 1882 showed that peas acquired large quantities of nitrogen from the air, and later suggested that they might 'favour the action of nitrogen-fixing organisms'.[6] But he was too busily engaged to follow the matter up, and once again an investigation in agricultural chemistry had been brought to a standstill for want of new methods of attack.

1 *Ann. Chim. Phys.*, 1838 (II), **67**, 5; **69**, 353; 1856 (III), **46**, 5.
2 J. B. A. Dumas and Boussingault, *Essai de Statique Chimique des Êtres Organisés*, Paris, 1841.
3 *Phil. Trans.*, 1861, **151**, 431; 1889, **180**A, 1; *J. Roy. Agric. Soc. Eng.*, 1891, ser. 3, **2**, 657.
4 A summary of the voluminous literature is contained in Löhnis's *Handbuch der landw. Bakteriologie*, pp. 646 *et seq.*
5 *Mitt. Landw. Lehranst.*, Poppelsdorf, 1858, **1**. Reprinted in *Zbl. Agrik. Chem.*, 1891, **20**, 837.
6 *Amer. Chem. J.*, 1885, **6**, 365; **8**, 327.

The beginnings of soil bacteriology

It had been a maxim with the older agricultural chemists that 'corruption is the mother of vegetation'. Animal and vegetable matter had long been known to decompose with formation of nitrates: indeed nitre beds made up from such decaying matter were the recognised source of nitrates for the manufacture of gunpowder during the European wars of the seventeenth and eighteenth centuries.[1] No satisfactory explanation of the process had been offered, although the discussion of rival hypotheses continued up till 1860, but the conditions under which it worked were known and on the whole fairly accurately described.

No connection was at first observed between nitrate formation and soil productiveness. Liebig[2] rather diverted attention from the possibility of tracing what now seems an obvious relationship by regarding ammonia as the essential nitrogenous plant nutrient, though he admitted the possible suitability of nitrates. Way came much nearer to the truth. In 1856 he showed that nitrates were formed in soils to which nitrogenous fertilisers were added. Unfortunately he failed to realise the significance of this discovery. He was still obsessed with the idea that ammonia was essential to the plant, and he believed that ammonia, unlike other nitrogen compounds, could not change to nitrate in the soil, but was absorbed by the soil by the change he had already described (p. 13). But he only narrowly missed making an important advance in the subject, for after pointing out that nitrates are comparable with ammonium salts as fertilisers he writes: 'Indeed the French chemists are going further, several of them now advocating the view that it is in the form of nitric acid that plants make use of compounds of nitrogen. With this view I do not myself at present concur: and it is sufficient here to admit that nitric acid in the form of nitrates has at least a very high value as a manure.'

It was not till ten years later, and as a result of work by plant physiologists, that the French view prevailed over Liebig's, and agricultural investigators recognised the importance of nitrates to the plant and of nitrification to soil fertility. It then became necessary to discover the cause of nitrification.

During the 'sixties and 'seventies great advances were being made in bacteriology, and it was definitely established that bacteria bring about putrefaction, decomposition and other changes; it was therefore conceivable that they were the active agents in the soil, and that the process of decomposition there taking place was not the purely chemical 'eremacausis' Liebig had postulated. Pasteur himself had expressed the opinion that nitrification was a bacterial process. The new knowledge was first brought to bear on agri-

---

1 *Instructions sur l'Établissement des Nitrières, Publié par les Régisseurs Généraux des Poudres et Salpêtre*, Paris, 1777.
2 *Principles of Agricultural Chemistry with Special Reference to the Late Researches Made in England*, London, 1855.

cultural problems by Th. Schloesing and A. Müntz[1] during a study of the purification of sewage water by land filters. A continuous stream of sewage was allowed to trickle down a column of sand and limestone so slowly that it took eight days to pass. For the first twenty days the ammonia in the sewage was not affected, then it began to be converted into nitrate; finally all the ammonia was converted during its passage through the column, and nitrates alone were found in the issuing liquid. Why, asked the authors, was there a delay of twenty days before nitrification began? If the process were simply chemical, oxidation should begin at once. They therefore examined the possibility of bacterial action and found that the process was entirely stopped by a little chloroform vapour, but could be started again after the chloroform was removed by adding a little turbid extract of dry soil. Nitrification was thus shown to be due to micro-organisms—'organised ferments', to use their own expression.

R. Warington[2] had been investigating the nitrates in the Rothamsted soils, and at once applied the new discovery to soil processes. He showed that nitrification in the soil is stopped by chloroform and carbon disulphide; further, that solutions of ammonium salts could be nitrified by adding a trace of soil. By a careful series of experiments described in his four papers to the Chemical Society he found that there were two stages in the process and two distinct organisms: the ammonia was first converted into nitrite and then to nitrate. But he failed altogether to obtain the organisms, in spite of some years of study, by the gelatin methods then in vogue. However, S. Winogradsky,[3] in a brilliant investigation, isolated these two groups of organisms, showing they were bacteria. He succeeded where Warington failed because he realised that carbon dioxide should be a sufficient source of carbon for them, so that they ought to grow on silica gel plates carefully freed from all organic matter; and it was on this medium that he isolated them in 1890.

Warington also established definitely the fact that nitrogen compounds rapidly change to nitrates in the soil, so that whatever compound is supplied as manure, plants get practically nothing but nitrate as food. This closed the long discussion as to the nitrogenous food of non-leguminous plants; in natural conditions they take up nitrates only (or at any rate chiefly), because the activities of the nitrifying organisms leave them no option. The view that plants assimilate gaseous nitrogen has from time to time been revived,[4] but it is not generally accepted.

The apparently hopeless problem of the nitrogen nutrition of leguminous plants was soon to be solved. In a striking series of sand cultures H. Hellriegel

1 *Comp. Rend.*, 1877, **84**, 301; **85**, 1018; 1878, **86**, 892.
2 *J. Chem. Soc.*, 1878, **33**, 44; 1879, **35**, 429; 1884, **45**, 637; 1891, **59**, 484.
3 *Ann. Inst. Pasteur*, 1890, **4**, 213, 257, 760.
4 For example, Th. Pfeiffer and E. Franke, *Landw. Vers.-Stat.*, 1896, **46**, 117; Thos. Jamieson, *Aberdeen Res. Assoc. Repts.*, 1905–08; C. B. Lipman and J. K. Taylor, *J. Franklin Inst., Calif.*, 1924, p. 475.

and H. Wilfarth[1] showed that the growth of non-leguminous plants, barley, oats, etc., was directly proportional to the amount of nitrate supplied, the duplicate pots agreeing satisfactorily; while in the case of leguminous plants no sort of relationship existed and duplicate pots failed to agree. After the seedling stage was passed the leguminous plants grown without nitrate made no further progress for a time, then some of them started to grow and did well, while others failed. This stagnant period was not seen where nitrate was supplied. Two of their experiments are given in Table 1.3.

TABLE 1.3   Relation between nitrogen supply and plant growth

Nitrogen in the calcium nitrate supplied per pot, g	none	0·056	0·112	0·168	0·224	0·336
Weight of oats obtained (grain and straw)	0·361 0·419	5·902 5·851 5·287	10·981 10·941	15·997	21·273 21·441	30·175
Weight of peas obtained (grain and straw)	0·551 3·496 5·233	0·978 1·304 4·128	4·915 9·767 8·497	5·619	9·725 6·646	11·352

Analysis showed that the nitrogen contained in the oat crop and sand at the end of the experiment was always a little less than that originally supplied, but was distinctly greater in the case of peas; the gain in three cases amounted to 0·910, 1·242 and 0·789 g per pot respectively. They drew two conclusions: (1) the peas took their nitrogen from the air; (2) the process of nitrogen assimilation was conditioned by some factor that did not come into their experiment except by chance. In trying to frame an explanation they connected two facts that were already known. M. Berthelot[2] had made experiments to show that certain micro-organisms in the soil can assimilate gaseous nitrogen. It was known to botanists that the nodules on the roots of Leguminosae contained bacteria.[3] Hellriegel and Wilfarth, therefore, supposed that the bacteria in the nodules assimilated gaseous nitrogen, and then handed on some of the resulting nitrogenous compounds to the plant. This hypothesis was shown to be well founded by the following facts:

1 In absence of nitrates peas made only small growth and developed no nodules in sterilised sand; when calcium nitrate was added they behaved like oats and barley, giving regular increases in crop for each increment of nitrates (the discordant results of Table 1.3 were obtained on unsterilised sand).

1 *Ztschr. Rübenzucker-Ind.*, Beilageheft, 1888.
2 *Comp. Rend.*, 1885, **101**, 775.
3 This had been demonstrated by Lachmann (p. 17) and by M. Woronin (*Mem. Acad. Sci.*, St. Petersburg, 1866, ser. 7, **10**, No. 6). J. Eriksson in 1874 (Doctor's dissertation, abs. in *Botan. Ztg.*, 1874, **32**, 381) carried on the investigation, while G. Brunchorst in 1885 (*Ber. Deut. Bot. Ges.*, **3**, 241) gave the name 'bacteroids'.

2 They grew well and developed nodules in sterilised sand watered with an extract of arable soil.

3 They sometimes did well and sometimes failed when grown without soil extract and without nitrate in *unsterilised* sand, which might or might not contain the necessary organisms. An extract that worked well for peas might be without effect on lupins or serradella. In other words, the organism is specific.

Hellriegel and Wilfarth read their paper and exhibited some of their plants at the Naturforscher-Versammlung at Berlin in 1886. Gilbert was present at the meeting, and on returning to Rothamsted repeated and confirmed the experiments. At a later date Th. Schloesing *fils* and E. Laurent[1] showed that the weight of nitrogen absorbed from the air was approximately equal to the gain by the plant and the soil, and thus finally clinched the evidence.

	Control	Peas	Mustard	Cress	Spurge
Nitrogen lost from the air, mg	1·0	134·6	−2·6	−3·8	−2·4
Nitrogen gained by crop and soil, mg	4·0	142·4	−2·5	2·0	3·2

The organism was isolated by M. W. Beijerinck[2] and called *Bacillus radicicola*, but is now known as *Rhizobium*.

Thus another great controversy came to an end, and the discrepancy between the field trials and the laboratory experiments of Lawes, Gilbert and Pugh was cleared up. The laboratory experiments gave the correct conclusion that leguminous plants, like non-leguminous plants, have themselves no power of assimilating gaseous nitrogen; this power belongs to the bacteria associated with them. But so carefully was all organic matter removed from the soil, the apparatus and the air in endeavouring to exclude all trace of ammonia, that there was no chance of infection with the necessary bacteria. Hence no assimilation could go on. In the field trials the bacteria were active, and here there was a gain of nitrogen.

The general conclusion that bacteria are the real makers of plant food in the soil, and are, therefore, essential to the growth of all plants, was developed by E. Wollny[3] and M. Berthelot.[4] It was supposed to be proved by E. Laurent's[5] experiments. He grew buckwheat on humus obtained from well-rotted dung, and found that plants grew well on the untreated humus, but only badly on the humus sterilised by heat. When, however, soil bacteria

1 *Ann. Inst. Pasteur*, 1892, **6**, 65.
2 *Bot. Ztg.*, 1888, **46**, 725, 741, 757; 1890, **48**, 837.
3 *Bied. Zbl. Agric. Chem.*, 1884, **13**, 796.
4 *Comp. Rend.*, 1888, **106**, 569.
5 *Bull. Acad. Roy. Belgique*, 1886, **2**, 128. See also E. Duclaux, *Comp. Rend.*, 1885, **100**, 66.

were added to the sterilised humus (by adding an aqueous extract of unsterilised soil) good growth took place. The experiment looks convincing,
but is really unsound. When a rich soil is heated some substance is formed
toxic to plants. The failure of the plants on the sterilised humus was, therefore,
not due to absence of bacteria, but to the presence of a toxin. No one has
yet succeeded in carrying out this fundamental experiment of growing plants
in two soils differing only in that one contains bacteria while the other does
not.

The rise of modern knowledge of the soil, and the return to field studies

Further investigation of soil problems has shown that they are more complex
than was at first supposed. Soils can no longer satisfactorily be divided into a
few simple groups: sands, clays, loams, etc., according to their particle size;
nor can attention be confined to the surface layer. It is necessary to take
account of their history. The properties of a soil depend not only on its
parent material but also, as shown by the Russian investigator V. V.
Dokuchaev[1] in particular, on the climatic, vegetation and other factors to
which it has been subjected.

The relations of the plant to the soil are also recognised as highly complex.
The older workers had thought of soil fertility as a simple chemical problem;
the early bacteriologists thought of it as bacteriological. E. Wollny[2] and
F. H. King[3] showed that the physical properties of the soil already studied by
Davy and Schübler play a fundamental part in soil fertility. Van Bemmelen
showed that soil has colloidal properties, and presentday workers have observed in the soil many of the phenomena investigated in laboratories
devoted to the study of colloids. Whitney and Cameron at Washington
greatly widened the subject by revealing the importance of the soil solution
and introducing the methods and principles of physical chemistry. Russell
and Hutchinson at Rothamsted showed that bacterial action alone would
not account for the biological phenomena in the soil, but that other organisms are also concerned, and subsequent work in the Rothamsted laboratories and elsewhere has revealed the presence of a complex soil population,
the various members of which react on one another and on the growing plant.

The nature of the subject necessitates a departure from the usual procedure. In purely laboratory investigations it is customary to adopt the
Baconian method, in which factors are studied one at a time, all others being
kept constant except the particular one under investigation. In dealing with
soils in natural conditions, however, it is impossible to proceed in this way:

1 *Tchernozéme de la Russie d'Europe*, St. Petersburg, 1883.
2 Papers by himself and his students in *Forschungen auf dem Gebiete der Agrikultur-Physik*, 1878–98.
3 *The Soil,* New York, 1899.

climatic factors will not be kept constant, and however careful the effort to ensure equality of conditions there is always the probability, and sometimes the certainty, that the variable factor under investigation is interacting with climatic factors and exerting indirect effects which modify or even obscure the direct effects it is desired to study. Hence, in recent years, statisticians have had to devise methods for dealing with cases where several factors are varying simultaneously.

This increased interest in the soil has shown itself in two directions. The development of soil surveys has encouraged an enormous development of soil studies *in situ*; and the introduction of modern statistical methods has given to field experiments a new value they completely lacked before. In the past, field experiments were always weakened by the unknown errors due to the circumstances that the soil of one plot was never strictly comparable with the soil of another. Modern methods of field plot technique have overcome this difficulty and yield results to which a definite value can be assigned so that the data can be utilised in further investigations.

2

The food of plants

Green plants synthesise their food from simple substances taken out of the air and the soil. It is common to speak of these substances as the actual foods: in reality they are the raw materials out of which the food is made. Plants, like all other organisms, have their tissues built out of carbohydrates, fats, proteins and nucleoproteins, and need for the functioning of their tissues a host of enzymes. Hence the plant needs large quantities of carbon, oxygen, hydrogen, nitrogen, phosphorus and sulphur for building up its tissues; it needs small quantities of at least iron, manganese, zinc, copper, boron and molybdenum and sometimes cobalt for building up its enzymes; and it needs potassium, magnesium and calcium, and sometimes sodium and chlorine, and often other electrolytes for these or other purposes. Other elements, such as silicon and aluminium may be necessary, and are certainly present in the tissues of all plants grown in the field, though they have not been shown to have essential specific effects on the growth and development of the crop. Carbon dioxide and water are probably the sole source of carbon and hydrogen for most plants; ammonium and nitrate ions are an adequate source of nitrogen, though some leguminous and other plants can supplement these with nitrogen from the air; and the other elements are usually taken up from the soil as simple inorganic ions.

Plants take up different amounts and different proportions of nutrients from a soil according to their species. The exact amounts taken up depend on the soil conditions, but certain broad generalisations can be made. Table 2.1 shows typical quantities of various substances English crops remove from the soil,[1] and E. Wolff[2] and H. Wilfarth and his co-workers[3] have published other tables. They all show that cereals absorb less mineral nutrients per acre than any other crop studied, although they give a large amount of dry matter per acre. The animal fodder crops—grass and clover hay and the root crops—all take up large amounts of nutrients, and hence the need for good conservation of the nutrients in farmyard manure made

1 R. Warington, *The Chemistry of the Farm*, 4th edn, London, 1886.
2 *Aschen Analysen*, Berlin, 1871.
3 *Landw. Vers.-Stat.*, 1905, **63**, 1.

TABLE 2.1   Amounts of various elements taken up by the common agricultural crops of England

| | t/ha | | kg/ha | | | | | | | | | |
	At harvest*	Dry matter	Ash	Nitrogen	Sulphur	Potassium	Sodium	Calcium	Magnesium	Phosphorus	Chlorine	Silicon
Wheat												
grain	2·0	1·72	34	38	3·0	8·6	0·6	0·8	2·5	7·0	0·1	0·3
straw	3·5	2·97	159	18	5·7	18·2	1·7	6·6	2·4	3·4	2·7	50·4
total	5·5	4·69	193	56	8·7	26·8	2·3	7·4	4·9	10·4	2·8	50·7
Oats												
grain	2·1	1·83	57	39	3·6	8·4	0·7	1·4	2·5	6·4	0·6	10·4
straw	3·2	2·64	157	20	5·4	34·4	3·8	7·9	3·5	3·1	6·8	34·3
total	5·3	4·47	214	59	9·0	42·8	4·5	9·3	6·0	9·5	7·4	44·7
Beans												
grain	2·0	1·81	65	88	4·9	22·6	0·4	2·4	2·7	11·2	1·2	0·2
straw	2·2	2·08	111	32	5·5	39·8	1·5	21·1	3·8	3·0	4·8	3·6
total	4·2	3·89	176	120	10·4	62·4	1·9	23·5	6·5	14·2	6·0	3·8
Meadow hay	3·7	3·16	228	55	6·4	47·5	7·6	25·7	9·8	6·1	16·4	29·8
Red Clover hay	5·0	4·21	290	110	10·5	79·5	4·3	72·1	19·1	12·2	11·0	3·7
Turnips												
roots	42	3·51	244	68	17·0	101·1	14·1	20·4	3·8	11·0	12·2	1·3
leaf	13	1·72	164	55	6·4	37·4	6·3	38·8	2·6	5·3	12·6	2·7
total	55	5·23	408	123	23·4	138·5	20·4	59·2	6·4	16·3	24·8	4·0
Mangolds												
roots	55	6·64	478	110	5·5	207·5	57·6	12·7	12·3	17·3	47·6	4·7
leaf	20	1·85	285	57	10·2	72·5	41·1	21·6	16·4	8·5	45·5	4·8
total	75	8·49	763	167	15·7	280·0	98·7	34·3	28·7	25·8	93·1	9·5
Potatoes	15	3·76	142	52	3·0	71·1	3·1	2·7	4·3	10·5	4·9	1·3

* Weight of crop. The yields quoted are not averages, though they were commonly obtained on farms before 1940.

from these crops, otherwise the soil will rapidly become impoverished. A consequence of this differential absorption of nutrients by crops is that if the nutrient status of the soil is low, crops taking large amounts of nutrients from a soil leave it poorer than crops taking only small amounts.

Carbon dioxide from the air and simple inorganic ions in the soil can certainly supply all the nutrients needed by the plant. But this does not prove that plants in the field obtain all their nutrients from these sources or that their growth may not be improved if they can take up complex substances of vitamin or hormone-like nature. Claims have, in fact, repeatedly been made that other sources of food are necessary if the crop is to make optimum growth, which is sometimes measured by crop yield and sometimes by improvements in its feeding value. In some conditions plants in fact obtain some of their food from other sources. The outer cells of their roots may contain fungal hyphae which extend around or even through the cells and into the soil. This association between root and fungus can be very strongly developed—as, for example, on forest trees growing on poor soil—giving structures known as mycorrhizas, which are described in more detail in chapter 13. The fungus in this association transfers nutrients, in particular nitrogen and phosphate, from the soil to the root cells, handing them on to these cells as complex substances. There is, however, no critical evidence to show that this mechanism plays any important part in the nutrition of crops grown on normal arable soils.

Many experiments have been made to see if the performance of a crop can be improved by supplying it either with vitamins or with plant hormones, but the results so far have been inconclusive. It is well established that isolated organs of a plant—such as excised roots—can only be grown in artificial conditions if some of the B vitamins are present, but this is no evidence that plants are unable to synthesise all the vitamins they need. There is, in fact, no well-established evidence that adding any vitamin or growth promoting factor to the soil ever improves crop production,[1] or has any appreciable effect on the vitamin content of the plant. This is much more dependent on climate and possibly soil than on manuring. Vitamin C contents in particular seem to be high in years of bright sunshine.[2]

---

1 For some evidence that vitamin B_1 may be of value, see J. Bonner and J. Greene, *Bot. Gaz.*, 1938, **100**, 226, and 1939, **101**, 491; but D. I. Arnon, *Sci.*, 1940, **92**, 264; C. L. Hamner, *Bot. Gaz.*, 1940, **102**, 156; W. G. Templeman and M. Pollard, *Ann. Bot.*, 1941, **5**, 133; and D. B. Swartz, *Bot. Gaz.*, 1941, **103**, 366, were unable to confirm this. H. Lundergårdh, *Kgl. Lantbr. Akad. Tidskr.*, 1943, **82**, 99, found some evidence that vitamin B_1 in the presence of phosphate and magnesium might be able to increase yields. For a short review of the literature, see R. L. Starkey, *Soil Sci.*, 1944, **57**, 264. For some evidence that indolyl- or naphthyl-acetic acids may be of value, see G. P. McRostie *et al.*, *Canad. J. Res.*, 1938, **16** C, 510, for wheat, A. Dunez, *C.R. Acad. Agric.*, 1946, **32**, 736, and 1947, **33**, 548, for wheat and other farm crops, and H. L. Stier and H. G. du Bay, *Proc. Amer. Soc. Hort. Sci.*, 1939, **36**, 723, for tomatoes. H. L. Pearse reviewed this subject in *Imp. Bur. Hort., Tech. Comm.* 12, 1939.

2 For cow-peas, M. E. Reid, *Bull. Torrey Bot. Cl.*, 1942, **62**, 204; for tomatoes in England, F. Wokes *et al.*, *Nature*, 1947, **159**, 172.

The question whether farmyard manure owes its value as a plant food to any so far unrecognised nutrient, needed only in minute quantities, cannot yet be definitely answered, though the experimental evidence is against any such possibility. Provided the soil conditions around the plant roots—as measured by the air and water supply—are favourable, it appears that the value of farmyard manure—as measured by crop yield—depends only on the amount of nutrients it can supply in simple form to the crop. Naturally, the farmyard manure itself often has an important role in creating a favourable air and water régime around the plant roots, but this effect is not of relevance to this particular question. Farmyard manure can, however, be a very valuable carrier of some minor elements, and in particular additions of farmyard manure may be the easiest way of maintaining an adequate supply of available iron to the plant.[1]

The second question—to which extravagant answers have been given by some workers—is whether the feeding value of the crop, either for animals or humans, is affected by the presence of farmyard manure or composts in the soil. The earlier experimental evidence on this point was conflicting, possibly because of imperfections in experimental technique, but the few recent accurate experiments that have been made have given no indication that the presence of farmyard manure or compost in a soil has any specific action in increasing the vitamin content[2] or the nutritive value[3] of the plant.

This result appears to be general, for fertilisers also seem to have little effect on the vitamin content of the leaves and roots of crops. However, the carotene and chlorophyll content of grass can be increased a little by nitrogen manuring,[4] as is shown by its darker green colour, and the carotene content of lucerne by boron under some conditions,[5] and of soyabean leaves by phosphates.[6] On the other hand, the concentration of vitamin C in the tissues of a crop is often somewhat lowered by any manuring that increases the growth of the crop. Thus L. J. Harris and D. J. Watson[7] found that normal dressings of farmyard manure or sulphate of ammonia lowered the vitamin C content of potato tubers. Again, small potato tubers[8] tend to have higher vitamin C contents than large, and slow-growing leaves of vegetable crops than fast growing.[9]

The plant needs its nutrients for three distinct but overlapping purposes. It must build its protoplasm and form all the enzymes needed for its vital

1 J. Bonner, *Bot. Gaz.*, 1946, **108**, 267.
2 For a review of this subject, see L. A. Maynard and K. C. Beeson, *Nutr. Abstr.*, 1943, **13**, 155.
3 D. I. Arnon, H. D. Simms and A. F. Morgan, *Soil Sci.*, 1947, **63**, 129.
4 B. Thomas and F. E. Moon, *Emp. J. Expt. Agric.*, 1938, **7**, 235. E. V. Miller, T. J. Army and H. F. Krackenberger, *Proc. Soil Sci. Soc. Amer.*, 1956, **20**, 379, who also showed that the contents of two of the B vitamins were increased.
5 W. L. Powers, *Proc. Soil Sci. Soc. Amer.*, 1939, **4**, 290.
6 W. J. Peterson *et al.*, *Amer. Fert.*, 1948, No. 3, 24.
7 Unpublished observations.
8 Unpublished observations of J. Meiklejohn.
9 S. H. Wittiver, R. A. Schroeder and W. A. Albrecht, *Soil Sci.*, 1945, **59**, 329.

processes and growth, it must build tissues to support and protect its proto-plasm, and it must be able to transport nutrients from one organ to another.

The plant's supporting and protecting tissues are built out of polymerised sugar residues such as celluloses, hemicelluloses and pectins, on the one hand, and lignins on the other, though the latter are typically formed as the tissues mature. They also contain inorganic constituents whose functions and chemical combinations are uncertain. Thus graminaceous plants—the grasses and cereals—accumulate considerable quantities of silica in their tissues. Further, many plants accumulate sugars or polymerised sugar residues, for example, starch and inulin, as a food reserve in their tissues, but other sub-stances, such as oils and proteins, may also accumulate in them for the same reason. The enzyme systems in the plant are built up out of proteins and nucleoproteins, and thus contain large proportions of nitrogen and phos-phate, with some sulphur. They also require certain metals and other elements, and unless the particular metal needed is present, the enzyme cannot function: only very rarely can that particular metal be replaced by another without destroying, or at least greatly weakening, the effectiveness of the enzyme. The liquid that transports the various nutrients throughout the plant, either from the roots, the green leaves or the storage organs, also contains inorganic ions. These may be nutrient ions, but they are also regulators of the osmotic pressure and hydrogen-ion concentration in the protoplasm, and for these latter functions the particular anions or cations present are of minor importance.

These considerations allow one to understand something about the total nutrient demands of the plant. Rapid growth can only take place when there is an adequate quantity of enzymes present, and hence after the plant has absorbed adequate quantities of the minerals necessary for their functioning. The maximum demand for these minerals therefore occurs when the plant is young, and the supply can be reduced later in the season; for as the cells in which the enzymes are situated age, much or all of these minerals can be transferred to new growing points to build new enzymes there. Further, if some of the minerals needed by the enzymes are in short supply, the older cells containing the enzyme may die prematurely and the limiting element transferred to the growing point. This need not affect the yield of cereals or root crops since it is food reserve material that is harvested, but it can severely limit the yield of pasture and forage crops which are grown for their green, and therefore actively functioning, leaves. Hence, these green crops must be able to obtain an adequate supply of all essential minerals throughout their growing season.

These points are illustrated in Fig. 2.1, taken from some work of A. E. V. Richardson and H. C. Trumble[1] on the uptake of nutrients, and the rate of growth of barley. Nitrogen, phosphate, potassium and calcium are all taken

1   *J. Dept. Agric. S. Aust.*, 1928, **32**, 224.

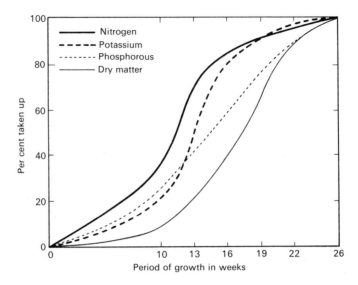

FIG. 2.1. Uptake of nutrients and production of dry matter in barley

up rapidly when the plant is small, as measured by the amount of dry matter present, but the rate of uptake falls when the plant is making its dry matter rapidly. The uptake of minerals, however, may continue even if they are not needed by the plant, for plants take up minerals because they are present in the soil solution, and this uptake goes on all through the active life of the roots.

The concentration of cations in some plant tissues, such as the actively functioning leaves and the fruits, tends to be a characteristic of the crop and fairly independent of the soil and manuring. Thus D. J. Watson[1] quotes the results of leaf analyses made on the Rothamsted permanent mangold experiment for the six years 1878–83, which showed that the leaf contained between 300 and 360 milli-equivalents of cations (sodium, potassium, magnesium and calcium) per 100 g of dry matter when the manuring, and, in consequence, the actual composition of the bases present, was varied within wide limits. D. R. Hoagland and J. C. Martin[2] found that tomato plants contained about 300 milli-equivalents of cations per 100 g of dry matter, although the potassium content could vary from 25 to 150 me. Similarly, T. B. van Itallie[3] found that Italian ryegrass contained about 200 me of cations per 100 g of dry matter in the leaf and F. E. Bear and A. L. Prince[4] found lucerne con-

1 *Emp. J. Expt. Agric.*, 1946, **14**, 57.
2 *Soil Sci.*, 1933, **36**, 1. For further examples, see J. T. Cope, R. Bradfield and M. Peech, *Soil Sci.*, 1953, **76**, 65.
3 *Soil Sci.*, 1938, **46**, 175.
4 *J. Amer. Soc. Agron.*, 1945, **37**, 217; 1948, **40**, 80. For additional data, F. E. Bear, *Soils and Fertilizers*, 4th edn.

tained between 150 and 200 me. But a more critical examination by R. K. Cunningham[1] of the cation content of ryegrass grown under a range of conditions has shown that its leaves do not have a constant content of cation, but it can vary from between 80 and 200 me, being higher the greater the total anion content of the leaf or its total nitrogen content. Further young tissues and leaves tend to have a higher ionic concentration in them than the older, and plants growing in a soil with an appreciable soluble salt content often accumulate considerable quantities of mineral salts in their tissues. As an extreme example some desert plants growing in saline soil can have between 20 and 50 per cent of their dry matter in the form of soluble salts, whereas most crops cannot accumulate more salts than 1–2 per cent of their dry matter.[2]

Plants not only take up nutrients from the soil, but also lose them to the soil. Rain washing over the leaves of vegetation carries away some nutrients from them, of which potassium is usually the element most strongly lost. This was first studied by foresters, and G. M. Will,[3] for example, working in a *Pinus radiata* forest at Rotorua, New Zealand, found that the rain washing through the canopy of the forest carried down annually with it more potassium and sodium than was present in the annual leaf fall, as is shown in Table 2.2. This table also shows that very little calcium or phosphate is

TABLE 2.2   Minerals in annual cycle in a New Zealand *Pinus radiata* forest. Quantities in m eq/m^2/year

	Ca	Mg	K	P
Rain in the open	1·3	1·4	2·2	0·0
Rain under canopy	4·7	12·6	21·5	4·9
In litter fall	61·6	21·2	16·5	24·6

washed out of the canopy, but nearly one-third of the magnesium in annual cycle appears in the rain wash. The same general result has been found for agricultural crops, so that many crops contain more potassium in their tissues during the initial stages of their maturity than later. There is still a lack of data on this point, but the total potassium content of cereals may be 50 per cent higher shortly after flowering than at harvest. This is illustrated in Table 2·3 for a winter wheat crop growing on a heavy Essex clay.[4]

The composition of the soil affects the mineral composition of plants growing on it; and to a limited extent the mineral composition of plant leaves taken under strictly standardised conditions can be used as a measure

1 *J. Agric. Sci.*, 1964, **63**, 97, 103, 109; 1965, **64**, 229.
2 For examples of such plants, see M. M. Shukevich, *Trans. Dokuchaev Inst.*, 1939, **19**, No. 2, 39.
3 *Nature*, 1955, **176**, 1180; see also *N.Z. J. Agric. Res.*, 1959, **2**, 719.
4 F. Knowles and J. E. Watkins, *J. Agric. Sci.*, 1931, **21**, 612. For another example, see E. K. Woodford and A. G. McCalla, *Canad. J. Res.*, 1936, **14**C, 245.

TABLE 2.3   Production of dry matter, and assimilation of nutrients from the soil by wheat growing in the open field. Weight in grams of substances in whole wheat plant grown in the field

	Before ear emergence					After ear emergence			
	1st Samp-ling (30.iv)	2nd Samp-ling (21.v)	3rd Samp-ling (4.vi)	4th Samp-ling (18.vi)	5th Samp-ling (2.vii)	6th Samp-ling (16.vii)	7th Samp-ling (23.vii)	8th Samp-ling (30.vii)	9th Samp-ling (6.viii)
Dry matter	770	2970	6190	9510	12270	14350	15060	14730	14210
Nitrogen	27	55	80	90	96	110	109	109	109
Potassium	26	79	140	149	121	103	94	86	75
Calcium	5·1	12·8	21·6	25·6	26·8	26·5	22·8	22·0	21·6
Phosphorus	3·2	10·5	18·0	24·4	24·8	27·0	27·8	27·8	27·8
Chlorine	6·0	15·1	24·2	28·8	33·9	30·5	27·4	23·6	19·3
Silicon	8	40	91	125	152	199	206	209	207

of the relative levels of the different plant nutrients in an available form in the soil within its root zone.[1] However, the leaves of different plant species and even different varieties within the same species will have different mineral compositions when growing in the same soil, and even the mineral composition of a leaf varies with its age. But soils low in phosphate, for example, will typically carry vegetation low in phosphate, so that, if the soil is very low in phosphate, ruminants grazing in these areas may suffer from severe phosphate starvation. Similarly, soils high in selenium will carry vegetation high in selenium and animals grazing in these areas may suffer from selenium toxicity.

The organic constituents of the plant, such as its protein content, are also affected to some extent by the soil. The amino acid distribution of the proteins within a plant is to a large extent genetically controlled, so that a low supply of soil sulphur will result in a low production of methionine which will result in a low production of protein. However, there is some evidence that appreciable changes in the ratios of different nutrient ions in the soil may only have a small effect on the ratios of some amino acids such as tryptophan and lysine.[2]

1 For the use of the mineral composition of sweet vernal grass *Anthoxanthum odoratum* in New Zealand for this purpose see N. Wells, *N.Z. J. Sci. Tech.*, 1956, **37B**, 473, and subsequent papers in that journal.
2 V. L. Sheldon, W. A. Albrecht with W. G. Blue, *Pl. Soil*, 1951, **3**, 33; with L. W. Reed, *Agron. J.*, 1960, **52**, 523.

3

The individual nutrients needed by plants

In the following discussion a brief account will be given of the effects of the various nutrients on crop growth. It will usually be assumed that all nutrients are present in adequate supply except the one under discussion. The field symptoms of the various deficiencies will not be described, as full descriptions with coloured plates will be found in several standard works.[1] Nor is it relevant to the general purpose of this book to discuss in any detail the physiological functions of these nutrients in the plant.

Nitrogen

Nitrogen is essential for plant growth as it is a constituent of all proteins and nucleic acids and hence of all protoplasm. It is generally taken up by plants either as ammonium or as nitrate ions, but the absorbed nitrate is rapidly reduced, probably to ammonium, through a molybdenum-containing enzyme. The ammonium ions and some of the carbohydrates synthesised in the leaves are converted into amino acids, mainly in the green leaf itself. Hence, as the level of the nitrogen supply increases compared with other nutrients, the extra protein produced allows the plant leaves to grow larger and hence to have a larger surface available for photosynthesis, and in fact, over a considerable range of nitrogen supply for many crops, the amount of leaf area available for photosynthesis is roughly proportional to the amount of nitrogen supplied.

This effect of nitrogen in increasing leaf growth is not its only effect on the leaf, for the higher the nitrogen supply the more rapidly the synthesised carbohydrates are converted to proteins and to protoplasm and the smaller the proportion left available for cell wall material, which is mainly nitrogen-free carbohydrates such as calcium pectate, cellulosans, cellulose and low-nitrogen lignins.

This effect of nitrogen in increasing the proportion of protoplasm to cell wall material has several consequences. It increases the size of the cells and

1 T. Wallace, *The Diagnosis of Mineral Deficiencies in Plants*, 1943, and Supplement, 1944, London, and *Hunger Signs in Crops*, ed. by G. Hambidge, Washington, 1941.

gives them a thinner wall, hence makes the leaves more succulent and less harsh. It also increases the proportion of water[1] and decreases that of calcium[2] to dry matter: the former because protoplasm has more water and the latter because it has less calcium than cell wall material. Excessive amounts of nitrogen give leaves with such large thin-walled cells that they are readily attacked by insect and fungus pests and harmed by unfavourable weather such as droughts and frosts. A very low nitrogen supply on the other hand gives leaves with small cells and thick walls, and the leaves are in consequence harsh and fibrous. The nitrogen supply has one other noticeable effect on the leaf: it darkens the green colour. The leaves of plants growing with a low level of nitrogen compared with other nutrients are pale yellowish to reddish green, which darkens rapidly as the nitrogen supply increases and become very dark green when it is excessive. Further increasing the nitrogen supply to the leaves tends to keep them green for a longer time, and in many cereals it increases the length of the growing season and delays the onset of maturity (see p. 35) presumably also through its effect in keeping the free carbohydrate content of the leaf low.

Crops grown for their carbohydrates, such as the root crops and the cereals, thus only benefit from nitrogen manuring through the increased leaf area brought about by the nitrogen, so that the additional yield of carbo-

TABLE 3.1   Relative proportions of roots and leaves at harvest under nitrogenous manuring (Rothamsted)*

Yields in t/ha							
White Turnips				Mangolds			
Nitrogen kg/ha	Roots	Leaves	Roots / Leaves	Nitrogen kg/ha	Roots	Leaves	Roots / Leaves
0	20·6	6·8	3·04	0	13·3	2·9	4·64
53	24·9	10·8	2·30	96	45·0	9·5	4·73
154	25·8	15·4	1·84	206	73·5	15·6	4·72

* This table must be interpreted with some caution, for the relative weight of leaves at harvest on the different treatments may not accurately reflect their relative weights during the growing season.

hydrate is usually less in proportion than the increase in leaf area. The root crops show this effect most strikingly when the length of the growing season is varied. Thus a high level of nitrogen manuring mainly affects the tops in a crop which is only in the ground for a short time, such as white turnips, but

1 For an example with mangolds at Rothamsted, see D. J. Watson, *Emp. J. Exp. Agric.,* 1946, **14**, 409.
2 For a review of the literature, see K. C. Beeson, *Bot. Rev.,* 1946, **12**, 424.

has a very considerable effect on both tops and roots of a crop that is in the ground for a long time, such as mangolds, as is shown in Table 3.1. Late sown sugar-beet also sometimes responds to nitrogen manuring by making a very large increase in leaf growth which is accompanied by a very disappointing yield of root.

In the past the cereal crops of the temperate regions—wheat, barley, oats and rye—grown for grain and not for fodder, only required a moderate supply of nitrogen, for too high a level leads to excessive straw, as shown in Table 3.2 and in Plate I. These show the yield and the appearance of the wheat crop grown on Broadbalk at Rothamsted—a field that has been in wheat almost continuously since 1843, and the manuring of the plots has been continued unaltered for a long time. The plots shown in Plate I have been manured as follows: Plot 2, 35 t/ha annually of farmyard manure and Plot 3 unmanured, both since 1843; Plots 5, 6, 7, 8 and 16 all have received annually 450 kg of superphosphate, 220 kg potassium sulphate, 110 kg of magnesium sulphate and 110 kg of sodium sulphate per hectare since 1852 (Plot 16 since 1885) and annual dressings of ammonium sulphate or sodium nitrate to give 48, 96, 145 and 96 kg per hectare of nitrogen since these dates. The excessive development of straw on some of these plots induces a liability of the crop to lodge, and the very high nitrogen dressing on Plot 8 may delay the time the wheat comes to maturity.

TABLE 3.2   Effect of increasing nitrogen supply on the growth of wheat, Rothamsted

Nitrogen in manure kg/ha	Yields in t/ha	
	Grain	Straw
0	1·19	2·09
48	1·88	3·40
96	2·44	4·80
145	2·54	5·35
193	2·58	5·85

In present day practice these harmful secondary effects have been minimised by using suitable stiff short-strawed varieties, for, under most English conditions, the amount of nitrogen a crop will stand is determined by the onset of these secondary effects rather than by the extra yield of grain due to the high manuring ceasing to be economic. These harmful effects, in the case of the cereals, can also be minimised to some extent by applying the high level of nitrogen as late as possible to the crop, for D. J. Watson[1] found that delaying the time at which the nitrogen top dressing was applied barely affected the increment in the grain yield due to the manure, but it appreciably reduced the increment in the straw yield.

1 *J. Agric. Sci.*, 1936, **26**, 391.

The results given for the effect of nitrogen manuring on the cereals of the temperate region—that high nitrogen delays maturity of the crop and encourages growth of the straw relative to the grain—is not necessarily true for some tropical cereals such as maize and the sorghums, for with them a high level of nitrogen manuring may have just the opposite effect: it hastens the time of flowering and maturity and increases the grain yield relative to the straw. This last point is illustrated in Table 3.3, derived from the results of

TABLE 3.3   The effect of increasing the nitrogen supply on the yield of maize, rice and wheat in the Nile delta (Abu Hammad, 1935 and 1936). Yields of grain and straw in tons per hectare

Nitrogen supplied kg/ha	*Maize*			*Rice*			*Wheat*		
	Grain	Straw	Grain / Straw	Grain	Straw	Grain / Straw	Grain	Straw	Grain / Straw
0	1·89	4·85	0·39	4·30	5·05	0·85	2·32	3·50	0·67
37	3·26	5·30	0·61	5·19	6·15	0·84	3·29	5·32	0·62
74	4·25	4·98	0·85	5·52	6·42	0·86	3·84	6·60	0·58
112	4·48	5·21	0·86						

some field experiments made on the same farm in the Nile delta by F. Crowther and his associates.[1] The yield of maize grain can be more than doubled by the nitrogen without the yield of straw being appreciably affected, whereas the result with wheat is comparable with that obtained at Rothamsted and illustrated in Table 3.2. Rice in this area is intermediate between these two, nitrogen manuring having almost no effect on the ratio of grain to straw, although in parts of India it behaves like wheat.[2] The sorghum dura in the Sudan Gezirah behaves like maize,[3] and other sorghums, the millets, cotton and sunflower have been stated to do likewise, whilst soyabeans and buckwheat may behave like rice.[4] This result is not, however, always found for maize. Thus H. V. Jordan[5] found that maize in Mississippi behaved in the same way as the small grains, in that increasing the level of nitrogen fertiliser increased the yield of stover more rapidly than that of grain.

The effect of nitrogen fertiliser on the nitrogen content of the crop depends both on the responsiveness of the crop and on the time the nitrogen is given relative to the development of the crop. If the nitrogen fertiliser greatly stimulates crop growth, the nitrogen content of the fertilised crop, expressed

1 *Roy. Agric. Soc. Egypt, Bull.* 28, 1937.
2 R. L. Sethi, K. Ramish and T. P. Abraham, *Ind. Counc. Agric. Res., Bull.* 38, 1952.
3 *Agriculture in the Sudan*, ed. J. D. Tothill, London, 1948, p. 474.
4 M. H. Cailahjan, *C.R. Acad. Sci. U.S.S.R.,* 1945, **47**, 146. F. Crowther (*Roy. Agric. Soc. Egypt, Bull.* 31, 1937) showed that nitrogen manuring hastened the time of flowering of cotton.
5 With K. D. Lavid and D. D. Ferguson, *Agron. J.,* 1950, **42**, 261.

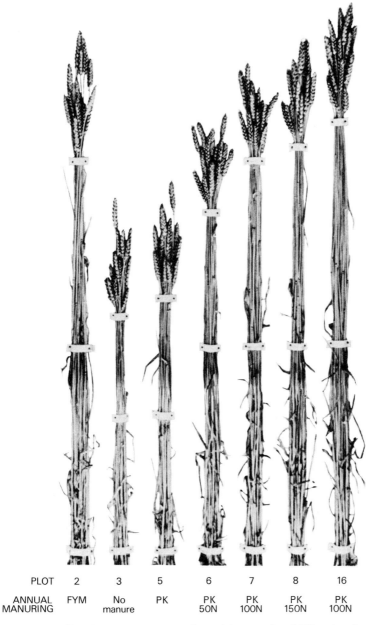

PLOT	2	3	5	6	7	8	16
ANNUAL MANURING	FYM	No manure	PK	PK 50N	PK 100N	PK 150N	PK 100N

N, in kg/ha, given as ammonium sulphate on plots 6,7,8 and sodium nitrate on plot 16.

PLATE 1. The effect of nitrogen fertilisers on the yield of wheat on Broadbalk field, Rothamsted in 1943, after 100 years in wheat

as a percentage of its dry or fresh weight, will often decrease, although the total uptake of nitrogen per acre will increase, as is shown in Fig. 3.1.[1] This is because the extra dry matter which the added nitrogen stimulates the plant to produce has a lower nitrogen content than that produced by a nitrogen-starved plant. As the level of nitrogen supply increases, however, this stimulating effect grows smaller, so the uptake of nitrogen is increasing relative to the additional production of dry matter due to it: the nitrogen content of the dry matter in the crop now begins to rise. But this rise only becomes appreciable when the growth of the crop ceases to respond economically to the additional fertiliser.

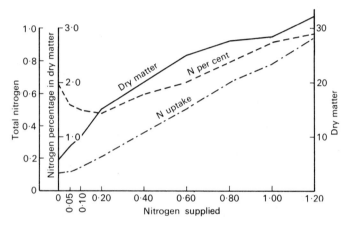

FIG. 3.1. The effect of increasing the nitrogen supply on the dry matter produced by mustard, and on its nitrogen content

Nitrogenous fertilisers can often be made to increase the nitrogen content of the crop more economically if they are given sufficiently near harvest for the crop to absorb much of the added nitrogen, but for the synthesised proteins not to have time to increase the growth of the crop appreciably. Thus, by adding a nitrogen fertiliser to a meadow one to three weeks before it is cut, the protein content of the hay will be increased though the yield of hay will barely be affected.[2]

The average response of crops to additional nitrogen under normal English agricultural conditions is shown in Table 3.4.[3] It shows that on the average 1 kg of nitrogen, as sulphate of ammonia, gives about 1 kg of protein

---

1 For a field example with barley, see E. J. Russell, *Min. Agric., Bull.* 28, 2nd edn, 1933.
2 H. B. Sprague and A. Hawkins, *New Jersey Agric. Expt. Sta., Bull.* 644, 1938; M. W. Evans, F. A. Welton and R. M. Salter, *Ohio Agric. Expt. Sta., Bull.* 603, 1939; A. H. Lewis, *J. Min. Agric.,* 1939, **46**, 77; *Emp. J. Expt. Agric.,* 1941, **9**, 43; W. S. Ferguson, *J. Agric. Sci.,* 1948, **38**, 33.
3 Based on E. M. Crowther and F. Yates, *Emp. J. Expt. Agric.,* 1941, **9**, 77. For other British data, see D. J. Halliday, *Jealotts Hill Res. Sta., Bull.* 6, 1948.

and 10 to 15 kg of starch equivalent.[1] These nitrogen responses differ from those to potash and phosphate in being relatively independent of climate if the rainfall lies between 50–100 cm, but they are reduced in years of considerable drought or excessive rain.

TABLE 3.4   Increase in the amount of crop and food value obtained by using 1 kg of nitrogen per ha. Average values for North European conditions in kg/ha

Crop	Crop	Protein equivalent	Starch equivalent
Potatoes	70	0·4	13
Kale	140	1·9	13
Swedes	170	1·1	12
Mangolds	220	0·8	13
Sugar-beet roots	70	0·4 ⎱ 1·2	11 ⎱ 17
Sugar-beet tops	80	0·8 ⎰	6 ⎰
Cereal grain	15	1·1 ⎱ 1·3	11 ⎱ 14
Cereal straw	25	0·2 ⎰	3 ⎰
Meadow hay	25	1·1	8

Plants can take up their nitrogen either as ammonium or as nitrate ions, and most plants probably can use either equally easily. The main difference between these two ions is that all the nitrate in the soil is dissolved in the soil solution, whilst if the soil contains much clay or humus, much of the ammonium will be present as an exchangeable cation and hence not in solution. Perhaps for this reason a nitrate fertiliser is more rapid acting than an ammonium as it would be present in a higher ionic concentration round the plants' roots. But in most arable soils, added ammonium ions are rapidly oxidised to nitrate, so no matter what form of nitrogen is given, nitrate is the only form present in appreciable concentration in the soil solution for the plant to take up.

Crops that respond to nitrogen manuring commonly take up and fix in their mature tissue between one-third and one-half of the nitrogen added as fertiliser; the remainder is lost to the crop and usually to the soil, probably being either denitrified or washed out into the subsoil during wet weather, though its fate has not been too well determined (see p. 346).

Phosphorus

Phosphorus, as ortho-phosphate, plays a fundamental role in the very large number of enzymic reactions that depend on phosphorylation. Possibly for this reason it is a constituent of the cell nucleus and is essential for cell division and for the development of meristem tissue. Its concentration in these

1 40 kg of nitrogen, the amount contained in 45 m³ of air—the volume of a fair-sized living-room —as sulphate of ammonia can provide enough additional food to feed one person for a year.

tissues can be demonstrated very beautifully if some radioactive phosphorus ^{32}P is mixed with the main phosphorus nutrient supply. This radioactive phosphorus behaves exactly like ordinary phosphorus in the plant, except in so far as its radiations may harm the plant tissues if it is present in too high a concentration. These radiations, however, will affect a photographic plate so that this phosphorus can be made to register its position. A radio-auto-graph of the phosphorus in a young barley plant is reproduced in Plate 2,

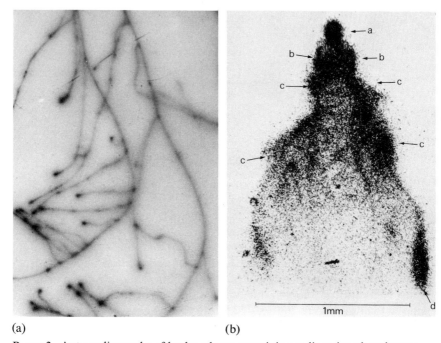

(a)   (b)

PLATE 2. Auto-radiographs of barley plants containing radioactive phosphorus:
(*a*) Showing the concentration of phosphorus in the tips of the growing roots (natural size).
(*b*) Showing the concentration of phosphorus in the apical meristem: (a) the leaf primordia; (b) the bases of leaf sheaves; (c) and (d) the initials of adventitious roots. ($\times 53$).

and it shows very clearly that the actively growing meristematic leaf and root cells contain far more phosphorus, in fact from several hundred to several thousand times more, than the cells that have ceased to divide.[1]

Phosphate deficiency is very widespread in the world, and in many countries such as Australia and South Africa crop production is limited over enormous areas by phosphate supply. In the British Isles phosphate deficiency can be very marked on many of the Jurassic clays of the English mid-

1 R. S. Russell and R. P. Martin, *Nature*, 1949, **163**, 71.

lands, on the acid Millstone grits of northern England and in large areas of Northern Ireland.[1]

Phosphate deficiency can be difficult to diagnose, and crops can be suffering from severe starvation without there being any obvious signs that lack of phosphate is the cause. By the time the deficiency has been recognised it may be too late to remedy it in annual crops. Thus wheat and barley take up much of their phosphate in the early stages of their growth, and starvation at this period cannot be rectified by a good supply later.[2]

Cereals suffering from phosphate starvation are retarded at every stage of their life-history, from the emergence of the second leaf to the time of ripening. They have a stunted root system and an even more stunted leaf and stem; the leaf colour is a dull greyish-green, a red pigment is often produced in the leaf bases and the dying leaves, and the tillering and the number of tillers bearing seed are depressed. On the other hand, except in extreme cases, the ratio of grain to straw is not affected. On soils badly deficient in phosphate, phosphatic fertilisers hasten the ripening processes, thus producing the same effect as a deficiency of water, but to a less extent. This ripening effect is well shown on the barley plots at Rothamsted: crops receiving phosphates are golden yellow in colour while those on the phosphate-starved plots are still green.

Certain indirect effects also follow: the ear of barley emerges from its ensheathing leaves a few days in advance of those receiving insufficient phosphate, and therefore has a better chance of escaping attack by the larvae of the gout fly (*Chlorops toeniopus* Meig.), which, hatching from their eggs on the top of the topmost leaf, crawl downwards seeking the ear for food.

Root crops suffering from severe phosphate shortage are also very stunted, and the effect of added phosphate can be spectacular. As a matter of history the early workers were so impressed with the great increase in the yield of roots obtained by phosphatic fertilisers that they assumed the phosphate had a specific action in encouraging root development, yet the permanent mangold field at Rothamsted (Barnfield) shows, in fact, that potassium increases the ratio of roots to tops far more than phosphate, and even the potassium effect is probably mainly a reflection of the early increase of leaf area.[3] Phosphate seems to increase leaf area without affecting the power of the leaves to transport carbohydrates to the roots, and it thus differs from nitrogen manuring, which also increases leaf growth but reduces their power of sending carbohydrates to the roots. Thus heavy nitrogen manuring of sugar-beet usually reduces the sugar content of the beet somewhat, although it may increase the amount of sugar produced per acre very considerably. All the experimental evidence available is in accord with this generalisation, and the

1 For an excellent description of cases occurring here, see Scott Robertson, *J. Min. Agric. Northern Ireland,* 1927, **1**, 7.
2 W. F. Gericke, *Bot. Gaz.,* 1925, **80**, 410; W. E. Brenchley, *Ann. Bot.,* 1929, **43**, 89.
3 D. J. Watson and E. J. Russell, *Emp. J. Expt. Agric.,* 1943, **11**, 49, 65.

statements made that crops respond especially well to dressings of phosphate on heavy soils is only a reflection of the low phosphate status of many English clay soils, and is not due to the power of phosphates to mitigate the effects of deficient aeration.

Excess of phosphate over the amount required by the crop sometimes depresses crop yield.[1] This usually occurs on light soils in dry years and has been attributed to the hastening of the maturation processes and consequent reduction of vegetative growth.

Plants take up their phosphorus almost exclusively as inorganic phosphate ions, probably principally as the $H_2PO_4^-$ ion,[2] for they may take this up more easily than the HPO_4^{2-}. Other phosphates besides ortho-phosphate act as phosphatic fertilisers, as for example meta- and pyro-phosphates and, though it has not yet been rigorously demonstrated, it is probable that these anions are hydrolysed to the ortho-phosphate before being absorbed.[3] Plants are, however, relatively inefficient users of phosphates in the field, for rarely more than 20 to 30 per cent of the amount supplied as fertiliser is taken up.

Potassium

Potassium is one of the essential elements in the nutrition of the plant, and one of the three that are commonly in sufficiently short supply in the soil to limit crop yield; hence it often needs to be added regularly in the fertiliser. Potassium differs from nitrogen and carbon, however, in not being a constituent of the plant fabric. It is important in the synthesis of amino acids and proteins from ammonium ions, for plants growing in solutions high in ammonium and low in potassium can have their tissues killed by the high concentration of the ammonium ions that accumulate in them under these conditions.[4] It is interesting to note that the only other element known that can replace potassium in this function is rubidium.[5] It is also probably important in the photosynthetic process, for potassium shortage in the leaf is commonly considered to lead to low rates of carbon dioxide assimilation,[6] though D. J. Watson[7] has been unable to confirm this on crops grown on the Rothamsted farm: it increased the leaf area and might put up the efficiency of assimilation slightly, but its effect was variable and always much smaller than that of nitrogen. But a discussion on the general effects of

1 For instances, see E. J. Russell, *J. Inst. Brew.*, 1923, **29**, 631 (barley); J. C. Wallace, *J. Min. Agric.*, 1926, **32**, 893; *Rothamsted Conf. Repts.*, No. 16, 1934 (potatoes).
2 C. A. Hagen and H. T. Hopkins, *Plant Physiol.*, 1955, **30**, 193.
3 W. T. McGeorge, *Arizona Agric. Expt. Sta., Tech. Bull.* 82, 1939.
4 T. W. Turtschin, *Ztschr. Pflanz. Düng.*, 1934, A34, 343; M. E. Wall, *Soil Sci.*, 1940, **49**, 393.
5 F. J. Richards, *Ann. Bot.*, 1941, **5**, 263, who used barley plants.
6 F. G. Gregory with F. J. Richards, *Ann. Bot.*, 1929, **43**, 119; with E. C. D. Baptiste, ibid., 1936, **50**, 579; O. Eckstein, *Plant Physiol.*, 1939, **14**, 113.
7 *Ann. Bot.*, 1947, **11**, 375.

potassium deficiency is complicated by the fact that they depend so much on the relative concentration of other elements, particularly sodium and calcium in the plant tissues.[1]

In the field the potassium supply in the soil may be adequate for crops growing under conditions of a low nitrogen and phosphorus supply, but become inadequate if these are increased. Hence, signs of potassium starvation are often seen when only nitrogenous and phosphatic fertilisers are given to a crop; and the most characteristic sign is the premature death of the leaves.

When nitrogen and potassium are simultaneously in short supply, the plants are stunted, their leaves are small and rather ashy-grey in colour, dying prematurely, first at the tips and then along the outer edges, and the fruit and seed is small in quantity, size and weight. These effects are general, and are seen on all soils, but best on light sandy or chalky soils and on certain peaty soils; and it is on these soils that potassic fertilisers are most likely to act on all crops. With large supplies of nitrogen relative to potassium, on the other hand, the leaves are large but relatively inefficient photosynthesisers; hence an abnormal concentration of nitrogen compounds compared to carbohydrate occurs in the leaf, leading to various undesirable effects, such as the greater liability of these leaves to fungus and bacterial diseases, and to a reduced resistance against damage by drought compared with those receiving more adequate dressings of a potassium fertiliser. Thus potassium acts as a corrective to the harmful effects of nitrogen, and is therefore often required for crops receiving high levels of nitrogen manures.

TABLE 3.5   Action of potassium salts on the yield of mangolds. Barnfield, Rothamsted, 50 years, 1876–1928.[2] Yields in tons per hectare

	Roots			*Leaves*		
Nitrogen supplied in manure, kg/ha	none	86	184	none	86	184
Series[3]	O	A	AC	O	A	AC
No potassium salts[4]	11·2	16·8	23·9	2·64	6·56	8·27
With potassium sulphate, 560 kg/ha	10·1	33·8	56·7	2·34	7·06	13·1

An adequate supply of potassium in the leaf is probably essential for the photosynthetic process to go on efficiently, and results, such as those given in Table 3.5, have been used to demonstrate the truth of this statement. This table gives the weight of mangold tops and roots at harvest obtained on certain plots at Rothamsted, some of which receive potassium, but all of which

1 For a discussion of these effects, see F. G. Gregory, *Ann. Rev. Biochem.*, 1937, **6**, 557.
2 Excluding 1885 when nitrogenous fertilisers were not applied owing to poverty of crop, and 1908 and 1927 when the crop failed.
3 Series A as 450 kg ammonium salts to 1915; 465 kg sulphate of ammonia since. Series AC as Series A, but in addition 2250 kg rape cake.
4 Series 5 and 6 respectively. Each receives 420 kg of superphosphate each year.

receive phosphate. It shows that with moderate dressings of nitrogen, given as sulphate of ammonia, potassium doubles the weight of roots, but has only a small effect on the yield of tops. One cannot argue, however, that this increase was necessarily due to the greater efficiency of the leaves, for D. J. Watson[1] has shown that the potassium increased the size of the leaves in the early part of the growing season, though this effect had disappeared by harvest, and that this initial increase was sufficient to account for the differences in the yield of roots without having to assume any effect of potassium on the efficiency of photosynthesis.

A result entirely similar to mangolds is found with sugar-beet. The results of a large series of manurial experiments carried out over the sugar-beet areas of Great Britain are given in Table 3.6.[2] Nitrogen alone depresses the sugar content, potassium increases it and phosphate has no effect, though these experiments, like that in Barnfield, cannot furnish the information required for their correct physiological interpretation.

TABLE 3.6   The effect of fertilisers on the yield of sugar-beet. Great Britain, 1934–48 (359 experiments). Yields and responses in tons per hectare

	Mean yield	*Sulphate of ammonia* 500 *kg/ha*	*Response to Superphosphate* 750 *kg/ha*	*Muriate of potash* 310 *kg/ha*
Roots	28·2	3·0	1·0	1·2
Sugar, per cent	17·3	−0·4	0·0	0·2
Sugar	4·87	0·44	0·19	0·26
Tops	23·9	6·8	0·8	0·8

Crops differ greatly in their responsiveness to potassium. Many fruit trees —apples, gooseberries, red currants—need ample supplies of potassium for good cropping; beans and potatoes among British field crops, and tomatoes among the glasshouse crops, are all very responsive. Leguminous pasture plants—clovers and lucerne—also seem to need adequate supplies of potassium, particularly if they are to compete successfully with grasses;[3] and for lucerne in addition potassium increases its winter hardiness, possibly because it encourages the plant to store more carbohydrate and protein in its root system.[4]

Excess potassium in the soil, as brought about by too high a level of potassium manuring, for example, will reduce very considerably the amount

1 *Ann. Bot.,* 1947, **11**, 375.
2 E. M. Crowther, *Roth. Expt. Sta. Rept.,* 1939–45, revised to include later data.
3 For the Rothamsted experiments on this point, see J. B. Lawes and J. H. Gilbert, *Phil. Trans.,* 1900, **192** B, 156; W. E. Brenchley, *Manuring of Grassland for Hay*, London, 1924.
4 L. F. Graber *et al., Wisconsin Agric. Expt. Sta., Res. Bull.* 80, 1927.

of other cations the crop can take up (see p. 28), and this may lead to crop growth being badly upset by these induced deficiencies of other cations.

Other elements needed in moderate quantities

Calcium appears to be essential for the growth of meristems and particularly for the proper growth and functioning of root tips. It is also present as calcium pectate, which is a constituent of the middle lamellae of the cell walls, and possibly for this reason it tends to accumulate in the leaf. Calcium deficiency appears to have two effects on the plant: it causes a stunting of the root system and it gives a fairly characteristic appearance to the leaf. Calcium deficiency also may have an indirect effect on the plant by allowing other substances to accumulate in the tissues so much that they may either lower the vigour or actually harm the plant. Thus a good calcium supply helps to neutralise the undesirable effects of an unbalanced distribution in the soil of nutrients and other compounds that can be taken up by the plant.

Magnesium is needed by all green plants as it is a constituent of chlorophyll. It also seems to play an important role in the transport of phosphate in the plant, and possibly as a consequence of this it accumulates in the seeds of plants rich in oil, for the oil is also accompanied by an accumulation of lecithin, a phosphate-containing fat.[1] Thus the phosphate content of a crop can sometimes be increased to a higher level by adding a magnesium rather than a phosphatic fertiliser, and it is for this reason that magnesium silicates, such as finely ground serpentine[2] or olivine,[3] are sometimes added to superphosphate to increase its effectiveness.

Sodium does not seem to be an essential element for any crop, even for salt marsh plants, yet certain crops undoubtedly grow better in the presence of available sodium supplies than in their absence, the sodium in these cases appearing to carry out some of the functions that potassium usually fulfils.

Crops can be divided into four groups with respect to their relative needs of sodium compared to potassium: some need sodium for optimum growth, some benefit if available sodium is present, some can tolerate part of their potassium supply being replaced by sodium and some can make no use of sodium even if the potassium supply is restricted. Table 3.7[4] shows the groups into which various agricultural crops are believed to fall, though the grouping may depend somewhat on climatic conditions.

The role of sodium in the nutrition of plants that need this element for optimum growth is not fully known, though one of its effects is to increase the succulence of the plant, that is, the amount of water held by unit dry weight of leaf tissue. This may be the reason why it appears to increase the

1 For a symposium on the role of magnesium in plant nutrition, see *Soil Sci.*, 1947, **63**, 1–28.
2 See, for example, H. O. Askew, *New Zealand J. Sci. Tech.*, 1942, **24** B, 79, 128.
3 D. U. Druzhinin, *Ztschr. Pflanz. Düng.*, 1936, **45**, 303.
4 Taken from P. M. Harmer and E. J. Benne, *Soil Sci.*, 1945, **60**, 137.

TABLE 3.7   Effect of sodium applied as a nutrient on several crops

Degree of benefit in deficiency of potassium		*Degree of benefit in sufficiency of potassium*	
None to very slight	*Slight to medium*	*Slight to medium*	*Large*
Buckwheat	Barley	Cabbage	Celery
Lettuce	Broccoli	Kale	Mangold
Maize	Brussels sprouts	Kohlrabi	Sugar-beet
Potato	Carrot	Mustard	Swiss chard
Rye	Cotton	Radish	Table beet
Soyabean	Millet	Rape	Turnip
Spinach	Oat		
Strawberry	Pea		
Sunflowers	Tomato		
White bean	Wheat		

drought resistance of these plants. It also increases the leaf area of sugar-beet.[1] F. J. Richards[2] suggests another role of sodium in helping crops such as barley to grow in a potassium-deficient soil is that it prevents an accumulation of other cations that may be toxic to the plant, for a deficiency of one cation leads to an accumulation of others.

Sulphur is an essential constituent of many proteins, and it is also a constituent of the oils produced by certain plants such as the mustard oil of the Brassicae. Organic sulphates also appear to be essential constituents in the plant. A lack of sulphur frequently shows in a yellowing of the leaf, and it was in fact first recognised in a trouble known as 'tea yellows' in Malawi, but some of the most striking responses have been with leguminous crops such as lucerne, groundnuts and clovers.

Silicon is another element that is probably not essential for the growth of most plants, though the general vigour of many grasses and cereals, and particularly rice, is decreased if grown in soils low in available silicon. Its role in the nutrition of these crops is not yet fully understood. It is however an important element in the ash of many graminaceous plants probably because their roots have no mechanism for filtering out the silicic acid dissolved in the soil water.

Chlorides are taken up by most plants from the soil solution, and there is no evidence that any soils are so deficient in chlorides for loss of crop to occur for lack of it, though chlorides are needed in small quantities by most crops. Some crops, however, often respond to additional dressings of chloride, for example, barley, lucerne and tobacco. Thus small dressings of chlorides appear to increase the yield and improve the texture of the tobacco leaf

1 J. Lehr, *Soil Sci.*, 1942, **53**, 399.
2 *Ann. Rev. Biochem.*, 1944, **13**, 611.

without affecting its burning qualities, though larger dressings, which still increase the yield, lower this quality.[1]

There is no strong evidence that chloride has any specific effect on plant growth, though it may sometimes hasten maturity. Its main function, for which, however, it is not specific, is as an osmotic pressure regulator and a cation balancer in the cell sap and in the plant cells themselves.

Trace elements in plant nutrition

Plants need very small quantities of certain elements—the so-called trace or minor elements—for their nutrition, and these include iron, manganese, zinc, copper and boron, whilst molybdenum and cobalt are beneficial under some circumstances. These elements are called trace elements, because only very small quantities, ranging from a few grams to a few kilograms per hectare, are usually needed by the crop; and in fact C. S. Piper and A. Walkley[2] estimate that a full crop of oats only removes about 20 g of copper, 100 g of zinc and 500 g of manganese, compared with 8 kg of phosphorus per hectare. But it must be realised that there is no sharp distinction between elements needed in large and small quantities, magnesium and sulphur being two good examples of intermediate elements for many crops. The literature on trace elements is now very extensive, and the reader should refer to recent books and reviews for more detailed discussion on their role in plant nutrition.

A shortage of one or more of these elements usually, but not always, affects the appearance of the plant, giving the leaves a chlorotic, bronzed or mottled colour, or altering its habit of growth, or causing the death of the growing points, so giving the plant a rosette appearance. Plants suffering from trace-element deficiencies need not show any symptoms of the deficiency at all—except in so far as growth is not as good as it might be, or they may only display symptoms for a short period in the growing season. Sometimes the symptoms are sufficiently characteristic for the deficiency to be diagnosed visually,[3] but often they are so indefinite, or even suppressed altogether, that they can only be diagnosed by chemical tests, either by analysing the minerals in the leaf tissue or by applying small quantities of any element suspected of being deficient to selected leaves or shoots of the plant.

On many soils, plants suffer from a deficiency of a number of elements simultaneously, though they may only show the symptoms of one, or even of none of them. An observation of W. A. Roach[4] illustrates the complexities arising in such a soil. Potatoes growing on a marsh soil in Kent showed symptoms of manganese deficiency which were cured by spraying with manganese,

1 *J. Amer. Soc. Agron.*, 1929, **21**, 113.
2 *Aust. J. Counc. Sci. Indust. Res.*, 1943, **16**, 217.
3 For colour reproduction of some of these symptoms, see *Hunger Signs in Crops*, ed. G. Hambidge, Washington, 1941; T. Wallace, *The Diagnosis of Mineral Deficiencies in Plants*, London, 1943, and *Supplement*, 1944.
4 *E. Malling Res. Sta. Ann. Rept.*, 1945, **29**, 83.

though this did not affect the yield of the crop. The crop was, however, also suffering from zinc deficiency although no symptoms whatever of this were visible; spraying with zinc improved the yield slightly, but spraying with zinc and manganese gave a very considerable increase in yield and obviously rectified the main cause of the low yields. Roach considers that the yields of many of our orchards are unnecessarily low because of a combined deficiency of several of these elements, none of which is severe enough to cause any visible symptoms. Again, plants can have their vitality lowered by shortages of trace elements without any symptoms of deficiency showing.

Trace elements have another characteristic; namely, that they are normally all very poisonous if present in the soil in an available form in more than very small amounts. Hence, it is very dangerous to apply them indiscriminately at more than very light dressings; though if concealed trace-element deficiencies are found to be widespread it will obviously become important to discover just how high a concentration of a mixture of them can be applied indiscriminately without harming the crop.

Trace elements in animal nutrition

Animals have not quite the same trace-element needs as plants.[1] Thus all animals require iodine, though iodine-deficiency diseases have rarely been reported among livestock. They need iron, manganese, copper and zinc, as do plants, but do not seem to require boron. Cattle and sheep, however, need cobalt, an element probably only required in very small amounts by legumes if they are to have active nodules; and some young animals, including man, need very small quantities of selenium[2] and chromium,[3] elements not required by plants. Normally, if pastures cannot adequately supply the trace element needs of the grazing animals, these can be met either by giving the animals licks or drenches containing the missing trace element, or by mixing these in with the main fertilisers, usually with the superphosphate given to the pastures.

The difference between the trace-element requirements of plants and animals can be seen in the effects of different hill grazings on the thriftiness of the sheep they carry. Thus, it is probable that sheep do better on some of the heather grazings than on the improved leys because the heather herbage is richer in trace elements.[4] The substitution of improved leys for the unimproved heather may require either the sheep or the land being given additional amounts of the more important trace elements, if any benefit is to be reaped from the increased amount of herbage produced.

1 For a review, see F. C. Russell, *Imp. Bur. Anim. Nutr., Tech. Comm.* 15, 144.
2 K. Schwarz and C. M. Folz, *J. Amer. Chem. Soc.,* 1957, **79**, 3292.
3 H. A. Schroeder, *J. Nutr.,* 1968, **94**, 475; 1969, **97**, 237; W. Mertz, *Physiol. Rev.,* 1969, **49**, 163.
4 B. Thomas *et al., Emp. J. Expt. Agric.,* 1945, **13**, 93.

The two main trace-element deficiency diseases in animals are due to copper and cobalt. Copper deficiency can occur in cattle grazing on copper-deficient pastures; but it can also occur on pastures apparently adequately supplied with copper for reasons which are not fully known, although in part the copper in these forages appears to be unavailable to the animal, and in part to be induced by a high sulphate or molybdenum content.[1] Thus teartness in pastures appears to be a molybdenum-induced copper deficiency, for the pastures giving this trouble have a herbage abnormally high in molybdenum. These troubles are ameliorated by feeding the animals extra copper.

Cobalt is needed by ruminants in much larger quantities than by non-ruminants, probably to help with the fermentation of food in the rumen, for the cobalt must be given by the mouth or as a pellet in the rumen—it is no use injecting it into their bodies. Many poor pastures all over the world are now known to have too low a cobalt content for sheep or cattle to thrive on them. In Great Britain these are mainly found on some hill pastures, and the sheep suffer from a disease known as 'pine'. Dressings of $\frac{1}{2}$ to 2 kg/ha of a cobalt salt are applied to the pastures and the effect lasts about three years. In New Zealand, where the soils are very deficient in phosphate also, it is usually applied mixed with superphosphate.

Animals can suffer from other deficiency diseases besides those due to trace elements. Thus, in parts of South Africa the herbage has such a low content of phosphate that the animals grazing on it suffer from acute phosphate starvation. Again, sheep need a fairly high content of available sulphur in their herbage to produce wool, as wool contains a high proportion of cystine, a sulphur-containing amino acid. The spring flush of good pastures on soils low in available sulphur can give a fodder too low in sulphur for the sheep, with a consequence that the quality of the wool falls—the wool becomes brittle and breaks easily.

Pasture plants can, however, contain sufficiently high concentrations of some elements in their tissues to be definitely harmful to animals feeding on them. The best known example is probably selenium in the United States. Prairie and range plants growing on seleniferous soils are not affected by the selenium, and some species can accumulate very large amounts; but animals feeding on these prairies may suffer severe selenium poisoning. Similarly, pasture grasses and clovers on some of the heavy Lower Lias clays of southern England may accumulate concentrations of molybdenum[2] that can seriously harm calves and young cattle grazing them. These are the teart pastures already mentioned in the paragraph dealing with copper-deficiency diseases.

The relation between the minerals composing the soil and the health of livestock grazing on the pasture or range can thus be very close. Thus the cobalt content of a rock tends to follow the magnesium content—their ions

1 See, for example, C. F. Mills, *J. Sci. Food Agric.*, 1957, **8**, S 88.
2 W. S. Ferguson, A. H. Lewis and S. J. Watson, *J. Agric. Sci.*, 1943, **33**, 44.

are of a similar size, hence hill grazings on granitic rocks, low in magnesium, tend to induce pine in sheep, while those on rocks well supplied with serpentine will be healthy.[1] Again, since some plants are much better extractors of these elements than others, the botanical composition of the herbage,[2] and because of this the way the grazing is managed, can affect the health of the stock markedly.

The effect of the level of trace elements in a soil on the nutritive value of the food grown on that soil for human nutrition is of minor importance for all communities which draw their food supplies from a wide area, but the more the community is dependent on the soils in a restricted locality, the greater the likelihood that any large trace element imbalance in the soil may have an effect on its health. Very few examples are available of such effects, but an interesting example may be furnished in the neighbouring Hastings and Napier areas in New Zealand. Dental caries is much more prevalent in Hastings than in Napier children, and T. G. Ludwig[3] has produced evidence that this greater resistance to decay is associated with the higher molybdenum content of the Napier vegetables, which, in turn, is a reflection of the higher level of available molybdenum in the parent marine sediments the Napier soils are derived from.

1 For examples, see E. B. Kidson, *J. Soc. Chem. Indust.*, 1938, **57**, 95; R. L. Mitchell, *Proc. Nutr. Soc.,* 1944, **1**, 183.
2 For examples of the power of plants to concentrate rare earths, see W. O. Robinson, *Soil Sci.,* 1943, **56**, 1.
3 With W. B. Healy and R. S. Malthus, *Int. Soil Conf. (N.Z.)*, 1962, 895.

4

Quantitative studies on plant growth

Attempts have been made in the past to find mathematical expressions for the relationship between the quantity or concentration of plant nutrients present and the growth of the crop, and in particular its yield. Two types of solution of this problem are possible:

1 A hypothesis is set up which seems to fit the facts; it is expressed by an equation and this is then applied to the experimental data.
2 The experimental data are studied by statistical methods and an empirical equation or regression formula is fitted thereto, with no assumption or hypothesis as to the underlying causes.

Owing to the complexity of the problem—plant growth is dependent on so many factors—no general solution can be expected by either method. The first procedure has, in fact, been far more widely followed, and has been of more practical utility than the second.

A number of different hypotheses have been put forward on the relationship between the amount of plant nutrient or other factors affecting plant growth and the growth or yield of the plant. One of the earliest was due to Liebig, who expressed it as the Law of the Minimum: the amount of plant

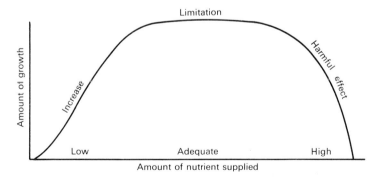

FIG. 4.1. General relation between any particular nutrient or growth factor and the amount of growth made by the plant

growth is regulated by the factor present in minimum amount and rises or falls according as this is increased or decreased in amount. The relation between plant growth and the amount of the limiting nutrient present in the soil can, therefore, be represented as a curve such as is shown in Fig. 4.1: growth increases with additions of the limiting factor until it ceases to be limiting, then plant growth becomes independent of this factor as it increases still more until a point is reached when it is becoming toxic and causes plant growth to decrease. This law unfortunately has only a very limited validity, for if several factors are low, but none too low, increasing any one will increase the yield, as will be shown in the section on the interaction of nutrients.

The relation between growth and nutrient supply

The first good experimental investigation was made by Hellriegel in the eighties of the last century. Barley was grown in pots of sand, all necessary factors were amply provided, excepting only one nutrient salt, the amount of which varied in the different pots. Table 4.1 gives an example of his results: it shows the effect of increasing the nitrogen supply on the total dry matter produced and on other aspects of growth, and Fig. 4.2 shows the dry matter

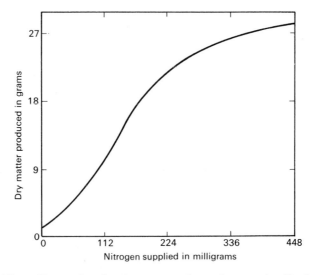

FIG. 4.2. Effect of increasing the nitrogen supply on the growth of barley

produced plotted against the amount of nitrogen supplied. The first increment of nitrogen produces a certain increase in yield, but the second and third increments produce proportionately more, thus giving a greater return than is expected if, as Liebig assumed, the effect be simply proportional to the amount present. The fourth and fifth increments, however, produce less effect. The curve, therefore, resembles an S and is described as sigmoid; this is a common shape for curves showing total growth made after the lapse of a definite period of time, though perhaps less common for curves relating to environmental factors.

TABLE 4.1   Effect of increasing the nitrogen supply on the growth of barley in sand cultures: H. Hellriegel and H. Wilfarth.[1] Nitrogen given as calcium nitrate

Mg of nitrogen supplied	—	56	112	168	280	420
Dry matter in crop, g	0·74	4·86	10·80	17·53	21·29	28·73
Increased yield for each extra 56 mg nitrogen	—	4·12	5·94	6·73	1·88	2·97
Grain, per cent of dry matter in crop	11·9	37·9	38·0	42·6	38·6	43·4
Weight of one grain, mg	19	30	33	32	21	30

Field experiments can lead to similar results if the factor under consideration is at a sufficiently low level. The effect of increasing dressings of sulphate of ammonia on the yield of wheat on the permanent wheat field at Rothamsted is given in Table 4.2, both for the twenty-year periods, 1852–71 and

TABLE 4.2   Response of wheat to sulphate of ammonia, Broadbalk field[2]

	Grain yield				*Straw yield*			
Nitrogen supplied annually in manure, kg/ha	0	48	96	144	0	48	96	144
Period, 1852–71								
Yield in t/ha	1·19	1·87	2·50	2·72	1·87	3·08	4·44	5·18
Increase for each 48 kg of nitrogen	—	0·68	0·63	0·22	—	1·21	1·36	0·74
Period, 1902–21								
Yield in t/ha	0·88	1·39	2·03	2·40	1·32	2·38	3·84	4·95
Increase for each 48 kg of nitrogen	—	0·51	0·64	0·37	—	1·06	1·46	1·11

1902–21. In both cases the second dose of nitrogen gives the largest response in the straw, and the same is true for the grain response in the latter period, but in the first twenty years, before the fertility of the unmanured had fallen very much, the response to the first dressing was rather larger than to the second. Thus the sigmoid curve does not always occur in field experiments, but perhaps only when the factor considered is at a very low level.

1 *Ztschr. Rübenzucker-Ind.*, Beilageheft, 1888.
2 E. J. Russell and D. J. Watson, *Imp. Bur. Soil Sci., Tech. Comm.* 40, 1940. The plot receiving no nitrogen has received none since 1839 at the latest.

The assumed relation between growth and nutrient supply

The smoothness of the curves found by experiment suggests that they can be expressed by a mathematical equation. E. A. Mitscherlich was among the first to do this, and his equation is certainly the best known and most widely used. He assumed that a plant or crop should produce a certain maximum yield if all conditions were ideal, but in so far as any essential factor is deficient there is a corresponding shortage in the yield. Further, he assumed that the increase of crop produced by unit increment of the lacking factor is proportional to the decrement from the maximum, or expressed mathematically:

$$\frac{dy}{dx} = (A - y)C,$$

where y is the yield obtained when x is the amount of the factor present, A is the maximum yield obtainable if the factor was present in excess, this being calculated from the equation, and C is a constant. On integration, and assuming that $y = 0$ when $x = 0$,

$$y = A(1 - e^{-Cx}).$$

This curve is not sigmoid in shape, but everywhere concave to the axis representing the nutrient supply. Mitscherlich's experiments were made with plants grown in sand cultures supplied with excess of all nutrients excepting the one under investigation. Table 4.3 shows the results obtained with oats

TABLE 4.3   Yield of oats with different dressings of phosphates. Mitscherlich*

P_2O_5 in manure	Dry matter produced	Crop calculated from formula	Difference	Difference expressed in terms of probable error†
Grams	Grams	Grams	Grams	
0·00	9·8 ± 0·50	9·80	—	—
0·05	19·3 ± 0·52	18·91	−0·39	−0·8
0·10	27·2 ± 2·00	26·64	−0·56	−0·3
0·20	41·0 ± 0·85	38·63	−2·37	−2·8
0·30	43·9 ± 1·12	47·12	+3·22	+2·9
0·50	54·9 ± 3·66	57·39	+2·49	+0·7
2·00	61·0 ± 2·24	67·64	+6·64	+3·0

* *Landw. Jahrb.,* 1909, **38**, 537.
† If this figure is less than 3, the agreement is considered satisfactory.

and monocalcium phosphate. Mitscherlich claimed to show by experiment that the proportionality factor C (called *Wirkungswert*, or *faktor* in Mitscherlich's papers) is a constant for each fertiliser, independent of the crop, the soil or other conditions. If this were so an experimenter knowing its value could, from a single field trial, predict the yields obtainable from any given quantities of the fertiliser, a result of great practical value. Further, it would be possible to estimate by direct pot experiment the amount of available plant food in a soil, one of the most difficult of all soil problems.

Mitscherlich has, indeed, used his formula for this purpose,[1] and in his very interesting book[2] he applies the expression in a variety of ways. Some work of E. M. Crowther and F. Yates[3] furnishes a later example of its use. By its aid they could put together in convenient tables all the results of fertiliser experiments that have been made on the various crops in Great Britain; and from these they formulated a suitable national war-time fertiliser policy for the country.

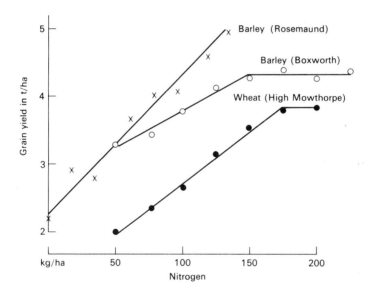

FIG. 4.3. Responses of wheat and barley to nitrogen, on soils low in nitrogen

Mitscherlich's work was extraordinarily stimulating and caused a veritable flood of controversy when it was first developed. His equation has been of great practical value though it is certainly not exact. Thus, the *Wirkungswert* for a particular nutrient is not a constant but depends somewhat on the other conditions of growth.[4] Further, the response curve is often sigmoid, fertiliser in excess decreases the crop yield, and the calculated maximum yield of the crop is sometimes far in excess of anything that can be obtained.

1 *Landw. Jahrb.*, 1923, **58**, 601.
2 *Bodenkunde für Land- und Forstwirte*, Berlin, 1913, and subsequent editions.
3 *Emp. J. Expt. Agric.*, 1941, **9**, 77. For another example, see O. W. Willcox, *Soil Sci.*, 1955, **79**, 467; **80**, 175; 1956, **81**, 57; **82**, 287.
4 See, for example, E. R. Bullen and W. J. Lessels, *J. agric. Sci.*, 1957, **49**, 319.

Other workers[1] have tried to improve on Mitscherlich's equation, and have had some success. But the interest in more exact equations has died down partly because response curves cannot be determined accurately from field experiments because of inherent soil and crop variability and partly

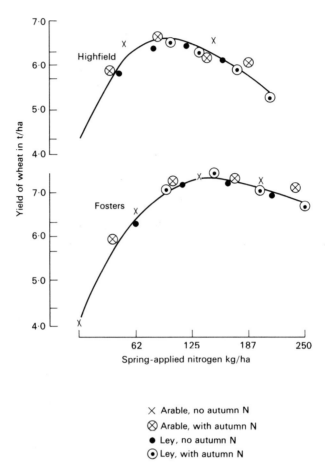

Fig. 4.4. Response of wheat to nitrogen on soils high in nitrogen

because the response of the crop to the fertiliser depends on so many extraneous factors. D. A. Boyd[2] has in fact shown that for practical purposes the response curve can often be broken down into two parts—a linear response to increasing levels up to a certain point followed by a compara-

1 For a recent review see G. W. Cooke, *Fertilizing for Maximum Yield*, Crosby Lockwood, 1972, and for an earlier discussion on the variability of the Mitscherlich *C faktor*, D. A. Boyd, *J. Sci. Fd. Agric.*, 1961, **12**, 493.
2 *Proc. 9th Congr. Int. Potash Inst.*, 1971, 461. For examples with the nitrogen response of sugar-beet see D. A. Boyd, P. B. H. Tinker *et al.*, *J. agric. Sci.*, 1970, **74**, 37.

tively sharp break after which there is either no effect on yield of additional fertiliser or only a small increase or sometimes a small decrease in yield. This is illustrated in Fig. 4.3, which shows the response of wheat and barley to nitrogen on three fields of low nitrogen status at three experimental husbandry farms, and in Fig. 4.4, for the response of winter wheat on two fields of high nitrogen status at Rothamsted. Boyd also showed that the response of potatoes to phosphate, on a series of soils low in phosphate, behaved in the same way as the cereals to nitrogen on the low-nitrogen soils. These figures justify the policy of many farmers in intensive agriculture of using more fertiliser than is strictly necessary, for this ensures maximum yields, since this yield is not very dependent on the exact level of fertiliser given; but they also show that if fertilisers are in short supply, it is more economic to use small dressings over as wide an area as possible rather than large dressings over a restricted area.

These curves illustrate another point of great practical importance. Nitrogen fertilisers should give a fairly predictable response in yield if lack of nitrogen is the principal factor limiting yield. If an experiment with nitrogen fertilisers is made on a crop giving rather a poor yield, and the nitrogen response is appreciably less than the expected response, this proves that nitrogen is either not limiting yield in this particular experiment, or that if it is, other factors limiting yield are also operative.

The interaction between nutrients

The Law of the Minimum, namely, that the amount of plant growth is regulated by the factor present in minimum amount, and rises or falls according as this is increased or decreased in amount, implies that if two factors are limiting, or nearly limiting growth, adding only one of them will have little effect on growth, whilst adding both together will have a very considerable effect. Two such factors are said to have a large positive interaction in such circumstances, for the response of the crop to both together is larger than the sum of the responses to each separately. If the crop response to the two factors together equalled the sum of its responses to each separately, we would say the two factors showed no interaction, or worked entirely independently of each other; and if the response to the two factors together was less than the sum of the responses to each factor separately, they are said to have a negative interaction with each other.

Two examples can serve as illustrations of such joint effects. The first is taken from a series of fertiliser experiments made at Rothamsted between 1938 and 1945, in which the response of potatoes to farmyard manure, sulphate of ammonia, muriate of potash and superphosphate is determined in all possible combinations of these fertilisers. Table 4.4 shows some of these responses for the three years 1939, 1943 and 1945, in which the crop only gave a low yield in the absence of fertiliser and manure.

TABLE 4.4   Response of potatoes to nitrogen and potassium.* Yield of potatoes in tons per hectare

Fertiliser	*No N No K*	*N, No K*	*No N, K*	*N, K*
No farmyard manure	10·0	10·9	16·1	20·6
With farmyard manure	23·1	27·6	23·6	28·5
Response to farmyard manure	13·1	16·7	7·5	7·9

	Response to nitrogen		*Response to potassium*		*N K inter-action*
	K absent	*K present*	*N absent*	*N present*	
No farmyard manure	0·9	4·5	6·1	9·7	3·6
With farmyard manure	4·5	4·9	0·5	0·9	0·4

	In presence of adequate potassium			
Response to	*FYM alone*	*N alone*	*FYM + N*	*Interaction*
	7·5	4·5	12·4	0·4

	In presence of adequate nitrogen			
Response to	*FYM alone*	*K alone*	*FYM + K*	*Interaction*
	16·7	9·7	17·6	−8·8

* Dressings given:

	1939	1943 and 1945
FYM	38	Mean of 20 and 40 tons per hectare.
N	100	75 kg/ha N, given as sulphate of ammonia.
K	170	105 kg/ha K, given as muriate of potash.

This table shows that in the absence of farmyard manure, potatoes give a bigger response to nitrogen if the potassium supply is adequate than if it is short, and, similarly, a bigger response to potassium if the nitrogen supply is adequate. Thus, in the example given there is a positive interaction between these two fertilisers: nitrogen in the absence of potassium increases the yield by 0·9 ton, potassium in the absence of nitrogen increases the yield by 6·1 tons, and if given together the yield is increased by 10·6 tons, which is 3·6 tons larger than the sum of the nitrogen alone and potassium alone responses (0·9 + 6·1 = 7·0 tons). On the other hand, provided adequate potassium was given, nitrogen gave an increase of between 4·5 and 4·9 tons whether farmyard manure was given or not, and farmyard manure gave an increase of between 7·5 and 7·9 tons whether nitrogen was given or not. These two materials thus act independently of each other, so that there is no interaction between them.[1] Potassium and farmyard manure provide an example of a negative interaction: in the presence of adequate nitrogen, potassium alone increases the yield by 9·7 tons, and farmyard manure alone by 16·7 tons, but the two together only increased it by 17·6 tons. Thus the response to potassium in the presence of the farmyard manure was only 0·9 ton, and the negative interaction between the two is 8·8 tons; in fact, in these experiments farmyard manure supplies all the potassium needed by the potatoes.

The second example illustrates the relationship between the water supply and the responsiveness of the crop to fertiliser, and in particular to nitrogen. Table 4.5 gives the result of a nitrogen-irrigation experiment in the Sudan Gezirah[2] and shows that cotton can only make efficient use of a nitrogen fertiliser if given an adequate water supply.

TABLE 4.5   Influence of water and nitrogen supply on the growth of cotton. Yield of seed cotton in kg/ha

Amount of nitrogen added as sulphate of ammonia kg/ha	Rate of watering		
	Light	*Medium*	*Heavy*
None	465	520	533
65	666	825	945
130	769	1023	1277

There is a further aspect of the Law of the Minimum that may be extremely important when one is rectifying deficiencies. If crop growth is being severely limited by a few factors, and if, as is usual for most crops, one is interested in the yield of a part of the crop only, such as the grain for example, it will often happen that the maximum yield of this part of the crop is obtained by allowing the individual plants a large amount of space, so the root system of each plant can tap as large a volume of soil as possible. If one now rectifies the limiting deficiencies, it may also be necessary to increase the number of plants per hectare to get the maximum possible increase in yield from this amelioration, for the individual plants will now need less volume of soil from which to draw their nutrients. Further, it may also be necessary to change the variety of the crop grown, for it often happens that the variety best adapted to poor conditions has not the potentiality to make really good use of a liberal supply of nutrients, whilst varieties that can make very good growth in optimum conditions may be relatively less satisfactory when growing in

1 For other examples of this, see E. M. Crowther and F. Yates, *Emp. J. Expt. Agric.,* 1941, **9**, 77; A. W. Oldershaw and H. V. Garner, *J. Roy. Agric. Soc. Eng.,* 1944, **105**, 98. For a possible explanation of this effect, see D. A. Boyd, *J. agric. Sci.,* 1959, **52**, 384.
2 F. G. Gregory, F. Crowther and A. R. Lambert, *J. agric. Sci.,* 1932, **22**, 617. See also F. Crowther, *Roy. Agric. Soc. Egypt, Tech. Bull.* 24, 1936.

poor conditions.[1] Table 4.6, taken from some work of B. A. Krantz[2] on the manuring of maize in North Carolina, illustrates the need for increasing the number of plants per hectare if one is to get the maximum benefit from a high level of nitrogen manuring. Incidentally, this closer spacing has the desirable consequence that it helps to suppress weed growth which otherwise becomes troublesome when generous fertiliser dressings are used with wide spacing.

TABLE 4.6   The influence of the number of maize plants per hectare on their response to a nitrogen fertiliser. Yield of grain in tons per hectare

Number of plants thousands/ha	Kg of nitrogen per hectare			
	22	79	135	190
10	2·4	3·4	3·7	3·6
17	2·4	3·6	4·2	4·1
25	2·6	4·1	4·3	4·5
32	2·3	3·7	4·6	4·6

Least significant difference 0·5 t/ha.
Plant numbers obtained by using spacings of 95 cm, 54 cm, 38 cm, 29 cm in rows 107 cm apart.

An important group of interactions between nutrients concerns those between the different cations in the soil and in the plant tissue. It has already been stated (see p. 28) that the leaves of plants tend to have a fairly constant total cation concentration in their dry matter, dependent mainly on the type of plant and to a much less extent on the soil and manuring. One important function of these cations is that of osmotic pressure and pH regulation, and for this purpose the composition of the cations present is largely irrelevant. But the leaves can only function properly if they contain a certain minimum concentration of the essential cations potassium, magnesium, calcium and the trace elements. Provided these minimal concentrations are exceeded, the relative concentrations of the different ions in the leaf are of little importance.[3]

The relative proportions of the cations in the leaf depend on the relative proportions of their available forms in the soil. The point of immediate importance is that increasing the relative concentration of any one cation in the soil, as, for example, by adding a potassic fertiliser, will increase its concentration in the leaf, and in consequence decrease that of the other cations. If the plant is growing in a soil rather short of available magnesium, so that the magnesium concentration in the leaf is approaching the minimum required for healthy growth, then adding a potassium fertiliser may cause the

1 For an example of this effect with sugar-cane, see J. A. Potter, *Trop. Agric. Trin.,* 1947, **24**, 94.
2 *Better Crops with Plant Food*, 1947, No. 26.
3 For a review and discussion of this, see F. J. Richards, *Ann. Rev. Biochem.,* 1944, **13**, 611. For an example with lucerne, see A. S. Hunter, S. J. Toth and F. E. Bear, *Soil Sci.,* 1943, **55**, 61.

magnesium concentration to fall below the minimum, so inducing a magnesium deficiency in the crop, which can be very severe if the crop growth is appreciably increased by the potassic fertiliser.[1] This phenomenon is sometimes called an ionic antagonism, and in this particular example one would say that the potassium was antagonistic to the magnesium. These potassium-induced magnesium deficiencies can often be seen on fruit trees and market-garden and glasshouse crops growing on soils low in exchangeable magnesium, for all these crops commonly receive generous dressings of potassic fertilisers. In the same way salt added to a sugar-beet crop may induce magnesium deficiency in the crop.[2] It is important to realise that this is no reason for not using a potassic fertiliser or salt, if the crop responds to either, but it is a very good reason for using in addition a magnesium fertiliser, such as a dolomitic limestone if the soil is rather acid, or otherwise magnesium sulphate.

Large dressings of lime may also induce magnesium or potassium deficiencies in the crop if the soil is low in either. These effects for potassium are, however, most noticeable for some crops on calcareous soils. Thus C. A. Bower and W. H. Pierre[3] found that maize and some of the sorghums, for example, need a fairly high level of potassium in their leaves compared with calcium and magnesium, so usually respond to potassic fertilisers even if the level of available potassium in the soil is fairly high. On the other hand, sweet clover and buckwheat need a much higher level of calcium compared with potassium in their leaves, so are much less responsive to potassic fertilisers on those soils, whilst flax, oats and soyabeans came intermediate between these two groups.

A further example is furnished by the manuring of lucerne on soils rather low in calcium. If one manures lucerne generously with a potassic fertiliser, it will take up potassium in preference to calcium, so the potassium content of the leaves will rise and their calcium content fall:[4] a tendency which may be undesirable for feeding purposes as a calcium-rich fodder is preferable to a potassium-rich one. Hence lucerne growing on a non-calcareous soil rather low in exchangeable calcium should be given small and frequent dressings of the potassic fertiliser rather than a single large one.

A detailed discussion on the actual level of fertilisers needed for maximum economic yields of different crops under different farming conditions is outside the scope of this book, and the reader is referred to specialist books such as G. W. Cooke's *The Control of Soil Fertility*[5] and *Fertilizing for Maximum Yield*[6] for British conditions, and S. L. Tisdale and W. L. Nelson's *Soil Fertility and Fertilizers*[7] for American.

1 For an example with potatoes, see T. Walsh and T. F. O'Donohoe, *J. agric. Sci.*, 1945, **35**, 254.
2 J. B. Hale, M. A. Watson and R. Hull, *Ann. Appl. Biol.*, 1946, **33**, 13.
3 *J. Amer. Soc. Agron.*, 1944, **36**, 608.
4 For an example of this, see F. E. Bear and S. J. Toth, *Soil Sci.*, 1948, **65**, 69.
5 Crosby Lockwood, 1967.
6 Crosby Lockwood, 1972.
7 Macmillan, 2nd edn, 1966.

5

The composition of the soil

Soils predominantly contain particles initially derived from the disintegration or decomposition of primary igneous rocks. But these particles may have been produced at an earlier geological period, moved by wind, water or ice over considerable distances, and been through one or more cycles of erosion and deposition. The processes that affect or alter these particles will be discussed in subsequent chapters, but one of the most fundamental is brought about by the plants that grow in the soil, for when the plants die they add to the soil energy-containing organic substances which they synthesised during their lifetime, and these serve as a food supply for a vast population of micro-organisms and even animals, which can in consequence inhabit the soil.

The soil mass is permeated with channels between the individual particles composing it, which are filled with water or air. The water, however, contains both dissolved gases and salts.

Soils may, therefore, consist of four parts:

1 Mineral matter derived from the rocks, but more or less altered by de-composition.
2 Calcium carbonate and resistant organic compounds derived from plants or organisms present at an earlier period.
3 Residues of plants and micro-organisms recently added to the soil.
4 The soil water, which is a solution of the various soluble and partially soluble salts present in the soil. Under temperate humid conditions this solution is dilute, but under some arid conditions, particularly in areas with poor drainage, it may become very concentrated. During dry periods such soils may contain appreciable quantities of soluble salts as crystals.

The soil has so far been pictured as consisting of a mass of particles, and this will be the aspect adopted in the rest of this chapter. But it has also been pictured as a network of channels filled with air and water and bounded by solid surfaces, and in which the roots of plants grow; its fundamental properties depend on the geometry of this interconnected network, called the 'pore space', and on the behaviour of water in the pore space, on the properties of the bounding surfaces, and on the mechanisms which supply plant nutrients both to the water in these channels and also to the solid surfaces.

The soil in the field is not a homogeneous material, for the individual soil particles are typically clustered together in the form of aggregates which vary widely in size, those visible to the naked eye being called crumbs and clods; and they are separated from one another over much of their surface by wider pores or cracks, each side of which may have a buckled or twisted surface. The phrase 'soil structure' is used to describe both the sizes and shapes of these aggregates and also the distribution of pores brought about by this clustering. A microscopic examination of the soil in its natural condition gives only limited information of these structural properties, but much more detail can be seen if the natural soil is impregnated with a resin, which is then hardened, so thin sections can be cut. An examination of these sections allows a study not only of the size, shape and distribution of the particles and pores large enough to be resolved by the optical microscope, but also of the way the individual particles forming the aggregates are arranged and oriented, even if they are too small to be resolved individually. The phrase 'soil fabric' was used by W. Kubiena,[1] who first developed this technique, to describe the types of spatial arrangement and orientation of the soil particles and soil pores as seen in these thin sections. Thus fabric, used in this sense, can be considered a particular aspect of structure, as defined above. This technique has been very helpful in the study of the activity of the small soil-inhabiting animals and of the role they play in the degradation of plant residues in the soil. It also allows the recognition of two important types of micro-structure in a soil, that in which the sand grains are comparatively clean and the finer particles are present in diffuse clusters between them, and that in which the sand grains have the finer particles cemented on their surface.

The mineral matter itself can be divided into two parts—particles possessing a definite crystal structure, and inorganic material that cannot be so described. The latter is composed of ill-defined precipitates of mixed hydrated oxides of iron, aluminium, manganese and silicon or of ill-defined mixed amorphous silicates; the former consist of crystals derived from the original rock, which tend to be the larger particles, and crystals formed from the products of weathering, which are usually the smaller particles. For these reasons mere chemical composition of the soil in bulk rarely gives information of much value: one wants to separate out the composition of the ill-defined deposits from the mineral grains and, as will be shown later, of the finer from the coarser mineral particles.

The size distribution of soil particles

Mere inspection of soil shows that it is composed of particles of different sizes and very irregular shapes, possessing a greater or less tendency to stick to-

1 *Micropedology*, Ames, Iowa, 1938. For a more recent description see R. Brewer, *Fabric and Mineral Analysis of Soils*, New York, 1964.

gether and form larger aggregates. These can be put into three groups, though the separation is not sharp or exact; clods and crumbs which are large and can be broken down by gentle mechanical means; granules, which are smaller, but in which the particles are more tenaciously held, so that some gentle chemical treatment is necessary to separate them; and concretions in which the fine material is bound still more firmly by a cement containing inorganic colloids. The power of forming clods and crumbs resides partly in the organic matter and partly in the mineral particles of smallest size, of which a relatively small percentage suffices.

The process of separating a soil into its component particles, and then estimating the proportion of particles in the various size ranges, is usually referred to as 'mechanical analysis', and the methods for doing this are now fairly standardised.[1] The first step is to crush the clods carefully so as much of the soil as possible will go through a 2 mm sieve, leaving only stones on the sieves. This operation introduces an arbitrary element into the results for those soils which contain rather soft, porous concretionary material, for the harder one crushes the soil the more the concretions are broken down into finer particles. Examples of such materials are, first, chalk, weathered limestone and even sometimes weathered granite, and second, soil particles bound together by weak inorganic cements, such as hydrated iron oxides or calcium carbonate. The material passing the 2 mm sieve is customarily referred to as 'fine earth', and the material left on the sieve as gravel.

The second stage is the dispersion of the crumbs passing the 2 mm sieve into their constituent particles. This dispersion is done in water, and involves separating particles that have a relatively strong attraction for each other when separated by distances of a few Ångström units, and keeping them separated afterwards. This can be done by subjecting the crumbs to a sufficiently strong mechanical force to tear them apart, such as is generated by passing strong ultrasonic vibrations through a water dispersion of the crumbs[2]; but it is more commonly achieved chemically, by altering the surface properties of the individual particles so that they repel each other, which is done by replacing the exchangeable calcium, magnesium and aluminium ions they contain with sodium. In some soils it is only necessary to shake the soil with a suitable sodium-saturated exchange resin, but for many the negative charge on the soil surfaces must be increased, by shaking in a Calgon (sodium hexametaphosphate) solution, for example, or by washing out the exchangeable cations with dilute acid and dispersing in sodium hydroxide. In soils containing appreciable amounts of positively charged particles such as allophanes or colloidal aluminium or ferric hydroxide, it may be necessary

1 For the method in use in Great Britain see G. W. Robinson, *Imp. Bur. Soil Sci., Tech. Common.* **26**, 1933; in America see P. R. Day, in *Methods of Soil Analysis*, Agronomy, 1965, Vol. 9, Chapter 43.
2 A. P. Edwards and J. M. Bremner, *J. Soil Sci.*, 1967, **18**, 47; W. W. Emerson, *J. Soil Sci.*, 1971, **22**, 50; and for a review J. R. Watson, *Soils. Fert.*, 1971, **34**, 127.

to increase their positive charge, for N. Ahmad and S. Prashad[1] found some of these soils dispersed best in a solution of zirconium nitrate of the correct concentration. The Zr^{4+} ion is small, replaces other cations easily, and probably increases the positive charges on the clay surfaces.

The standard International method consists of destroying the organic matter either by boiling the soil with hydrogen peroxide, or treating it with sodium hypobromite,[2] to remove the organic matter which binds soil particles together; then washing out the exchangeable cations with dilute acid, washing out the acid, adding sodium hydroxide and shaking. But a very common method of dispersion is to add a Calgon solution to a soil-water suspension and disperse the soil using a drink-mixer. The International method has the great disadvantage that it involves dissolving out any calcium carbonate present in the soil, and so gives results of limited significance for those calcareous soils in which calcium carbonate particles form an appreciable proportion of the finer particles.

The next problem in mechanical analysis is to find methods for specifying the size distribution of the soil particles. Since they have irregular shapes, their size cannot be described adequately by any one single measurement, though both their mass and their volume are theoretically definable, except possibly for the smallest particles which absorb a certain amount of any gas or liquid which comes in contact with their surfaces. There are, however, no methods for determining either the mass or volume distribution of the particles, so less direct size properties must be used. In practice the size of the larger particles is specified by the standard sieve mesh which just retains them, or just allows them to pass through; but this can only be used for sizes down to about 0·05 mm, and the size of smaller particles is measured by their velocity of sedimentation in a vertical column of water at a standard temperature, usually 20°C. But since it is difficult to think of particle sizes in terms of their settling velocities, these are usually converted into the diameter of a sphere of density 2·6 that would have this settling velocity, making use of Stokes' equation

$$v = \frac{g(\sigma - \rho)d^2}{18\eta}$$

where v is the settling velocity in cm/sec, g is the acceleration due to gravity (about 980 cm/sec²), σ is the density of the settling particle and ρ of the water ($\sigma - \rho$ is taken as 1·6), η is the viscosity of water (about 0·010 at 20°C) and d is the diameter of the sphere, commonly called the 'equivalent diameter' of the soil particle. The fact that many particles have a density different from 2·6 is no more disturbing to the use of this equivalent diameter than the fact that they are not spherical.

1 *J. Soil Sci.*, 1970, **21**, 63.
2 E. Troell, *J. agric. Sci.*, 1931, **21**, 476; S. J. Bourget and C. B. Tanner, *Canad. J. Agric. Sci.*, 1953, **33**, 579.

The size of particles which constitute the fine earth are divided into a few conventional groups. Atterburg originally proposed that they be divided into four groups, coarse sand 2–0·2 mm, fine sand 0·2–0·02 mm, silt 20–2 μ and clay finer than 2 μ; and sometimes additional divisions at 0·6, 0·06 mm and 6μ have been inserted. But for reasons which are discussed in the section on soil texture, in practice it is more convenient to make these divisions to give fractions with appreciably different properties. There is general agreement that 2 μ is a convenient division between silt and clay, and a division at about 50 μ between silt and sand has been widely accepted. It has not been found necessary to subdivide the silt fraction, but the sand fraction is always subdivided, though there has been little agreement on where the subdivisions should come. The American practice has been to use coarse sand 2–0·5 mm, medium sand 0·5–0·25, fine sand 0·25–0·10, and very fine sand 0·10–0·05 mm. But the upper limit of 2 μ for the clay particles typically contain many silt-like particles, so it would probably be desirable to subdivide the clay somewhere between 0·5 and 0·2 μ.

The coarse sand, on the International method, and all the sand fractions on the American scale, are separated from the finer particles on standard sieves; and the finer particles are separated by sedimentation in water, either by measuring the density of the suspension at a given depth with a hydrometer[1] or by taking samples of the suspension with a pipette from a standard depth after standing for a definite period.[2] This pipette technique can be refined so very shallow sampling depths can be used, which allows a ready fractionation of the clay, certainly to 0·5 μ and perhaps to 0·2 μ.[3] Table 5.1 gives examples of results obtained by these methods, where the size of the clay particles is expressed in terms of the negative logarithm of the observed settling velocity in cm/sec ($pv = -\log_{10} v$), as well as of the conventional equivalent diameter.

The distribution of particles finer than about 0·2 μ cannot be made by sedimentation under gravity, because such particles diffuse slowly from the part of the suspension where they are concentrated to that part where they are more dilute, due to their Brownian motion. They are therefore diffusing against the direction of settling, and this blurs the sharpness of the separation between different particle sizes. But this complication becomes less important as the gravitational field is increased, so sedimentation in a centrifugal field must be used for finer fractionation.[4]

Fuller information about the size distribution of soil particles would be given by continuous distribution curves instead of a discrete number of fractions, and methods for obtaining such curves have been devised, but unfortu-

1 G. J. Bouyoucos, *Soil Sci.*, 1927, **23**, 343; 1928, **25**, 365; 1932, **33**, 21. R. G. Downes, *Aust. J. Counc. Sci. Indust. Res.*, 1944, **17**, 197. I. A. Black, *J. Soil Sci.*, 1951, **2**, 118. P. R. Day, *Soil Sci.*, 1950, **70**, 362 and 1953, **75**, 181.
2 For a discussion on the accuracy of this method see M. Köhn, *Landw. Jahrb;*, 1928, **67**, 495.
3 A. N. and B. R. Puri, *J. agric. Sci.*, 1941, **31**, 171. E. W. Russell, *J. agric. Sci.*, 1943, **33**, 147.
4 For methods, see C. E. Marshall, *Proc. Roy. Soc.*, 1930, **126** A, 427, and *Proc. Soil Sci. Amer.*, 1939, **4**, 100, and C. Brown, *J. Phys. Chem.*, 1944, **48**, 246.

TABLE 5.1   An extended mechanical analysis of some clay soils

	Sand	Silt	Clay	*per cent of the clay present in the fraction*		
				pv 3·5–4·5 d 1·9–0·6 μ	4·5–5·5 0·6–0·19 μ	< 5·5 > 0·19 μ
Sudan Gezira	16·6	14·1	54·9	9·6	9·1	81·2
Malayan rubber soils						
Clay type	10·1	5·0	81·0	6·5	62·1	31·4
Silt type	10·2	27·0	62·0	30·3	44·8	24 8
English soils						
Oxford Clay	15·6	31·3	47·7	16·7	17·4	66·0
Weald Clay	18·3	33·5	48·3	13·5	37·9	48·6
Hereford fruit	23·7	38·5	31·2	32·0	23·1	44·9

nately most methods contain inherent errors[1] though it is possible that a new method described by C. L. Bascombe[2] may have overcome them. But for agricultural purposes accurate mechanical analyses have not proved of great value, nor does there seem to be any demand for continuous distribution curves, so little attention has been given to this problem in recent years.

Soil texture

Soil texture is a phrase that has no universally accepted definition, because two different concepts of how soil texture is defined are current. Soil texture was concerned originally with a property which referred to how open a soil was, or how it worked in the field, and a farmer might refer to the texture of his soil as being 'light' or 'heavy', or sometimes he would describe it by the number of horses needed to plough the land. Thus he would talk of 'two-horse' or 'four-horse land'. But he would also describe the texture of his soil by referring to it as a sandy, silty, loamy or clay texture; and a soil with a coarse sandy texture, for example, would have many of the properties of a coarse sand, a clay soil of a clay and a loam soil would be one in which the properties due to these three components were fairly evenly balanced. This concept of soil texture was taken up by many soil scientists who then defined texture with reference solely to the size distribution of the particles forming the soil. But they still kept the farmer's concept of workability implicit in their definition of textural class, so that the classes are not defined arbitrarily, but so as to be as consistent as possible with the farmer's description. This could be done because an experienced soil surveyor can, under certain conditions,

1 For a discussion of the older work see B. A. Keen, *Physical Properties of Soils*, London, 1931. For the more modern, two symposia arranged by the Institute of Chemical Engineers in 1947 and the Institute of Physics in 1954.
2 *J. Sedim. Petrol.*, 1968, **38**, 878.

estimate the proportion of sand, silt and clay particles in a soil sample by handling it when moist. This type of classification is illustrated in Fig. 5.1, in which the textural class of a soil is determined from its particle size distribution, but the boundaries between the classes are chosen so the classes fit as well as possible to the farmer's concept of what properties these classes should have. This particular classification is used by the American and the British Soil Surveys.

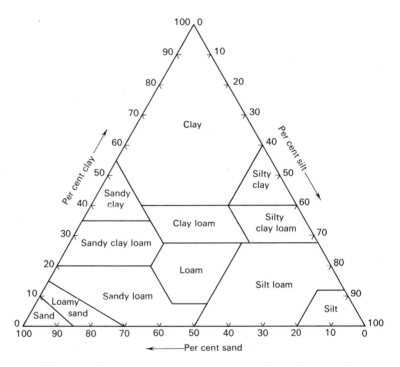

FIG. 5.1. The composition of the textural classes of soils used by the United States Soil Survey. (Sand, 2–0·05 mm; silt, 0·05–0·002 mm; clay, below 0·002 mm)

This use of the concept of soil texture as a measure of the particle size distribution in a soil is thus dependent on there being a close correlation between textural class as so defined and the farmer's appreciation of the workability of the soil. In general this correlation is good for most of the soils of western Europe and the north-eastern parts of the USA, although it breaks down very badly for many tropical and subtropical soils. But if this correlation is to be high, the particle size limits for silt and sand must be suitably chosen, and it was to make this correlation high that the American soil surveyors chose 50 μ instead of 20 μ for the upper limit of the silt, though they did not give this explicitly as the reason.

This definition of textural class means that there is no longer any essential connection between class as defined by Fig. 5.1, and class as defined by the openness of the soil or by its workability in the field. This is particularly true if comparisons are made between soils whose clays have different mineralogical compositions. Thus soils high in kaolinitic clay will have some typical clay soil properties weakly developed, while those high in montmorillonitic clay will have those properties very strongly developed; and the proportions of free iron oxides and of organic matter can have a marked influence on the feel and the workability of a soil. Thus some properties that have been considered textural must now be put into categories which embrace structure and consistency properties, the latter including plasticity and cohesion. These will be discussed on page 480. It is most desirable that, as a consequence of this separation of structural and consistency from textural properties, textural words be used correctly, and that the use of the words 'heavy' and 'light', in the sense of a heavy loam for a clay loam, for example, be discontinued. If these distinctions are borne in mind, some of the confusion that exists on whether one should use mechanical analysis or the feel of the soil for textural descriptions could disappear, because different properties are being assessed. Unfortunately there is still a divergence between field texture as assessed by some soil advisers and texture based on particle size distribution, due to the advisers purposely not breaking down the more stable soil micro-aggregates when working the soil between their fingers to assess its texture; so they will underestimate the amount of clay and overestimate the amount of silt, giving the soil a coarser texture than corresponds to its true particle size distribution.

The mineralogical composition of the soil particles

Sand and silt fractions

These particles can be divided into two main groups: crystalline mineral particles derived from primary rock and rock fragments; and micro-crystalline aggregates or amorphous deposits composed, for example, of calcium carbonate, ferric or aluminium hydroxides, or silica, which have been formed either from products of weathering or from residues of plant and animal life. There may also be present crystals formed in the soil such as calcite and possibly quartz, as well as minerals formed during the weathering process, and perhaps non-crystalline or poorly crystalline residues of the weathering of rock minerals.

The principal minerals found in the silt and sand fraction of the soil[1] are:

1 Quartz.

---

1 For details, see for example H. B. Milner, *Sedimentary Petrography*, 3rd edn, London, 1936; A. N. Winchell, *Elements of Optical Mineralogy*, 5th edn, New York, 1937.

2 Felspars.
   (a) Microcline and orthoclase[1] $KAlSi_3O_8$, which are potash felspars. Both are resistant to weathering, microcline being the more resistant.
   (b) Plagioclase, a series of mixed crystals having albite or soda felspar $NaAlSi_3O_8$ and anorthite or calcium felspar $CaAl_2Si_2O_8$ as end members. The sodium-rich members are about as resistant to weathering as orthoclase, while the calcium-rich members weather more easily.

3 Micas.
   (a) Dioctahedral micas usually containing no divalent metals; muscovite, a potassium aluminium silicate, $H_2KAl_3(SiO_4)_3$, which is fairly stable.
   (b) Trioctahedral micas, usually containing divalent metals; biotite, a potassium magnesium iron aluminium silicate not very resistant to decomposition. No true micas containing calcium are known, and soda mica is very rare.
   (c) Glauconite: a potassium mica, relatively low in aluminium but high in ferric iron and containing some magnesium and ferrous iron.[2]

4 The ferromagnesian minerals, which are low in aluminium and are divided into the pyroxenes $(MgFe)SiO_3$, the amphiboles $(MgFe)_7(Si_4O_{11})_2(OH)_2$ and the olivines $(MgFe)_2SiO_4$, where $(MgFe)$ refers to one ion only. Magnesium and ferrous iron are completely interchangeable in these minerals, and in general there is a certain amount of replacement of either by calcium; while in the amphiboles, and particularly in hornblende, sodium, potassium, calcium and aluminium may replace part of the magnesium or iron, and aluminium may replace part of the silicon. These minerals are usually not very resistant to decomposition.

5 Various minerals, such as zircon, garnet, apatite, ilmenite $FeTiO_3$, the iron oxides haematite Fe_2O_3, and magnetite Fe_3O_4, and the hydrated oxide limonite $FeO(OH).xH_2O$.

Igneous or primary rocks are composed of these minerals, and Table 5.2 gives a rough classification of the principal rocks and their mineral constitution.

6 Certain clay minerals which may be present in large particles, such as vermiculites and chlorites, occur in the sand fractions of some soils derived from certain basic igneous rocks. Kaolins are also often present in the silt fraction of soils, particularly if derived from granites. These clay minerals may be cemented by hydrous iron or aluminium oxides or dehydrated iron oxides.[3]

1 The following formulae are given only to show the type of constitution. The composition of actual specimens may differ considerably from the type formula.
2 S. B. Hendricks and C. S. Ross, *Amer. Mineral.*, 1941, **26**, 683.
3 J. S. Hosking, M. E. Neilson and A. R. Carthew (*Aust. J. Agric. Res.*, 1957, **8**, 45) give examples of these. See also a series of papers by D. M. McAleese and coworkers, *J. Soil Sci.*, 1957, **8**, 135; 1958, **9**, 66, 81, 289.

TABLE 5.2　The mineral constitution of the principal igneous rocks

Size of mineral crystals			Constitution					
Coarse	*Medium*	*Fine*	*Quartz*	*Alkali felspars*	*Plagio-clases*	*Micas*	*Pyroxenes amphi-boles*	*Olivines*
Granite	Quartz porphyry ⎱ Felsite ⎰	Rhyolite	×	×	+	+	O	−
Syenite	Microsyenite	Trachyte	O	×	O	+	O	−
Diorite	Microdiorite	Andesite	O	+	×	+	×	−
Gabbro	Dolerite (diabase)	Basalt	−	−	×	−	×	+
Peridotite ⎱ Serpentine ⎰		Picrite-basalt	−	−	O	−	+	×

× plentiful　　+ less plentiful　　O rare　　− absent

Quartz is by far the commonest mineral in most soils, and also the most resistant to decomposition. In soils derived from sedimentary deposits it often makes up 90 to 95 per cent of all the sand and silt particles.[1] Soils directly derived from primary rock contain much less, the actual quantity depending on the quartz content of the rock itself and the amount of weathering it has been subjected to. J. Hendrick and G. Newlands[2] have, for example, found

TABLE 5.3　Distribution of quartz and felspar in the sand and silt fraction of a soil derived from granite

Depth in cm	*0–10*	*10–25*	*25–45*	*45–60*	*Rotten rock*	*Un-weathered rock*
Quartz in fraction (%)	50	45	30	25	25	32
Felspar in fraction (%)	28	30	27	36	65	53

some Aberdeenshire soils derived from a basic igneous olivine-gabbro which are practically quartz-free, and the other soils in their paper illustrate the rapid rise in quartz content to over 70 per cent if the soil has been derived from rock which has already been subjected to weathering. However, the sand and silt particles in soils derived from basic igneous rocks can differ from those in

1 This is in marked contrast with the content in the average parent rock material, which according to F. W. Clarke is:

Felspars	60%	Micas	4%
Amphiboles and Pyroxene	17%	Other minerals	7%
Quartz	12%		

(analyses of 700 igneous rocks)

He further estimates that the lithosphere is composed of 95 per cent igneous rocks and 5 per cent sediments down to a depth of half a mile (*US Geol. Surv., Bull.* 770, 1924).

2 *J. Agric. Sci.*, 1923, **13**, 151.

nearly all other types of soil in that they contain large particles of silicate minerals classified as clay minerals, such as vermiculite and chlorite, for example, which possess cation exchange capacity.[1] Table 5.3, which is taken from some results of R. P. Humbert and C. E. Marshall,[2] illustrates this effect by giving the distribution of quartz and felspar in the sand and silt fraction down the profile of a soil derived from granite.

The clay fraction

The clay fraction is typically differentiated mineralogically from the silt fraction by being composed predominantly of minerals which are formed as products of weathering and which are not found in unweathered rocks. These minerals rarely ever occur in particles larger than 2 μ, and are usually present as particles smaller than this. They are much more resistant to weathering in the soil than are rock minerals ground to a comparable particle size, and they comprise the particles that carry the physical and chemical properties characteristic of clays.

The coarser clay fractions, particularly those larger than 0·5 μ in diameter, may contain appreciable proportions of quartz and sometimes of mica, but the fractions finer than 0·1 μ are almost entirely clay minerals or other products of weathering, such as hydrated ferric, aluminium, titanium, and manganese oxides.

The great difficulties in making a mineralogical analysis of clay particles are that X-ray photographs cannot be taken of single crystals as they are too small. Originally photographs of clay powders, with the clay particles in random orientation, were used and, though this method is still in use, much additional information is obtained by forming oriented flakes of clay particles, and treating these with various reagents, such as glycol, whose molecules will take up standard positions between the individual crystals of some clays.[3] Other types of method are also available. These minerals are all hydrated and lose their water at different temperatures, hence they can to some extent be differentiated by their dehydration curves. The energy change involved as this water is lost, and other changes which occur in the clay mineral, are most marked in certain critical temperature ranges, which are characteristic for different minerals. Again, some clay mineral particles have characteristic shapes: they may be flat with sharp edges, or have blurred edges, or be rod-shaped, and these can be determined by photographing the particles in the electron microscope.

1 See, for example, D. M. McAleese and S. M. McConaghy, *J. Soil Sci.*, 1957, **8**, 135, for basaltic soils in Northern Ireland.
2 *Missouri Agric. Expt. Sta., Res. Bull.* 359, 1943.
3 For an account of these and other methods for recognising and distinguishing between the different clay minerals see *X-ray Identification and Crystal Structures of Clay Minerals*, ed. G. W. Brindley, London, 1951, and *Differential Thermal Analysis of Clay Minerals*, ed. R. C. Mackenzie, London, 1955.

The result of all this work has been to show that for many soils the clay particles do not all belong to one definite and clearly defined class; they appear either to be a mixture of particles belonging to different mineralogical groups, or even for each particle itself to be a mixture of groups. The main minerals present are kaolinites, illites and hydrous micas, montmorillonites, vermiculites and chlorites; but because the soil clay particles are usually poorly crystallised, much work has been done on the simpler type minerals, which are only found in the pure state in certain very restricted localities. For this reason the composition and properties of these type minerals will be discussed in the next chapter.

Other inorganic components

Most soils contain small quantities of inorganic particles which have such a highly disordered structure that they are almost amorphous, and are often called the 'amorphous constituents'.[1] They consist of hydroxides, hydrous oxides and oxides of iron, aluminium and manganese, and for brevity will be called hydrous oxides when no particular form is specified. The most widespread of these hydrous oxides are the ferric, which impart the yellow, brown and red colours to soils, and they may appear as uniformly dispersed over the soil particles, or present as localised streaks or stains. In some soils they are present as definite particles, whose size varies from submicroscopic up to the size of gravels, often with a smooth rounded surface, and which are usually very hard. But they may be present as cements which bond soil particles together into concretions of indefinite shape, or into a pan; and these vary in strength from being soft enough to be crushed between the fingers up to massive ironstone.

These hydrous oxides probably are adsorbed on the surface or edge of clay particles as discrete colloidal particles, which are often micro-crystalline, although aluminium hydroxide may form a continuous film, perhaps a monolayer over these surfaces. The evidence from electron microscope photographs is that if iron hydroxide is deposited on a clay surface, the particles grow larger as they age,[2] possibly up to a size of 50–200 Å,[3] the actual size depending on the impurities either in the crystal lattice or adsorbed on the crystal surface.[4] But in some examples the hydrous oxide appears to remain amorphous both to X-ray and electron diffraction, though these may be intimately intermixed with an appreciable proportion of aluminium hydroxide

1 For a review of these see B. D. Mitchell, V. C. Farmer and W. J. McHardy, *Adv. Agron.*, 1964, **16**, 327.

2 For an example of a photograph of coated kaolinite see E. A. C. Follett, *J. Soil Sci.*, 1965, **16**, 334, and from changes in cation exchange capacity in bentonite, J. S. Clark and W. E. Nichol, *Canad. J. Soil Sci.*, 1968, **48**, 173.

3 Obtained in some Australian soils by D. J. Greenland, J. M. Oades and T. W. Sherwin, *J. Soil Sci.*, 1968, **19**, 123.

4 K. Norrish and R. M. Taylor, *J. Soil Sci.*, 1961, **12**, 294.

(a)

(b)

PLATE 3. Coatings on the surface of clay particles of Urrbrae loam soil:
(a) After treatment with sodium dithionite and sodium acetate at pH 5:
(b) After treatment with 5% sodium carbonate.
These photographs are electron micrographs of carbon replicas of the clay surfaces.

or silica.[1] These particles may be adsorbed on the edges or sometimes on the surfaces of the clay, plates[2] particularly of kaolinite particles, as is illustrated in Plate 3. A. K. M. Habibullah and D. J. Greenland[3] have obtained evidence from some East Pakistan soils that poorly drained and seasonally waterlogged soils contain stable amorphous gels of ferric oxide, and the better-drained ones crystalline hydrated oxide. These gels contain a considerable admixture of aluminium hydroxide.

Many methods have been introduced for determining the amount of these hydrous oxides in soils, all based on treating the soil with solvents which either bring into solution all, or else a particular fraction, of these oxides, without dissolving any material from the crystalline clay particles. Free aluminium hydroxide or disordered mixed amorphous alumina and silica gels can be removed by treating the soil with a hot 5 per cent sodium carbonate solution,[4] but thin films of hydrated alumina can be dissolved by treating the soil with an acidified neutral salt solution, such as a calcium chloride solution acidified to pH 1·5 with hydrochloric acid.[5] A number of methods have been used to dissolve out the free iron oxides. The most powerful method is to treat the soil with a suitable reducing agent, and sodium dithionite $Na_2S_2O_4$ is one in common use,[6] which reduces the ferric ions to ferrous, and these are kept in solution either by adjusting its pH or by adding a chelating agent such as sodium citrate or tartrate. Another widely used method, first introduced by O. Tamm,[7] is to dissolve the ferric oxides in a solution of sodium or ammonium oxalate and oxalic acid whose pH is about 3·2; but this solution only dissolves very fine ferric oxide particles.[8]

There is still little reliable information on the chemical composition of these hydrous oxides in soils because, before the advent of the electron-probe analyser, the individual crystals were too small to analyse, and the larger concretions could contain other soil particles occluded in them. They are probably never pure, for aluminium hydroxides usually contain small quantities of iron and silicon, and ferric hydrous oxides probably always contain some aluminium. There is, in fact, evidence that aluminium can replace up to one quarter of the iron in the crystal,[9] though the replacement is usually considerably less than this. These precipitates also contain many other ions, probably adsorbed on their surface, such as silicate, phosphate, molybdate, and man-

1 E. A. C. Follett, W. J. McHardy *et al*, *Clay Minerals*, 1965, **6**, 35.
2 D. J. Greenland and G. K. Wilkinson, *Proc. 3rd Int. Clay Conf.*, 1969, **1**, 861.
3 *J. Soil Sci.*, 1971, **22**, 179.
4 B. D. Mitchell and V. C. Farmer, *Clay Miner. Bull.* 1962, **5**, 128 and E. A. C. Follett, W. J. McHardy *et al.*, *Clay Minerals*, 1965, **6**, 23.
5 C. K. Twencboah, D. J. Greenland and J. M. Oades, *Aust. J. Soil Res.*, 1967, **5**, 247.
6 See, for example, R. C. Mackenzie, *J. Soil Sci.*, 1954, **5**, 167, and with B. D. Mitchell, *Soil Sci.*, 1954, **77**, 173; V. J. Kilmer, *Proc. Soil Sci. Soc. Amer.*, 1960, **24**, 420.
7 *Medd. Skögsförsokanst.*, 1922, **19**, 385, and 1934, **27**, 1.
8 B. J. Anderson and E. A. Jenne, *Soil Sci.*, 1970, **109**, 163; J. A. McKeague, J. E. Brydon and N. M. Miles, *Proc. Soil Sci. Soc. Amer.*, 1971, **35**, 33.
9 K. Norrish and R. M. Taylor, *J. Soil Sci.*, 1961, **12**, 294.

ganese; and most of the phosphate and molybdate present in soils high in free ferric oxides may be locked up in these precipitates. If the hydrated oxides are being formed in the presence of humic colloids, these may be adsorbed on the oxide surface, and cause the precipitate to consist of very fine particles with a strongly disordered structure.

As already noted, ferric hydroxide is not a normal constituent of aerated soils. The iron oxides occur either in the α-form as limonite or goethite Fe O(OH) and haematite Fe_2O_3, or else in the γ-form as lepidocrocite and maghemite. Limonite is usually considered to differ from goethite by being almost amorphous; and the γ forms are probably characteristic of ferric hydrous oxides produced from the oxidation of ferrous iron in the presence of organic matter.[1] The presence of maghemite, which is weakly magnetic so its presence can easily be demonstrated in a soil, is therefore a good indication that the soil in the past has been under reducing conditions in the presence of organic matter. The conditions controlling the dehydration of goethite and lepidocrocite to haematite and maghemite are not fully known, though a pronounced hot dry season is conducive to the formation of haematite. The ferric oxides give soils their bright colours; goethite gives colours varying from yellow through brown to red, possibly dependent on its state of sub-division, lepidocrocite is usually orange, haematite red and maghemite light brown.

Aluminium hydroxide $Al(OH)_3$ probably having the gibbsite structure is a common constituent in many soils, but is usually only present in very small amounts. It is possible that it is an unstable constituent in the presence of silicic acid, being converted to a crystalline clay. Some soils may also contain the hydrous oxide form, either with the boehmite or diaspore structure. A highly disordered aluminosilicate occurs in some soils which is known as 'allophane'. It is typically formed as a weathering product of a disordered primary silicate mineral such as volcanic glass and pumice,[2] and its presence in a soil is often good evidence of the earlier presence of volcanic ash in that soil.

The aluminium hydroxide precipitates usually occur as films on the surface of clay particles, whereas the hydrous ferric oxides are usually present as very small discrete concretions up to 100 Å in size.[3] Plate 3 gives an example of this for the clay particles of the Urrbrae red-brown earth soil at the Waite Institute, Adelaide, S. Australia. The upper photograph shows that after the soil has had the free iron oxides removed with dithionite, the clay particles are still coated with a precipitate that is non-crystalline at the magnification used. It is however removed if the clay is treated with sodium carbonate, for it is composed of a mixture of dehydrated alumina and silica gels.

1 S. Henin and E. Le Borgne, *Ann. Agron.*, 1955, **5**, 11, and H. W. van der Marel, *J. Sediment. Petrol.*, 1951, **21**, 12.
2 See, for example, M. Fieldes, *NZ J. Sci.*, 1966, **9**, 599, 608.
3 D. J. Greenland, J. M. Oades and T. W. Sherwin, *J. Soil Sci.* 1968, **19**, 123.

Silica can also occur in amorphous forms, both as a consequence of weathering and as a product of vegetation. Under some conditions, which do not seem to have been investigated in detail, silica can be deposited between soil particles to give a 'silica' pan (see p. 735) as happens in some sandstones, in which silica binds the sand particles together. Also all plants contain some silica, which is returned to the soil as the plant dies. Grasses, in particular, contain appreciable amounts, often in the form of particles of opal which can be clearly seen using a low power microscope; these can be found in grassland soils,[1] and probably also in tropical grassland and savannah soils subjected to regular firing.

1 F. Smithson, *J. Soil Sci.*, 1958, **9**, 148.

6

The constitution of clay minerals

The clay minerals occurring in the soil are built up out of sheets of oxygen or hydroxyl ions, and so belong to the class of layer-lattice minerals.[1] Oxygen and hydroxyl ions have been singled out since they are much the largest ions in the crystal and the size of the unit cell out of which the crystal is built is controlled by their position and packing in the lattice. The word sheet will be used for the group of ions lying on a plane, and a layer is built up out of several sheets stacked usually in a definite orientation one above the other. Most clay minerals are built out of layers containing either three or four sheets of oxygen or hydroxyl ions, though a few such as sepiolite and polygorskite are built out of amphibole-type chain lattices instead of layer lattices.

Only two types of oxygen sheets occur in layer lattices, which are illustrated in Plate 4. One type consists of oxygens arranged in regular rows and columns so each oxygen touches six others in the sheet, and the second type has a similar arrangement except that one quarter of the oxygens needed for a complete sheet are lacking, so that each oxygen only touches four neighbours instead of six, and the sheet is a honeycomb with hexagonal holes. Some or all of the oxygens in a complete sheet contain a proton so are hydroxyls, a substitution not found in the hexagonal sheet, and the distance apart between the oxygens in the complete sheet is larger than between the oxygens in the hexagonal sheet, since each sheet contains the same number per unit area, as shown in the Plate 4. If there was no distortion the ratio of the spacings between oxygen centres would be $\sqrt{3}$ to 2 for the hexagonal and complete sheets.

If a hexagonal sheet is placed on a complete sheet, each oxygen of the hexagonal sheet will touch three oxygens of the complete sheet, and between these four oxygens is what is known as a tetrahedral hole because the centres of the four oxygens can be pictured as forming the corners of a tetrahedron. All these holes contain a cation, usually silicon but in many clays a proportion may be filled with aluminium. These cations are thus in four-coordination with oxygen, and all the silicon present in rocks or soils is coordinated with four oxygen ions in this way. If two complete sheets are placed one above the

1 For a more complete discussion of this subject see R. E. Grim, *Clay Mineralogy*, 2nd Edn, McGraw-Hill, 1968.

other, so each oxygen in one sheet touches three oxygens of the other, there is an additional space formed between three oxygens of one sheet and three of the other which can be pictured as an octahedron with its corners at the centres of the six oxygen ions. In clay minerals, none of the tetrahedral spaces between these two sheets is occupied but a proportion of these octahedral spaces, varying from between about two-thirds to nearly all, contain a cation, usually predominantly aluminium if only about two-thirds of the holes are filled and predominantly magnesium if most are filled. These ions are thus in six-coordination with oxygen, which is the same as for the free ion in water or in gibbsite $Al(OH)_3$ and brucite $Mg(OH)_2$.

In soils the layer-lattice clay minerals fall into two groups, those built out of three and those out of four oxygen sheets. Those built out of three sheets—the kaolin group—consist of a hexagonal oxygen sheet, a complete oxygen sheet in which about two-thirds of the oxygens are hydroxyls, and a second complete oxygen sheet in which all the oxygens are hydroxyls. This group is often known as the 1:1 group of clay minerals, because the two complete oxygen sheets are thought of as constituting a single unit. The other group—often known as the 2:1 group—has a symmetrical structure and is built up of two outer hexagonal oxygen sheets enclosing the two complete oxygen sheets, in which two-thirds of the oxygens in each sheet are hydroxyls.

The type mineral of the kaolin group is kaolinite, and it typically contains enough silicon, aluminium and hydrogen ions to balance almost completely the negative charges due to the oxygen ions in the sheets, so that its composition rarely departs far from a silicon in every tetrahedral space between the hexagonal and complete sheets, and from aluminium in two-thirds of the octahedral spaces. The composition of the unit cell is thus

$$6\,O \quad (4\,Si) \quad 2\,O+4\,OH \quad (4\,Al) \quad 6\,OH$$

where Si and Al have been put in parentheses to show they occupy spaces between the oxygen ions. A unit cell with this structure carries no electric charge. The corresponding uncharged structure for the 2:1 group is

$$6\,O \quad (4\,Si) \quad 2\,O+4\,OH \quad (4\,Al) \quad 2\,O+4\,OH \quad (4\,Si) \quad 6\,O$$

which corresponds to the mineral pyrophyllite; though clays based on

$$6\,O \quad (4\,Si) \quad 2\,O+4\,OH \quad (6\,Mg) \quad 2\,O+4\,OH \quad (4\,Si) \quad 6\,O$$

which corresponds to the mineral talc are also found. The thickness of the kaolin layer is 7·2 Å and of the pyrophyllite 9·3 Å.[1]

Isomorphous replacement

Soil clays usually have cation compositions different from the model uncharged structures just illustrated, but the range in the cation compositions is

1 1 Å (Ångström unit) $= 10^{-10}$ m or 10^{-8} cm.

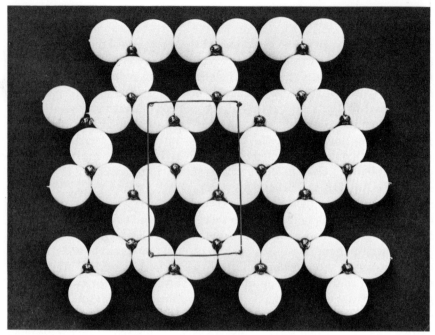

(*a*) The oxygen layer containing silicon in 4-coordination.

(*b*) The oxygen and hydroxyl layer containing aluminium in 6-coordination.

PLATE 4. The structure of the oxygen and hydroxyl layers in clay minerals and the structure of pyrophyllite.

(*c*) The hydroxyl layer in close packing.

(*d*) The arrangement of the layers to form a sheet of mineral pyrophyllite.

very strictly limited because the clay structure is controlled by the packing of the oxygen-hydroxyl sheets, which, in turn, controls the size of the holes available for the cations. Thus, the radius of the largest sphere that can just fit in the tetrahedral hole between four spheres of radius r that are touching each other is $0.225\ r$, and since the radius of an oxygen ion in four-coordination is 1.28 Å, the radius of the largest cation that would just fit into this hole is 0.29 Å. Similarly, the radius of the largest sphere that can just fit into an octahedral space when the six spheres are touching each other is $0.412\ r$, and the radius of the oxygen ions in these sheets is 1.40 Å, so the radius of the largest cation is 0.61 Å. However, the hexagonal oxygen sheet can only fit over its associated complete sheet if the positions of the oxygens are moved a little from their ideal position, or if the four oxygens forming the tetrahedral holes are twisted from their positions in the sheets, so that the tetrahedral and octahedral spaces are somewhat different from those given above.[1]

Silicon is the only commonly occurring cation that can fit into the tetrahedral space without causing further distortion of the sheets, but aluminium—the second smallest common cation when in four-coordination with oxygen—can replace a limited proportion of the silicons without causing such a severe distortion that the lattice loses stability. It is possible that clays formed in the soil do not have more than 5 per cent of the silicon replaced by aluminium, although micas, which are formed at high temperatures, have up to 25 per cent replacement. Clays are found in the soil with up to 15 per cent replacement, but these may be residual products of weathering of micas.

A wide range of cations are found in the larger octahedral spaces. Table 6.1 gives the radius of the more commonly occurring cations when in six-co-ordination with oxygen, and shows that all the other cations are appreciably

TABLE 6.1   The radii in Å of certain ions in six-coordination*

4+		3+		2+		1+	
V	0.63	Al	0.51	Mg	0.66	Li	0.68
Ti	0.68	Cr	0.63	Ni	0.69	Cu	0.96
		Co	0.63	Cu	0.72	Na	0.97
		Fe	0.64	Co	0.73	K	1.33
		Mn	0.66	Fe	0.74	NH_4	1.48
				Mn	0.80		
				Ca	0.99		

* Taken from L. H. Ahrens, *Geochem. Cosmochem. Acta*, 1952, **2**, 155.

larger than aluminium. This again has the consequence that only a limited proportion of the aluminium can be replaced by other ions before the distortion of the lattice becomes so severe that it has lost stability. However, the spacing between the two complete sheets can be increased somewhat without

1 For a description of the distortions this brings about see E. W. Radoslovich, *Trans. 9th Int. Congr. Soil Sci.*, 1968, **3**, 11.

the layer losing stability. Thus, all the octahedral holes can be filled with magnesium, instead of only two-thirds being filled with aluminium, and the layer will still be stable if some of the magnesium is replaced by other cations, including aluminium on the one hand and divalent cations not much larger than magnesium on the other.[1]

The replacement of one cation by another of about the same size is known as 'isomorphous', because it hardly alters the shape of the lattice. It takes place when the clay is being formed, and once the ion is in position it is difficult to get out from the clay without destroying the lattice. Isomorphous replacement typically leads to the layer acquiring a negative charge, for every silicon replaced by an aluminium, and every aluminium in six-coordination replaced by a divalent cation, will cause the layer to acquire a negative charge, a topic which will be discussed in more detail on p. 86.

The charge a clay lattice can carry appears to be limited, so that isomorphous replacement is to some extent a self-compensating process. Thus, one ion of aluminium in the octahedral spaces is likely to be replaced by more than one, but by less than one and a half magnesium ions, on the average; so that while there is an increase in negative charge due to this replacement, it is less than one unit. However, in the dioctahedral clays the proportion of octahedral holes filled usually lies between 2·0 and 2·22, and in the trioctahedral between 2·88 and 3·0, with a very clear separation between the two groups.[2] Further, some of the increase in charge that is brought about by isomorphous replacement is compensated by an increase in the proportion of oxygen ions in the sheets forming the octahedral spaces that are hydroxyls. The actual level of charge a clay can have depends on the way the external cations neutralise it. The charge can be as high as 1·9 units per unit cell when it is neutralised by potassium ions bound very close to the surface, but only about 1·3 units when neutralised by aluminium or magnesium ions separated from the surface by a mono-molecular layer of water, to about 0·7 units when neutralised by individual cations not held close to the surface in a regular arrangement. Thus as a micaceous clay loses potassium by weathering, ionic re-arrangements take place in the sheets, resulting in a lowering of the lattice charge, and some oxygens in the hexagonal layer may accept protons to become hydroxyls.[3]

The structure of the kaolin clay minerals

These are built out of layers, each of which consists of a hexagonal oxygen sheet and two complete sheets of oxygen or hydroxyls, as already described.

1 For a review of this subject, see C. S. Ross and S. B. Hendricks, *US Geol. Surv.*, *Prof. Paper* 205B, 1945, and *Proc. Soil Sci. Soc. Amer.*, 1942, **6**, 58, and J. W. Earley, B. B. Osthaus and I. H. Milne, *Amer. Mineral.*, 1953, **38**, 707. These papers give examples of analyses of clay minerals to illustrate the range of replacements found.
2 R. E. Grim, *Clay Mineralogy*, 2nd Edn, McGraw-Hill, 1968.
3 A. C. D. Newman, *Clay Miner*, 1967, **7**, 215.

This group includes a number of different minerals, such as kaolinite, which often occurs as roughly hexagonal plates several microns in size, or as halloysite which occurs as hollow tubes, or in less well-characterised and usually small particles as is shown in Plate 5. Kaolinite particles only carry a small negative charge on their surface, when expressed in terms of charge

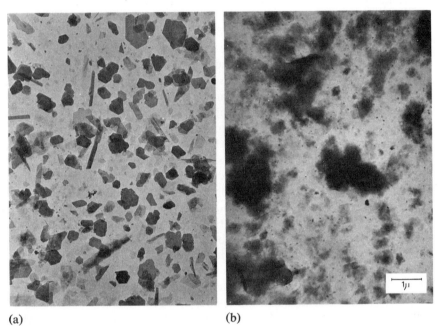

(a)   (b)

PLATE 5. (*a*) Electron microscope photographs of kaolinite particles between 0·1 and 0·2 μ. (×15 000).
(*b*) Electron microscope photograph of bentonite particles (×16 000). These particles are too small and too thin to be resolved, and are mainly aggregated together in the photograph.

per unit weight, compared with most other clays, and consist of a number of uncharged layers stacked one upon the other in definite positions relative to each other with the outer layer or layers carrying the negative charge. The layers are held together by hydrogen bonds being formed between the oxygens in the outer hexagonal sheet of one layer and the hydroxyls in the outer complete sheet of the adjacent layer. Hydrogen bonding can only take place when the sheets are close together, and uncharged.

Halloysite consists of individual layers separated from each other by a mono-molecular layer of water. Typically the layers carry a small negative charge, due to isomorphous replacement, and the exchangeable cations which neutralise it are situated in this water layer. The layers are held together by weak hydrogen bonding between oriented water molecules and the oxygen and hydroxyl ions on adjacent layers. The particles are tubular with an internal diameter and a wall thickness each of about 200 Å, but both may be larger than this. The curvature is probably due to the spacings between the oxygens in the hexagonal sheet being greater than that needed for the oxygens in the hydroxyl sheet of gibbsite, so the outer sheet of each curved layer is the open hexagonal sheet.[1]

The other kaolin minerals occurring in soil clays have smaller particles than typical kaolinites, with less definite shape and with a higher negative charge per unit weight. They also contain some water. Some of the particles may consist of layers with irregular distortions interleaved between layers of kaolinite.[2]

The 2:1 group of clays

The layer lattices of these clays typically carry an appreciably higher negative charge per unit weight than do the kaolin group, and this negative charge is neutralised by cations external to the oxygen layers. They fall into three fairly well separated classes, each characterised by possessing a negative charge in a different range: those with the highest charge being the micaceous clays, those with intermediate the vermiculite clays, and those with the lowest the smectite or montmorillonite clays.

The micaceous clays, as their name suggests, have a structure similar to that of the micas. There are two principal groups of micas—the dioctahedral in which aluminium occupies about two-thirds of the octahedral spaces, and the trioctahedral in which magnesium and other divalent ions occupy nearly all of these spaces. The typical dioctahedral mica is muscovite, which has very little replacement of aluminium in the octahedral spaces with other cations, and the typical trioctahedral is biotite, which has an appreciable proportion of the magnesium cations replaced by ferrous, manganous and other divalent cations of about the same size. All the micas have about one-quarter of the silicon ions replaced by aluminium, so carry a negative charge of two units per unit cell. This charge is neutralised by potassium ions which are the right size to fit into the hexagonal holes on the lattice surface, and adjacent layers are so arranged that each potassium is partly in a hole in one layer and partly in a hole of the adjacent layer. It therefore should touch twelve oxygen ions, six in each layer, and be close to two hydroxyls which form the floor of the holes, and which belong to the oxygen-hydroxyl sheet below the oxygen sheet. This bond between adjacent layers is relatively strong, for much of the negative charge on the lattice is located on the oxygen ions on the outer layer, since it is due to replacement of silicon by aluminium in the tetrahedral spaces, so that the positively charged potassium ions are very close to the source of the

1 T. F. Bates, F. A. Hildebrand and A. Swineford, *Amer. Mineral.*, 1950, **35**, 463.
2 L. Bramao, J. G. Cady, *et al.*, *Soil Sci.*, 1952, **73**, 273.

negative charge. The potassium ions, however, bind adjacent sheets of di-octahedral micas together more strongly than they do those of the triocta-hedral, possibly because the direction of the dipole of the hydroxyl ion below the hexagonal hole is vertically upwards in the trioctahedral, but twisted from the vertical in the dioctahedral. This may be the reason biotite micas lose their potassium more easily than muscovite micas. The micaceous clays differ from the micas in having a smaller electric charge, due to a smaller proportion of aluminium ions in the tetrahedral spaces, and in having a higher water content, or at least losing more water when heated, than a mica. These clays are often known as illites, but this name is not universally accepted as the typical illites probably consist of a mixture of clay minerals and are not pure micaceous clays, as defined here.

The vermiculite clays differ from the micaceous in having a lower net nega-tive charge on the layer lattice, and adjacent layers are held together by mag-nesium or aluminium ions in six-coordination with water, giving a water layer two molecules thick between the clay layers, which will also help to hold the layers together through hydrogen bonding. The bond between adjacent layers is therefore considerably weaker than in micaceous clays, because the posi-tively charged ions are much further away from the source of the negative charges in the lattice. This is reflected in the vermiculite clays having a basal spacing of 14–15 Å compared with the 10 Å of a typical mica.

The smectites or montmorillonites have a still lower negative charge on the clay lattice, almost all of which is due to isomorphous replacement in the octa-hedral spaces. The bond between adjacent layers, when the clay is dispersed in water, is still weaker than in vermiculite, so that they are still further separated, and any cation can enter between them. It is, in consequence, rela-tively easy to disperse these clays into small particles one layer thick. The particles are usually very small, as can be seen in Plate 5, which gives an elec-tron microscope photograph of a dried bentonite suspension. It shows that most of the particles are too small to be resolved at the magnification used. If all the negative charge is neutralised by sodium ions, and the water content of the clay is gradually increased, the water will penetrate between the in-dividual layers causing the distance between adjacent parallel layers to in-crease. This causes the clay to swell, and it will continue to swell until the forces holding the particles parallel to each other become weaker than those generated by Brownian movement, when the clay will deflocculate. Thus con-tinuous swelling of the sodium-saturated clay in water is the characteristic property of the smectite group, and they are in consequence often spoken of as the swelling clays.

There is another group of clay minerals, the chlorites, which are related to the vermiculites. A typical chlorite consists of alternate layers of a 2:1 lattice with a brucite sheet, and differs from a vermiculite in that the layer between the 2:1 layers is a double sheet of hydroxyls instead of a double sheet of water molecules. Chlorites normally only carry a small charge per unit weight, but the 2:1 lattice usually carries an appreciable negative charge, as up to 50 per cent of the silicon ions are replaced by aluminium, and the brucite layer carries a positive charge due to replacement of magnesium ions by aluminium, to counterbalance it.

Soil clays

All the principal types of clay minerals are found in a relatively pure form in geological deposits, and most of the work on clay minerals has been done on these materials. The clay particles present in soils often differ appreciably in properties from the pure type minerals, particularly if they have been produced by a weathering process in the soil profile. They usually have a less well-ordered structure and are smaller in size. If a dispersion of soil clay is examined at moderately powerful magnification in an electron microscope, particles of all sizes can be seen, many of which are very small, are flat or plate-shaped, usually with fairly clearly defined edges,[1] though this cannot always be seen on photographs. If the clay contains much kaolinite, a certain proportion of the particles are usually present as discrete plates of roughly hexagonal shape.

Many clay particles are composed of only a few lattice layers and, although their broken edges look fairly sharp on high-quality electron microscope photographs, there is evidence from the adsorption of non-polar molecules by soil clays that the layers may have varying amounts of overlap with their neighbours at the broken edge of the particle, so the face of this broken edge may contain channels one, two or three layers wide of variable length.[2] Both the layers themselves and the interlayer ions may differ in constitution in a single clay particle. Further, it is unusual for all the clay particles to have a similar composition, and in the past it was difficult to recognise the various constituents in a soil clay fraction.[3] In consequence names were given to clays which were thought to be uniform in composition but are now known to be mixtures. The names illite and beidillite, in particular, have been widely used in the past, neither of which are now accepted as valid, for both were given to clays now known to be mineralogically mixed.[4]

The surfaces of clay particles in many soils are not clean, but have films of iron and aluminium hydrated oxides, often associated with silica deposited on them as is shown in Plate 3. Many important soil properties, particularly those associated with positively charged surfaces, are due to these films, and

1 See for example E. A. C. Follett, W. J. McHardy *et al.*, *Clay Minerals*, 1965, **6**, 23. Also H. Beutelspacher and H. W. van der Marel, *Atlas of Electronmicroscopy of Clay Minerals and Their Admixtures*, Amsterdam 1966.
2 L. A. G. Aylmore and J. P. Quirk, *J. Soil Sci.*, 1967, **18**, 1.
3 For a discussion see H. W. van der Marel, *Contrib. Miner. Petrol.*, 1966, **12**, 96.
4 For illite see K. V. Raman and M. L. Jackson, *Clays Clay Miner.*, 1966, 53; for beidellite, R. E. Grim, *Clay Mineralogy*, McGraw-Hill, 1953.

it is often necessary to remove them to obtain more precise information on the structure, composition and properties of the crystalline clay particle.[1]

The surface area of clay particles

The surface area of a clay particle is usually defined as the area of the particle that is accessible to ions or molecules when the clay is in an aqueous solution. In so far as the particles do not show the phenomena of interlamellar swelling, they have a definite size; and provided the particles are dispersed in the solution, they possess a definite surface area. Even particles showing interlamellar swelling possess a definite surface area, provided the ions or molecules being considered can enter freely between the individual lamellae. If the clay particles or individual lamellae are so tightly bound that ions or water cannot freely enter, the apparent surface area of the particle becomes the area of its external surface. The difference between these two areas can be very large for montmorillonites, but is usually small for other clays.

There are a number of methods available for determining surface areas of clay particles based on measuring the amount of a suitable substance that must be absorbed by the clay to cover the clay surface with a monolayer. A substance that has been widely used is nitrogen at $78°K$, when it is a liquid, but the technique involves outgassing the clay under high vacuum and usually high temperature. The nitrogen molecules cannot enter between the layers of montmorillonite, for example, so this method underestimates the surface area of such clays. The use of cetyl pyridinium bromide dissolved in water allows the determination of the surface area of clays without the need for drying them,[2] and so can be used for fully dispersed clay suspensions.

The surface areas that are found for clays vary from 700–800 m^2/g for a well-dispersed sodium montmorillonite, 300–500 for vermiculites and some mixed layer clays, 100–300 for micaceous clays and 5–100 for kaolinitic clays. Lower figures than these have been found for soil clays, possibly those containing relatively unweathered micaceous particles. Thus D. M. Farrer[3] obtained a figure of 30–50 m^2/g for typical English Jurassic clays consisting predominantly of micaceous and montmorillonitic clays. The amorphous constituents, aluminium and ferric hydroxide gels and silicic acid gels, have surface areas between 100 and 500 m^2/g.

The charge on soil clay particles

A clear distinction must be made between the net charge on the individual alumino-silicate layers forming a clay particle and the net charge on the particle itself. Thus, micaceous clays carry a high lattice charge but a low particle

1 For methods of doing this, see K. Wada and D. J. Greenland. *Clay Miner.*, 1970, **8**, 241.
2 D. J. Greenland and J. P. Quirk, *J. Soil Sci.*, 1964, **18**, 178 and *Int. Soil Confr, NZ*, 1962, 69.
3 *J. Soil Sci.*, 1963, **14**, 303.

charge, whereas for a well-dispersed montmorillonite clay the two charges are identical. Micaceous clay particles have a lattice charge of 1·3–1·5 units per unit cell, and about five-sixths of this charge is neutralised by potassium, so that the actual charge on the particles is of the order of 30–40 m eq/100 g clay, whereas a montmorillonite will have a charge between 80–120 m eq and a well-dispersed vermiculite up to 150 m eq. On the other hand, chloritic clays have their sheets bound together more tightly than do vermiculite clays, and so behave like micaceous clays except that the particles carry a somewhat lower charge. Soil kaolins and halloysites may have a wide range of charges, from 3–5 up to possibly 20–30 m eq though sometimes at least part of the apparent charge may be due to very small montmorillonite particles being adsorbed on the broken edges of the kaolinite plate.

It is interesting to translate these negative charges into the mean surface density of charge. Taking montmorillonite as an example and using a molecular weight of a unit cell as 720, a surface area of $46·1 \text{ Å}^2$ on each face, and a negative charge of 100 m eq/100 g there is 0·72 unit charge per unit cell, equal to a charge density of one negative charge per 130 Å^2 or about $1·5 \times 10^{-6}$ eq/m^2 of surface. This calculation is only approximate, since isomorphous replacement affects the molecular weight and area of the unit cell. Further, 80 m eq is probably a more common charge, giving a surface area of 160 Å^2 per unit charge. This area is large enough to allow large organic cations to neutralise this charge and still to form only a mono-ionic layer on the surface.

Greenland and Quirk[1] have determined the surface area and charge on a number of clay minerals and soil clays and find values of 80–180 Å per charge for expanding lattice clays, $50–60 \text{ Å}^2$ for micaceous clays and $100–300 \text{ Å}^2$ for kaolinites. It is worth noting that kaolinites have surface densities of charge similar to montmorillonite, although their exchange capacity per gram is much lower due to their lower surface area per gram.

So far we have been concerned with the negative charges carried on the surface of the layers forming a clay particle, but these particles can also carry electric charges on their broken edges. At the broken edge the open packed hexagonal oxygens can only touch one instead of two silicon ions, so only one of their negative charges is neutralised by a silicon and the other is neutralised in acid conditions, by a hydrogen ion, giving a silanol or silicic acid-like group $\equiv$Si—OH, which dissociates the hydrogen ion in alkaline conditions. This can be seen in the buffer curve of the montmorillonite given in Fig. 7.11 on page 110. The charge on the particle is due solely to its permanent negative charge up to pH 6, and it then increases by about 50 m eq/ 100 g up to pH 11–12. A rough calculation shows that this is of the order to be expected, for a typical montmorillonite particle is one layer thick and can be considered a circular disc about 200 Å diameter. There will be a hydroxyl attached to a silicon at about every 5 Å around the periphery, so such a

1 *Soil Conf. NZ*, 1962, 79.

particle would contain 340 unit cells and 125 broken bond hydroxyls, which is equivalent to 50 m eq/100 g of clay. Naturally the larger the clay crystal, and most micaceous clay particles are larger, the smaller is the increase in negative charge with increasing pH.

Kaolinite particles also show this effect, though the ratio of the pH dependent charge to the permanent negative charge is much higher than in the micaceous clay minerals. They can also acquire a positive charge in acid conditions, Schofield and Samson[1] have shown that the conditions on their broken edges under very acid and very alkaline conditions can be represented as follows:

$$
\begin{array}{cc}
\text{charge} & \text{charge} \\
\text{Si}\!\!<\!\!\begin{array}{l}\text{OH}\\ \text{OH}\end{array} & \text{Si}\!\!<\!\!\begin{array}{l}\text{O} \quad -1\\ \text{O} \quad -\frac{1}{2}\end{array} \\
\text{Al}\!\!<\!\!\begin{array}{l} \quad +\frac{1}{2}\\ \text{OH} \quad +\frac{1}{2}\\ \text{H}\end{array} & \text{Al}\!\!<\!\!\begin{array}{l}\text{OH} \quad -\frac{1}{2}\end{array} \\
\text{acid conditions} & \text{alkaline conditions}
\end{array}
$$

This requires that the surface of each unit cell on the broken edge, which has an area of 33 $Å^2$, should acquire a positive charge of one unit in acid and a negative charge of two units in alkaline conditions. These positive charges are neutralised either by simple anions, or sometimes in part by very small clay particles, such as montmorillonite particles, sticking to the broken edge.

1 R. K. Schofield and H. R. Samson, *Disc. Faraday. Soc.*, 1954, **18**, 135.

7

The cation and anion holding powers of soils

We have seen in the previous chapter that soil clays carry negative charges on their surfaces due to isomorphous replacement of cations in their crystal lattices, which are balanced by simple cations, typically calcium, magnesium, potassium, sodium and, in some soils, aluminium and manganese, situated on the external surfaces. We will be seeing in Chapter 15, that soils also contain humus which contains true weak acids, and, in the pH range found in most cultivated soils, these acids have dissociated a part of their hydrogen ions so also carry a negative charge. In addition soils may carry some positive charges, particularly in acid conditions, if they have sesquioxide films on their surfaces, due to the adsorption of hydrogen ions by hydroxyls on these films; but the positive charge is small compared with the negative, if the pH is over 6 in most agricultural soils.

These cations which are outside the clay particle can be quantitatively replaced by other cations if a suitable salt solution is percolated through the soil. Thus, if an ammonium salt, say, ammonium chloride, is percolated through a soil in which calcium is the principal neutralising cation, the leachate will initially be found to contain both calcium chloride and ammonium chloride and, if the leaching is continued long enough, most of the calcium on the clay will have been replaced by ammonium. This is the phenomenon of cation exchange, for one cation has been exchanged quantitatively, that is equivalent for equivalent, by another; and because these cations are exchangeable they are said to be the exchangeable cations held by the soil.

The composition of the exchangeable cations held by a given soil depends on the composition of any weatherable minerals the soil may contain and their rate of weathering, possibly also on the mineral composition of the rain water and of any dust blown in from outside the area and, for agricultural soils, on the amount of cations applied in fertilisers, lime or manures to the land. Typically, in well-farmed temperate soils, calcium constitutes over 80 per cent of the exchangeable cations, magnesium up to 15 per cent and sodium and potassium together rarely more than 5 per cent. Soils derived from magnesium-rich rocks will have a higher proportion of magnesium, and soils containing free sodium salts a higher proportion of sodium. Natural leached non-calcareous soils often have as much or more exchangeable magnesium as calcium, but in these soils aluminium ions usually constitute an appreciable proportion of the exchangeable cations.

Cation exchange

The Ratio Law

The proportions of the different cations held by a soil are largely dependent on their concentrations or, to be more exact, on their activities in the solution bathing the soil particles. If a soil holds only two species of cation, each with the same valency, as, for example, sodium and potassium, or magnesium and calcium, then the ratios of the amounts held depends primarily on the ratio of their activities in the bathing solution. But if one of the cations is monovalent and the other divalent then the ratio of the amounts held by the soil is dependent on the ratio of the activity of the monovalent to the square root of the divalent in the solution. Similarly, if one is monovalent and the other trivalent, then the ratio of the two on the soil depends on the ratio of the activity of the monovalent to the cube root of the activity of the trivalent. In the general case, if a soil is in equilibrium with a large volume of a dilute solution, the equilibrium will not be upset if the activities of the monovalent ions are altered in one ratio, of the divalents in the square root and the tri-valents in the cube root of that ratio. This statement is often called the Ratio Law. It has the important consequence that if a soil containing both mono-valent and divalent cations is in equilibrium with its bathing solution, then on diluting the solution by adding water, the equilibrium will be upset and a proportion of divalent cations in the solution will exchange with an equiva-lent number of monovalents from the soil. Conversely, if the solution is concentrated by water being removed, as, for example, by plants taking up water from the solution, some monovalent ions from the solution will ex-change with some divalents from the soil. If the monovalents are principally sodium, this may seriously complicate the maintenance of the permeability of the soil, as will be discussed on p. 137.

The Ratio Law is reasonably valid for a wide range of soil conditions, but its validity depends on certain conditions being fulfilled. The first essential condition is that all the exchangeable cations are allowed for. Thus, if the soil is acid, the concentration of the aluminium ions in the soil solution must not be ignored. The second condition is that the relative concentration of the cations close to the fixed negative charges on the clay surface is high in rela-tion to the external solution, and that anions from the solution do not penetrate into the Stern layer or the inner part of the diffuse Gouy layer (see p. 138). This condition is obviously not fulfilled if there are positive charges on the soil surfaces close to the negative charges, or if any anions in the solution can be absorbed on these surfaces. The Natal soil, whose buffer curve is given in Fig. 7.13 is an example of a soil for which the Ratio Law will not in general, be valid.

The Ratio Law will also fail if a particular species of cation in the solution is present in a sufficiently high concentration to cause an increase in the amount of cations held by the clay that becomes exchangeable. Thus, a high calcium concentration in the solution may cause an opening-up of the clay lattices which, in turn, may increase the number of interlayer potassium ions that become exchangeable. It will also be difficult to test when the composition of the bathing solution is sufficiently concentrated for appreciable errors to arise in the determination of the activity ratios of the ions. The Ratio Law has been tested by a number of workers and has been found valid within these limitations. Thus, it has been shown to be valid by R. K. Schofield and A. W. Taylor[1] for a Rothamsted soil when one cation was hydrogen and the other sodium, potassium, calcium, calcium plus magnesium or aluminium; by A. W. Taylor[2] for the potassium–calcium pair on some Rothamsted soils low in potassium; and by P. H. T. Beckett[3] for potassium–calcium and potassium–calcium plus magnesium for a Lower Greensand soil, as is shown in Table 7.1. M. E. Sumner and J. M. Marques[4] tested it

TABLE 7.1   The validity of the Ratio Law for a Lower Greensand soil. Exchange system potassium–calcium and potassium–calcium plus magnesium

Effect of varying the concentration of $CaCl_2$ in the solution		Effect of varying the ratio of calcium to magnesium in the solution	
Concentration of Ca + Mg in the solution *mol/l*	*Equilibrium activity ratio* $a_K/\sqrt{a_{Ca+Mg}}$ $(mol/l)^{\frac{1}{2}}$	*Relative activity of magnesium in the solution* a_{Mg}/a_{Ca}	*Equilibrium activity ratio* $a_K/\sqrt{a_{Ca+Mg}}$ $(mol/l)^{\frac{1}{2}}$
0·0603	0·0138	0·0695	0·0141
0·0194	0·0136	0·1675	0·0137
0·0062	0·0128	0·375	0·0137
0·0021	0·0131	0·557	0·0135
0·00063	0·0138	0·746	0·0139

for the potassium–calcium system on a Natal soil similar to the one whose buffer curve is given in Fig. 7.13, but whose positive charges only equalled 20 per cent of its negative, and found it was valid provided the electrolyte concentration did not exceed 10^{-2} M.

1 *J. Soil Sci.*, 1955, **6**, 137; *Proc. Soil Sci. Soc. Amer.*, 1955, **19**, 164.
2 *Proc. Soil Sci. Soc. Amer.*, 1958, **22**, 511.
3 *Soil Sci.*, 1965, **100**, 118.
4 *Agrochimica*, 1968, **12**, 191.

Relative strengths of cation adsorption by clays

Soils hold different species of cations with different tightnesses of binding, and correspondingly cations in solution have different powers of displacing a given exchangeable cation. Thus, if a calcium-saturated soil is shaken up in a series of chlorides of constant normality, sodium chloride will displace less calcium than will potassium or ammonium chlorides, which will displace less than caesium chloride; and magnesium chloride will displace less calcium than will strontium chloride, which will displace less than barium chloride. Correspondingly, if samples of a soil are saturated with different cations, and shaken up in, say, a standard potassium chloride solution, the potassium will displace more sodium from the sodium-saturated soil than ammonium from the ammonium-saturated, and it will displace a little more magnesium than calcium, and more calcium than barium from the corresponding saturated soils. Table 7.2, due to J. E. Gieseking and H. Jenny[1]

TABLE 7.2   Displacing power of different cations. Putnam subsoil clay holding 4·5 m eq of exchangeable NH_4 or Ca. Chloride solution contains 4·5 m eq of cation in 500 ml of solution.

| | Per cent of exchangeable cation displaced | |
Chloride	NH_4–clay	Ca–clay
NaCl	35·3	12·7
KCl	51·3	28·8
RbCl	62·6	43·8
CsCl	68·8	50·8
$MgCl_2$	65·4	47·5
$CaCl_2$	63·6	—
$BaCl_2$	71·7	53·0
$ThCl_4$	80·9	80·2
HCl	84·9	77·8

| | Clay holding 4·5 m eq of exchangeable cation | | | |
	Mg-clay	Ba-clay	La-clay	Th-clay
KCl	31·3	26·7	14·0	1·8

illustrate some of these points for a Putnam subsoil clay, which is a montmorillonite though, for this clay, potassium seems to displace calcium slightly more easily than magnesium.

The relative tightness of binding of cations to clays depends to some extent on the type of clay, particularly if potassium is one of the cations concerned, for it tends to be held strongly on the surface of micaceous clays.

1 *Soil Sci.*, 1936, **42**, 273.

Some old results of P. Schachtschabel[1] illustrate this. He equilibrated a humic acid, a montmorillonite, a kaolinite and a muscovite in an acetate solution 0·05 N in both calcium and ammonium, and found that calcium formed 92, 63, 54 and 6 per cent of the exchangeable ions on these four materials, showing the strong preference for calcium by the humic acid and ammonium by the muscovite; and potassium would be expected to behave very similarly to ammonium.

The ability of clays to hold barium ions strongly is illustrated by R. Bradfield's[2] observation that, if a sodium-saturated bentonite is shaken with a barium sulphate suspension, the clay is converted to a barium-clay leaving sodium sulphate in solution. Again, the ability of clays to hold calcium more tightly than sodium is illustrated by some observations of G. H. Chaudhry and P. Warkentin.[3] If they leached soils high in exchangeable sodium slowly with a solution that was half-saturated in calcium sulphate, the leachate was almost pure sodium sulphate, until the exchangeable sodium was reduced to a low level; but if they leached a soil high in exchangeable magnesium with this calcium sulphate solution, the leachate always contained appreciable amounts of calcium sulphate, showing that the magnesium is held about as strongly as calcium.

The relative tightness of binding for large ions has not been worked out in detail, but they tend to be bound more tightly than small ones. Thus, diamines are more tightly held than the corresponding simple amines, and the tetra-substituted ammonium ion is still more strongly held; and very large cations, such as methylene blue and brucine, are so strongly held that it is very difficult to displace them, except with other equally large cations.[4] As a consequence of this very tight binding, clay particles dispersed in an acid medium can be made positively charged by adding just sufficient of one of these large cations to neutralise the negative charges on the clay surface. But clay particles can absorb more milli-equivalents of some large organic cations than corresponds to their negative charge and some, such as the cetyl pyridinium ion, will form an oriented monolayer covering the clay surface provided the balancing anion is bromide, so it can be used to determine the surface area of soil clays.[5]

The quantitative laws of cation exchange

The relative tightness of binding of cations to a clay, or their relative displacing power, can be measured quantitatively. If a soil holds only two species of cations, M_1 and M_2, and $(Ex\,M_1)$ and $(Ex\,M_2)$ represent the

1 *Kolloid-Beih.*, 1940, **51**, 199.
2 *J. Phys. Chem.*, 1932, **36**, 340.
3 *Soil Sci.*, 1968, **105**, 190.
4 J. E. Gieseking, *Soil Sci.*, 1939, **47**, 1.
5 D. J. Greenland and J. P. Quirk, *J. Soil Sci.*, 1964, **18**, 178.

amounts of M_1 and M_2 held as exchangeable cations on the soil and AR_{1-2} is the activity ratio of these ions in the bathing solution when in equilibrium, then a quantity k can be defined by the equation

$$AR_{1-2} = k(\text{Ex } M_1)/(\text{Ex } M_2)$$

which gives a measure of the relative strength with which the soil holds M_1 compared with M_2. This equation assumes the validity of the Ratio Law, and the activity ratio is defined as the ratio of the activity of the two ions in the solution, if they have the same valency; but as $\sqrt[n]{a_1}/\sqrt[m]{a_2}$ where n and m are the valencies of the cations, if they are different. The value of k is independent of the units in which one expresses the activities if the ions have the same valency; but if M_1 is monovalent and M_2 divalent, k is 31·6 (equals $\sqrt{1000}$) times larger if the activities are expressed in milli-moles per litre instead of in moles per litre. In this chapter, activities will be expressed in moles per litre unless stated to the contrary. This relation between the activity ratio and the ratio of the exchangeable ions is known as Gapon's equation, and Gapon, who introduced it, considered that k should be a constant independent of the activity ratio, an assumption which has been found to be reasonably correct for a fairly wide range of activity ratios.

Gapon's equation is the simplest of numerous equations that have been proposed linking the relative activities of the ions in solution to their relative proportions as exchangeable ions. A second simple equation that has been widely used is due to W. H. Kerr[1], and has a better theoretical justification. It is the same as Gapon's equation if the cations have the same valency, but if ion M_1 has a valency of m and M_2 of n, and a_1 and a_2 are their activities in the bathing solution, the equation is

$$(a_1)^n/(a_2)^m = k(\text{Ex } M_1)^n/(\text{Ex } M_2)^m$$

which in the case of exchange between mono- and divalent ions can be put in the form

$$AR_{1-2} = k(\text{Ex } M_1)/\sqrt{(\text{Ex } M_2)}$$

which is now directly comparable with the Gapon equation. B. C. Coulter and O. Talibudeen[2] have given examples of the validity of the general equation for the Ca:Al and K:Al exchange with some clay minerals and two soils. It is possible to calculate forms of these equations, based on the Gouy diffuse double layer theory, which have a much wider range of validity than the Gapon and Kerr equations, but their particular form depends on the simplifying assumptions that must be made to allow the necessary equations to be formulated and solved.[3] Figure 7.1 illustrates the important and surprising result that, for a number of soils, the two equations sometimes

1 *Soil Sci.*, 1928, **26**, 385; A. P. Vanselow, *Soil Sci.*, 1932, **33**, 95.
2 *J. Soil Sci.*, 1968, **19**, 237; 1969, **20**, 72.
3 For a review of these, see G. H. Bolt, *Neth. J. agric. Sci.*, 1967, **15**, 81.

fit the same experimental data almost equally well over an appreciable concentration range,[1] which is usually smaller for the Gapon. However, the
Gapon equation is often only valid for the potassium–calcium exchange
when the proportion of exchangeable potassium to calcium is small.

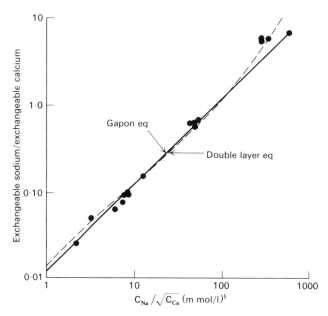

FIG. 7.1. Comparison of the Gapon Equation with a more exact equation based on
diffuse double layer theory for sodium–calcium exchange with an illite

These equations assume that the ratio of the exchangeable cations on the
clay surface is dependent solely on their activity ratio in the bathing solution,
and that it does not depend on the past history of the clay. This reversibility
is rarely found for actual soils or clays when they are present as crumbs or
aggregates, because a proportion of the exchangeable ions do not appear to
be readily accessible to the bathing solution. Thus, if dried crumbs of a
sodium-saturated and a calcium-saturated soil are equilibrated with a
sodium and calcium chloride solution of a given activity ratio, the soil that
was originally calcium-saturated is likely to hold more calcium than the soil
that was originally sodium-saturated, because the clay plates that are held
together by the calcium ions have only a very limited power of swelling in
solutions; while sodium soils do not suffer from this restriction (see p. 135),
so are more likely to give the true reversible equilibrium value.[2]

1 Taken from G. H. Bolt, *Soil Sci.*, 1955, **79**, 267. See also J. V. Langerwerff and G. H. Bolt,
  ibid., 1959, **87**, 217.
2 For an example with a NH_4-Ca system see R. van Bladel and H. Laudelout, *Soil Sci.*, 1967,
  104, 134.

The Gapon constant k gives a measure of the relative tightness of binding of cations to a clay; it varies from 1·7 to 3·1 for the sodium–potassium exchange on montmorillonites,[1] being greater the higher the charge density on the clay surface, and it is as high as 7 for micaceous clays,[2] which is a reflection of the strength with which these clay surfaces hold potassium ions. The corresponding value of k for the potassium–calcium exchange on a montmorillonite is 0·075, showing that calcium is held the more strongly.[3] R. C. Salmon[4] found a value of k for the magnesium–calcium exchange on a montmorillonite and on a micaceous clay to be 1·22 for magnesium saturations up to 80 per cent,[5] but the value was 3·3 for a peat at low magnesium content rising to 5·3 when 40 per cent of the exchangeable cations were magnesium. His values for forty British soils fell between the value for the clays and the peat. However, soils high in vermiculite probably have a greater affinity for magnesium, and soils containing much hydroxy-aluminium ions a greater affinity for calcium.[6] The US salinity laboratory find $k = 2·14$ for the sodium–calcium exchange for a large number of soils, presumably micaceous, because they also find a value of $k = 0·30$ for the potassium–calcium exchange. Another pair of cations of different valencies of relevance to field soils is calcium and aluminium. B. C. Coulter and O. Talibudeen[7] showed that the value of k for this exchange favoured the Al^{3+} ion most strongly for vermiculite clays, least strongly for montmorillonite, with the micaceous intermediate.

Soils never, in fact, contain only two species of cation, but the Gapon equation is still of practical value for two species when other species are present, and its validity is not affected by treating magnesium and calcium as equivalent ions over a quite wide range of magnesium–calcium ratios. Thus, the Gapon equation relating a monovalent ion M to the divalent ions in a soil is commonly written

$$\frac{\text{Activity of ion M in solution}}{\sqrt{(\text{activity of } Ca^{2+} + Mg^{2+} \text{ in solution})}} = \frac{k \, Ex \, (M)}{Ex(Ca + Mg)}$$

Its validity for a given value of exchangeable potassium to exchangeable calcium plus magnesium has already been illustrated in Table 7.1, for a Lower Greensand soil, and it was also found valid by R. K. Schofield and A. W. Taylor[8] for the Rothamsted soil. C. A. Bower[9] further showed that,

1 A. A. Tabikh, I. Barshad and R. Overstreet, *Soil Sci.*, 1960, **90**, 219.
2 Based on data given in the *US Dept. Agric. Handbk. 60*, 1954.
3 I. Shainberg and W. D. Kemper, *Soil Sci.*, 1967, **103**, 4.
4 *J. Soil Sci.*, 1964, **15**, 273.
5 J. S. Clark (*Canad. J. Soil Sci.*, 1966, **46**, 271) found a value of 1·06 for the montmorillonite he used, which was independent of pH in the range pH 3·5 to 9·2.
6 V. E. Hunsaker and P. F. Pratt, *Proc. Soil Sci. Soc. Amer.*, 1971, **35**, 151.
7 *J. Soil Sci.*, 1968, **19**, 237.
8 *Proc. Soil Sci. Soc. Amer.*, 1955, **19**, 164.
9 *Soil Sci.*, 1959, **88**, 32.

for a Chino soil, the ratio of the exchangeable sodium to the sum of the exchangeable calcium plus magnesium was independent of the amount of potassium in the solution, even if the potassium constituted up to 30 per cent of the exchangeable cations. This result is illustrated in Fig. 7·2, which also shows that for this soil the sodium–calcium curve is the same as the sodium–calcium plus magnesium up to an activity ratio of unity, that is between 30 and 50 when the concentrations are expressed in milli-moles per litre, above

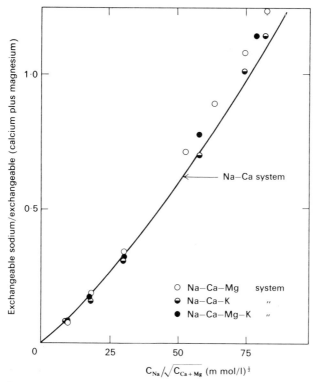

FIG. 7.2. Influence of other cations on the applicability of the Gapon equation for the sodium–calcium plus magnesium exchange

which there is a small differential effect due to magnesium. This result would not be found if calcium plus magnesium only form a small proportion of the total exchangeable ions. Thus. P. B. Tinker[1] found it did not hold for some acid southern Nigerian sands in which the exchangeable aluminium was appreciably greater than the exchangeable calcium plus magnesium; nor does the Ratio Law hold for such soils if the aluminium ion concentration in the solution is ignored.

1 *J. Soil Sci.*, 1964, **15**, 24.

The quantity k in the Gapon equation does not remain constant as one of the ions becomes present in a very low proportion. This is illustrated in Fig. 7.3, taken from some work of P. H. T. Beckett[1] for a Lower Greensand soil. In this figure the activity ratio of the equilibrium solution is plotted on one axis, and the amount of potassium taken up or given up by the soil is plotted on the other.

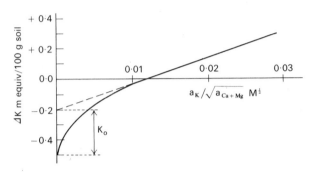

FIG. 7.3. The exchangeable potassium–activity ratio—Q/I—curve for a Lower Greensand soil

Beckett called such a curve a Q/I curve, Q representing the change in the quantity of the exchangeable cation, in this case potassium, held by the soil, and I the intensity of the potassium ion activity relative to the dominant cations in the bathing solution, in this case calcium and magnesium. The soil contains just over 8 m eq/100 g of exchangeable cations, of which 6·9 m eq are calcium, 0·7 magnesium and 0·5 potassium, so the sum of the exchangeable calcium plus magnesium is effectively constant in the range of activity ratios covered in the figure. Figure 7.3 also shows that, for activity ratios in excess of 0·01, there is a linear relation between increase in the amount of exchangeable potassium held by the soil and the increase in activity ratio, as one would expect if the Gapon equation is valid. This linearity above a certain activity ratio is not found for all soils, however, although many workers have made this assumption. Thus the curve does not become linear for either the Rothamsted or Woburn soils in the activity ratio range found even from samples taken from plots which have received a high level of fertiliser over a long period of years.[2] Below an activity ratio of 0·01, this Lower Greensand soil releases more potassium for a given drop in activity ratio than would be found if the linear relation continued to hold. This amount is shown as K_0 in Fig. 7.3, and for this particular soil equals about 0·3 m eq. G. H. Bolt[3]

1 *J. Soil Sci.*, 1964, **15**, 9; *Soil Sci.*, 1964, **97**, 376.
2 T. M. Addiscott, *J. agric. Sci.*, 1970, **74**, 131.
3 *Soil Sci.*, 1964, **97**, 382. See also J. C. van Schouwenburg and A. C. Schuffelen, *Neth. J. agric. Sci.*, 1963, **11**, 13 and G. H. Bolt, M. E. Sumner and A. Kamphorst, *Proc. Soil Sci. Soc. Amer.*, 1963, **27**, 294.

calculated that this potassium was held about 200 times more tightly to the soil than the remainder. It also differs from the remainder in that equilibrium between solution and soil particles is almost instantaneous for changes in activity ratios above 0·01, but is slow if the new equilibrium involves the tightly bound potassium.

This tightly bound potassium is probably held in holes or wedge-shaped spaces situated on the broken edge of clay sheets, into which the potassium ion can just fit, for neither a well-dispersed montmorillonite nor humic matter holds any of this tightly bound potassium, though a poorly dispersed montmorillonite will.[1] In soils, these spots can be blocked by humic matter, and by material, probably aluminium ions, which can be removed either by raising the pH or by treatment with sodium hexametaphosphate. This type of spot which holds potassium very firmly is a normal feature of nearly all the soils that have been examined, and these spots typically account for between 2 and 4 per cent of the exchangeable ions held by a soil.[2] This property is not confined to potassium, although it is most strongly shown by potassium, for it is also shown by sodium,[3] and in some soils all the exchangeable sodium is held on such sites; it is sometimes[4] but not always[5] shown by calcium or magnesium. Thus, clays, but not peats, possess specific sites for certain ions, probably depending on the size of the hole, crack or imperfection in the surface layer of the clay crystal in which the site is situated.

The potassium uptake-release curve for a soil, such as is shown in Fig. 7.1, can be described by three parameters—the activity ratio of the solution with which the soil is in equilibrium AR_0, the slope of the linear part of the curve, and the amount of specifically adsorbed potassium K_0. AR_0 can therefore be used to define the potassium status of the soil relative to its magnesium and calcium status, and the slope of the line, which is the reciprocal of the Gapon constant for the exchange, specifies the amount of potassium that must be added to a soil to give a unit increase in the activity ratio. Beckett called this slope the instantaneous potential buffer capacity of the soil, but the word potential is best omitted, and will be so in this book. It is largely dependent on its clay and humus content, though for a given clay content it is higher for a kaolinitic than an illitic clay. Figure 7.4 also due to Beckett,[6] gives the Δ K-AR curves for four contrasting soils. The value of the buffer capacity per unit of clay is between 2·0 and 2·4 for the Lower Greensand,

1 P. H. T. Beckett and M. H. M. Nafady, *J. Soil Sci.*, 1967, **18**, 263.
2 P. H. T. Beckett, *J. Soil Sci.*, 1964, **15**, 9. R. C. Salmon, *Rhod. J. agric. Res.*, 1965, **3**, 15, gives similar curves for thirty Rhodesian soils.
3 P. B. Tinker and J. Bolton, *Nature*, 1966, **212**, 548, and *Trans. 2nd and 4th Comm. Int. Soc. Soil Sci. (Aberdeen)*, 1966, 233. Also R. Levy and D. Hillel, *Soil Sci.*, 1968, **106**, 393. J. Bolton, *J. Soil Sci.*, 1971, **22**, 417.
4 P. H. T. Beckett, *Soil Sci.*, 1965, **100**, 118.
5 R. C. Salmon, *J. Soil Sci.*, 1964, **15**, 273.
6 *J. Soil Sci.*, 1964, **15**, 9.

Coral Rag and Middle Lias soils, all of whose clays are largely micaceous, and 3·1 for the basalt, much of whose clay is kaolinitic. The value for the sand is only 0·7, probably due to the humus being much more important than the clay, for Beckett and Nafady have shown that oxidising the humus

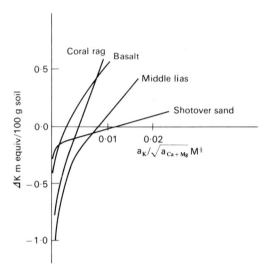

FIG. 7.4. The Q/I curves for four contrasting British soils

with hydrogen peroxide increases this buffering capacity. As already noted, for many soils the buffer capacity is not a constant in the range of exchangeable potassium contents of field soils, but increases with decrease in the exchangeable potassium content, expressed as a proportion of the cation exchange capacity.[1]

The intensity of exchangeable potassium, at a given level of exchangeable potassium relative to the other cations, can also be expressed in terms of the change in free energy of the system ΔG, when a small amount of potassium exchanges for a small amount of, say, calcium, for $\Delta G = RT \ln$ (potassium–calcium activity ratio), where R is the gas constant (8·31 J/mole/°K), and T the temperature of the system in °K. This reduces to $\Delta G = -5\cdot5(pK - \frac{1}{2}pCa)$ kJ/mole at 15°C, where pK and pCa are the negative logarithms, to base 10, of the potassium and calcium ion activities in the solution. The expression $(pK - \frac{1}{2}pCa)$ is sometimes called the potassium potential in the solution (relative to calcium), and some authors find it is more convenient to use this potential than the activity ratio.

The change in free energy when a calcium-saturated soil is converted to a potassium-saturated soil is calculated by integrating these values of ΔG over

1 See T. M. Addiscott, *J. agric. Sci.*, 1970, **74**, 131, for the Rothamsted and Woburn soils.

the whole exchange isotherm, that is from the condition where all the exchangeable ions are calcium to where they are all potassium. Thus, J. Deist and O. Talibudeen[1] calculated this change for a number of British soils and found that the free energy decreased by between 4 and 16 kJ/eq when calcium was replaced by potassium, and for a smaller selection of soils, by between 4 and 6 kJ/eq when sodium was replaced by potassium.

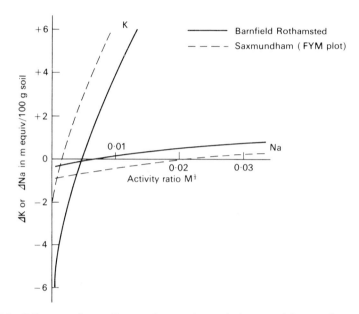

FIG. 7.5. Q/I curves for sodium and potassium relative to calcium and magnesium for two British soils.

This difference in the strength with which soils hold exchangeable potassium and sodium can be seen also in the exchange isotherm curves, as is illustrated in Fig. 7.5 for two English soils.[2] The soils are very weakly buffered for the sodium ions, compared with the potassium, and in both these soils, the sodium activity ratio in the soil is higher than the potassium activity ratio—a result Bolton found for all the soils he examined.

The fixation of potassium by soils

If a mica is put in a suitable chloride solution very low in potassium ions, potassium will slowly be released from between adjacent layers of the mica at its broken edge, and the spacing between the layers increased, so they slowly

1 *J. Soil Sci.*, 1967, **18**, 125. See also J. Robeyns, R. van Bladel and H. Laudelout, *J. Soil Sci.*, 1971, **22**, 336, for an example of the alkali metal ions and bentonite.
2 Taken from J. Bolton, *J. Soil Sci.*, 1971, **22**, 419.

open up to allow the larger hydrated ion to enter.[1] The stronger the binding between the potassium ions and the adjacent mica layers, the lower must be the concentration of potassium in the bathing solution before potassium begins to be released from between the layers. Thus if the bathing solution is 1 N NaCl, the potassium concentration in the solution must be below 10^{-5} M for potassium to be released from a muscovite mica, but below 2×10^{-4} M from a biotite.[2] There is a critical potassium concentration in the bathing solution for any particular mica sample, below which the mica will slowly release potassium until the concentration has risen to the critical value. The release is slow, because the potassium ions between the mica layers are not readily accessible to the outside solution, that is, they need an activation energy, which is about 50 kJ/mole for a biotite, to become accessible for exchange; and the depth of penetration of the replacing cation increases with the square root of the time of contact.

The micaceous clays found in soils are less ordered than are well-crystalline micas, and the forces between the interlayer potassium ions and the adjacent layers weaker. Thus, the critical potassium concentration in the soil solution to prevent interlayer potassium being lost may be as high as 5×10^{-3} M, about 500 times greater than for a muscovite. It is likely that in these micaceous clays the potassium ions do not occupy all the interlayer positions, but that there are channels occupied by other, larger, ions which cause the 2:1 layer to be buckled and so increases the rate of diffusion of ions into and out of the clay particle,[3] making some of the potassium ions more accessible to exchange. Further a proportion of the interlayer spaces in soil clays are occupied by OH_3^+ ions, which have about the same size as the potassium, and since the proton is very mobile, the $(OH_3)^+$ can easily become an uncharged water molecule, in which case the adjacent layers will repel each other and open up, allowing a larger cation to enter.

This process of slow exchange in a solution of low potassium concentration is only partly a reversible process, for if a clay that has been depleted of non-exchangeable potassium is put into a more concentrated potassium solution, a proportion of the potassium will be converted into a non-exchangeable form in a matter of minutes,[4] whereas desorption of this potassium is always a slow process.

Figure 7.6 illustrates some of these points for two Broadbalk soils,[5] one of which has received potassium fertilisers for most years since 1843 and the other which has received none; and it shows that if the soil is put in a solution whose potassium concentration is either larger or smaller than its equi-

1 For illustrations of this see J. A. Rausell-Colom, T. R. Sweatman *et al.*, in *Experimental Pedology*, Butterworth, 1965.
2 A. C. D. Newman, *J. Soil Sci.*, 1969, **20**, 357.
3 J. C. Martin, R. Overstreet and D. R. Hoagland, *Proc. Soil Sci. Soc. Amer.*, 1946, **10**, 94; I. Barshad, *Soil Sci.*, 1954, **27**, 463.
4 Quoted by P. W. Arnold, *Fertiliser Soc., Proc.* 115, 1970.
5 B. C. Matthews and P. H. T. Beckett, *J. agric. Sci.*, 1962, **58**, 59.

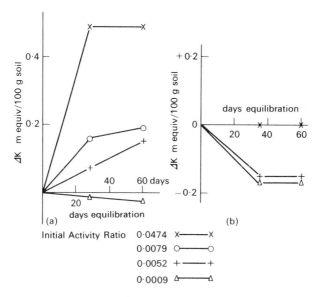

FIG. 7.6. The fixation and release of potassium from two Broadbalk soils when shaken in solutions of different activity ratios.

    (a) Plot 11, N P fertiliser, no K; $AR_0 = 0.0015$
    (b) Plot 13, N P K fertiliser; $AR_0 = 0.0145$

librium value in the soil, potassium will either be taken up from the solution or given up to the solution. In the graph the potassium–calcium activity ratios in the solution have been plotted rather than potassium concentrations, though it is the potassium concentration rather than the activity ratio that is important.

Soils in the field do not have a constant potassium ion concentration throughout the year, for the crop growing in the soil typically removes potassium continuously during its active period of growth, so will be reducing the exchangeable potassium in the soil; but since the rate of release of fixed potassium is appreciably slower than the rate of uptake, the exchangeable potassium falls during the growing season and then rises after crop growth has ceased,[1] which is typically in the winter in temperate regions. Again if a potassium fertiliser is added to a soil, a part of the added potassium will be converted to the fixed form, which will be greater, the greater the proportion that is not taken up by the crop.[2] If a soil regularly receives more potassium than is used by the crop, the strength of the 10·4 Å mica line in X-ray photographs of the clay may increase.[3]

1 See for example R. H. Bray and E. E. De Turk, *Proc. Soil Sci. Soc. Amer.*, 1939, **3**, 101, and E. E. De Turk, L. K. Wood and R. H. Bray, *Soil Sci.*, 1943, **55**, 1.
2 N. J. Volk, *Soil Sci.*, 1934, **37**, 267; W. E. Chambers, *J. agric. Sci.*, 1953, **43**, 479.
3 N. J. Volk, *Soil Sci.*, 1934, **37**, 267; C. I. Rich and J. A. Lutz, *Proc. Soil Sci. Soc. Amer.*, 1965, **29**, 167.

The amount of potassium fixed by a clay would not be expected to alter the relative tightness with which the accessible surfaces of the clay bind their exchangeable cations. This implies that the Gapon constant k for the potassium–calcium exchange, or Beckett's buffering capacity, should be independent of the amount of fixed potassium held by a soil. This has been demonstrated for the Rothamsted and Woburn soil,[1] and four other English soils,[2] and a group of West Indian soils.[3]

Various methods have been proposed for determining the amount of fixed potassium that a soil holds which can be fairly readily converted to the exchangeable form; but the fixed potassium is likely to be held in a wide range of positions, from those from which it can fairly readily be removed to those held as tightly as between muscovite sheets. Further, crops vary very greatly in their ability to reduce the potassium ion concentration in the soil solution (see p. 616), so no useful definition can be given for the crop available fixed potassium a soil holds. However, hydrogen-saturated exchange resins can be a useful means of measuring the less strongly fixed potassium,[4] particularly since resins of different bonding strengths are available. Sodium tetraphenyl boron, which forms an insoluble compound with potassium, can also be used for this purpose.

The proportion of potassium ions added to a soil that is converted to the fixed form is increased by drying the soil, so that alternate wetting and drying in the root zone of a crop will tend to increase the amount of fertiliser potassium that is converted to the fixed form. This effect of drying has the consequence that there need be no close connection between the amount of potassium fixed and the proportion of the exchangeable ions that are potassium, for this will depend on the intensity and frequency of the drying of the soil. Thus there can be no reversible equilibrium between the amounts of fixed and exchangeable potassium in a soil.

The fixation of fertiliser potassium takes place more readily in neutral than in acid soils, and liming an acid soil increases its ability to fix potassium.[5] The negative charge on acid clays is neutralised, in part, by aluminium ions which will bond the sheets together as in vermiculite if the sheets carry a sufficiently high negative charge. Potassium ions will only be able to diffuse very slowly into these interlamellar spaces to displace the aluminium, unless their concentration in the external solution is high.[6] But if the soil is limed, the aluminium ions will lose their charge, the interlamellar spacing will increase and calcium ions will be able to enter freely. Potassium ions will then

1 T. M. Addiscott, *J. agric. Sci.*, 1970, **74**, 131.
2 P. H. T. Beckett and M. H. M. Nafady, *J. Soil Sci.*, 1967, **18**, 244.
3 P. Moss, *Soil Sci.*, 1967, **103**, 196.
4 T. Haagsma and M. H. Miller, *Proc. Soil Sci. Soc. Amer.*, 1963, **27**, 153. T. E. Brown and B. C. Matthews, *Canad. J. Soil Sci.*, 1962, **42**, 266.
5 E. T. York, R. Bradfield and M. Peech, *Soil Sci.*, 1953, **76**, 379, 481; 1954, **77**, 53.
6 A. L. Page and T. J. Ganje, *Proc. Soil Sci. Soc. Amer.*, 1964, **28**, 199. D. L. Carter, M. E. Harward and J. L. Young, *Proc. Soil Sci. Soc. Amer.*, 1963, **27**, 283.

be able to enter these wider spaces more easily, displace the calcium and form islands which will pull the sheets together into a spacing approaching that of mica.

Potassium is not the only ion which shows this property of fixation and release by clays. Rubidium and ammonium, two monovalent ions of about the same size as potassiums, behave in exactly similar a manner; but whereas rubidium is a rare ion in the soil of little agricultural consequence, ammonium can occur in sufficient concentration to become fixed and so inaccessible for rapid exchange or for oxidation by bacteria. Fixed ammonium is the cause of the very low apparent carbon–nitrogen ratios found in subsoils (see p. 305), for it is non-exchangeable but is released during acid Kjeldahl digestion, when the organic nitrogen is being determined as ammonium. This fixed ammonium behaves just as fixed potassium. Thus, it is protected against displacement by calcium or sodium ions, for example, by a sufficient concentration of potassium ions in the outside solution.[1]

Potassium fixation complicates the concept of exchangeable cations because, during the process of fixation, some other cations become entrapped within the interlayer spaces that have been collapsed by the potassium ions. The exchangeable potassium can be determined if the potassium ions are replaced by ammonium, from ammonium acetate, for the high concentration of ammonium will prevent any fixed potassium ions being displaced, though some potassium may be fixed if the extraction is allowed to go over a long period of time. If strong solutions of sodium, calcium or barium salts are used to extract the exchangeable cations they will normally extract some fixed potassium as well,[2] unless the extraction is done quickly. But this difficulty of determining exchangeable potassium can be overcome by the use of a potassium isotope, for only exchangeable potassium equilibrates with it if the potassium activity ratio used corresponds to that in which the soil is in equilibrium.[3]

Acid clays

The discussion so far has been principally concerned with clays whose negative charges have been neutralised by the so-called exchangeable bases, to use an archaic but still extremely useful word, namely, calcium, magnesium, sodium, potassium and ammonium. If a calcium clay is leached with ammonium chloride, an ammonium clay is produced, and if this is then leached again with calcium chloride, the ammonium appears in the leachate as ammonium chloride. But if a calcium clay is leached with dilute hydrochloric acid, and this clay is then leached with calcium chloride, a proportion

1 G. E. Leggett and C. D. Moodie, *Proc. Soil Sci. Soc. Amer.*, 1963, **27**, 645. A. D. Scott, A. P. Edwards and J. M. Bremner, *Nature*, 1960, **185**, 792.
2 J. Deist and O. Talibudeen, *J. Soil Sci.*, 1967, **18**, 125.
3 E. R. Mercer and A. R. Gibbs, *Trans. Comm. II and IV, Int. Soc. Soil Sci. (Aberdeen)*, 1966, 233; M. E. Sumner and G. H. Bolt, *Proc. Soil Sci. Soc. Amer.*, 1962, **26**, 541.

of the displaced cations will be aluminium for most clays, but a mixture of aluminium and magnesium in those clays containing appreciable magnesium in the octahedral sheet. A hydrogen-saturated clay is, in fact, unstable for such a clay decomposes spontaneously to release sufficient aluminium or other metallic cations from the clay crystal to neutralise the negative charge.[1] In general this decomposition also releases silica from the clay, some of which appears in the solution as silicic acid, H_4SiO_4. Clays are stable in acid solutions provided most of the negative charges are neutralised by metallic cations, and it is for this reason one can extract free aluminium hydroxide from a clay by treating it with 0·05 N, or even 0·1 N HCl provided the solution also contains an adequate concentration of a salt, such as 1 N KCl or $CaCl_2$.[2] The reason for the instability of the hydrogen clay is probably that a proton from an exchangeable hydrion H_3O^+, situated near the broken edge of the clay plate, becomes transferred to an oxygen ion coordinated with aluminium in either the tetrahedral or octahedral sheet, or with magnesium in the octahedral. This reduces the negative charge around these cations which, being at the broken edge, will then escape into the solution.[3]

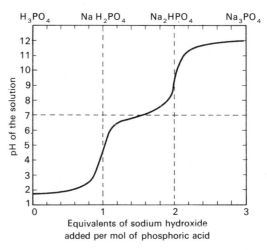

FIG. 7.7. Titration curve of phosphoric acid with sodium hydroxide

The electrochemistry of acid clays has had an interesting history,[4] because for a long time many soil chemists thought of acid soils as true acids, that is as hydrogen soils in which the hydrogen had definite dissociation constants. Further, the concept of a clay as a weak acid had a certain experimental

1 I. Barshad, *7th Int. Congr. Soil Sci.* (*Madison*), 1960, **2**, 435, and N. T. Coleman and D. Craig, *Soil Sci.*, 1961, **91**, 14.
2 Used by C. K. Tweneboah, D. J. Greenland and J. M. Oades, *Aust. J. Soil Res.*, 1967, **5**, 247.
3 R. J. Miller, *Proc. Soil Sci. Soc. Amer.*, 1965, **29**, 36.
4 For a short summary see H. Jenny, *Proc. Soil Sci. Soc. Amer.*, 1961, **25**, 428.

justification, for there are similarities between the neutralisation or buffer curve of a weak acid and an acid soil, if determined electrometrically. As an example, Fig. 7.7 gives the buffer curve for phosphoric acid, a weak polybasic acid whose three hydrogens have very different dissociation constants, and Fig. 7.8 gives buffer curves for a Rothamsted soil. This soil had been under

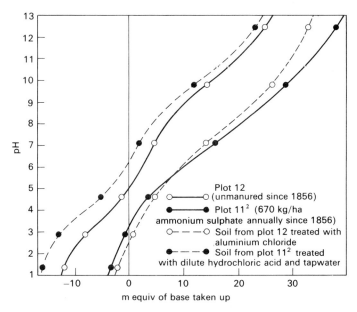

FIG. 7.8. Influence of aluminium on the buffer curve of Rothamsted soil (R. K. Schofield)

grass for several centuries, and plot 12 received no manure for at least a century while plot 11^2 has been manured annually with ammonium sulphate, and is in consequence very acid. The curve for plot 12 has similarities to that for a weak dibasic acid, with pKs about 4 and 9, while plot 11^2 is similar to a weak monobasic acid with a pK about 7, although the curves are too smooth to conform to the standard Henderson curves for simple weak acids.

There are several other points of interest shown in Fig. 7·8. First of all, the buffer curve for the soil of plot 12, if treated with aluminium chloride, is very similar to that of plot 11^2, indicating that the acid soil is probably an aluminium soil; and if the aluminium is removed from the soil of plot 11^2 by leaching with dilute acid, and the soil immediately leached with a calcium bicarbonate solution (tapwater), its buffer curve becomes similar to that of plot 12, except that it starts with a higher pH. Another feature, commonly found in many actual soils, is shown by the soil of plot 11^2 for its buffer curve is almost a straight line over the range normally found in agricultural soils from pH 4·5 to pH 8·5.

Soil and clay buffer curves are not easy to interpret. In the first place, the shape of the curve is very dependent on the salt concentration in the suspension, a point whose significance has often been overlooked, because the pH of a soil suspension is itself very dependent on the salt concentration if measured electrometrically (see p. 124). The shape of the curve depends on the hydroxide used, as is illustrated in Fig. 7.9, which shows the curves for a given soil and calcium and sodium hydroxides.

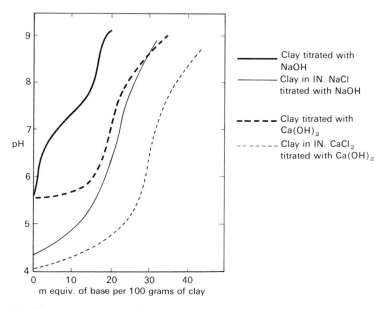

FIG. 7.9. Titration curve of a clay separated from a Bengal soil, showing the effect of the hydroxide used and the salt concentration

Neutralisation curves for acids, however, can be determined in other ways, two of the most direct being conductometric and thermometric titrations. The difference between a conductometric and electrometric determination is that changes in conductivity of the suspension are measured instead of changes in pH, and in thermometric the heat evolved per mole of hydroxide neutralised is determined.

Figure 7.10 gives an example of the conductometric titration curve for a freshly prepared H-montmorillonite, and for the same clay after it had aged in water for nine weeks; and it also gives the potentiometric curve for the fresh clay.[1] The conductometric curve for the fresh clay shows breaks in its slope at 64, 74 and 88 m eq, and that for the aged clay at 29, 50, 73 and 90 m eq. It is probable that the breaks at 64 and 29 m eq in the two curves are due to

1 R. P. Mitra and B. S. Kapoor, *Soil Sci.*, 1969, **108**, 11; for similar results see M. E. Harward and N. T. Coleman, *Soil Sci.*, 1954, **78**, 181.

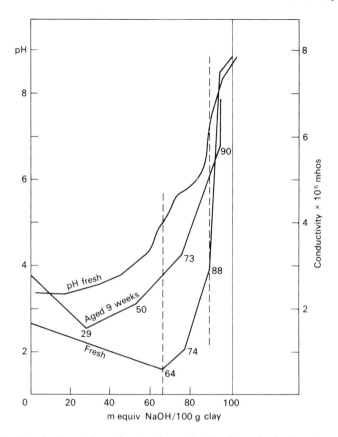

Fig. 7.10. Conductometric and potentiometric titration of a fresh and an aged acid montmorillonite clay

the neutralisation of exchangeable hydrogen ions, at 74 and 73 m eq to the neutralisation of hydroxy aluminium ions (see p. 116) and at 88 and 90 m eq to the neutralisation of hydrogen ions that have dissociated from the silanol or silicic acid groups at the broken edges of the clay lattices. The break at 50 m eq in the aged curve may be due to the neutralisation of aluminium ions. On this interpretation, the fresh clay contains 64 m eq of exchangeable hydrogen, 10 m eq of neutralisable hydroxy aluminium ions and 14 m eq of silanol hydrogens, while the aged clay contains 29 m eq of exchangeable hydrogen, 21 m eq of exchangeable aluminium, 23 m eq of neutralisable hydroxy aluminium and 17 m eq of silanol hydrogens.

Figure 7.11 gives the thermometric and potentiometric neutralisation curves for a somewhat similar montmorillonite, both when genuinely present as a hydrogen clay and after it had been allowed to age.[1] The heat evolved

1 W. H. Slabaugh, *J. Amer. Chem. Soc.*, 1952, **74**, 4462.

for the true hydrogen montmorillonite was 57 kJ/eq for the first 80 m eq of NaOH added, which corresponds almost exactly to the heat of neutralisation of hydrogen ions $H^+ + OH^- \rightarrow H_2O$. This is interpreted as showing that this sample contained 80 m eq of exchangeable hydrogen. After these

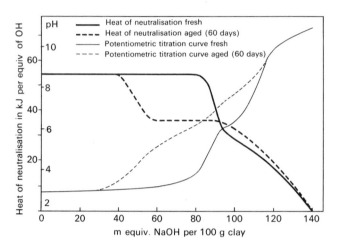

FIG. 7.11. The heat of neutralisation and the buffer curve of a freshly prepared and an aged acid montmorillonite

hydrogen ions had been neutralised, the heat evolved falls from 30 kJ/eq to zero almost linearly with the amount of hydroxide added, as it increases from 90 to 140 m eq. This probably represents the net heat of neutralisation of the hydrogen ions dissociating from the silanol groups, and so shows that it decreases approximately linearly with the increase in the proportion of hydrogens that have dissociated. The thermometric curve for the aged clay shows that it contains about 40 m eq of exchangeable hydrogen and about 50 m eq of presumably some form of aluminium ions having a heat of neutralisation of 38 kJ/eq. This technique does not allow the recognition of any sharp breaks in the curve, but if there is a sharp break between the ions having a heat of neutralisation of 57 kJ/eq and of 38 kJ/eq, the latter figure probably represents the heat of neutralisation of simple aluminium ions. This is because M. E. Harward and N. T. Coleman,[1] who also used this technique, found that their aged bentonite had a heat of neutralisation of only 21 kJ/eq in this part of the curve, which could be due to the aluminium on their clay being in the form of partially neutralised hydroxy ions.

Films of aluminium and ferric hydroxides and hydrous oxides carry positive charges, and the number depends on the isoelectric point of the surface, the pH and the salt concentration in the solution. The isoelectric

1 *Soil Sci.*, 1954, **78**, 181.

point of these films depends on the conditions prevailing when they were being precipitated, and is up to pH 9 for aluminium hydroxide and pH 8 for ferric.[1] The isoelectric point often appears to be lower than this in soils due to strongly adsorbed anion contaminants, such as silicates, phosphates and humates, present on their surface. Thus the number of positive charges found by these titration methods is increased if the humic matter is oxidised under alkaline conditions, with sodium hypobromite, for example. This is not found if the oxidation is done with hydrogen peroxide, for this produces some oxalates which are strongly adsorbed on the positively charged spots.

It is probable that the aluminium hydroxide films are usually a more important source of positive charges than the ferric in soils.[2]

The determination of the electrical charges on soil particles

Potentiometric buffer curves of soils and clays, as usually determined, are of the general form expected if the clay was genuinely a weak acid, and appear to be inconsistent with the picture of a clay particle carrying a definite negative charge due to isomorphous replacements. R. K. Schofield,[3] in a classical paper published in 1949, showed clearly under what conditions the information obtained from buffer curves could be properly interpreted in terms of the electrical charges carried by soil particles. Two conditions must be met: the soil must be free from exchangeable aluminium cations, and the conditions must be suitable to determine the number of positive charges present at the same time as the negative.

Positive charges will be most active in acid conditions whether they are due to the adsorption of hydrogen ions or the dissociation of hydroxyl. A clay dispersed in an acid salt solution will therefore adsorb some of the anions of the salt to neutralise the positive charges it contains. But the positive charges need not be neutralised by simple anions, for if they are very close to negative charges and the salt concentration is not too high, the double layers around the positive and negative spots will interpenetrate, with a consequent decrease in the cation concentration in the interpenetrating layers. Hence, such a system can only show its maximum adsorption for anions if the electrolyte concentration is sufficiently high to compress the double layers so that there is no interpretation. Schofield found that, in many soils, these spots were so close together that the chlorides of sodium, potassium or ammonium would not compress them sufficiently to prevent interpenetration, and he had to disperse the soil in alcohol, which suppressed the double layers almost completely, and then determine the adsorption of chloride from alcoholic hydrochloric acid, before he could

1 G. A. Parks, *Chem. Rev.*, 1965, **65**, 177.
2 C. K. Tweneboah, D. J. Greenland and J. M. Oades, *Aust. J. Soil Res.*, 1967, **5**, 247.
3 R. K. Schofield, *J. Soil Sci.*, 1949, **1**, 1.

measure the total number of positive charges the soil carried. He thus showed how buffer curves, properly used, could determine the relation between the pH of a soil and the number of positive charges it carries, the negative charge due to isomorphous replacement and the number of negative charges that can be created by the dissociation of hydrogen ions as the pH increased. The application of these ideas to actual buffer curves can be illustrated in Fig. 7.12, which gives the buffer curve of montmorillonite clay

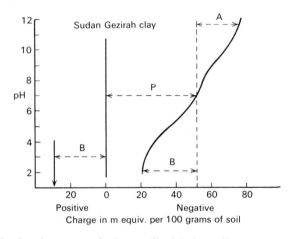

Fig. 7.12. The titration curve of a heavy alluvial clay soil.
P is the permanent negative charge on the clay.
A is the additional negative charge developed at high pH.
B is the positive charge developed at low pH.
The arrow shows the maximum uptake of chloride from alcoholic HCl.

soil from the Sudan Gezira. In this soil, the number of positive charges being discharged as the pH rises from 2 to 7 is almost exactly equal to the increase in apparent negative charge over this pH range, showing that the cause of the rise in negative charge is the suppression of the positive charges, and the actual negative charge remains constant up to pH 7. This is the charge due to isomorphous replacement. Only after the pH ran above 7, did additional 30 m eq of negative charges per 100 g soil arise. This is probably because the dissociation of hydrogen ions from the silanol groups on the broken edges of the clay plates, which is the amount expected if the clay, which is a montmorillonite, consisted of particles about 200 Å in size. Table 7.3 also illustrates this in more detail for a Rothamsted subsoil, carefully freed from exchangeable aluminium. The total positive charge, as determined in alcoholic HCl was 2 m eq, though the chloride adsorption from the $N/5 \, NH_4Cl$ solution was only 1·5, and the table shows that the negative charge was constant from pH 2·3–3·8, and then it began to rise. In this case, the cause of the rise between pH 3·8 and 6·2 is not known, but is probably

TABLE 7.3   Uptake of ammonium and chloride by a Rothamsted subsoil from an N/5 ammonium chloride solution at different pHs. Uptake in milli-equivalents per 100 grams of oven-dry soil

pH	Chloride adsorbed	Positive charge	Excess of NH_4 over Cl adsorbed	Negative charge
2·05	1·5	2·0	19·5	21·5
2·3	1·3	1·8	21·3	23·1
2·6	1·2	1·7	21·7	23·4
3·1	1·2	1·7	21·6	23·3
3·3	1·1	1·6	21 7	23·3
3·8	0·6	1·1	22·3	23·4
5·5	0·0	0·5	24·0	24·5
6·2	−0·4	0·1	25·7	25·8
7·15	−0·6	−0·1	27·0	26·9
7·4	−0·5	0·0	28·2	28·2

Samples treated with acid ammonium oxalate. Uptake in milli-equivalents per 100 grams of untreated oven-dry soil

pH	Chloride adsorbed	Positive charge	Excess of NH_4 over Cl adsorbed	Negative charge
2·5	−0·3	0·2	23·2	23·4
3·65	−0·5	0·0	23·5	23·5
4·15	−0·6	−0·1	23·3	23·2
7·5	−0·4	0·1	26·8	26·9

due to the presence of hydroxy-aluminium ions that were not removed when the exchangeable aluminium was.

In general, the negative charges due to isomorphous replacement are greater than the positive charges carried by soils, but there are a few soils in which the positive charges can be greater, so that below a certain pH, the soil will carry a net positive charge instead of a negative. The buffer curve of such a soil is illustrated in Fig. 7.13. It is a soil derived from a basalt, probably high in kaolinite clay and in iron and aluminium hydrated oxides. Such a soil is said to be amphoteric and to possess an isoelectric point, which for this soil occurs at pH 4 under the conditions for which this buffer curve was obtained. Amphoteric soils are rare in nature, and are probably only found in moderately humid climates for soils derived from rocks high in iron and aluminium.[1]

The discussion has been oversimplified for soils high in active aluminium hydroxide films, for the number of free positive and negative charges such a soil possesses depends on the salt concentration in the bathing solution. This is because the charges are determined by the number of hydrogen ions these surfaces possess, and at a given pH, these increase with the salt con-

[1] M. E. Sumner, *Clay Miner. Bull.*, 1963, **5**, 218, gives further examples of such soils from the same region.

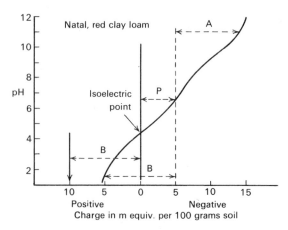

FIG. 7.13. The titration curve of a tropical red loam.
P is the permanent negative charge on the soil particles.
A is the additional negative charge developed at high pH.
B is the positive charge developed at low pH.
The arrow shows the maximum uptake of chloride from alcoholic HCl

centration (see p. 130). Thus R. G. Barber and D. L. Rowell[1] showed that a red earth soil from Nachingwea, Tanzania, carried an appreciable positive charge at pH 7·5 in $N.NH_4Cl$, but none in 0·02 N. Some of their results are given in Table 7.4. By making certain likely assumptions, they could interpret them as being due partly to free positive charges, partly to positive and negative charges being so close that their double layers overlapped, partly to negative charges on the hydroxide and partly to the soil clay minerals possessing a permanent negative charge of about 3 m eq. The distribution of charge at two values of the soil pH are given in Table 7.5. Table 7.4 also gives the apparent cation exchange capacity of the soil as determined by the standard ammonium acetate method, except that the concentration of the acetate is varied. This result, that the apparent exchange capacity increases with the concentration of the acetate, is well known for soils high in aluminium hydroxide gels, such as those containing allophane. It is probable that the surfaces which have these variable positive and negative charges are usually composed of aluminium hydroxide because, although the soils possessing this property are typically red and high in iron, careful removal of the iron only has a small effect on the results.[2] This is due to the greater tendency of ferric hydrated oxides to form crystals of relatively low surface area compared to the aluminium hydroxide films.

1 *J. Soil Sci.*, 1972, **23**, 135. For another example see N. G. Reeve and M. E. Sumner, *Proc. Soil Sci. Soc. Amer.*, 1971, **35**, 38.
2 T. L. Deshpande, D. J. Greenland and J. P. Quirk, *Trans 8th Int. Congr. Soil Sci.*, 1964, **3**, 1213; but M. E. Sumner, *Agrochim.*, 1962, **6**, 183 gives an example where the iron is more important.

TABLE 7.4   Absorption of ammonium and chloride by Nachingwea soil. Amounts of ammonium and chloride adsorbed in m eq/100 g soil

pH of solution (approx.)	Concentration of the ammonium chloride solution					
	1·0 N		0·2 N		0·02 N	
	NH_4	Cl	NH_4	Cl	NH_4	Cl
3	11·4	10·7	5·1	1·4	3·5	1·0
5	12·6	9·5	5·7	0·8	4·4	0·4
6·5	13·4	9·2	6·1	−0·1	5·0	0·1
7·5	16·5	9·5	7·8	1 1	6·3	0·3

Cation exchange capacity determined in ammonium acetate at pH 7
Concentration of acetate   1·0   0·2   0·02   0·004   N
Cation exchange capacity   9·68   7·91   6·45   5·39   m eq/100 g
Soil pH 4·78 in 1:1 water.

TABLE 7.5   Distribution of charges on a Nachingwea soil. Charges in m eq/100 g soil

Normality of NH_4Cl	pH 3			pH 7·5		
	Positive	Mixed	Negative	Positive	Mixed	Negative
1·0 N	14·2	0·0	14·1	11·7	0·0	20·2
0·2	6·4	4·8	10·0	5·2	3·6	11·9
0·02	2·0	1·0	4·5	1·6	1·4	7·9
0·002	0·6	0·3	3·2	0·5	0·4	4·8

The behaviour of exchangeable aluminium ions in acid soils

If an acid solution of aluminium chloride is titrated with a base, say, sodium hydroxide, the aluminium ions remain in solution until the pH has risen to about 4, and they then begin to precipitate out as $Al(OH)_3$, and when the pH has risen to about 4·5, the concentration of aluminium ions in solution has fallen to a low level; above pH 5, most of the ions will be present as the $Al(OH)^{2+}$. The negative logarithm of the dissociation constant of aluminium hydroxide is about 33·8, and the neutralisation reaction of the aluminium ions involves the loss of three protons from the six water molecules surrounding each ion, the elimination of the other three water molecules, and the sharing of each of the hydroxyls so produced between two aluminium ions. Thus, each aluminium ion remains in six-coordination, but with hydroxyls instead of water. The precipitate has the gibbsite structure, with each aluminium ion surrounded by six hydroxyls, and each hydroxyl shared by two aluminiums.

However, if the titration is stopped when only 80 per cent of the sodium hydroxide needed to neutralise the aluminium ions has been added, the solution may remain clear without any precipitate forming for many months, but if the solution is dialysed or diluted, a precipitate of $Al(OH)_3$ will form.[1] This behaviour is due to aluminium ions forming complex hydroxy ions, whose charge per aluminium ion is only 0·6 units instead of 3 for the simple ion. The constitution of these ions cannot yet be determined definitely, but R. A. Weissmiller[2] suggests they are composed of hexagonal rings of aluminium ions in which only a proportion of the oxygen ions in coordination with the aluminium are hydroxyls, so the complex is charged; and these rings may join up by sharing two aluminium ions. These would have compositions such as $Al_6(OH)_{12}(H_2O)_{12}^{6+}$ or $Al_{10}(OH)_{22}(H_2O)_{18}^{8+}$ for the single and for the double ring.

If an acid soil, containing aluminium, is titrated with a base, such as sodium hydroxide, in the presence of a salt solution, the aluminium ions are not fully neutralised until a pH much higher than 4–4·5 is reached. The Rothamsted soil, whose buffer curves are given in Fig. 7.8, illustrates this, for the curve for the high-aluminium soil of plot 11[2], is almost identical with the curve for the low-aluminium soil of plot 12 above pH 7.5, as can be seen by displacing one relative to the other so they coincide at this pH. The reason why the full neutralisation of aluminium takes place at a much higher pH when adsorbed on a clay surface than when in solution is that the apparent dissociation constant of water adsorbed on a clay surface is considerably lower than for free water, so that when a clay is in a solution of pH 7, for example, the dissociation of hydrogen ions from this adsorbed water may be 100 times greater than from free water (see p. 122). This means that a soil at pH 6 could still contain some Al^{3+} ions on its surface; and it could contain complex ions at a still higher pH. It is not yet certain what conditions favour the formation of these complex ions, but slow and only partial neutralisation of the negative charge on the aluminium ions, such as is obtained in the field by liming an acid soil, is probably one.[3]

These complex aluminium ions, when adsorbed on a clay surface, behave differently from the simple aluminium ions, for the latter can take part in cation exchange, and be exchanged against potassium ions, for example. But these complex ions are too strongly adsorbed on the clay surface so cannot take part in cation exchange, due to the large decrease in the entropy of the system (see p. 148), but they reduce the effective negative charge on a clay particle, in so far as this is determined by leaching the soil with a neutral salt, such as ammonium or barium chloride, and then determining the amount of ammonium or barium held by the soil. They can only be removed by pro-

1 P. H. Hsu, *Proc. Soil Sci. Soc. Amer.*, 1966, **30**, 173; and with T. F. Bates, *Mineral. Mag.*, 1964, **33**, 749. See also R. C. Turner, *Soil Sci.*, 1968, **106**, 291, 338, where $Ca(OH)_2$ was used as the base.
2 R. A. Weissmiller, J. L. Ahlrichs and J. L. White, *Proc. Soil Sci. Soc. Amer.*, 1967, **31**, 459, and see J. S. Richburg and F. Adams, *Proc. Soil Sci. Soc. Amer.*, 1970, **34**, 728.
3 N. T. Coleman, *Proc. Soil Sci. Soc. Amer.*, 1964, **28**, 35 and R. C. Turner, *Soil Sci.*, 1968, **106**, 291.

longed treatment of the soil with acidified neutral salt solution, or be completely neutralised by raising the pH above 7·5.

It is not yet clear where these large partially neutralised ions sit relative to the negative charges on the clay surface, but it is likely they can never grow to extensive films because it is unlikely they can cover many negatively charged spots when the film itself is unchanged. It is more likely that they will remain as a mosaic of spots each containing only a few, possibly only one or two hexagonal rings, so their composition may vary from $(Al_6(OH)_{12}(H_2O)_{12})^{6+}$ to $(Al_6(OH)_{18}(H_2O)^6)^0$, for example, and they will be situated on the surface in such a way that they cover the minimum number of negative charges.[1] This would allow simple cations to be fairly close to each negative charge when the soil pH is sufficiently high for the aluminium spots to be uncharged. If these spots contained some ferric ions as an impurity, they might not be dissolved by acidified potassium chloride solution at pH 3, which is the solution Schofield used to obtain the buffer curves for the Gezira clay shown in Fig. 7.12, In acid conditions, they can form a mosaic of positive charges in sufficiently close proximity to negative charges on the clay surface for their diffuse double layers to interpenetrate, unless these were almost completely suppressed by dispersion of the clay in alcohol.

It is also possible that these complex aluminium ions may form between contiguous clay sheets, binding them together as in a vermiculite. Potassium fixation takes place more actively in a limed than an acid soil, presumably due to the potassium ions having more difficulty in replacing interlamellar aluminium than interlamellar calcium ions, although these clay surfaces may hold potassium more tightly than aluminium. This behaviour would be expected if some of the interlamellar aluminium ions were present as complex ions.

We can now summarise the interpretation of electrometric buffer curves below pH 7 for mineral soils, particularly subsoils low in organic matter. The curves are due partly to polymerised hydrated aluminium ions dissociating hydrogen ions as the soil becomes more neutral. Under some conditions these partially neutralised aluminium ions may form sufficiently stable polymers for their removal to require strong acid treatment. Typically, the buffering capacity of these aluminium ions or polymers is approximately uniform in the pH range met with in soils from about pH 5 to pH 7·5, that is the buffer curve is approximately a straight line in this pH range. Soils also contain films of mixed iron and aluminium hydrated oxides which are positively charged in acid conditions, and which may have an isoelectric point above pH 7. Soils containing humic matter have an additional buffering mechanism, for humic matter contains genuine weakly acid groups, the strongest of which are carboxylic acids whose pK is in the range pH 4–5 (see p. 294).

1 For a review see C. I. Rich, *Clays Clay Miner.*, 1968, **16**, 15.

It is unlikely that ionic ferric iron behaves in the same way as aluminium in soil. The behaviour of its hydroxide is of less importance because ferric hydroxide forms at a much lower pH, so even if the iron could form ionic polymers comparable to the aluminium, they would not be expected in agricultural soils.

Cation exchange capacity of a soil

The early soil chemists thought of a soil as a weak acid possessing a definite base-holding capacity, and the acidity was thought to be neutralised by liming the soil. A neutral soil was thus saturated with calcium, as all its acidity was neutralised. An acid soil was unsaturated, and its degree of unsaturation was measured by the amount of exchangeable bases, that is calcium, magnesium, sodium and potassium that it held expressed as a percentage of the amount of these cations it held when saturated. This simple concept is obviously quite inadequate to explain the cation exchange and cation-holding power of soils, but the concepts of exchange capacity and base saturation are still current because of their supposed practical value.

The cation exchange capacity of a soil, as it is now called, is an arbitrary concept arbitrarily defined. A historically interesting definition is the amount of exchangeable calcium a soil will hold when in equilibrium with a calcium carbonate suspension through which air containing 0·03 per cent CO_2, the standard concentration in the atmosphere, is bubbled.[1] This should correspond to the total amount of cations the well-aerated soil should hold when in equilibrium with free calcium carbonate. But the most commonly used definitions are based on determining the amount of simple cations, such as ammonium or barium that a soil can hold when a salt solution buffered at some arbitrarily agreed pH is leached through the soil. The most widely used salt is ammonium acetate buffered at pH 7, but A. Mehlich's[2] method using barium chloride buffered at pH 8·1 with triethanolamine is also widely used. These two methods often give about the same value of the exchange capacity, though this is not always the case, and this exchange capacity is not necessarily the same as the permanent negative charge on the clay particles plus the cations held by the organic matter at that pH, although for many soils these two do not differ widely.

An alternative definition of exchange capacity, which may be of as much value as the previous ones, is the total amount of exchangeable cations the soil is actually holding, as determined by leaching with a neutral salt such as IN KCl or 0·1 N $BaCl_2$. The solutions are meant to replace not only the exchangeable calcium and magnesium, but the exchangeable aluminium as well, and they will replace any other exchangeable cations held by the soil,

---

1 R. Bradfield and M. Peech, *Trans. 2nd Comm. Int. Soc. Soil Sci.*, 1933A, 63, and *Proc. Soil Sci. Soc. Amer.*, 1942, **6**, 8.
2 *Proc. Soil Sci. Soc. Amer.*, 1939, **3**, 162, and *Soil Sci.*, 1948, **66**, 429.

such as divalent manganese. But unless these methods are properly standard-ised they will not displace all the simple aluminium ions, and probably none of the polymerised ions.[1] But if the methods are properly standardised, the difference between the exchange capacities so determined should measure the amount of complex aluminium ions held by the soil plus the additional amount of cations the soil organic matter holds between the actual pH of the soil and the pH of the buffered solution.

The concept of the exchange capacity of a soil has always been linked with that of the percentage saturation of this capacity with 'bases', that is calcium, magnesium, potassium and sodium; and as there are a number of different definitions of exchange capacity, so the percentage saturation of a soil with these alkali and alkaline earth cations will depend on the particular defini-tion used. In the past it was assumed that from the point of view of general crop production a neutral soil was the ideal to be aimed at, that is a soil in which all the acidity was neutralised, and this was thought of as a soil with a pH of 7. A soil at pH 7 would therefore by definition be saturated.

However, it is increasingly being found experimentally that crop pro-duction on many acid soils is limited by the aluminium ion concentration in the soil solution (see p. 660), which, in turn, depends on the concentra-tion of other cations in the solution, since the two processes that affect its concentration are the solubility of any aluminium hydroxide precipitates in the soil, and the cation exchange reactions between the soil solution and the exchangeable cations on the soil. This has two consequences. First, that the most useful agronomic definition of the degree of base unsaturation of a soil is the ratio of the exchangeable aluminium to the total exchangeable cations that the soil is actually holding. The second consequence is that in soils in which the sum of the exchangeable calcium, magnesium and aluminium constitute the bulk of the exchangeable cations, and the Gapon exchange equation is valid, the percentage unsaturation, as so defined, controls the concentration of the aluminium ions in solution through the relation:

$$(a_{Al})^{1/3}/(a_{Ca+Mg})^{1/2} = k \, Al_{exch}/(Ca + Mg)_{exch}$$

where a_{Al} and a_{Ca+Mg} are the activities of the aluminium and calcium plus magnesium ions in solution, and Al_{exch} and $(Ca + Mg)_{exch}$ are the amounts of exchangeable aluminium and calcium plus magnesium in the soil.

R. C. Turner[2] has examined the validity of this equation for a large number of soils, and found a unique relationship between these two quanti-ties. This is shown in Fig. 7.14,[3] in which $\frac{1}{3}p \, Al - \frac{1}{2}p(Ca + Mg) - 2.73$, called the corrected lime potential in the figure, is plotted against the percentage saturation of the soil with calcium plus magnesium, that is the percentage of the cations exchangeable with 2N NaCl that are calcium plus magnesium.

1 B. L. Sawhney, *Clays Clay Miner.*, 1968, **16**, 157.
2 With W. E. Nichol, *Soil Sci.*, 1962, **93**, 374; **94**, 58, with J. S. Clark, *Soil Sci.*, 1965, **99**, 194.
3 R. C. Turner and J. S. Clark, *Trans. Comm. II and IV Int. Soc. Soil Sci.*, 1966, 207.

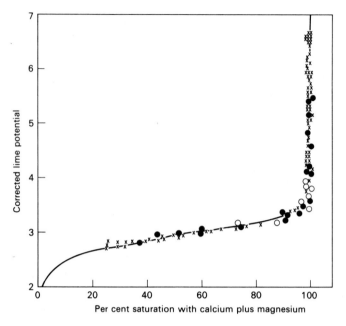

Fig. 7.14. The relation between the corrected lime potential of a soil and its per cent saturation with calcium plus magnesium

The reason the constant 2·73 is introduced is to allow for the effect of the solubility of gibbsite on the aluminium ion activity in the solution.

Clark[1] also finds that some soils are saturated at pH 5, and this is probably the reason that acid-sensitive crops can be grown on some acid tropical soils whose pH is about 5·5 and that liming these soils does not increase their yield (see p. 664).

It is still not quite certain if all the polymerised aluminium ions have lost their charge at pH 8·1, that is when the soil is in equilibrium with Mehlich's buffered barium chloride solution. There is some evidence that either simple or complex aluminium ions can exist between adjacent clay lattice layers sufficiently strongly held together for them to be very slowly neutralised in alkaline conditions. Thus McLean[2] found that if an acid soil was incubated with calcium carbonate, 1·3 equivalents of calcium as calcium carbonate had to be added for the acidity of the soils, as measured in Mehlich's buffer to be reduced by one equivalent. This is presumably because charged sites previously inaccessible to the buffer became opened up as the lime slowly discharges the polymerised aluminium ions, and allows calcium ions to enter the interlamellar space and neutralise the charges they were covering.

1 *Proc. Soil Sci. Soc. Amer.*, 1966, **30**, 93.
2 With W. R. Hourigan *et al.*, *Soil Sci.*, 1964, **97**, 119.

There are a group of soils which do not possess a definite cation exchange capacity, as it is usually defined, in that the apparent capacity depends strongly on the concentration of the salt solution used for saturating the soil with a cation. This behaviour is probably characteristic of certain types of allophane minerals,[1] which are very disordered alumino-silicates, and the number of positive and negative charges on the colloidal surfaces depends on the electrolyte concentration, as is shown in Tables 7.4 and 7.5.

In the discussion so far it has been assumed that the buffering of soils resides in either the clay fraction or the organic matter, but there are soils in which the sand or silt fraction may possess an appreciable cation exchange capacity. This is usually due to those fractions containing some vermiculites or chlorites, and they are probably most commonly found in soils derived from basic igneous rocks that are not too strongly weathered.[2]

The pH of a soil

The pH of an aqueous solution is defined as the negative logarithm of the hydrogen ion activity in the solution, which is the same as the hydrogen ion concentration if the solution is very dilute, but is increasingly smaller than this concentration as the salt content of the solution increases. This concept appears to be quite definite so long as one is dealing with volumes that are large compared with molecular dimensions, for the individual molecules and ions composing the solution are uniformly dispersed throughout it within the limits set by variations due to molecular and ionic thermal movements. These volumes may, however, have to be very large, in terms of molecular dimensions, at the end of the pH range of importance in soils. Thus a solution at pH 7 will, on the average, contain one hydrogen ion in a cube of side $0.25\,\mu$—just about the minimum-sized volume fairly clearly visible in a microscope fitted with a one-twelfth inch objective. But the pH of a soil dispersed in water is not a simple concept like this; for the soil particles, which carry ions attached to them, are very large compared to molecular dimensions, and the ions are therefore not uniformly distributed throughout the solution. The concept of the pH of a soil, or rather a soil suspension, can therefore only be discussed in relation to the properties of the ionic atmosphere around the soil particles.

Consider a negatively charged soil particle dispersed in water. The negative charge is neutralised by cations, some of which sit firmly on the clay surface forming the Stern layer, and some dissociate into the dispersion medium to form the Gouy diffuse double layer (see p. 139). The thickness of this double layer depends in part on the ions dissolved in the dispersion medium, and

---

1 For New Zealand soils see K. S. Birrell and M. Gradwell, *J. Soil Sci.*, 1956, **7**, 130; 1961, **12**, 307.
2 For an example for a soil from basalt in N. Ireland see D. M. McAleese *et al.*, *J. Soil Sci.*, 1957, **8**, 135; 1959, **9**, 66, 81, 289.

is thicker the more dilute the solution and the greater the hydration and the lower the valency of the cations. Conversely, the more concentrated the salt solution, the thinner the double layer, and the more nearly is the cation concentration just outside the Stern layer the same as in the solution in bulk.

This effect of salts is directly relevant to the limitations inherent in the concept of the pH of a soil. The hydrogen-ion concentration in the solution surrounding the soil particles in less than or the pH of the solution is higher than that close to the soil particles themselves, due to the hydrogen ion concentration gradient in the double layer. And as the double layer is made more compact by adding an electrolyte to the soil water system, so the hydrogen ion concentration gradient across the double layer is reduced, and the pH in the solution falls, to become more nearly equal to that close to the surface of the soil particle. The pH of the solution is thus greater than the pH just outside the Stern layer unless the salt concentration in the solution is high.

In some soils, the activity of the hydrogen ions in the Stern layer may be much higher than of those just outside it. This can be proved by comparing the rate of a pH-dependent enzyme activity when in free solution and when adsorbed on a clay surface. A. D. McLaren and E. F. Estermann,[1] for example, showed that the hydrolysis of denatured lysozyme by the enzyme chymotrypsin reached its maximum value at pH 7 when in solution, but at pH 9 when the enzyme was adsorbed on the surface of a kaolinite, showing that the hydrogen ion activity in the Stern layer appeared to be about a hundredfold higher than in the solution.

Soil pH is commonly measured either colorimetrically or electrometric-ally. A colorimetric determination is based on the use of an organic acid whose undissociated acid has a different colour from its anion, so that its colour is largely controlled by the pH of the solution in which it is dissolved. However, the colour can only be measured at all accurately in a solution from which the soil particles have been removed, and the colour also depends, to some extent, on the concentration of the various salts present in the solution as well as its pH.

Soil pH can also be measured electrometrically by measuring the potential difference between an electrode whose potential is determined by the hydrogen ion concentration of the solution in contact with it and a reference electrode whose potential should be independent of the pH. The normal hydrogen electrode is the glass electrode, and the reference electrode is a calomel electrode in which calomel is bathed in a solution of potassium chloride. Unfortunately, there is still some uncertainty about the exact behaviour of both the glass and the calomel electrodes when in colloidal dispersions, and of the calomel electrode in many solutions. Thus very appreciable junction potentials can be set up between a clay suspension or

1 *Arch. Biochem. Biophys.*, 1957, **68**, 157.

paste and the calomel electrode,[1] so that the measured potential difference between the two electrodes cannot be interpreted solely in terms of the hydrogen ion concentration outside the glass electrode. Further, it is not quite certain to what extent the glass electrode itself is affected by the hydrogen ions in the double layers around the clay particles, when they are fairly thin. The consequences of these uncertainties are that the least objectionable method for measuring soil pH is to measure the pH of a solution that is in equilibrium with the soil in its natural condition, but naturally this is a very inconvenient method because the composition of this solution has to be determined empirically. In practice, the pH of a soil has been commonly determined by shaking it up in distilled water, using either 2·5 or 5 times the weight of water to soil, and keeping the calomel electrode in the supernatant liquid. Since soils always contain some soluble salts, such as nitrates and bicarbonates, the apparent pH of the soil will depend on the amount of water added to the soil to make the suspension or dispersion, as is illustrated in Table 7.6.[2] Further the amount of soluble salts in a soil varies continuously

TABLE 7.6   The effect of soil-water ratio on the soil's pH. Calcareous alkaline soils from Arizona

Volume of water added cm³ per 100 g dry soil	pH of soil suspension or paste		
	Soil 1	Soil 2	Soil 3
10	7·45	9·10	7·95
25	7·60	9·40	8·00
100	7·70	9·85	8·20
1000	8·15	9·90	9·20

throughout the year, depending on such factors as the amount of rain percolating through the soil and the rate of nitrification in the soil, so that the apparent pH of a soil is often higher in wet weather or winter than in dry weather or summer, and under crop than under fallow.[3] However, soils in the field are very rarely uniform, and the seasonal variations are usually smaller than the variations from site to site in the same field, and variations of up to 1·0 to 1·5 units of pH may easily be found between samples taken on the same day.[4]

1 H. Jenny, T. R. Nielsen *et al.*, *Science*, 1950, **112**, 164.
2 W. T. McGeorge, *J. Amer. Soc. Agron.*, 1937, **29**, 841. For further examples see M. R. Huberty and A. R. C. Haas, *Soil Sci.*, 1940, **49**, 455 and 1941, **51**, 17, and L. E. Davis, *Soil Sci.*, 1943, **56**, 405.
3 See, for example, L. D. Bowen, *Soil Sci.*, 1927, **23**, 399; A. M. Smith and I. M. Robertson, *J. agric. Sci.*, 1931, **21**, 822; M. Raupach, *Aust. J. agric. Sci.*, 1951, **2**, 73, and Y. Kanehiro, Y. Matsusaka and G. D. Sherman, *Hawaii Agric. Exp. Sta., Tech. Bull.* 14, 1951.
4 See, for example, G. M. Robertson and K. Simpson, *E. Scot. Coll. Agric. (Edinburgh), Tech. Bull.* 8, 1954, and M. Raupach, *Aust. J. Agric. Res.*, 1951, **2**, 83.

This dependence of measured soil pH on salt concentration, which is continuously varying in natural soils, can be reduced by making all the measurements in a salt solution that is sufficiently strong to swamp the effects of the changes that occur naturally. But adding a salt solution to the soil will cause cation exchange to take place, and, in particular, exchange of hydrogen ions from the clay surface with the cation of the added salt may take place, which will give a pH that is lower than the soil pH under normal conditions. The added salt solution should therefore cause as little exchange as possible, and for this reason R. K. Schofield and A. W. Taylor[1] proposed the use of 0·01 M calcium chloride solution for temperate soils, on the grounds that it approximates to the calcium concentration in the soil solution (see p. 542), that normal changes in the natural salt concentration in the soil will have little additional effect, that it will usually cause little cation exchange, and that it does not give appreciable junction potential with the calomel electrode. The pH of a soil measured in this solution is more constant and is much less dependent on the soil-solution ratio than is the pH measured in a water suspension and it is also presumably closer to the pH of the solution around the plant roots. It is typically lower by between 0·5 to 0·9 units, the difference tending to be greater in near-neutral than in very acid soils. However, if some soils whose pH is within the range 5·0–6·0 are put in this calcium chloride solution, their pH rises slowly, possibly due to polymerised aluminium ions taking up a proton from the solution, as a consequence of a change in the calcium–aluminium activity ratio in the solution.[2]

The pH of a soil is influenced by two other factors besides the pH rise from the surface of the soil particles to the soil solution in bulk. In the first place, soils may contain substances capable of changing their degree of oxidation and reduction, with consequent fall or rise in pH (see p. 680). Since many actual soils are often a constantly changing mosaic of pockets of good and poor aeration, this means that the pH of a soil may vary quite appreciably from point to point; but it also means that one cannot talk about the pH of soils seasonally waterlogged, for this often varies through two units of pH during the year.

The pH of a soil is also influenced by the carbon dioxide concentration in the soil air. The higher this concentration the lower the pH, and the pH of a neutral or calcareous soil is very sensitive to small changes in the carbon dioxide concentration when the concentration itself is low, that is, when it is not very different from the atmospheric value of 0·03 per cent. Table 7.7 illustrates the magnitude of this effect for some neutral non-calcareous soils.[3] In calcareous salt-free soils, its magnitude can be calculated approxi-

1 *J. Soil Sci.*, 1955, **6**, 137, and *Proc. Soil Sci. Soc. Amer.*, 1955, **19**, 164. See also R. C. Turner and W. E. Nichol, *Canad. J. Soil Sci.*, 1958, **38**, 63, and J. B. Collins, E. P. Whiteside and C. E. Cress, *Proc. Soil Sci. Soc. Amer.*, 1970, **34**, 56 for other examples.
2 B. W. Bache, *J. Soil Sci.*, 1970, **21**, 28.
3 W. E. Nichol and R. C. Turner, *Canad. J. Soil Sci.*, 1957, **37**, 96. For further examples, see R. S. Whitney and R. Gardner, *Soil Sci.*, 1943, **55**, 127, and **56**, 63.

TABLE 7.7   Effect of carbon dioxide concentration on the pH of neutral soils

	pH measured in water			*pH measured in* 10^{-3} *M CaCl$_2$*		
	Pressure of CO$_2$ atmospheres					
	0·0004	0·001	0·05	0·0004	0·001	0·05
Soil 1	7·01	6·92	6·53	6·46	6·37	6·22
Soil 2	7·42	7·20	6·70	6·77	6·75	6·38
Soil 3	8·09	7·44	6·98	7·52	7·05	6·69

mately from the solubility of calcium carbonate and carbon dioxide in water and the dissociation constants of carbonic acid, because in such soils the pH is basically determined by that of the system: calcium carbonate–calcium bicarbonate–carbon dioxide–water which is given approximately by the equation

$$2\,pH = K + pCa + pCO_2$$

where pCO_2 is the negative logarithm of the partial pressure of carbon dioxide, in atmospheres, in equilibrium with the solution, pCa the negative logarithm of the activity of the calcium ions, and K is a constant whose value lies between 10 and 10·5, depending on the value used for the solubility constant of calcium carbonate.[1] The reduction of pH is therefore approxi-

TABLE 7.8   The pH and calcium ion concentration of a calcium carbonate and a calcareous clay suspension in equilibrium with carbon dioxide at varying pressures

Carbon dioxide pressure in atm.	*pH of calcium carbonate suspension*	*pH of clay suspension*	*Calcium held by the clay m eq per 100 g*	*Calcium ion conc. in solution in m eq/l*
0·00033	8·42	8·57	71·3	0·53
0·001	8·00	8·30	68·0	0·75
0·003	7·77	7·95	62·4	1·14
0·01	7·33	7·62	52·2	1·70
0·03	7·00	7·30	37·9	2·52
0·1	6·65	6·95	—	3·84

mately proportional to the logarithm of the partial pressure of carbon dioxide, a result which has been confirmed experimentally. Table 7.8 gives the pH of a clay–calcium carbonate–carbon dioxide–water system, and the concentration of calcium in the soil solution.

1   For a more exact treatment see F. S. Nakayama, *Soil Sci.*, 1968, **106**, 429.

The pH of the suspension depends somewhat on the presence of the clay, for F. Simmons,[1] from whose work this part of the table has been derived, found that the pH was about 0·3 units lower in the absence than in the presence of the clay. The last column of the table gives the solubility of calcite in water in the absence of clay, and is calculated from some results of G. L. Frear and J. Johnston.[2] Hence the two sets of figures are not quite comparable, but they give the orders of magnitude of the effects observed. Further, if the soil contains a soluble calcium salt dissolved in the soil solution, this will lower the pH still further, as is shown in the second part of Table 7.7.

The calculations which have been made in the foregoing paragraphs are based on the assumption that the calcium carbonate in the soil is present as calcite. But S. R. Olsen and F. S. Watanabe,[3] by comparing observed with calculated values of pH, have shown that, in some soils, at least part of the calcium carbonate is in a form more soluble than calcite, probably due to it being precipitated in the presence of magnesium or sulphate ions,[4] so that, for a given carbon dioxide concentration, the pH is higher than expected: and this can be particularly noticeable in clay soils in which much of the carbonate is present in the clay fraction. Crops growing on such soils are liable to show iron chlorosis[5] (see p. 646), possibly because of the higher HCO_3^- ion concentration around their roots.

One consequence of the soil being a fairly well buffered system is that if a fresh sample of soil is taken from a field and quickly shaken up in say a 0·01 M $CaCl_2$ and its pH measured almost immediately, this will correspond reasonably to the pH of that soil in equilibrium with the mean field carbon dioxide concentration in that sample.

The pH of a well-aerated calcareous soil containing few exchangeable sodium or magnesium ions thus cannot exceed about 8·5 when in equilibrium with the atmosphere; but if the CO_2 concentration rises to 0·1 per cent, a low figure for most soils, the pH has already dropped to 8, and if it rises to 1 per cent, a common figure for pastures, the pH is under 7·5. Now, as will be shown later (p. 659), many plants suffer from phosphate and minor element deficiencies if the pH rises much above 8, and hence they can only thrive on calcareous soils if they can maintain an adequate concentration of carbon dioxide around their roots. This is more easily achieved in soils of low porosity, such as clays, then in soils of high porosity, such as well-drained sands, and in soils rich in organic matter than in poor, and in soils under pasture than under arable.

A calcareous subsoil can have a pH greater than 8·5 because if water,

1 *J. Amer. Soc. Agron.*, 1939, **31**, 638.
2 *J. Amer. Chem. Soc.*, 1929, **51**, 2082.
3 *Soil Sci.*, 1959, **88**, 123.
4 H. E. Doner and P. F. Pratt, *Proc. Soil Sci. Soc. Amer.*, 1969, **33**, 691.
5 P. H. Yaalon, *Plant Soil*, 1957, **8**, 275.

initially in equilibrium with the carbon dioxide in the atmosphere, percolates through a calcareous soil in which no carbon dioxide is being produced, the solution loses some bicarbonate or carbonate ions to calcium ions on its way down. As a consequence, the solution is effectively in equilibrium with a carbon dioxide concentration of 10^{-6} bar, the calcium concentration is about 1.4×10^{-4} M, and the pH about 9.9.[1] R. C. Turner and his coworkers have in fact measured pHs between 8.8 and 9.7 for the equilibrium solution of different soils in the laboratory.

Soils normally only have a pH over 8.5 if they contain enough exchangeable sodium for sodium carbonate to be present in the soil solution. These soils are known as alkali soils and are discussed on p. 761, but R. S. Whitney and R. Gardner[2] showed that even if their pH is above 9.5, when measured in equilibrium with the atmosphere, it may be below 8 when in equilibrium with air containing 1 per cent of CO_2. Hence, in calcareous and alkali soils the effective pH around the plant roots can be considerably lower than the pH determined by shaking the soil up in water, because of the greater concentration of carbon dioxide the roots can maintain in their immediate neighbourhood, and this effect can be still further increased due to the electrolyte concentration in the solution.

Soils do not normally have pHs below about 4.0 to 4.5, unless they contain free acids such as sulphuric. This limit is set by the pH of an aluminium soil; for clays containing much replaceable hydrogen are unstable and decompose to release sufficient aluminium ions to neutralise the negative charges the hydrogen ions were neutralising (see p. 106). Thus in Assam, India, some acid tea soils have received very large dressings of ammonium sulphate over the years, yet their pH has remained at about 4.5.[3]

This discussion has therefore brought out the following points concerning the pH of a soil:

1 The correct interpretation of the measured pH of a soil may be very difficult.
2 There is no such thing as the pH of a soil-water system when dispersed in pure water, because the pH rises across the diffuse electrical double layers.
3 The pH of a soil, as usually measured, depends on the salt concentration in the soil solution and the CO_2 concentration in the soil air, which are constantly changing.
4 Typically the pH of a soil in the field varies appreciably over the field, even when all the samples are taken the same day.

These conclusions mean that the pH of a soil can have no precise significance in agricultural practice, a conclusion which is amply justified by

1 R. C. Turner, *Soil Sci.*, 1958, **86**, 32, and with W. E. Nichol and K. E. Miles, *Canad. J. Soil Sci.*, 1958, **38**, 94.
2 *Soil Sci.*, 1943, **55**, 127 and **56**, 63.
3 N. G. Gokhale and N. G. Bhattacharyya, *Emp. J. Exp. Agric.*, 1958, **86**, 309.

experience, as will be discussed in chapter 24. They also mean that very rarely can one make any use of a really accurate measurement of the pH of a field soil, even if one measures it in a suitable salt solution; and it is probable that little information of value would be lost if the pH of field samples was measured only to the nearest 0·2 of a unit.

Anion adsorption by soils

Soils may contain free positive charges, as has already been noted, and these charges must be balanced by anions in the soil solution; and as the number of these charges varies, due to changes in pH or salt concentration, so will the number of anions balancing them. The seat of these charges can be aluminium ions on the broken edges of clay particles, as shown on p. 88 for kaolinite or they can be on the surfaces of aluminium hydroxide films or of ferric hydroxide, goethite or other iron oxide surfaces. These balancing anions are in the solution outside the solid surface, and are readily exchangeable with other anions introduced into the solution.

The aluminium and iron hydroxide or oxide surfaces can hold anions by another process, known as ligand adsorption. Their surfaces consist of water molecules or hydroxyls forming part of the coordination shell of the six oxygens around each cation, and a proportion of them are in the shell of one cation only, unlike the hydroxyls or oxygens in the crystal itself, which are shared between two cations. Thus, each hydroxyl which is coordinated with only one aluminium ion, for example, will carry an effective charge of $-\frac{1}{2}$ unit, and each water molecule a charge of $+\frac{1}{2}$ unit; so the net charge on the surface is determined by the number of protons associated with the oxygens on the outside of the crystal lattice. This can be pictured as:

$$
\text{crystal} \left| \begin{array}{l} OH_2^{+\frac{1}{2}} \\ OH_2^{+\frac{1}{2}} \end{array} \right|^{+1} \quad \xrightarrow{-H^+} \quad \text{crystal} \left| \begin{array}{l} OH_2^{+\frac{1}{2}} \\ OH^{-\frac{1}{2}} \end{array} \right|^{0} \quad \xrightarrow{-H^+} \quad \text{crystal} \left| \begin{array}{l} OH^{-\frac{1}{2}} \\ OH^{-\frac{1}{2}} \end{array} \right|^{-1}
$$

$$
\quad\quad\;\; \text{surface} \qquad\qquad\quad\;\; \text{surface} \qquad\qquad\quad\;\; \text{surface}
$$

$$
\text{low pH} \qquad\qquad\qquad \text{medium pH} \qquad\qquad\qquad \text{high pH}
$$

In these diagrams the solid vertical line represents the plane of the surface cations (aluminium or ferric ions), the oxygens outside this line are in six-coordination with the cation and the number of protons they contain determines the charge; and the dotted vertical line represents the outer surface of the crystal and the net charge on the surface is given. In this example, the proton is the ion that determines the charge on the surface, so hydrogen ions are called the potential determining ion; and when the surface carries no net charge, it is at its isoelectric point, which is specified by the pH of the solution.[1]

1 For examples with goethite and haematite, see R. J. Atkinson, A. M. Posner and J. P. Quirk, *J. Phys. Chem.*, 1967, **71**, 550. For a fuller account of these properties see C. J. B. Mott in *Sorption and Transport Processes in Soils*, Soc. Chem. Ind., 1970, Monog. 37, 40.

The net charge on the surface must be neutralised by anions or cations in the solution, which are readily exchangeable for other ions. They form the outer layer of the surface electric double layer, or the Helmholtz double layer as it is often called, and the oxygen ions with their associated protons form the inner layer. Although at the isoelectric point the surface consists of a mosaic of positive and negative spots, they are in general so close together, in that they are on contiguous ions, that the surface behaves as if it were uncharged.

This picture is unfortunately a little simplified because the crystal surface exerts a slow buffering action if the pH of the bathing solution is altered. If the pH is raised, the surface slowly releases hydrogen ions, and if lowered slowly absorbs them. This is probably due in part to protons on the crystal surface slowly migrating into or out of the superficial layers due to the change in their concentration on the surface, and in part to the existence on the surface of micro-cracks of molecular dimensions, through which the protons diffuse only very slowly; and these two may in fact be effectively the same process. This characteristic is also shown by soils containing much free iron or aluminium hydroxides, for if they are shaken up in a dilute salt solution of different pHs, the pH slowly drifts back towards its original value, and acid or alkali must be added at intervals over periods measured in months rather than days before the soil is in equilibrium with the solution at its new pH.

The oxygen ions on the crystal surface can be replaced under suitable conditions by any other ion that can enter into six-coordination with the aluminium or ferric ion, such as many oxyacids, and fluoride. This exchange is known as ligand exchange, for it takes place in the inner Helmholtz layer, and its principal characteristics have been described in detail by F. J. Hingston.[1] It always involves an increase in the negative charge or a decrease in positive charge on the surface, which is less than the charge carried by the anion being adsorbed, with the consequence that the pH of the solution rises. Thus ligand exchange always involves the adsorption of hydrogen ions, and only takes place with weak acids in the pH range in which there are both undissociated molecules and free anions of the acid present, presumably because only in this range can the acid provide the hydrogen ions that must be absorbed. Ligand exchange is, in fact, at a maximum when the pH of the solution is equal to or a little lower than the pK of the acid.

The process of ligand exchange can be pictured as

$$\text{crystal surface} \begin{vmatrix} OH_2^{+\frac{1}{2}} & 0 \\ OH^{-\frac{1}{2}} & \end{vmatrix} + H^+ + OA^- \longrightarrow \text{crystal surface} \begin{vmatrix} OA^{-\frac{1}{2}} & -1 \\ OH^{-\frac{1}{2}} & \end{vmatrix} + H^+ + H_2O$$

1 With R. J. Atkinson *et al.*, *Nature*, 1967, **215**, 1459, *Trans. 9th Int. Congr. Soil Sci.*, 1968, **1**, 669.

followed by a proportion of the hydrogen ions being absorbed by hydroxyls. In this particular example the initial surface has been given no net charge but ligand exchange can take place equally well on surfaces carrying a net positive or a net negative charge. It is thus quite different from non-specific anion adsorption, of chloride or nitrate for example; for this can only occur when the surface carries a positive charge, and the pH is in consequence below the isoelectric pH for that surface. The actual increase in negative charge due to ligand exchange depends on many factors, which have not been worked out in detail. Figure 7.15 gives the actual increase in charge

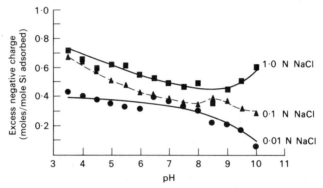

FIG. 7.15. The increase in negative charge on a goethite surface due to the adsorption of silicate

per silicate anion adsorbed for the particular goethite sample used by Hingston. Once an anion has been adsorbed by this exchange, it can only be replaced by another anion that can increase the negative charge on the surface still further, that is more than one equivalent of the replacing anion must be adsorbed for each equivalent of the anion being desorbed. Thus,

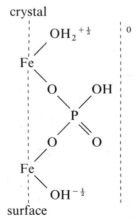

FIG. 7.16. Diagram to illustrate a possible mechanism for strong ligand adsorption of phosphate

anions held by ligand adsorption cannot be displaced by washing with a solution of sodium chloride, for example, though they can by washing with sodium hydroxide, for the hydroxyl ions can replace or remove protons from the water molecules on the surface, so increase its negative charge. Anions held by ligand adsorption can usually exchange freely with isotopically labelled anions of the same species, if introduced into the solution. However, phosphate may be adsorbed so strongly by ligand exchange that it takes an appreciable time for it to become equilibrated with labelled phosphate added to the solution, perhaps because two of its oxygens are coordinated with adjacent ferric or aluminium ions, as shown in Fig. 7.16, though this diagram is not meant to imply that there is no net charge on the inner Helmholtz layer.[1]

The maximum amount of any anion that can be held by ligand adsorption depends on the concentration of the anion in the solution, on the pH of the solution, and on the specific affinity of the ion for the surface. At a given pH, the relation between the concentration of the anion in solution and the amount adsorbed follows the Langmuir equation approximately, as is shown in Fig. 23.3 on p. 569 for phosphate so that there is a definable maximum adsorption for the anion at that pH. The curve of maximum adsorption against pH is known as the adsorption envelope for that anion. Adsorption only begins when the weak acid begins to dissociate, it is at a maximum at about its pK, and it falls to a low value when it is fully dissociated. If the anion is polybasic, as, for example, phosphoric acid, adsorption is at a maximum in acid conditions, and there are definite changes in the slope of the envelope at the pH values corresponding to the pK for the $H_2PO_4^-$ and $H PO_4^{2+}$ anions. Figure 7.17 gives the adsorption envelope for phosphate and silicate for an acid soil which was known to absorb or fix phosphate strongly.[2] The maximum adsorption for silicate occurs at pH 9·2, which is a little below its pK of 9·6; the phosphate envelopes has breaks at pH 6·4 and 11·6, that is, slightly lower than the pK of the two anions, at 7·2 and 12·7. Figure 7.17 shows that silicic acid is taking part in ligand exchange at a pH well below that at which it dissociates an appreciable number of hydrogen ions. This is due to both goethite and gibbsite surfaces inducing dissociation of silicic acid under acid conditions; and acid soils formed from basic igneous rocks under conditions giving appreciable amounts of iron or aluminium hydroxides will adsorb large amounts of silicic acid, which can be displaced by phosphate, for example.

The maximum amount of an acid that is adsorbed is not the same for all acids, though in the graph the amounts of phosphate and silicate at their maximum adsorption are about the same. Thus, Hingston finds that, at maximum adsorption on goethite, there is one $H PO_4^{2-}$ anion per 66 Å², one molybdate per 35 Å², and one fluoride per 19 Å², though the dibasic phosphate anion may be in coordination with two ferric ions. There is some

1 U. Kafkafi, A. M. Posner and J. P. Quirk, *Proc. Soil Sci. Soc. Amer.*, 1967, **31**, 348.
2 C. H. Obihara and E. W. Russell, *J. Soil Sci.*, 1972, **23**, 105.

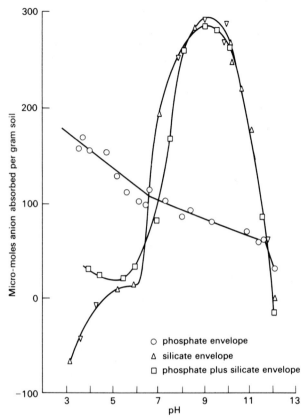

FIG. 7.17. Adsorption envelopes for phosphate, silicate and phosphate plus silicate for a phosphate-fixing soil

evidence that there are spots on the surface that are specific for certain anions, though many of the spots appear to be common for most.[1] Again, in the graph, the adsorption envelope is given for adsorption from a solution of phosphate and silicate, sufficiently concentrated in each to ensure maximum adsorption if alone; and, for this soil, the curve for total adsorption from the mixed solution either lies a little below or falls on the higher of the adsorption envelopes for the two acids. In this particular example, the soil only adsorbed silicate from the solution after the pH had risen above 7, when silicate adsorption is greater than phosphate adsorption. Thus, the presence of silicate in the phosphate solution did not affect phosphate adsorption if the pH was below 7. On the other hand, above pH 7, the presence of phosphate in the silicate solution reduced the amount of silicate adsorbed, and the presence of silicate reduced the amount of phosphate adsorbed, so the soil adsorbs phosphate more strongly than silicate, and the

1 F. J. Hingston, A. M. Posner and J. P. Quirk, *Faraday Soc. Disc.*, 1971, **52**.

silicate cannot compete with the phosphate, and the silicate can be adsorbed in the presence of phosphate only when it can increase the negative charge on the clay surface, that is when the pH of the solution exceeds 7.

The agricultural significance of this result must be clearly understood, for the results as given refer to solutions well supplied with both phosphate and silicate. If a soil is low in phosphate, and the phosphate concentration in the soil solution is low, adding silicic acid, in the form of a soluble silicate, may easily cause some phosphate to be displaced, particularly if the soil is neutral in reaction; for the concentration of silicic acid in the soil solution may then be much higher than that of the phosphate, and the phosphate concentration be far below that necessary for the soil to be saturated with phosphate at that pH. It is therefore quite explicable why adding silicate to a soil should increase the availability of the soil phosphate (see p. 638). In the same way adding phosphate to an acid soil very low in molybdate may increase the availability of the molybdate by displacing some from the soil surface.

The adsorption of sulphate by soils has not been examined in detail. Soils high in active aluminium or ferric hydroxide will adsorb sulphates in acid conditions,[1] but these are conditions where the surfaces contain free positive charges, so the adsorption could be due to simple neutralisation of these charges. There is no evidence that the sulphate anion can enter the co-ordination shell of the aluminium or ferric ion, nor would it be expected to on Hingston's theory, as sulphuric acid is a strong acid.

A further point of interest in ligand exchange is the use of a sodium fluoride solution for picking out soils high in active aluminium hydroxide. The fluoride ion will displace hydroxyls from the coordination shell of aluminium, and probably gives a large increase in negative charge on the surface, as the pH of the solution rises. M. Fieldes and K. W. Perrott[2] used this to devise a rapid test for allophane in soils, for they observed that if a crumb of an allophane soil was put on phenolphthalein paper and a few drops of a sodium fluoride solution put on the soil, the paper turned pink. However, as found by J..E. Brydon and J. H. Day,[3] this test is not specific for allophane soils; it gives the same result for the B-horizon of a podsol, or indeed for any soil that contains aluminium soluble in Tamm's acid oxalate solution.

1 M. E. Harward and S. C. Fang with T. T. Chao, *Proc. Soil Sci. Soc. Amer.*, 1964, **28**, 632, and with T. C. Tsun, *Soil Sci.*, 1962, **94**, 276.
2 *N.Z. Jl. Sci.*, 1966, **9**, 623.
3 *Canad. J. Soil Sci.*, 1970, **50**, 35.

8

The interaction of clay with water and organic compounds

When a dry clay particle is wetted with water, there is an interaction between the water molecules and the clay surface. Some water molecules may be adsorbed on the clay surface through hydrogen bonding, and some may be absorbed by the exchangeable ions becoming hydrated, which may result in some of them dissociating from the surface into the water. The effect of the cation on the water molecules is greater the greater its charge and the smaller its size, so the greater its surface charge density; and these effects are shown by the relative moisture content of the clay when in equilibrium with water vapour at low relative humidities,[1] by the heat evolved when the clays are wetted,[2] and by the greater apparent density of the clays in water than in an inert liquid.[3] Thus, a magnesium-saturated clay absorbs more water vapour from an atmosphere of low relative humidity, has a higher heat of wetting and a higher apparent density in water than has a calcium, and these than a sodium, and these than a potassium clay. These ionic affects are not confined to the system clay–water, but are also shown, usually to a smaller extent, with clay and any other liquid whose molecules possess dipole moments, such as the alcohols for example. Measurements of the heat evolved when a dry clay is wet—its heat of wetting—show that dry calcium-saturated clays liberate about 100 J/100 g clay for each milli-equivalent of exchangeable calcium in the clay and a potassium clay about 60 J/100 g for each milli-equivalent of potassium, the figures being somewhat higher for micaceous and somewhat lower for montmorillonite clays.[4] The heat of wetting drops rapidly as the vapour pressure of the moist clay rises.

The swelling of clays

When a dry montmorillonite clay is slowly wetted by allowing it to come into equilibrium with water vapour at gradually increasing pressure, the clay

1 M. D. Thomas, *Soil Sci.*, 1928, **25**, 485; H. Kuron, *Koll. Chem. Beih.*, 1932, **36**, 178.
2 W. W. Pate, *Soil Sci.*, 1925, **20**, 329; L. D. Baver, *J. Amer. Soc. Agron.*, 1928, **20**, 921; M. S. Anderson, *J. agric. Res.*, 1929, **38**, 565; H. Janert, *J. agric. Sci.*, 1934, **24**, 136.
3 E. W. Russell, *Phil. Trans.*, 1934, **233 A**, 361.
4 H. W. van der Marel, *Z. PflErnähr.*, 1966, **114**, 161.

begins to swell due to water condensing between the clay plates, forcing them apart. S. B. Hendricks[1] showed that for magnesium- and calcium-saturated montmorillonite, the first process was that the ions adsorbed six water molecules of hydration. This increases the spacing between adjacent clay layers from 10 to 15 Å. When the relative humidity has reached about 50 per cent, the spacing between the layers has jumped to 19 Å, due to a water monolayer being intercalated on each side of this hydration double layer and the clay surface, giving a water layer four molecules thick. These two clays swell no further as the relative humidity approaches 100 per cent. Sodium and potassium montmorillonites behave like the calcium and magnesium clays at low relative humidities, except that a much higher relative humidity is needed for the jump from 10 to 15 Å. The potassium clay does not swell to the 19 Å spacing as the relative humidity rises towards saturation, although the X-ray line corresponding to the 15 Å spacing becomes more diffuse. The sodium clay swells first to the 15 Å, then to the 19 Å spacing; but at a higher relative humidity the spacing jumps to 40 Å, though the X-ray line is diffuse, and the spacing then continues to increase as the relative humidity increases. If the clay is suitably dispersed in dilute sodium chloride solutions, spacings of up to 350 Å have been found, provided the clay was carefully freed from all iron and aluminium impurities, or 120 Å if used in its natural condition.[2]

K. Norrish also showed that this swelling behaviour at low moisture contents depends on the surface density of charge on the clay lattice, for vermiculite and illite clays do not swell as much as montmorillonite. Thus the potassium and ammonium vermiculites only form a water monolayer and the calcium and magnesium a layer two water molecules thick.

These results must be interpreted with caution, however, for a calcium-saturated clay can swell in water to a volume greater than corresponds to the interlamellar water films. This is because when the dry clay is wetted, the heat evolved heats up the water or bathing solution, setting up convection currents that create pores or channels or cracks between groups of clay plates. This additional swelling is therefore intercrystalline not interlamellar, where the word crystal signifies packets of oriented clay plates separated by two or three layers of water molecules.

It should be noted also that interlamellar water films, when only three or four molecules thick, exclude anions of salts, so that if a dry calcium clay, for example, is wetted in a solution that is 4 M in calcium chloride, the clay will remove sufficient water from this solution to form its interlamellar water film, and it will not take up any salt in the process.[3]

The water molecules held in films between the clay lattice have a degree of ordered structure. The molecules in contact with the lattice surface form

1 With R. A. Nelson and L. T. Alexander, *J. Amer. Chem. Soc.*, 1940, **62**, 1457.
2 With J. A. Rausell-Colom, *Proc. 10th Int. Conf. Clays, Clay Min.*, 1963, 123, and see B. P. Warkentin, G. H. Bolt and R. D. Miller, *Proc. Soil Sci. Soc. Amer.*, 1957, **21**, 495.
3 A. M. Posner and J. P. Quirk, *Proc. Roy. Soc.* 1964, **278A**, 35.

covalent hydrogen bonds with oxygens in the tetrahedral layer. Their exact ordering on the surface is not known, but it probably consists of an open hexagonal network, similar to that of ice, which is controlled by the hexagonal network of oxygen ions in the tetrahedral layer of the lattice; and it is probable that only every other water molecule forms a covalent bond with a lattice oxygen.[1] The covalently-bonded water molecules are strongly polarised so that, if there are no external restrictions, a second layer of water molecules can form over the first, having the same structure as the first, with alternate water molecules having covalent hydrogen bonds to the oxygen of the strongly polarised water molecules; and in this way thick films of ordered water molecules can be built up. This probably represents the condition of the water between the sheets of a sodium clay when the films are over 30 Å thick.

This structure will be disrupted, to a greater or less extent, by the hydration shells around the exchangeable cations, and in particular by their energies of hydration or the tightness with which they bind the water molecules to their surface. Magnesium and calcium ions, for example, bind six water molecules very strongly to their surface, the ion being then in six-coordination with the oxygens of the water, so that the position of these cations on the clay surface determines the position of the water molecules in their neighbourhood. Sodium ions, on the other hand, have a much smaller energy of hydration, so bind water molecules only weakly to their surface, and so that the positions of the water molecules close to the surface will be more nearly determined by the surface oxygens.

The swelling of a sodium-montmorillonite can be measured by another method which is to place the clay in sodium chloride solutions of decreasing concentration. An example of the type of result obtained is given in Fig. 8.1, due to D. L. Rowell.[2] There is at first no swelling of the clay in a concentrated solution, then below a certain concentration it begins to swell, and the swelling volume increases rapidly as the concentration falls, until there is a region in which the swelling volume increases linearly with the decrease in the negative logarithm of the concentration, during which time the clay has the appearance of a uniform gel. At a certain concentration the clay breaks up into loosely bound small flocs and at a still lower concentration it disperses into individual clay plates. The stage before this is spoken of as disaggregation, for the clay disperses into units formed out of a stack of a few clay layers held parallel to one another. Figure 8.1 also shows the swelling that would be expected if it was due solely to the clay layers moving apart but remaining parallel to one another with the spacing given by X-ray analysis. Quite clearly most of the swelling is not due to this uniform separation be-

1 For a discussion of some proposed structures see R. E. Grim, *Clay Mineralogy*, 2nd Edn, 1968, McGraw-Hill, and of some properties of these films see P. F. Low, *Adv. Agron.*, 1961, **13**, 269.
2 *Soil Sci.*, 1965, **100**, 340.

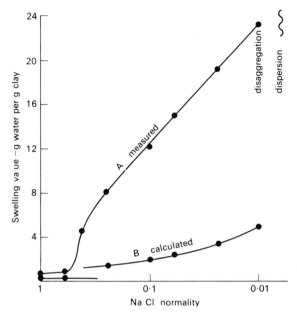

F IG. 8.1. Measured and calculated swelling of a Na-montmorillonite in NaCl solutions

tween individual clay layers, but to the separation of clay aggregates or domains; and this interaggregate swelling, where the aggregates have random orientations relative to each other, is the major cause of the swelling.

Figure 8.2 shows another common feature of the behaviour of soils containing sufficient exchangeable sodium for them to deflocculate in a dilute solution.[1] Chloride solutions of constant sodium–calcium activity ratio, but decreasing molarity, were percolated through a bed of soil crumbs, where the soil had different contents of exchangeable sodium; and Fig. 8.2 shows the concentration at which the swelling of the crumbs became sufficient for the permeability of the soil bed to drop by 15 per cent in 5 hours–the threshold curve; the concentration at which the percolation liquid started to be turbid, that is when soil particles began to disperse in the solution; and the concentration at which the soil crumbs had deflocculated and the bed became impermeable. These three points are of great practical importance in the field, although they will typically occur at more dilute concentrations in the field due to external disruptive forces.

There is some evidence from the swelling behaviour of a montmorillonite clay containing both exchangeable sodium and calcium ions, that the individual interlayers between two clay layers contain either sodium or calcium ions, but not a mixture of both.[2]

1 D. L. Rowell, D. Payne and N. Ahmad, *J. Soil Sci.*, 1969, **20**, 176.
2 B. L. McNeal, *Proc. Soil Sci. Soc. Amer.*, 1970, **34**, 201.

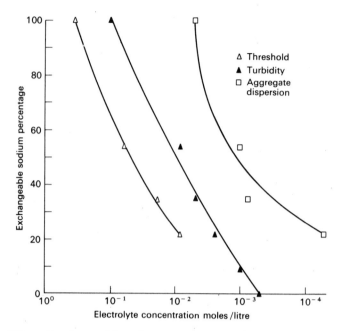

FIG. 8.2. Effect of exchangeable sodium percentage on the electrolyte concentration at which different types of structure collapse occur

This swelling is dependent on the electrical charges on the broken edges of the clay aggregates or crystals, for these will contain positive charges unless the pH of the dispersion liquid is high. This can be shown by converting these positive to negative charges by adding some hexametaphosphate to the clay. This does not affect the chloride concentration at which swelling begins, but below this critical concentration, it increases the swelling at a given chloride concentration, and it increases the chloride concentration at which disaggregation and dispersion occur. Increasing the positive charges by precipitating ferric or aluminium hydroxide on the particles has the opposite effect, of reducing swelling in a solution of a given concentration and reducing the concentration at which disaggregation and dispersion occur.[1]

The diffuse electrical double layer around clay particles

The ionic conditions on the outside surface of a clay particle or a packet of clay particles dispersed in water or an electrolyte solution are controlled by the proportion of the exchangeable cations that disperse into the solution. A clay surface probably behaves as an effectively uncharged surface if the negative charge on the lattice is neutralised by monovalent cations that are

1 H. M. E. El Rayah and D. L. Rowell, *J. Soil Sci.*, 1972, **23**.

tightly bound to the surface, such as caesium or many large organic cations. But if the cations are hydrated a proportion tend to dissociate from the surface and will cause an electrical potential gradient to be set up near the surface. The system clay lattice-exchangeable cations-solution can be looked upon as forming a complex electrical double layer, known as the Helmholtz double layer, the inner layer being the surface of the lattice carrying the negative charge and the outer layer being composed of two parts—a positive layer due to the cations bound to the lattice surface known as the Stern layer, and a positive layer diffused in the solution close to the lattice surface, known as the Gouy diffuse layer. The proportion of the cations that dis· sociate from the Stern into the Gouy layer depends partly on the concentration of salts dissolved in the water, and partly on the tightness with which they are held to the surface.

The distribution of ions in the diffuse layer, when the solution contains a salt can be calculated, under certain assumptions, from well established physical principles; and Fig. 8.3 illustrates a typical distribution curve for

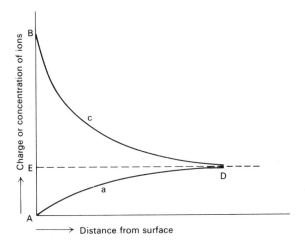

FIG. 8.3. The ionic concentrations in a diffuse electric double layer. AB: negative charge on the surface. BD: concentration of cations in the solution. AD: concentration of anions in the solution

the anions and cations outside a negatively charged surface. But the relevance of these calculations to clay systems depends on the appropriateness of the assumptions made; the most important of which, and the least generally valid, is that the outer surface of the inner Helmholtz layer can be considered to have a uniform density of negative charge, on the molecular scale. This assumption is reasonable when a fairly large proportion of the exchangeable cations have dissociated into the solution and the double layer is thick, but it becomes increasingly invalid as the proportion of cations dissociating from the surface drops, so the distance apart between the free negative charges on the surface becomes large. If, for example, the clay lattice possesses one negative charge per 100 $Å^2$, and only 5 per cent of the exchangeable cations dissociate, the average area occupied per free negative charge is 2000 $Å^2$, or the mean distance between negative charges is 45 Å, or about twenty times the crystallographic diameter of the ion. At distances of the order of 10 diameters, or 20 Å, from the surface, the negative charge must be considered as a mosaic of isolated charged spots rather than as uniformly distributed. In practice it is found that sodium-saturated clays in dilute salt solutions behave much as predicted from the simple theory, but that calcium- and aluminium-saturated clays do not.

There are four properties of this double layer of importance. First, its thickness decreases as the electrolyte concentration in the water increases; and at equal equivalent concentrations of electrolyte, divalent cations decrease the thickness of the double layer more strongly than monovalent, and trivalent more strongly than divalent. As a corollary to this, if clay particles are suspended in water containing very little electrolyte, the double layer is thicker if the exchangeable ions are sodium and potassium than if they are calcium or magnesium, and it is probably thicker for sodium than for potassium.

Second, the cation concentration gradient in this diffuse double layer sets up an electrical potential difference between the surface of the clay particle and the bulk of the solution, that is, work must be done to transfer a cation from the surface of the particle to the bulk of the solution. Hence, this double layer hinders the free interchange of cations between these two regions. A consequence of this effect has already been mentioned on p. 122. Hydrogen ions cannot have the same activity throughout these regions, and the thicker the double layer, the smaller will be the contribution of the hydrogen ions that can dissociate from the clay surface to the hydrogen ion concentration in the bulk of the solution. The more the diffuse double layer is compressed on to the clay surface, the more freely will the hydrogen ions be able to move from the clay surface into the solution, and the lower will be its apparent pH. Also, since at a given concentration, calcium ions compress the double layer more strongly than potassium, a soil will have a lower pH in a dilute calcium chloride solution than in a potassium chloride solution of the same normality. This result is not confined to hydrogen ions. If, say, a sodium clay is dispersed in pure water, adding a dilute electrolyte to the solution will increase the proportion of sodium ions that dissociate from the clay surface or Stern layer by reducing the potential difference across the double layer, though the effective thickness of the diffuse or Gouy layer will decrease.

This is the third important characteristic of the double layer. The most direct way of estimating the proportion of cations that dissociate into the Gouy layer should be by measuring the activities of the cations around the clay particles using suitable membrane electrodes, in the same way as the hydrogen ion concentration around the clay particle is measured with a glass electrode. C. E. Marshall pioneered this approach,[1] and showed that the results obtained with it are concordant with one depending on determining the contribution of the dissociated exchangeable cations to the electrical conductivity of a clay suspension;[2] but it is not certain how relevant are results obtained on dilute suspensions to the conditions prevailing in clay pastes. Marshall's results are that between 20 and 40 per cent of the exchangeable sodium ions, but only 0·5 to 3 per cent of the exchangeable calcium or magnesium ions dissociate in his conditions, the actual amount depending on the surface density of the charge on the clay lattice and on the proportion of other ions present on the clay. D. G. Edwards and his coworkers,[3] working with much more concentrated clay pastes, and using a technique to be described in the next paragraph, found that the effect of salt concentration was very large, an effect not studied by Marshall, and that only about 1 per cent of sodium ions dissociated from a sodium clay in 0·01 M NaCl, but about 15 per cent did in 0·1 M NaCl.

Fourth, the concentration of the anions in the solution is lower in the diffuse double layer than in the bulk of the solution. Hence, if a dilute sodium chloride solution is added to a dry sodium clay, and if the clay does not absorb any chloride, the concentration of chloride in the solution will rise because chloride will be partially expelled from the diffuse double layer. This phenomenon is known as negative adsorption, and it creates great difficulties when adsorption of anions by soils is being measured; for the more dilute the solution, the more important is this exclusion from the double layer. Measurements of the negative adsorption of chloride, when a clay is dispersed in a chloride solution, allows the calculation of the effective volume of solution from which the chloride has been excluded, so if the effective thickness of the double layer is known, the area of the surfaces on which the diffuse layer has developed can be calculated.[4] Edwards and his workers showed that the area so determined is less than the probable area of clay surface bathed by the water because of the failure of the assumption of the uniformity of charge density on its surface.

The diffuse layers around clay particles in a paste are not independent of each other if the average distance between the particles is less than twice the thickness of the layers. The properties of interpenetrating diffuse layers between parallel plates can be studied mathematically and the results compared with the observed behaviour of clays. Two properties can be easily studied: negative adsorption, and the swelling pressure exerted by the system if it is in a confined space but is in contact with the solution through a rigid

1 See, for example, *Trans. 4th Int. Congr. Soil Sci.*, 1950, **1**, 71.
2 W. T. Higdon and C. E. Marshall, *J. Phys. Chem.*, 1958, **62**, 1204; see also I. Shainberg and W. D. Kemper, *Proc. Soil Sci. Soc. Amer.*, 1966, **30**, 707.
3 *Trans. Faraday Soc.*, 1965, **61**, 2808.
4 See R. K. Schofield, *Nature*, 1947, **160**, 408; F. A. M. de Haan, *J. Phys. Chem.*, 1964, **68**, 2970; *Proc. Soil Sci. Soc. Amer.*, 1963, **27**, 636.

permeable membrane. Again, it is found that, for sodium-montmorillonite clays, the assumption of the uniformity of charge on the clay surface is not too unrealistic,[1] while it also gives pressures of the correct order of magnitude for calcium montmorillonites if the suspension is pictured as consisting of crystals built up of four or five clay sheets surrounded by a diffuse layer.[2]

Deflocculation and flocculation of clay suspensions and soils

A clay suspension is said to be deflocculated when the individual particles of the clay, once dispersed in the suspension, remain dispersed, which happens if a force of repulsion builds up sufficiently strongly as any two particles approach each other to prevent them coming close enough together for van der Waals forces of attraction to come into play.[3] If the salt concentration in a deflocculated suspension is increased to a certain concentration, the particles will begin to stick together when they come close to each other, and at a somewhat higher concentration all the particles will stick together in loose flocs which settle through the clear solution. The suspension is then said to be flocculated. The sediment is loose and voluminous, and if shaken appears to redisperse into individual particles, which rapidly re-form flocs. This sediment is quite different from any sediment formed by the coarser deflocculated particles settling under gravity, for this is so compact that it is difficult to redisperse the particles; but once they are dispersed they remain deflocculated. If the suspension is more concentrated, nearer to a paste, the deflocculated suspension will behave more or less like a liquid, and when shaken will show characteristic dark and light striations due to the plate-shaped particles setting themselves along the lines of flow. If sufficient salt is added to this suspension, it will set to a semi-solid mud; and if the paste is not too concentrated, some clear liquid will separate from it and rise up through channels to the surface. If the clay is a sodium clay, there is a range of clay and sodium chloride concentrations in the suspension when the clay forms a thixotropic suspension, that is the whole suspension sets to a semi-solid mass which becomes liquid on shaking and re-forms the semi-solid mass again on standing.

The phenomena of deflocculation and flocculation are dependent on the interactions between two sets of double layers on the clay particles. So far we have only considered the double layer on the surface of the clay plates

1 See, for example, B. P. Warkentin with G. H. Bolt and R. D. Miller, *Proc. Soil Sci. Soc. Amer.*, 1957, **21**, 495, and with R. K. Schofield, *J. Soil Sci.*, 1962, **13**, 98.
2 See, for example, A. V. Blackmore and R. D. Miller, *Proc. Soil Sci. Soc. Amer.*, 1961, **25**, 169. W. W. Emerson, *J. Soil Sci.*, 1962, **13**, 31, 40 and I. Shainberg, *Trans. 9th Int. Cong. Soil Sci.*, 1968, **1**, 577.
3 For a more detailed account of these phenomena see H. van Olphen, *An Introduction to Clay Colloid Chemistry*, Interscience Pub., 1963.

whose inner layer only carries a negative charge, that is when there are no iron or aluminium hydroxides on the surface; but there is also a double layer around the broken edges of the clay plates, and this may have quite different properties for, as Schofield and Samson showed for kaolinite (see p. 88) the broken face may carry a positive charge due to the unbalanced charges on the aluminium ions in the octahedral positions and, if the pH is high enough, a negative charge due to the unbalanced charges on the silicon ion in the tetrahedral position. The broken edge may also have a negative charge if the break comes where there is an aluminium ion in the tetrahedral position. The exact electrical conditions in this double layer for any particular clay suspension cannot be forecast, but the higher the pH of the suspension, the lower will be the positive charges and the lower the pH, the greater their number.

The interaction between clay particles dispersed in a suspension depends on the interactions between these two sets of double layers. A sodium montmorillonite, in a sufficiently dilute salt solution whose pH is not too low, will have effectively sodium ions in both double layers, because the sodium ions close to the broken edge will form a sufficiently diffuse double layer to blanket the positive charges. Under these conditions, when two particles approach each other, sufficiently large forces of repulsion will be set up to keep them apart. If now the salt concentration is raised sufficiently, the diffuse layer due to the sodium ions will be compacted to such an extent that it no longer overlaps the positive charges on the broken edges. When two particles approach each other under these conditions, a force of attraction will be set up between the positive charges on the broken edge of one particle and the negative charge on the face of another, and they will join up edge to face. If the clay concentration is suitable, which is about 2 per cent for a Wyoming bentonite, this structure of edge-to-face packing can be so open that the whole suspension forms a semi-solid mass; but the forces of attraction between the particles are still so small that a simple shaking of the suspension will break them. This is the thixotropic condition that has been mentioned. If the sodium chloride concentration becomes greater, the forces of attraction edge to face become greater, the structure becomes more compact, but, since it is still edge-to-face like a house of cards, it remains open, so the flocs that form can be easily dispersed.

It is now possible to state the conditions under which a clay can form a stable deflocculated suspension or paste: it is that the positive charges on the broken edges of the clay plates are suppressed. As already mentioned, this can be done by raising the pH of the suspension, with sodium hydroxide, for example. But it can also be done by using a polyphosphate, such as sodium hexametaphosphate or Calgon, for some of the phosphate oxygens will enter the coordination shell of the aluminium and the sodium ions will dissociate from others, giving a negative charge to the whole of the broken edge.

A calcium-saturated clay differs from a sodium in that the calcium ions tend to remain in the Stern layer, so that the broken edges maintain their positive charge. A calcium clay can only be deflocculated in a much more dilute solution than a sodium clay. For this reason a calcium clay can only be deflocculated in a solution that is very dilute; in fact, the maximum concentration of calcium chloride in which a calcium clay will deflocculate is over a hundred times smaller than for the sodium clay in sodium chloride.

Face-to-face orientation of clay particles is typical of settling or slow drying from a deflocculated suspension, for this is the pattern produced when plates that tend to repel each other are forced close together by the removal of water. Face-to-face orientation is also a stable configuration for the micro-aggregates which compose a normal calcium-saturated montmorillonite, for these cannot be dispersed by shaking the clay up in distilled water, for example. The individual clay aggregates or crystals are built up out of clay layers separated by water layers four molecules thick which contain the exchangeable calcium ions, and these layers can only be separated if they are pulled apart by an external force, as can be done by applying sufficiently strong ultrasonic vibrations to them. But once they have been separated they will remain apart, if the solution is sufficiently low in salts, so they can form a stable deflocculated suspension.

Figure 8.4 gives a diagrammatic representation of the various types of structure clay particles can display in flocculation phenomena.[1] In (a) the clay particles are fully dispersed and have random orientation, in (b) they still possess random orientation, but form an open network enclosing a large volume of solution. This gel structure is characterised by having spaces sufficiently wide not to affect appreciably the rate of diffusion of solutes that are not adsorbed by the clay. In (c) and (d) there is a partial collapse of the gel structure, and a proportion of the clay particles have collapsed to give face-to-face aggregates. These aggregates have been called clay domains by Quirk, and are discussed further on p. 485. In (e) the solution bathing the clay domains is sufficiently dilute to allow them to swell, but still to maintain their relative orientations; but if the clay contains sufficient exchangeable sodium to allow it to deflocculate, this swollen structure will collapse under stress to give the random orientations shown in (a).

Aluminium and iron hydroxide films precipitated on the surface of clay layers will also make it more difficult to deflocculate the clay, except at a very high pH, because their positive charges will increase the strength of the edge-to-face attractive forces, and this will be noticeable with sodium saturated soils as well as with calcium. This stabilisation of edge-to-face bonds can be of great importance in soils in areas where sodium salts are present in the soil or in the irrigation water, for these hydroxides reduce the effect of sodium replacing an increasing proportion of exchangeable

1 I am indebted to Professor Quirk for this diagram.

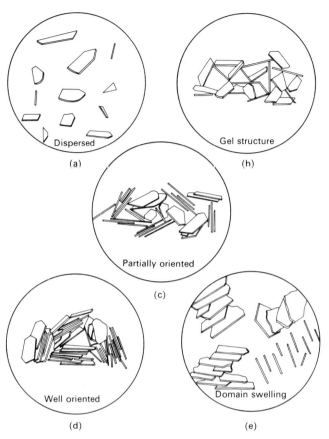

FIG. 8.4. Diagrammatic representation of a deflocculated clay suspension (a), a clay gel (b) and clay flocs (c) and (d)

calcium ions on the swelling of the soil and the loss of stability of its structure[1] (see p. 138).

A deflocculated soil in the field has very undesirable properties. A wet deflocculated soil paste dries into large hard clods separated by a few wide and deep cracks, whereas a flocculated soil dries into smaller clods, not so hard and separated by narrower cracks; this difference in behaviour is illustrated in Plate 6. A deflocculated soil mud is mouldable, while the corresponding flocculated paste is more granular and crumbly. Correspondingly, when a dry deflocculated clod is wet in water or a very dilute salt solution, it swells, loses its shape and develops a very fuzzy zone between the clod and the solution, in which particles will disperse. On the other hand,

1 B. G. Davey and P. F. Low, *Trans. 9th Int. Congr. Soil Sci.*, 1968, **1**, 607; J. d'Hoore, *Trans. 6th Int. Congr. Soil Sci.*, 1956, B, 365.

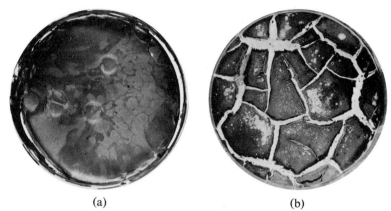

PLATE 6. (*a*) Dried paste of a deflocculated sodium saturated Lower Lias clay. (*b*) Dried paste of a sodium saturated Lower Lias clay flocculated with sodium chloride.

a flocculated clod will typically keep its shape, and maintain a sharp transition zone between the clod and the solution. However, as will be explained in chapter 21, the clay clod will usually slake down into smaller discrete crumbs if it is wetted in air.

Serious deflocculation in soils is usually only found in the field when more than a certain critical percentage of the exchangeable cations are sodium and the soil is wetted by rain or irrigation water of a low salt content.

But deflocculation can be shown by non-sodic soils. Thus, an acid or calcium clay soil wetted with water will suffer a degree of deflocculation if the wet soil is sheared, by puddling or by a tractor tyre slipping over the surface of the soil. This deflocculation of calcium clays is usually suppressed if the calcium ion concentration exceeds about 10^{-3} M, and clay soils containing free calcium carbonate will usually contain sufficient carbon dioxide in the soil air to maintain this concentration. E. C. Childs[1] noted, for example, that liming a clay soil of probably rather unstable structure resulted in the drainage water from the field being clear instead of cloudy. There had been a tradition among many clayland farmers in England that heavy dressings of calcium carbonate improved the workability of their soils, but E. W. Russell and J. J. Basinksi[2] were unable to show that they had any appreciable effect. Gypsum has also been applied to many clay soils to improve their flocculation and consistency properties, but again there are very few reports of appreciable improvements unless very heavy dressings have been used.

1 *J. agric. Sci.*, 1943, **33**, 136.
2 *Trans. 5th Int. Congr. Soil Sci.*, 1954, **2**, 166.

Adsorption of organic compounds by clays and soils

Soils and clays can adsorb organic compounds through a number of different mechanisms.[1] Organic cations will take part in cation exchange and, in general, the larger the cation the more strongly it will be held. Thus ions such as methylene blue and quarternary ammonium ions are held very strongly, while ions such as mono-methyl-ammonium are held not much more strongly than the ammonium. Ammonium ions containing a long chain, such as cetyl tri-methyl-ammonium, are held very strongly,[2] up to the cation exchange capacity of the soil,[3] much more so than tetra-methyl-ammonium, for example. If only a small proportion of the exchangeable cations are organic cations containing a long aliphatic chain, the chain may lie on the clay surface and be held there by weak van der Waals forces, but if the cations are in excess, the chain may be held at right-angles to the surface. Thus, if excess cetyl pyridinium bromide is added to a soil, the pyridinium groups will form a close packed arrangement on the clay surface, with the surplus bromide ions probably held in the aromatic ring, and the aliphatic groups being held together by weak hydrogen bonding or van der Waals forces.[4] It is for this reason it can be used to measure the surface area of clays.

Water molecules on the surface of a clay particle have an effective dissociation constant greater than 10^{-14}, so will dissociate hydrogen ions at lower pH than will free water. This has the consequence that uncharged organic compounds, such as amines, which can accept a proton to become a cation will be held as such in their protonated state. Strongly adsorbed or polarised water molecules around a cation will also donate protons, and this is more marked for aluminium than for magnesium and calcium, and is little shown by sodium or potassium. This proton-donating power of water increases as the soil becomes drier.

Organic anions can be adsorbed by di- or tri-valent exchangeable cations under conditions in which the cation has one of its charges neutralising a negative charge on the clay surface and another a charge on the anion.[5] The cation is the bridging mechanism between the clay and the organic anion, and both divalent cations, such as calcium and magnesium, and trivalent or multivalent ions such as aluminium or hydroxy-aluminium, will bind humic colloids to clay by this mechanism.

1 For an older review see D. J. Greenland, *Soils Fert.*, 1965, **28**, 415, 521; for more recent, M. M. Mortland, *Adv. Agron.*, 1970, **22**, 75; and D. J. Greenland in *Sorption and Transport Processes in Soils, Soc. Chem. Ind. Monog.* 37, 79, 1970, and in *Soil Chemistry* ed. J. M. Bremner and G. Chesters. I have drawn on all of these extensively for this section.
2 H. van Olphen, *Clay Miner. Bull.*, 1951, **1**, 169.
3 S. B. Hendricks, *J. Phys. Chem.*, 1941, **45**, 65.
4 D. J. Greenland and J. P. Quirk, *Clays Clay Miner.*, 1962, **9**, 484; *Int. Soil Conf. (N.Z.)*, 1962, 79.
5 For the adsorption of humic and fulvic acid by soils, see D. J. Greenland, *Soil Sci.*, 1971, **111**, 34.

Some uncharged organic molecules will form hydrogen bonds with water molecules that form part of the hydration shell of an exchangeable cation, and the more tightly held the water molecule to the cation, the more easily will it form a hydrogen bond. The typical compounds that form these bonds are aliphatic alcohol groups, such as are present in polysaccharides or in synthetic polyvinyl alcohols. These bonds are weak if much water is present, but they are additive, so that a long flexible molecule with a large number of hydroxyl groups along its length may be bound strongly to a clay surface because of the large number of hydrogen bonds formed. If the soil is dried, the hydrogen bonds will increasingly be between water molecules directly coordinated to the exchangeable cation and the hydroxyls on the molecule, and the bonding will, in consequence, become stronger. In this bond, water molecules are the link between the molecule and the cation.

Polymers held by this bond can, however, be extremely difficult to desorb from the clay surface because of the very great increase in the translational entropy of the system when wetted, compared with the unabsorbed system; for the polymer will have displaced a large number of water molecules from the neighbourhood of the exchangeable cations, and so increased the randomness of their movement, and so their translation entropy. Thus, the free energy G of the system can be decreased very considerably by this adsorption, for

$$\Delta G = \Delta H - T\Delta S$$

where ΔH is the heat of adsorption, T the temperature in $°K$ and ΔS the change in entropy. It is not necessary for ΔH to have a large negative value for the free energy to be decreased considerably. Further, since ΔG is principally determined by the entropy term, raising the temperature of the system does not reduce the adsorption, for this increases the randomness of molecular movements and, hence, the entropy. This adsorption process is sufficiently strong to cause coiled molecules, such as some naturally occurring polysaccharide gums, to uncoil for this will increase the number of hydrogen bonds and hence the entropy of the system.

Large molecules can also be held on the surface of clay particles by van der Waals forces, which come into play when the atoms of the molecule come sufficiently close to those on the clay surface to be within their zone of interaction. These are normally weak forces, unless the atoms are very close together, so that these forces are increased if after adding the molecules to a moist soil, it is then dried. It may then be extremely difficult to displace these adsorbed molecules after rewetting the soil for, once again, they will have taken the place of a number of water molecules that would normally form a monomolecular layer on the clay surface, so the entropy of the system will be increased very considerably if the molecule is large. Many large natural polymers, such as polysaccharide gums, are built out of aliphatic chains and planar rings which contain hydroxyls, so that once these have been added to a moist soil, and the soil dried, they can only be displaced from the soil by chemical treatments which decomposes them.

Strong hydrogen bonding also takes place between silanol or silicic acid groups $\equiv$Si—OH and water molecules. These groups occur on the broken edges of clay lattices, on the surface of quartz and other silica surfaces, and possibly on the silicic acid adsorbed on sesquioxide surfaces. Aliphatic hydroxyls will form hydrogen bonds with these water molecules, and if the surface is dried, direct with the silanol hydroxyl. There is no evidence that

the oxygens of the siloxane group $$\begin{matrix} \equiv\text{Si} \\ \diagdown \\ \text{O} \\ \diagup \\ \equiv\text{Si} \end{matrix}$$ ever form hydrogen bonds with water.

9

The physiology of the microbial population

The microbial population of the soil

Soils differ from a heap of inert rock particles in many ways, but one of the more important is that they have a population of micro-organisms living in them which derives its energy by oxidising organic residues left behind by the plants growing on the soil, or by the animals feeding on these plants. In the final analysis, the plants growing on the soil subsist on the products of microbial activity, for the micro-organisms are continually oxidising the dead plant remains and leaving behind, in a form available to the plant, the nitrogenous and mineral compounds needed by the plants for their growth. On this concept, a fertile soil is one which contains either an adequate supply of plant food in an available form, or a microbial population which is releasing nutrients fast enough to maintain rapid plant growth; an infertile soil is one in which this does not happen, as, for example, if the micro-organisms are removing and locking up available plant nutrients from the soil.

The soil micro-organisms can be classified into major divisions, such as the bacteria, actinomycetes, fungi and algae—the microflora, and the protozoa, worms and arthropods—the microfauna and fauna. But these divisions are not always clear-cut, because some organisms have a number of properties that are characteristic of both plants and animals. For this reason some biologists have used the word 'protista' to cover all micro-organisms.

The classification of the soil micro-organisms into orders and genera also presents very great difficulties. The standard Linnean criteria of sexual and reproductive organs is usually inapplicable, because these are either absent or very poorly developed in many groups, and there are still no accepted criteria for deciding which of the very large number of properties of these organisms are most suitable to form a satisfactory basis for their classification. The bacteria illustrate these difficulties very clearly. Their resting-stage shapes, where they occur, are too similar for most of them to be classified by this means, and it seems unlikely that they will be found to possess sufficiently numerous surface characteristics, even when examined with the electron microscope (see Plates 7b and 11a) for any detailed morphological classification[1] to be possible. Their early classification was developed by pathologists, who were impressed by their specificity in being the cause of an identifiable disease, for example; but the outstanding characteristic of the common soil bacteria is their adaptability rather than their specificity.

The methods of studying what the micro-organisms do in the soil are again inadequately developed.[2] There are still no satisfactory ways of following what the smaller micro-organisms do, though there are a variety of methods giving partial pictures of their activities. The method of examining a natural soil directly with a microscope, using incident light illumination, has been developed by W. L. Kubiena,[3] and though of considerable value for the study of the larger organisms such as some fungi and nematodes, it is still limited by great technical difficulties in the study of bacteria and other organisms that have almost transparent bodies.

J. Rossi and S. Riccardo[4] and N. Cholodny[5] independently devised a method for the direct examination of the soil organisms. Microscope slides or cover-slips are buried in the soil for a few days, and carefully dug up, the main lumps of soil removed, and the micro-organisms on the slide fixed and stained. Such preparations often give an excellent idea of the methods of growth and of the morphology of the different types, and one can sometimes even see that one organism has been feeding on another, but the results obtained by this method must be carefully interpreted. In the first place, some organisms will grow better on a continuous solid surface than in between the soil particles, hence will develop more strongly on these slides than in the soil. To overcome this limitation J. S. Waid and M. J. Woodman[6] suggested burying nylon net in the soil and then studying the fungal flora that grew on or round the fibres. In the second place, it has been customary to examine these slides after the organisms have been killed, fixed and stained, so that only their gross morphology, i.e. whether they are cocci, rods, fine hyphae, etc., could be noted, and only rarely were fruiting bodies found. Techniques such as phase contrast microscopy, however, allow direct examination of the living organisms to be made, but their value for studying where the micro-organisms are actually living in the soil is still very limited.

Another group of methods gives a limited picture of where the organisms are living in the soil, but the means used to make them visible involves killing them. C. Rouschel and S. Strugger[7] and E. Burrichter[8] have developed tech-

---

1 See for example S. T. Cowan, *Symp. Soc. gen. Microbiol.*, 1962, **12**, 433, for a discussion of these problems.
2 For a review of this subject see R. D. Durbin, *Bot. Rev.*, 1961, **27**, 522.
3 *Micropedology*, Ames, 1938.
4 *Nuovi Ann. dell' Agricolt.*, 1927, **7**, 92, 457. For a summary of Rossi's work in English, see *Soil Sci.*, 1936, **41**, 53.
5 *Arch. Mikrobiol.*, 1930, **1**, 610.
6 *Pédologie Gand.*, 1957, **3**, 1.
7 *Naturwiss.*, 1943, **31**, 300. *Canad. J. Res.*, 1948, **26C**, 188. See also A. Stockli, *Schweiz., landw. Mh.*, 1959, no. **4**/5, 162.
8 *Ztschr. PflErnähr*, 1953, **63**, 154.

niques which allow one to see bacteria and other organisms in a fluorescence microscope and P. C. T. Jones and J. E. Mollison[1] and Y. T. Tchan and J. S. Bunt[2] used dyes that were intended to stain living but not dead organisms. Rouschel and Strugger found that the dye, acridine orange, when absorbed by living (or recently dead) organisms fluoresces green, but when absorbed by dead cells or organic matter fluoresces red to red-brown, and L. A. Babiuk and E. A. Paul[3] more recently have introduced fluorescein isothiocyanate, which gives better resolution between living and dead cells. Jones and Mollison found that living (or recently dead) organisms would absorb aniline blue and hold it against a solvent which would remove it from organic matter and dead cells. Bacteria, actinomycetes and living fungal mycelium all stain an intense blue colour by this technique, and can be clearly distinguished from any soil or organic particles they may be on: some of their photographs are reproduced in Plate 10. However, these dyes do not distinguish between living and dead organisms as clearly as had been hoped, for usually many microbial cells are found that have an intermediate colour between the fully alive and the completely dead.

A better picture of where soil organisms are living in the soil can be got by staining the organisms in a block of soil with one of these dyes, drying the block, impregnating it under vacuum with a thermosetting resin and cutting a section 20–30 μ thick, which is thin enough to be viewed with transmitted light,[4] and a typical picture obtained by this technique is reproduced in Plate 10c (p. 216). Fungal hyphae can be seen in these thin sections of soil without the use of a stain.[5] Micro-organisms can also be seen in suitably prepared soil preparations using an electron microscope, particularly if it has a considerable depth of focus and Plate 11 gives three photographs of soil micro-organisms taken with a Stereoscan electron microscope by T. R. G. Gray.[6] The bacteria show up as small spheres, most of which are separated one from another, but the resolution is not yet adequate to show any great detail of the surface structure of the organisms.

The characteristic of all these methods which allow direct observation of where the micro-organisms are actually living in the soil is that the organisms are all dead at the time of observation, so most cannot be closely identified. However, certain species or strains of bacteria and fungi can be recognised under suitable conditions using the fluorescent-antibody technique. The antibody of a selected bacterium or fungus is obtained, usually from rabbit serum following a series of injections, which is then conjugated with a fluor-

1 *J. Gen. Microbiol.*, 1948, **2**, 54.
2 *Nature*, 1954, **174**, 656.
3 *Canad. J. Microbiol.*, 1970, **16**, 57.
4 For illustrations see D. Jones and E. Griffiths, *Pl. Soil*, 1964, **20**, 232.
5 S. T. Williams and D. Parkinson, *J. Soil Sci.*, 1964, **15**, 331, and A. Burges and D. P. Nicholas, *Soil Sci.*, 1961, **92**, 25.
6 *Science*, 1967, **155**, 1668; for other illustrations see C. A. Hagen, E. J. Hawrylewicz and B. T. Anderson, *Appl. Microbiol.*, 1968, **16**, 932.

escent dye, such as fluorescein or rhodamine isothiocyanate. A soil smear is then made on a glass slide, washed with this suspension, and all bacterial or fungal cells which absorb the antibody will fluoresce when the smear is viewed either by incident or transmitted ultraviolet light.[1]

Another method for studying the soil organisms is based on the pure culture plating technique. The soil is dispersed in water or suitable diluent and a sample of the soil suspension is diluted, mixed with a suitable culture medium, and poured into a petri dish containing nutrient agar. Organisms from the individual colonies can then be picked out and their metabolism and behaviour studied in isolation. This method selects from among the common soil organisms those that can be brought into suspension, and which will grow on the particular nutrient medium used; it can also be used to pick out less common organisms that possess any easily recognisable characteristic; but it suffers from the defect that usually only a small proportion of the soil micro-organisms appear to develop on these plates (see p. 162), and, hence, one cannot conclude that the organisms so isolated are a representative sample of the microbial population. Further, it only gives limited information about the reactions which the various bacteria actually carry out in the soil, for bacteria can grow on organic substances in pure culture which they would be unable to use in the very competitive environment of the soil.

Beijerinck and Winogradsky developed another group of methods of great value in isolating organisms having an 'unusual' metabolism, based on altering the nutrient conditions either in the soil, or more usually in the culture plate, so that they are especially favourable to organisms capable of this metabolism, but unfavourable to other types. Thus, Winogradsky was able to encourage bacteria which could derive their energy from the oxidation of ammonium or nitrite by supplying the soil with ammonium or nitrite as the sole source of energy.

These various methods supplement each other. The methods of direct observation show something of what the micro-organisms look like in the soil, and often where they live, but only rarely do they show what they are doing; while the pure culture methods usually give some indication of what the micro-organisms can do, but not what they look like, or where they are, in the soil.

The nutrition of the microflora: Autotrophic and heterotrophic organisms

Organisms need food for two distinct purposes: to supply energy for their necessary vital processes and to build up their body tissues. Some of the

---

1 See for example E. L. Schmidt and R. O. Bankole, *Science*, 1962, **136**, 776; 1968, **162**, 1012. Also J. Eren and D. Pramer, *Soil Sci.*, 1966, **101**, 39, and I. R. Hill and T. R. G. Gray, *J. Bact.*, 1967, **93**, 1888. For another method of labelling see M. A. Darken, *Appl. Microbiol*, 1962, **10**, 387, and P. H. Tsao, *Soil Biol. Biochem.*, 1970, **2**, 247 (optical brightness method).

microflora can use entirely different sources of food for those two purposes, whereas for others and for most animals the same food serves both purposes equally. The microflora use three different methods for obtaining energy: most algae and a few kinds of bacteria can use direct sunlight in much the same way that green plants can, though most photosynthetic bacteria do not evolve oxygen during the process; some bacteria can use, and some can only use, energy set free by the oxidation of certain reduced inorganic compounds, such as hydrogen, sulphides, sulphur and ammonia, for example; while most of the microflora can only use energy set free when organic substances, such as sugars or simple fatty acids, are being oxidised or degraded.

The microflora have in the past been classified into autotrophs which were considered to need no organic substances for their energy or growth, and heterotrophs which needed organic substances for both, but this classification is inadequate.[1] There are bacteria that can be classified as obligate autotrophs, which can only get their energy from a specific inorganic oxidation and their carbon from bicarbonate, and, further, their growth may be inhibited by quite low concentrations of certain amino and carboxylic acids that are nutrients for most other organisms. But there are also bacteria which can function autotrophically in the above sense, but which can also get their energy from organic oxidations, and may only give maximum growth when they are not solely dependent on an inorganic oxidation for their energy supply. There are also bacteria which can only use organic sources of carbon for growth, but can, though need not, get a part of their energy from an inorganic oxidation. Moreover, some of the bacteria using bicarbonate for their body carbon may have their growth stimulated if given a suitable growth factor such as biotin or a component of a complex nutrient such as yeast extract. Examples of all these bacteria are found in soils.

Fully autotrophic organisms require much more energy for growth than many heterotrops, since much more energy is needed to convert the carbon of bicarbonate than of organic substances such as sugars into protoplasm and cell wall constituents. A number of the inorganic oxidations used by these organisms release less, and sometimes considerably less, energy per mole of oxygen taken up than does the oxidation of sugar, for example. Thus, the oxidation of glucose releases about 530 kJ/mol of oxygen used,[2] the oxidation of ammonium to nitrite releases 160, nitrite to nitrate 150, and ferrous to ferric iron 180. On the other hand, some oxidations release about the same amount. Thus the oxidation of hydrogen sulphide to sulphur releases 670 kJ, sulphur to sulphate 500 and hydrogen gas to water 470. Autotrophic bacteria using the first group of oxidations are likely to make relatively little growth

1 For a review see S. C. Rittenberg, *Adv. Microbiol Physiol.*, 1969, **3**, 159.
2 1 kJ = 1000 joules = 239 calories. The figures quoted here are taken from R. E. Buchanan and E. I. Fulmer, *The Physiology and Biochemistry of Bacteria*, Vol. I. Ballière, Tyndall and Cox, 1928, except those for ammonium and nitrite oxidation which were kindly supplied by Dr J. E. Prue and Dr C. J. B. Mott.

per unit of oxygen consumed, compared with a heterotroph using sugar, and in consequence will appear to be slow-growers in culture solutions.

There are a number of photosynthetic micro-organisms in soils. The most widely distributed are the algae, which possess chlorophylls either similar or closely related to the chlorophylls of green plants, so evolving oxygen gas as a consequence of photosynthesis. Under some conditions photosynthetic bacteria are found in soils, but these cannot evolve free oxygen, so can only grow in reducing conditions containing suitable oxidisable substrates, such as hydrogen sulphide, ferrous ions or methane. In so far as these bacteria are found in soils, they are usually only present in soils that are regularly water-logged, such as some padi soils.

Autotrophic organisms can obtain their nitrogen supply from ammonium or nitrate salts, and these are usually the preferred sources, although the majority can use certain amino acids. A few autotrophic organisms can use or fix atmospheric nitrogen as their primary source, for example, all photo-synthetic bacteria growing under suitable anaerobic conditions and some blue-green algae under aerobic conditions.

Heterotrophic organisms are usually classified either by their nutritional requirements or by the biochemical changes they bring about in various nutrient media. Most members of the microflora can use simple sugars as their main source of body fuel, but they vary in the types of sugar they use. The majority can use glucose (dextrose) as their primary source, and produce, sometimes only if necessary, enzymes capable of converting a wide range of carbohydrates into this sugar.

Heterotrophic organisms have a wide variety of demands in their food supply. The unspecialised members can obtain all the carbon for their body tissues from simple sugars or fatty acids and their nitrogen from inorganic nitrogen compounds, or, for some bacteria, from gaseous nitrogen. These processes are possible because the organisms can synthesise all the enzymes needed for their life and growth. Other members need more complex food supplies and one can consider that this is due to the organism losing its power to synthesise one or more of these enzymes. Carbon dioxide is also an essential growth factor for many heterotrophic bacteria and fungi, as it is for all auto-trophs. Werkman and Wood[1] have been able to follow a little of the chemistry of this carbon dioxide need by using radioactive carbon and have shown that it is needed to convert pyruvic acid   a product of sugar fermentation—into oxalacetic acid, which plays an important role as a hydrogen carrier in the respiratory oxidation of sugars.

This loss of power of synthesising critical substances is scattered over many groups of the microflora in such an erratic way that it has no systematic significance. On the whole, the fastidiousness of the organism is a reflection of its usual food supply: the more specialised the supply, as, for example, in a

1 For a review of this work, see C. H. Werkman and H. G. Wood, *Adv. Enzymol.*, 1942, **2**, 135; and H. A. Krebs, *Ann. Rev. Biochem.*, 1943, **12**, 519.

well-adapted parasite, the more likely the organism is to have lost the power of synthesising some of the enzymes produced by related organisms utilising a less specialised food supply. It has always been assumed in the past that autotrophic organisms, that is, organisms deriving the carbon of their protoplasm from carbon dioxide, could synthesise all the enzymes and amino acids needed for their growth. This assumption is now known to be false for some of the photosynthetic sulphur bacteria, as some need one or more constituents of the B vitamin complex.[1] Hence, there may also be other members of the autotrophic bacteria, in the sense defined above, that need some of the growth-promoting substances—a possibility that has not yet been properly investigated.

All micro-organisms need minerals as well as sources of carbon and nitrogen, though the relative amount needed varies widely with different types. As with plants, potassium and phosphate are needed in considerable quantities; sodium, magnesium, calcium, iron and sulphur, often in relatively high quantities, though the demand for these varies widely as between different organisms; magnanese, copper, zinc, boron and cobalt usually in small quantities by most organisms; while molybdenum or vanadium are probably necessary if the organisms are either using nitrate or atmospheric nitrogen as their source of nitrogen.[2] The function of most of these metals is to form part of the prosthetic group of an enzyme; thus, iron and copper are needed for various porphyrins taking part in hydrogen transfer during respiration. Azotobacter forms a good example of an organism needing a trace metal only when certain enzymes are required; it can only fix atmospheric nitrogen when supplied with traces of molybdenum or vanadium, but it can grow perfectly well in the absence of these metals if an available supply of ammonium ions is present.

The respiration of the microflora: Aerobic and anaerobic organisms

All organisms must respire to live, but they have developed different methods of respiration. Some—the obligate aerobes—must have access to free oxygen, while others either do not need such access—the facultative anaerobes—or else can respire only in the absence of free oxygen—the obligate anaerobes. Most soil members of the fungi and actinomycetes require conditions of good aeration, and few, if any, members outside the yeasts have been recognised as anaerobes. The soil bacteria, on the other hand, include groups which appear to have varying degrees of tolerance to progressive oxygen deficiencies, ranging from strict obligate anaerobes through numerous groups relatively insensitive to the oxygen supply to strict obligate aerobes. Many soil aerobic bacteria have a very efficient terminal oxygen acceptor, probably a cyto-

1 S. H. Hutner, *J. Bact.*, 1946, **52**, 213.
2 For a review of the nutrition of fungi, see R. A. Steinberg, *Bot. Rev.*, 1939, **5**, 327.

chrome oxidase, so can satisfy their oxygen requirements if the concentration of dissolved oxygen in the soil solution exceeds about 5 μM, which is about 1 per cent of the concentration of dissolved oxygen in equilibrium with the atmosphere.[1] If this is generally true for the soil bacteria, their apparent relative tolerance to poor aeration is probably due to their tolerance to the various reduced substances produced by other bacteria using an anaerobic system of respiration. It is, however, possible to separate out those able to grow in the absence of free oxygen from those requiring it at a certain minimum concentration.

Respiration can be defined as the process of carrying out a chemical reaction which liberates energy with the transfer of part of the energy so liberated into energy available for the vital needs of the organism. The typical sources of energy in the heterotrophic cell are sugars, mainly or perhaps only glucose, and the process of respiration normally consists in breaking down the sugar molecule into simpler ones having smaller total heats of formation. The mechanism of this breakdown can be most simply pictured as transfers of hydrogen atoms from the sugar to a hydrogen acceptor coupled with the residues taking up either water, or the hydroxyl groups from water, when necessary. Thus, the complete oxidation of glucose to carbon dioxide and water involves the tacking on of at least six hydroxyl groups to the six carbon atoms and the removal of at least twenty-four hydrogen atoms, twelve from the sugar itself and twelve from the six water molecules dissociated in the process, and these reactions are brought about by a battery of respiratory enzymes coupled with their appropriate prosthetic groups or coenzymes that function as hydrogen carriers. Some of these prosthetic groups or coenzymes contain a member of the vitamin B complex in their constitution, some are metal porphyrins, such as cytochrome, and some are simple four-carbon dicarboxylic acids, such as oxalacetic.

Oxygen is the final hydrogen acceptor in aerobic respiration, and combined oxygen can play this role under anaerobic conditions. Thus, many bacteria can reduce nitrates to nitrogen gas during respiration, a process involving the transfer of six hydrogen atoms from the sugar for each molecule of nitrate so reduced; other bacteria can reduce sulphates to sulphides, a process involving the transfer of eight hydrogen atoms per molecule of sulphate reduced. Even the oxygen of carbon dioxide can be so utilised, and again eight hydrogen atoms are transferred for each molecule reduced to methane, and this is probably the principal source of the methane produced in anaerobic decompositions.

Oxygen sources are not necessary for anaerobic respiration. The simplest type, in which one molecule of sugar is split up into two of lactic acid:

$$C_6H_{12}O_6 \quad \rightarrow \quad 2C_3H_6O_3,$$

1 For a summary see D. J. Greenwood, in *Ecology of Soil Bacteria*, ed. T. R. G. Gray and D. Parkinson, Liverpool UP, 1967.

is a reaction brought about by a number of bacteria, and involves no net transfer of hydrogen at all. But, in general, when an external oxygen supply is not available, respiration proceeds by part of the products of respiration acting as hydrogen acceptors and the remainder as hydrogen donors. The extreme case of this is the typical end product of the anaerobic fermentation of carbohydrates by a mixed bacterial flora which produces only methane and carbon dioxide:

$$C_6H_{12}O_6 \rightarrow 3CO_2 + 3CH_4.$$

Here the carbon of the methane has accepted the maximum number of hydrogen atoms a carbon atom can accept, and the carbons of the CO_2 have donated the maximum number. This process need not go to completion, thus alcohol yeast converts glucose into alcohol and carbon dioxide:

$$C_6H_{12}O_6 \rightarrow 2C_2H_5OH + 2CO_2.$$

In this case only two instead of three of the carbon atoms have lost all their hydrogen.

The outstanding difference between the anaerobic and aerobic fermentation of sugar lies in the energy liberated. Oxidising glucose to carbon dioxide and water liberates about 2800 kJ/mol of sugar oxidised; but converting glucose to lactic acid liberates about 88 kJ, or to alcohol and carbon dioxide about 75 kJ, or to methane and carbon dioxide about 180 kJ/mol of sugar decomposed. Thus, anaerobic organisms need to decompose far more organic material to derive a given amount of energy than do aerobic organisms; alternatively, given additions of organic matter will furnish far greater opportunity for increase of microbial activity in aerobic than in anaerobic conditions.

The byproducts of microbial metabolism: Microbial excretions

Micro-organisms, like larger organisms, take in food and excrete byproducts, which are either products of respiration or components in the food supply that cannot be assimilated. Autotrophic non-photosynthetic organisms must naturally always be excreting the oxidised product of the substrate from which they have derived their energy, while heterotrophic organisms must always be excreting carbon dioxide.

Many micro-organisms can multiply in a well-aerated carbohydrate solution and excrete little else except carbon dioxide if the solution contains adequate quantities of ammonium or nitrate. If the conditions are altered, for example, if the nitrogen supply or the degree of aeration is greatly reduced, these same organisms will begin to excrete a variety of energy-rich compounds instead of the fully oxidised carbon dioxide. Thus, if growth under aerobic conditions is being limited by a reduction in the nitrogen supply, the micro-organisms may keep up their carbohydrate intake but make less efficient use of it. They will then excrete partially oxidised carbohydrate of relatively high energy content to carry away the energy they have not needed to use.[1]

The excreted products of respiration by heterotrophic organisms growing under limiting conditions consist of two distinct types of compounds: complex carbohydrate gums typically produced by many groups of bacteria under aerobic conditions, and a variety of simple soluble compounds, such as the aliphatic acids from butyric to formic or the derived aldehydes, ketones or alcohols, the simple dibasic acids succinic and oxalic, and the simple hydroxy acids such as citric, tartaric and lactic—though citric, tartaric and oxalic acids are mainly excreted by certain groups of fungi, such as the Aspergilli.

Micro-organisms usually excrete the nitrogen of originally combined nitrogen which is surplus to their requirements as ammonium ions under aerobic conditions, though some of the larger may excrete urea or uric acid. But if the aeration is reduced, or anaerobic conditions set in, complex and usually foul-smelling amines will also be produced such as the aliphatic amines cadaverine and putrescine and the aromatics indole and skatole.

Micro-organisms also excrete small quantities of compounds that are often fairly specific to a given group. Some of these are antibiotics which will affect the microbial population growing in the immediate neighbourhood of the organism excreting it. Penicillin, which the fungus *Penicillium notatum* excretes under suitable cultural conditions, is an example. Other compounds that are excreted have not yet been isolated, such as those which cause the medium in which the micro-organism is growing to become 'stale', so that the organism will no longer grow in it, even though there is an adequate supply of nutrients present. Some others are growth substances, such as various vitamins and amino acids which are needed by other members of the population because they have lost the ability to synthesise them.

Micro-organisms can only use insoluble substances, such as cellulose and other polysaccharides and insoluble proteins as sources of nutrients if they can produce enzymes capable of converting these materials into soluble substances such as simple sugars or amino acids; for only soluble substances can diffuse or be transported through the cell wall into the cytoplasm. Most soil micro-organisms must therefore either excrete the enzymes necessary for solubilising their nutrient substrate either into the soil solution or onto the outside of their cell walls.

1 See, for example, J. W. Foster, *Chemical Activities of Fungi*, Academic Press, 1949.

10

The organisms composing the population

Bacteria and actinomycetes

Historically, bacteria have been far the most important group of soil organisms; they were the first to be studied intensively, and for long were regarded as the only organisms important in normal conditions. Laboratories and departments were set up for their study in most agricultural research institutions, and for many years they received more attention than all the other soil organisms put together, although they had not been shown to play such a predominant role in the vital processes of the soil as was implied by this one-sided development. However, during the 1950s the soil actinomycetes became the centre of interest since some of them are important sources of antibiotics. These two groups of organisms are being discussed together because there appears to be no clear dividing line between them, and the most commonly occurring soil micro-organisms fall into groups that are intermediate between typical bacteria and typical actinomycetes. An important distinction between typical members is that when a bacterial cell multiplies it tends to form clumps of cells, while an actinomycetes cell will form a chain of cells. The individual cells fall in the same size range, and often have similar cell wall constituents.

The number of bacteria in the soil

Bacteria and actinomycete cells cannot usually be seen in a natural soil when it is examined under a microscope, but techniques have been developed, based on the use of dyes, that either stain living cells in the soil, but not the soil colloids or organic matter[1] or stain the living cells so they have a different colour from the organic matter when examined in a fluorescent microscope (see p. 151). Results obtained using these techniques are consistent with the generalisation that the bacteria mainly live on the surface of the soil and humus particles, and that they hold on to these surfaces very tightly.

Techniques for counting bacterial and actinomycetes cells present in the

1 H. J. Conn, *New York State Agric. Expt. Sta., Tech. Bull.* 64, 1918. S. Winogradsky, *Ann. Inst. Pasteur*, 1925, **39**, 299.

soil have been developed (p. 152), of which the first to be used extensively was that of H. G. Thornton and P. H. H. Gray.[1] One difficulty of their technique is that the dye they used, erythrosin, stains moribund as well as newly dead cells the same colour as active ones. Dyes are now available that distinguish rather more sharply between living and dead cells but, no matter what dye is used, cells are found which either take up the dye weakly or give a colour intermediate between the typical active and definitely dead cell. These methods therefore tend to overestimate the number of active cells in the soil. The number of organisms counted by these techniques will be referred to in this book as the total cell count in the soil; and, for brevity, the word bacteria will be used to include actinomycete cells, except when it is necessary to make the distinction.

The method that has normally been used to count the number of bacteria in the soil is the so-called 'plating' technique and is much less direct. A known weight of soil is shaken up with a diluent, and some of this suspension, after dilution, is poured in to Petri dishes or plates and nutrient agar added. After incubation the number of bacterial colonies developed on the plate is counted, and so the number present per gram of soil can be calculated. But the number so calculated can only be considered the number present in the soil if:

1 each colony on the plate developed from one and only one bacterium;
2 all the bacteria in the soil sample are brought into suspension;
3 all the bacteria in the suspension can grow on the nutrient medium used.

None of these conditions, in fact, holds. Some bacteria regularly seem to consist of colonies held together by gummy substances, and so are counted as one organism instead of many. Most bacteria live on the surface of clay or organic matter particles[2] and it may be very difficult to dislodge them without killing them.[3] Further, when the soil is shaken up in the dispersion solution, bacteria in the solution may be absorbed by small soil crumbs settling through the solution, for clay crumbs will absorb some bacteria quite strongly.[4] Some bacteria will not grow close to their neighbours, so, in general, the greater the dilution of the bacterial suspension used on the plates, the greater the number of bacteria counted per gram of soil.[5] Finally, there can be no one single nutrient medium on which all soil bacteria will develop, though the most suitable medium for maximum count is one that is poor in readily assimilable nutrients, such as a water extract from a fertile soil.[6] It is also

1 *Proc. Roy. Soc.*, 1934, **115** B, 522.
2 F. E. S. Alexander and R. M. Jackson, *Nature*, 1954, **174**, 750; M. L. Jackson, W. Z. Mackie and R. P. Pennington, *Proc. Soil Sci. Soc. Amer.*, 1947, **11**, 57.
3 For the possible use of ultrasonics see I. L. Stevenson, *Pl. Soil*, 1958, **10**, 1.
4 For an example see N. Lahav, *Pl. Soil*, 1962, **17**, 191 and K. C. Marshall, *J. gen. Microbiol.*, 1969, **56**, 301.
5 See a series of papers by N. James and M. L. Sutherland, *Canad. J. Res.*, 1940–43, 18C–21C, and J. Meiklejohn, *J. Soil Sci.*, 1957, **8**, 240.
6 L. E. Casida, D. A. Klein and T. Santoro, *Soil Sci.*, 1964, **98**, 371.

possible that a proportion of soil bacteria, possibly in a resting stage brought about either by shortage of food or by the presence of antibiotics, take an appreciable time before they come out of that stage and can begin to grow in the medium.[1] Hence, the plating method will underestimate the number of soil bacteria.[2]

The direct and the plating method thus over- and underestimate the number of bacteria in the soil, but the difference between the numbers is surprisingly large, as is shown in Table 10.1,[3] which refers to some arable plots cropped every year with mangolds (Barnfield) and to some grass plots from which one or two hay crops are taken annually (Park Grass).

TABLE 10.1   Bacterial numbers in Rothamsted field soils

Plot	Manuring	pH	Numbers in millions per g		Ratio total count to plate count
			Total cell count (direct method)	Plate count	
	Barnfield				
1–0	Farmyard manure	7·6	3730	28·9	129
4–A	Complete minerals + ammonium sulphate	7·2	1770	15·10	117
8–0	No manure	8·0	1010	7·55	133
	Park Grass				
13	Farmyard manure	4·6	2390	2·25	1064
11–1	Complete minerals + ammonium sulphate	3·8	2400	1·35	1780
12	No manure	5·6	3040	7·50	405

The direct method is seen to give over a hundred times as many bacteria on the arable and a thousand times as many on the grass plots as the plating method. All other workers who have used the direct methods of counting have obtained numbers of the order of 10^9 bacteria per gram of soil,[4] while their plate counts are usually about a hundred times smaller, although F. A. Skinner, P. C. T. Jones and J. E. Mollison,[5] taking all precautions to get as high a plate count as possible, found that the total cell count on the Broadbalk

1 Y. T. Tchan, *Proc. Linn. Soc. N.S.W.*, 1952, **77**, 89; L. E. Casida, *Appl. Microbiol.*, 1965, **13**, 327.
2 For a discussion on the plate count technique see L. E. Casida (p. 97) and V. Jensen (p. 158) in *Ecology of Soil Bacteria*, ed. T. R. G. Gray and D. Parkinson, Liverpool UP, 1967.
3 Unpublished data of H. G. Thornton and P. H. H. Gray.
4 Thus, S. Strugger, *Canad. J. Res.*, 1948, **26**C, 188, using his fluorescent microscope technique found between 1 and 8·6 × 10^9 per gram soil.
5 *J. gen. Microbiol.*, 1952, **6**, 261. For a comparison with the Muguga (Kenya) soils see J. Meiklejohn, *J. Soil Sci.*, 1957, **8**, 240, who found the ratio varied from 4 in the dry season to 80 in the wet.

plots were only about fifteen times higher than the plate, the mean values of the two being 5146×10^6 and 337×10^6 per gram of soil. It is interesting to note, however, that T. R. G. Gray and his workers[1] found that the two methods gave very similar results for a dune-sand soil under forest, the direct count giving between 4–6 and the plate 1–2 $\times 10^8$ cells per gram of soil.

The full interpretation of this difference cannot yet be given, but it is probably due to only a proportion of the bacterial and actinomycete cells being active at any one time, the remainder being in a resting state. This would imply that the plate count gives a good estimate of the metabolically active bacteria at the time of sampling, and this is borne out from an estimate of the rate of production of carbon dioxide from the soil. This is commonly about 1 kg/ha/hr, which is the rate expected for 50×10^6 bacteria per gram of soil[2] if growing actively, assuming they produce one half of the CO_2, are evenly distributed throughout the top 15 cm of soil, and respire about 10^{-7} g CO_2 per million bacteria per day.

The figures given in Table 10.1 can be used to estimate the weight of the bacteria in an acre of soil. The typical soil bacteria are very small. Their mean volume is not known, but is probably between 0·5–0·2 μ^3 or 0·5–0·2 $\times 10^{-12}$ cm³, so if there are 3 $\times 10^9$ bacteria per gram of soil, they will weigh between 1·5 and 0·6 mg/g of soil, or, assuming they average this number throughout the top 15 cm of soil and the top 15 cm of soil weighs about 2500 t/ha, the bacteria will weigh between 3·5 and 1·5 t/ha live weight or say between 700 and 300 kg dry weight. This soil would probably contain about 3 per cent of organic matter by dry weight, so that there would be about 75 t/ha of dry organic matter, of which the bacteria would constitute less than 1 per cent.

The fluctuations in the number of soil bacteria

The numbers of bacteria in the soil are never stationary. D. W. Cutler, L. M. Crump and H. Sandon,[3] as long ago as 1920 and 1921, showed that the numbers counted by their plating technique fluctuated from day to day during the course of the year, and C. B. Taylor[4] showed that these fluctuations occurred from hour to hour, whether they were counted by Thornton and Gray's total cell count method or were counted on mannite or soil extract plates. He found that the fluctuations in numbers determined by these three methods were uncorrelated, as is shown in Fig. 10.1,[5] where the plate count was made on soil extract agar, suggesting that the fluctuation in total numbers is made up of a series of fluctuations occurring independently in different groups of bacteria.

1 In *Ecology of Soil Bacteria*, Ed. T. R. G. Gray and D. Parkinson, Liverpool UP, 1967.
2 F. E. Clark, *Soil Biology*, Ed. A. Burges and F. Raw, Academic Press, 1967.
3 *Phil. Trans.*, 1922, **211** B, 317.
4 *Proc. Roy. Soc.*, 1936, **119** B, 269.
5 Taken from H. G. Thornton and C. B. Taylor, *Trans. 3rd Int. Cong. Soil Sci.*, (*Oxford*), 1935, **1**, 175.

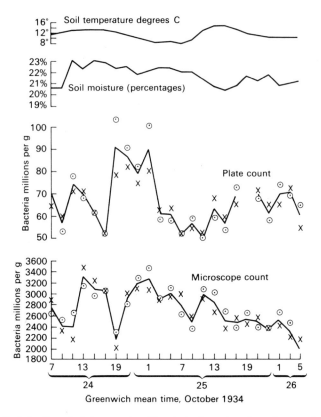

FIG. 10.1. Total cell and plate counts of bacteria from two-hourly samples of fallow garden plot soil. In the curves shown here, the dots and crosses represent numbers in duplicate samples; all curves show mean values

The causes of these variations are not known, nor is it even certain whether field variations from sample to sample may not be the most important cause. However, Taylor has shown that these variations occur even if the soils are kept at a constant temperature and moisture content.

The types of soil bacteria

The prime difficulty in the classification of the soil bacteria can now be appreciated. The only certain way the bacteria can be classified is by their behaviour on different nutrient media, but it is possible that the majority of the types of bacteria present in the soil have never yet been grown on any medium; in all the discussions of bacterial classification which involve platings this limitation must be remembered.

The soil bacteria can be classified by their nutritional requirements. The division between autotrophs and heterotrophs has already been mentioned

(p. 153). The heterotrophs can be classified according to the complexity of their nutritional requirements, as has been attempted by Lochhead and Chase[1]. They isolated soil bacteria on as non-selective a medium as possible, and then tested the proportion that can grow on media of increasing simplicity. Thus, they have suggested seven groups as follows:

1 Those growing in a simple glucose-nitrate medium containing mineral salts.
2 Those that can only grow when ten amino acids are added to this medium.
3 Those that can only grow when cysteine and seven growth factors (aneurin, biotin, etc.) are added.
4 Those needing both the amino acids and the growth factors.
5 Those needing yeast extract.
6 Those needing soil extract.
7 Those needing both yeast and soil extract.

They found that fertile soils contain a greater proportion of bacteria with complex requirements, i.e. belonging to groups 5, 6 and 7, than infertile; that the soil extract from a fertile soil has a greater growth-promoting power than from an infertile; and that the bacteria in the immediate vicinity of plant roots fall more into groups 2, 3 and 4, than into the more complex groups.[2] Katznelson and Chase[3] found that adding generous dressings of farmyard manure to a soil increased the proportion of bacteria having complex growth requirements at the expense of those having very simple ones, and V. Garcia, working with soils from some of the Rothamsted plots, confirmed this result and found that soils which had large dressings of fertilisers had the same distribution of bacterial types as the unmanured.

This classification can, however, be criticised on the ground that, as the various growth factors in yeast and soil extract are recognised, some may be simple substances that should be included in the simpler media. Thus, A. G. Lochhead and R. H. Thexton[4] subsequently showed that vitamin B_{12} is an important constituent in some soil extract media, and there is no fundamental reason why this should not have been included in the growth factor mixture used for groups 3 and 4.

The morphological classification of soil bacteria is also difficult, as some groups can have several forms. A usual classification for the soil bacteria has been into six groups:

1 Small cocci, usually about 0·5 μ in diameter.
2 Short straight rods, about 0·5 μ in diameter and 1–3 μ long; the shorter often being called coccoid rods.
3 Short curved rods—the Vibrios.

1 *Soil Sci.*, 1943, **55**, 185.
2 P. M. West and A. G. Lochhead, *Canad. J. Res.*, 1940, **18** C, 129. See also H. Katznelson and L. T. Richardson, ibid., 1943, **21** C, 249.
3 *Soil Sci.*, 1944, **58**, 473.
4 *Nature*, 1951, **167**, 1034.

4 Long rods, usually about 1 μ in diameter.
5 Rods sometimes showing branching.
6 Thin flexible rods, with very thin cell walls, usually under 0·5 μ in diameter and 2–10 μ long.

There also appears to be some very small bacteria in the soil, existing as cocci or coccoid rods with diameters down to 0·15 μ, but some of these may be stages in the life of larger bacteria.[1]

This morphological classification has, however, little systematic significance, for most of the soil bacteria are strongly pleomorphic, that is their shape depends on the culture medium in which they are growing. In particular, the majority of group 1, the small cocci seen in the soil, may give rise to forms in groups 2, 4 and 5 in suitable media. Nor is there any correlation between the morphological and the nutritional classification, for members in any morphological group may occur in each of the nutritional groups. It is because of difficulties such as these that much interest is being taken in the application of numerical taxonomy to bacteria and actinomycetes.[2] The basis of these methods is to specify quantitatively a selected, usually large, number of different characteristics of the organism, and group together those showing the maximum number of similarities.

Direct observation of the bacterial flora in soils shows up the overwhelming predominance of cocci and rods in most soils, as is shown in Plates 10 and 11, though cocci, which may principally be actinomycete spores, predominate in a few. Clumps of cocci or coccoid rods can sometimes be seen embedded in slime and some of the organisms are surrounded by gummy capsules, as is also shown in Plate 10a and c. Large rods are relatively rare and occur either in chains or singly. Large cocci sometimes occur, commonly in groups of two, three or four together in a colony. Plate 7 shows typical bacterial colonies of these kinds.

The plating technique, using a nutrient-poor non-selective medium picks out, as Conn showed several years ago, many small slow-growing short rods. C. B. Taylor and A. G. Lochhead,[3] for example, found that about 90 per cent of all the bacteria picked out were short rods. The largest proportion of these are pleomorphic, and cover the range between the ill-defined bacterial genus *Arthrobacter*[4] through typical *Corynebacter* and *Mycobacter* to the ill defined but undoubted actinomycete genus *Nocardia*. Most of the rest of this group are included in the *Pseudomonas-Achromobacter* group of bacteria and a very small proportion are included in the *Radiobacter-Rhizobia* group. The most

1 See P. P. Laidlaw and W. J. Elford, *Proc. Roy. Soc.*, 1936, **120** B, 292, for a possible example of small soil bacteria in sewage, and S. W. Orenski *et al.*, *Nature*, 1966, **210**, 221.
2 For example, see a group of papers in *The Soil Ecosystem*, The Systematics Assoc. Publ. 8, 1969, by T. R. G. Gray, 73; M. Goodfellow, 83; S. T. Williams *et al.*, 107.
3 *Canad. J. Res.*, 1938, **16** C, 162.
4 For a description of *Arthrobacter* strains in soils see E. G. Mulder and J. Antheunisse. *Ann. Inst. Pasteur*, 1963, **105**, 46.

common larger cells are actinomycetes typically belonging to the genera *Streptomyces* and *Micromonospora*, and it is species of *Streptomyces* which produce a volatile substance with the odour of freshly turned earth, which N. W. Gerber[1] has named geosmin. Since the relative importance of these organisms depends on the soil pH, some experienced field workers can often estimate the pH of a soil to about half a unit by its smell.[2]

S. Winogradsky[3] was one of the first bacteriologists to make a direct examination of the bacteria in a soil, and he introduced the hypothesis that the small cocci and coccoid rods characterised soils containing no readily fermentable material, while the rod-shaped and spore-forming bacteria characterised soils which still contain readily decomposable material. He called the first population autochthonous or indigenous, and he pictured it as feeding on soil humus and consequently having a low biological activity; and the second zymogenous which was capable of rapid multiplication, and therefore of high biological activity in the presence of its food supply and then rapidly sinking into resting spores when the food supply is exhausted. The autochthonous group also contains a higher proportion of forms having complex nutritional requirements than the zymogenous. It is unlikely, however, that the soil population can be sharply divided into these two groups, although no critical examination of the two populations based on modern techniques has been made.

Some bacterial cells develop flagellae in certain stages of their growth or under suitable external soil conditions, and this allows them to move in the soil water films. Not much is known about their rate of movement in the soil, or the distance they can move, nor of the effect of the thickness of the water films on their movement (see p. 222), but some early work of H. G. Thornton and N. Gangulee[4] suggested that it could be as high as 1 mm/hr under very favourable conditions.

There are also bacteria in the soil that feed on some of the common species of soil bacteria and possibly some other microbial cells. They do this by excreting a substance that dissolves, or lyses, the bacterial membrane that encloses the living protoplasm, and then absorbing the liquid cell contents.[5] Some belong to a group of very small and motile organisms known as *Bdellovibrio* and others to the Myxobacteria or slime bacteria. Although many species of myxobacteria are known, the soil only seems to contain very few. Thus, B. N. Singh[6] only found three species commonly present and a fourth more rarely. He found them in all the British arable soils and in about two-

1 Quoted by D. Pramer, *Ecology of Soil Bacteria*, Ed. T. R. G. Gray and D. Parkinson, Liverpool UP, 1967.
2 This was brought to my attention by Dr P. J. Harris.
3 *Comp. Rend.*, 1924, **178**, 1236.
4 *Proc. Roy. Soc.*, 1926, **99** B, 428.
5 See, for example, J. M. Beebe, *Iowa St. Coll., J. Sci.*, 1941, **15**, 307, 319; and A. E. Oxford and B. N. Singh, *Nature*, 1946, **158**, 745.
6 *J. Gen. Microbiol.*, 1947, **1**, 1.

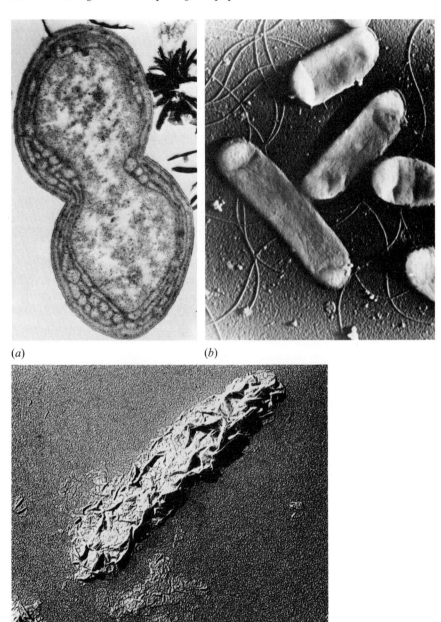

(a)   (b)

(c)

PLATE 7. Electron micrographs of soil bacteria:
(a) *Nitrosococcus nitrosus* dividing. (× 47 000).
(b) Rod-shaped bacteria with flagellae (Au shadowed, × 25 000).
(c) Illite clay particles adsorbed on the surface of the bacterium *Rhizobium trifolii*. (× 34 500).

thirds of the pasture soils he examined, and in the single soil in which he tried to count them there were between 2000 and 80 000 per g.

These higher predaceous bacteria are not the only members of the Myxobacteria present in the soil, as many of the cellulose and chitin[1] decomposers, such as the *Sporocytophaga* and the *Cytophaga*, belong to this group,[2] but it has not yet been conclusively shown that they can ever form the characteristic fruiting bodies of the predaceous species.

Adsorption of clay particles by bacteria

Some, possibly most, species of bacteria will adsorb clay particles on their surface. Plate 7c is an electron microscope photograph of a bacterial cell which has adsorbed particles of a sodium-saturated montmorillonite clay on its surface[3]. K. C. Marshall[4] has also obtained evidence that some strains of *Rhizobia* have a surface which carries a fairly uniform negative charge, due to the dissociation of hydrogen ions form surface carboxylic acids, and these bacteria adsorb about 0·5–0·6 μg of an illitic clay per 1 μ^2 of bacterial surface, and this weight of clay has a surface area of about 200 μ^2. N. Lahav[5] obtained a similar figure for the amount of montmorillonite clay that a strain of *Bacillus subtilis* would absorb. Marshall considers the clay particles are adsorbed through positive charges on their broken edges being attracted to the negative charges on the bacterial cell wall, so the clay particles are perpendicular to this surface. He also found that other rhizobial strains carry some positive charges on their surface, due to the presence of amino groups, as well as negative, and these only adsorb about one-half the amount of clay. The photograph also indicates that the clay film around the bacterial cell is more than one particle thick.

No corresponding work has been done on the adsorption of humus particles by bacteria. In the soil, the bacteria seem to be concentrated on clumps of humus particles rather than on mineral particles,[6] but this could simply be because the humus formed the nutrient substrate for these bacteria rather than there being any specific adsorption of humus by the bacterial cell wall.

Bacteriophages

Some groups of bacteria and actinomycetes can be attacked by bacteriophages, which appear to be bodies analogous to plant and animal viruses. They enter the living cell and multiply in it apparently by reorganising the cell contents and then lyse the cell wall and so diffuse into the surrounding

1 For a review of this subject, see R. Y. Stanier, *Bact. Rev.*, 1942, **6**, 143.
2 R. Y. Stanier, *J. Bact.*, 1947, **53**, 297.
3 M. M. Rope and K. C. Marshall, *Microbial Ecol.*, 1974, **1**.
4 *Aust. J. biol. Sci.*, 1967, **20** 429; *Biochem. biophys. Acta*, 1968, **156**, 179. *J. gen. Microbiol.*, 1969, **56**, 301, and see also M. L. Jackson, W. Z. Mackie and R. P. Pennington, *Proc. Soil Sci. Soc. Amer.*, 1947, **11**, 57.
5 *Pl. Soil*, 1962, **17**, 191.
6 P. C. T. Jones and J. E. Mollison, *J. gen. Microbiol.*, 1948, **2**, 54.

medium. Some phages can grow in bacteria without causing lysis of the cell, so appear to be well-adjusted parasites. Bacteria and actinomycetes belonging to all the principal widely distributed groups in the soil can be attacked, but soil fungi rarely appear to be. The phages of greatest practical interest from the point of view of crop growth are those attacking the nodule bacteria, but very little is yet known about their distribution in nature, nor even about how often they are of any importance in causing 'clover sickness' or a reduction in the nitrogen-fixing ability or number of nodule bacteria in soil.

Actinomycetes

The typical filamentous actinomycetes belong predominantly to the Streptomyces and Micromonospora groups.[1] They form very fine often much-branched hyphae when growing, which break up into spores, either by the tip of the hypha producing one or two spores or else by a length of hypha, often a length of hypha twisted into a coil, breaking up into a line of spores. In the soil these characteristic hyphal coils are only produced if they have free access to air, so that they are typically produced in the surface pores of a moist soil which are shaded from the direct sun.[2] However, in the soil there are groups of actinomycetes that come intermediate between the Nocardia and the hyphal forms, so the separation between soil Nocardia and Streptomyces can be difficult. In fact, a large proportion of the soil actinomycetes are difficult to classify as their morphology probably often depends on their nutritional history.

The soil actinomycetes are nutritionally a very adaptable group; they are probably without exception heterotrophic, and can use a wide range of carbon and nitrogen compounds, such as celluloses, hemicelluloses, proteins and possibly lignin.[3] The Nocardia also contain members that can be adapted to decompose many unusual compounds or synthetic chemicals. They do not seem to need any growth-promoting substances, though many, particularly members of the Streptomyces, may excrete a range of vitamins, growth substances and antibiotics into the surrounding medium. Most members are aerobic, and some may have a limited ability to reduce nitrates. Only a few actinomycetes are parasitic on plants and then usually on their roots.

Soil actinomycetes are typically aerobic organisms and, like the fungi, are commoner in dry than wet soils. They are also commoner in warm than cool soils; thus H. L. Jensen[4] found that they can be the dominant members of the microflora at temperatures around 28°C if the water supply is restricted, and many of the thermophilic types can thrive in the aerated parts of a compost heap even when the temperature reaches 60° to 65°C. They also appear to be

1 For a review of the soil actinomycetes, see E. Küster in *Soil Biology*, Ed. A. Burges and F. Raw, Academic Press, 1967.
2 D. Erikson, *J. gen. Microbiol.*, 1947, **1**, 45.
3 S. A. Waksman, *Antonie van Leeuwenh.*, 1947, **12**, 49.
4 *Proc. Linn. Soc. N.S.W.*, 1943, **68**, 67.

very active under pastures, and may be the dominant micro-organisms in the surface layers of grassland in the United Kingdom if the soil is not too acid.

Fungi

The fungi form the second of the two great groups of soil micro-organisms, and whether they or the bacteria-actinomycetes group predominate depends on local conditions.[1] Soil fungi cannot be seen with either the naked eye or with a magnifying glass in normal arable soils, but in some woodland and forest soils white mycelial strands, consisting of a rope of hyphae, can sometimes be seen in the humus layer. Individual mycelia and fruiting bodies can sometimes be seen when a soil is directly examined in the field using Kubiena's microscope technique, and both spores and hyphae can be seen, when thin sections of soil, hardened with a suitable resin, are examined under a microscope.[2]

These methods of observing the soil fungi show that the hyphae are commonly between 5–20 µ in diameter, with a few being as fine as 2 µ and some up to 20–30 µ; and some fungal resting stages such as some chlamydospores, sclerotia and vesicles may be over 100 µ in size. The hyphae are often of irregular shape, possibly due to growing through irregular shaped soil pores. Fungal spores are commonly seen, but fungal fructifications only rarely, because they can only occur where there is space for them to form, such as on the surface of worm burrows, cracks, the under-surface of clods in arable fields or on the soil surface itself. Also, in many species, they only form at irregular intervals when all the environmental conditions are favourable, and Basidiomycetes can be active in soils yet only form their fruiting bodies at intervals of several years. Fungal spores are, however, fairly readily carried through the coarser pore space of the soil by members of the soil fauna.

Fungal hyphae are nearly always seen when a soil is stained and examined for bacteria, but the hyphae show all gradations of staining from deep to almost unstained, and many of the hyaline or transparent hyphae that take up the stain strongly only do so at one end, which is presumably the site of metabolic activity and growth, while the other end of the hypha looks empty of cell contents and may sometimes carry a bacterial population presumably feeding on the dead cell wall. Little is known for certain about the condition of the unstained hyaline hyphae; many are merely cell walls containing no protoplasm, but they cannot always be distinguished from those containing protoplasm.[3] J. H. Warcup[4] developed a technique for removing individual

---

1 For recent reviews on soil fungi see *The Ecology of Soil Fungi*, Ed. D. Parkinson and J. S. Waid, Liverpool UP, 1960, and J. H. Warcup in *Soil Biology*, Ed. A. Burges and F. Raw, Academic Press, 1967.
2 See, for example, D. P. Nicholas, D. Parkinson and N. A. Burges, *J. Soil Sci.*, 1965, **16**, 258, and D. Jones and E. Griffiths, *Pl. Soil*, 1964, **20**, 232.
3 See, for example, P. C. T. Jones and J. E. Mollison, *J. gen. Microbiol.*, 1948, **2**, 54.
4 *Trans. Brit. Mycol. Soc.*, 1957, **40**, 237.

pieces of hypha from a soil and inoculating them on agar plates, and found that in a wheatfield in South Australia 75 per cent of those removed from the soil in autumn, after the stubble had been ploughed in and the rains had started, grew in the medium compared with 3 to 15 per cent of those removed in summer when the soil had become dried out. For a soil under grass in a neighbouring field, a still smaller proportion grew, varying from 1 to 25 per cent. Unfortunately it is not possible to be certain what proportion did not grow because they were dead and what proportion did not do so because the medium into which they were inoculated was unsuitable to induce growth. His evidence also suggested that most hyaline hyphae were short-lived in moist soils and soon lost their cell contents if not growing actively, but that dark brown to black hyphae were longer-lived. Mycelial threads of Basidiomycetes can certainly be very long-lived in forest litter. These observations all refer to isolated fungal hyphae in the soil, but soils normally have plants growing on them with their roots growing in them, and the root whether alive or dead, can have a large fungal population growing between it and the soil or in it and the soil, and these fungal hyphae are usually active. It could be that a proportion of the non-viable hyphae found in the soil are the remains of those fungi after the root structure has completely disappeared.

This discussion shows that it is still very difficult to obtain a quantitative estimation of the amounts of viable mycelium and fungal spores in a soil, though techniques have been developed which separate fungal spores from a soil by a flotation technique based on their property of not wetting in water.[1]

The soil fungi predominantly belong to the groups that form filaments or mycelia,[2] though some species of Myxomycetes (slime fungi), yeasts and Chytridiales (Chytrids) are also found. The filamentous fungi belong to the Phycomycetes, which have branched unseptate mycelia; to the Moniliaceae or, where the perfect form is known, to the Ascomycetes, which have branched septate mycelia; and to the Basidiomycetes. The common soil Phycomycetes include members of the Saprolegniales including the genus *Pythium*, and of the Mucoraceae including the genera *Mucor*, *Rhizopus* and *Zygorrhynchus*. The common soil Moniliaceae include the genera *Trichoderma*, *Aspergillus*, *Penicillium*, *Cephalosporium* and *Fusarium*. These are probably usually present as spores in soils to which fresh organic matter has not recently been added. The soil Basidiomycetes are not so well identified,[3] but species of *Rhizoctonia* are common and pasture and forest soils contain members of the Hymenomycetes, or mushroom-like fungi, and the Gasteromycetes or puffball-like.

---

1 J. H. Warcup, *Nature*, 1955, **175**, 953; D. Parkinson and S. T. Williams, *Pl. Soil*, 1961, **13**, 347 and 1965, **22**, 167.
2 For a description of typical soil fungi see J. C. Gilman, *A Manual of Soil Fungi*, 2nd edn, Ames, 1957.
3 For lists of Basidiomycetes found in soils, see J. H. Warcup, *Trans. Brit. Mycol. Soc.*, 1959, **42**, 45, and with P. H. B. Talbot, ibid., 1962, **45**, 495; 1963, **46**, 465; 1965, **48**, 249. For a summary of his work see his Chapter in *Soil Biology*, Ed. A. Burges and F. Raw, Academic Press, 1967.

They also include most of the fungi forming mycorrhiza on the roots of forest trees, and those responsible for the 'fairy rings' of old pastures. These are probably usually present as viable pieces of mycelium or hyphae rather than as spores. In addition, there are fungi which can catch and feed on nematodes, mainly belonging to the Hyphomycetes, of which the species *Arthrobotrys oligospora* appears to be the most common, and others which feed on amoebae, mainly belonging to the Phycomycetes and put in the family Zoopagaceae, though some of the former feed on amoebae and other protozoa and some of the latter on nematodes.[1]

Direct observation of fungi in soils shows that there are a number of typical patterns of mycelial growth. S. Hepple and A. Burges[2] have described five such patterns.

1 The Penicillium pattern, in which the hyphae densely colonise a piece of substrate but they do not grow out into the surrounding medium, so spore formation occurs on its surface.
2 The Mucor ramannianus pattern, in which the hyphae colonise the substrate, but they also grow out into the surrounding medium forming chlamydospores well away from its surface.
3 The Basidiomycete pattern in which the long-lived mycelium will grow and colonise a substrate and then grow to another substrate well separated from the first. It produces fruiting bodies from rhizomorphs or well-developed mycelial strands.
4 The Zygorrhynchus pattern in which the fungus seems to grow at random through the soil as single hyphae and not be associated with any particular substrate. It may be growing on organic matter dissolved in the soil solution.
5 The fairy ring pattern in which the fungus migrates in a well defined mycelial zone, but the mycelia do not appear to be connected with any particular substrate. The mycelia cause profound microbiological changes in the zone in which they are active, for at least two-thirds of the normal fungal population appears to be killed and most of the fungi that are active in this zone are not commonly active away from it.[3] These fairy ring fungi are Basidiomycetes of the family Agaricales and belong to the genera *Marasmius*, *Psalliota* and *Tricholoma*.

A number of different types of hyphae can be seen in soils. Thus D. P. Nicholas and his colleagues[4] considered they could recognise a number of morphological types which included dematiaceous hyphae, thin hyaline septate hyphae, broad aseptate hyaline and broad septate brown-stained hyphae, and sparsely septate fragments of purple black hyphae.

1 C. Dreschler, *Biol. Rev.*, 1941, **16**, 265, and for a list of common English species see C. L. Duddington, *Nature*, 1954, **173**, 500.
2 *Nature*, 1956, **177**, 1186.
3 See, for example, J. H. Warcup, *Ann. Bot.*, 1951, **15**, 305.
4 *J. Soil Sci.*, 1965, **16**, 258.

It is not known what types of compounds activate fungal spores or pieces of hyphae in the soil to make them begin growth, although in pure culture it is only necessary to put them in a nutrient solution. But for some species spore germination may be prevented by the presence of fairly ubiquitous chemical compounds which inhibit germination, for these spores will germinate when transferred from the soil to distilled water.[1] Many plant roots produce soluble exudates which stimulate certain specific fungi to start growing, and hyphal extension may be as rapid as 40 μ per hour.

There are not yet any reliable methods available for estimating quantitatively the activity of the soil fungi as measured, for example, by their contribution to the oxygen uptake by the soil, or the carbon dioxide production in it, though qualitative observation indicates that it is greatest shortly after fresh organic matter has been incorporated. The older work on fungal numbers and activity was based on counting fungal colonies on plates, in the same way that bacterial numbers are counted. By making conditions in the plates unsuitable for bacteria, and by preventing any of the fungi from growing quickly, for example, by making the medium acid and by adding rose bengal and streptomycin,[2] it is possible to get a considerable proportion of the spores or pieces of mycelium to grow; and the numbers counted vary from a few thousands to over a million per gram of soil.

The soil fungi are probably all heterotrophic, but the species present have a wide variety of food requirements, ranging from those which can utilise simple carbohydrates, alcohols and organic acids, and nitrates or ammonia as their source of nitrogen, through those which can use celluloses and lignins, and those which require either growth factors, such as members of the B group of vitamins, to those which can only grow in competition with the general soil population as parasites of living plant roots and parasites and predators of living soil animals. Nearly all soil fungi need to be supplied with either inorganic nitrogen salts or organic nitrogen compounds, though some yeasts, a *Saccharomyces* and a *Rhodotorula*, which are mainly subsoil inhabitants, can fix atmospheric nitrogen.[3]

The saprophytic fungi can be very efficient converters of food into microbial tissues; some can synthesise 30 to 50 per cent of the carbon in the food into their cell substance,[4] which is higher than the corresponding figure for most bacteria when growing in the presence of an abundant food supply, though possibly not so much higher than for the autochthonous flora (see p. 167). This high efficiency of conversion has the corollary that rapidly growing fungi make very high demands on the available nitrogen of the soil, much of which is subsequently only slowly released in a form available to plants. Some

1 Quoted by J. H. Warcup, *Soil Organisms*, Eds J. Doeksen and J. van der Drift, North Holland, 1963, chapter 3.
2 J. P. Martin, *Soil Sci.*, 1950, **69**, 215.
3 G. Metcalfe and S. Chayen, *Nature*, 1954, **174**, 841, and E. R. Roberts and T. G. G. Wilson, ibid., 842.
4 S. A. Waksman, *Principles of Soil Microbiology*, 2nd edn, London, 1931, p. 244.

of these fungi synthesise humic-like substances or their precursors, and they may contribute appreciably to the humic matter in the soil.

The filamentous fungi generally need aerobic conditions to flourish, though they do not need aerobic conditions all along their filaments: they are capable of sending filaments into poorly aerated pockets of soil, but only if much of the filaments are growing in well-aerated conditions, and only these well-aerated parts ever produce spores. Species differ among themselves, however, in their tolerance to poor aeration. Thus P. Burges and E. Fenton[1] found that *Penicillium nigricans*, a fungus usually restricted to the upper 5 cm of the surface soil, is less tolerant of a high CO_2 concentration than *Zygorrhynchus vuillemini*, a species which is usually more abundant below 10 cm. However, typically, fungi are more common near the surface of the soil than lower down, and H. L. Jensen[2] in Denmark found them more common on light well-aerated soils than in heavier, as shown in Table 10.2. They can tolerate a wide

TABLE 10.2 Effect of $CaCO_3$ on numbers of fungi, actinomycetes and bacteria in soils as found by plating methods. Number per gram of soil

	Heath soil				Sand soil				Light loam			
	pH	Fungi, thousands	Bacteria + Actino-mycetes, millions	Actinomycetes, per cent	pH	Fungi, thousands	Bacteria + Actino-mycetes, millions	Actinomycetes, per cent	pH	Fungi, thousands	Bacteria + Actino-mycetes, millions	Actinomycetes, per cent
Untreated soil	3·7	610	0·84	0	4·7	341	5	61	5·8	127	8	36
CaCO₃ added	7·5	393	398	21	7·6	365	23	35	7·6	120	17	20

pH range, though typically they flourish under acid conditions such as occur in heaths and forests,[3] probably because few bacteria are really active in such acid soils. However, J. H. Warcup[4] showed that, for a group of soils at Laken-heath, Suffolk, all under grass and where the pH varied from 3·9 to 8·0, most of the Phycomycete and Ascomycete species inhabiting the acid sandy soils were distinct from those inhabiting the neutral and calcareous soils. He also noted that the dominant species in the acid soils belonged to the *Penicillia* which sporulated very freely, and the high numbers found on these soils may simply reflect the type of fungus present rather than the fungal activity.

1 *Trans. Brit. Mycol. Soc.*, 1953, **36**, 104.
2 *Soil Sci.*, 1931, **31**, 123.
3 H. L. Jensen, *J. Agric. Sci.*, 1931, **21**, 38; J. H. Warcup, *Trans. Brit. Mycol. Soc.*, 1951, **34**, 376 and S. L. Jansson and F. E. Clark, *Proc. Soil Sci. Soc. Amer.*, 1952, **16**, 330.
4 *Trans. Brit. Mycol. Soc.*, 1951, **34**, 379.

S. D. Garrett[1] has developed an ecological classification of the soil fungi based on the principal food supply they use in the highly competitive environment of the soil. The true saprophytes range from the 'sugar' fungi, which can only use relatively simple and easily decomposable organic matter, but not cellulose or lignins, through the cellulose decomposers to the lignin decomposers. There are two other important groups in the soil: the specialised root-inhabiting fungi, which will be discussed further in chapter 13, and the fungi which are predaceous either on other fungi[2] or on other members of the soil population, including the soil fauna.

The sugar fungi are typically Phycomycetes, and since there are a very large number of different organisms in the soil which can use the same simple organic compounds for food, they must be adapted to exploit these food supplies ahead of their competitors. Their primary source of food is injured, moribund or recently dead plant tissues. They are widely distributed through the soil typically as spores, though sometimes as short lengths of mycelium; the spores germinate immediately a suitable food source comes near, and the hyphae or mycelia grow very rapidly, so dominating the population there before other organisms have begun to multiply. In addition many of them produce antibiotics, and the growth of their hyphae is often only prevented by a relatively high antibiotic concentration.

The cellulose decomposing fungi, most of which belong to the Ascomycetes, Fungi Imperfecti and Basidiomycetes, come intermediate between the sugar and the lignin fungi in their rate of growth. Many of them are widely distributed throughout the soil as spores; and many of them, like many of the sugar fungi, can be copious producers of antibiotics; a number are relatively tolerant to antibiotics, in that a higher concentration is needed to reduce their growth rates than for the other groups of fungi. There is, however, no sharp distinction between them and the lignin fungi, for many of them will decompose small quantities of lignin if well supplied with cellulose or other more easily decomposable material.

The typical lignin fungi are species of the higher Basidiomycetes and are characterised by very slow growth rates. Slow-growing is not an ecological disadvantage for them because there is little competition for their food supplies as no other organisms are known to decompose lignins in lignin-rich material. They decompose lignins more easily if there is a reasonably high cellulose content associated with the lignin which is not too readily accessible to other micro-organisms. Once established, the fungus seeks new sources of food by sending out thick mycelial strands or rhizomorphs into the soil, apparently to allow the point on the strand at which it begins its attack on another piece of lignified material, usually the apex, to be relatively well supplied with food; the fungus appears to expend considerable energy in initiating the decomposition of such material. Garrett himself illustrates this by the crude analogy

1 *New Phytol.*, 1951, **50**, 149.
2 For a review, see J. E. de Vay, *Ann. Rev. Microbiol.*, 1956, **10**, 115.

that one cannot start a coke fire with paper and matches only. This behaviour is also shown by the higher Basidiomycetes which are specialised root-inhabiting parasitic fungi of tree roots; their rhizomorphs also can usually only gain entry to a root if they arise from a root which is already heavily attacked (see p. 249). The typical field conditions in which these fungi are dominant are the surface litter of forest soils and the mat in old turf soils, since it is in these sites that mycelial migration from one food source to another only need take place over short distances. Typically these Basidiomycetes do not produce antibiotics,[1] nor can their mycelia grow in regions where other fungi, such as *Trichoderma viride*, are producing them (see p. 246).

Algae

The soil algae are microscopic chlorophyll-containing organisms, and belong mainly to the Cyanophyceae (Myxophyceae) or blue-green algae, the Xanthophyceae or yellow-green algae, the Bacillariaceae or diatoms, and the Chlorophyceae or green algae. The soil forms typically comprise smaller and simpler species than the aquatic forms, and consist either of species which only occur as small organisms or of dwarfed forms of species that can occur as large organisms.[2] The morphology of the soil forms is also simple: they occur either as simple unicellular organisms or simple filaments or colonies. Many of the soil algae have their cell walls covered with a thick layer of a gummy substance, while the cell walls of most diatoms are partially silicified.

The soil algae are found not only on the surface and just under the surface, where sunlight or diffused light may be able to penetrate, but also several centimetres below the surface where no light can penetrate. The surface and immediate sub-surface forms presumably function as green plants, converting the carbon dioxide of the air into the protoplasm and taking up nitrates or ammonia from the soil. What activity those in the dark can display is still undecided. Earlier workers, such as B. M. Bristol Roach,[3] considered that the algae grew heterotrophically below the soil surface, and she found on arable land that there were often more algae at a depth of 10 cm than at the surface. The common soil algae can certainly grow in the dark on a medium containing sugars, but it is very doubtful how far they can compete with the bacteria or fungi for such readily assimilable compounds which are nearly always in extremely short supply. J. L. Stokes[4] showed in some laboratory experiments that adding such substances as sugars, lucerne meal, straw or

1 P. W. Brian, *Bot. Rev.*, 1951, **17**, 357, gives examples of some which do and J. H. Warcup, *Ann. Bot.*, 1951, **15**, 305, of some which probably do.
2 See J. W. G. Lund, *New Phytol.*, 1945, **44**, 196 and 1946, **45**, 56, for a discussion on the validity and interpretation of this generalisation for the diatoms; he gives the references to the earlier work on which it was based.
3 *J. Agric. Sci.*, 1927, **17**, 563.
4 *Soil Sci.*, 1940, **49**, 171.

manure to a soil, and incubating in the dark, increased the numbers of bacteria very considerably, but had little effect on the numbers of algae. Y. T. Tchan and J. A. Whitehouse,[1] using a fluorescent-microscopy technique, found that algae appeared to multiply only in the top few millimetres of the soil surface under natural conditions, and those cells found lower down were probably washed down from the surface or, as suggested by F. E. Fritsch several years earlier, carried down by the soil fauna or by tillage operations. J. W. G. Lund[2] has noted that most of the soil algae are either motile themselves, such as the diatoms, or have motile zoospores, so they can presumably migrate from the subsurface soil back into the surface soil.

The relative importance of the different groups of algae in different soils has not yet been worked out. It seems to be generally true that in temperate soils green algae and diatoms are probably about equally common and the blue-green less common, though the relative differences between the blue-green and the rest may be larger in infertile than in fertile soils.[3] In many tropical soils, however, the blue-green appear to be the dominant algal group.

Algae develop most readily in damp soils exposed to the sun, hence usually have their maximum development in spring and autumn when the soil is damp, the sun not too hot and other vegetation sparse. They develop most freely on fertile soils well supplied with nutrients, and tend to be less numerous on light, infertile, acid soils. The green algae are the dominant group of algae in acid soils, but as the soils become more neutral the blue-green algae and the diatoms become equally important, and on fertile soils the blue-green may be the dominant group.[4]

The numbers of algae occurring in a gram of soil vary widely according to conditions. Both Bristol Roach and Stokes record numbers up to 100 000 or 200 000, though up to 800 000 per gram have been reported in some Utah[5] and Hungarian[6] soils and up to 3 000 000 in some Danish soils.[7] One can make a rough estimate of the volume of this algal protoplasm, as it is a reasonable approximation to take it as equal to a sphere of 10 μ radius, and this gives volumes of the order of a few tenths of a cubic millimetre per gram of soil.

The soil algae probably affect plant growth in four ways: they may add some organic matter to the soil, help bind the soil particles on the surface together, improve the aeration of swamp soils, and fix atmospheric nitrogen.

The amount, and the significance, of the organic matter algae add to normal soils is not known, but the part played by this organic matter in recolonising

1 *Proc. Linn. Soc. N.S.W.*, 1953, **78**, 160.
2 In *Soil Biology*, Ed. A. Burges and F. Raw, Academic Press, 1967.
3 E. W. Fenton, *Trans. Bot. Soc. Edinburgh*, 1943, **33**, 407.
4 J. W. G. Lund, *New Phytol.*, 1945, **44**, 196; 1946, **45**, 56; 1947, **67**, 35.
5 T. L. Martin, *Proc. 3rd Int. Cong. Microbiol.*, New York, 1940, 697.
6 D. Fehér, *Arch. Mikrobiol.*, 1936, **7**, 439.
7 J. B. Petersen, *Dansk. bot. Arch.*, 1935, **8**, No 9.

burnt or barren land is fundamental. F. E. Fritsch and E. J. Salisbury[1] found slimy green algae as the primary colonisers of burnt-over heathland in England. M. Traub[2] found blue-green algae were the primary colonisers on the barren mineral layer created by the eruption of Krakatoa in 1883; W. E. Booth[3] found them growing actively on bare eroded soil in Oklahoma, and he made the interesting additional observation that this algal film over the surface did not reduce the infiltration capacity of the soil; and N. N. Bolyshev and T. I. Evdokimova[4] found they produced the slippery surface crust or *takyr* on saline lake bed soils in Central Asia.

Some of the blue-green algae of the family Nostococcaceae, including members of the genera *Nostoc*, *Anabaena*, *Aulosira* and *Cylindrospermum*, as well as a few belonging to the families Rivulariaceae, Stigonemataceae and Scytonemataceae,[5] have been shown to possess the power of fixing nitrogen from the atmosphere,[6] and thus have simpler food requirements than any other organisms, since they can obtain both their carbon and nitrogen from the air. This power of nitrogen fixation is not dependent on light, for these algae can fix nitrogen readily in the dark if supplied with sugar, but it is lost if the algae are supplied with nitrates, ammonium or asparagine. They need, however, small quantities of molybdenum,[7] and there seems little reason to doubt that the enzyme system responsible is similar to that in Azotobacter. Allison and Singh found the optimum pH for fixation was on the alkaline side of neutrality, from about pH 7 to 8·5, and fixation occurred in the range pH 6–9.

These nitrogen-fixing blue-green algae are probably of great importance in swamp rice or padi soils (see chapter 25), in which they are almost universally present. It has always been rather difficult to understand how many tropical rice soils can carry rice crops almost indefinitely without showing signs of nitrogen starvation. Singh found the algal film that developed on paddy rice soils in the United Provinces and in Bihar in India was composed of blue-green algae, all of which were active nitrogen fixers. He found that his species of *Anabaena*, which came from Indian ricefields, could excrete over 40 per cent of the nitrogen fixed as soluble organic compounds, and although these compounds may not be immediately assimilable by the rice, they are probably rapidly rendered so. A. Watanabe[8] has shown that species of *Toly-*

1 *New Phytol.*, 1915, **14**, 116.
2 *Ann. Jard. Bot. Buitenzorg*, 1888, **7**, 221. This has since been somewhat qualified by C. A. Barker, *The Problem of Krakatoa as seen by a Botanist*, The Hague, 1930.
3 *Ecology*, 1941, **22**, 38.
4 *Pedology*, 1944, Nos 7–8, 345. This observation has been confirmed for the algal crust on desert soils in Arizona by J. E. Fletcher and W. P. Martin, *Ecology*, 1948, **29**, 95.
5 A. Watanabe, S. Nishigaki and C. Konishi, *Nature*, 1951, **168**, 748, and A. E. Williams and R. H. Burris, *Am. J. Bot.*, 1952, **39**, 340.
6 K. Drewes, *Zbl. Bakt. II*, 1928, **76**, 88; F. E. Allison, S. R. Hoover and H. J. Morris, *Bot. Gaz.*, 1937, **98**, 433; P. K. De, *Proc. Roy. Soc.*, 1939, **127** B, 121, G. E. Fogg, *J. Expt. Biol.*, 1942, **19**, 78; and R. N. Singh, *Indian J. Agric. Sci.*, 1942, **12**, 743.
7 H. Bortels, *Arch. Mikrobiol.*, 1940, **11**, 155.
8 With S. Nishigaki and C. Konishi, *Nature*, 1951, **168**, 748.

pothrix, selected for high nitrogen fixing ability, appreciably improved the yields of rice when inoculated into soils from which they were absent, probably by increasing the nitrogen supply to the plant. Nitrogen-fixing blue-green algae are often the dominant organisms in the algal crusts which form on desert soils and these crusts may be high in nitrogen.[1] There is no evidence yet that algae play any significant role in enriching the soils of the temperate regions with nitrogen, although species belonging to genera containing nitrogen-fixing forms are fairly widespread.

One can summarise our presentday knowledge of the importance of algae in the economy of the soil by saying that they have been shown to be of great importance in colonising bare soil or soil devoid of organic matter, but that only the blue-green algae have proved important agriculturally, and then only in hot climates. It is possible that they are of prime importance in the cultivation of rice, which is grown under waterlogged conditions, by supplying the rice roots with free oxygen and by fixing atmospheric nitrogen which they can use either directly or indirectly; and, since rice forms the staple diet of nearly half the human race, this role is, on a world agricultural basis, by no means negligible.

Protozoa

The soil protozoa are mostly rhizopods and flagellates, though a few ciliates can usually be found. The rhizopods include the amoebae, of which *Naegleria gruberi* and *Hartmanella hyalina* are typical representatives, and the testaceous rhizopods, which are similar organisms, but which have a hard shell covering parts of their body, of which *Difflugia* and *Euglypha* are common soil genera. The soil amoebae vary greatly in size, the common small ones being from 10 to 40 μ, but they go up to giant forms several tenths of a millimetre across. The flagellates have one or more flagellae to help them move, and are usually small organisms 5 to 20 μ in length. The commonest soil forms are species of *Cercomonas*, *Oicomonas* and *Heteromita*. The ciliates, which have many short cilia covering the whole or part of their bodies, are most commonly represented by *Colpoda cucullus* and *C. steinii*, and are usually 20 to 80 μ in length.

The soil protozoa are typically considerably smaller than the normal protozoa of stagnant water for example, for whereas the water forms have a large volume of space in which they can move, the soil forms are restricted to moving in the soil pores, and then only in those containing some water; for they cannot move in dry soil. Furthermore, they can only be active when living in a water film. The majority form cysts during their life cycle, and in this state they can withstand desiccation, though whether desiccation itself encourages encystment has not been definitely established. Others can apparently go into a state of suspended animation when the soil dries out, reviving immediately

1 J. E. Fletcher and W. P. Martin, *Ecology*, 1948, **29**, 95.

it becomes damp again.[1] Encystment is a definite phase in the life cycle of some soil amoebae, so that the number of active amoebae of these species present in the soil depends on the rate of encystment and of hatching. Further, the rate of hatching seems to be dependent on the type of bacteria growing in the neighbourhood of the cysts, presumably because hatching is stimulated by products of excretion specific to these groups of bacteria.[2]

Protozoa, or protozoa-like organisms, can feed in three ways: some possess chlorophyll and are autotrophic, but these are confined to a few genera of flagellates such as *Euglena*; some, again mainly flagellates, can feed saprophytically, absorbing nutrients from solution in pure cultures, but what proportion of these do this in the highly competitive environment of the soil is not known; most can feed, and normally only feed, by capturing and digesting solid particles such as bacteria.

Bacteria form the staple food supply of the protozoal population, though small algae, yeasts, flagellates and amoebae are also ingested. But not all bacteria can serve as food. Amoebae, for example, will eat some species of bacteria voraciously, other species only if there is no more acceptable source of bacteria available, and other species not at all.[3] Singh has found that these inedible species include practically all the bacteria producing red, green, blue or fluorescent pigments so far tested, and some of these seem to excrete substances that are definitely toxic to the amoebae. Thus, the protozoa must exert a strong selective influence on the composition of the bacterial population.

The effect of protozoa used to be considered entirely harmful to bacterial life, it being thought that the protozoa reduced the beneficent action of the bacteria in maintaining soil fertility, but it has now been established that some bacteria, in pure culture, work more efficiently in the presence than in the absence of predaceous protozoa. For example, D. W. Cutler and D. V. Bal[4] showed that *Azotobacter* fixed more nitrogen when mixed with the ciliate *Colpidium colpoda*, and Table 10.3, taken from D. W. Cutler and L. M. Crump's book,[5] shows the influence of amoebae on the rate of oxidation of sugars by bacteria. The bacterial efficiency is here defined as the micrograms of carbon dioxide produced by 1000 million bacteria per twenty-four hours, when the bacteria and amoebae are growing in an aerated solution.

Protozoa cannot yet be counted directly in the soil. The older methods for determining their numbers were based on letting them feed on an uncontrolled bacterial population, and the numbers so determined varied rapidly from day to day, from a few hundred to as many as a hundred thousand, and

1 J. M. Watson, *Nature*, 1943, **152**, 694.
2 I am indebted to Miss L. M. Crump for these unpublished observations of hers.
3 B. N. Singh, *Ann. Appl. Biol.*, 1941, **28**, 52, 65; 1942, **29**, 18. Also *Brit. J. Expt. Path.*, 1945, **26**, 316.
4 *Ann. Appl. Biol.*, 1926, **18**, 516.
5 *Problems in Soil Microbiology*, Longman, 1935.

TABLE 10.3   Average efficiencies per 1000 million bacteria at different bacterial densities in the presence and absence of amoebae

Density of bacteria in millions per cubic centimetre of solution	Weight of CO_2 produced in 10^{-6} g per 1000 million bacteria in 24 hours	
	Bacteria + amoebae	Bacteria alone
0–100	73·2	—
100–200	15·6	9·3
200–300	13·4	7·9
300–400	10·6	8·3
400–500	10·1	3·7
500–600	6·2	3·5

the proportion encysted from 0 to 100 per cent, within twenty-four hours.[1] But it is possible that these large fluctuations are spurious and are due to an unsatisfactory counting technique, for B. N. Singh introduced an improved method, based on counting the protozoa on a controlled population of an edible bacteria, and found very much smaller fluctuations.

These short-period fluctuations make it difficult to study the effects of external conditions on the protozoal fauna, as only factors having a very large effect on numbers are sufficient to override them. The effect of season, however, can be demonstrated, for at Rothamsted protozoal numbers are higher in spring and autumn than in summer or winter. On the other hand, protozoal numbers do not seem to be very sensitive to changes in the moisture content, aeration or pH of the soil, nor even to its organic matter content, although the bacterial numbers are. Thus, on the permanent mangold field (Barnfield) at Rothamsted, B. N. Singh[2] found that a plot which had received no manure for nearly ninety years contained between 400 and 11 000 active amoebae on nine different sampling dates from April 1945 to August 1946; a plot which had received many annual dressings of 35 tons per hectare of farmyard manure contained between 6000 and 45 000 per gram of soil on these same sampling dates: and a plot that had received for many years a heavy dressing of fertiliser without any farmyard manure contained between 9000 and 31 000. The organic carbon in the two soils receiving no farmyard manure was about 0·8 per cent, while in the farmyard manure plot it was about 2·5 per cent, and the bacterial numbers, which were determined by total cell count on only two of these occasions, were between 5000 and 8000 million on the farmyard manure and between 2000 and 3000 million on the other two. A similar example from Broadbalk Field at Rothamsted is given in Table 12.2 on p. 221.

Only very rough estimates can be given of the biomass of the soil protozoa. B. N. Singh[3] quotes as an example of a Rothamsted soil, a population of

1 See for example, D. W. Cutler, L. M. Crump and H. Sandon, *Phil. Trans.*, 1922, **211** B, 317.
2 *J. gen. Microbiol.*, 1949, **3**, 204.
3 *Ann. Appl. Biol.*, 1946, **33**, 112.

70 500 flagellates, 41 400 amoebae and 377 ciliates per gram of soil; and J. D. Stout and O. W. Heal[1] estimate that on an assumed mean volume of 50 μ^3 for a flagellate, 400 μ^3 for an amoebae and 3000 μ^3 for a ciliate, this population would only weigh about 2 $\times$ 10^{-2} mg/g soil, or about 5 g/m^2, which is about one hundred times smaller than the weight of bacteria or fungi, though it could be as a much as one-fifth to one-tenth the weight of the soil meso- and macro-fauna. They also estimate that a forest soil may contain 20 g/m^2 or even more, due to a much higher number of the larger organisms in forest than in arable soils, and an important reason for this is probably the greater proportion of large pores in the forest soils kept moist by the forest litter in which they can live.

Stout and Heal also estimated the volume of oxygen consumed annually by the protozoal population on the unmanured, the complete fertiliser and the farmyard manure plot on Broadbalk using Singh's figures, which worked out at 20, 80 and 100 litres/m^2 on these three plots, or at approximately 1–2 per cent of the total oxygen used by the soil (see p. 405). They also calculated that the protozoa would consume annually about eighty times the weight of the standing crop of bacteria, which they defined as the bacteria which would grow on dilution plates.

Amoeboid and flagellate stages of other organisms

A number of organisms which look like flagellates and amoebae may not be protozoa at all, but a stage in the life cycle of organisms belonging to other groups. K. B. Raper has described in some detail certain species of Acrasieae,[2] which have an amoeboid stage termed Myxamoebae. During this stage these organisms live in the soil as amoebae feeding on bacteria, and reproduce by simple fission, but at some stage in their life history they congregate together, forming characteristic structures that produce spores, which grow into amoebae when conditions for growth become favourable again. B. N. Singh[3] has shown that British soils only contain two species of these organisms, both belonging to the genus *Dictyostelium*, and he isolated these two from thirty-three out of the thirty-eight arable soils, but only from three out of the twenty-nine pasture soils he examined. On the other hand, Raper[4] found a wider range of species in some American forest soils.

In the same way organisms which are flagellates during part of their life cycle may belong to the algae or fungi. In particular, there are species of soil Myxomycetes, or slime fungi, that exist as flagellates during part of their life cycle, feed on bacteria and divide by simple fission, but at another stage pairs of flagellates will fuse and grow into a large multinucleate plasmodium, or

1 In *Soil Biology*, Ed. A. Burges and F. Raw, Academic Press, 1967.
2 See, for example, *Amer. J. Bot.*, 1940, **27**, 436, for pictures of these organisms.
3 *J. Gen. Microbiol.*, 1947, **1**, 11, 361.
4 *Quart. Rev. Biol.*, 1951, **26**, 169, and see M. Sussman, *Ann. Rev. Microbiol.*, 1956, **10**, 21.

mass of naked protoplasm, behaving as a large amoeba and continuing to feed on bacteria, yeast, fungal mycelium and small protozoa. Under certain conditions, not yet known, it breaks up into spores which in due course become flagellates again.

There are also large, amoeboid-like organisms in the soil that feed on bacteria and small protozoa and are multinucleate like myxomycete plasmodia, but that do not appear to give rise to flagellates. T. Goodey[1] isolated two species of such multinucleate organisms, which he put in the genus *Leptomyxa*, and B. N. Singh[2] found that one of these species is widespread in British soils. He isolated it from all the arable soils he examined, where it numbered about 1000 per gram of soil, but he only found it present in about one-third of the pasture soils.

1 *Arch. Protistenk.*, 1914, **35**, 80.
2 *J. Gen. Microbiol.*, 1948, **2**, 8, 89.

11

The soil fauna other than protozoa

The invertebrate fauna of the soil has only recently been studied in any detail,[1] and the reason for its relative neglect compared with microbial studies has been partly the difficulty of quantitatively isolating the animals from the soil mass, and partly the very great problems of their systematic classification when they have been separated out. Again, until recently very little was known about the actual food of many members of this population, and even now our knowledge is very incomplete largely due to lack of any adequate techniques for studying what they are feeding on in the soil. But techniques are now available for fixing and sectioning the soil without disturbing its structure,[2] so one can see where the small members of the population are, which, in turn, gives a more complete picture of the activity of this population than we have had before.

The size of the soil invertebrate population depends both on the food supply and also on the physical condition of the soil. The soil invertebrates need a fairly well-aerated soil for active growth: they cannot thrive in waterlogged soils nor in wet soils that have been puddled by the trampling of cattle. However, if the soil is fairly open, such as the surface of a forest or old pasture soil, and the soil becomes temporarily waterlogged by heavy rain or flooding from above, enough air bubbles are likely to be entrapped in the soil pores for the smaller animals to survive for considerable periods of time. This entrapping of air bubbles does not take place to anything like the same extent if the water table rises to the soil surface, and this has a far more serious consequence for the more aerobic members of the population. The population is, on the whole, more tolerant of dry conditions, so a large proportion of the animals tend to concentrate in the top 2 to 5 cm of the soil surface, and in many soils only a few groups, such as the earthworms, are common in the next 5 to 7 cm.

The food supply determines the size of the population in normal soils, and

1 For an excellent account of our knowledge on this subject up to 1955, see *Soil Zoology*, London, 1955, and in particular the papers by W. Kühnelt, pp. 3 and 29, and D. K. McE. Kevan, pp. 23 and 452.
2 See for example, N. Haarlov and T. Weis-Fogh, *Oikos*, 1953, **4**, 44, and *Soil Zoology*, 1955, p. 429. Also J. F. Newman, *Crop Loss Assessment Methods*, F.A.O., 1970.

it consists primarily of dead plant tissue, but it also includes living plants and the soil microflora. This combination of need for good aeration and a very superficial food supply has two important consequences. First, it is very difficult to define what is meant by a soil animal, for many of them are living in the litter as well as in the soil. Second, if the litter is distributed unevenly on the soil, as, for example, in a tussocky pasture or meadow, many of these animals will also be very unevenly distributed over the soil surface. In such a pasture there are a number of quite different environments in and above the soil, each with its characteristic population, so it is misleading to lump them all together as if they were a uniform population. It has been found, in all the investigations in which the effect has been looked for, that in fact the animals of a given species are not distributed uniformly or at random throughout the soil but always show tendency to congregate in some places and be rare in others.[1]

The food supply of the invertebrate population is not restricted to dead or living plant material, although this is the most important. In addition there is a population feeding on the excreta of the primary population and another that is feeding on the population itself. Animals feeding on dead plant tissues or their associated microflora are commonly called saprophagous, on living plants phytophagous, on animal excreta coprophagous, and on other animals predaceous. The higher and the larger members of the community may use all these methods of obtaining their food though, generally speaking, only specialised genera are predaceous on the larger invertebrates.

The foresters have studied the sources of food more intensively than other workers. They have shown that one bit of primary organic matter often passes through the gut of several different groups of animals before being degraded to resistant humus. Thus A. P. Jacot[2] showed that the process of decomposition of leaf litter or dead plant roots could be as follows: first, the more resistant material is softened by fungi, then it is eaten by saprophagous animals, particularly mites, which continue the decomposition of the plant material, possibly with the aid of a microbial population in their gut; and finally the undigested material is excreted and forms the food supply for a whole chain of other animals until it is converted to resistant humus-like material. In fact, if soils poor in earthworms are examined under the microscope[3] a great part of the 'humus' present is recognisable as the excreta of soil invertebrates.

Very little is yet known about the types of material in the plant debris that are available to the different species of the saprophagous fauna. But it appears probable, for example, that some woodlice and millipedes can only digest the sugars, starches and accessible proteins in the plant tissues; some termites,

1 See, for example, G. Salt and F. S. J. Hollick, *J. Exp. Biol.*, 1946, **23**, 1, for wireworms, and
2 *J. Animal Ecol.*, 1948, **17**, 139, for animals in pasture soils.
   Ecology, 1936, **17**, 359.
3 W. L. Kubiena, *Bodenk. PflErnähr.*, 1942, **29**, 108. See also P. E. Müller, *Studien über die natürlichen Humusformen*, Berlin, 1887.

snails and earthworms can digest part of the celluloses and hemicelluloses contained in the plant tissues, and some nematodes and mites can live symbiotically with, or predaceously on, the cellulose-decomposing bacteria attacking the tissues.[1] It is also possible in many forest and pasture soils that fungal mycelium forms the transition compounds between plant and soil faunal tissue, for the majority of the soil fauna appears to have no power of digesting the celluloses, hemicelluloses or lignins in the dead plant tissues: this is attacked by fungi and converted into their protoplasm, and it is their mycelia that forms the principal food supply of a large proportion of the mites and springtails, as is shown by a study of the contents of their gut.[2] This has the consequence that the soil fauna living on plant remains only oxidise a small fraction of the material they ingest, and it is probable that only rarely do they use more than 10 per cent of the energy in the food they consume. Further, they differ from most soil micro-organisms in using most of this energy for their vital processes, and only converting a small fraction into their own tissues. In so far as most of these animals will probably be the primary source of food for the predatory population in the soil, little if any of the carbon they assimilate will contribute to the humus supply of the soil.

The soil fauna can for many purposes be divided into two groups: the meso- or meiofauna which are too small when mature to affect the pore size in the soil, and so are without mechanical influence on it, and the macrofauna, which increase the size of the pores in which they move. The former group includes the rotifers and nematodes, which live in the soil water films, and the micro-arthropods—the mites, springtails and other small insects—which live in the air spaces. These latter animals may help stimulate fungal attack on plant debris, for they may carry fungal spores on their bodies, so distribute them over new volumes of soil and organic remains.

The larger soil fauna have three very important effects in the soil; they assist the aeration and drainage of the soil through the channels and burrows they leave behind them as they move through the soil in search of food; they macerate and grind up the plant litter they eat, excreting it in a form more readily available to the attack of the soil micro-organisms; and they distribute this macerated plant debris together with some of the microflora throughout the volume of soil in which they are working. The saprophytic soil fauna can have two other effects on the soil. In the first place, their excreta compose the typical humus form in the soil (see p. 707), a fact emphasised by W. L. Kubiena[3] as a result of his direct microscopic observations on the forms of humus actually present in undisturbed surface soils. The gut of these animals is therefore an important site for humus formation, and this is particularly true for many species of earthworms which ingest soil along with their food and excrete mull humus in the form of wormcasts. In the second place, many

1 H. Franz, *ForschDienst.*, 1942, **13**, 320.
2 K. H. Forsslund, *Medd. Skogsförsöksanst.*, 1945, **34**, 1.
3 See, for example, his book *The Soils of Europe*, Madrid and London, 1953.

of the larger saprophytic animals attack freshly dead plant tissues, and if for any reason the supply of suitable food is reduced, they will start attacking the corresponding plant tissue before it is dead, thus sometimes becoming a serious pest to farmers and gardeners. This typically happens on the farm when the rate of supply of organic matter to the soil is suddenly reduced, as, for example, when a long ley or an old pasture is ploughed out.

The effect of the soil fauna in moving soil is sometimes very considerable. Earthworms, ants, termites and moles may all put appreciable amounts of subsurface or subsoil on the soil surface, and in the semi-arid steppes there are a number of other mammals that burrow into the soil, down to 3 metres at times, bringing up soil from this depth and leaving it as mounds on the surface. These burrows often get filled with surface soil giving the typical channels of black soil, or *crotovinas* to use the Russian word, in the light-coloured parent material in these areas. On poorly drained soils other burrowing animals, such as crayfish, can be active. Thus J. Thorp[1] has described how these animals will make burrows down from the surface to the water table, if it is within the top 3 metres, and he found them active on many soils having a pronounced clay pan. Thus these animals are not only bringing subsoil up to the surface, but also by burrowing through a clay pan assisting in the mass movement of water.

The soil fauna can be of indirect importance in agriculture through some of them being alternative hosts to certain animal parasites, and this particularly concerns the fauna of pastures, for the grazing animal can easily pick up infected soil invertebrates with its fodder. Thus, the liver fluke and some of the lungworms of sheep must pass a part of their life cycle in certain gasteropods (slugs and snails); so that good drainage of pastures, by reducing the gasteropod population, reduces the incidence of these diseases. Again, some sheep and horse tapeworms have, as alternative hosts, species of orebatid mites which are common in old matted pasture but which do not flourish in young leys. But it is not always possible or desirable to control the alternative host; thus some lungworms of pigs have certain common species of earthworms as alternative hosts, and it is both difficult, and probably also undesirable, to try to reduce the incidence of this disease by eliminating the earthworm host.

The invertebrates that have been most frequently recognised in soils include among the mesofauna nematodes, acarine mites and collembola, pauropoda, symphyla, thysanura and protura; and among the macrofauna enchytraeid worms and earthworms, millepedes and centipedes, many insects, mainly dipterous flies and beetles, and a few crustacea and gasteropods. Photographs of some typical members of the fauna of English soils are given in Plates 8 and 9.

The determination of the numbers of soil animals, their liveweight and their rate of respiration in a soil is difficult. Many of the animals are fairly

1 *Sci. Mon.*, 1949, **68**, 180. He gives field descriptions of the soil-moving effects of many of these animals with estimates of the weight of soil they move.

PLATE 8. Typical members of the soil fauna (natural size).
Chilopoda: Centipedes 4, 5, 28, 40. *Diplopoda:* Millepedes 19, 22, 41.
Arachnida: (*a*) Araneae: Spiders 31, 33; (*b*) Acarina, Mites: Gamasidae 2, 14, 36; Oribatidae 24, 32, 37, 45; Tyroglyphidae 42.
Insecta: (*a*) Collembola: Springtails 3, 6, 10, 13, 16, 20, 50. (*b*) Lepidoptera (larva) 8. (*c*) Coleoptera: Staphylinoidea (adults) 7, 9, 17, 30, 35, 38, 44; Staphylinoidea (larvae) 15, 23, 27, 29, 34, 41; Carabidae (adult, small species) 25; Carabidae (larva, large species) 12; Elateridae (larvae): Wireworms 21, 26, 39, 45. (*d*) Diptera: Bibionidae (larvae) 1, 43, 48, 49; Cyclorrapha (larva) 11; Anthomyidae (adult) 18.

large, so their numbers can only be accurately determined if a large volume of soil is sampled. But this sampling problem is particularly difficult because the animals are not uniformly distributed throughout the soil, but tend to be more concentrated in some spots in a field than in others. Thus, J. E. Satchell[1] found on a Park Grass plot at Rothamsted that the population of the earthworm *Lumbricus castaneus* varied from 0 to 14 in neighbouring quadrants of a square yard (0.84 m^2). Further, some animals, such as ants, live in the soil but move very considerable distances over the soil surface. The second problem is that the mere process of taking a sample can cause some animals to

1 In *Soil Biology*, Ed. A. Burges and F. Raw, Academic Press, 1967.

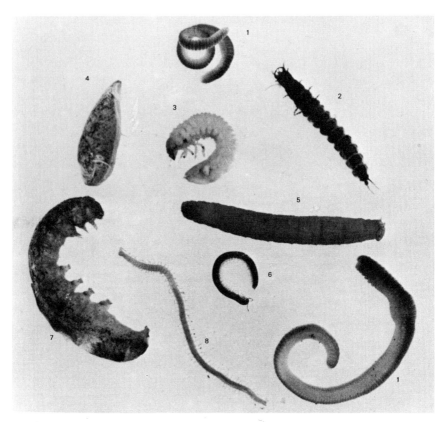

PLATE 9. 1. Earthworm, *Lumbricus* sp. (Oligochaeta); 2. Beetle larva, Carabidae (Coleoptera); 3. Chafer grub, *Phyllopertha* sp. (Coleoptera); 4. Slug, *Agriolimax* sp. (Mollusca); 5. Leatherjacket, Tipulidae (Diptera); 6. Millepede (Diplopoda); 7. Cutworm, Agrotidae (Lepidoptera); 8. Centipede (Chilopoda).

move out of the sampled volume. Burrowing species of earthworms, for example, will move down to the bottom of their burrows, so some will be missed. It is also very difficult to separate out quantitatively the smaller members of the population from the soil. This can result in the age distribution of the population being wrongly determined; for immatures, being smaller than the mature form are more likely to be missed. Table 11.1 illustrates this relation between the size of the population and the size of the individuals as determined for the population of wireworms in two Cambridge pastures.[1]

A number of analyses of the soil invertebrate populations have been published, and the results show that the relative importance of the different

1 G. Salt and F. S. J. Hollick, *Ann. Appl. Biol.*, 1944, **31**, 52.

TABLE 11.1   The size distribution of wireworms
(*Agriotes* spp.) in two Cambridge pastures

Length of wireworm in mm	Per cent by numbers in the size class	
	Field 1	Field 2
2–6	59	52
6·1–10	26	23
10·1–14	9	12
14·1 24	6	13

groups depends on the soil conditions and on the type of plant residues being returned to the soil. Table 11.2 gives some old results of C. H. Bornebusch[1] for two forest soils, one under mull carrying a flourishing ground vegetation and the other under mor, almost bare of ground vegetation; the table gives estimates of the numbers, weights and metabolic activity of the different groups. It is quite certain that the numbers of many of the smaller animals are under estimated, but the general results given are representative of much later work.

Mites and springtails abound in these soils and, on the basis of numbers alone, form almost the whole population, though they make a negligible contribution to its total weight. But their contribution to the total oxygen demand, or to the total quantity of organic matter oxidised away by the population may become appreciable, particularly in the raw humus or mor soils. Table 11.2 also illustrates the important point that the oxygen demand of the soil fauna, or the rate at which it oxidises the soil organic matter, does not necessarily increase with increases in the total number of animals present.

These results also show that where earthworms thrive they form the major portion of the total weight of the animal population. Thus, Bornebusch found that the weight of all animals excluding earthworms varied from 40 to 190 kg/ha for the ten different soils he studied, but in the two good mull soils of his group the $1\frac{1}{4}$ million earthworms present weighed about 550 kg/ha. Further, in two forest sites, on which he only took a few samples, $2\frac{1}{2}$ to $3\frac{1}{2}$ million per hectare weighed 1700 to 2000 kg, which as he points out, equals the weight of livestock carried per hectare on firstclass Danish pastures, although the earthworms will only be respiring about one-tenth the amount of carbon dioxide that the livestock would.[2] The relation between the numbers of earthworms and their weight is not, however, simple for it depends on the species present. Thus, in Table 11.2, the average weight of an earthworm in the mull is about 300 mg—they were mainly the medium-sized *Allolobophora*

1 *Forstl. Forsoegsv., Danmark*, 1930, **11**, 1.
2 K. Mellanby, *Soils and Fert.*, 1960, **23**, 8.

TABLE 11.2   The numbers, weights and oxygen consumption of various groups of animals in Danish forest soils

| | *Mull soil under beech pH 6·1–5·8* | | | *Raw humus under beech pH 5·6–3·6* | | |
| | | *Per cent contribution to the* | | | *Per cent contribution to the* | |
	Numbers in millions per hectare	*total weight of the animals present*	*total oxygen consumption of the animals present*	*Numbers in millions per hectare*	*total weight of the animals present*	*total oxygen consumption of the animals present*
Earthworms	1·78	75·1	56·2	0·82	22·4	12·2
Enchytraeid worms	5·35	1·5	6·0	7·81	6·5	14·4
Gasteropods	1·04	7·0	7·3	0·52	13·4	7·5
Millepedes	1·78	10·6	15·0	0·40	4·7	3·9
Centipedes	0·79	1·8	3·5	0·20	2·1	2·0
Mites and springtails	44·3	0·4	4·5	112·4	2·3	16·1
Diptera and Elateridae larvae	2·44	2·4	3·7	13·1	43·8	35·2
Other insects, isopods and spiders	4·71	1·2	3·8	6·50	4·7	8·7
Total number of animals	61·7 million/ha			14·1 million/ha		
Total weight of animals	705 kg/ha			240 kg/ha		
Oxygen used per m² at 13°C	0·33 litre/day			0·20 litre/day		

caliginosa—while on the raw humus soil they are mainly the small *Dendrobaena octoedra*, which only weigh about 67 mg.

One other use can be made of the respiration figures given in Table 11.2. Bornebusch estimated the annual oxygen consumption of the animal population in the various soils he studied; it was about 90 and 60 $g/m^2/year$ in the two soils given in the table, and this is approximately the weight of food oxidised by the soil animals, for about 1 g of oxygen is needed to oxidise 1 g of organic matter to carbon dioxide. He further estimated that the annual leaf fall returns about 400 g/m^2 of organic matter to the soil, so that, ignoring the weight of plant roots being decomposed, about which nothing is known, the animal population is responsible for the oxidation of about 15 to 20 per cent of the plant remains. These estimates must be accepted with great caution, however, because of the many assumptions that have had to be made in converting laboratory determined respiration rates of selected species of the fauna to field rates for the population as a whole.

An example of the numbers of arthropods present in a pasture soil is given in Table 11.3. This field, which is on a Gault clay, had been under grass for at least ten years and been used principally for grazing.[1] In addition to those animals given in the table, the following occurred in numbers less than $2\frac{1}{2}$ million per hectare: Psocoptera, Lepidoptera and Hymenoptera among the insects, and Chelonethida. The table also shows that a number of arthropods lived in the subsoil, which was on the whole well supplied with roots and was stained with humus down to 15 to 20 cm.

TABLE 11.3 Numbers of arthropods collected from surface and subsoil on a Cambridgeshire pasture, 23 November 1943

	Numbers in millions per hectare	
	Surface soil 0–15 cm	*Subsoil 15–30 cm*
Thysanura	37·1	29·2
*Protura	2·7	10·9
*Collembola	432	182
Thysanoptera	10·9	0·5
Hemiptera	15·3	25·2
Coleoptera	28·2	15·8
Diptera	5·2	1·5
Total insects	670	266
Araneida	1·4	0·0
*Acarina	1200	448
*Pauropoda	0·2	5·9
Symphyla	2·7	36·1
Diplopoda	1·6	2·3
Chilopoda	2·2	4·3
Total Arthropods	1880	761

(* Numbers of animals so marked are known to be underestimated.)

The number of soil arthropods in the surface soil works out at about one per cubic centimetre of soil, rising to a maximum of three per cubic centimetre in the samples with the highest number. The authors consider the modal volume of the mesofauna as about 5×10^{-5} cm^3 per individual, so their volume to the volume of soil is about 1 to 20000, and if the surface soil contains 5 per cent of air when wet but well drained, they would occupy about 0·1 per cent of the air space. P. W. Murphy[2] has collected the corresponding figures for forest mor, which came out to 2 to 15 per cubic centimetre with an animal-to-mor ratio by volume of about 1 to 30000.

Unfortunately Salt and his coworkers did not determine the number of other soil animals in the pasture they sampled. J. B. Cragg[3] has collected data for the populations on some northern Pennine hill pastures in Yorkshire. On

1 G. Salt, F. S. J. Hollick *et al.*, *J. animal Ecol.*, 1948, **17**, 139; H. Franz, *Bodenk. PflErnähr.*, 1943, **32**, 336, found similar total numbers in the top 10 cm of Austrian pasture soils.
2 *J. Soil Sci.*, 1953, **4**, 155.
3 *J. Ecol.*, 1961, **49**, 477.

a pasture on the Carboniferous limestone, he found about the same total numbers of mites and collembola as Salt, although on his pasture there were twice as many collembola as mites. They, however, only formed 0·5 per cent of the liveweight of the soil fauna. The total liveweight was 1·9 t/ha, made up of 1·4 tons of lumbricids, 0·15 tons of enchytraeids and 0·36 tons of tipulid (cranefly) larvae. He estimated this population would use about 55 mg/m²/hr of oxygen at 13°C, and that the demands of these four groups of animals would be 4, 44, 30 and 20 per cent of the total. He also comments that the weight of lambricids on this pasture corresponds to about thirty hill sheep per hectare, on land that normally only carries about one, but which if improved could carry between five and seven.

Nematodes

The nematodes, or eelworms, are non-segmented worms with thin spindle-shaped bodies, many of which when adult are between 0·5 and 1·5 mm in length and 10 to 30 µ thick. They can be considered to fall into three classes nutritionally: those feeding on bacteria and other small cells, those feeding on the cell contents of other members of the soil population such as fungi, protozoa, nematodes and oligochaetes, and those feeding on the cell contents and juices in plant roots and which are, in consequence, plant pathogens. Some of the fungal feeders, such as *Aphelenchus avenae* may have a preference for fungi attacking plant roots such as *Fusarium* spp., *Rhizoctonia solani* and *Armillarea mellea*.[1] There are a very large number of soil species, and their systematic classification poses very great problems,[2] and different species vary very widely in the specificity of their food requirements. Probably all feed on liquid food only, for even the bacterial feeders may only feed on the liquid cell contents and not on the cell wall.

Nematodes live in the soil water films, and they have very specific space requirements if they are to be numerous. Most are strictly aerobic,[3] so although they may be able to stay alive in flooded soils, they are only active when it is drained. They move in water films, but since they are fairly large organisms they are confined to fairly coarse pores. Some important ecto-parasites of plant roots are relatively large nematodes, such as species of *Xiphinema*, *Longidorus* and *Trichodorus*, living in the soil and sucking the root juices out through long stylets, and some of these are important vectors of plant virus diseases. The adults need pores at least 60 µ in size to pass through, though their larval stages can pass through pores only 20 µ in size, so they can only live in the spaces between soil crumbs or in the pores between sand grains.

1 R. and S. K. Mankau in *Soil Organisms*, Ed. J. Doeksen and J. van der Drift, North Holland, 1963.
2 For an accepted classification see T. Goodey, *Soil and Freshwater Nematodes*, 2nd edn, Methuen, 1963.
3 S. D. Van Gundy, L. H. Stolzy *et al.*, *Phytopath.*, 1962, **52**, 628.

They are, in consequence, most active in coarse sandy soils in wet weather. In eastern England, the roots of young sugar-beet plants growing in these soils can suffer severely from their attacks in wet late springs, which causes the beet to show the symptoms of 'docking disorder'; but since these nematodes cannot remain active in dry conditions, the beet can grow out of this disorder during dry spells that are sufficiently long for the crop to grow a new root system.[1] Many of these ectoparasites can feed on a wide range of roots, so they cannot easily be controlled by normal crop rotations, though they can be severely checked with nematicides. Pore space requirements are not so important for the gall-forming or root-inhabiting nematodes, for they are returned to the soil in the channels made by the roots, and can move through these root channels to attack any susceptible roots growing in them. They can therefore be serious pests in all soils, but since most only attack a relatively narrow range of crops, they can be controlled by suitable crop rotations fairly easily when they do not form cysts, and with much greater difficulty when they can remain as cysts in the soil for a number of years. Many of the nematodes that are not parasitic on plants are small organisms that can move in finer textured soils, but even they are probably only common in well-structured soils such as pastures and forest soils and forest litter.

Nematodes are preyed upon both by other nematodes and by some species of fungi which have developed trapping mechanisms which immobilise the nematode and allow their hyphae to grow inside its body emptying it of its contents. Unfortunately there are not yet any methods available for stimulating these nematode predators to help control the numbers of plant parasitic nematodes in soils.

It is possible that in some soils, particularly some pasture soils, nematodes parasitic on plant roots may constitute about half the nematode population, while in forest soils nearly all will feed on soil organisms. The liveweight of nematodes may be as high as 100 to 200 kg/ha in mull soils under deciduous forest and in some pastures.[2] However, they have a high metabolic rate when active and will then have an oxygen demand per unit of liveweight appreciably higher than other members of the soil fauna.

Enchytraeid worms

These worms have not received much attention, probably because they are of greatest relative importance in soils of low agricultural value.[3] They have been studied by C. O. Nielsen[4] in some Danish pastures, by F. B. O'Connor[5]

1 F. G. W. Jones, D. W. Larbey and D. M. Parrott, *Soil Biol. Biochem.*, 1969, **1**, 153.
2 See, for example, C. O. Nielsen, *Natura Jutlandica*, 1949, **2**, 1 and A. Stöckli, *Z. PflErnähr.*, 1952, **59**, 97.
3 For a summary of our knowledge up to 1965 see F. B. O'Connor in *Soil Biology*, Ed. A. Burges and F. Raw, Academic Press, 1967.
4 *Oikos*, 1953, **4**, 187; 1955, **6**, 153.
5 *Oikos*, 1957, **8**, 161; 1958, **9**, 271.

in a Welsh coniferous forest, and by J. E. Peachey[1] in some Pennine moor-lands. These soils contain between 10 000 and 300 000 per square metre, with a liveweight of between 30 and 100 kg/ha. It is not certain what they feed on, but they consume rotting material and may feed on the bacteria and fungi which it contains. They also appear to prey on pathogenic eelworms by feed-ing on their early larval stages.

Earthworms

The importance of earthworms was first stressed by Charles Darwin in his book *The Formation of Vegetable Mould through the Action of Worms*[2]—one of the classics of soil science—and also by V. Henson.[3] But in spite of this early start, and of their obvious importance in soils, for a long time little work was done on their ecology. Yet both C. H. Bornebusch,[4] working with Danish forest soils, and H. M. Morris,[5] working with Rothamsted arable soils, found that where earthworms flourished they constituted between 50 and 75 per cent of the total weight of the animals present, although on a further calculation of Bornebusch (see p. 192) their effectiveness in decom-posing organic matter was not quite as high as this figure might suggest. Further, foresters such as P. E. Müller[6] and W. Kubiena[7] have remarked on the fact that the only obvious animal excreta in soils containing earthworms is that of the earthworm, while on soils not containing earthworms the excreta of a large number of other animals becomes obvious. The conclusion seems to be justified, therefore, that in north-western Europe where earthworms flourish they dominate the whole soil fauna.

From the agricultural point of view earthworms can be classified into two groups: the family Lumbricidae, and the other families belonging to the suborder of the order Oligochaeta. The earthworms of northwestern Europe are all lumbricids, but the native species in India, Africa, the Americas and Australasia are not. However, as the native vegetation is replaced by Euro-pean or Eurasian species, e.g. the ryegrass/white clover pastures of New Zealand, the European species of earthworms seem to have come along with the plants, and dominate these soils—the native species being unable to live in this entirely new soil habitat.

The general problems of earthworm zoology have been described in a monograph, now rather out of date, by J. Stephenson.[8] Their separation into

1 In *Progress in Soil Zoology*, Ed. P. W. Murphy, Butterworth, 1962; *Pedobiologia*, 1963, **2**, 81.
2 Published by J. Murray (London) in 1881, and reissued by Faber and Faber in 1945 under the title *Darwin on Humus and the Earthworm*.
3 *Ztschr. wiss. Zool.*, 1877, **28**, 354; *Landw. Jahrb.*, 1882, **11**, 661.
4 *Forstl. Forsoegsv. Danmark*, 1930, **11**, 1.
5 *Ann. Appl. Biol.*, 1922, **9**, 282; 1927, **14**, 442.
6 *Studien über die natürlichen Humusformen*, Berlin 1887.
7 *Bodenk. PflErnähr.*, 1942, **29**, 108.
8 *The Oligochaeta*, Oxford, 1930. For a more recent account of their physiology see M. S. Laverack, *The Physiology of Earthworms*, Pergamon, 1963.

genera and species presents great taxonomic difficulties, particularly for sexually immature individuals, so many of the earlier workers either did not identify the species they were studying, or worse, gave incorrect identifications. Some species can only reproduce sexually, some parthenogenetically and some can use either mode.[1] They lay their eggs in cocoons, and the young earthworm takes from six to eighteen months to reach maturity, depending on species and food supply.[2] L. Cernosvitos and A. C. Evans[3] have published a key for the identification of the twenty-five British species belonging to eight genera, of which seventeen are widespread. They vary in size from the large *Lumbricus terrestris*, which may have a length exceeding 25 cm and weighs between 2 and 7 g, to small species with lengths about 2·5 cm and weighing about 50 mg. The factors controlling which species will be dominant and which be almost lacking in a soil are not yet known in any detail. In British pastures one or more of the larger worms such as *Allolobophora longa*, *A. nocturna*[4] and *L. terrestris* are usually dominant; though in leys and in highly productive pastures, such as some of those in New Zealand, *A. caliginosa* and *Lumbricus rubellus* may be the dominants. In British arable soils *A. caliginosa*, *A. chlorotica* and *Eisenia rosea* are usually important, and the first two are often the dominants in leys of several years' duration which come in arable rotations. In addition another fairly large worm, *Octolasium cyaneum*, and the smaller *L. castaneus* and *Dendrobaena rubida* are fairly common in British pastures, and *E. foetida* in farmyard manure heaps.

The principal food of earthworms is dead or decaying plant remains, including both leaf litter and dead roots, but there are large differences in the acceptability of litter from different plant species. W. Wittich[5] found that forest leaf litter with a nitrogen content exceeding 1·4 per cent, such as from *Alnus glutinosa*, *Fraxinus excelsior* or *Sambucus nigra* were taken by *L. terrestris* more readily than litter from *Fagus sylvatica* or species of *Betula* and *Quercus*, whose nitrogen content was under 1 per cent, though since there was a close correlation between nitrogen content and soluble carbohydrates in the litters, it was not possible to decide which of these two constituents was the more important. C. H. Bornebusch[6] obtained roughly similar results with forest trees, but in addition showed that the typical ground flora of mull forest soils—*Mercurialis perennis*, *Urtica dioica* and *Oxalis acetosella*—were even more strongly preferred. Litter high in polyphenols, such as the needles of coniferous trees and fresh oak leaves, are relatively little eaten.

The level of available nitrogen supply is probably an important factor controlling the size of the earthworm population. Animal dung is an attractive

1 S. Muldal, *Heredity*, 1952, **6**, 55.
2 For a discussion of the factors affecting the rate of cocoon production in British soils see A. C. Evans and W. J. McL. Guild, *Ann. Appl. Biol.*, 1948, **35**, 471.
3 *Synopses of the British Fauna*, Linnean Soc. No. 6, 1947, and revised by B. M. Gerard, 1964.
4 These two species are classed together as *A. terrestris* by some authors.
5 *Schr. Reihe forstl. Fak. Univ. Gottingen*, 1953, **9**, 5.
6 *Dansk. Skovforen. Tidsskr.*, 1953, **38**, 557.

food for many species of earthworm, and large populations are frequently found under cow pats in pastures. Faecal pellets of soil arthropods and fragments of their cuticles are also probably eaten. Additions of farmyard manure to soils will increase the size of the population. At Rothamsted, the Barnfield plots receiving 35 t/ha of manure annually carry about 900 kg/ha of earthworms, which is comparable to the weight in many pastures, while the corresponding plots not receiving the manure carry about only one-sixth this weight.[1] Even when applied to old grass plots, as on Park Grass field at Rothamsted, it increases the weight of worms.

J. E. Satchell[2] has attempted to draw up a nitrogen balance sheet for *Lumbricus terresticus* in an ash–oak woodland at Merlewood in the English Lake District. He estimated the annual return of nitrogen in the leaf litter at about 50 kg/ha, the amount of nitrogen returned as dead earthworm tissue at about 60 to 70 kg/ha, and the amount of metabolic nitrogen excreted at about 30 kg/ha, making a total annual turnover by this one species of worm alone at about 100 kg/ha or twice the amount of nitrogen in the leaf fall. It is unlikely that these estimates are accurate, but they do emphasise what a severe competition there must be both among the worms themselves and between the worms and other members of the soil population for nitrogen.

It is still very difficult to determine what are the immediate sources of food for earthworms, and little is known about the biochemistry of their digestive system. They can presumably use many soluble carbohydrates and accessible proteins, and their gut contains cellulases, so they can use some celluloses, though they are inefficient users, since it is still present in worm casts; this may be because the residence time of food in their gut is only about 20 hours. The gut also contains chitinases, so presumably they can also use chitins. However, J. N. Parle[3] also found that the numbers of bacteria and actinomycetes, as determined on soil extract agar, increased in the gut, but he considered it more probable that they were benefiting from the worm's digestive enzymes rather than contributing to its nutrition.

The habits of life of the different species of earthworm differ appreciably. Many feed on surface litter, some coming to the surface to collect it and then drag it down into their burrows, and others only pulling down the litter that is directly above their burrow. They typically consume soil along with the plant debris, for reasons that are not known, though H. B. Miles[4] found that the worm *Eisenia foetida* needed to consume protozoa if it was to grow normally. Their excreta, in the form of wormcasts, therefore typically contains a high proportion of soil. Most species excrete this in the body of the soil, often in their burrows, but some make casts on the soil surface. At Rotham-

1 I am indebted to Dr C. A. Edwards for these data.
2 In *Soil Biology*, Ed. A. Burges and F. Raw, Academic Press, 1967.
3 *J. gen. Microbiol.*, 1963, **31**, 1, 13.
4 *Soil Sci.*, 1963, **95**, 407.

sted only *Allolobophora longa* and *A. nocturna* have been proved to make casts regularly on the surface, but other species such as *Lumbricus terrestris* sometimes do.

The shape of the burrow, and its depth, also depends on species. Some are always making new burrows and live entirely in the surface layer of the soil; others, particularly the larger ones, have fairly permanent burrows, which may be 1·5 to 2 m deep for *L. terrestris*. On old pastures on clay-with-flints at Rothamsted, *L. terrestris* forms burrows to about 100 cm, *Octolasium cyaneum* to about 55 cm, *A. nocturna* to 33 cm, and the other species all lived in the top 23 cm of soil which was the depth of soil which had all been formed from wormcasts.[1]

Earthworms can only thrive in soils under certain specific conditions. They are intolerant of drought and frost,[2] and hence dry sandy soils, and thin soils overlying rock are not usually favourable environments for them.[3] They need a reasonably aerated soil, hence heavy clays or undrained soils are also un- favourable as are pastures whose surface is puddled by overgrazing in wet weather. Thus, under given conditions of management, they will be most numerous in loams and less numerous in sands, gravels and clays.[4] Many can, however, survive up to a year in water if it is reasonably aerated,[5] and the author has found them alive in late autumn in undrained English clay pastures in burrows already filled with water due to the late autumn rains. Earthworms are much more active in spring and autumn, when the surface soil is moist and warm, than in summer when the surface is usually dry, and in winter when it is too cold. A. Stockli[6] has given an example of the effect of spring rainfall on the weight of earthworm casts produced on a Swiss pasture: 7 t/ha for the three months April to June in a year when the rainfall during the period was 265 mm, and 25 t/ha the next year, when the rainfall was 420 mm.

Most earthworms, including all the larger species, need a continuous supply of calcium, and if they are feeding on a calcium-rich material will excrete calcium surplus to their requirement as calcite from special glands in their digestive tract. Earthworms are, therefore, absent on soils low in calcium. They are also absent from acid soils, and in fact if they are put on the surface of a soil that is too acid they are unable to burrow into it and very soon die. M. S. Laverack[7] showed that *Allolobophora longa* reacted violently when bathed in a buffer solution below pH 4·6–4·4, *Lumbricus terrestris* below pH 4·3–4·1 and *L. rubellus* below pH 3·8; and these correspond very well to the lower limit of soil pH in which they are found. A soil pH of about 4·5 is in

1 J. E. Satchell in *Soil Biology*, Ed. A. Burges and F. Raw, Academic Press, 1967.
2 For examples of the effect of frost limiting earthworm numbers, see H. Hopp, *Proc. Soil Sci. Soc. Amer.*, 1948, **12**, 503, 508.
3 W. Kubiena, *Bodenk. PflErnähr.*, 1942, **29**, 108.
4 W. J. McL. Guild, *Ann. Appl. Biol.*, 1948, **35**, 181.
5 B. L. Roots, *J. Exp. Biol.*, 1956, **33**, 29.
6 *Landw. Jahrb. Schweiz.*, 1928, **42**, 1.
7 *Comp. Biochem. Physiol.*, 1961, **2**, 22.

fact fairly critical for earthworm activity; for in woodland and pasture soils with a pH above it, earthworms account for a high proportion of the weight of the soil fauna and below it for a relatively low proportion, as is shown in Table 11.4. C. H. Bornebusch divided up the earthworms in Danish forest soils into two groups: acid-tolerant litter-dwelling species, such as *Bimastus eiseni*, *Dendrobaena octaedra* and *D. rubida*, which can be found in litter with a pH below 4 and are not usually found in soils or litter above pH 5, and an acid-intolerant group of soil-burrowing species, which are normally present in soils with a pH above 4·5. But J. E. Satchell[1] has found in some English Lake District woodlands a third group which are present both in very acid and in neutral soils, two of which are mainly surface- or litter-dwellers and three, *L. terrestris*, *L. rubellus* and *Octolasium cyaneum*, which are soil-dwellers. It is not known how commonly these last three are found in acid soils, for he also found that at Rothamsted *L. terrestris* and *O. cyaneum* are absent or almost absent from pastures below pH 4·5 although present in considerable numbers in soils above that pH.[2]

The weight of earthworms present in soils can be impressive. Bornebusch's figures for the numbers and weights of earthworms in two Danish forest soils has already been referred to (see p. 191), though he also found on two productive forest sites $2\frac{1}{2}$ to $3\frac{1}{2}$ million per hectare weighing 1700 to 2000 kg. A. G. Davis and M. M. Cooper,[3] on productive four-year-old rye grass/white clover leys on a deep brick earth at Wye, Kent, found about 6 million per hectare weighing nearly 1700 kg and P. D. Sears and L. T. Evans[4] and R. A. S. Waters[5] at the Grasslands Station, Palmerston North, New Zealand, found on their most productive pastures numbers up to 7 million per hectare and weighing 2400 kg; and they noted a close correlation between the productivity of the pasture and the weight of earthworms, there being about 170 kg of worms for every 1000 kg of annual dry-matter production, averaged over the year.

The weights and numbers for typical British pastures are usually lower than these figures, but so is their carrying capacity for sheep or stock. Numbers of the order of $1\frac{1}{4}$ million per hectare, weighing up to 650–1100 kg are common, though J. B. Cragg[6] has given an example of an upland pasture in Yorkshire of relatively low carrying capacity for sheep, yet carrying 1400 kg/ha of earthworms. Arable soils usually contain a much smaller weight, often of the order of 100 kg/ha, and typically composed of smaller species, unless the land is given large regular treatments of farmyard manure. Ploughing out a pasture may cause a very rapid drop in earthworm numbers though the

1 In *Soil Zoology*, Ed. A. Burges and F. Raw, Academic Press, 1967.
2 J. E. Satchell in *Soil Zoology*, Ed. D. K. McE. Kevan, Butterworth, 1955.
3 *J. Brit. Grassland Soc.*, 1953, **8**, 115.
4 *N.Z. J. Sci. Tech.*, 1953, **31** A, Suppl. 42.
5 *N.Z. J. Sci. Tech.*, 1955, **36** A, 516. A. Finck, *Z. PflErnähr.*, 1952, **58**, 120, compared numbers and weights on productive German pastures.
6 *J. Ecol.*, 1961, **49**, 477.

numbers sometimes remain high during the first year of arable.[1] Killing pasture off with a herbicide such as paraquat and slit-seeding a cereal crop without further cultivation gave a much higher population of earthworms in the first year out of pasture than did traditional ploughing, as is shown in Table 11.4.[2]

TABLE 11.4 Effect of contact herbicide and traditional cultivation on earthworm weights when a pasture is replaced by wheat. Herbicide: paraquat. Site: Woburn (England) sandy loam. Weights in g/m^2

Species	Pasture replaced	Paraquat treated	Ploughed
Lumbricus terrestris	Autumn 1967	56·2	1·6
Others		14·7	0·4
Lumbricus terrestris	Autumn 1968	71·8	17·6
Others		36·7	26·1

It is important to note that shallow cultivation of orchard soils does not necessarily give low earthworm numbers, indicating again the important role food supply may play in determining their activity. Thus F. Raw[3] found at least as high numbers of *L. terrestris* in apple orchard soils that received regular shallow cultivations as under a gang-mown grass sward, but most of these earthworms would have been below the layer of soil that was loosened by cultivation when this operation was performed. The two cultivated orchards he examined in fact contained nearly 2 t/ha of *L. terrestris* which was as high as the highest grassed orchard, and these worms buried over 90 per cent of the leaf fall, which contained just over 1 t/ha dry matter, during late autumn and winter. It is interesting to note that orchards heavily sprayed with bordeau or other copper sprays contained very few earthworms and had peaty surface mats and poor structure,[4] but that regular spraying with normal levels of DDT did not affect the number or species of earthworms present, although they accumulated DDT in their body tissues.[5]

Earthworms, where they flourish, as in virgin soils and pastures, are the principal agents in mixing the dead surface litter with the main body of the soil, so making it more accessible to attack by the soil micro-organisms. Peat

1 See, for example, A. C. Evans and W. J. McL. Guild, *Ann. Appl. Biol.*, 1947, **34**, 307; 1948, **35**, 485, and W. J. McL. Guild, *J. Animal Ecol.*, 1951, **20**, 88; 1952, **21**, 169.
2 From Rothamsted Exp. Stn., *Ann. Rept.*, 1968, 218. For a similar example from Hohenheim, Germany, see F. Schwerdtle, *Z. PflKrankh.*, 1969, **76**, 635.
3 *Ann. Appl. Biol.*, 1962, **50**, 389.
4 J. M. Hurst, H. H. Le Riche and C. L. Bascombe, *Plant Path.*, 1961, **10**, 105.
5 A. Stringer and J. A. Pickard, *Long Ashton Res. Sta., Rept.* 1963, 127.

formation is, in fact, typical of uncultivated land too acid for earthworms; thus at Rothamsted a mat of dead vegetation accumulates on all the pasture plots whose pH is below 4, apparently solely because there are no earthworms or other animals capable of mixing this debris into the body of the soil.

Earthworms improve the aeration of a soil in two ways. They are very extensive channellers and burrowers, and these channels serve to improve the drainage of the surface soil, to give regions of good aeration and to loosen the whole soil throughout the zone in which they are working. They are most active in the surface layers of the soil, rarely working the soil intensively below 15 to 20 cm, though a few species will make burrows to much greater depths and these often serve as passages for plant roots to penetrate into the subsoil. They can also play an important role in irrigated fields, for N. A. Dimo[1] found up to 15 million channels per hectare coming to the soil surface in fields of irrigated lucerne; and these allow the irrigation water to penetrate rapidly through the surface layers of the soil into the subsoil. C. H. Edelman[2] gives an example from Holland of gang-mowing in a grass orchard on a clay loam giving 5 million large worm channels and many more smaller ones per hectare, and A. Finck[3] in Germany found up to 10 million in the subsoil of well-manured arable fields and 1 million common on normal arable. On the other hand F. Raw[4] only counted about half a million channels which reached to the soil surface in some Wisbech apple orchards.

They also improve the aeration and at the same time the water-holding capacity of the soil crumbs by improving the soil structure. Surface soil under pastures or forests in which earthworms are active has a structure which used to be known as a mould or mull, which is characteristic of earthworm activity and is admirably suited for plant growth. It is only brought about by those species producing definite wormcasts, for A. C. Evans[5] found at Rothamsted that the top few inches of pasture soils had a pore space of 67 per cent where these species were present but only of 40 per cent where they were absent, although both groups of pastures had about the same weight of worms. The form of the food supply may also be important, for A. K. Dutt[6] and R. J. Swaby[7] found that wormcasts only possessed this stable structure if produced on pasture soils; they had this property to a much less extent if produced on arable soils and subsoils.

The cause of this improvement in structure is not yet fully understood and is more fully discussed on pp. 500–501. J. N. Parle[8] found the stability of the structure of newly formed casts increased for the first two weeks and then

1 *Pedology*, 1938, No. 4, 494.
2 *Trans. 5th Int. Congr. Soil Sci.* (*Leopoldville*), 1954, **1**, 119.
3 *Z. PflErnähr.*, 1952, **58**, 120.
4 *Nature*, 1959, **184**, 1661.
5 *Ann. Appl. Biol.*, 1948, **35**, 1.
6 *J. Amer. Soc. Agron.*, 1948, **40**, 407.
7 *J. Soil Sci.*, 1950, **1**, 195.
8 *J. gen. Microbiol.*, 1963, **31**, 13.

began to decrease in phase with an increase and subsequent decrease in the length of fungal hyphae in the casts. He also found that the cast contained a considerably higher proportion of polysaccharides than the soil, but their level dropped fairly rapidly as the cast aged. There is as yet no definite evidence that the earthworm gut is the site for the production of genuine humic acids, as distinct from polysaccharide gums, but this is a technical problem that is difficult to investigate critically. Earthworms may also help to stabilise the structure of soils not in wormcasts in another way altogether, for A. E. Needham[1] found that half the nitrogen excreted by *Lumbricus terrestris* was in the form of mucoproteins secreted by cells within the epidermis, which constitute the slimy film on their bodies, and these may stabilise the pore space distribution of the soil in contact with the worm's body as it moves through the soil.

TABLE 11.5   The chemical composition of wormcasts compared with that of the whole soil (Connecticut)

Depth of sample	Arable soil			Forest soil (mean of 4 soils)			
	Cast	0–15 cm	20–40 cm	Cast	A1	A3	B
			In parts per million of soil				
Exchangeable calcium	2790	1990	481	3940	747	155	171
Exchangeable magnesium	492	162	69	418	140	43	59
Exchangeable potassium	358	32	27	230	138	32	25
Available phosphorus	67	9	8	9	7	3	3
Nitrate nitrogen	21·9	4·7	1·7	—	—	—	—
Total carbon in per cent	5·17	3·35	1·11	15·6	5·9	2·1	1·0
Total nitrogen in per cent	0·353	0·246	0·081	0·625	0·327	0·130	0·064
C/N	14·7	13·8	13·8	25·1	18·0	16·3	15·0
pH	7·0	6·4	6·0	5·3	4·6	4·6	4·7
Per cent of base saturation	93	74	55	63	32	18	12

The chemical composition of wormcasts differs from that of the soil mass because the worms feed selectively on the material in the soil, concentrating in the cast most of the nitrogen and mineral constituents present in the organic matter ingested. Thus, the casts are richer in available plant nutrients than the soil, as is shown by some results of H. A. Lunt and H. G. M. Jacobson[2] given in Table 11.5.

Since the cast contains the waste products of the worm's metabolism, a part of the nitrogen present is in the form of ammonium ions, urea, and possibly uric acid and allantoin,[3] and these are either readily available for

1 *J. exp. Biol.*, 1957, **34**, 425.
2 *Soil Sci.*, 1944, **58**, 367.
3 A. E. Needham, *J. exp. Biol.*, 1957, **34**, 425.

uptake by plants or are rapidly ammonified. Since the cast soon becomes aerated, these are all nitrified, so that the cast is higher in mineral nitrogen than the ingested soil, and K. P. Barley and A. C. Jennings[1] found that in some of their experiments with *Allolobophora caliginosa* about 6·4 per cent of the insoluble nitrogen ingested was converted to these soluble compounds. The body tissue of earthworms, which may consist of up to 70 per cent protein, decomposes rapidly at death, and this will also contribute first to the ammonium and then the nitrate content of the soil. Soils in which earthworm activity is high therefore will tend to have a high rate of mineral nitrogen production. This does not necessarily imply that such soils are better supplied with mineral nitrogen than soils in which earthworm activity is less, but it is likely to be so whenever this activity increases the rate of decomposition of added organic matter. This effect of earthworms is less likely to be important in cultivated soils, when the cultivation operations bury and incorporate all dead stubbles and plant remains with the soil, than in pastures, but this earthworm effect should become increasingly important as minimal cultivation techniques are introduced. There is, however, still no firm evidence that the presence of earthworms in cultivated soils has any effect on crop yields.[2]

Darwin first drew attention to the great quantity of soil earthworms can move per year; he estimated that the earthworms on some pastures outside his house could form a new layer of soil 18 cm thick in thirty years, or that they brought up about 50 t/ha of soil annually, enough to form a layer 5 mm deep, a figure which represents fairly well the amount of earth the casting species of earthworms can put on the soil surface every year. Furthermore, A. Stöckli found a similar figure in some Swiss and A. C. Evans in some Rothamsted pastures. It is not yet possible to estimate from what depth the soil cast on the surface has been derived, but most of it has probably come from the top 10 to 15 cm. Some species of tropical earthworms may cast considerably greater quantities of soil on the surface, for D. S. Madge[3] found two species produced about 170 t/ha of casts during the two to six month wet season in some western Nigerian pastures in Ibadan.

This weight however, does not represent the total weight of soil that passes through the gut of the earthworms each year, for this has been derived purely from the casting species. No attempt has yet been made to measure this accurately, but A. C. Evans[4] has made a rough estimate by assuming the weight of wormcasts produced by worms is proportional to the weight of the worms. His results for some Rothamsted fields with different cropping histories are given in Table 11.6, and show that some fields have high earthworm activity although few casting species are present, so only very few wormcasts are

1 *Aust. J. agric. Res.*, 1959, **10**, 364.
2 For a review of this subject see K. P. Barley, *Adv. Agron.*, 1961, **13**, 249, and J. E. Satchell, *Soils Fert.*, 1958, **21**, 209.
3 *Pedobiologia*, 1969, **9**, 188.
4 *Ann. Appl. Biol.*, 1948, **35**, 1.

produced. *Allolobophora nocturna* and *A. longa* are the casting species present, so their numbers are given separately. The table also shows that normal arable soils have a low earthworm activity compared with established leys and pastures.

TABLE 11.6   Estimated consumption of soil by worms at Rothamsted

| Cropping of field | Number of worms thousands/hectare | | Weight of worms | Worm-casts | Soil excreted below ground | Total soil consumption |
	A. nocturna and A. longa	Others	kg/ha	tons/ hectare	tons/ hectare	tons/ hectare
Old pasture (1)	850	1070	990	62	30	89
(2)	350	420	610	27	23	50
Established leys:						
18 years old	300	320	560	30	25	52
8 years old	75	820	740	5	52	57
5 years old out of woodland	120	470	730	5	42	47
5 years arable after old pasture	75	320	130	7	5	12

Casting species of earthworms, by bringing up soil from below as wormcasts, alter the texture of the surface soil, for they avoid sand particles, particularly coarse sand particles, when they feed on soil. Thus on old Rothamsted pastures, the top 10–15 cm of soil is almost stone-free and Evans showed it has an appreciably lower coarse sand content than the subsoil, and it is separated from the subsoil by a layer of soil that is appreciably more stony than the soil immediately below this layer. An aspect of earthworm activity that has received little attention is the consequence of casting on the distribution of plant nutrients in the soil, for as the earthworms move soil particles and granules from one part of the soil profile to another, they also move the less soluble plant nutrients.

One can summarise our present knowledge on the importance of earthworms in agriculture as follows. They are of very great importance in undisturbed forest and pasture soils above pH 4·5, for they are the principal agents in mixing dead plant debris on the surface of the soil with the soil itself, and in doing this they help to keep the surface soil loose and well aerated. They may also play an important role in the conversion of plant litter into humus, as their gut may form a favourable environment for the necessary chemical changes to take place, but this has not yet been rigorously proved. Only three British species produce wormcasts on the surface of the soil, all the rest excrete in the soil itself.

Arthropods

Mesofauna

The soil-inhabiting acarine mites vary from about 0·1 to 1 mm in size and are present in very considerable numbers, belonging to a large number of genera and species. Each species has probably fairly specific food requirements, but the mites as a whole feed on a wide range of materials such as decomposing plant remains and fungi, though perhaps many of these appearing to feed on plant remains are in reality feeding on the micro-organisms carrying out the decomposition, and a number are predaceous on other members of the meso-fauna. A. P. Jacot[1] has described how they will feed on the inside of a dead root leaving a channel with its outer corky layer. The springtails or collembola are rather larger, 0·5 to 2 mm in size, and are minute wingless insects. There are probably considerably fewer species in the soil than there are of mites, and each species may be less specific in their food. They feed on decaying plant tissue and the micro-organisms decomposing it, on dead insects, and pre-sumably on the remains of other members of the soil fauna. The species present in the soil vary with the depth at which they live, the soil surface in-habitants being pigmented, and have well-developed eyes and springing organs, all of which are lacking or poorly developed in the subsoil dwellers. The mites and springtails have been extensively studied in forest soils, where they concentrate in the F and H layers of forest litter. These layers form a very suitable habitat for them, since they provide a supply of decomposing leaf litter all through the year and a habitat that is usually moist but not water-logged. The numbers of mites and springtails are lower in arable than in un-disturbed soils. Ploughing-out grassland therefore reduces their numbers, but the actual number in an arable soil depends on the rate that organic matter is being added. The numbers are, for example, higher in soils receiving regular dressings of farmyard manure than in those only receiving fertilisers.[2] The saprophytic forms can only utilise a small fraction of the energy and nutrients in the plant debris they consume so their principal effect in the litter layer is to macerate a relatively large amount of debris and distribute it as faecal pellets.[3]

Larvae of beetles and dipterous flies

The larvae of several beetles and dipterous flies live in the soil and can be im-portant burrowers and channellers in it. They occur in considerable numbers in some forest soils, particularly in the mor layer, as can be seen from Table

1 *Ecology*, 1936, **17**, 359, and *Quart. Rev. Biol.*, 1940, **15**, 28.
2 C. A. Edwards and J. R. Lofty, *The Soil Ecosystem*, Systematics Assoc. Publ. 8, 1969.
3 For a detailed description of the micro-arthropods in some Danish soils see N. Haarlov, *Oikos*, 1960, Suppl. 3.

11.2, and in some grassland soils. They are normally saprophagous, but they will attack either the roots of agricultural crops or the young plant just above ground level. Among the principal agricultural pests on living crops are wireworms, which are the larvae of elaterid beetles, predominantly of the genus *Agriotes*, and leatherjackets, which are the larvae of craneflies (Tipulidae).

Ants and termites

Ants and termites have many features in common. They are social insects living in nests, and for many genera these nests are in the soil. Some genera fill the soil with passages and chambers for their brood without making any mound, while others make mounds either out of surface or of subsoil without, however, mixing in any humic matter in the way earthworms do. This channelling and burrowing in the soil can have a very appreciable effect on the aeration and ease of drainage of soils, and H. Hopp and C. S. Slater showed that they could be as efficient as earthworms in improving plant growth in this way.[1] They are very active insects and can forage over very considerable distances for their food, so are common in land carrying a sparse vegetation, such as the fringes of deserts. There are genera in each group which harvest green leaves, often of grasses, and bring them back to their nests to be used as a food base on which they cultivate fungi in special gardens or combs on which they feed. Some of the ants also collect the seeds of the grasses at the same time. In semi-arid country, particularly if the grass is scarce due to overgrazing, these insects can remove almost all the remaining grass, complicating the problem of rehabilitating the range enormously, and leaving it exposed to water run-off and soil erosion. J. Thorp[2] has described this harmful effect of ants in the Great Plains area in the centre of the USA, where he finds 25 to 50 ant hills per hectare each surrounded by a circle of bare soil from 2 to 12 m in diameter, and W. G. H. Coaton[3] describes the corresponding trouble with termites on similar rangeland in Zululand, South Africa.

Ants differ from termites in that they are not confined to equatorial and subtropical regions, but are common inhabitants of temperate soils. They can also differ in their feeding habits, for many genera of ants are predaceous, feeding on the mesofauna and smaller or more immature members of the soil macrofauna, including termites, of which they are one of the principal predators. They are not restricted to the soil fauna, but prey on insects feeding on plants and crops. A much quoted example of the control of an insect pest by a species of ant is that of the corrid bug *Pseudotherapsis wayi* causing premature nut drop of coconuts in Zanzibar and the coastal area of East Africa, which can be fairly effectively controlled by the ant *Oecophylla longinoda*.

1 *Soil Sci.*, 1948, **66**, 421.
2 *Sci. Monthly*, 1949, **68**, 180.
3 *Farming in S. Africa*, 1954, **29**, 243.

Unfortunately this ant is itself preyed upon by three other species[1] which do not feed on this bug, and we have yet to learn how to control the fortunes of battle between warring species of ants.

Ants also differ from termites in that some genera cultivate aphis or coccids in their nests and feed on their exudates. These 'milch cows' may either be cultivated on roots of plants, particularly of grasses, or may be tended on the leaves of trees or other plants, and the tending includes protecting the aphis or coccid colonies against attacks by fungi and predators which would otherwise keep them in check.[2] The size of the ant colony may be controlled by the size of the colonies it can tend, and the damage done to crops by these colonies may be greater than any benefit that may accrue to the crop from the incidental reduction of insect numbers brought about by these same ants.

Termites differ from ants in that many genera have developed the use of fungal gardens to a much greater extent, and use wood as the base on which to grow the fungi. They are in consequence the dominant insect in many of the woodland and forest soils of the equatorial and subequatorial regions, the ants in these areas being primarily predaceous on them.

Naturally, the detailed habits of the different species of ants and termites differ very considerably, even if using the same general method of obtaining their food supply and living in the same kind of nest. In spite of the importance of these animals in the decomposition of plant residues and in the control of the soil population, little is known about their biology, feeding habits or digestive processes. It is very difficult, for example, to count the number of these insects on a hectare of land, or estimate the weight or kind of food they consume per year. Hence in the following account of termite activity in soils, there must be many gaps, and probably a good deal of faulty deductions from observations, which must be borne in mind.

Termites are the dominant animals in many tropical soils, in the sense that they probably consume more organic matter per hectare per year than any other group. H. Drummond,[3] who was one of the first commentators on their activity, described them as the topical analogue of the earthworm, a phrase that has often been quoted. The termites can be classified into a number of groups, such as the dry-wood termites, which live in trees and therefore have no effect on the soil, the wood-feeding termites which are probably of more importance in the damage they do to buildings than in their effect on the soil, the fungus-growing termites and the humus-feeding ones.[4] They can also be classified by the mounds they build, some building very large ones, others small ones, and others burrowing in the soil but building none, but having their nests at different depths. In Africa, Thailand and Malaya all the large

1 M. J. Way, *Bull. ent. Res.*, 1953, **44**, 669.
2 M. J. Way, *Bull. ent. Res.*, 1954, **45**, 93 and 113, describes the behaviour of the ant *Oecophylla longinoda* in protecting colonies of the scale insect *Saissetia zanzibarensis* on clove trees.
3 *Tropical Africa*, London, 1888.
4 See, for example, W. V. Harris, *E. Afr. Agric. J.*, 1949, **14**, 151, and *Soil Zoology*, London, 1956.

mound builders are fungus growers, but in Australia they are not. In Africa some harvester termites collect green leaves, and carry them into nests 3 m or more below the soil surface, with the consequence that all the mineral nutrients in the leaves not required by the termites will be left far below the topsoil.

Little is known of the effect of the so-called humus-eating termites on the soil organic matter. They pass humus-rich topsoil through their gut, though whether they break down the humus in the soil or partially decayed plant remains or any fungi and other micro-organisms in the soil is not known. They do not, however, enrich the soil they excrete in humic matter as earthworms do; they are said to impoverish the topsoil, but it is not certain what is meant by this phrase.

Some species of termites have very considerable ability to penetrate through hard pans and possibly into laterite crusts, and termite activity may have been a factor in causing the particular structure of the vesicular or vermicular laterite crusts so common on the Miocene peneplains of central Africa (see p. 730).[1]

The termites with the most spectacular effect on the soil, the large mound-builders, belong to the genus *Macrotermes* in Africa. They are probably confined to 'miombo', that is, *Brachystegia-Isoberlinia* woodland, which is characterised by having an adequate rainy season and a long dry one; and they cover hundreds of thousands of square kilometres of equatorial and southern Africa. The mounds can be 12 to 18 m in diameter and up to 7·5 m high, and their remains may persist long after a change of climate has changed the vegetation. The mounds are not inhabited all the time by the termite which built them, but serve as a shelter for other species of ants and termites, and some of these may inhabit the mound even when the building species is in occupation. These mounds may contain hundreds of tons of earth above soil level, and commonly occur at a spacing of about two per hectare, which gives a measure of the distance around the mound that these termites will forage. The important factor for most of these mound-building termites is an adequate wood supply. Soil pH, soil type and depth, and topographical position are unimportant, except that they avoid wet sites though they may build on soil that has impeded drainage so is temporarily waterlogged in the rains.

The exact shape and architecture of the mounds depends somewhat on the species present and other local conditions, but no detailed description of these variations has been given. The mound consists of two parts, a compact outer casing and an inner nest. The soil forming the mound is subsoil, which may have come up from as much as 3 m, though most is derived from depths of 30 to 90 cm, but the channels from which the subsoil is derived can extend far beyond the mound itself. The mound consists of sand grains, usually about 2 mm in size, cemented with finer soil particles, below 0·2 mm, which the termite carries in its crop. If the soil is sandy the mound will usually have more

1 H. Erhart (*Comp. Rend.*, 1951, **233**, 804, 966; 1953, **237**, 431) has discussed this problem with reference to the laterite crusts in French West Africa.

clay in it than the surface soil, and this would also happen if the subsoil is heavier than the surface soil. The casing is continually eroding away as it carries no vegetation if the mound is inhabited, though it gets grassed over or protected by trees if it is deserted. There is thus over a period of years a very considerable movement of subsoil to the surface which is liable to be washed down hill in the rains, so giving soil creep. This has the consequence that all particles larger than about 2 to 4 mm accumulate in a fairly definite 'stone-line' at the bottom of the zone of maximum termite activity, though this is not the only mechanism which gives this characteristic feature of many tropical soils. Little is known about the average rate of transport of subsoil to the surface in these mound areas, but P. H. Nye,[1] working in an area of rather small mounds, estimated it at just over 1 t/ha/year. This figure should be compared with Nye's estimate of 50 t/ha of wormcasts produced during the six months of rain by the worm *Hippopera nigeriae* on a neighbouring area under a bush which shades the ground and has an abundant leaf fall. The soil in the cast is surface soil and is all finer than 0·5 mm and mostly finer than 0·2 mm.

Inside the casing is the nest, which is full of channels and contains a number of chambers filled with the fungus gardens. The soil is again subsoil but it tends to have a higher pH and be more saturated with exchangeable bases than the original subsoil, presumably because at least part of the minerals present in the wood brought in are left behind in the soil when the wood has disappeared.[2] The decomposition of the wood by the fungi is so efficient that this soil is hardly enriched in organic matter. One would expect the soil in the nest to contain more phosphate as well as more bases, but the evidence for this is inconclusive, P. R. Hesse[3] finding no extra phosphate, but R. L. Pendleton[4] finding extra in some of the mounds he investigated. The fate of the minerals in the wood that is introduced has not yet been worked out.

This lack of knowledge complicates the interpretation of another feature found in some mounds, which is a layer of concretionary or nodular calcium carbonate at the base of the mound. Hesse only found this in calcareous soils or on land with impeded drainage which collected calcium salts from higher land during the rainy season. But Pendleton, and C. G. Trapnell have found examples of accumulation sometimes on acid soils, where the only reasonable explanation seems to be that the calcium was derived from the wood brought in.[5]

There has been much discussion on the fertility of soils in a termite mound compared with the soil around it, but much of the inconsistencies in the litera-ture seem to be due to ignoring the fact that soil fertility is made up of many

1 *J. Soil Sci.*, 1955, **6**, 73.
2 This has also been noted by J. B. D. Robinson, *J. Soil Sci.*, 1958, **9**, 58.
3 *J. Ecol.*, 1955, **43**, 449 (East Africa).
4 *Thai Sci. Bull.*, 1941, **3**, 29.
5 For an example from Rhodesia, where calcium carbonate accumulation could be noted to 6 m depth on a slightly acid soil, see J. P. Watson, *J. Soil Sci.*, 1962, **13**, 46.

different factors. If the soil around the mound is acid and low in calcium, and the crop being grown is calcium-demanding, such as sisal, one would expect to find better sisal on the mounds, and even on mounds that have been more or less levelled, than on the soil around; and this effect can be seen quite strikingly on some Tanzanian sisal estates. On the other hand if the subsoil is very poor in nutrients, most of the soil in the mound will also be poor, particularly in available nitrogen, and levelling the mound can result in a nutrient-poor subsoil being distributed over a considerable area of land.[1]

Areas where active mound-building *Macrotermes* are common can be difficult to cultivate both because the mounds themselves are difficult to cultivate, and also because, if knocked down with a bulldozer, they may be built up again surprisingly quickly. Since these termites are primarily wood-eaters, they would be expected to die out quickly when forest is cleared for cultivation, and this is seen in some areas of East Africa, but some will attack annual crops and then cause serious trouble. It is also difficult to maintain mulches on the surface of the soil in areas where soil-inhabiting termites of any kind are active.

Ants affect the soil in ways very similar to termites, particularly those species which are mound-builders, although anthills are much smaller than most termite mounds. In and close to the mound, the soil is full of their burrows, the bulk density of the soil is reduced, the potash and phosphate status is frequently higher, and much of the soil forming the mound may have come from the subsoil. Some species of ants only use their mound for a limited number of years, and these may then slowly become levelled as in tropical areas, but they are not normally used by other species of ant.[2]

Myriapods and Isopods

Both millepedes (Diplopoda) and centipedes (Chilopoda) are present in soils, often in considerable numbers.[3] In general the food of the millepedes is decaying plant litter, and the centipedes are predaceous on the immature forms of the larger soil fauna. Some millepedes at least, and possibly all, are unable to digest cellulose, so must consume a very large volume of litter to extract sufficient sugars and other simple carbohydrates for their requirements, hence can be very important as mechanical comminuters and mixers of litter with soil. K. L. Bocock,[4] for example, estimated that the millepede *Glomeris marginata* feeding on ash litter only utilised 6 to 10 per cent of the dry matter that it ingested and only converted about 0·3 to 0·5 per cent into its body tissues, but it is possible the proportion so converted is determined by the level

1 For an example of this, see J. A. Meyer, *Bull. Agric. Congo*, 1960, **51**, 1047.
2 For a description from American prairie and forest soils see F. D. Hole with F. P. Baxter, *Proc. Soil Sci. Soc. Amer.*, 1967, **31**, 425, with M. Z. Salem, ibid., 1968, **32**, 563; for an English example, N. Waloff and R. E. Blackith, *J. anim. Ecol.*, 1962, **31**, 421.
3 For a general account, see J. G. Blower, *Soil Zoology*, London, 1955, p. 138.
4 In *Soil Organisms*, Ed. J. Doeksen and J. van der Drift, North Holland, 1963.

of protein in the litter. Some consume a certain amount of soil, possibly only soil on the litter, but do not appear to excrete a mull humus but a mull-like moder in Kubiena's terminology.[1] In both the millepedes and centipedes there are species adapted for burrowing in the soil and species that mainly feed and live above the soil surface.

The humidity of the air controls the suitability of the environment for both these groups of animals. Many species have little ability to control water loss from their body in a dry atmosphere, and many cannot control water intake if placed in water. A good forest mull, or the surface layers of forest soils with leaf litter forms a very favourable environment for most species. Hence forest soils that are not too acid but are for some reason not suitable for high earthworm populations, such as acid sandy soils, usually have a high millepede population.

In agriculture millepedes are found in some pastures, and may occur in considerable numbers in arable soils well supplied with farmyard manure, and these may be so high that they can be a pest for crops such as potatoes.[2]

The woodlice—isopods—are predominantly saprophagous or phytophagous, but their relative importance in the decay of organic matter and in altering the soil structure is unknown, though their gut may be a favourable environment for the humification of plant residues to take place.[3] They probably often thrive in sites too dry for earthworms. Thus, N. A. Dimo[4] had shown that they may be very active channellers and burrowers during the summer in semi-desert soils when earthworms are aestivating. He found them burrowing down to 60 or 90 cm and bringing up about five tons of soil per hectare during the season in a condition of small crumbs with a mellow structure.

Gasteropods

Slugs and snails are the two soil representatives of the Gasteropods, and so far little is known about their distribution or numbers. Most are surface-feeders, usually active in damp conditions, hence during the day either burrow into the soil or into a dense, shady spot, such as under stones or leaves or in thick grass tussocks. Some species of snails contain high concentrations of the enzyme cellulase in their gut, so obviously must digest cellulose, but it does not seem to be known if any of the principal soil-inhabiting slugs have this enzyme or if their food is confined to simple sugars. The forest floor is the well-adapted environment for them, and, as B. Lindquist[5] has shown, litter and fungi appear to be their principal sources of food.

1 J. G. Blower, *Trans. 6th Int. Congr. Soil Sci.* (*Paris*), 1956, **C**, 168.
2 For a discussion of the cause of high numbers, see J. L. Cloudsley-Thompson, *Proc. Zool. Soc.*, 1951, **121**, 253.
3 H. Franz, *Bodenk. PflErnähr.*, 1943, **32**, 336.
4 *Pedology*, 1945, No. 2, 115.
5 *Kungl. Fysiograp. Sällsk. i Lund Forkandl.*, 1941, **11**, Nr. 16.

In agricultural soils, H. F. Barnes and J. W. Weil[1] have also shown that the food of slugs is usually dying vegetation, such as freshly fallen leaves, old grass, and so on, and that they usually attack the actively growing parts of plants only when other food is scarce. They are therefore typically scavengers rendering old plant tissues more usable by the smaller members of the animal population by excreting them in a macerated and partially digested form. Some slugs, such as species of *Testacella*, are predaceous on other slugs, worms and centipedes. Normally, the numbers of slugs on arable land is not very high, though in some areas they may damage potatoes, and D. C. Thomas[2] has given an example of a wheatfield carrying $1\frac{1}{2}$ million slugs per hectare and weighing about 400 kg.

The soil-inhabiting mammals

Animals of the mouse family—mice, voles and shrews—and moles are present in appreciable numbers in some undisturbed soils, such as forest and prairie soils, and to a less extent in some pastures. Though their total weight per hectare is only small, probably under 5 kg in forest soils[3] and up to 10 kg on open ranges,[4] they can cause a very important loosening of the surface layers of the soil by honeycombing it with their burrows and nests, and many of them also transport an appreciable amount of subsurface or subsoil and leave it on the soil surface as mounds.[5] Some semi-arid soils can also carry a population of mammals that burrow into the deeper subsoil, and these burrows may be filled in course of time with surface soil rich in humus. Thus, these soils have numerous channels of dark soil running in the lighter humus-poor subsoil.

1 *J. Animal Ecol.*, 1944, **13**, 140; 1945, **14**, 71.
2 *Ann. Appl. Biol.*, 1944, **31**, 163; 1947, **34**, 246.
3 W. J. Hamilton and D. B. Cook, *J. Forestry*, 1940, **38**, 468.
4 W. P. Taylor, *Ecology*, 1935, **16**, 127.
5 For detailed examples, see W. J. Hamilton, *Bull. New York Zool. Soc.*, 1940, **43**, 171, and J. Thorp, *Sci. Mon.*, 1949, **68**, 180.

12

The general ecology of the soil population

The various groups of soil organisms do not live independently of each other, but form an interlocked system more or less in equilibrium with the environment. This is obviously true for the predators, but it is equally true for the saprophytic species, for they are all competing for the available food supply, and each has developed a series of symbiotic and antibiotic relations with its neighbours for increasing its ability to get its share of the available food. Thus the composition of the population in any soil achieves a certain equilibrium which depends on the environment; and for any particular soil farmed on a definite system the equilibrium is fairly stable. The soil, in fact, usually appears to be fairly 'well-buffered' biologically. Hence, if any particular soil-inhabiting organism is absent from a soil it cannot usually be made an active component of the soil population merely by inoculating it into the soil;[1] it can only be introduced by altering the environment either through the crops grown on the soil, the energy-containing material added to the soil, or the general environmental conditions of life in the soil itself. This limitation naturally applies with much less force if the organism has a close relationship with any of the crops growing on the soil. D. Parks[2] showed this stability of population applied to some extent even if a foreign fungus was introduced with the food supply. He buried pieces of rotting grass, in which various saprophytic fungi were actively growing, in a soil and found that after a short time the soil fungi had displaced most of the foreign fungi from the rotting grass. On the other hand, because a soil organism can decompose a substrate in pure culture it need not be stimulated into activity if that substrate is added to the soil, for it may be unable to compete with some other soil organisms also able to carry out this decomposition. Thus E. Griffiths and D. Jones[3] isolated twenty species of fungi from a soil which would decompose cellulose in pure culture, but when they added cellulose to the soil only one or two appeared to be active, though the particular species that were

1 For an example of an attempt to do this, see S. A. Waksman and H. B. Woodruff, *Soil Sci.*, 1940, **50**, 421.
2 *Trans. Brit. Mycol. Soc.*, 1955, **38**, 130 and 1957, **40**, 283.
3 *Trans. Brit. Mycol. Soc.*, 1963, **46**, 285.

active could be altered by altering the pH or the nutrient status of the soil.

This stability of the population naturally refers to the conditions holding under normal conditions: it may be upset when new sources of energy-material, such as plant debris, composts or farmyard manures are added to the soil, for these additions can cause a rapid, though temporary, rise in certain groups, particularly some of the fungi.

The distribution of micro-organisms through the soil space

The soil micro-organisms live on the surface of the soil particles, but they are not evenly distributed. In spite of their very large numbers, a large proportion of the surface area of the soil particles is unoccupied. Part of the reason for this is that a considerable proportion of the surfaces are inaccessible to the organisms themselves. Thus, there are few bacteria smaller than 0·5 μ, and most are about 1 μ in size, but pores of this size need suctions of 6 and 3 bar respectively to empty them of water and medium to fine textured hold considerable amounts of water more strongly than this. It is also possible that micro-organisms can only be really active if in pores considerably larger than 1 μ, so that except for very sandy soils or soils in very good tilth, most of the soil surface is unsuited for high microbial activity. Yet even allowing for these inaccessible surfaces there are relatively large surfaces of the soil, on the microscopic scale, which appear to be unoccupied yet available for colonisation. A. Burges,[1] who has stressed the importance of this fact, suggests that one of the major problems in soil microbiology is to obtain a proper understanding of the size and extent of the soil microhabitats.

Direct microscopic observation shows that the fungi tend to congregate round particles of decaying plant and animal debris, and the bacteria live on these as well as on the surface of soil crumbs.[2]

A proportion of these bacteria, as well as some living in the crumb pore spaces, are in colonies, often embedded in gum; but the techniques for direct examination of where bacteria are actually living in soils are not well developed. Plate 10c gives an example of a bacterial colony inside a crumb, as seen in a thin soil section prepared by E. Griffiths and D. Jones.[3] It is not certain what proportion of bacteria are dispersed in the soil as single cells or colonies of only a few cells. Techniques based on crushing soil crumbs in sterile water, and staining, such as that developed by P. C. T. Jones and J. E. Mollison,[4] show about half of the population is present in colonies of three

1 *Micro-organisms in the Soil*, London, 1958.
2 C. Rouschel and S. Strugger, *Naturwiss.*, 1943, **31**, 300; *Canad. J. Res.*, 1948, **26** C, 188.
3 *Pl. Soil*, 1964, **20**, 232.
4 *J. gen. Microbiol.*, 1948, **2**, 54.

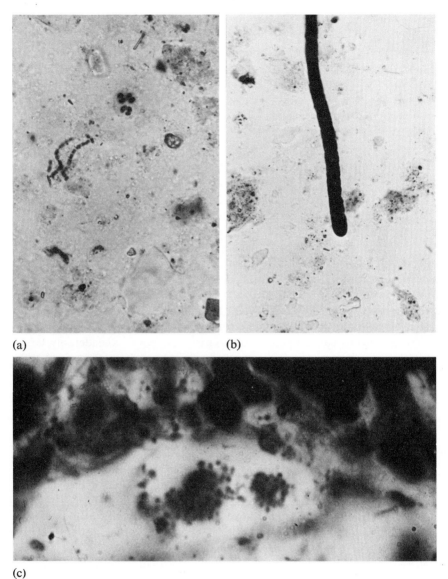

(a)   (b)

(c)

PLATE 10. Micro-organisms in the soil:
(*a*) Bacteria, a bacterial colony and actinomycetes.
(*b*) Fungal hypha, displayed by Jones' and Mollison's technique. (× 1000).
(*c*) Bacteria in the pore of a Kikuyu red loam crumb, thin soil section; Jones' and Griffiths' technique. (× 2400).

cells or less; but this could be a reflection of the break-up of larger colonies during crushing rather than of what exists in the undisturbed soil. One of their photographs is reproduced in Plate 10a and shows bacteria present both as colonies of bacteria and as single cells as well as two chains of actinomycete spores. The stereoscan electron microscope photographs of soil micro-organisms taken by T. R. G. Gray[1] and reproduced in Plate 11 show most of the bacteria present as individual cells. In all these photographs the bacteria are present as small cocci. Plate 11 also gives photographs of actinomycete and fungal hyphae, and these appear to lie on the surface of the soil particles, but this could be an artifact due to the method of preparing the samples for examination in the electron microscope.

The Rossi–Cholodny slide technique has been the method most extensively used in the past for investigating the distribution of soil micro-organisms. This technique is known to have severe limitations (see p. 151), but the results obtained are concordant with those obtained by other methods, and they will be described for three conditions—in fallow soil, in soil carrying a crop, and in soil to which fermentable organic matter has been added.

In a fallow soil[2] only a sparse population develops on the slips, extensive development only taking place near pieces of organic matter. The bacteria seem to be mainly small organisms in small colonies, which appear to be pure cultures of the organisms and which may be compact or loose and composed of cocci or short rods. A few of the looser colonies often contain rather longer rods, all lying more or less parallel to each other. There are a number of isolated bacteria, commonly short rods, scattered over the slide, long rods are scarce and usually occur near pieces of organic matter, and the larger cells commonly occur in clusters of two to four. Actinomycetes are fairly common, usually occurring as a line of spores, or conidia, each about the size of small coccus or coccoid rod, and hence indistinguishable from them if occurring individually. Fungal mycelium is rare, and this often is attacked by bacteria.

In cropped land the population does not differ in kind but becomes more numerous. Extensive bacterial development occurs around some plant root-lets, the colonies of coccoid or short rods becoming so dense that they cannot be resolved into their constituent organisms under the microscope. Actino-mycetes in particular, and fungal mycelium to some extent, share in this general development, though they are not conspicuous in the immediate neighbourhood of the rootlets.

The effect of adding fermentable organic matter can easily be seen. The population becomes much more dense, and typical groups of bacteria develop. Thus, adding sugars can give many Azotobacter-like colonies appearing on the slide, and adding cellulose many typical Cytophaga-like bacteria. This initial bacterial attack is often followed by development of fungi and

1 For the technique used, see *Science*, 1967, **155**, 1668.
2 See, for example, R. L. Starkey, *Soil Sci.*, 1938, **45**, 207.

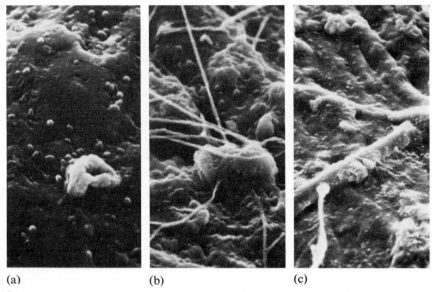

(a)   (b)   (c)

PLATE 11. Scanning electron microscope photographs of: (*a*) bacteria (*b*) actino-mycetes and (*c*) fungal hyphae, growing in a chitin enriched sand dune soil under *Pinus nigra laricio*. (*a*) and (*b*) × 2000, (*c*) × 1000. Bacterial cells are also visible in both (*b*) and (*c*).

actinomycetes, which are later attacked by other bacteria.[1] In acid soils, or when some types of organic material resistant to decomposition are added, fungi develop initially, but decomposition is slower than when bacterial development begins. Thus, fungal mycelia can often be seen radiating from pieces of decomposing organic matter, though after a few weeks' decomposition the fungi may be mainly represented by spores.[2] But fungi, being large micro-organisms, can be seen directly in the soil with a suitable microscope and their development followed. W. Kubiena and C. E. Renn,[3] for example, followed the development of different groups of fungi when various substances were added to the soil and showed that the species developing depended on the air supply, so that the fungal flora was not the same inside as on the surface of the soil crumbs.

W. Kubiena[4] has stressed the control which the size of the soil pores has on the development of the soil fungi; typically their spores and fruiting bodies are smaller, and sometimes much smaller, and their hyphae may be finer, in the soil where space is restricted than in pure cultures where there is no such restriction. This need of fungi for relatively large pores in which to grow

1 J. Ziemiecka, *Zbl. Bakt.* II, 1935, **91**, 379.
2 H. L. Jensen, *Proc. Linn. Soc. N.S.W.*, 1934, **59**, 200.
3 *Zbl. Bakt.* II, 1935, **91**, 267.
4 *Micropedology*, Ames, 1938.

should make coarse sandy soils a particularly suitable habitat for them, and in many acid sands they constitute far the greater proportion of micro-organisms present.[1] Also micro-fungi can only form their fruiting bodies in the larger pore spaces and typically cracks, channels and spaces between crumbs.[2]

This dwarfing effect also affects protozoa and diatoms; thus, both he and M. Koffmann[3] found the largest ciliates present in the soil were only 18 to 22 μ in size, whereas they often attain 60 μ in cultures, and it also controls the size of some of the arthropod population, for W. Kühnelt[4] found that soils with large pores contained larger mites than soils with small pores. This control is naturally not effective for those organisms possessing the power of increasing the pore size, such as the burrowing and channelling invertebrates.

The weight of the different organisms in the soil cannot yet be determined accurately, largely because of the great uncertainty in the weight of the individual organisms. A Rothamsted arable soil containing 3 per cent organic matter, or, say, 65 to 70 t/ha of dry organic matter in the top 15 cm of soil, will contain about 3000 million bacteria or similar-sized cells and about 30 000 flagellates and 20 000 amoebae per gram. The dry weight of the bacteria was estimated at up to 700 kg/ha on p. 163, and this will include most of the actinomycetes. The amount of dry matter in the fungal hyphae and spores is not known, but if one assumes that there are about 40 metres of mycelium per gram of soil, as suggested in Table 12.2, and this has an average diameter of 5 μ, there will be about 1700 kg/ha of fungal mycelium, if it is all alive. However, a proportion of this is undoubtedly dead, so the organic matter content of the mycelium will probably be less. But a dry weight of 350 kg/ha is possible, which is comparable to that of the bacterial. A rough calculation gives the weight of protozoa as about 170 kg/ha live weight, or 35 kg/ha dry matter, and Table 11.2, p. 192, suggests that the soil fauna, excluding the protozoa and nematodes, have a dry weight of 100–200 kg/ha. Therefore a very rough calculation indicates that the weight of the dry matter in the soil population in an arable soil may be of the order of 1400 kg/ha, or something between 1 and 2 per cent of the organic matter.

The effect of the energy supply

The general requirements of the soil organisms are the same as those of plants: energy, nutrients, water, suitable temperatures and the absence of harmful conditions. The great difference between them is the source of energy: green plants derive their energy directly from sunlight, and the soil

1 For a forested podsol soil see A. Burges in *Soil Organisms*, 151, Ed. J. Doeksen and J. van der Drift, North Holland, 1963.
2 For a review see D. M. Griffin, *Ann. Rev. Phytopath.*, 1969, **7**, 289.
3 *Arch. Mikrobiol.*, 1934, **5**, 246.
4 Quoted by P. W. Murphy, *J. Soil Sci.*, 1953, **4**, 155.

organisms—apart from the photosynthetic forms—obtain theirs either directly or indirectly from the products of plant metabolism. The size of the soil population is thus controlled by the rate at which energy-containing material synthesised by plants is added to the soil.

The organisms derive their energy in different ways from the added plant tissues; some attack the dead or dying plant tissues, or even the tissues of the living plant, and these form the primary source of energy for the remainder of the population, which is either predaceous on them while alive or else lives on the waste products of their metabolism or on their dead bodies. The number, or the activity, of the micro-organisms is controlled by the amount of energy that can be released by the decomposition of the added organic matter and, no matter how many stages or what organisms are involved in its degradation, only a certain definite amount of energy can be extracted; and this amount cannot exceed the energy set free when the organic matter is completely oxidised.

The energy in the organic matter differs fundamentally from the nutrients in it. Nutrients can be used over and over again by an unending succession of organisms. An atom of nitrogen never loses its value; it might in the course of a single day form part of a fungus, a bacillus which decomposed it, an amoeba which ate the bacillus, and a bacterium which decomposed the dead amoeba; for all these organisms one and the same atom of nitrogen would be a perfectly good nutrient. But energy cannot be used in this way. It is as indestructible as matter, but once transformed to heat, it cannot be used by micro-organisms or any other living things; whatever energy is dissipated by one organism becomes out of reach of the others. It follows, therefore, that no factor affecting the trophic life in the soil which does not add to the stock of energy material can permanently increase the numbers of all the groups; if one group increases, others necessarily decrease.

The supply of energy and the turnover of nutrients thus become the chief factors determining the kinds and numbers of organisms in the soil. An exact measure of the amount of energy available to the soil population in a given weight of soil cannot be given, for this depends on the accessibility of the energy sources to the soil population, their ease of decomposibility, and the soil aeration. For a well-aerated soil, the amount of energy dissipated by the population can be estimated from the total quantities in the organic matter at the beginning and the end of the period, and in any organic matter added during this period. Estimates obtained from analytical data for two of the Broadbalk plots are given for illustration only in Table 12.1.

This great increase in the amount of energy dissipated in the two plots is barely reflected in the numbers of micro-organisms present in the soil, for whereas the farmyard manure plot is dissipating fifteen times as much energy each year as the unmanured, or as the neighbouring plots receiving fertilisers only, yet it has scarcely twice as many bacteria or protozoa, and only about the same number of fungi, as these plots, as can be seen in Table 12.2. Only

TABLE 12.1   Annual energy changes in soil: Broadbalk. Approximate estimates only. (10^9 J/ha/year) .

	Farmyard manure added	No manure added
Added in manure	15	Nil
Added in stubble	2	0·3
Total added	17	0·3
Net loss from soil	Nil	0·5–1
Stored in soil	0·5–1	Nil
Dissipated per annum	16	1
Per day: MJ (10^6 J)	103	6·8
Equivalent to the requirements of	30 men	2 men
The human food grown provides for	5 men	1 man

the protozoa seem to be really dependent on the fertility of the soil, in the sense that their active numbers roughly follow the average wheat yield.

Clearly, the organisms on the farmyard manure plot must be living much more actively than on the other plots, presumably because they pass a smaller proportion of their time in resting stages.

So far as present knowledge goes, the soil organisms are living right up to their income in the matter of nutrients and energy supply. Any increase in the

TABLE 12.2   The effect of long-continued applications of farmyard manure and fertilisers on the numbers of micro-organisms in some Broadbalk plots.[1] Numbers per gram of air-dry soil. Mean of six determinations made at monthly intervals from 20-1-48 to 23-6-48

Treatment	Unmanured	Complete fertiliser	Farmyard manure
Plot number	3	7	2
Number of bacteria			
Total cells in thousand millions	1·6	1·6	2·9
On plates in millions	50	47	67
Number of Fungi			
Pieces of mycelium in millions	0·85	0·94	1·01
Length of mycelia in metres	38	41	47
On plates in millions	0·16	0·26	0·23
Number of Protozoa			
Total in thousands	17	48	72
Active in thousands	10	40	52

1 Unpublished observations of P. C. T. Jones, J. E. Mollison, and F. A. Skinner. I am much indebted to these workers for permission to use their data. The data for protozoa have been published by B. N. Singh, *J. Gen. Microbiol.*, 1949, **3**, 204.

available organic matter capable of supplying energy at once increases the numbers of micro-organisms. Further, it appears to be a fairly general rule that, under natural conditions, the greater the number of soil-inhabiting organisms in a soil, the greater the number of species present, but this generalisation has not yet been adequately tested.

Activity of the soil population

The numbers of micro-organisms in a soil give no direct measure of the activity of the microbial population. The activity of the population is not a concept that can be given a quantitative definition, but for many purposes it can be measured by the amounts of either CO_2 or heat evolved by the population, that is, by the rate at which either the oxidisable carbon compounds or the energy available for organic growth are being dissipated. Neither of these definitions gives a perfect measure of the activity, for there is no necessary connection between the heat and CO_2 evolution by an organism, for the amount of CO_2 produced per joule evolved increases as the oxygen tension decreases because energy-rich materials are then being either stored in or lost from the soil.

In general, the CO_2 evolution increases as the number of bacteria increases. D. Fehér[1] found a close correspondence between the bacterial numbers, as determined by the plating technique, and the CO_2 evolution at different seasons of the year for different soils, and concluded that the activity of the whole population rose and fell with the numbers of bacteria so determined. But H. L. Jensen[2] found the CO_2 evolution was poorly correlated with the bacterial numbers so determined, but was sometimes well correlated with them when determined by the total cell count. The only conditions when the fungi seemed to contribute appreciably to the CO_2 evolution was when there was a vigorous growth of mycelium on buried Cholodny slides, and this occurred for only one or two weeks after fermentable organic matter was added to the soil.

The effect of moisture content

Moisture content affects the activity of the soil population in two different ways: through the thickness of the water films in the soil and consequently its aeration, and through the reduction in free energy of the water as the films become thinner and the soil drier. Bacteria can only move in water films, and although many are smaller than 1 μ in size they only appear to be readily mobile in films appreciably thicker than this. D. M. Griffin and G. Quail,[3] for example, found little movement if the suction of the soil water exceeded

1 *Arch. Mikrobiol.*, 1934, **5**, 421; 1938, **9**, 193.
2 *Proc. Linn. Soc. N.S.W.*, 1936, **61**, 27.
3 *Aust. J. Biol. Sci.*, 1968, **21**, 579.

300 cm corresponding to pores about 5 μ in size. The rate of movement, however, is dependent on the moisture content of the soil at a given water suction, and hence on the soil texture, and becomes slower as the soil becomes drier, presumably because the water film pathway between two points in the soil becomes more tortuous.[1] Filamentous fungi differ from bacteria in that their hyphae need not grow in a continuous water film but can grow across an air space, so we would expect these fungi to be able to grow in drier soils than bacteria; a conclusion reached by Y. Dommergues[2] experimentally.

Soil fungi and bacteria must take up their nutrients from the water films in which they are growing, but these films can be thinner than the organism, even if the organism must maintain a film around its cell walls. The effect of the thickness of the film on the metabolic rate of the organism depends on the source of its food supply. If the supply is from the hydrolysis of a solid insoluble substance, such as a bit of dead root, for example, the water requirement of the organism may be very low, as the humidity of the soil air is high until the soil becomes dry, so its metabolic rate may remain high until the suction of the soil water has reached levels appreciably drier than 15 bar. On the other hand if the nutrient supply is a soluble substance, such as ammonium or nitrite, for example, the rate of diffusion of these ions to the bacterial surface through their water films will become smaller as the water films become thinner, which will automatically reduce their rate of oxidation. Thus, one would expect to find that the production of ammonium from solid organic residues—the process of ammonification—should continue as the soil becomes drier even after the rate of oxidation of ammonium to nitrate has dropped to a very low level—a result that has been observed.[3]

D. M. Griffin[4] reviewed the literature on the effect of soil moisture content on fungal activity, and concluded that the suction of the soil water has little direct effect until it exceeds 15 bar, and many fungal processes are continuing, though at a slower rate, to suctions exceeding 100 bar. A point that does not appear to have received attention, however, is that microbial activity always involve the excretion of byproducts that are toxic, or at least inhibiting to the growth of the organism, and as the soil becomes drier and the water films thinner, these will diffuse away from the organism increasingly slowly. This may be the reason why metabolic activity often decreases with increasing suction, even before the suction has reached very high values.

The effect of high soil moisture contents on microbial activity is through the effect of thick water films on the oxygen supply to the organisms, and the carbon dioxide concentration around them. D. J. Greenwood[5] has emphasised that the micro-environment of a bacterium is either aerobic, if the oxygen

1 Y. A. Hamdi, *Soil Biol. Biochem.*, 1971, **3**, 121.
2 *Ann. Agron.*, 1962, **13**, 265, 391.
3 See, for example, J. B. D. Robinson, *J. agric. Sci.*, 1957, **49**, 100.
4 *Biol. Rev.*, 1963, **38**, 141; *Ann. Rev. Phytopath.*, 1969, **7**, 289.
5 In *The Ecology of Soil Bacteria*, Ed. T. R. G. Gray and D. Parkinson, 1967, Liverpool UP, p. 138.

concentration in the soil atmosphere exceeds 5 μM, corresponding to 1 per cent of the oxygen present in a fully aerobic soil, or anaerobic, due to the efficient terminal oxidases most possess. Many soil fungi also have a cytochrome oxidase system so would not be expected to be sensitive to moderate oxygen deficiency, yet many appear to be so, but this may be due to their sensitivity to CO_2 concentrations, for many aerobic fungi have their activity appreciably affected by quite small increases in the CO_2 concentration in the air or bathing solution;[1] this may also be the reason why many of them are confined to the superficial layers of the soil or to the surface of cracks or channels in the soil.

The effect of temperature

For a given organism, its metabolic activity tends to increase with temperatures up to a certain value and then decrease as the temperature increases beyond this optimum, and in the complex environment of the soil, the number of active organisms also decreases when the temperature has exceeded a certain value, the exact value depending upon local conditions. Different organisms tend to have their maximum activity at different temperatures, and the temperature at which any organism has its maximum metabolic rate may be different from the temperature at which it is most abundant. Also different organisms react differently to given changes of temperature.

The effect of the normal variations of soil temperature on the numbers of bacteria in agricultural soils to which decomposable organic matter has not recently been added is uncertain. H. L. Jensen[2] working with a pasture soil in New South Wales, W. G. E. Eggleton[3] with an English pasture soil and N. James and M. L. Sutherland[4] with a fallow soil at Winnipeg, all failed to show any effect of temperature on bacterial numbers if the associated moisture effects were eliminated. On the other hand, D. W. Cutler and L. M. Crump[5] using a fallow soil at Rothamsted found the number of bacteria tended to decrease with increasing temperatures whilst D. Fehér and M. Franck[6] working with Hungarian pasture and forest soils found they increased.

H. L. Jensen[7] also investigated the effect of soil temperature on the relation between the rate of CO_2 evolution by a soil and the number of bacteria it contained, and showed that as the temperature rose the CO_2 evolved per 1000 million bacteria, as determined by the direct cell count, increased from about 16 per cent of their dry weight per day at 15°C to 30 per cent at 28°C.

1 See, for example, A. Burges and E. Fenton, *Trans. Brit. mycol. Soc.*, 1953, **36**, 104.
2 *Proc. Linn. Soc. N.S.W.*, 1934, **59**, 101.
3 *Soil Sci.*, 1938, **46**, 35.
4 *Canad. J. Res.*, 1940, **18** C, 435.
5 *Problems in Soil Microbiology*, Longman, 1935.
6 *Arch. Mikrobiol.*, 1933, **4**, 447; 1937, **8**, 249; 1938, **9**, 193.
7 *Proc. Linn. Soc. N.S.W.*, 1936, **61**, 27.

and to 50 per cent at 37°C; while if plate counts were used as the basis there was no clearly marked temperature dependence, and for some soils the CO_2 evolution per day would be over sixty times the dry weight of the organism, which would correspond to higher rates than observed for young cultures of active bacteria growing under optimal condition. He considered the fungi were inactive under these conditions.

When decomposable organic matter was added to the soil, Jensen found that the fungi, as measured by the abundance of mycelia on Cholodny slides, came into prominence, and the density of mycelium was correlated with the CO_2 production; in fact, the multiple correlation of this and the total cell count for bacteria with CO_2 production was about 0·8. At low temperatures, 4° to 7°C, the importance of the fungi was small compared with the bacteria, at 14° to 16°C it was about equal to that of the bacteria, at 28°C it was considerably larger, and at 37°C it was equal or less; but in all these cases this was only true for the first week of decomposition, after that the fungal activity fell off rapidly compared with the bacterial.

Biochemical processes brought about by soil micro-organisms

It is often important to study biochemical reactions taking place in the soil under conditions when one is not concerned with the actual species of micro-organisms responsible for them. Thus it is of great importance to study the rate at which compounds potentially toxic to the soil population or to plants are decomposed in the soil, and these include many present-day herbicides and insecticides. Again, the actual compounds in which many inorganic elements occur is of great importance for plant growth, such as those containing inorganic nitrogen, sulphur, manganese and iron, and these also are closely controlled by the soil micro-organisms.

Two general techniques have been developed to study the ability of the soil population to carry out certain chemical and biochemical reactions, and the detailed chemical or enzyme processes involved. The more powerful of the two, and particularly useful if the substances concerned are water-soluble, is the perfusion technique, in which a solution is repeatedly percolated through a column of soil, usually under well-aerated conditions, and the solution is analysed at suitable intervals to study the rate of progress of the relevant chemical reactions. This is very suitable, for example, for studying the rate of oxidation of an ammonium compound to nitrate, or the rate of detoxification of a herbicide. The other technique is the Warburg respirometer, in which a small volume of soil is placed in a respirometer and its rate of oxygen uptake and carbon dioxide production is measured.

A study of the rate of oxidation of ammonium to nitrate in the soil perfusion apparatus will illustrate the type of information that is obtained using these techniques. If a dilute ammonium solution is percolated through a soil, the rate at which nitrate is produced usually increases for a period of days, and then reaches a maximum value, and this rate will be maintained provided other soil conditions, such as the calcium status in this example, are maintained. If now a small amount of a substance which inhibits this oxidation is introduced, such as ethyl urethane, oxidation is much reduced or ceases, and will not begin for a period of days, weeks or months, but it will then increase again to its original value. If more urethane is added, still keeping it dilute, its effect will be short-lived, or even absent, and it is now possible to increase slowly the concentration of added urethane without the ammonium oxidation being inhibited, and it thus becomes possible to detoxify a relatively high concentration of urethane, which, if used initially, would have inhibited the oxidation of ammonium for a very long time.[1] Instead of urethane, nitrite could have been used to show this last effect, for soils can be conditioned to oxidise concentrations of nitrite which would have been toxic to the bacteria oxidising them if used initially at that concentration.

An experiment such as this brings out three points of importance: there is a certain maximum rate at which a given amount of soil can carry out a chemical change; it takes the soil a period of time to decompose an inhibitor of a biochemical reaction, but later on its addition has little if any effect on the reaction; and soils can be adapted to decompose toxic substances at concentrations, which if used initially would entirely prevent the reaction from taking place.

Quastel interpreted the first of these results as showing that the microorganisms can only occupy a limited number of spots on the surfaces of the soil particles, and once they have multiplied up and occupied these spots, they cease to multiply, except to make good losses due to other organisms grazing on them. The organisms are fairly firmly anchored to these spots, as their cells do not appear in the percolating solution, unless the soil is low in clay, such as a sand, for example,[2] and the spots are fairly specific to the organism. Alternatively, the maximum rate of oxidation could be determined by some other factor, such as the rate that oxygen can diffuse to the surface of the bacteria.[3]

Quastel[4] also considers that in the enriched soil the organisms carrying out the oxidation are in a resting not a multiplying phase, for once the maximum rate of oxidation has been reached, it is not affected by adding a growth inhibitor, such as sulphanilamide, to the solution. Unfortunately it is not certain that his explanation of the results is valid. Thus, D. J. Greenwood and H. Lees[5] found that if they percolated an amino acid through a soil, only

1 J. H. Quastel and P. G. Scholefield, *Appl. Microbiol.*, 1953, **1**, 282.
2 J. H. Quastel and P. G. Scholefield, *Soil Sci.*, 1953, **75**, 279.
3 D. J. Greenwood and H. Lees, *Pl. Soil*, 1960, **12**, 175.
4 *Proc. Roy. Soc.*, 1955, **143** B, 159.
5 *Pl. Soil*, 1960, **12**, 175.

about 40 per cent of the carbon appeared as CO_2 when the amino acid was being oxidised at the maximum rate, the remainder being converted probably to a protein which accumulated in the soil, and which could be microbial protoplasm.

The interpretation of the second result is that when an enzyme inhibitor is introduced into the soil, it may take weeks or months for a large population of an organism which can decompose it to build up. What is not known is whether these organisms were initially present in the soil in very low numbers as resting cells, so it takes a fairly long time for them to multiply up, or whether a micro-organism already present in the soil undergoes a mutation which enables it to carry out this decomposition. Under some conditions it would seem that the latter is probably correct, because of the very long lag period that is sometimes found. What is known is that the new organisms may be extremely specific for the particular chemical introduced. Thus in the example given above, the soil which will decompose ethyl urethane after an interval of time will not decompose methyl or propyl urethane. The interpretation of the third result is that organisms can be trained to become tolerant of a poison by conditioning them to relatively low concentrations which are not fully toxic, and gradually increasing the concentration. This effect is well known in conditioning bacteria to become tolerant to antibiotics and insects to insecticides.

The perfusion technique has proved very valuable for studying the biochemical pathways of microbial oxidations,[1] such as for example the intermediates involved in the oxidation of sulphides to sulphates through thiosulphates,[2] and for showing that oxidations which could in theory be brought about by chemical processes are in fact brought about by micro-organisms, such as the oxidation of manganous ions to manganese dioxide and related compounds.[3] It can also show how far a given oxidation is specific to a given group of organisms. Thus, D. H. Greenwood and H. Lees[4] showed that a soil enriched to oxidise any one amino acid at its maximum rate was not enriched to oxidise any other at its maximum rate.

As a general rule, soils contain micro-organisms capable of converting all the normal inorganic constituents to that state of oxidation or reduction which would be predicted if they are to be in equilibrium with the oxidation-reduction potential of that soil. Typical reversible oxidation-reduction processes involving inorganic substances in the soil are the system nitrate-nitrite, sulphate-sulphide, manganese dioxide-manganic ions, and ferric-ferrous ions. There are naturally other oxidation and reduction processes going on, such as the oxidation of ammonium to nitrite, the reduction of nitrite to nitrous oxide or nitrogen gas, and the production of hydrogen gas and

1 See, for example, J. H. Quastel, *Proc. Roy. Soc.* B, 1955, **143**, 159.
2 H. Gleen and J. H. Quastel, *Appl. Microbiol.*, 1953, **1**, 70.
3 P. J. G. Mann and J. H. Quastel, *Nature*, 1946, **158**, 154.
4 *Pl. Soil*, 1960, **12**, 175.

methane in very poorly aerated soils; and some of the oxidations and reductions may not be straightforward because of the production of organic substances which themselves may bring about a reduction process, and this
applies particularly to the ferric-ferrous system. These reactions are discussed in more detail in Chapter 25.

The organisms responsible for these oxidations and reductions are usually
bacteria, and certainly in some examples the species carrying out the oxidation are different from these carrying out the reductions. In general the
oxidations are carried out by specialised autotrophs whilst the reductions are
brought about by unspecialised heterotrophs. Correspondingly an oxidation
may involve several organisms: the oxidation of ammonium to nitrate involves two, and the oxidation of sulphur or even sulphides to sulphates
several.

Soil enzymes

It is a well-known fact that soils contain enzymes capable of carrying out a
number of simple but important biochemical reactions. They contain, for
example, urease which hydrolyses urea to ammonium carbonates, catalases
and peroxidases which prevent the build-up of peroxides, phosphatases which
hydrolyse polyphosphates to ortho-phosphate and sulphatases which hydrolyse organic sulphate esters to an alcohol and inorganic sulphate.[1] Unfortunately, it has not proved possible to separate the enzymes from the soil, so
the whole subject has been complicated by the difficulty of separating chemical reactions brought about by soil micro-organisms from those brought
about by enzymes not forming part of the living microbial cells. Free enzymes
could be present either on the cell walls of micro-organisms or adsorbed on
to the surfaces of soil particles, for clays will absorb them very strongly since
they are proteins. Very recently R. G. Burns, M. H. El-Sayed and A. D.
McLaren[2] have separated a urease-active organic fraction containing no clay
particles from a soil, which has the interesting property that it is not inactivated by the proteolytic enzyme pronase; and they interpret this as showing
that the enzyme occurs in porous organic particles whose pores are wide
enough to admit urea and ammonium ions freely, but are too fine to admit
pronase.

K. N. Paulson and L. T. Kurtz,[3] however, have provided indirect evidence
that in the case of urease activity, up to 90 per cent may be due to urease
adsorbed on clay particles, but that the microbial proportion could be temporarily increased by added additional urea to the soil. In so far as the soil
enzymes are adsorbed on clay particles, the environment of the adsorbed

---

1 For a list of enzymes that have been recognised in the soil, see J. J. Skujins in *Soil Biochemistry*,
  Ed. A. D. McLaren and G. H. Peterson, Arnold, 1967.
2 *Soil Biol. Biochem.*, 1972, **4**, 107.
3 *Proc. Soil Sci. Soc. Amer.*, 1969, **33**, 897; 1970, **34**, 70.

enzyme on the molecular scale is affected by the clay, as described on p. 122. Thus, the effective pH close to the enzyme can be quite different from the pH of the soil as usually measured, so the apparent pH dependence of a soil enzyme reaction can differ appreciably from that of the enzyme in solution.

Soil enzymes are probably mostly produced by the soil micro-organisms and may be excreted into the soil solution by the living organism or be released into the solution upon the death and lysis of the microbial cell. The more active the soil population, the higher the level of the principal enzymes in the soil. Thus regular additions of farmyard manure to a soil, or the use of rotations that raise its humus level, will also increase the amount of the various enzymes in it.[1] It is possible that some members of the soil fauna and some plant roots may release enzymes into the soil, but it is difficult to be certain that these enzymes have not been produced by micro-organisms in the animal's gut or in the root's rhizosphere.

Soil enzyme activity is measured when the soil has been sterilised in such a way that the enzymes have not been inactivated; but this raises great difficulties in practice. Most enzymological studies have been made on soil treated with toluene, as this certainly reduces microbial activity very strongly, though it is unlikely to kill all the soil organisms. Other methods have included the use of heat, because the soil enzymes tend to be more stable to heat than the soil micro-organisms, and the use of gamma irradiation. But none of these methods can be considered absolutely satisfactory.

Dissolution of soil minerals by microbial activity

Microbial activity in a soil can cause the decomposition of some soil minerals, and for a long time it has been considered probable that some bacteria in the rhizosphere of certain crops can solubilise rock phosphate,[2] so allowing the crop to use this source of phosphate which it would be unable to do in the absence of the organism. Again, many research workers have obtained evidence that some organisms can decompose certain silicate minerals, releasing soluble silica into the soil solution. The mechanism which causes these dissolutions are organic acids which are excreted by the organisms and which can chelate calcium and other polyvalent metals. Citric and tartaric acids are probably the more important acids excreted by fungi, and lactic and 2-keto-gluconic acids by bacteria and actinomycetes; and certainly substrates rich in sugars or polysaccharides encourage the excretion of these acids by bacteria.

D. M. Webley and his colleagues[3] have shown that some strains of the

---

1 For an example see S. U. Khan, *Soil Biol. Biochem.*, 1970, **2**, 137.
2 F. C. Gerretsen, *Pl. Soil*, 1948, **1**, 51; R. J. Swaby and J. Sherber in *Nutrition of Legumes*, Butterworth, 1958, p. 289.
3 With R. B. Duff, *Pl. Soil*, 1965, **22**, 307, with R. B. Duff and R. O. Scott, *Soil Sci.*, 1963, **95**, 105, with M. E. K. Henderson and I. F. Taylor, *J. Soil Sci.*, 1963, **14**, 102.

bacterium *Pseudomonas fluorescens* will produce appreciable amounts of 2-ketogluconic acid, and that it appears to be commonly found in the rhizosphere and on the surface of the roots of crops. Thus, they found 1 per cent of the bacteria in these two regions of barley roots were of this type, whereas only about 0·02 per cent of the soil bacteria were. They also found that a large proportion of the micro-organisms colonising weathering rock surfaces were able to liberate silica from the minerals. Some of the organisms liberating phosphate from mineral phosphates appear to set free more inorganic phosphates than they need for their own use, as these authors found that usually less than 20 per cent of the phosphate solubilised was present as an organic, possibly as a sugar, phosphate.

The break-down of toxic chemicals in the soil

It has long been known that soils can acquire the ability to decompose compounds not normally present in them, but which may be toxic to the microbial population if used in high enough concentration. Thus soils can acquire the ability to oxidise phenol, cresols and hydrocarbons.[1] But interest has recently centred round the ability of the soil to decompose the whole host of new synthetic substances used as herbicides and insecticides. These substances have been selected because of their ability to affect plant or animal life at very great dilutions, and their continued use in agriculture would be extremely dangerous if they persisted in the soil for long periods unchanged. The kind of result found for the decomposition of the urethane applies equally to the herbicides. Thus, a soil which has acquired the ability to decompose the herbicide 2-4-D (2, 4, dichlorophenoxyacetic acid) will usually decompose MCPA (2, methyl 4, chlorophenoxyacetic acid) but not 2, 4, 5 T (2, 4, 5 trichlorophenoxyacetic acid).[2] Again some of the organisms which decompose DNOC (2, 4 dinitro-ortho-cresol) decompose any nitrophenol in which the nitro group is in the para-position, such as p-nitrophenol, 2, 4 dinitrophenol or 2, 4, 6 trinitrophenol, but they cannot decompose m-nitrophenols or 2, 5- or 2, 3 dinitrophenols.[3] This is a partial example of the general finding that benzene rings with substitution in the ortho- or para-position are more readily degraded than those with substitution in the meta-.

Some compounds have a constitution that is so resistant to microbial attack that they will last for very long periods in the soil. Thus, the herbicide CMU ($Cl.C_6H_4.NH.CO.N(CH_3)_2$) and other substituted ureas are fairly persistent, and chlorinated hydrocarbon insecticides such as DDT and BHC

1 See, for example, R. Wagner, *Ztschr. Gärungsphysiol.*, 1914, **4**, 289, and P. H. H. Gray and H. G. Thornton, *Zbl. Bakt. Abt.* II, 1928, **73**, 74.
2 L. J. Audus, *Pl. Soil*, 1951, **3**, 170.
3 H. L. Jensen and K. Gundersen, *Nature*, 1955, **175**, 341.

(γ-hexachlorobenzene) are very persistent and may remain in appreciable amounts in the soil for four to six years after commercial dressings,[1] and in the case of BHC it may cause a taint in potatoes, tobacco and other crops grown in the soil throughout this period. However, a substance may be very resistant to decomposition and so persistent in the soil provided it remains aerobic, but may lose its persistence if the soil becomes anaerobic periodically.

In all the examples which have been studied of the decomposition of organic compounds which are toxic to most of the soil organisms when they are added to a soil, the same general conclusion has been reached on the build-up of ability to decompose these active substances. The organisms responsible for the decomposition, in so far as they have been isolated, are sometimes bacteria and sometimes fungi. The substance must initially be present at a low enough concentration not to kill the majority of the soil organisms. It may remain in the soil undecomposed for a period of months or years, which is usually shorter in soils having a high than a low microbial activity, for example, in surface soils than in subsoils.[2] The substance then begins to decompose, sometimes to another substance also very difficult to decompose but not necessarily toxic and after a time it, too, decomposes.[3] Further additions decompose more quickly, and the concentration of the active substance can be slowly increased without affecting the rate of decomposition, until a concentration is reached which would have been definitely toxic to the soil population if it had been added initially.

Symbiotic and antibiotic relations between the microflora

The various groups of soil organisms do not live independently of each other, and many have developed a series of symbiotic relations with their neighbours. It seems to be a characteristic of all micro-organisms that they excrete complex organic compounds when growing actively, usually only in very small amounts, and much of the symbiotic and antibiotic relations take place through the action of these substances on other organisms in their neighbourhood. These relations are, however, rarely very specific, though it is possible certain Basidiomycetes need the exudate of a particular bacterium if they are to produce fruiting bodies. Thus, W. A. Hayes[4] has shown the cultivated mushroom *Agaricus bisporus* will only form fruiting bodies in the presence of a bacterium or a group of bacteria similar to *Pseudomonas putida*,

1 N. Allen, R. L. Walker *et al.*, *U.S.D.A. Tech. Bull.* 1090, 1954; L. W. Jones, *Soil Sci.*, 1952, **73**, 237, and J. E. Dudley, J. B. Landis and W. A. Shands, *U.S.D.A. Farmers Bull.* 2040, 1952.
2 A. S. Newman, J. R. Thomas and R. L. Walker, *Proc. Soil Sci. Soc. Amer.*, 1952, **16**, 21. For a more recent discussion see C. A. Edwards, *Soils Fert.*, 1964, **27**, 451, and I. J. Graham Bryce, *Roy. Inst. Chem. Rev.*, 1970, **3**, 87.
3 L. J. Audus and K. V. Symonds (*Ann. Appl. Biol.*, 1955, **42**, 174) give an example where the intermediate is phytotoxic.
4 With P. E. Randle and F. T. Last, *Ann. appl. Biol.*, 1969, **64**, 177.

and that the function of the casing that must be spread over the mushroom compost is to allow these bacteria to multiply in it. The chemical agent that the bacteria excrete has not yet been identified, though it may be a steroid.

There are several probably non-specific types of beneficial interactions which may occur in soils. In the first place, a natural soil contains a wide range of micro-habitats, having a range of oxidation-reduction potentials when wet, even if well drained. A number of aerobic bacteria are known to produce peroxidases and catalases which prevent the build-up of peroxides, which are strong inhibitors of anaerobic bacteria. It is therefore likely that some anaerobic or micro-aerophilic bacteria can grow in the neighbourhood of these aerobes. In the second place, two organisms may be living together because one needs an excretion product such as an amino acid or growth factor of another. Again, two organisms living together can effect a decomposition neither can do when growing alone, and although there are few if any examples of this in the soil, there are numerous examples of two soil organisms carrying out a decomposition more quickly when growing together than when alone.[1]

Other soil organisms develop antibiotic relations with selected species,[2] which are brought about by their excreting substances into the soil solution that prevent the affected organisms from growing or that kill them by lysis, that is, by dissolving their outer membrane. The fungus *Penicillium notatum*, which excretes penicillin, is such an organism, but in the soil the fungus *Trichoderma viride* is a much more important example, for it excretes two antibiotics—gliotoxin and viridin—having very powerful effects on the soil population and particularly on plant pathogens living in the soil.

The ability to produce antibiotics is not confined to fungi, for it is also strongly developed by many actinomycetes and bacteria. At least half the soil fungi and actinomycetes probably produce such substances in the laboratory under suitable conditions,[3] but the proportion of bacteria producing them is probably much smaller. The substances so far recognised usually act only on other members of these three groups, but there are some bacteria, such as *Pseudomonas aeruginosa* and *Ps. pyocyanea*, which produce substances that can kill species of soil protozoa belonging to the amoebae, flagellates and ciliates.[4] These substances have a very varied chemical composition and are often named after the organism producing it, as, for example, penicillin, mentioned above. But just because these organisms can

1 See, for example, A. G. Norman, *Ann. appl. Biol.*, 1930, **17**, 575, and S. A. Waksman and I. J. Hutchings, *Soil Sci.*, 1937, **43**, 77. For an example when one bacterium is a cellulose decomposer, see L. Enebo, *Nature*, 1949, **163**, 805.
2 S. A. Waksman, *Soil Sci.*, 1937, **43**, 51, and *Bact. Rev.*, 1941, **5**, 231, has given full historical reviews of the development of these ideas and has described many examples of this antibiosis. See also his book *Microbial Antagonisms and Antibiotic Substances*, 2nd edn., New York, 1947.
3 R. L. Emerson *et al.*, *J. Bact.*, 1946, **52**, 357.
4 B. N. Singh, *Nature*, 1942, **149**, 168; *Brit. J. Expt. Path.*, 1945, **26**, 316.

produce antibiotics when grown on suitable media in the laboratory, it does not follow either that they produce them in the soil, or if they do, that their concentration in the soil solution is high enough to have any pronounced effect on the neighbouring micro-organisms, particularly since there are many soil organisms which decompose them in dilute solution. The zone of action of these antibiotics is therefore confined to a region very close to the actively excreting organisms, and since these are often confined to pockets containing pieces of decaying organic material which forms a readily available source of energy for fungi, it is likely that concentrations of antibiotics sufficiently high to affect an appreciable volume of soil only occur in such pockets.

It has been found very difficult to assay the amount of any particular antibiotic in the soil, so little is known about the conditions favouring their production, but it is probable that production by an organism is greatest when it is growing rapidly on a suitable food supply. These antibiotic substances may collect in the subsoil, for A. S. Newman and A. G. Norman[1] found that the microflora of an arable soil below the depth of cultivation responded to additions of fermentable material much more slowly than did those in the surface soil, and this response was not increased by inoculating the subsoil with some surface soil.

Crops may perhaps sometimes suffer from these substances. Thus, R. A. Steinberg[2] has shown that tobacco plants show morphological deformities if grown near some strains of non-pathogenic bacteria in otherwise aseptic conditions, and this was due to soluble organic compounds diffusing from the bacterial cells; and he could reproduce these symptoms by allowing a very low concentration of some organic compounds, such as the amino acid isoleucine, to build up around the roots.[3] Now these symptoms are typical of 'frenching' in tobacco, a disease not apparently due either to a nutrient deficiency or to a pathogenic organism. Hence, he suggested that the cause of frenching in the field is due to too high a concentration of diffusible microbial products accumulating in the soil near the tobacco roots.

Bacteria can also be parasitic on soil fungi. Cholodny slides buried in soil repeatedly show portions of fungal mycelia being attacked by bacteria: one end of the mycelium appears free from bacteria, then comes a large concentration of small rods and cocci on what appears to be almost empty mycelium, and behind this the original line of the mycelium can be followed by colonies of bacteria outlining the old wall.[4] The interpretation of this observation is not clear. P. C. T. Jones[5] showed that even if the fungi were growing in the absence of bacteria the main fungal protoplasm appeared to keep in the

1 *Soil Sci.*, 1943, **55**, 377.
2 *J. Agric. Res.*, 1947, **75**, 199.
3 *Science*, 1946, **103**, 329; *J. Agric. Res.*, 1947, **75**, 81.
4 See, for example, D. M. Novogrudsky, *Mikrobiologia*, 1948, **17**, 28.
5 I am indebted to Dr Jones for this information.

growing parts of the mycelium, leaving the back parts almost empty, and it may be that the bacteria are simply attacking these nearly empty regions. On the other hand, R. L. Starkey[1] considered that the fungal tissue on which the bacteria were living could not provide enough food for the development of all these bacteria, and assumed they can extract nutrients from the living mycelium before it is killed. Bacteria may, therefore, play an important role in limiting the amount of fungal mycelium in a soil, though their importance cannot yet be assessed as there are so many organisms in the soil that can feed on fungal mycelium; and, in fact, it seems that the life of any long piece of mycelium in the soil must be fairly short. These observations only apply to Phycomycete and other fast-growing hyphae and not to the slow-growing rhizomorphs of basidiomycetes.

These interrelations between the various members of the microflora may be of considerable importance in the control of plant pathogens in the soil, for the growth of some pathogenic fungi can be inhibited by some common soil fungi and bacteria. R. Weindling[2] has claimed that the soil fungus *Trichoderma viride* can control the damping-off of citrus seedlings by a *Rhizoctonia*, and A. Lal[3] that it can control *Ophiobolus graminis*, the fungus causing 'take-all' in wheat and barley. J. P. Chadiakov[4] has also claimed that bacteria of the genera *Pseudomonas* and *Achromobacter* can control some *Fusaria*, causing a wilt of flax.

The great practical difficulty is to find means of encouraging these desirable fungi or bacteria to multiply, for they can only control the pathogenic organisms effectively if they are sufficiently active in the soil, that is, if they have an adequate food supply. It cannot usually be done by inoculating the desirable organism into the soil, as it will not be able to compete with the indigenous population for food, nor, in general, will adding an inoculum of a desirable organism already present help, for again the extra numbers of the organism will not usually be able to find enough food to keep them active. However, C. H. Meredith[5] has reported what appears to be an exception to this rule, for he claimed to have reduced the severity of panama disease in bananas, caused by the fungus *Fusarium oxysporum* var. *cubense* by adding to the soil an inoculum of an Actinomycete derived from the soil itself which is antagonistic to this fungus.

The only way desirable organisms can have their activity increased is by altering the soil conditions. This can be done in two ways: either by adding organic matter and increasing the activity of the whole saprophytic population, and this will usually increase the activity of organisms that are antago-

1 *Soil Sci.*, 1938, **45**, 207.
2 *Phytopath.*, 1932, **22**, 837; 1934, **24**, 1153; 1936, **26**, 1068; with H. S. Fawcett, *Hilgardia*, 1936, **10**, 1. He called the fungus *T. lignorum*.
3 *Ann. Appl. Biol.*, 1939, **26**, 247.
4 *Mikrobiologia*, 1935, **4**, 193.
5 *Phytopath.*, 1946, **36**, 983.

nistic to specialised plant pathogens; or by altering the soil conditions in such a way that the desirable organisms can obtain a larger share of the available food supply. Thus, the fungus causing root rot in cotton, *Phymatotrichum omnivorum*, can be kept in control in irrigated Arizona fields by adding 25 to 40 tons per hectare of manure or other decomposable organic matter annually to the soil well before the cotton is planted.[1] Weindling and Fawcett showed that the *Rhizoctonia* fungus causing damping-off of citrus seedlings could be controlled in the field by *Trichoderma viride* if the soil was made sufficiently acid for the *Trichoderma* and similar fungi to multiply at the expense of the majority of the soil population. And the cotton root rot fungus can be attacked by saprophytic fungi in the roots of the cotton plant after harvest if the cotton plant is then cut below the crown; for this injury allows saprophytic fungi to enter the roots rapidly and destroy the root rot fungus in the process of attacking the root cells themselves.[2] Some further methods of controlling fungi attacking plant roots will be described later on pp. 246 *et seq.*

The problem is in some ways easier when the pathogenic fungus can only live in plant tissue, and is carried over in the soil in infected tissue, for anything that hastens the decomposition of the plant remains in the soil hastens the killing-out of the fungus. However, although the problem is easy in theory, it is difficult to find suitable methods in practice because the infected plant remains are usually resistant to decomposition. As an example, *Ophiobolus graminis*, the cause of take-all in wheat and barley, is carried in the soil in infected straw, and it is not possible to get all the stubble decomposed between harvest-time in August and the beginning of the next growing season in March. But it may be possible to help the soil organisms in their work of destroying the pathogen if its resistance can be weakened in any way. In the particular case of *Ophiobolus*, S. D. Garrett[3] has shown this can be done by depleting the soil of available nitrogen directly after harvest; for the fungus, which is growing in the nitrogen-poor straw, appears to need some soil nitrogen during this period. Hence, any device which does this, such as under-sowing the corn with Italian rye-grass or trefoil (*Medicago lupulina*), for young trefoil takes up soil nitrogen in autumn, will weaken the fungus.[4] When the rye-grass or trefoil is ploughed in in the spring, the readily available nitrogen in the green manure is released as nitrate which the young barley crop takes up. Since the resistance of corn to damage by *Ophiobolus* is increased as the general level of fertility, and of available nitrogen in particular, is increased, trefoil is preferable to Italian rye-grass as the under-sown crop.

1 C. J. King, *U.S. Dept. Agric.*, *Circ.* 425, 1937; for corresponding results in Texas, see R. B. Mitchell *et al.*, *J. Agric. Res.*, 1941, **63**, 535.
2 F. E. Clark, *U.S. Dept. Agric.*, *Tech. Bull.* 835, 1942.
3 *Ann. Appl. Biol.*, 1944, **31**, 186; 1948, **35**, 435.
4 For an agricultural example from the English chalk lands, see R. Sylvester, *Agric.*, 1947, **54**, 422, and S. D. Garrett, ibid., 425.

Interactions between the soil microflora and fauna

The standard example of the relation between the soil's fauna and microflora
is that between the protozoa and bacteria, though B. N. Singh's work has
shown that other protozoal-like organisms, such as myxobacteria, Acrasieae,
giant amoeboid organisms and presumably Myxomycetes all feed on bacteria
as well. Bacteria, however, vary in their edibility to these different organisms:
some bacteria appear to be readily eaten by them all, some by only a few, and
a few appear to be inedible to all the organisms tested—these latter mainly
being pigmented bacteria which may secrete strong antibiotics. Thus, out of
eighty-seven different bacterial species of various origin tested for edibility
by five groups of these organisms, a soil amoeba, a myxamoeba, a giant
amoeba and two myxobacteria, twelve were eaten by all five groups, twenty-
five were eaten by four, seventeen by three, thirteen by two, thirteen by one
group only, and seven, confined to coloured bacteria most of which produced a red or a pink pigment, were not eaten by any.[1] Thus, the great
majority of soil bacteria can be eaten by at least one group of these bacterial
feeders known to be present in the soil.[2]

The soil fauna can affect the microflora in several ways: the larger saprophytic soil animals, by being motile, distribute both the decaying organic
matter and some of the microbial population throughout the soil layer in
which they are working; for they will be ingesting food at one place, mixing
it with bacteria in their gut, and excreting it at another. In fact, this process of
comminuting and distributing dead plant litter throughout the surface layers
of the soil is the predominant action of the larger soil invertebrates on the
microbial population.

Animals feeding on decaying organic matter, and particularly those that
ingest soil at the same time, also affect the composition of the microbial
population, for these animals will always be taking into their gut the micro-
organisms attached to the plant residues or soil ingested, and be subjecting
them to the action of their digestive juices, though whether they can in fact
affect the composition of the microbial population is not known. It is possible that earthworms can depress fungal development in the soil by feeding
on fungal mycelia, for it is an interesting fact, of unproven generality, that
fungal mycelia only seem to develop abundantly in those soils of the temperate regions where earthworms are scarce, though more experimental work
is needed to establish the validity and interpretation of this generalisation.
On the other hand, mites graze on fungal spores and are probably the principal organisms removing such spores from the humus horizons in coniferous
forests.

There appears to be another interaction of importance, this time between
the soil invertebrates and the protozoa, for some of the saprophytic inverte-

1 F. J. Anscombe and B. N. Singh, *Nature*, 1948, **161**, 140.
2 For a review, see H. G. Thornton and L. M. Crump, *Rothamsted Ann. Rept.* 1952, 164.

brates carry a protozoal population in their gut to help digest some of the more resistant plant products, such as cellulose, which the animal appears to feed on. Thus, the digestive tracts or organs of some soil invertebrates constitute an important environment of the soil protozoa—an environment of presumably great activity.[1]

Effect of drying, heating or sterilising a soil

It has been known from time immemorial in India that heating a soil, either by lightly burning it or by burning stubble on it, will increase the yield of the following crop. This practice is known as *rab* and is mentioned in the Vedas,[2] and also by Vergil,[3] who was aware that the benefit of this treatment was very short lived. A similar practice in areas having an intense dry season is to leave the bare soil exposed to the sun, so it gets baked out. This is again practised in India and in Egypt, where it is known as *Sheraqi*.[4] Some investigations on these effects were made in the early 1920s,[5] but only recently have they received the attention they deserve because of their great importance in tropical agriculture.[6]

Drying a soil, or heating it and re-wetting it, results in a flush of decomposition of the soil humus, and a flush of ammonium and then of nitrates, provided the soil has been inoculated with a little fresh soil before re-wetting; and this flush lasts for five to ten days. The more intensely the soil is dried, the higher the temperature it is heated to and the longer the soil is kept dry, and the higher the organic matter content of the soil, the larger is the magnitude of the flush. Further, a soil can be dried and wetted a number of times, and on each re-wetting another flush occurs of only slightly less magnitude than the previous one, except that the first flush is usually appreciably greater than the subsequent ones. As a rough guide, if a soil is dried to 100°C about 1 to 2 per cent of the organic matter is decomposed at each flush. Freezing and thawing a soil can also give a flush of decomposition[7] as can killing all or most of the soil organisms by treating the soil with a sterilising agent such as chloroform or methyl bromide.[8] The organic matter oxidised in each flush is derived partly from the oxidation of humic substances desorbed from the

1 For an account of this, see E. A. Steinhaus, *Insect Microbiology*, Ithaca, N.Y., 1946.
2 'Like to a tender plant whose roots are fed,
  On soil o'er which devouring flames have spread '.
  Stories of the Buddha's Former Births, trans. H. L. Francis.
3 *Georgics*, Bk. 1, 84–93.
4 See, for example, J. A. Prescott, *J. agric. Sci.*, 1919, **9**, 216; 1920, **10**, 177.
5 See, for example, A. F. Gustafson, *Soil Sci.*, 1922, **13**, 173, and A. N. Lebendiantzev, *Soil Sci.*, 1924, **18**, 419.
6 See, for example, H. F. Birch, *Pl. Soil*, 1958, **10**, 9; 1959, **11**, 262; 1960, **12**, 81; and *Trop. Agric.* (*Trin.*), 1960, **37**, 3, from whose papers the data in the next paragraph are taken.
7 See, for example, A. R. Mack, *Canad. J. Soil Sci.*, 1963, **43**, 316, and D. E. Harding and D. J. Ross, *J. Sci. Fd. Agric.*, 1964, **15**, 829.
8 D. S. Jenkinson, *J. Soil Sci.*, 1966, **17**, 280.

soil and rendered soluble by the drying process[1] and partly from the microbial cells killed by the drying process or the chemical sterilant.

Partial sterilisation of soils

It is common practice for horticulturalists growing crops intensively in glasshouses to partially sterilise their soils at frequent intervals to kill any pathogens in the soil that can attack the roots of their crops. The commonest method in commercial use is to blow low pressure steam into the soil for a short time, then leave the soil for a few weeks, often after a good flooding, because if a crop is planted directly after steaming it either grows very badly or not at all, while after the short rest it makes excellent growth. Partial sterilisation is also carried out by incorporating volatile fumigants such as methyl bromide into the soil, covering the soil surface with polythene sheeting for a while, and then allowing the fumigant to volatilise away completely before planting the next crop.

Not all the consequences of partial sterilisation are understood. If steam sterilisation is used, there is a production of ammonium and of soluble organic matter including root toxins,[2] of which the chemical constitution has not yet been determined, a flush of organic matter decomposition giving a further production of ammonium ions and, if the soil is at all acid, of readily extractable manganese[3]—a result sometimes found when a soil is dried and then wetted. Most of the root pathogenic fungi are fairly sensitive to moist heat and are killed if the soil temperature exceeds 60°C, but many glasshouse crops make poorer growth in soil heated to 80° than to 60°, possibly due in part to the higher manganese and higher nitrite found in soils heated above 65 to 70°.[4] Its effect on the microbial population is an initial reduction in all groups, but later a rapid build-up of bacteria, sometimes including nitrifiers, and a slower build-up of protozoa; but the fungal population remains depressed for up to twelve or eighteen months and it cannot be built up more quickly by inoculating fungi into the soil.[5] The cause of this is not yet known.

Chemical sterilisation or fumigation differs from steam sterilisation in many ways. It initially reduces the microbial population, and afterwards the bacterial population builds up much more quickly than fungal. The effect on the nitrifiers depends on the chemical—methyl bromide, for example, markedly reducing bacterial activity, while ethylene dibromide has little

1 H. F. Birch, *Pl. Soil*, 1959, **11**, 262.
2 A. D. Rovira and G. D. Bowen, *Pl. Soil*, 1966, **25**, 129.
3 See, for example, J. H. L. Messeng, *Pl. Soil*, 1965, **23**, 1, who also finds that extractable aluminium may also be increased (*Nature*, 1965, **207**, 439).
4 J. R. Dawson, R. A. H. Johnson *et al.*, *Ann. appl. Biol.*, 1965, **56**, 243.
5 J. E. Brind, *Forestry Commission Bull.* 37, 1965, 206; B. N. Singh and L. M. Crump, *J. gen. Microbiol.*, 1953, **8**, 421. For the effect on nitrifiers, see also J. N. Davies and O. Owen, *J. Sci. Fd. Agric.*, 1953, **4**, 248; 1951, **2**, 268 and C. D. Oxley and E. A. Gray, *J. agric. Sci.*, 1952, **42**, 353.

effect, although both cause a marked flush of ammonium production.[1] The nematicide DD also reduces the rate of nitrification, which means that ammonium tends to accumulate in the soil, and since this cannot be easily leached from the soil, acts as a good source of nitrogen to crops in wet springs,[2] and for this reason helps to reduce the effect of the damage done to the root system by the residual nematodes. Chemical fumigation tends to encourage the rapid multiplication of the fungus *Trichoderma viride* and related species,[3] which are important antagonists to many pathogenic soil-borne fungi such as *Armillaria mellea* and *Fomes annosus*.[4] Most research workers have also found that these chemicals are more effective and last longer in sandy than in heavier soils.

The effect of chemical sterilants which only partially sterilise a soil can have a depressing residual effect on crop production. Table 12.3 gives two examples of this both on spring wheat after formalin has been applied to the soil.[5] The Woburn site is an old arable sandy loam soil liable to drought, so the results are given for irrigated plots where water did not limit the response to nitrogen, and the Rothamsted site, on a clay loam, had nineteen cereal crops in the previous twenty-one years. The table shows that treating the soil with formalin gave a very large increase of yield at both Woburn and Rothamsted in the year of application, but the yield in the second year after application was actually lower than on the untreated soil, and there is no evidence that applications in two successive years are having any cumulative beneficial effect.

These experiments show that sterilisation has little effect on the responsiveness of the crop to nitrogen in spite of the very great difference of yield, and therefore of nitrogen uptake, by the crop, particularly at Woburn. The experiments included plots having a still higher dressing of nitrogen, but this either gave only a small response or an actual depression of yield compared with the nitrogen dressings given in the table, except that in 1964 at Woburn the wheat on the unfumigated land gave a further response of 650 kg/ha to an additional 75 kg/ha N, and at Rothamsted in 1966 the plots sterilised in 1965 gave a further 550 kg/ha to an additional 62 kg/ha N. Thus, although the chemical treatment must have liberated ammonium nitrogen, and so increased the nitrogen supply to the crop, it did not appreciably affect its responsiveness to additional nitrogen fertiliser.

The reduction in yield in the year succeeding sterilisation is due to the natural antagonists of the plant pathogens being seriously reduced by the

1 E. R. Tillett, *Rhod. J. agric. Res.*, 1964, **2**, 13.
2 For an example on sugar-beet in East Anglia, see A. P. Draycott and P. J. Last, *J. Soil Sci.*, 1971, **22**, 152.
3 S. B. Saksena, *Trans. Brit. Mycol. Soc.*, 1960, **43**, 111; J. E. Mollison, ibid., 1953, **36**, 215; J. H. Warcup, ibid., 1952, **35**, 248, and 1951, **34**, 519.
4 S. D. Garrett, *Canad. J. Microbiol.*, 1957, **3**, 135, and R. Weindling, *Phytopath.*, 1934, **24**, 1153; 1932, **22**, 837.
5 F. V. Widdowson and A. Penny, *Rothamsted Rept.* 1969, Part II, 113.

TABLE 12.3   The effect of partial sterilisation with formalin on the yield of spring wheat. Yields and responses in t/ha

WOBURN (*crop irrigated*)					
1964		1965			
Formalin treatment		*Formalin treatment*			
None	*With*	*None*	*In 1964*	*In 1965*	*In 1964 and 65*
Yield with 75 kg/ha N					
1·45	3·76	0·90	0·75	3·64	2·46
Response to an additional 75 kg/ha N					
0·84	0·92	0·48	1·17	1·04	1·28
ROTHAMSTED					
1965		1966			
None	*With*	*None*	*In 1965*	*In 1966*	*In 1965 and 66*
Yield with no additional nitrogen					
2·13	3·00	2·50	1·63	3·66	3·46
Response to 62 kg/ha N					
1·42	1·38	1·38	1·12	0·88	1·35
Response to an additional 62 kg/ha N					
−0·21	0·06	1·16	0·79	0·79	−0·04

sterilising treatment, whereas the pathogens themselves could multiply on the roots or stubble of the crop. The two principal pathogens were the cereal cyst eelworm *Heterodora avenae* and the take-all fungus *Ophiobolus graminis*, and both were more severe on the second crop after treatment than on the untreated.[1] There is nothing peculiar to partial sterilisation in this effect, for if a crop resistant to take-all is grown between a run of successive susceptible cereal crops, the first cereal crop after the resistant crop is usually much less seriously attacked, and the following one more seriously, than if the resistant crop had not been taken.

The effect of the formalin treatment on the microbial population both on the Rothamsted and the Woburn soil is long-lasting; for D. S. Jenkinson and D. S. Powlson[2] found that, although the microbial population on the treated and untreated soils respired at the same rate five and a half years after treatment, yet if the soils were then partially sterilised with chloroform vapour, the subsequent rate of respiration and of nitrogen mineralisation was less on the treated soil, indicating that it still had a lower microbial biomass.

1 G. A. Salt, *Rothamsted Rept.* 1970, Part II, 138; T. D. Williams, *Ann. Appl. Biol.*, 1969, **64**, 325.
2 *Soil Biol. Biochem.*, 1970, **2**, 99.

13

The association between plants and micro-organisms

The rhizosphere population

The influence of the plant roots on the soil population in their neighbourhood is considerable, for plant roots, rootlets and root hairs may carry large concentrations of micro-organisms on their surface. Not all the rootlets or root hairs carry these concentrations, and those that do so are not necessarily uniformly covered. The interfacial volume between the plant rootlets and the bulk of the soil, in which this population lives, is often called the rhizosphere;[1] but this volume has no exact bounds, for some fungal hyphae live both in the soil and penetrate into the outer cortical cells in the root surface on the one hand, and there is no sharp boundary between the rhizosphere and the bulk of the soil on the other. It is possible that the layer may be up to 1 to 2 mm thick,[2] and if so, nearly all the surface soil may be in the rhizosphere for crops such as some pasture grasses which have a very extensive root system in the surface soil. Some workers have also introduced the concept of the rhizoplane population, which is defined as the bacterial and possibly the actinomycete population living on the outer surface of the root. There is, in fact, a continuous gradation in the characteristics of the microbial population living on the root surface, in the soil but close to the root, and in the soil well away from the root.[3]

The picture obtained by direct microscopical examination of plant roots taken from a soil is that the growing root tip is typically free of micro-organisms, but that the zone of root elongation behind the tip carries a bacterial population, often of sparse clumps of bacteria near the tip becoming almost a sheath further back. The older part of the root may carry a considerable fungal and bacterial population, the fungal population becoming of

1 L. Hiltner, *Arb. deut. landw. Ges.*, 1904, **98**, 59, first used this word.
2 A. D. Rovira, *Aust. J. agric. Res.*, 1961, **12**, 77.
3 For reviews of this subject see J. Macura in *The Ecology of Soil Bacteria*, Ed. T. R. G. Gray and D. Parkinson, Liverpool UP, 1967; A. D. Rovira and B. M. McDougall in *Soil Biochemistry*, Ed. A. D. McLaren and G. H. Peterson, Arnold, 1967; A. D. Rovira, *Ann. Rev. Microbiol.*, 1965, **19**, 241.

increasing importance as root cells become moribund.[1] There also tends to be a concentration of protozoa and nematodes in the rhizosphere which are predatory on the rhizosphere population,[2] but some of the nematodes are parasitic on the root itself.

The size of the bacterial population has usually been investigated through the use of the plating technique, which is a technique known to underestimate bacterial numbers. The results obtained by a number of workers for a range of crops in a range of soils indicate that the soil that adheres fairly tightly to the root surface may contain up to 200 times as many bacteria as in the body of the soil, with ratios in the region between 10 and 50 being common.[3] On the other hand, H. A. Louw and D. M. Webley[4] found that for the oat rhizospheres they examined, the ratio of bacterial numbers as determined by the plating technique, using a non-selective medium and the direct cell count was between 1 and 5 in the rhizosphere and between 8 and 11 in the soil—a ratio that is much smaller than is usually found (see p. 162). Thus, a part of the apparent concentration of bacteria reported by most workers in the rhizosphere compared with the body of the soil, could be due to the counting technique they have used. A few workers[5] have tried to estimate the activity of the rhizosphere population from the rate of microbial respiration in the rhizosphere soil compared with the soil in bulk, and the results they have obtained show that it may be between two to four times as high, but the reliability of the techniques used and the interpretation of the results are uncertain.

The bacterial population differs from that in the soil as a whole in that it contains larger proportions of gram-negative non-spore-formers such as *Agrobacterium radiobacter*, of a group of bacteria which are classified as similar to nodule bacteria, of *Pseudomonas* spp., and of *Mycobacteria* and *Corynebacteria*; while most *Bacilli* and gram-positive cocci are reduced, although *B. polymyxa*, *B. circulans* and *B. brevis* may be stimulated.[6] The population also contains a higher proportion of chromogenic and motile forms and more ammonifiers, nitrifiers, denitrifiers and aerobic cellulose decomposers, but a lower proportion capable of using aromatic acids such as benzoic.[7]

1 D. Parkinson, G. S. Taylor and R. Pearson, *Pl. Soil*, 1963, **19**, 332.
2 See V. E. Henderson and H. Katznelson, *Canad. J. Microbiol.*, 1961, 7, 163, and *Soil Sci.*, 1946, **62**, 343.
3 For a review see F. E. Clark, *Adv. Agron.*, 1949, **1**, 241, and R. L. Starkey, *Bact. Rev.*, 1958, **22**, 154.
4 *J. appl. Bact.*, 1959, **22**, 216.
5 See H. Katznelson and J. W. Rouatt, *Canad. J. Microbiol.*, 1957, 3, 673, and H. W. Reuszer, *Proc. Soil Sci. Soc. Amer.*, 1949, **14**, 175.
6 F. E. Clark, *Trans. Kansas Acad. Sci.*, 1940, **43**, 75.
7 See, for example, J. W. Rouatt and H. Katznelson, *J. app. Bact.*, 1961, **24**, 164; *Proc. Soil Sci. Soc. Amer.*, 1960, **24**, 271, and J. I. Sperber and A. D. Rovira, *J. app. Bact.*, 1959, **22**, 85. For older work see R. L. Starkey, *Soil Sci.*, 1929, **27**, 355 and A. G. Lochhead, *Canad. J. Res.*, 1940, **18** C, 42.

The essential nutrient requirements of the rhizosphere bacterial population also differ from that for the general soil population in that it contains a much lower proportion requiring complex growth factors such as are present in soil and yeast extract, but a much higher proportion needing simple amino acids.[1] This is what would be expected if the principal sources of nutrients for this population are root exudates, sloughed off cells from the growing root cap, moribund root hairs and moribund cortical cells in the older root region. It is still technically difficult or not yet possible to make any quantitative estimate of the relative importance of these sources, but root excretions are probably the principal source in the younger regions of the root, and moribund tissues become of increasing importance as the root ages. Root excretions contain a wide variety of simple substances; sugars, amino acids, organic acids, nucleotides and enzymes, and the zone of elongation behind the root cap is a zone where excretion and particularly amino acid excretion[2] appears to be active;[3] which could account for the bacterial population in this region being characterised by simple nutritional requirements and high metabolic activity. A further source of nutrients of unknown importance, is the mucilagenous sheath or mucigel which may surround the rootlets of some plants under some conditions (see p. 521) and in which bacteria can sometimes be seen embedded.

There is no question that the rhizosphere population is dependent on the crop for its principal source of energy and nutrients, but it is much less certain how far it affects the growth of the plant under normal conditions. There is no evidence that it weakens a healthy plant by causing it to excrete such large amounts of energy and amino acids that its growth is affected, though under conditions of very low nutrient supply in the soil the rhizosphere population may compete for any nutrient in short supply so reducing the supply available to the crop. Thus, D. A. Barber and A. C. Loughman[4] showed that the rhizosphere population around barley roots growing in a medium very low in phosphate, reduces the amount of phosphate available to the plant compared with a sterile root system (see p. 526). Under some conditions also the rhizosphere may be a zone of active denitrification, even if the soil as a whole is aerated.[5]

The population may also affect the availability of a nutrient to a crop by converting it to an insoluble form just outside the root surface. Thus, M. I. Timonin[6] found that this population could induce manganese deficiency in

1 A. G. Lochhead and J. W. Rouatt, *Proc. Soil Sci. Soc. Amer.*, 1955, **19**, 48.
2 R. Pearson and D. Parkinson, *Pl. Soil*, 1961, **13**, 391; M. N. Schroth and W. C. Snyder, *Phytopath.*, 1961, **51**, 389; B. Frenzel, *Planta*, 1960, **55**, 169.
3 For reviews see A. D. Rovira, *Soils Fert.*, 1962, **25**, 167, and with B. M. McDougall in *Soil Biochemistry*, Ed. A. D. McLaren and G. H. Peterson, Arnold, 1967.
4 *J. exp. Bot.*, 1967, **18**, 170, and for a review see D. A. Barber, *Ann. Rev. Pl. Physiol.*, 1968, **19**, 71.
5 J. W. Woldendorp, *Meded. Landb. Hoogesch., Wageningen*, 1963, **63**, 1.
6 *Proc. Soil Sci. Soc. Amer.*, 1947, **11**, 284; and see also F. C. Gerretsen, *Ann. Bot.*, 1937, **1**, 207.

some varieties of oats by oxidising the manganous ions to manganese dioxide, and he also found that this harmful effect of the population could be aggravated by applying a straw mulch. M. W. Loutit and her colleagues[1] have also shown that this population can influence the molybdenum uptake by a crop through a mechanism that has not yet been characterised.

The rhizosphere population may, under some conditions, increase the supply of nutrients to a crop. F. C. Gerretsen,[2] for example, showed that roots carrying a rhizosphere population took up more phosphate than sterile roots from insoluble calcium phosphate, which is probably due to some members excreting 2 keto-gluconic acid.[3] However, the actual mechanism may be more complex because a proportion of the soluble phosphate may be present as organic phosphates,[4] though this need not affect the availability of the phosphate to the roots, both because roots can take up many soluble organic phosphates and also because the rhizosphere population possesses considerable phosphatase activity.[5] Some of the micro-organisms may also produce soluble organic substances which chelate iron and manganese,[6] probably after reduction of the ferric to the ferrous state and the tetravalent manganese to manganous ions, which the crop roots can take up. This may be the principal source of iron for crops in most soils and of manganese in neutral and calcareous soils.

One particular aspect of the rhizosphere–plant interaction has received a great deal of attention. Russian agronomists have for several decades been inoculating the seed of wheat with a culture of Azotobacter before drilling, and have found that on the average this practice may increase yields up to about 10 per cent; and they also find some evidence that inoculating the seed with *Bacillus megatherium* may increase the phosphate supply to the crop, though the evidence for this is less convincing.[7] Azotobacter is not an organism that is normally concentrated in the rhizosphere, but M. E. Brown and her colleagues[8] at Rothamsted have shown that if the wheat seed has a sufficiently large population of Azotobacter, these organisms grow along or are carried along the root surface as it grows into the soil, and give a number of large colonies on the roots. They also showed that their beneficial effect was due not to an increase of the nitrogen supply to the crop but to the excretion of growth substances, such as some gibberellins and β-indolyl acetic acid[9]

1 *Pl. Soil*, 1967, **27**, 335; *N.Z.J. agric. Res.*, 1968, **11**, 420; *Soil Biol. Biochem.*, 1970, **2**, 131.
2 *Pl. Soil*, 1948, **1**, 51.
3 H. A. Louw and D. M. Webley, *J. appl. Bact.*, 1959, **22**, 227.
4 R. B. Duff, D. M. Webley and R. O. Scott, *Soil Sci.*, 1963, **95**, 105.
5 M. P. Greaves and D. M. Webley, *J. appl. Bact.*, 1965, **28**, 454.
6 A.R.C. Letcombe Laboratory, Ann. Rept. 1971, 79; ibid. Ann. Rept. 1969, 35.
7 For a review of the Russian work see R. Cooper, *Soils Fert.*, 1959, **22**, 327.
8 R. M. Jackson and M. E. Brown, *Ann. Inst. Pasteur*, 1966, **111**, Suppl. 112. For a photo of this development on barley roots see A. D. Rovira and B. M. McDougall in *Soil Biochemistry*, Ed. A. D. McLaren and G. H. Peterson, Arnold, 1967.
9 M. E. Brown and S. K. Burlingham, *J. gen. Microbiol.*, 1968, **53**, 135; *J. exp. Bot.*, 1968, **19**, 544, with N. Walker, *Pl. Soil*, 1970, **32**, 250.

which, when taken up by the young plant, modifies its growth and subsequent yield.

This effect is not confined to seed inoculation with Azotobacter, for A. D. Rovira[1] found wheat and several other crops responded in the same way to inoculation with a Clostridium and *Bacillus polymyxa* though he did not determine the cause. But A. G. Norman[2] found that under some conditions they could excrete antibiotics which depressed the growth of the root or injured its cells.

Association of fungi with plant roots

A plant root, during its life in the soil, will exist in four conditions: it begins as an actively growing organism, it then becomes mature and ceases growth, later becoming senescent, and finally it becomes diseased and dies; and in each of these stages it usually has a characteristic fungal flora. Further, the rhizosphere fungi may be growing on the root surface, or they may penetrate the outer cortical cells, or the inner, or the stele, and each of these zones has its characteristic fungal flora. The characteristic species on the healthy root behind the zone of active growth belong to the genera *Fusarium*, principally *F. oxysporum*, *Gliocladium*, *Penicillium*, often the *P. lilacium* group, *Cylindrocarpon radicola*, *Rhizoctonia* (*Corticum* or *Pellicularia*), *Mortierella*, and sometimes *Trichoderma viride*. In addition, there are a number of both dark and hyaline hyphae which cannot yet be identified. None of these fungi, with the exception of *T. viride*, are common soil forms nor are their hyphae normally found in the soil.[3]

The relative proportions of the species present on or in the root depends on the plant and the soil conditions. Thus *Fusarium* tends to be more common on acid soils and *Cylindrocarpon* on neutral. Further, *Rhizoctonia*, *Gliocladium*, *Trichoderma*, *Penicillium* and *Mortierella* are usually confined to the root surface, or if they penetrate the root, to the outer cortical cells; while *Fusarium* and *Cylindrocarpon*, as well as the sterile hyphae, have a greater ability to penetrate into the inner cortical cells or even the outer cells of the stele.[4] As the root ages, the fungal hyphae tend to penetrate deeper into the root tissues,[5] and many mature or senescent roots have lost their cortical cells leaving the stele much more exposed to infection. The activity of the fungi, as measured by their respiration, or by the length of mycelia in

1 *Pl. Soil*, 1963, **19**, 304 and with E. H. Ridge, *Trans. 9th Int. Congr. Soil Sci.*, 1968, **3**, 473.
2 *Trans. 7th Int. Congr. Soil Sci.*, 1960, **2**, 531.
3 D. Parkinson, G. S. Taylor and R. Pearson, *Pl. Soil*, 1963, **12**, 332. For examples of the fungi found on wheat roots see P. M. Simmonds and R. J. Ledingham, *Sci. Agric.*, 1937, **18**, 49 (Canada), and M. D. Glynne, *Trans. Brit. Mycol. Soc.*, 1939, **23**, 210. For examples in grass and clover roots R. H. Thornton, *N.Z. J. agric. Res.*, 1965, **8**, 417.
4 G. S. Taylor and D. Parkinson, *Pl. Soil*, 1965, **22**, 1; P. D. Gadgil, *Pl. Soil*, 1965, **22**, 239, and J. S. Waid, *Trans. Brit. Mycol. Soc.*, 1957, **40**, 391.
5 J. S. Waid, *Trans. Brit. Mycol. Soc.*, 1962, **45**, 479.

the soil, increases as the young plant grows; it reaches a maximum when the vegetative growth rate of the crop is at its maximum.[1]

Root-rot fungi

Many of the fungi which develop on the young root can be weak unspecialised parasites, for they can enter and kill juvenile but not mature roots. The attack is usually only serious if seedling growth receives a check so the rootlet remains juvenile for an extended period. This can occur if the soil temperature is high enough to allow the seed to germinate, but sufficiently low to give very slow root growth. Thus, they are the cause of the characteristic damping-off of seedlings growing under unfavourable conditions.[2] The specialised *Fusarium* and *Verticillium* vascular wilt fungi fall into the same group, for they are long-lived in the soil and can only attack juvenile roots; but once in the root they escape into the vascular system, and initially are confined there by the active resistance of the living cells to infection, though once a cell dies the fungus enters. Plant resistance to these vascular wilts thus occurs entirely in the root system.

The ability of a plant root to withstand attack by these specialised root-rot fungi depends partly on the health and vigour of the root at the point of attack and partly on the vigour of the fungus growing in the soil or on the root surface before it penetrates into the root cells, and this may depend on the conditions in the rhizosphere. The rhizosphere may contain fungi, such as *Trichoderma viride*, which are antagonists to the root infecting fungi, so that altering the root environment to encourage these antagonists reduces the susceptibility of the crop's roots to attack. *Trichoderma viride*, for example, is more resistant to chemical sterilants than most pathogenic fungi, so treating a soil with one of these chemicals encourages its development (see p. 239). It grows more strongly in acid than calcareous soils, as has been shown in detail by J. Rishbeth[3] in his study of the susceptibility of pine roots to attack by *Fomes annosus* on some sandy East Anglian soils; for the *Fomes* attack was much more serious, and the growth of *Trichoderma* was much poorer on the calcareous than on the acid soils. Garrett has also suggested that this effect of soil acidity encouraging the growth of *Trichoderma* on root surfaces may be the explanation of some observations made by A. A. Hildebrand and P. M. West.[4] They found that root-rots of strawberries and tobacco caused by a complex of weak unspecialised root-parasitic fungi could be reduced by ploughing in soyabean tops, which reduced the soil pH, whereas

1 D. Parkinson and A. Thomas, *Canad. J. Microbiol.*, 1969, **15**, 875.
2 For a fuller discussion of this see L. D. Leach, *J. agric. Res.*, 1947, **75**, 161. For an example with maize see J. L. Harper, *Ann. appl. Biol.*, 1955, **43**, 696, and *New Phytol.*, 1955, **54**, 107, 119; 1956, **55**, 35.
3 *Ann. Bot.*, 1950, **14**, 365; 1951, **15**, 1, 221.
4 *Canad. J. Res.*, 1941, **14** C, 183, 199. See also H. Katznelson and F. E. Chase, *Soil Sci.*, 1944, **58**, 473.

doing the same with red clover tops was ineffectual; they also found that any other soil treatment which reduced the soil pH was beneficial.

Other organisms which are good producers of powerful antibiotics may protect plant roots against the root-rot fungi in the same way as *T. viride*. Thus, F. M. Eaton and N. E. Rigler[1] found that varieties of cotton that were resistant to *Phymatotrichum* root-rot had roots carrying a much higher number of blue-green fluorescent bacteria of the *Pseudomonas-Phytomonas* group in their rhizosphere than had the roots of susceptible varieties.

It is possible that the antagonistic action of the bacteria may be indirect, for Z. Krezel[2] found that certain members of the rhizosphere population produced antibiotics which were taken up by the roots and were distributed throughout the plant in the plant sap, so protecting the plant from attack. The example he worked on was the protection of the bean *Phaseolus vulgaris* from attack by *Xanthomonas phaseoli*.

The rhizosphere environment can be altered by adding different types of decomposable organic matter to the soil. Thus, R. Mitchell and M. Alexander[3] found that adding 560 kg/ha of chitin to a soil two weeks before planting a *P. vulgaris* bean reduced the severity of attack by the root-rot *Fusarium solani.* f. *phaseoli* markedly; and Z. A. Patrick and L. W. Koch[4] found that adding the soluble products from decomposing rye or grass to a soil increased the severity of attack of tobacco roots by *Thielaviopsis basicola*. The inoculation potential of a root-rot fungus may also be markedly affected by changes in the oxygen or carbon dioxide content of the soil air. Thus, G. C. Papavizas and C. B. Davey[5] found that the pathogenic activity of *Rhizoctonia solani* on radish and sugar-beet seedlings was reduced by raising the CO_2 concentration in the soil air, and Canadian workers[6] found a similar result for the Fusarium wilt in flax (*Fusarium oxysporum* f. *lini*), for its virulence was much reduced if the soil aeration was poor. D. A. Lapwood and his associates[7] have shown that *Streptomyces scabies*, the cause of potato scab, can only cause appreciable infection of potato tubers if it can gain entry to the tubers during the three weeks after tuber initiation, but it can only do this if the soil is dry. By maintaining the soil moist during this critical period, by irrigation if available, a good quality marketable crop can be obtained on moderately infected land.

It is sometimes possible to weaken the root-rot fungus during its resting phase in the soil. For instance, some parasitic strains of *Fusarium oxysporum* can be killed or severely weakened if subject to anaerobic conditions, so

1 *J. agric. Res.*, 1946, **72**, 137.
2 *Ann. Inst. Pasteur*, 1964, **107**, 168.
3 *Proc. Soil Sci. Soc. Amer.*, 1962, **26**, 556.
4 *Canad. J. Bot.*, 1963, **41**, 747.
5 *Phytopath.*, 1962, **52**, 759.
6 R. O. Lachance and C. Perrault, *Canad. J. Bot.*, 1953, **31**, 515. R. H. Stover, ibid., 693.
7 *Ann. appl. Biol.*, 1966, **58**, 447; 1970, **66**, 397; *Plant Path.*, 1967, **16**, 131; B. G. Lewis, *Ann. appl. Biol.*, 1970, **66**, 83.

infected land can be much improved by being flooded in warm weather, particularly if it contains much decomposable organic matter. Moreover, land heavily infected with banana wilt, *Fusarium oxysporum* f. *cubense*, can have its infection very considerably reduced by waterlogging the soil for several months, although this method has the limitation that the surface layer of the soil will remain aerobic, due to algal growth (see p. 687), so the fungus can maintain itself in this layer.[1]

Again, the take-all fungus—*Ophiobolus graminis*—which attacks the wheat and barley roots can be weakened when it is in the cereal stubble if the level of available nitrogen in the soil can be kept low during the autumn, as has been described on p. 235.

The specialised root-rot fungi have considerable power to decompose cellulose and to attack mature root cells, with which they show a degree of symbiosis, for they do not initially kill or seriously disorganise the cells they are attacking. They differ from the mycorrhizal fungi, which will be considered in the next section, for these enter only juvenile roots or the growing points of laterals and they form a stable symbiosis with the cells they enter. The root-rot fungi typically grow on the root surface before they penetrate into the root cells, and for fungi which attack mature tree roots such as *Armillaria mellea, Fomes lignosus* and *F. annosus*, their mycelia may run along a root for several metres before they send down branches into root cells themselves. Further, in undisturbed equatorial forests, *A. mellea* grows over the surface of the roots of some species of trees as a weft without penetrating into the roots properly, provided the tree is healthy. Only if the forest tree is weakened, by a very severe drought, for example, or if the tree is cut down, can the fungus enter the root and start growing actively inside it. Even a plantation crop such as tea, which is usually considered to be susceptible to *A. mellea*, may carry rhizomorphs on the surface of its roots without suffering any serious attack if it is growing vigorously.[2]

Plant roots have a certain resistance to attack by these fungi, which the fungus must overcome before it can enter. Garrett has introduced the concept of the inoculation potential of the fungus in this context. A fungus with a high inoculation potential will have the energy or ability to overcome a high resistance to attack by the host, and one with a low inoculation potential will only be able to enter a root with a very low resistance. For a given fungus, the better the food supply at the point of attack, the greater is its inoculation potential.

This concept of inoculation potential can be used to illustrate the type of attack caused by three quite different fungi. The fungus of 'take-all' in wheat, *Ophiobolus graminis*, only sends out very fine hyphae into the soil and can only infect the fine roots of susceptible Gramineae. The fungus of cotton root-rot, *Phymatotrichum omnivorum*, sends out mycelial strands from the

1 R. H. Stover, *Soil Sci.*, 1953, **76**, 223; 1954, **77**, 401; 1955, **80**, 397.
2 I am indebted to Dr I. A. S. Gibson for this observation.

root in which it is living and it can infect the thicker roots of herbaceous plants, provided it starts from a fair-sized piece of root. Fungi causing the root-rot of trees, such as *Armillaria mellea* and *Fomes lignosus*, send out rhizomorphs starting from a large piece of tree root; and if the root of the tree is relatively resistant to attack, its resistance can only be broken down if it grows fairly close to the source of infection, so the point of attack is not too far away from the food source, but only after the fungus has built up a weft of mycelia on the root which then send a large number of branches down into the root more or less simultaneously—a method which obviously increases the inoculation potential of the fungus very considerably.

A point of great importance in the control of the specialised root-disease fungi on herbaceous and tree crops in the field is that when the plant is cut down, its roots do not die immediately, so for a considerable time the root will maintain its immunity to saprophytic and weakly parasitic fungi, while becoming a better food source for the specialised parasite; and roots of some tree species that were resistant to infection by these specialised parasites when the tree was alive can lose their resistance and become attacked once it has been felled. Such roots will carry the parasite in the soil for a considerable time, which may be measured in years in the case of trees, and are a source of infection for any new susceptible roots growing in their neighbourhood for this period. This source of infection is of great importance when one is converting natural forest, whose roots may carry a heavy population of *Armillaria mellea* or *Fomes lignosus*, to plantations of susceptible crops such as tea or rubber.

The current method of control, which was worked out by R. Leach[1] for the control of *A. mellea* in tea, is to make the roots of the trees an unfavourable environment for the specialised fungus but a more favourable one for the general soil saprophytes. Leach showed that this could be done if the roots were dead by the time the tree was felled, and this could be brought about by ring-barking the tree down to the cambium leaving the outer xylem layers intact. This causes the starch reserves in the roots to become depleted, for the roots must use energy to translocate water and mineral nutrients from the soil to the crown of the tree through the outer xylem layers, but prevents any of the carbohydrates synthesised by the leaves being translocated to the roots, for this translocation is through the phloem layers which have been cut by the ring barking. When the starch reserves in the roots are depleted the root dies, and when all the roots are dead, so is the tree, though this process may take at least three years for a mature forest tree. When this dead tree is felled, its roots, being dead, are no longer a favourable habitat for the fungus.

1 *Proc. Roy. Soc.*, 1937, **121** B, 561, and *Trans. Brit. Mycol. Soc.*, 1939, **23**, 320.

The mycorrhizal association between fungi and plant roots

Fungi have probably been living symbiotically with the roots of land plants ever since they evolved. When growing in association with algae, the combined organism is a lichen, but they occur in association with the roots of bryophytes, pteridophytes and higher plants, and their associations with the roots of the latter are known as mycorrhizas.[1] They are the fungal equivalents of the bacteria forming nodules on leguminous plants and the actinomycetes forming nodules on a range of other plants. The mycorrhizal associations can conveniently be grouped into four classes: those mainly between some higher Basidiomycetes and Ascomycetes with the roots of many forest trees; those between some Basidiomycetes and orchids, and those between a group of fungi and ericaceous plants, neither of which will be considered here; and those between fungi placed in the genus *Endogone*, which is either an Ascomycete or a Phycomycete, and the roots of many higher plants.

Mycorrhiza of forest trees

This association between a Basidiomycete fungus and a tree root is in many ways similar to that between the specialised root-rot fungi, such as *Armillaria* and *Fomes*, and tree roots. They are typically sugar fungi and ecologically obligate root inhabitants; the attack of a new root commonly begins by their hyphae running over the root surface; their hyphae, which enter the root, do not initially kill any cortical cells in which they enter; the viable mycelium of at least some of the mycorrhizal and root-rot fungi can remain in the soil in old rotten bits of root for many years,[2] and then infect new susceptible roots growing in their neighbourhood; and they may be more sensitive than related saprophytes to antibiotics. They differ in that the mycorrhizal fungi appear to enter susceptible roots more easily than can the root-rot fungi, and to remain a symbiont for a very much longer time. It is possible that there are some fungi which are normally symbiotic but which can become mildly parasitic if conditions for the growth of the tree become unfavourable.

There are two distinct types of mycorrhizal associations on tree roots with some intermediate types. Some fungi cause the formation of short many-branched structures on the sublateral tree roots which take the place of root hairs, as is shown in Plates 12 and 13 for Scots pine and the date palm. This structure is composed of both a mantle or sheath of fungal mycelium enclosing the roots whose cortical cells are enlarged, and also of fungal hyphae

1 For a general review of mycorrhizal associations, see J. L. Harley, *The Biology of Mycorrhiza*, London, 1959, and *Ann. Rev. Microbiol.*, 1952, **6**, 367; E. Melin, *Ann. Rev. Plant Phys.*, 1953, **4**, 325.

2 For an example of *Boletus scaber* and *Amanita muscaria* remaining viable in old rotten pine roots and giving mycorrhiza on birch roots, see G. W. Dimbleby, *Forestry*, 1953, **26**, 41.

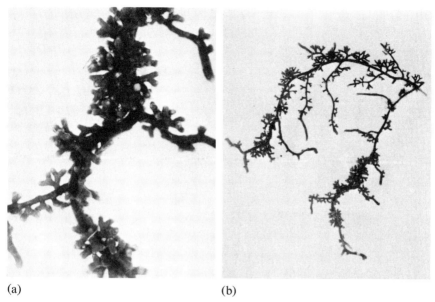

(a)   (b)

PLATE 12. Part of the root system of Scots pine (*Pinus sylvestris*), showing differentiation into long roots and short forked mycorrhizas: (*a*) × 6; (*b*) natural size.

filling the intercellular spaces, as can be seen from the cross-section of the mycorrhiza reproduced in Plate 13. The mantle is often coloured, the colours ranging from white through golden brown to black, and usually has a smooth surface, though it may be loose and have many hyphae radiating from it into the soil;[1] and it is these structures that were originally called mycorrhiza, or fungus roots, by B. Frank.[2] This type of mycorrhiza is known as ectotrophic. The other extreme type, the endotrophic, does not normally form a smooth mantle and its hyphae invade the cortical cells of the roots without killing them; and there are transitional types in which some hyphae are within and some between the cells, known as ect-endotrophic.

Little is known of the length of life of a mycorrhiza. On some northern pines, they last a single season only, dying off in the autumn; the fungus probably rests in the soil over winter and reinfects the short roots of the sub-laterals when growth starts in the spring. In other trees it is probable that the fungus is always present in the root, and it infects new short roots by growing epiphytically over the root surface.

Most species of tree have been found to carry mycorrhiza if grown under

1 For descriptions and photographs of different types of mycorrhiza, see M. C. Rayner, *Trees and Toadstools*, London, 1945, and J. L. Harley, *The Biology of Mycorrhiza*, London, 1959. For a histological study of a mycorrhiza on beech see F. A. L. Clowes, *New Phytol.*, 1950, **49**, 248; 1951, **50**, 1.
2 *Ber. deut. bot. Ges.*, 1885, **3**, 128. Frank spelt the word mycorhiza.

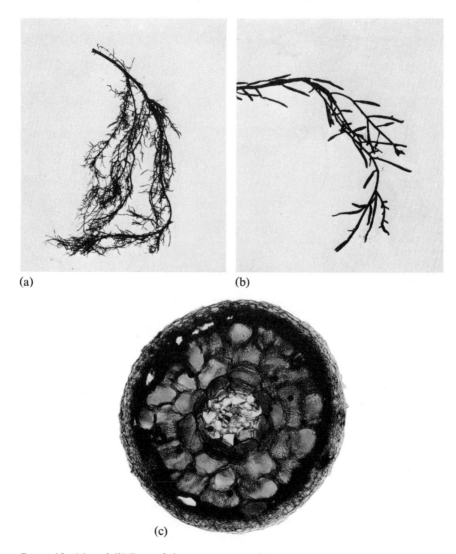

PLATE 13. (*a*) and (*b*) Part of the root system of the date palm (*Phoenix dactylifera*). The fine feeding roots are all mycorrhizas. (*a*) half natural size, (*b*) ×7.
(*c*) Transverse section of a mycorrhiza of Scots pine formed by association with *Boletus bovinus* (×130).

suitable conditions, and it is probably true to say that in most existing natural forests and in many plantations on suitable soils, mycorrhiza are normal features of the root system, whatever the tree. The fungi are usually Hymenomycetes, such as species of *Boletus*, *Amanita* and *Lactarius*, but they include at least one member of the Gasteromyceres (puffballs) *Rhizopogon* and, by inference, truffles and truffle-like fungi, such as some *Tuberales* and *Elaphomyces* belonging to the Ascomycetes, as well as several fungi whose systematic position is not known because their fruiting bodies have not yet been observed.[1] A given species of fungus will usually form mycorrhiza on a number of tree species, and a given tree can carry mycorrhiza formed from a number of fungal species; and it is possible that most trees and fungi can be classified as carriers and producers either of ectotrophic or of endotrophic mycorrhizas.

Mycorrhizas only develop freely on trees under three fairly definite soil conditions. They need a supply of organic matter in the soil, so develop most freely in the surface litter of the forest floor, and they can often be encouraged in a mineral soil by adding leaf litter or compost to it. They need well-aerated soils and are inhibited by water logging and reducing conditions, so sandy soils are usually more conducive to mycorrhizal formation than clays or peat bogs; and loosening a heavy clay, or draining a bog, often increases mycorrhizal development. They also need a restricted, but not too restricted, supply of nutrients, and in fact mycorrhizal formation appears to be the tree's response to low availability of nutrients. Thus mycorrhiza are more common on tree roots growing in a mor or raw humus layer, when practically every organ carried by the shallower sublateral roots is a mycorrhiza, than in a mull layer when mycorrhiza may be relatively few, which is presumably connected with the higher availability of nutrients in the mull. On the other hand there are soils so low in phosphate or available nitrogen that adding a phosphate fertiliser on the one hand,[2] or a nitrogen on the other, will result in a strong development of mycorrhizas and increase in tree growth and vigour.

S. D. Garrett has discussed the ecology of these mycorrhizal fungi in the soil. They typically need simple carbohydrates such as sugars as their source of energy, while closely related non-symbiotic fungi can use cellulose and often lignin. Thus, in the field they obtain all their carbohydrates from the roots they are living in and, as E. Bjorkman[3] has shown, they most frequently occur on roots high in carbohydrates, which normally only occurs if protein synthesis is lagging behind carbohydrate synthesis in the tree, and this is likely to be a consequence of a moderate shortage of available soil nitrogen and phosphate. Although most of the fungi probably draw their carbohydrate supply entirely from the tree, yet some may obtain a part of theirs by decomposing some of the cellulose or even lignin in the leaf litter in competition with the saprophytes, because they can draw on a readily available source of energy to increase their inoculation potential

1 E. Melin, *Handb. d. biolog. Arbeitsmeth.*, Abt. XI, Teil 4, pp. 1015–1108 (1936), gives lists of fungi forming mycorrhiza on various trees.
2 For an example with phosphate, see A. L. McComb, *Iowa Agric. Expt. Sta., Res. Bull.* 314, 1943; and with nitrogen, H. Hesselman, *Medd. Skogsförsöksanst.*, 1937, **30**, 529.
3 *Medd. Skogsförsöksanst.*, 1941, **32**, 23; *Symb. Bot. Upsaliens.*, 1942, **6**, 191; *Svensk. Bot. Tidskr.*, 1949, **43**, 223.

which is denied to the saprophytes. Some of the fungi are ecologically obligate root inhabitants and can only remain viable in a suitable tree root, either living or dead; but it is very difficult to prove in what form others remain viable in the soil, and whether they can grow saprophytically there. It is, however, probable that most of them can only grow actively in the soil if they are attached to a tree root, and their fruiting bodies must be attached to a root by a rhizomorph.

The role played by the mycorrhizal fungus in the nutrition of the tree on Garrett's picture is as follows: the fungus is well supplied with energy from the tree, so that part of it which is in the soil can compete strongly for whatever nutrients are in short supply with the soil-inhabiting fungi that are growing slowly on the difficultly decomposable leaf litter. Ecologically this means that the mycorrhiza association will be most strongly developed on soils suitable for the growth of the particular tree except that the nitrogen, phosphate, or some other nutrient element is in rather short supply, which is what is observed. If any nutrient is in too short supply, the tree cannot grow strongly enough to supply the fungus adequately with carbohydrate; and if the nutrients needed for protein synthesis are in good supply, the level of simple carbohydrates in the roots will also be too low for strong fungal growth.

It is very difficult to estimate the quantities of simple carbohydrates the mycorrhizal fungus withdraws from the tree, or to compare this with the carbohydrate requirements of the additional feeding roots which would be necessary if the fungus was not there. There is, however, no need to assume that the effective root system which the fungal hyphae in the soil provides normally makes any larger energy demands on the tree than the root system which it replaces, though this is probably not true on those occasions when the fungus produces a large crop of fruiting bodies (sporophores or toadstools). The amount of fruiting bodies formed in any one year, and the proportion of years in which any are formed, depends very much on the climate and other factors which have not been worked out; but L. G. Romell[1] quotes figures for a typical crop under spruce in Denmark in a favourable year. The fruiting bodies contained about 180 kg/ha dry matter and 5 kg/ha of nitrogen, and he estimated that the fungus would need to withdraw at least 390 kg/ha of carbohydrate from the trees to produce this crop, which is equivalent to one quarter of the organic matter in the annual increment of timber in the trees.

Table 13.1 illustrates how the mycorrhiza increase the nutrient uptake of pine seedlings compared with uninfected roots growing under comparable conditions; and it also shows that if the soil is well supplied with nutrients, the uninfected roots will take them up quite easily. This last conclusion may not apply to the root systems of all trees, for some have a relatively poor

1 *Svensk. SkogsvFören. Tidskr.*, 1939, **37**, 348.

TABLE 13.1   Effect of mycorrhiza on the mineral nutrition of pine growing in poor soil

Per seedling	White pine[1]					Virginia pine[2] on Iowan prairie soil	
	Sampled 10/6/36		Sampled 20/8/36				
	Non-mycor-rhizal	Mycor-rhizal	Non-mycor-rhizal	Mycor-rhizal	Non-mycor-rhizal on soil with complete fertiliser	Non-mycor-rhizal	Mycor-rhizal-
Dry weight in mg	160	208	181	337	550	152	323
Nitrogen content as per cent of dry weight	1·23	1·48	1·20	1·60	2·50	1·88	1·78
in mg	1·98	3·09	2·17	5·39	13·75	2·87	5·75
Phosphorus content as per cent of dry weight	0·08	0·13	0·07	0·21	0·24	0·097	0·184
in mg	0·13	0·27	0·13	0·72	1·32	0·15	0·60
Potassium content as per cent of dry weight	0·55	0·50	0·45	0·63	1·10	0·63	0·66
in mg	0·88	1·03	0·81	2·12	6·05	0·96	2·17
Number of mycorrhizal roots	—	—	—	—	—	7	350
Number of non-mycorrhizal roots	—	—	—	—	—	279	321

ability to take up phosphates even from fairly readily available sources, and many have a poorer ability than agricultural crops. Thus A. L. McComb[3] found that by adding phosphate to an Iowan prairie soil the growth of jack pine seedlings, as measured by their weight, was increased twenty-eightfold, while that of oats in the same soil by only 80 per cent; and this increase in the growth of the pine was brought about through a strong development of mycorrhiza. R. O. Rosendahl[4] likewise found that pine mycorrhiza could extract more potash from orthoclase than could uninfected roots.

A final point on the role of mycorrhiza in tree nutrition concerns legumin-ous trees in closed equatorial rain forests. C. Bonnier[5] has noted that, in the rain forests around Yangambi in the Congo (Zaire), the roots of leguminous trees are covered with mycorrhizas but have no nodules, while in the open forest the roots are nodulated but do not carry mycorrhiza. This point deserves more careful study, but if it is found to be correct, the interpretation

1 H. L. Mitchell, R. F. Finn and R. O. Rosendahl, *Black Rock Forest Pap.* No. 10, 1937.
2 A. L. McComb, *J. Forestry*, 1938, **36**, 1148.
3 *Iowa Agric. Expt. Sta., Res. Bull.* 314, 1943.
4 *Proc. Soil Sci. Soc. Amer.*, 1943, **7**, 477.
5 *I.N.E.A.C., Sci. Ser. No.* 72, 1957.

may be that in the intense competition for all nutrients from the forest floor, only mycorrhiza can compete; while in the open forest, where competition will be less severe, but where the level of available nitrogen will remain low, the competition is principally for nitrogen, which leguminous trees can bypass through nodulation.

Endotrophic mycorrhiza

Endotrophic mycorrhizas are far more widespread among plants than ectotrophic, and the roots of almost every plant species growing on some soils appear to be mycorrhizal.[1] These mycorrhizas rarely have a characteristic root form, so their presence can only be recognised from a microscopic examination of root sections. There is, however, no sharp distinction between ectotrophic and endotrophic mycorrhizas on the one hand or between endotrophic mycorrhizas and infection by rhizosphere fungi on the other. Roots will typically contain a number of fungal hyphae, some of which are entirely or predominantly intercellular growing in and sometimes filling the spaces between the cortical cells, and some of which are predominantly intracellular, growing within the cells and usually forming typical intracellular inclusions. These can be of four types—arbuscules or cauliflower-like organs composed of tufts of hyphae, peletons or skeins of hyphae and oidia both of which usually almost fill the cell, and vesicles which are usually intercellular but may be intracellular. These vesicles are usually less than 0·1 mm in size, and a proportion of them will be returned to the soil as a resting spore when the mycorrhiza decomposes. These are all illustrated in Plates 14 and 15. Arbuscules are typically found in the inner cortical cells and are short-lived, the cell apparently dissolving the inclusion. Pelotons and vesicles may be found in the outer cortical cells, and then are either short-lived and dissolved by the cell, or may develop into resting spores and be released into the soil on the death and decomposition of the outer cortical cells.

The characteristic mycorrhizal fungi are put in the genus *Endogone*, which is possibly a Mucorales though it may be an Ascomycete rather than a Phycomycete. It is not yet possible to put these fungi into species, and so far they can only be classified on the basis of the morphology of their sporocarps with their containing spores, which they produce in the soil. These are large organs, the sporocarps being up to 1 mm in size and containing up to ten spores usually yellow or honey-coloured and themselves up to 0·3–0·4 mm; and typically soils contain one or two of these spores per cubic centimetre. Nine different shapes of spore have been separated from soils from different parts of the world, only seven of which appear to give mycorrhizal *Endogone*.[2]

---

1 For reviews see B. Mosse, *13th Symp. Soc. gen. Microbiol.*, 1963, 146; J. L. Harley, *Trans. Brit. Mycol. Soc.*, 1968, **51**, 1; *New Phytol.*, 1950, **49**, 213.
2 B. Mosse and G. D. Bowen, *Trans. Brit. Mycol. Soc.*, 1968, **51**, 469, and J. W. Gerdemann and T. H. Nicolson, ibid., 1963, **46**, 235.

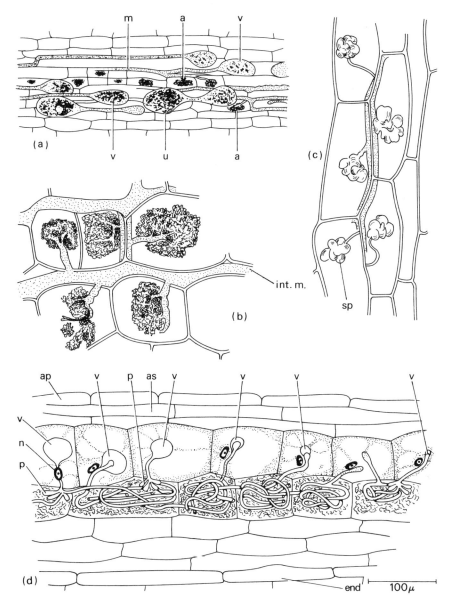

PLATE 14. (*a*) Vesicles in a strawberry rootlet (×60); (*b*) and (*c*) Arbuscules in strawberry rootlets (×600); (*b*) well developed; (*c*) partially decomposed; (*d*) Vesicles and pelotons in autumn crocus roots (×210). m = mycelium; int. m = intercellular mycelium; v = vesicles; a = arbuscules; sp = sporangioles; end = endodermis; ap = as = epidermis.

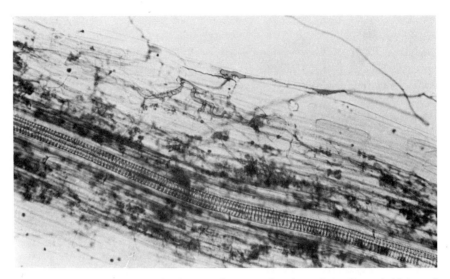

PLATE 15. *Endogene* mycorrhiza in an onion root showing arbuscules and thickened hyphae (appressoria) within the root, and mycelium outside the root.

As far as is known, although the distribution of the different shaped spores differs in different soils, even in neighbouring wheat fields at Rothamsted, the specificity of the fungus for plant roots does not appear to depend on the shape of the spore.

These spores only germinate when a root grows near them, and they appear to need both the presence of the root and of a rhizosphere pseudomonad for germination. After germination the hypha may penetrate into only one epidermal cell or it may grow along the surface of the root before it sends branches into several. Entry only causes a slight yellowing of the cell that is penetrated, in contrast to the browning and necrosis of the cell produced by the entry of most other fungi. The hypha then grows down into the deeper cortical cells and soon produces short-lived arbuscules in them. Later they produce vesicles on or between the outer cortical cells, and sporocarps in the soil; but they do not send hyphae either into the vascular system or the endodermal cells, so cease to infect the plant once the cortical cells have died. These mycorrhizas are therefore short-lived structures unless the root remains juvenile for a long time, as may happen to roots of some trees and shrubs over winter. Each new rootlet is infected afresh from the soil. Although epidermal penetration is the normal mode of entry, in some grasses entry is through root hairs and almost every root hair may have had an infection.[1]

1 T. H. Nicolson, *Trans. Brit. Mycol. Soc.*, 1959, **42**, 421.

The hyphae probably only grow actively in the soil if they are attached to a plant root. The main hypha is dark and thick-walled and has a diameter of 20 to 27 μ, with branches having diameters of 7·5 to 10 μ, and there are thin-walled ephemeral laterals 3 to 5 μ in diameter. These ephemeral hyphae ramify as a haustorial form on many bits of decomposing plant tissue in the soil,[1] which presumably indicates that they are feeding organs. The mycelia normally extend up to 1 cm from the root, they become more developed as the cortical cells age, and they produce the sporocarps already mentioned.

Endogone mycorrhiza are more common on soils low in nutrients than on soils well supplied. Thus, plants growing on soils low in available nitrogen,[2] phosphate,[3] manganese[4] or zinc[5] may have up to twenty infection points per millimetre of root, while there may be under one per millimetre if the plant is growing on a well-manured soil.[6] Correspondingly, if soils are sterilised by

TABLE 13.2   Response of onions to inoculation with *Endogone*. Harvested after 10 weeks

	Soil sterilised by gamma radiation		Unsterilised soil		Standard error
	Inoculated	*Control*	*Inoculated*	*Control*	
Height of plant, cm	29·8	13·3	26·7	19·9	1·16
Bulb diameter, cm	1·89	0·38	1·34	0·48	0·11
Fresh weight, g	13·2	2·3	9·4	3·9	0·78

gamma radiation, plants inoculated with *Endogone* grow much better than uninoculated if the soil is low in any nutrient than if it is well supplied; and even in an unsterilised soil that is low in phosphate, for example, a planted seedling that already contains mycorrhiza will initially grow much faster than a corresponding one that is free of any; though later on in the season, after these have become infected, they also will start growing more rapidly. This effect is shown in Table 13.2 taken from the work of B. Mosse and her associates.[7] Similar results have been obtained by C. D. Holevas[8] for the response of strawberries to inoculation on low phosphate but not high phosphate soils at East Malling. On the other hand, some observations of Mosse

1 T. H. Nicolson, *Trans. Brit. Mycol. Soc.*, 1959, **42**, 421 and B. Mosse, *13th Symp. Soc. Gen. Microbiol.*, 1963, 146.
2 For wheat at Rothamsted see D. S. Hayman, *Trans. Brit. Mycol. Soc.*, 1970, **54**, 53.
3 D. S. Hayman, *Trans. Brit. Mycol. Soc.*, 1970, **54**, 53.
4 For an example from S. Australia see G. Samuel, *Trans. Roy. Soc. S. Aust.*, 1926, **50**, 245.
5 A. E. Gilmore, *J. Amer. Soc. hort. Sci.*, 1971, **96**, 35.
6 T. H. Nicolson, *Trans. Brit. Mycol. Soc.*, 1959, **42**, 421; 1960, **43**, 132. For an example with strawberries and apples at East Malling, Kent, see B. Mosse, ibid., 1959, **42**, 439.
7 *Nature*, 1969, **224**, 1031; *New Phytol.*, 1971, **70**, 19, 29.
8 *J. Hort. Sci.*, 1966, **41**, 57.

and Hayman[1] suggest that the fungus is intolerant of strongly acid soils, for mycorrhiza was absent from plants growing on these soils, nor would plants become infected if the soil was inoculated with fungal spores.

Endogone is not the only fungus that forms these mycorrhizas, although it is much the most common. Under some conditions, probably rare, a *Pythium*, related to *P. ultimum*, forms vesicles and arbuscules in root cortical cells.[2] Some Rhizoctonias commonly form pelotons or coils of hyphae in the outer cortical cells, and this type of infection is often as common as that producing arbuscules. When both are found in the cortex, it is assumed that the double infection of both an *Endogone* and a *Rhizoctonia* has taken place. Under some conditions Rhizoctonias will also fill three or four outer cortical cells with oidia or oval beaded cells 6 to 10 μ in size joined by short lengths of hyphae 1 to 2 μ in diameter.[3] G. Stevenson[4] has also shown that a number of fungi which live on the root surface, which included species of *Mortierella*, *Moniliopsis* as well as of *Pythium* and *Rhizoctonia*, may act as weak mycorrhizal fungi, although any intracellular intrusions they produce only have a small surface area. Her actual observations were that a number of pioneer species of plants growing on wasteland of low nutrient status had their growth markedly improved if their roots were infected by any of these fungi, but she did not show if this improvement was brought about by fungal excretions into the rhizosphere which were taken up by the root, or by inter- or intracellular hyphae excreting nutrients into the root cells.

These root-inhabiting fungi will become parasitic on the root if the plant loses vigour for any reason, and the same result appears to hold for *Endogone*. Thus, H. S. Reed and T. Fremont[5] found that citrus trees in some Californian orchards carried numerous mycorrhizas which were perfectly normal in healthy trees, but which became parasitic in trees growing under unfavourable conditions; A. A. Hildebrand[6] found the same results for strawberries.

1 *Rothamsted Ann. Rept.*, 1968, 85; 1969, 95.
2 L. E. Hawker, A. M. Han, *et al.*, *Nature*, 1957, **180**, 998; *Trans. Brit. Mycol. Soc.*, 1957, **40**, 375, and B. Mosse, *13th Symp. Soc. Gen. Microbiol.*, 1963, 146.
3 T. H. Nicolson, *Trans. Brit. Mycol. Soc.*, 1959, **42**, 421.
4 *Trans. Brit. Mycol. Soc.*, 1964, **47**, 331.
5 *Phytopath.*, 1935, **25**, 645; *Rev. d. Cytologie*, 1935, **1**, 327.
6 *Canad. J. Res.*, 1934, **11**, 18; 1936, **14** C, 11.

14

The decomposition of plant material

The plant constituents

These form the primary material both for the food of the soil organisms and for the production of soil organic matter. They can be divided into three main groups of material: the cell contents, the reserve food supply, and the cell wall and structural material. The first group is rich in proteins and sugars, and the second in starches, fats and proteins. The third group consists of two fairly distinct materials, the skeletal framework and cementing and encrusting substances.

The skeletal framework is built up from cellulose and cellulosan fibres. Pure cotton cellulose consists of chains of glucose residues linked at the $1:4\beta$ positions, as shown in Fig. 14.1, but structural cellulose also contains some xylose and glucuronic acid residues similarly linked in the glucose residue chain.[1] The cellulosans of the angiosperms are shorter chains, mainly of xylose and a few uronic acid residues similarly linked, as in Fig. 14.1, and in the gymnosperms there are also chains of mannose residues, but their linkage has not been definitely established. The architecture of the skeletal framework is still under discussion; it is probably built up of micelles containing about a hundred oriented cellulose chains, each containing several hundred glucose residues with up to 20 to 30 per cent by weight of shorter cellulosan chains similarly oriented.[2]

The cementing and encrusting material consists of two distinct groups of substances: polymerised sugar units and the lignins. The former predominates in the young shoot, but the latter in the mature. Initially the lignin only covers the cell walls and the outside of the cellulose micelles, but as the tissue matures it may encrust the fibres in the micelle also. The carbohydrates are normally arbitrarily divided into those substances soluble in water or

1 A. G. Norman, *Biochem. J.*, 1936, **30**, 2054; with W. H. Fuller, *J. Bact.*, 1943, **46**, 281; S. P. Saric and R. K. Schofield, *Proc. Roy. Sci.*, 1946, **185** A, 431.
2 See K. H. Meyer, *Natural and Synthetic High Polymers*, New York, 1942, for a description of the constitution of cellulose, and A. G. Norman, *The Biochemistry of Cellulose, the Polyuronides, Lignin, etc.*, Oxford, 1937, for a general discussion of these problems. For a review of the chemistry of the polysaccharides, see A. G. Norman, *Ann. Rev. Biochem.*, 1941, **10**, 65.

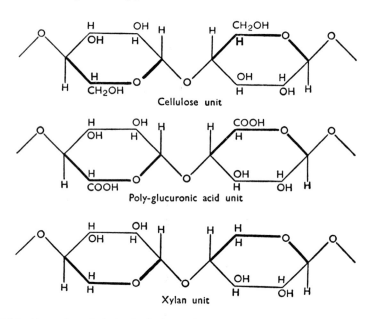

Fig. 14.1. Formulae of chains built from glucose units through $1:4\beta$ linkages

dilute acids—the pectins, gums and mucilages—and those soluble in dilute alkali—the hemicelluloses—though this group contains some of the cellulosans already mentioned. These substances are built up of either glucose and the corresponding glucuronic acid and xylose residues, or else galactose and the corresponding galacturonic acid and arabinose residues, though mannose and other residues sometimes occur. The gum molecules probably have short main chains with many shorter side-chains[1]; pectin is possibly long chains of galacturonic acid residues linked through the $1:4\alpha$ positions, as in starch; and the hemicelluloses short chain molecules containing an appreciable proportion of pentose and uronic acid residues. But our knowledge of these substances is imperfect, and there are not yet any accepted methods in general use for separating and purifying the individual hemicelluloses, so that the interpretation of experimental results is complicated by the possibility that other molecules may be present besides the type under investigation.

Lignins are richer in carbon and poorer in oxygen than cellulose, their typical analyses being

	Carbon	*Hydrogen*	*Oxygen*
lignin	61–64%	5–6%	30%
cellulose	44·5%	6·2%	49·3%

1 For a review of the constitution of plant gums, see E. L. Hirst, *J. Chem. Soc.*, 1942, 70.

but the lignins of most agricultural crops differ from carbohydrates in containing nitrogen.[1] Thus the lignins of leguminous plants may contain up to 3 per cent, and while those of the grasses are usually lower, from 1·5 to 2·0 per cent, they overlap those of the legumes. The lignin of cereal straw has, however, a much lower nitrogen content, 0·5 per cent being perhaps a typical figure. The lignins of wood have a still lower nitrogen content.

Lignins are composed of substituted phenyl-propane groups linked together.[2] The phenyl group has a hydroxyl in the *para*-position and often a methoxyl on the *meta*, in which case it is known as the guaiacyl radicle, and perhaps sometimes with a second methoxyl, as in the syringyl radicle. The propane may have a double bond, i.e. be a propene, and also a hydroxyl.

Phenyl propane   Guaiacyl radicle   Syringyl radicle

The position of the nitrogen is unknown, but it is probably present as a tertiary amine in a linkage similar to that in pyridine. Bondi and Meyer claim to have shown that the lignins of grasses, cereals and legumes are built out of three of these units and hence have a molecular weight of about 650. Further, they showed that grass lignins contained two methoxyls and leguminous lignins one methoxyl per molecule, while each contained two phenolic and one aliphatic hydroxyls. The third phenolic hydroxyl that should have been present presumably takes some part in the polymerisation process.

The decomposition of plant residues

The biochemistry of the rotting of plant residues in the soil has not been followed in any detail because of the lack of suitable analytical methods, though the rotting or composting of plant residues, and particularly cereal straws, outside the soil has been studied in more detail. There is first of all the difficulty of specifying the composition of the plant tissue, and it may also be necessary to specify the actual pattern of distribution of the various components in the cell walls of the plant tissue. Then decomposition involves microbial activity and growth, so that associated with the disappearance of

1 A. Bondi and H. Meyer, *Biochem. J.*, 1948, **43**, 258; E. R. Armitage, R. de B. Ashworth and W. S. Ferguson, *J. Soc. Chem. Indust.*, 1948, **67**, 241.
2 For a review of the structure of lignins, see H. Hibbert, *Ann. Rev. Biochem.*, 1942, **11**, 183; E. G. V. Percival, *Ann. Rep. Chem. Soc.*, 1942, **39**, 142.

some components of the plant tissue, there is an associated appearance of microbial protoplasm and cell wall constituents, which in some respects are similar to the corresponding plant cell contents and wall.[1] In addition, some plant tissues may undergo biochemical changes brought about by the micro-organisms without the material being absorbed by the organism, and it is likely that some of the plant lignins are involved in this type of reaction.

The general course of the decomposition of plant tissue, such as cereal straw, outside the soil under moist aerobic conditions is reasonably under-stood. Initially, the hemicelluloses encrusting the cellulose fibres are strongly attacked, mainly by fungi, and this attack is accompanied by a considerable evolution of heat and carbon dioxide. The most prominent feature of the attack, once this initial stage is over, is the loss of cellulose, which accounts for the major part of the loss of organic matter. Thus, Norman,[2] working

TABLE 14.1   The disappearance of hemicelluloses and celluloses from rye straw during aerobic decomposition. Temperature of decomposition 35°C. Extra nitrogen added as ammonium carbonate. All weights calculated on the quantities initially present in 100 g organic matter

Duration of decomposition in days	0	4	8	24	84
Organic matter	100	97	78	62	49
Cellulose	48	47	33	25	20
Associated xylan	9·0	9·1	8·9	5·4	4·2
Groups in the hemicellulose fraction:					
Uronic acid	4·5	4·2	3·0	2·3	1·8
Anhydropentose	17·0	13·8	7·4	5·5	4·2
Pectin as calcium pectate	0·35	0·35	0·41	0·62	0·85

with oat straw rotting aerobically under favourable conditions, found that 100 g of oat straw, containing 23 per cent of free hemicelluloses and 44 per cent of pure cellulose, lost 10 g of hemicellulose and under 1 g of cellulose in the first four days, an additional 1·5 g of hemicellulose and 4 g of cellulose in the next four, and an additional 1·25 g of hemicellulose but 11 g of cellulose in the next eight days; the maximum temperature was reached on the seventh day, but the maximum rate of heat production occurred between the second and sixth days. The reason for the slowing down in the rate of hemicellulose decomposition when only half has disappeared is presumably due to the inhomogeneity of this fraction. Finally, the products of microbial activity appear to be the principal products undergoing decomposition. Table 14.1 gives a further illustration of these results.[3]

1 For an account of the composition of the body substances of bacteria, see J. R. Porter, *Bacterial Chemistry and Physiology*, New York, 1946.
2 *Ann. Appl. Biol.*, 1930, **17**, 575.
3 A. G. Norman, *Biochem. J.*, 1929, **23**, 1367.

The course of lignin decomposition is difficult to follow for analytical reasons. Lignin is resistant to microbial attack and relatively few species can decompose it. Since it encrusts some of the cellulose fibres in the plant cell wall, it affects the decomposition of plant residues by rendering the fibres inaccessible to the microflora. Thus 15 per cent of lignin seriously reduces the rate of decomposition of cellulose; 20 to 30 per cent, as is common in woods, slows up the decomposition so much that they cease to have any agricultural value as a source of humus; and a content of around 40 per cent, as occurs in coir, renders the fibre extremely resistant to decomposition.[1]

Lignin decomposition can differ from that of the polysaccharides in that the micro-organisms attacking the fibres of the latter typically hydrolyse the whole fibre and absorb the products into their protoplasm, although they may excrete saccharides surplus to their requirements as polysaccharide gums. But organisms which decompose lignins may only absorb a portion of the products of hydrolysis, the unabsorbed part undergoing chemical reactions outside the living cell, possibly catalysed by enzymes excreted by the decomposing micro-organisms, to give some of the components present in soil humus.

The course of the decomposition depends on the relative rates at which the heat is produced and dissipated in the initial stages, for these determine the temperatures at which the decomposition takes place, and hence the composition and activity of the microflora. These relative rates are controlled by the decomposability of the substances, the water supply and the aeration; a moist, loose, well-aerated heap decomposes and produces heat more rapidly than a compacted badly aerated heap, but if the heap is too loose it will also lose its heat rapidly.

The effect of the temperature at which decomposition is taking place has not yet been worked out in detail. The general results seem to be that in the temperature range 5° to 30°C,[2] the lower the temperature of decomposition the slower is the process, and after it has slowed down the smaller is the loss of organic matter and the higher is the organic nitrogen content of the compost. Low temperatures, in fact, as H. L. Jensen[3] has been able to show, directly encourage the accumulation of microbial protoplasm. At very high temperatures,[4] in the range 45° to 75°C, both the loss of organic matter and the organic nitrogen content of the compost decrease with increasing temperature, probably due to the increasing restriction of the microflora to specialised thermophilic organisms, and to their tendency to attack proteins rather than cellulose. In the range 30° to 45°C,[5] increasing temperature may

1 See, for example, W. H. Fuller and A. G. Norman, *J. Bact.*, 1943, **46**, 291; W. J. Peevy and A. G. Norman, *Soil Sci.*, 1948, **65**, 209.
2 S. A. Waksman and F. C. Gerretsen, *Ecology*, 1931, **12**, 33.
3 *Proc. Linn. Soc. N.S.W.*, 1939, **54**, 601.
4 S. A. Waksman, T. C. Corden and N. Hulpoi, *Soi Sci.*, 1939, **47**, 83.
5 A. G. Norman, *Ann. Appl. Biol.*, 1931, **18**, 244.

not have any marked effect on the loss of organic matter, but may allow the organic nitrogen content of the compost to increase.

The nitrogen demands of the microbial population during decomposition are of obvious practical importance. On the whole, material containing less than 1·2 to 1·3 per cent nitrogen, on the dry weight basis, when rotting in the presence of ammonium salts, will cause some of the ammonium to be taken up and converted into organic nitrogen compounds. If the material contains more than 1·8 per cent nitrogen, some of this nitrogen will be converted into ammonium ions during the rotting process, though if the nitrogen content is not much in excess of this, ammonium may be taken up and fixed in the initial stages of the decomposition. Materials with intermediate nitrogen contents, between 1·2 and 1·8 per cent, tend to have no net effect on the level of ammonium during the decomposition, though they also often take up ammonium initially and release it again later.[1] Figure 14.2, taken from some

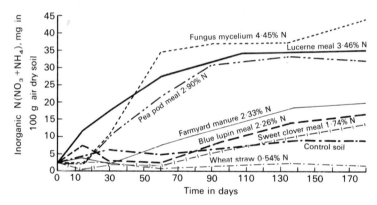

FIG. 14.2. Rate of accumulation of ammonium and nitrate in a loam soil to which 5 per cent of various organic substances have been added

of H. L. Jensen's work,[2] illustrates these points. Thus, the rotting of nitrogen-poor plant or animal remains in a soil will lower its content of mineral nitrogen, i.e. ammonium and nitrate ions, whilst the rotting of a nitrogen-rich material will increase it. But these effects depend somewhat on the type of material being rotted, and neither the nitrogen content alone, nor the ratio of the carbon in the material to the nitrogen, is a safe guide to the effect of the rotting process on the mineral nitrogen level in the soil.

1 For an example of this, see N. H. Parberry and R. J. Swaby, *Agric. Gaz., N.S.W.*, 1942, **53**, 357. I am indebted to Dr R. J. Swaby for showing me a fuller report of these experiments containing the nitrogen contents of the various materials used. C. N. Acharya *et al.* (*Indian J. Agric. Sci.*, 1946, **16**, 178) obtained somewhat similar results.
2 *J. Agric. Sci.*, 1929, **19**, 71.

An experiment of Waksman and F. G. Tenney[1] illustrates these effects well. They added ground-up plant roots of varying nitrogen content to a soil and measured its nitrate nitrogen content during decomposition. They added to 13 kg of soil sufficient material to contain 600 mg of nitrogen, incubated it for three months, then leached the soil well and determined the nitrate nitrogen in the leachates. The results, which are given in Table 14.2, show that ground clover roots, with a nitrogen content of 1·7 per cent, contains just about enough nitrogen for its own decomposition, but the other plant roots remove nitrogen from the soil during this process.

The nitrogen demands during the rotting of a material can be specified by its nitrogen factor, which is defined as the number of grams of nitrogen,

TABLE 14.2   Relation between the nitrogen content of various plant roots and the nitrate nitrogen in the soil

Material used	Nitrogen content of material Per cent	Weight of material added Grams	Nitrate nitrogen in leachings Milligrams	Gain or loss of nitrogen Milligrams
Control soil	—	—	947	—
Oat roots	0·45	133	207	− 740
Timothy roots	0·62	97	398	− 549
Maize roots	0·79	76	511	− 436
Clover roots	1·71	35	925	− 22
Dried blood	10·71	5·6	1750	+ 803

in the form of ammonium ions, immobilised during the decomposition of 100 g of the material. The nitrogen factor, however, is not a constant for a given material, but depends somewhat on the conditions prevailing during the decomposition. However, E. H. Richards and A. G. Norman,[2] who introduced it to specify the amount of nitrogen that must be added to heaps of vegetable waste in the field to allow rapid decomposition, found that in farm practice it was sufficiently constant, both during the course of a rot and for a given material rotting in compost heaps under natural conditions, to be of practical value in farm advisory work. In the first place the nitrogen factor is not quite constant during the course of a rot, as is shown both in Table 14.3, taken from Richards and Norman's work, and also by the fact, already mentioned on p. 266, that materials with a small negative nitrogen factor may remove ammonium and nitrate nitrogen, if available, during the early stages of decomposition and release it later on. In the second place, the nitrogen factor of a given material cannot be predicted from its nitrogen

1 *Soil Sci.*, 1927, **24**, 317.
2 *Biochem. J.*, 1931, **25**, 1769.

TABLE 14.3   Effect of duration of decomposition on the nitrogen factor of oat straw. Nitrogen content of straw, 0·30 per cent

Duration of decomposition in days	8	16	24	48
Per cent loss of organic matter	13	25	37	50
Nitrogen factor	0·75	0·67	0·75	0·80

content. It is usually higher for natural materials, such as straw, than for pure substances, such as cellulose or straw treated to remove some of its constituents, as is shown in Table 14.4. The nitrogen in a material is not all equally available to micro-organisms: they may use ammonium salts, if

TABLE 14.4   Effect of treating straw on its decomposability and nitrogen factor

Oat straw extracted with	Composition of straw					Decomposition for 48 days	
	Cellulose and associated cellulosans	Hemicellulose as uronic acid anhydride	anhydro-pentose xylose	Lignin	Per cent nitrogen	Loss of organic matter	Nitrogen factor
Control	49	5·4	11·5	18·5	0·30	50	0·80
Hot water	51	6·2	13·3	—	0·27	18	0·51
4 per cent NaOH	84	1·0	1·5	11·1	0·12	21	0·16
4 per cent alcoholic NaOH	66	3·2	7·5	8·2	0·11	21	0·36
2 per cent HCl	62	2·4	1·0	25·6	0·34	9	0·27

present, in preference to difficultly available nitrogen compounds in the material. Thus, Richards and Norman found that bean husks, which had a nitrogen content of 3 per cent, rot to liberate ammonium in the absence of readily available nitrogen, but immobilise some nitrogen in its presence as is shown in Table 14.5.

TABLE 14.5   Effect of added nitrogen on the nitrogen factor of rotting bean husks. Nitrogen content of bean husks, 3 per cent. Duration of decomposition: 48 days

	Loss of organic matter %	Nitrogen content of compost %	Nitrogen factor
No added nitrogen	48	4·7	−0·52
With added ammonia	41	5·4	+0·15

The rate of decomposition of a material with a low nitrogen content can be increased considerably if a suitable source of nitrogen is added, for the nitrogen supply obviously limits the maximum amount of protoplasm that can be present, and, if it is too low, it limits biological activity. The added nitrogen may also allow a larger proportion of the carbonaceous organic matter to be converted into microbial protoplasm or its byproducts, hence it may increase the proportion of the added organic matter that is converted into humic substances.[1] This forms the principle underlying all processes for rapidly obtaining a compost or manure from straw without passing it through a stockyard.[2] Since its nitrogen factor is about 0·8, at least 8 kg of nitrogen should be added per ton of straw to ensure its rate of decomposition is not limited by nitrogen shortage. This is commonly added as ammonium sulphate, but any other material rich in nitrogen which yields ammonium ions, such as urine, animal droppings, dried blood or young green plant material is equally suitable. Nitrate–nitrogen is not used as readily as ammonium–nitrogen,[3] and it is also more liable to loss by denitrification. It has been claimed that organic nitrogen compounds give a mellow and friable compost in contrast to inorganic nitrogen compounds, which give a more sticky compost, possibly because the former encourage a greater development of fungi compared to bacteria, but there is little critical evidence for this statement.

The actual amount of mineral nitrogen locked up when material with a low nitrogen content is decomposing depends on several factors, not all of which have been well established. For a given carbon and nitrogen content in the decomposing material, the amount of soil nitrogen locked up during the decomposition increases as the speed of decomposition increases. Hence, materials low in lignins typically immobilise more nitrogen than those high in lignins,[4] and decomposition in warm soils immobilises more nitrogen than in cool soils. Thus L. A. Pinck, F. E. Allison and V. L. Gaddy[5] found that under Maryland conditions a ton of wheat straw immobilised about 7 to 8 kg of nitrogen if decomposing during the winter, but between 10 and 13 kg during the summer. Again, decomposition under acid conditions appears to lock up more nitrogen than under neutral conditions,[6] though the significance of this result is not known.

H. F. Birch[7] found that rotting Kikuyu grass (*Pennisetum clandestinum*) containing 2 per cent nitrogen released no mineral nitrogen over a period of 33 days if the heap was kept moist, but it released some if it was dried and rewetted during this period. Correspondingly, a sample containing 1·5 per cent nitrogen released some mineral nitrogen if dried and wetted three times

1 F. J. Salter, *Soil Sci.*, 1931, **31**, 413.
2 H. B. Hutchinson and E. H. Richards, *J. Min. Agric.*, 1921, **28**, 398.
3 S. L. Jansson, M. J. Hallam and W. V. Bartholomew, *Pl. Soil*, 1955, **6**, 382.
4 E. H. Richards and A. G. Norman, *Biochem. J.*, 1931, **25**, 1769; E. J. Rubins and F. E. Bear, *Soil Sci.*, 1942, **54**, 411.
5 *J. Amer. Soc. Agron.*, 1946, **38**, 410.
6 H. L. Jensen, *J. Agric. Sci.*, 1929, **19**, 71.
7 *Pl. Soil*, 1964, **20**, 43.

during this period. The cause of this effect was not determined, but it is presumably due in part to the increase in the CO_2 production brought about by the drying treatment, and possibly also to the decomposition of some of the microbial protoplasm produced during the wetting periods and killed by the drying treatment.

The process of decomposition may lock up other nutrients besides nitrogen, or be retarded by their lack. The ash of microbial protoplasm may contain up to 25 per cent of phosphorus and 10 per cent of potassium,[1] so that the micro-organisms may compete with growing crops for the available supplies of these nutrients if the soil or the decomposing organic matter is very low in them. I. P. Mamchenko[2] has brought forward some evidence that increasing the phosphate supply of material well supplied with nitrogen, such as clover residues, increases the proportion of carbonaceous material converted to humic substances by reducing the proportion respired as carbon dioxide. There is also some evidence that the rate of decomposition of tree leaves depends on the cation content of the leaves,[3] though whether this is due simply to the effect of the bases in controlling the pH of the rotting material and therefore the composition of the microflora, or to their nutrient effect on the micro-organisms has not been determined.

The discussion so far has assumed that the decomposition was taking place under conditions of good aeration. If the aeration becomes poor, the rate of decomposition becomes smaller, the proportion of the soil population taking part in the decomposition becomes less, and a larger proportion of the organic carbon that is taken up by the organisms is excreted as high energy compounds and a smaller proportion is given off as carbon dioxide. These changes do not begin to be important until the oxygen content of the water bathing the micro-organisms has fallen to about 5×10^{-6} M, which corresponds to about 1 per cent of the partial pressure of oxygen in the atmosphere (see p. 415).[4] Pockets of poor aeration in a mass of decomposing plant tissue can accumulate products of anaerobic decomposition which will be rapidly decomposed if the aeration is improved.[5] This often happens when a compost heap is turned, for any volumes that had become anaerobic will typically become aerobic on turning and allow rapid decomposition of these products with a rapid evolution of heat.

In conclusion, the conditions favouring rapid decomposition of plant products are:

1 The material should have a low lignin, and probably also a low wax, content.

2 The material should be in as fine a state of comminution as possible; and

1 S. A. Waksman, *Principles of Soil Microbiology*, 2nd edn, London, 1931, p. 367.
2 *Vest. Udob. Agrotekh. Agropoch.*, 1941, No. 3, 5.
3 W. M. Broadfoot and W. H. Pierre, *Soil Sci.*, 1939, **48**, 329.
4 D. J. Greenwood, *Pl. Soil*, 1961, **14**, 360.
5 J. F. Parr and H. W. Reuszer, *Proc. Soil Sci. Soc. Amer.*, 1959, **23**, 214.

it is probable that the principal natural agents bringing this about are the larger soil animals.

3 There should be an adequate supply of available nitrogen.
4 The pH should not be allowed to fall too low, otherwise the microbial population becomes unduly restricted and the larger soil animals die out.
5 The aeration should remain good and the moisture supply adequate. Anaerobic conditions or waterlogging lead to a restricted population, mainly bacterial, though the nitrogen demands of the population fall.
6 The temperature should be fairly high. Temperatures above 45°C or below 30°C give lower rates of decomposition than temperatures in this range.
7 For substances rather difficult to decompose, mixed groups of materials decompose quicker than single groups. Thus, in forest floors, mixed litter from several species of trees usually decomposes quicker than litter from a single type.[1]

Composting

Composting is the term usually applied to the rotting down of plant and animal remains in heaps before the residue is applied to the soil. The chemical processes of decomposition in the compost heap and the soil differ in two respects: the compost heap is usually not so well aerated, and the temperature of the decomposition can be far higher in it than in the soil.

The points of importance in composting are ensuring that the water content and the aeration of the heap are within certain limits: in practice the rate and type of decomposition is controlled by varying the aeration conditions rather than the water content of the heap. If the heap is too dry, too compacted, or too waterlogged, little decomposition takes place, and if it is kept moist and loose, decomposition is at its maximum.

There appears to be no virtue in rotting down materials in the compost heap compared with letting them rot down in the soil itself as far as adding plant nutrients to the soil or increasing its humus content are concerned. The relative value of the two methods depends on circumstances. One cannot always add farmyard manure or waste organic material to the soil when it is available: one must often leave it until it is convenient to put it on the field, and composting is a way of storing these materials until they are required. Again, some soils are naturally very open and free draining, and adding undecomposed material will often exaggerate the harmful consequences of this open structure. Under these conditions it is much preferable to add well-rotted humic material, for this will tend to reduce the openness of the soil. Again, an important difference between the unrotted plant material and the compost is in physical condition. The unrotted material typically holds little water and is coarse and fibrous; the compost holds much water and is friable.

1 See, for example, J. T. Auten, *J. Forestry*, 1940, **38**, 229; F. G. Gustafson, *Plant Physiol.*, 1943, **18**, 704.

On light, sandy soils, which in any case are too open for optimum crop growth, a compost is obviously preferable on this score, while on some heavy soils the uncomposted material may be better.

A third factor, the importance of which has not yet been established, is that rotting a given amount of material in the soil will usually give a larger addition to the humus and nutrient content of the soil than if it is first composted and then added to the soil. Experiments have been made at Rothamsted for a number of years in which a given dressing of straw and nitrogen ploughed into the soil is compared with the same quantity of straw rotted with the same quantity of nitrogen in a compost heap and then applied to the land. The ploughed-in straw has always given a better yield than the compost, as is shown in Table 14.6.

TABLE 14.6   Comparison of straw composted with straw ploughed in on a three-course rotation at Rothamsted.* Yield in t/ha

| | 18 years, 1934–51 | | | |
| | Year of applying straw | | Year after application | |
	Compost	Straw ploughed in	Compost	Straw ploughed in
Potatoes	20·2	23·7	19·6	20·3
Sugar-beet: sugar	4·63	5·15	4·53	4·76
Barley: grain	3·45	3·86	3·30	3·46

* *Rothamsted Ann. Rept.* 1951, 135.

The cause of the better crop response to the rotting in the soil compared with rotting in the compost heap may be due to losses of nitrogen and potassium during the composting process. Thus C. N. Acharya and his coworkers[1] found that nearly one-quarter of the nitrogen in a mixed organic refuse, whose C/N ratio was initially 32, was lost during the first twenty weeks of composting, probably mainly by volatilisation as ammonia, but there was no net loss if some soil was added to the refuse. Presumably the ammonia set free in the decomposition is absorbed by the clay in the soil and then nitrified. Potassium is lost if any liquid drains out of the compost heap, and straw compost can be a useful source of potassium for the crop.

Composting has, however, a considerable farm health aspect. It allows many parasitic organisms in the plant and animal material to be destroyed before the residues are put back on the land, and this destruction is more complete the higher the maximum temperature attained and the better the compost has been made. High temperatures have the further advantage that all the weed seeds present are also killed and decomposed. Hence, if possible

1 *Indian J. Agric. Sci.*, 1945, **15**, 214; 1946, **16**, 90.

trouble from either of these sources is suspected, good high-temperature composting may be well worth while.

Finally, the effect of composting on the content of members of the vitamin B complex[1] and other growth-promoting substances, such as the auxins,[2] has been investigated by many workers, who all find that the amount of these substances in the heap decreases during the course of the decomposition. Composts are therefore not rich sources of these substances, but there is no evidence that any farm crop benefits in any way by increasing the levels of these compounds in the soil.

The products of microbial synthesis formed during decomposition

Many species of micro-organisms can decompose plant celluloses and hemi-celluloses in pure culture, but most of these are unable to compete with the well-adapted species in the very competitive environment of the soil or compost heap. Under moist warm aerobic conditions the initial active population contains a large proportion of fungi, most of them belonging to Garrett's sugar fungi (see p. 176), for these multiply the most rapidly while simple plant cell contents are available for decomposition. The next stage in the soil depends on the soil conditions and the plant material added. If the soil is acid, these fungi are replaced by cellulose-decomposing fungi, while if it is neutral a general bacterial population appears to become dominant, which often seems to begin on the hyphae of the sugar- and cellulose-decomposing fungi. But if the plant material contains strongly lignified tissue, fungi remain an important part of the population, for there are few bacteria which will decompose lignins. If the soil is poorly aerated bacteria are dominant from the beginning, with the consequence that the lignified tissues accumulate as peat.

There are a considerable number of species of fungi and actinomycetes which can decompose cellulosic plant material under aerobic conditions in pure culture, and a considerable number of bacterial species which can do this under conditions of poor aeration, but the number of bacterial species which decompose these tissues in aerobic conditions in the soil appear to be rather limited, although they form a very important part of the microflora in subacid and neutral soils.[3]

The products of microbial metabolism include microbial cell-wall components such as polyuronide gums and other polysaccharides, as well as humic-like substances. Many bacteria, including some of the cellulose decomposers, also produce extra-cellular polyuronide gums, particularly if they

1 R. L. Starkey, *Soil Sci.*, 1944, **57**, 247.
2 J. H. Hamence, *J. Soc. Chem. Indust.*, 1945, **64**, 147.
3 For reviews of these, see A. G. Norman and W. H. Fuller, *Adv. Enzymol.*, 1942, **2**, 239, and R. Y. Stanier, *Bact. Rev.*, 1942, **6**, 143.

are working under conditions of limited nitrogen supply;[1] for they may have to attack a large amount of polysaccharide to extract sufficient nitrogen for their growth; and since it is nitrogen and not energy supply that is limiting their growth, they have to excrete energy-rich byproducts. Some fungi produce dark-coloured humus-like products when growing on cellulose and other lignin-free plant products[2] (see p. 313). Some bacteria may also produce such products, both from the easily decomposable components in recently dead plant tissue,[3] and also from the mycelia of the fungi that originate the attack on the more resistant plant remains.[4]

The introduction of techniques based on adding decomposable organic matter to the soil labelled with ^{14}C has shown that it takes a long time for all the added carbon to become fully humified. Some results obtained at Rothamsted by D. S. Jenkinson[5] are typical of the kind of information that these techniques will give. He grew ryegrass labelled with ^{14}C, cut it when succulent and before flowering, and mixed either the tops or the roots with soil and followed the decomposition over a period of four years. There was a rapid loss of ^{14}C, as carbon dioxide, during the first six months, during which two-thirds of the labelled carbon added to the soil had been lost. During the next three and a half years a further 14 per cent of the added ^{14}C was lost, and during this period the rate of loss followed a first-order reaction with a half life of about four years. This compared with a half life of twenty-five years for the organic matter previously in the soil; this is illustrated in Fig. 14.3. In addition Jenkinson showed that although the ryegrass tops decomposed faster than the ryegrass roots, after six months the differences had disappeared. He also found that the rate of loss of labelled carbon was the same if the ryegrass had been mixed with a soil low in humus or high in humus, and that doubling the amount of ryegrass added did not affect the proportion of labelled carbon lost at different times.

Present evidence is consistent with the picture that most of the labelled carbon left in the soil has been through microbial protoplasm, and that even after four years much of it is still in an appreciably different form from the bulk of the soil humus. This can be seen from the kinds of treatment that attack heavily labelled fractions, for they typically remove only small amounts of humus. Thus, the fraction solubilised by boiling with water, or by treating with cold dilute HCl is very heavily labelled, as is the fraction that is mineralised on killing the micro-organisms with chloroform, for

1 See, for example, S. Winogradsky, *Ann. Inst. Pasteur*, 1929, **43**, 549; A. G. Norman and W. V. Bartholomew, *Proc. Soil Sci. Soc. Amer.*, 1940, **5**, 242.
2 H. L. Jensen, *J. agric. Sci.*, 1931, **21**, 81; L. A. Pinck and F. E. Allison, *Soil Sci.*, 1944, **57**, 155.
3 M. M. Kononova, *Pedology*, 1943, No. 6, 27; No. 7, 18; 1944, No. 10, 456; with N. P. Belchikova, ibid., 1946, 529.
4 F. Y. Geltser, *Sovet. Agron.*, 1940, No. 11–12, 22; also see her book *The Significance of Micro-organisms in the Formation of Humus*, Moscow, 1940; J. G. Shrikhande, *Biochem. J.*, 1933, **27**, 1551.
5 *J. Soil Sci.*, 1965, **16**, 104; 1966, **17**, 280; 1968, **19**, 25.

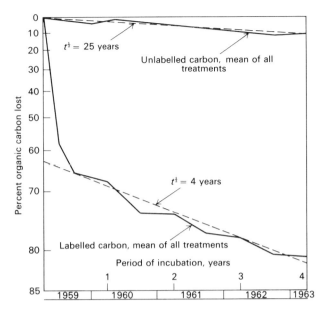

F<small>IG</small>. 14.3. Loss of labelled ryegrass carbon and unlabelled soil carbon from a Broad-balk soil

example, although the proportion of the labelled carbon in the soil that is in these fractions is small: for one soil, extraction with cold 0·1 N HCl removed only 4 per cent of the labelled carbon in the soil, but the percentage of ^{14}C in the extract was 7·0, while in the soil as a whole it was only 1·6. The interpretation put on these results is that although the biomass at any one moment only contains a small proportion of the total labelled carbon, yet labelled carbon is concentrated in the biomass as the residues of the ryegrass added several years previously are still a more important food supply for the soil population than is the mass of the soil humus.

Green manuring

Plant residues can be applied direct to the soil instead of via the compost or manure heap, and the usual method of doing this is to plough in a growing crop. This method of green manuring can have several effects on the soil, depending on conditions: it may increase the organic matter content of, or the available nitrogen in, the soil; it may reduce the loss of mineral nitrogen by leaching; and it may concentrate nutrients likely to be deficient in the surface soil and leave them there in a readily available form. Green manuring normally cannot confer all these benefits on the soil simultaneously, and it may confer none of them.

Green manuring, when properly used, can either increase the humus content or else the supply of available nitrogen in the soil, but rarely can it do both at the same time, and therein it differs from well-made farmyard manure, for the humus content only appears to be increased appreciably if material fairly resistant to decomposition is added to the soil, and resistant plant material is typically low in nitrogen; and the available nitrogen supply is only increased if readily decomposable material high in nitrogen, such as young green plants, are decomposing. Thus, the effect of a given crop as a green manure depends on its maturity when ploughed under. Table 14.7 illustrates this point for rye.[1] Young rye decomposes rapidly with a large production of carbon dioxide, and consequently leaves little residue but much nitrate nitrogen, whilst mature rye decomposes much more slowly, leaving a large residue and making demands on the soil's supply of available nitrogen.

TABLE 14.7   The effect of maturity of rye on its composition and rate of decomposition. Fresh material, containing 2 g dry matter, added to 100 g soil and incubated at 25° to 28°C for 27 days

Stage of maturity	Moisture content of material	Per cent of dry matter soluble in cold water	Nitrogen content of dry matter	Carbon dioxide evolved, mg	Mineral nitrogen liberated (+) or absorbed (−) mg
Plants 25–35 cm high	80	34	2·5	287	+ 22·2
Just before heads form	79	23	1·8	280	+ 3·0
Just before flowering, leaves and stalks	57	18	1·0	200	− 7·5
Grain in milk stage, leaves and stalks	15	10	0·24	188	− 8·9

Green manures are not as effective as farmyard manure, per unit of carbon, in increasing the organic-matter content of the soil, because of their greater decomposability. M. Chater and J. K. R. Gasser[2] found that on the Woburn sandy loam soil 25 per cent of the carbon in nineteen annual dressings of 25 t/ha of farmyard manure remained in the soil at the end of this period, but only 14 per cent of the carbon added as green manure. Similarly, C. A. Mooers[3] found that in a twenty-year experiment in which cowpeas were grown as an autumn catch crop in a continuous winter wheat experiment, the organic matter content of the soil decreased by 0·11 per cent, or by about 2·5 t/ha over the twenty-year period on the plots in which the cowpeas,

1 S. A. Waksman and F. G. Tenney, *Soil Sci.*, 1927, **24**, 317.
2 *J. Soil Sci.*, 1970, **21**, 127.
3 *Tennessee Agric. Expt. Sta.*, *Bull.* 135, 1926.

yielding about 2·5 t/ha of dry matter, had been ploughed under, and increased by about 0·11 per cent on the plots which received 10 t/ha of farmyard manure, also containing about 2·5 t/ha of dry matter, each year. The 50 t of cowpea dry matter, however, reduced the loss of humus from the soil, for the humus content fell by 0·24 per cent on other plots that did not receive either farmyard manure or green manure. Hence, about 50 t of cowpea dry matter increased the humus content of the soil by 2·7 t, and 50 t of farmyard manure dry matter by 8 t compared with the untreated.

Green manuring has been suspected of hastening the decomposition of soil humus. The evidence for this has come from studies on the effect of adding decomposable organic matter labelled with radio-carbon or heavy nitrogen to a soil, and determining the effect of this decomposition on the rate of decomposition of the organic matter already present in the soil. Table 14.8 gives the result of such an experiment made in 1948 by F. E. Broadbent,[1] who used sudan grass labelled with ^{13}C and ^{15}N as the green manure. This experiment shows that the active decomposition of the sudan grass has increased the rate of production of carbon dioxide from the humic matter in the soil threefold and of mineral nitrogen twofold, but subsequent work has not confirmed the universality of this effect, and, in fact, several workers[2] have suspected that if it does occur, it is very small.

TABLE 14.8   The effect of additions of decomposable matter in increasing the rate of decomposition of the soil humus. Amounts liberated in milligrams

	Soil plus sudan grass	*Soil alone*		*Soil plus sudan grass*	*Soil alone*
Total CO_2 evolved	283	19	Total mineral nitrogen released	4·83	1·40
Part derived from sudan grass	225	—	Part derived from sudan grass	1·93	—
Part derived from soil	58	19	Part derived from soil	2·89	1·40

Green manuring has, in most parts of the world, been applied more successfully to increasing the available nitrogen supply than the humus content of soils. It is, however, only widely practised under certain definite conditions. The crop must be grown as a catch crop between the main cash crops, and consequently the practice is commonly commended in regions where the winters are sufficiently mild and moist to allow the crop to grow during this period. The crop must not compete with the main crops in any

1 *Proc. Soil Sci. Soc. Amer.*, 1948, **12**, 246.
2 See, for example, papers by D. S. Jenkinson and by D. Sauerbeck in *The Uses of Isotopes in Soil Organic Matter Studies*, Pergamon, 1966.

way, and in particular not for water, so that this system is not used under dry-farming conditions.

Hence, regions having a long growing season, such as regions with mild winters and with a well distributed rainfall, or else an adequate irrigation system, are well suited to green manure crops. The practice is prevalent in the humid tropics and subtropics, and in orchards when the growing season extends into the dormant period of the trees; it is also extensively advocated in the south-eastern states of the USA, where maize and cotton, both summer crops, are the two main cash crops, and where legumes will grow during the winter. Green manure crops can be particularly valuable in saline soils, because in comparison with fallow, they reduce the evaporation from, and hence the salt content in, the surface soil, for they take their water from the subsoil and shade the surface. Further, when they are ploughed in, their residues help to increase the availability of phosphates and trace elements to the succeeding crop due to the lowering of the soil pH brought about by the carbon dioxide produced in the process of decomposition.

Leguminous crops, such as some vetches, peas and clovers, are commonly used for green manuring as they increase the soil's supply of nitrogen. But these crops will normally only make adequate growth and fix enough nitrogen to make their cultivation worth while if the soil contains an adequate supply of calcium, potassium and phosphates. An example of the need for this is given in Table 14.9,[1] taken from some experiments made on a soil with pH 5·5 in Mississippi. In this example 31 and 62 kg/ha N in the green manure gave rather more cotton than 20 and 40 kg/ha N in a mixed fertiliser.

The nitrogen liberated during the decomposition of the green manure crop can only benefit the subsequent crop if the latter is sufficiently developed to take up the nitrogen soon after it is released and before the nitrate produced is leached out of the soil; and this period is fairly short, particularly in light soils under warm moist conditions, as the protein of the living green plant decomposes more rapidly than does that in dried or dead plant material. Hence, a long, wet period between the ploughing-in of a green manure crop and the establishment of the following crop, particularly on light soils, can result in much of the nitrifiable nitrogen of the green crop being leached out of the soil with the consequence that the following crop obtains little, if any, benefit from the green manure.[2] On the other hand, the main crop should not be sown too soon after the green crop has been turned in, as the period covering the first flush of decomposition is very unfavourable for germination and the growth of very young plants.

Green manure crops can confer other benefits on the land: by growing during wet off-seasons they reduce the loss of nitrogen and other nutrients

1 C. D. Hoover, *Proc. Soil Sci. Soc. Amer.*, 1942, **7**, 283.
2 See E. M. Crowther in *Fifty Years of Field Experiments at the Woburn Experimental Station*, by E. J. Russell and J. A. Voelcker, for a discussion of the causes of the failure of the Woburn Green Manuring Experiment to benefit the soil or the succeeding crop.

TABLE 14.9   Influence of fertilisers for green manure crop on the yield of cotton. Experiment at Holly Springs, Mississippi. Winter green manure crop: Hairy vetch

	Yield of seed cotton t/ha	Yield of vetch	
		Dry matter t/ha	Nitrogen kg/ha
(a) No green manure before cotton			
Fertiliser applied to cotton:			
335 kg/ha 6–8–4	1·??	—	—
Ditto, plus 112 kg/ha nitrate of soda	1·61	—	—
(b) With green manure, but no fertilisers applied to cotton			
Fertiliser applied to vetch:			
None	1·05	0·52	19
Phosphate	1·31	0·81	31
Dolomitic limestone	1·66	1·31	63
Limestone and phosphate	1·74	1·36	64

by leaching; they can often utilise less available forms of phosphate[1] and zinc than the main crop and, hence, increase the availability of these for the crop; and by decomposing rapidly they liberate large quantities of carbon dioxide in the soil which can increase the availability of phosphates in alkaline calcareous soils. These benefits may be of particular value in deciduous orchards, for they have a long dormant season. The practical difficulties in growing these off-season cover crops for green manuring are to get them established without interfering with harvesting or competing with the trees for nutrients before harvest, and to manage them so they are not killed prematurely by the winter sprays and washes used.[2]

Green manure crops can ease problems of soil management in several ways. R. Fauck[3] found that ploughing in a suitable green manure crop on some Senegal soils of unstable structure helped to keep the surface open, and allowed a greater proportion of the rainfall to penetrate into the soil. The cash crop, in this case, was responding to the additional water in the soil.

Ploughing in a green crop can have another function that has barely been examined yet. It is possible that the best way of ploughing in straw is to undersow the corn crop with a green manure crop, probably a legume, and under these conditions the benefit conferred by the legume on the

1 For examples from S. Africa see E. R. Orchard and A. L. Greenstein, *Un. S. Africa, Dept. Agric., Sci. Bull.* 290, 1949.
2 For an example of their management in English orchards, see W. S. Rogers and Th. Raptopoulos, *J. Pomol.*, 1946, **22**, 92.
3 *Trans. 5th Int. Congr. Soil Sci.*, 1954, **3**, 155.

succeeding crop appears to be enhanced. F. C. Bauer[1] has given the results of an experiment in Illinois on the three-course rotation, maize-oats-wheat, in which the effect of undersowing the wheat with sweet clover and returning the straw to the wheat stubble was investigated. The results for the first twelve years of the experiment were:

	Yield of maize in t/ha		
Residue returned to wheat stubble	*No sweet clover*	*Sweet clover under-sown in wheat*	*Benefit of sweet clover*
None	3·6	4·2	0·6
Wheat straw and maize	3·7	5·0	1·3

and they show that undersowing the wheat with sweet clover benefited the following maize crop, but this benefit was almost doubled if the wheat and maize straw were put on the sweet clover ley, while the straw alone had no effect.

It is a common practice for farmers in several parts of the world who crop their land fairly continuously with a corn or maize crop to undersow the crop with a legume. Thus, wheat growers in Missouri commonly undersow their wheat with lespedeza, which can be made both to seed itself for the following year and to provide a hay crop; and Iowan farmers often sow sweet clover or vetch in their maize.[2]

Again, farmers on some English chalk and light soils growing a large acreage of barley will undersow the crop with trefoil (*Medicago lupulina*). The barley straw, left by the combine, is pressed into the trefoil with a ring roll, and wherever it touches the ground, will begin to decompose as soon as the soil becomes moist. This weakens the straw so that it can be ploughed in in late autumn or winter quite easily. The trefoil not only helps the straw to decompose, possibly by keeping the soil surface damp, but keeps the nitrate nitrogen in the soil low by taking it up, and in consequence helps to control 'take-all' (*Ophiobolus graminis*) in the following cereal crop (see p. 235), but it also supplies nitrogen for the decomposition of the straw when it is finally ploughed in.

Micro-organisms can produce plant toxins during the initial rapid stages of decomposition of plant residues. In general, these are of no practical significance, either because they are present in too dilute a solution to have any effect on plant roots or because they are absorbed by clays. But they may have importance under conditions when the decomposing matter is unevenly distributed in a soil, so an appreciable mass of organic matter is decomposing in a heap. Just as under some conditions coloured solutions will drain out of a compost heap,[1] so when straw is worked into the soil surface, as in stubble mulch farming, phenolic acids, such as p-hydroxy-benzoic, vanillic or proto-catechuic, may be produced in sufficient concentration to harm young wheat seedlings[2] (see p. 799).

1 For a note on some properties of these liquids, see C. Bloomfield, *Chem. Ind.*, 1969, 1633.
2 See, for example, W. D. Guenzi and T. M. McCalla, *Proc. Soil Sci. Soc. Amer.*, 1966, **30**, 214, and for a review T. M. McCalla and F. A. Hoskins, *Bact. Rev.*, 1964, **28**, 181.

15

The composition of the soil organic matter

The separation of humus from the soil particles

The composition of the soil organic matter, and in particular of the soil humus, has exercised the minds of soil chemists from the very beginning. The soil organic matter consists of a whole series of products which range from undecayed plant and animal tissues through ephemeral products of decomposition to fairly stable amorphous brown to black material bearing no trace of the anatomical structure of the material from which it was derived; and it is this latter material that is normally defined as the soil humus. The great difficulty in all investigations on the composition of soil humus has been that it can neither be separated from unhumified matter nor from the mineral constituents of the soil. The methods used separate out only a part of the humus, and some may alter the original constitution of the humus into forms that disperse more easily. Further, the composition and properties of humus cannot yet be determined precisely when it is mixed with the mineral matter of a soil in its field condition, for it is not the only organic material in the soil: the soil also contains a large number of soil organisms, though they may only constitute a few per cent of the soil carbon (see p. 219), and it also contains plant debris in various stages of decomposition.

Techniques have recently been developed, based on ultrasonic dispersion, which allow a good separation of unhumified or partially humified material from typical humus. The soil is dispersed in a heavy liquid, with a specific gravity of about 2, to which a suitable surfactant has been added and the fully humified material which is adsorbed on the soil mineral particles sinks, and the unhumified or partially humified floats. Much of the partially humified material is associated with silica in the form of phytoliths, or simply with amorphous silica, and hence has a density of up to 1·8; and this is the reason that one needs to separate out this material in a heavy liquid. G. W. Ford[1] found that most of the Australian soils he examined had

---

1 With D. J. Greenland and J. M. Oades, *J. Soil Sci.*, 1969, **20**, 291.

between 20 to 30 per cent of their carbon in this light fraction. It contains many recognisable bits of plant debris, has a carbon content between 14 and 28 per cent, and most of its ash is silica; but it probably contains some dispersed humus as it is usually coloured brown.

Humic material primarily consists of negatively charged colloids. Some of it is bound to the clay through polyvalent cations, such as calcium or aluminium, one of whose charges neutralise a negative charge on the clay and the other on the humus. Some is bound to positively charged surfaces of iron or aluminium hydrated oxides, either through electrostatic bonding or by specific anion adsorption (see p. 128), and some is held on clay surfaces through hydrogen bonding, or by van der Waals forces (see p. 147).

The classical way of separating a humus fraction from the soil, and the one that is still most widely used, is to extract the soil with sodium hydroxide, a concentration between 0·1 and 0·5 N being common, and the extraction is usually done at room temperature. This replaces the polyvalent cations by sodium and precipitates them as their hydroxides, or converts the aluminium to the aluminate anion, it increases the negative charge on the humus colloids and it decreases the positive charge on any sesquioxide surfaces. In so far as the humus is bonded to the clay through polyvalent cations or through specific adsorption, treatment with a chelating agent such as sodium pyrophosphate or acetyl acetone will be an effective dispersing agent; and J. W. Parsons and J. Tinsley[1] have shown that anhydrous formic acid is a valuable dispersing agent for certain humic fractions, including the soil polysaccharides. But no dispersing agent yet available will disperse all the humus from an agricultural soil, although some will disperse most of that present in the B horizon of a podsol. However, as more is known about the kind of substances present in humus, and the mechanisms by which they are adsorbed on soil particles, methods are being developed which will extract them more or less quantitatively.

The total quantities of certain components of the soil organic matter can be determined without separating out all the humus from the soil quantitatively. Thus, by suitable hydrolytic methods, the total content of pentose and hexose sugars can be determined, and so can the total content of a number of amino acids, provided no complicating polymerisation reactions take place between any of the products of hydrolysis.

The fractionation of the humus dispersion

The classical method for fractionating the humic colloids that disperse in the sodium hydroxide extract is to acidify the suspension with sulphuric or hydrochloric acid, which causes a part of the dispersed organic matter to precipitate. The part that stays in solution is known as fulvic acid, that which

1 *Soil Sci.*, 1961, **92**, 46; *Proc. Soil Sci. Soc. Amer.*, 1960, **24**, 198.

precipitates out as humic acid, and that part of the organic matter which does not disperse in the alkali but remains in the soil as humin. The humic acid is sometimes fractionated by treatment with ethanol, that part which disperses in the alcohol being hymanomelanic acid. The divisions thus made are shown in the following diagram.

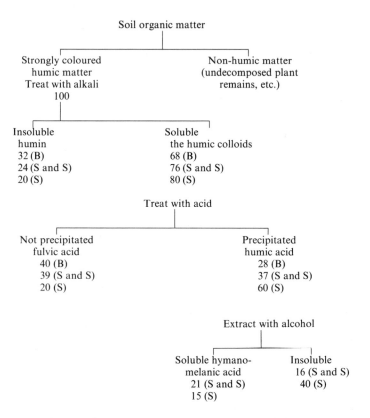

The figures under the fractions show how the carbon is divided among them, as determined by M. Berthelot and G. André (B).[1] O. Schreiner and E. C. Shorey (S and S)[2] and A. Schmuck (S).[3] Although these fractions are given definite names, they are not homogeneous. Each contains particles with a wide range of molecular weights and constitution, nor is there any clear-cut division between the fractions. Thus the proportion of the humic matter that is precipitated by the acid depends on the type of acid used and its strength; and the fractionation could equally easily be made by adding a salt instead of an acid, particularly if the alkali is first removed with a suitable H^+-saturated exchange resin. Again, there is no clear distinction

1 *Ann. Chim. Phys.*, 1892, ser. 6, **25**, 364.
2 *U.S. Dept. Agric. Bur. Soil, Bull.* 74, 1910.
3 *Pedology*, 1930, No. 3.

between the humin fraction and that which disperses, for altering the concentration of the alkali, the temperature of extraction, or pretreating the soil with acid before alkali extraction all alter the amount of humus that disperses. Further, a part of the humin fraction may be partially decomposed plant and microbial debris that does not disperse in sodium hydroxide.

Soil chemists have been criticised for maintaining this archaic terminology, since we know that there is not a simple fulvic, or humic or hymanomelanic acid; but the terms have continued to be useful, and should not be misleading. However, the term humic acid does not always have a precise meaning. Some authors refer to what are here called the humic colloids as humic acid, and there is no accepted name for the residual humic acid fraction after the soil polysaccharides or the hymanomelanic acid have been removed from the humic acid, strictly defined. The tendency has been to call this residual fraction humic acid also, making it very necessary for the reader to be quite certain what any given author means by the term.

Constitution of humic colloids

The humic colloids are built up of carbon, oxygen, hydrogen, nitrogen, sulphur and phosphorus; and as far as is known, no other element forms an integral part. The elemental composition of the humus from different soils varies within quite wide limits, but for many agricultural soils the ratio of carbon to nitrogen to sulphur to phosphorus is of the order 100:10:1:2 on a weight basis (see p. 305). It is not yet possible to determine the oxygen and hydrogen content of the humus in the soil, but this has been done by a few workers on the fulvic and humic acid fractions. M. M. Kononova's results for some Russian soils, however, are typical.[1] She found mean carbon, oxygen, hydrogen and nitrogen contents of 61, 31, 3·7 and 4·1 per cent for the humic and 46, 48, 3·5 and 2·4 per cent for the fulvic. The characteristic difference between these two fractions is the much higher oxygen and lower carbon content of the fulvic than the humic. The nitrogen content of the fulvic is also lower, though its C/N ratio is not always higher.

The dispersed humic and fulvic particles have a wide range of molecular weight, but there are probably still some difficult problems of technique before the humic suspensions can be separated into fractions each with particles having a molecular weight in a given range. The technique currently in use is known as gel filtration or gel chromatography, and is based on adding a small amount of the humic fraction to the top of a column of granules of Sephadex—a synthetic cross-linked polydextran in which the pores are all smaller than a given size—and leaching the fraction through this column with a suitable solution. The principle of the method is that, if the column is long enough, all particles smaller than the pore size of the

1 *Trans. 6th Int. Congr. Soil Sci.*, 1956, **B**, 557.

granules will be held back for a time, since they can enter these pores, but all particles larger than the pores will leach through the coarse pores and not be held back. Unfortunately, the method has been subject to an error, for some of the larger particles become adsorbed on the surface of the gel, so are held back and appear as smaller particles; and only recently have modifications in the technique been proposed which are likely to minimise this error.[1] All the published results quoted here, however, are subject to this error, which results in an underestimate of the proportion of the larger particles in a suspension.

The molecular weights of fulvic acid, as determined by these techniques are typically below 10 000, and the humic above 5000 and going up to several million,[2] though it is difficult to carry out fractionations above a molecular weight of 100 000. Figure 15.1 illustrates the particle size distribution of a

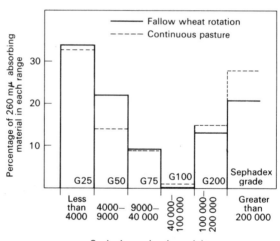

FIG. 15.1. Distribution of Sephadex molecular weights in two humic acid samples

fulvic plus humic acid dispersion from a permanent pasture plot and from a wheat-fallow rotation plot on the Urrbrae silt loam soil at the Waite Institute, Adelaide.[3] This shows that the dispersed humic material falls into two groups, one with a molecular weight less than 40 000 and the other with a weight in excess of 100 000. Unfortunately, it is not yet known if this result is an artefact, if it is of general validity, or if it is peculiar to the Waite soil.

1 R. S. Swift and A. M. Posner, *J. Soil Sci.*, 1971, **22**, 237.
2 See N. C. Mehta, P. Dubach and H. Deuel, *Z. PflErnähr.*, 1963, **102**, 128; A. M. Posner, *Nature*, 1963, **198**, 1161; J. H. A. Butler and J. N. Ladd, *Aust. J. Soil Res.*, 1969, **7**, 229 (gel filtration). See also E. L. Piret, R. G. White *et al.*, *Sci. Proc. Roy. Soc. Dublin*, 1960 A **1**, 69; F. J. Stevenson, Q. van Winkle and W. P. Martin, *Proc. Soil Sci. Soc. Amer.*, 1953, **17**, 31 (ultra-centrifuge).
3 A. M. Posner, *Nature*, 1963, **198**, 1161.

The fraction with a molecular weight in excess of 100000 may itself fall into two groups, one with a weight of about 200000 in which the particles are more or less spherical with a diameter of about 40 Å, and the other with a molecular weight over 1 million and with ellipsoid particles about 100 Å in length.[1]

Humic colloids contain a range of polysaccharides, proteins or polypeptides, and substances of uncertain composition but rich in aromatic rings, as well as a large range of substances present in small or very small quantities. These include waxes and asphalts on the one hand,[2] and substances likely to be present in the living cells of the soil organisms, such as purine and pyrimidine bases and nucleic acids[3] on the other. There is no reason to suppose that these play an important role in the properties of humus as they affect the soil or the plant.

The humic fractions in the different molecular weight ranges have different properties and compositions, but there are not yet any extensive studies on the effect of particle size. It is possible that the largest particles are polysaccharide polymers, and that the lower molecular weight fractions are more oxidised with a higher exchange capacity than the polysaccharides, and that a proportion of the larger non-polysaccharide particles have properties closer to the lignins than to typical humic colloids.[4]

The soil polysaccharides

The greater part of the polysaccharide fraction is probably present in the soil as particles separate from the remainder of the humus, for they can be separated from the remainder quite easily. G. D. Swincer and his colleagues,[5] for example, made the separation by first treating the soil with dilute acid, then extracting the humic colloids in alkali, passing this extract first through a sufficiently long column of a suitable H-resin which held back the humic acids, and then through a pad of Polyclar, a polyvinyl pyrolidone plastic, which absorbed the remainder of the non-polysaccharide material. This gave an extract containing about 40 per cent of the total soil polysaccharides almost uncontaminated with other material. A further 40 per cent of the soil polysaccharides could then be dispersed from this soil sample by acetylating the residual polysaccharides with acidified acetic anhydride.

The proportion of the organic carbon that is present as polysaccharide or carbohydrate material varies from 5 to 25 per cent.[6] Suitable techniques

1 R. L. Wershaw, P. J. Burcar *et al.*, *Science*, 1967, **157**, 1429.
2 See, for example, R. I. Morrison and W. Bick, *J. Sci. Fd. Agric.*, 1967, **18**, 351.
3 For a review see G. Anderson, in *Soil Biochemistry*, Ed. A. D. McLaren and G. H. Peterson, Dekker, 1967.
4 For an example from Australia see R. S. Swift, B. K. Thornton and A. M. Posner, *Soil Sci.*, 1970, **110**, 93.
5 With J. M. Oades and D. J. Greenland, *Aust. J. Soil Res.*, 1968, **6**, 211.
6 For a review, see G. D. Swincer, J. M. Oades and D. J. Greenland, *Adv. Agron.*, 1969, **21**, 195.

will extract between 55 to 95 per cent of this material present in the soil, though some of the extracted material will have had its composition altered to bring it into solution. These polysaccharides are composed of the hexose sugars glucose, galactose and mannose, of the pentose sugars arabinose and xylose, of the de-oxy-hexoses fucose and rhamnose, of the amino sugars galactosamine and glucosamine, and of galacturonic and glucuronic acids. It is likely that much of the glucosamine is present as N-acetyl-glucosamine, the principal constituent of chitin, which is an important component of fungal cell walls. In addition, as improved methods of hydrolysing poly-saccharides are being developed, an increasing number of other sugars and amino sugars are being recognised in small yields, many of which are characteristic of bacterial or fungal metabolism. The general picture is, however, that these polysaccharides constitute an important fraction of the soil humus, and they are derived from microbial cell wall material or exocellular gums.

The dispersed polysaccharide suspensions have as wide a range of particle sizes as have the humic acid dispersions as a whole. As an example, the polysaccharide fraction separated from the Urrbrae soil at the Waite Insti-tute[1] has the following distribution of material: with molecular weights between 4000 to 20000, 29 per cent; between 20000 and 40000, 25 per cent; between 40000 and 100000, 9 per cent; between 100000 and 200000, 11 per cent; and over 200000, 26 per cent. The larger particles are ellipsoidal or long rods that are not strongly hydrated.[2]

It is not possible to obtain a polysaccharide fraction completely free from amino acids, and much of this is probably present as protein intimately inter-mixed with the polysaccharides, probably as a glycoprotein, such as are known to be present in microbial cell wall material. Nor can the poly-saccharides be entirely freed from some humic material. It is, in fact, possible that some of the polysaccharides form esters with some of the more typical humic acid colloids which can be saponified in $2 N . NaOH$.[3] Swincer[4] found the dispersed carbohydrate fraction from the Urrbrae soil to consist of about 55 per cent of neutral sugars, 6 per cent of amino sugars, at least 10 per cent of uronic acids, 6 per cent of amino acids and 2 per cent of nitrogen that was not accounted for. However, if the uronic acids were estimated from the carboxyls determined by titration, they would account for nearly 30 per cent, so that 98 per cent of the dispersed material would be accounted for. He also found that the fraction of low molecular weight, under 4000, has an appreciably higher amino acid content and a much lower uronic acid content than the higher molecular weight fractions.

1 G. D. Swincer, J. M. Oades and D. J. Greenland, *Aust. J. Soil Res.*, 1968, **6**, 211.
2 A. G. Ogston, *Biochem. J.*, 1928, **70**, 598.
3 F. E. Clark and K. H. Tan, *Soil Biol. Biochem.*, 1969, **1**, 75.
4 *Aust. J. Soil Res.*, 1968, **6**, 225.

The nitrogen compounds

A proportion of the nitrogen in soil humus is liberated as ammonium on acid hydrolysis and a proportion as amino acids. It is probable that most of the amino acids are derived from proteins, although it has not yet been possible to prove rigorously whether proteins or only smaller polypeptides are present. Soils do not contain either free α-amino nitrogen groups or free proteins, nor has any protein yet been liberated in reasonable yield or in pure form from a soil, although P. Simonart and his coworkers[1] obtained fractions that were probably very high in protein by treating a humic acid extract with phenol. It is possible to prove the existence of either polypeptides or proteins in the soil by treating a humus preparation with the proteolytic enzyme pronase,[2] which will release between 25 and 40 per cent of the amino acids that are released by acid hydrolysis.

Acid hydrolysis of soil organic matter releases between 20 and 30 per cent of the soil nitrogen as amino acids, and over 30 different acids have been identified although only 14 are usually present in appreciable yield.[3] The distribution of nitrogen between these acids is relatively constant for different soils implying that the humus proteins from different soils have a fairly constant composition,[4] which is not dissimilar to the composition of microbial protein. The distribution of nitrogen between these acids is roughly 18 per cent for the basic, 13 per cent for the acidic and 60 per cent for the neutral amino acids with 7 per cent for a miscellaneous group.[5] Some of the amino acids found, usually only in low yield, are not constituents of proteins, but are typical products of microbial metabolism. Other amino acids are present as mucopeptides associated with the amino sugar muramic acid and some of the alanine is a constituent of the teichoic acids present in humus, which are derived from the cell walls of gram-positive bacteria.

An appreciable part of the nitrogen in soil organic matter is released as ammonium on acid hydrolysis. This is often called amide nitrogen, because many amides release ammonium on acid hydrolysis, and part of the ammonium released almost certainly derives from amides. Part probably also derives from the hydrolysis of hydroxy-amino acids, such as serine and threonine, and part from some easily hydrolysable amino sugars; but it is probable that a part, and perhaps a major part, is derived from functional groups at present unidentified.

Table 15.1 taken from the work of D. R. Keeney and J. M. Bremner[6]

1 P. Simonart, L. Batistic and J. Mayaudon, *Pl. Soil*, 1967, **27**, 153.
2 J. N. Ladd and P. G. Brisbane, *Aust. J. Soil Res.*, 1967, **5**, 161; *9th Int. Congr. Soil Sci.*, 1968, **3**, 309, 319.
3 For a review, see J. M. Bremner in *Soil Biochemistry*, Ed. A. D. McLaren and G. H. Peterson, Dekker, 1967.
4 See, for example, F. J. Sowden, *Soil Sci.*, 1956, **82**, 491.
5 This is for a Flanagan silt loam, calculated by Bremner.
6 *Proc. Soil Sci. Soc. Amer.*, 1964, **28**, 653.

gives a summary of the distribution of nitrogen in so far as it can be fraction-ated, for ten virgin Iowa soils and their cultivated counterparts. On the average about one-quarter of the soil nitrogen is amino acids, about 5 per cent is in amino sugars, about a quarter as unidentified amide nitrogen and nearly a half is unidentified. Results from other parts of the world broadly agree with these figures, though the amino sugar nitrogen can be as high as

TABLE 15.1   Distribution of organic nitrogen in some Iowa soils. Ten virgin soils with their cultivated counterparts. Per cent of total soil N in fraction

Nitrogen	Virgin soils		Cultivated soils	
	range	mean	range	mean
Non-hydrolysable	18·4–36·7	25·4	19·2–34·3	24·0
Hydrolysable				
ammonium	18·6–25·9	22·2	18·7–29·0	24·7
amino acids	19·4–34·3	26·5	17·8–31·0	23·4
hexosamines	3·3– 6·2	4·9	4·3– 7·1	5·4
unidentified	17·9–25·1	21·0	19·5–28·9	22·5

10 per cent, or perhaps higher,[1] and the amino acid nitrogen may be as high as 30–40 per cent; but the general result is that over half the nitrogen is in compounds that cannot yet be identified. J. H. A. Butler and J. N. Ladd[2] fractionated the humic acid from the Urrbrae red brown earth of the Waite Institute into a number of size fractions, using gel filtration, and found that in the molecular weight range from under 5000 to over 150 000 the amino-nitrogen content increased with molecular weight, which if it was largely present as protein would imply that the larger humus particles had the highest protein content.

The consitution of the unidentified part of the soil nitrogen is still quite unknown. It is possible a part is an artefact, for Bremner has pointed out that if proteins are being hydrolysed in the presence of furfural, which itself is produced from the hydrolysis of some polysaccharide gums, some of the amino acids released will react with the furfural to give complexes very resistant to hydrolysis. However, he considers that the amount of nitrogen so locked up is likely to be small. It is possible that a proportion is present in simple nitrogen containing compounds held so tightly by the clay, perhaps in between clay sheets, that they are protected from hydrolysis. J. M. Bremner,[3] for example, found that treating soils with a mixture of hydro-fluoric and hydrochloric acids released organic nitrogen compounds not dispersible by sodium hydroxide, and the proportion so released increased

1 F. J. Stevenson, *Proc. Soil Sci. Soc. Amer.*, 1960, **24**, 472.
2 *Aust. J. Soil Res.*, 1969, **7**, 229.
3 *J. Agric. Sci.*, 1959, **52**, 147.

down the soil profile, as did the amount of ammonium ions fixed by the soil in non-exchangeable form. It has commonly been assumed that much of the unhydrolysable nitrogen is present either in heterocyclic rings or as bonding either heterocyclic or aromatic rings together. Unfortunately, there are not yet any methods available for determining how much, if any, nitrogen is held in such positions.

A proportion of the soil nitrogen can be brought into solution by hydrolysing the soil under pressure in 0·01 M $CaCl_2$. This gives a solution which is only weakly coloured, which may contain up to 20 to 25 per cent of the soil nitrogen, some of which is present as ammonium ions, and which also contains hexose sugars, presumably derived from the hydrolysis of amino sugars.[1] It is not yet known what other nitrogen constituents are hydrolysed in this process.

Organic phosphates

Humus contains appreciable amounts of organic phosphates, which are present as complex organic esters. They differ from some of the carbon and nitrogen compounds in the soil in that they can be extracted almost quantitatively by relatively mild treatments.[2] Thus R. L. Halstead[3] was able to disperse almost all the organic phosphates from some soils using ultrasonic dispersion in acetyl acetone, and T. I. Omotoso and A. Wild[4] dispersed about 90 per cent in the soils they used by a pretreatment with acid followed by dispersion using a Na^+-saturated exchange resin and then extraction with acetyl acetone. Most of the organic phosphate remains in solution when an alkaline extract is acidified, so is present in the fulvic acid fraction, and that in the humic fraction appears to have a similar composition to that in the fulvic. The phosphate appears to be in the lower molecular weight range of particles, mainly below 10 000.[5]

The first group of phosphate compounds to be recognised in soil organic matter were inositol phosphates, and in particular phytic acid or inositol hexaphosphate. The introduction of chromatographic techniques has now shown that there are a whole range of inositol phosphates, from the mono to the hexa, present in humic compounds, and, further, that the inositol is not only present in the myo-form, the only form that has been recognised in plants and animals, but in the scyllo-, chiro-, dl- and neo-forms.[6] These

1 See, for example, G. Stanford, *Soil Sci.*, 1968, **106**, 345; 1969, **107**, 203, 323; 1970, **109**, 190.
2 For a review, see G. Anderson in *Soil Biochemistry*, Ed. A. D. McLaren and G. H. Peterson, Dekker, 1967.
3 With G. Anderson and N. M. Scott, *Nature*, 1966, **211**, 1430.
4 *J. Soil Sci.*, 1970, **21**, 224. J. D. Williams, J. K. Syers *et al.* (*Soil Sci.*, 1970, **110**, 13) using an improved method were able to disperse almost all the organic phosphate from some New Zealand soils.
5 J. R. Moyer and R. L. Thomas, *Proc. Soil Sci. Soc. Amer.*, 1970, **34**, 80, probably found rather more above 10000 than did Omotoso and Wild.
6 D. J. Cosgrave, *Aust. J. Soil Res.*, 1963, **1**, 203. *Nature*, 1963, **200**, 568. *Soil Sci.*, 1966, **102**, 42.

have only been recognised in the soil, but presumably they must be products of microbial metabolism.

The inositol phosphates do not occur free in the soil, but are probably associated with proteins and possibly polysaccharides,[1] and are set free by alkaline oxidation and hydrolysis, usually with sodium hypobromite. They can then be precipitated from the suspension with ferric chlorides, which specifically precipitates the inositol phosphates, and be purified and fractionated chromatographically.

The results published for the proportion of organic phosphorus that can be obtained as inositol phosphate varies very widely for different soils, the proportion varying from 10 to 50 per cent, with 15 to 30 per cent being common;[2] and J. K. Martin[3] found that, for the New Zealand grassland soils he examined, the proportion of the inositol to the other organic phosphates increases down the soil profile. The proportion of the phosphate in the different polyphosphates also is very variable, but most is usually in the hexaphosphate, and the amounts in mono- di- and triphosphate is usually small, being under 10 per cent of the total inositol phosphate. Omotoso and Wild found these phosphates only in the lower molecular weight fraction of the only extract they fractionated, while the hexa- and penta- were only found in a higher molecular weight fraction.

The ratio of hexa- to penta- is very variable, ranging from about equal amounts to over four, with tetra- usually being less than penta-. In the same way the proportions of the different inositols varies very widely from soil to soil. Myo- is usually the commonest with scyllo- the second most common, dl- rarely accounts for more than 10 per cent, and neo- is lacking in many soils, and when present is only present in small quantities. The ratio of myo- plus dl- to scyllo- varies from about unity to 8; and no reasons can yet be given for the very variable amounts of inositol phosphates or their compositions between different soils.

No other recognisable phosphate compounds have yet been isolated from the soil humus in appreciable amounts, though Omotoso and Wild found that 17 per cent of the organic phosphate isolated from the English pasture soil was present as two definite sugar phosphates which they could not identify, though each contained equal amounts of hexose sugar and phosphate.

Other organic phosphates have been found in soil extracts, but usually only in very small amounts. Thus G. Anderson[4] found about 1 or 2 per cent

1 G. Anderson and R. J. Hance, *Pl. Soil*, 1963, **19**, 296.
2 For some results for Australian soils see C. H. Williams and G. Anderson, *Aust. J. Soil Res.*, 1968, **6**, 121. For Scottish and Canadian soils see R. B. McKercher and G. Anderson, *J. Soil Sci.*, 1968, **19**, 47, 302. For some English and Nigerian soils T. I. Omotoso and A. Wild, *J. Soil Sci.*, 1970, **21**, 216. For a summary R. B. McKercher, *Trans. 9th Int. Congr. Soil Sci.*, 1968, **3**, 547.
3 *N.Z. J. Agric. Res.*, 1970, **13**, 522.
4 *Soil Sci.*, 1961, **91**, 156. *J. Soil Sci.*, 1970, **21**, 96.

of the organic phosphorus as nucleotides, but this could all be present in the RNA and DNA of the living micro-organisms. Small amounts of phospho-lipids have also been found.[1] In addition R. L. Halstead and Anderson[2] have found small amounts of ribitol- and glycerol-phosphate almost certainly derived from the teichoic acids of microbial cell walls.

The sulphur compounds

About as much organic sulphur as organic phosphorus is present in humus, but there is much less knowledge about the compounds in which it occurs. It is more closely associated with the organic nitrogen than is the phosphorus, and the organic N:S ratio in humus from a given group of soils is usually less variable than the organic N:P.[3]

No identifiable sulphur-containing compounds have been isolated from humus, though the sulphur-containing amino acids, cysteine, cystine, and methionine are probably constituents of the humic proteins. The organic sulphur is usually divided into two groups: that which can be reduced to hydrogen sulphide by hydriodic acid (HI), which is composed of sulphate esters and sulphated polysaccharides, and the remainder which is assumed to be compounds in which the sulphur is bonded direct to a carbon atom.[4] This group includes the sulphur-containing amino acids, but it must almost certainly contain other compounds as well, for up to two-thirds of the organic sulphur in surface soils can be in this form. Most of the organic sulphur in plant tissue is usually in this form also, so the sulphate esters are largely microbial products, and much is in the high molecular weight fraction of humic dispersions associated with polysaccharide material. At least some of these sulphate esters can be hydrolysed by soil sulphatase enzymes, and drying and then moistening a soil results in a rapid hydrolysis of some of these groups.[5]

Functional group analysis

Although it is not possible to give any structural formulae for the humic and fulvic acid colloids, it is possible to say something about the functional groups they contain. Unfortunately, the results of functional group analysis can be very difficult to interpret, partly because many of the results given in the literature are for preparations that have only been poorly described, and partly because the constitution of the colloids is so complex that the normal methods of functional group analysis do not give such clear-cut separations

1 R. J. Hance and G. Anderson, *Soil Sci.*, 1963, **96**, 94, 157.
2 *Canad. J. Soil Sci.*, 1970, **50**, 111.
3 C. H. Williams, E. G. Williams and N. M. Scott, *J. Soil Sci.*, 1960, **11**, 334.
4 For an example of the problems of fractionation see J. R. Freney, *J. Sci. Fd. Agric.*, 1969, **20**, 440.
5 I am indebted to Dr P. Cooper for this observation.

as they do for simple organic compounds. The principal active groups are carboxylic acids and phenolic hydroxyls, which give the colloids acidic properties, and ketones and aliphatic hydroxyls.

The kind of difficulty that arises in trying to make quantitative determinations of these groups is that not all the groups need be on the outer surface of the colloid, so that the amount determined will depend on the extent to which the groups inside the particle become accessible for determination, either due to the particle swelling in the bathing liquid so all the groups become accessible, or to the chemical treatment used for determining the groups itself opening up the framework. This, in turn, may depend on the amount of metallic ions, particularly aluminium and ferric, that are present in the colloid.

The determination of the carboxylic acid groups in the colloid illustrates many of the complications that arise. Their presence can be proved by infrared absorption spectroscopy, for both the undissociated carboxylic acid and the carboxylate anion give characteristic lines which are not too much affected by the presence of other groups. Infrared data show that humic acids possess carboxylic acids, most of which have dissociated at pH 7 to 8, but some of which do not dissociate until a pH of 10 to 11.[1] A titration curve which probably illustrates this point is given in Fig. 15.2, which was obtained by W. S. Gillam.[2] He methylated a humic acid preparation using a process that was meant to methylate only phenolic hydroxyls and not carboxylic hydroxyls, so the methylated curve should give the titration curve for the carboxylic groups, and the difference between the two curves the curve for the phenolic groups.

The shape of the titration curve can be interpreted as showing that the carboxylic acids are not uniformly distributed over the surface of the polymers, but have a tendency to occur in units of two or more sufficiently close together to affect the dissociation constants of the different hydrogen ions. Thus, A. M. Posner[3] found that for a series of humic fractions obtained from an Urrbrae red-brown earth, the more gentle the method of extraction, the higher the content of carboxylic acids, the greater the degree of clustering of the carboxylics, and the lower the mean pK of the acids, as one might expect. Thus, the extract obtained with pyrophosphate possessed 400 m eq of carboxylic acids per 100 g, with a mean pK = 5·2 and a mean cluster size of 3·5 carboxylics, while the sodium hydroxide extract contained 240 m eq of carboxylics with a mean pK = 6·0 and a mean cluster size of 2·5. This conclusion, that a proportion of the carboxylics occur in clusters, is confirmed by the ease with which a proportion of the carboxylics can be

1 See, for example, M. Schnitzer and S. I. M. Skinner, *Soil Sci.*, 1963, **96**, 86; G. H. Wagner and F. J. Stevenson, *Proc. Soil Sci. Soc. Amer.*, 1965, **29**, 43; B. K. G. Theng, J. R. H. Wake and A. M. Posner, *J. Soil Sci.*, 1967, **18**, 349.
2 *Soil Sci.*, 1940, **49**, 433; for a further example, see F. E. Broadbent and G. R. Bradford, ibid., 1952, **74**, 447.
3 *J. Soil Sci.*, 1966, **17**, 65; *Trans. 8th Int. Congr. Soil Sci.*, 1964, **3**, 161.

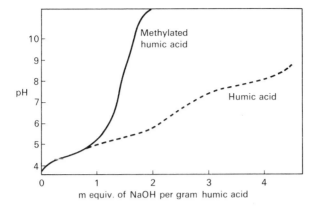

FIG. 15.2. The effect of methylating humic acid on its titration curve

converted into anhydrides, which is only possible if they occur on contiguous carbon atoms either on an aliphatic chain or aromatic ring.[1]

There are still a number of observations concerning the acidic properties of humic dispersions that cannot be given a definite explanation. As an example, it is still uncertain to what extent the buffer curve above pH 7 is a genuine equilibrium curve that is reversible. If a humus suspension is taken up to say pH 9 and left, its pH will usually drop;[2] and if the buffer curve is obtained for an acid humus suspension up to say pH 9, and acid is then added to obtain the buffer curve with decreasing pH, the new curve will typically lie below the first curve, showing that at the higher pH values, more alkali is needed to raise the pH by a given amount than acid is required to lower it by that amount;[3] no critical explanation of these results has been given. Part of the cause of these two phenomena is that many humic suspensions contain some complexed iron and aluminium ions, and these are slowly desorbed and precipitated at high pH, setting free carboxylate groups which dissociate hydrogen ions at much lower pH values (see p. 307). Most modern work, however, is done with humic acids which have had at least most of the iron or aluminium in the initial extract removed with strong cation exchange resins. Part of the cause of the fall of pH when it is raised rapidly could be due to the polymer swelling and uncovering carboxylic acids which had been sealed off from access to the solution; and this could be a simple explanation for those carboxylics appearing to need a pH between 10–11 before they dissociate, though the reversibility of this high pH dissociation has not been investigated. It is also possible that in part this drop in pH could be due to the hydrolysis of organic esters, or to keto groups

1 G. H. Wagner and F. J. Stevenson, *Proc. Soil Sci. Soc. Amer.*, 1965, **29**, 43.
2 See, for example, M. Schnitzer and U. C. Gupta, *Proc. Soil Sci. Soc. Amer.*, 1965, **29**, 274.
3 A. E. Martin and R. Reeve, *J. Soil Sci.*, 1958, **9**, 89.

becoming converted to enols which can dissociate a hydrogen ion. This transformation, which is a slow reaction in a number of compounds is

$$\underset{\underset{\text{O}}{\|}}{-\text{C}-\text{CH}_2-} \quad \text{going to} \quad \underset{\underset{\text{OH}}{|}}{-\text{C}=\text{CH}-}$$

Finally, if precautions are not taken, the humic extracts will absorb oxygen with the production of new acidic groups (see p. 298), so all critical work must be carried out either in nitrogen or with adequate precautions against access of oxygen.

Humic acids possess some phenolic hydroxyl groups. These can be determined by titration methods, if it is assumed that all the carboxylic acids in a humus have dissociated by the time the pH has risen to 8, and that the phenolic hydroxides do not begin to dissociate appreciably until the pH has risen above 8 and have all dissociated at a pH of 11 to 12. The alcoholic hydroxyls do not begin to dissociate until a still higher pH. Under these conditions, the carboxyls can be determined by using a calcium or barium acetate solution at pH 7 to 8 to saturate the humic acid, and a barium hydroxide solution to saturate both the carboxylic and the phenolic groups.[1]

However, there is some evidence that hydrogen ions dissociate from phenolic hydroxides on some humic acid fractions at an appreciably lower pH than from phenol itself, probably due to the phenolic hydroxyl and the carboxylic acid being on neighbouring carbon atoms.[2] This would give an additional cause of overlap between the dissociation of hydrogen ions from these two groups. It is theoretically possible to determine the phenolic hydroxyls by methylating the extract, which should convert these hydroxyls to methoxyls, and if this is done quantitatively, their amount can be measured by the reduction in cation exchange capacity. In general, both the humic and the fulvic acid fractions contain more, and usually appreciably more, carboxylic acid groups than phenolic hydroxyls, but there is still little quantitative data on this.

Humic colloids also contain aliphatic hydroxyls, but it is still uncertain how reliable are the methods used for their determination. They almost certainly behave as acidic groups if dispersed in a very strongly protonating solvent, such as di-methyl-formamide, for S. O. Thompson and G. Chesters[3] found exchange capacities of the order of 1100 to 1200 m eq per 100 g for humic fractions dispersed in this solvent, and the difference between this figure and the 300 or 400 usually found for the carboxylic acids plus phenolic hydroxyls is probably principally due to the aliphatic alcohols present.

Humic acids contain carbonyl groups in forms other than carboxylic acids, and the evidence is consistent with these being present as ketones,

1 F. Martin, P. Dubach *et al.*, *Z. PflErnähr*, 1963, **103**, 27.
2 See, for example, D. S. Gamble, *Canad. J. Chem.*, 1970, **48**, 2662.
3 *J. Soil Sci.*, 1969, **20**, 346.

some as diketones such as —CH_2—CO—CO—CH_2—, and some in stable lactone,[1] quinone, or pyrrone units.[2] There is inadequate experimental data to draw any conclusions about the relative proportion of these different carbonyls in humic acids.

This functional group analysis will account for a considerable amount of the oxygens present in some very mobile, and probably very oxidised fulvic acid fractions, but for only a part of those present in the less oxidised humic fractions. This is illustrated in Table 15.2 for fractions extracted from two horizons of a Canadian podsol.[3] The validity of this table naturally depends on the validity of the determination of the functional groups, and the possibility cannot be ruled out that some of them have been underestimated. But there is no reason to doubt the general conclusion. It is not known where the unaccounted for oxygens are situated. Only a very few are present as methoxyls, and it is likely that a number are present as ether linkages, but these cannot be yet determined quantitatively.

The acidities given in Table 15.2 can be expressed in terms of the average number of carbon atoms present in the polymers per acidic group. The fulvic

TABLE 15.2 Composition of fulvic and humic acids from a Canadian podsol

| | % Composition of organic matter | | | | Active Groups in m eq/g | | | | Per cent total oxygen in active groups |
	C	H	N	O	COOH	phen-olic OH	alco-holic OH	ketones	
Fulvic Acids									
A horizon	49·9	4·67	1·27	44·1	6·1	2·8	4·6	3·1	87
B horizon	48·4	3·28	0·60	47·7	9·1	2·7	4·9	1·1	94
Humic Acids									
A horizon	55·5	5·18	2·19	37·0	2·1	3·5	3·4	1·0	54
B horizon	56·7	4·94	2·47	35·9	3·7	2·9			
5 Canadian* surface soils	58·2	4·8	5·1	31·9	2·2				

* From R. E. Wildung, G. Chesters and D. E. Behmer, *Pl. Soil*, 1970, **32**, 221.

acids contain between 4·5 and 7 carbons per carboxylic and between 3·5 and 4·5 per total acid group, while the humic contains between 13 and 23 and between 7 and 9 for each of these two acidities. The fulvic also contains between two-thirds and three-quarters of an oxygen atom per carbon, and the humic half an oxygen per carbon. This also shows, in another way, how strongly oxygenated these polymers are.

1 See, for example, F. E. Broadbent and G. R. Bradford, *Soil Sci.*, 1952, **74**, 447.
2 B. K. G. Theng and A. M. Posner, *Soil Sci.*, 1967, **104**, 191.
3 Taken from J. R. Wright and M. Schnitzer, *Trans. 7th Int. Congr. Soil Sci.*, 1960, **2**, 120.

There are a number of reports in the literature of the exchange capacity of the humic fraction in the soil itself. Two methods have principally been used: determining the reduction in the exchange capacity of a soil when the organic matter is removed by treatment with hydrogen peroxide,[1] and doing a multiple regression between the exchange capacity of a series of soils and their clay and organic carbon contents.[2]

The exchange capacities of the humus determined by these methods for different soils varies widely, partly because the various authors have used different methods for determining the cation exchange capacity of the soils. But the general result is that the exchange capacity of humus at about pH 7 for neutral well-drained soils is over 200 m eq per 100 g humus, with many results between 250 and 300 m eq. The values are lower for more acid or for poorly drained soils and figures between 30 and 70 have been found for acid soils. The probable reason for such low values is that much of the organic carbon in these acid soils is not fully humified, for the exchange capacity of the fulvic and humic acid fractions from in them is about the same as from neutral soils. Since the humus is more strongly buffered than clay in the pH range of normal soils, for a given organic carbon and clay content, the humus becomes increasingly important as the seat of cation exchange as the pH of the soil rises.[3]

Two other properties of humic colloids depend on the existence and location of functional groups, but the experimental data cannot yet be interpreted in terms of these groups. Humic colloids in alkaline solution will take up atmospheric oxygen,[4] and this autoxidation results in an increase in the acidity of the polymer. Also if the autoxidation is taking place in the presence of ammonia, some ammonia is absorbed and some of the nitrogen converted into a form from which it cannot easily be released by acid hydrolysis. This property is shown more strongly by lignin and humic acids from peat[5] than by humic acids from agricultural soils. It is also shown by a number of polyphenols related to lignin, particularly those with hydroxyls in the 1:2, 1:2:3 and 1:4 positions, such as occur in catechol, tannic and gallic acid and hydroquinone. This, or a similar oxidation is also brought about by some polyphenoloxidases. There is little quantitative data on the factors controlling the amount of oxygen taken up or into what groups it becomes incorporated, nor is much known about the amount of ammonia that can be fixed or the groups into which the strongly fixed nitrogen goes. N. M. Atherton and his colleagues[6] have found that a number of humic

1 L. C. Olson and R. H. Bray, *Soil Sci.*, 1938, **45**, 483.
2 E. G. Hallsworth and G. H. Wilkinson, *J. Agric. Sci.*, 1958, **51**, 1; T. L. Yuan, N. Gammon and R. G. Leighty, *Soil Sci.*, 1967, **104**, 123.
3 See, for an example for Wisconsin soils, C. S. Helling, G. Chesters and R. B. Corey, *Proc. Soil Sci. Soc. Amer.*, 1964, **28**, 517.
4 J. M. Bremner, *J. Soil Sci.*, 1950, **1**, 198.
5 S. Mattson and E. K. Andersson, *Lantbr Högsk. Ann.*, 1943, **11**, 107.
6 Tetrahedron, 1967, **23**, 1653.

acids from peats and soils, from which proteins and carbohydrates were removed by prolonged acid hydrolysis, took up between 0·5 and 5 g of oxygen per 100 g humic acid. The humus of soils can, in fact, be characterised by the amount of oxygen they will take up per unit of organic carbon from mild oxidising agents such as hypoiodite.[1]

The second property, which has only recently received any attention, is the ability of humic acid to decompose nitrites under acid conditions. Nitrites are unstable in acid solutions, breaking down to NO and NO_2 or to nitric acid and NO; but in the presence of humic acids this reaction goes more rapidly and in addition nitrogen gas, with sometimes a small quantity of N_2O, is evolved[2] and a portion of the nitrite–nitrogen becomes fixed in the humic acid and converted into some groups resistant to hydrolysis. Table 15.3 taken from some work of J. M. Bremner and D. W. Nelson,[3] illustrates

TABLE 15.3 The decomposition of nitrite by organic matter. Decomposition in a 4 M sodium acetate buffer pH 5. Nitrite enriched with ^{15}N (8 mg N in 20 ml)*

| | Per cent recovery of nitrite N after 24 hours incubation | | | | | |
	As nitrite	As nitrate	As NO_2	As N_2	As N_2O	As fixed N
Buffer alone	82	3	15	0	0	0
With 20 g soil 8·9% C	44	3	15	21	1	15
4·3% C	59	4	14	12	1	9
2·3% C	68	3	15	7	<1	6
0·3% C	78	3	15	1	<1	2
1 g humic acid	49	6	16	19	1	9
1 g lignin	23	8	13	27	1	27

* Ignited soil, oxidised soil (with KOBr), quartz sand and clay mixed with the buffer had no effect on the decomposition.

these points. The groups on the humic acid responsible for the formation of nitrogen gas are probably phenols and polyphenols, for these were the only pure chemicals found to be active; and lignins contain more of these than humus and are more active. Bremner[4] suggests the first reaction of phenols with nitrite is the formation of a nitrosophenol, which reacts with additional nitrite giving a new hydroxyl group replacing the nitroso-group and nitrogen gas, with or without nitrous oxide.

Thus these two properties of autoxidation and of nitrite decomposition both imply the presence of phenols and polyphenols in humic matter. The

1 A. G. Norman and W. J. Peevy, *Proc. Soil Sci. Soc. Amer.*, 1939, **4**, 183; C. D. Moodie, *Soil Sci.*, 1951, **71**, 51.
2 F. J. Stevenson, R. M. Harrison *et al.*, *Proc. Soil Sci. Soc. Amer.*, 1970, **34**, 430.
3 *Trans. 9th Int. Congr. Soil Sci.*, 1968, **2**, 495.
4 *Pontif. Acad. Sci. Ser. Var.*, 1968, **32**, 143.

evidence for quinones present in humus in more than very small amounts is poor, which suggests that it is phenolic hydroxyls on neighbouring carbons that are the groups responsible for these properties. That such groups are probably an important constituent of humus can also be inferred from the carbon–hydrogen atom ratio, which is of the order of unity, while it would be expected to be nearer two if most of the carbon were in aliphatic chains.

The humic acid core

The identifiable compounds obtained by hydrolysis of soil humus usually only account for under half, and sometimes appreciably under half, of the total organic carbon and nitrogen; and very great problems have arisen in trying to identify the types of compounds present in the remainder due to lack of appropriate analytical techniques. Some authors refer to this residual humic material as the humic acid core, but since it is the residue left after hydrolysis there is always the possibility that the hydrolysis itself has altered the chemical nature of the residue, and this core material is not present as such in the soil. In so far as the protein or polypeptide moiety in the humic acid fraction is bonded to chemical groups in the core polymer, their removal or partial removal by hydrolysis could easily cause molecular rearrangements in the residual polymer.

It is often implicitly assumed that the humic acid core is a fairly homogeneous material chemically, in spite of the fact that the humic particles themselves have a wide range of size or molecular weight. This assumption is not strictly true, for this core can be separated into two fractions by electrophoresis, a major fraction that is dark brown in colour and does not fluoresce, and a minor fraction that is light coloured and is fluorescent.[1] Further, a green humic acid fraction can sometimes be separated out chromatographically, which is probably a fungal metabolite,[2] and the clearest way to show that it is in fact chemically heterogeneous would be by comparing the properties of the humic acids separated from the same soil by different extractants. Unfortunately, no comparisons of the humus core residues have been made, but the difference in properties of these different dispersions from which the carbohydrate fraction has been removed strongly suggests that the cores would differ in that those extracted by a weak extractant, such as sodium pyrophosphate, have less resemblance to lignin than have those extracted with sodium hydroxide.[3]

The chemical composition of this core material differs from that of the humic acid in its lower nitrogen content. This is to be expected because the hydrolysis removes protein-like material. The actual nitrogen contents vary with the source of the humus, but a content of about 2 per cent is probably

1 H. H. Johnston, *Proc. Soil Sci. Soc. Amer.*, 1959, **23**, 293.
2 K. Kumada and H. M. Hurst, *Nature*, 1967, **214**, 631.
3 A. M. Posner, B. K. G. Theng and J. R. H. Wake, *Trans. 9th Int. Congr. Soil Sci.*, 1968, **3**, 153.

representative for agricultural soils, falling to 0·4 per cent for acid peat. H. H. Johnston[1] found for the soil he was using that the material resistant to hydrolysis by 6 N. HCl had a composition of: 52·5 per cent C; 38·4 per cent O; 4·65 per cent H; 1·98 per cent N; 6·7 per cent MeO. It thus had a C/N ratio equal to 26, and for each atom of C there were: 0·55 atoms O, 0·9 atoms H, and 0·032 atoms N. It differs from typical unhydrolysed humic acids by having a rather lower carbon, a rather higher oxygen and much higher methoxyl content, and it differs considerably from lignins, which have a higher carbon and methoxyl, and lower oxygen and much lower nitrogen content.

A number of methods have been used to obtain information on the structure and composition of this core material, based on the principle of subjecting it to a range of degradative treatments, such as alkaline fusion, oxidative treatments of varying intensity, reduction with sodium amalgam, and hydrogenation under pressure. The objective is to break the polymer into smaller units which can be recognised, and from this to deduce the constituent monomers and their method of bonding. Differential thermal analysis and infrared spectroscopy have also been used to give additional information.[2]

These methods have shown that an appreciable proportion, probably of the order of one-half, of the carbon is present in aromatic rings which are relatively easily oxidised, probably because they are heavily substituted. Lignins are not present in appreciable quantity, but a proportion of the rings have substitutions typical of lignin polyphenols, and a further proportion have substitutions not found in lignins, such as on the 1, 3, 5 positions, typical of the flavenoid group of polyphenols which are characteristic of microbial metabolism.[3] A proportion of the lignin-derived groups still retain the propyl group attached to the ring, a feature that is probably not found in the flavenoid group.

Studies of the polymerisation of model substances which may be present in soils, under conditions which could exist in soils, have suggested detailed chemical reactions which yield substances similar to humic core material. Most studies have been concerned with the polymerisation of simple polyphenols which can be derived from lignin under oxidising conditions either in the presence or absence of ammonium ions and amino acids.[4] These oxidising reactions can be either autoxidation in alkaline medium or enzymatic oxidation by microbially produced polyphenoloxidases in neutral or acid conditions. Thus, a substituted 3,4 dihydroxybenzene can have its ring cleaved between the carbons in the 4 and 5 positions, the phenolic

1 *Proc. Soil Sci. Soc. Amer.*, 1959, **23**, 293.
2 For a review of these methods and interpretation of their results, see G. T. Felbeck, *Adv. Agron.*, 1967, **17**, 328.
3 N. A. Burges, H. M. Hurst and S. B. Walkden, *Geochim. cosmochim. Acta.*, 1964, **28**, 1547.
4 See, for example, W. Flaig, *Geochim. cosmochim. Acta*, 1964, and in *The Use of Isotopes in Soil Organic Matter Studies*, Pergamon, 1966.

hydroxyl oxidised to carboxylic acid, the nitrogen of ammonium incorporated between the carbon in position 3 and 5 to give a substituted pyridene carboxylic acid (α-picolinic acid).[1]

A further possible group of polymers in humus could be produced by the Maillard reaction between sugars and amino compounds which gives humus-like products when some foods are stored under unsuitable conditions.[2] The reaction is probably between a diketone, produced during microbial decomposition of carbohydrates and amino acids, although the reaction may require some of these to be in the enol form; and it results in the formation of substituted polymerised pyrazine rings. B. K. G. Theng and A. M. Posner[3] have, in fact, found evidence for the existence of such diketones in humic colloids on the basis of their infrared absorption spectra. Unfortunately the proportion of the carbon in soil humic polymers that has been derived from this type of reaction cannot yet be determined, because present-day methods of analysis yield so little precise evidence on their constitution.

The most probable picture of the composition of the noncarbohydrate fraction of the humus polymers is that each polymer is built up by stepwise additions of polyphenols and diketones which absorb some proteins or polypeptides during polymerisation; and each of these units is cross-linked to active groups on the polymer surface through linkages such as methylene or ether oxygens. Each particle is probably produced within a microbial cell either when it is alive or during autolysis of the protoplasm on its death; or it could be produced on the outer surface of the living cell. As will be shown later, this material appears to undergo further changes in the soil, for it becomes more stable to acid hydrolysis and less easily decomposed by soil organisms, but nothing is known about what processes are taking place during this transformation. This gives no explanation of the range of molecular weights or sizes of these particles, but they are small compared with cell size, for most of them have molecular weights less than 50 000; nor does it explain the relatively small variations in their chemical composition.

The elemental ratios in humus

Although humus is composed of carbon, nitrogen, oxygen, hydrogen, sulphur and phosphorus, most of the early studies were only concerned with the carbon and nitrogen contents, and later with the carbon, nitrogen, sulphur and phosphorus contents, because of the analytical difficulties of determining organic oxygen and hydrogen. But the chemical determination of each of these first four elements is subject to considerable error, if the determination

1 D. W. Ribbons and W. C. Evans, *Biochem. J.*, 1962, **83**, 482.
2 For short reviews, see J. M. Bremner in *Soil Biochemistry*, Ed. A. D. McLaren and G. H. Peterson, Dekker, 1967, and M. M. Mortland and A. R. Wolcott in *Soil Nitrogen*, Ed. W. V. Bartholomew and F. E. Clark (Agronomy Monog. 10), 1965.
3 *Soil Sci.*, 1967, **104**, 191.

is being made on the humus in the soil as distinct from humic acids separated from the soil.

Organic carbon determinations are subject to error as many soils contain charcoal, often produced by the burning of vegetation in the past, and it is not yet possible to determine all the former without including some of the latter. Further, a method of determining organic carbon used in many laboratories—the Walkley–Black method—is known to be very inefficient, although quick and easy to use.

Organic nitrogen is subject to error both because of the importance of using a suitable catalyst and of observing the correct conditions if the Kjeldahl digestion is to be efficient, and also because many soils contain ammonium ions fixed between the layers of the clay particles, and as these are released during the Kjeldahl digestion they will masquerade as organic nitrogen unless specially determined.

Organic phosphorus and organic sulphur determinations may also be subject to considerable experimental error.

A further point of importance concerns the definition of organic matter. The determinations are usually made on soil samples containing an unknown quantity of partially decomposed plant remains so the determinations refer not to the humus alone, but to the humus plus the non-humified material plus the soil biomass in the sample. These limitations must be constantly borne in mind in any interpretation of the composition of soil humus determined in the soil sample itself.

It has been known for a very long time that the organic carbon-organic nitrogen ratio, the C/N ratio expressed as a ratio of the content by weight, is relatively constant for a given soil under a wide range of management conditions. This is illustrated in Table 15.4 for soils from different fields at

TABLE 15.4   The carbon and nitrogen contents and C/N ratio of some Rothamsted soils

Situation	*Per cent C*	*Per cent N*	*C/N*
Old Woodland 12–17 cm	2·38	0·250	9·5
Park Grass, old pasture* 0–22 cm			
Unfertilised pH about 6	3·4	0·28	12·1
Fertilisers and lime pH about 7	3·7	0·32	11·5
no lime pH about 5	3·2	0·25	12·8
Subsoil 22–45 cm pH about 6·5	1·4	0·14	10·0
Broadbalk in 1966† 0–22 cm			
No manure since 1839	0·84	0·099	8·5
Complete fertilisers since 1843	1·00	0·115	8·7
35 t/ha farmyard manure since 1843	2·59	0·251	10·3

* *Rothamsted Rept.* 1963, 258.   † *Rothamsted Rept.* 1968, Part II, 97.

Rothamsted. The C/N ratio varies only between 8·5 and 12·8 in spite of a wide variation in the carbon content and the land use. This ratio of about 10 is very commonly found for many neutral well-drained agricultural soils throughout the world; though under wetter or more poorly drained conditions and under acid conditions this ratio is higher and may be as high as 17.[1] The effect of pH on the C/N ratio of humus can also be seen in peats and fens. Thus, H. Hesselman[2] found that the nitrogen content of forest humus tends to increase with increasing pH; and B. D. Wilson and E. V. Staker[3] found that a very definite correlation between the C/N ratio of some New York peats and their calcium content, and in fact the correlation is closer with calcium than with pH. Cultivating virgin soils with a high C/N ratio often lowers the ratio, for it encourages the decomposition of unhumified plant debris; and the mean C/N ratio of grassland soils under similar moisture conditions tends to decrease with increasing mean annual temperature, i.e. as the mean rate of decomposition increases.

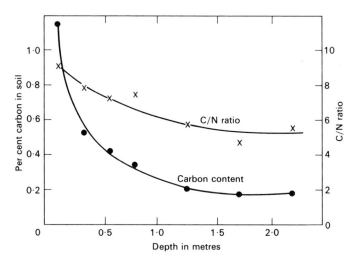

FIG. 15.3. The percentage of carbon and the C/N ratio in the first 2·1 m of Broadbalk soil

The constitution of the soil humus probably varies down the soil profile, for the C/N ratio decreases with depth, as is shown in Fig. 15.3 for Broadbalk,[4] and may apparently fall to as low as 4. A part of this apparent fall is due to the inclusion of fixed ammonium in the organic nitrogen figure, but even

1 See, for example, C. H. Williams *et al.*, *J. Soil Sci.*, 1960, **11**, 334 for some Scottish and *Aust. J. Agric. Res.*, 1961, **12**, 612 for some Australian soils.
2 *Medd. Skogsförsöksanst.*, 1926, **22**, 169.
3 *Cornell Agric. Exp. Stat.*, *Bull.* 537, 1932.
4 B. Dyer, *U.S. Dept. Agric.*, *Off. Exp. Sta. Bull.*, 106, 1902.

allowing for this, the fall can be quite appreciable, as is shown in Table 15.5.[1] Little is known about the changes in the composition of the humic colloids down the profile, though Bremner found that the C/N ratio of the humic matter that disperses in pyrophosphate or sodium hydroxide actually increases with depth, although the proportion of the organic fraction that disperses decreases. It is still therefore uncertain how far the apparent decrease with depth genuinely reflects the composition of the subsoil humus.

TABLE 15.5   Effect of adsorbed ammonium ions on the apparent fall in the C/N ratio with depth. Broadbalk

Depth in cm	*Per cent carbon*	*Per cent nitrogen*	*C/N uncorrected*	*Per cent organic nitrogen*	*C/N corrected*
2·5–23	1·04	0·122	8·5	0·113	9·2
46–69	0·37	0·055	6·7	0·044	8·4
117–137	0·14	0·031	4·5	0·024	5·8
185–205	0·14	0·031	4·5	0·024	5·8

The organic nitrogen-organic sulphur or N/S ratio also shows relatively little variation in well-managed agricultural surface soils, and is usually between 7 and 8, or as commonly written the N/S ratio is commonly about 10:1·2–1·5.[2] It is possible that the level of organic sulphur is more closely bound up with the level of organic nitrogen than of organic carbon.

The organic nitrogen-organic phosphorus ratios in soils are much more variable than the organic nitrogen-organic sulphur ratios, for the published ratios vary from 10:0·2 to 10:3, a fifteenfold range. It is still not certain how far this very large range genuinely refers to humus, or how far it is an artefact due to the inefficient methods used to determine organic phosphorus by the earlier workers, or to the inadequate separation of partially decomposed plant residues from the humic colloids; though there is little doubt that the organic phosphorus varies much more than the organic sulphur in comparison with organic nitrogen. This greater variability of the N/P ratio is what would be expected, since much of the organic phosphorus is not apparently associated with nitrogen, and since nearly all the organic phosphorus can be dispersed from a soil comparatively easily, while a proportion of the organic carbon, nitrogen and sulphur appears to be much more firmly adsorbed.

1   J. M. Bremner, *J. agric. Sci.*, 1959, **52**, 147. F. J. Stevenson (*Soil Sci.*, 1959, **88**, 201) found similar results for a variety of American soils.
2   For some Scottish soils, C. H. Williams with E. G. Williams and N. M. Scott, *J. Soil Sci.*, 1960, **11**, 334; for some Australian, with J. Lipsett, *Aust. J. agric. Res.*, 1961, **12**, 612; for some New Zealand, T. W. Walker and A. F. R. Adams, *Soil Sci.*, 1958, **85**, 307; for some Minnesota, C. A. Evans and C. O. Rost, *Soil Sci.*, 1945, **59**, 125.

C. H. Williams[1] using a number of non-calcareous Scottish soils and T. W. Walker and A. F. R. Adams[2] a number of New Zealand virgin and pasture soils found N/P ratios in the range of 10:2 to 3, though the ratio fell to 10:1·3 for the Scottish calcareous soils. On the other hand, R. W. Pearson and R. W. Simonson[3] found ratios between 0·6 and 1·2 for some Iowa soils, C. H. Williams and J. Lipsett of 10:0·7 for some Australian red-brown earths very low in available phosphate, and P. H. Nye and M. H. Bertheux[4] 10:0·4 to 0·2 for some Ghanaian forest and savanna soils on a variety of parent materials, some of which would certainly be very low.

A consequence of the relative constancy of the C/N/S ratio in a soil is that added organic matter decomposes to leave a residue having this ratio. Typically plant residues contain much more carbon than corresponds to these ratios. When they are incorporated into the soil and begin to decompose, a part of the carbon they contain will be liberated as carbon dioxide, and if there is no loss of nitrogen, the C/N ratio will fall during the decomposition, so there is no necessary correlation between the C/N ratio of the added residues and the final humus. Naturally, if the residues are high in nitrogen, the decomposition will release ammonium ions into the soil, and if low in nitrogen, will remove ammonium or nitrate ions from the soil, as described on p. 266, or else encourage the process of nitrogen fixation. Since neither sulphur nor phosphorus will be liberated as a gas, they will not be lost from the soil, and the N/P/S ratios in the plant residue and the humus need not change much, though if the residues are from crops that have been well manured with phosphate, some phosphate is likely to be liberated as inorganic phosphate during the decomposition.

Interaction of humic acid with metallic ions

Humic colloids form very stable compounds with a number of metallic ions, though in the past it was difficult to prove how far these were normal salts and how far coordination complexes. There are at least four types of interaction possible: a simple neutralisation of the negative charge on the humic acid, a coordination of a polar humic group with a water molecule forming part of the hydration shell around the cation, a coordination compound in which the carboxylate group is directly coordinated with an aluminium or iron ion, and a true chelate. With the introduction of infrared spectroscopy it is possible to distinguish between these interactions. It is likely that, for the divalent cations, coordination is usually through water of hydration; but for the trivalent, coordination is with the ion itself, as described for ligand exchange on p. 128.

1 With E. G. Williams and N. M. Scott, *J. Soil Sci.*, 1960, **11**, 334.
2 *Soil Sci.*, 1958, **85**, 307.
3 *Proc. Soil Sci. Soc. Amer.*, 1939, **4**, 162.
4 *J. agric. Sci.*, 1957, **49**, 141.

Humic colloids separated from soils, and particularly acid soils, often contain appreciable amounts of aluminium and probably ferric iron, and these ions affect the buffer curve of the extract appreciably. Some results of A. E. Martin and R. Reeve[1] illustrate this point very well. They extracted the humic fraction from the B horizon of a number of podsols using acetyl acetone as the extracting agent, for this will take out nearly 90 per cent of the organic carbon. This extract contained aluminium ions at a concentration of between one aluminium for 5 to 9 carbon atoms, which was reduced to 1 aluminium per 152 carbons by the use of a suitable strong cation-exchange resin. Figure 15.4 shows the effect of this removal of aluminium on the

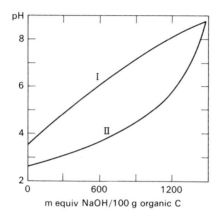

FIG. 15.4. The buffer curves of a dispersed humus with a carbon–aluminium ratio of 8·5 (curve I) and 152 (curve II) in N.KCl

buffer curve of the extracts, as determined with sodium hydroxide in a normal potassium chloride solution. The aluminium-containing humus had a buffer curve that was almost linear over the pH range 3·5 to 8, while the low-aluminium humus started off at a pH of 2·5, was strongly buffered up to pH 5, and then increasingly weakly buffered up to pH 9. Both extracts held the same amount of sodium at pH 9, but the low aluminium extract held 620 m eq at pH 7 compared with 380 for the high aluminium, and this increase of 240 m eq was brought about by the removal of 560 m eq of aluminium. This presumably implies that the aluminium is present as a hydroxylated cation, such as $Al(OH)^{2+}$ or $Al(OH)_2^{+}$. One can also interpret these curves as showing that the low aluminium acid humus had a pK of 4·1, while the high aluminium humus had a pK of 6·2. Thus the aluminium tended to make the humic acid weaker, without affecting the total amount of replaceable hydrogen at pH 9. The curve for the high aluminium fraction was not reversible, for if the suspension that had been brought up to pH 8·5

1 *J. Soil Sci.*, 1958, **9**, 89.

was back-titrated with acid, the new curve fell below the original curve and was more similar to the low aluminium curve. Since the low-aluminium curve was reversible, this implies that some of the complexed aluminium was extracted and precipitated as aluminium hydroxide in the alkaline solution. Alternatively, this effect could be due to an aluminium ion forming a coordinate bond with two carboxyl groups, on different small humic particles;[1] and on this picture the greater the amount of aluminium present, the greater the proportion of humic particles held together in these compound particles; but once the aluminium is precipitated as the hydroxide, the bonds would be broken and the individual humic particles set free.

Humic and fulvic acids will complex other ions besides aluminium: M. Schnitzer and S. I. M. Skinner[2] showed that a fulvic acid from a Canadian acid podsol, probably very similar to the Australian ones used by Martin and Reeve, would complex iron, up to one ferric ion per five atoms of carbon, and these complexes behaved in the same way as the aluminium complexes. The infrared absorption data showed that at saturation all the carboxylic acid had been converted to carboxylate and the ferric iron behaved as a monovalent cation, so was presumably $Fe(OH)_2^+$. They also found that one-third of the carboxylics dissociated easily, one-third not so freely and the last third with much greater difficulty. The other point, illustrated in Fig. 15.4, that is important to note, is that the complexed ion is not completely precipitated as the hydroxide until a high pH has been reached and the titration curve shows that this precipitation must take place gradually over a wide range of pH.

The humic polymers will complex other ions as well as iron and aluminium; in particular, they complex copper strongly. The strength with which different metallic ions are held can be measured by the drop in pH which occurs when the chloride or sulphate of the ion is added to the suspension;[3] and experiments of this type show that on the whole the order of the strength of adsorption follows the Irving-Williams order of the stability of these metal chelates, namely, that trivalent ions iron and aluminium are most strongly absorbed, followed by copper and then much more weakly zinc, nickel, cobalt and manganese, with calcium and barium often being comparable to some of these latter ions. The number of hydrogen ions released for each ion complexed is always less than would correspond to their replacement by the unhydroxylated ion, which could be due either by the ion being adsorbed as a hydroxylate, and therefore having a lower effective valency than the simple hydrated ion, or by the ion forming a coordination complex.

An interesting consequence of the greater strength of binding of some ions

---

1 S. Yariv, J. D. Russell and V. C. Farmer, *Israel J. Chem.*, 1966, **4**, 201, showed aluminium could do this with benzoic acid.
2 *Soil Sci.*, 1963, **96**, 86.
3 See, for example, M. Schnitzer, *Trans. 9th Int. Congr. Soil Sci.*, 1968, **1**, 635; S. U. Khan, *Proc. Soil Sci. Soc. Amer.*, 1969, **33**, 851.

than others is that if, for example, a little copper sulphate is added to a humic suspension containing exchangeable calcium whose pH is somewhat below neutrality, copper may be absorbed without any appreciable amount of calcium being exchanged, the copper displacing hydrogen ions probably from the carboxylic acids rather than calcium from the carboxylate. It also means that if one determines the exchange capacity of a humus extract at, say, pH 7, it is appreciably greater if copper is used as the saturating ion than if calcium is.[1] A further consequence is that the concept of cation exchange capacity has no unique value for these humic colloids, since it depends entirely on what saturating cation is being used and on the exact conditions under which it is being determined.

There is some evidence that at least a fraction of these absorbed ions are held by a phenolic hydroxyl and a carboxylic acid on adjacent carbons of a benzene ring, for the amount absorbed is appreciably reduced if the suspension is treated to remove either or both of these groups.[2] But it is unlikely that a large proportion of cations are held by a specific grouping like this because copper and zinc, for example, are held with a wide range of bond strengths,[3] presumably due to an element of randomness in the distribution of the carboxylic acid and phenolic hydroxyl groups on the polymer surface; for copper $=N-H$ groups of humic peptides are probably also involved.

There is still incomplete evidence on how far the absorption of these cations is due to the formation of coordination complexes or how far to their forming simple non-dissociating salts. About 1·6 hydrogen ions are released per divalent ion absorbed, which could be due to a proportion of the cations being present in the hydroxylated form, so having a mean valency of less than 2. On the other hand, this absorption probably causes the polymer to acquire some positive charges, for they will now absorb phosphate,[4] probably by one valency of the polyvalent cation absorbing a phosphate anion and the other neutralising a carboxylate.

The interaction of humus and clay

Humic material has for a long time been known to absorb, or be absorbed by, clay particles. Th. Schloesing[5] wrote in 1874 that 'L'argile possède une certaine tendance à s'unir aux humates du terreau pour former probablement une de ces combinaisons entre colloides signalées par Graham.' He attempted to separate the humus from the clay by suspending the mixed sol in ammonia and then adding ammonium chloride which flocculated the clay, but left

1 See, for example, S. O. Thompson and G. Chesters, *J. Soil Sci.*, 1969, **20**, 346; B. Chatterjee and S. Bose, *J. Coll. Sci.*, 1952, **7**, 414.
2 M. Schnitzer and S. I. M. Skinner, *Soil Sci.*, 1965, **99**, 278; *Proc. Soil Sci. Soc. Amer.*, 1969, **33**, 75; R. I. Davies, M. V. Cheshire and I. J. Graham Bryce, *J. Soil Sci.*, 1969, **20**, 65.
3 F. E. Broadbent, *Soil Sci.*, 1957, **84**, 127, and with N. S. Randhawa, *Soil Sci.*, 1965, **99**, 295.
4 M. Levesque, *Canad. J. Soil Sci.*, 1969, **49**, 365.
5 *Comp. Rend.*, 1874, **78**, 1276.

much of the humus in suspension. He also found that the quantity of chloride required to flocculate the clay increased with the amount of humus present, a phenomenon that was first investigated in any detail by E. Fickendey.[1]

Several lines of evidence have been advanced to show that much or most of the humus in a soil can be in close association with the clay particles. Direct microscopic examination of thin sections of most mull and arable soils, using the technique of W. Kubiena,[2] is consistent with a close association of humus and clay particles, for they constitute a plasma in which these two components cannot be distinguished; nor can most of the humus in these soils be seen in electron microscope photographs.[3] Again, if a soil is dispersed ultrasonically in a heavy liquid, with a specific gravity as high as 2, most of the humic fraction settles with the mineral fraction, for the organic matter that floats is primarily unhumified or partially humified and is only lightly contaminated with dark-coloured material.[4]

A second line of evidence for the existence of the clay-humus complex is derived from the methods by which humus is dispersed from a soil. Thus, if a soil is leached with sodium chloride, to convert it into a sodium soil, and is then deflocculated, a portion of the humus will disperse. If the soil is now treated with, say, sodium pyrophosphate or sodium hydroxide a further portion will disperse, but a portion will still be left in association with the clay. If the soil is now treated with a dilute solution of hydrofluoric acid, or as is more usual with a mixture of hydrofluoric and hydrochloric acids, and a concentration between 0·1 M and 1 M has been used by various workers,[5] and the soil again treated with sodium hydroxide, a further portion of the humus disperses; presumably due to the dissolution of silica from some components of the clay fraction.

Humic colloids carry a negative charge, as do pure clay minerals, so the clay mineral would not be expected to absorb these colloids, and there is in fact no evidence to suggest that when saturated with sodium ions, for example, they can absorb any appreciable amount of humic acid. There are probably three major mechanisms by which clays bond humus to their surface: through polyvalent exchangeable cations, through ligand exchange, and through London or van der Waals forces.[6]

Bonding through polyvalent cations is brought about partly by the cation, such as calcium or aluminium having one of its charges neutralised by a negative charge on the clay surface and one or more by the humic surface; partly by hydrogen bonding of a humic particle to water molecules forming part of the hydration envelope of the cation; and partly by an oxygen which

1 *J. Landw.*, 1906, **54**, 343.
2 See, for example, some of the photographs in his *Soils of Europe*, Madrid, 1953.
3 E. M. Wroth and J. B. Page, *Proc. Soil Sci. Soc. Amer.*, 1946, **11**, 27.
4 G. W. Ford, D. J. Greenland and J. M. Oades, *J. Soil Sci.*, 1969, **20**, 291.
5 See, for example, J. M. Bremner and T. Harada, *J. agric. Sci.*, 1959, **52**, 137.
6 For a review of this subject see D. J. Greenland, *Soil Sci.*, 1971, **111**, 34.

has lost a hydrion by dissociation in the humic particle entering the co-ordination envelope around an aluminium or ferric ion by ligand exchange. These bonds are strong enough to raise appreciably the pH at which the cation is precipitated as its hydroxide (see p. 307). Humus held by these bonds is dispersed by treating the soil with a solution that absorbs or chelates these cations more strongly than does the clay-humus bond. Sometimes some of the cations can be removed by simple exchange with, say, potassium or sodium from a chloride solution, but most of these bonding cations can only be removed by extraction with a solution such as sodium pyrophosphate, sodium ethylene diamine tetra-acetate or acetyl-acetone which can remove at least a proportion by chelation.

Surfaces of ferric and aluminium hydroxides or hydrated oxides will absorb humic acids by ligand exchange (see p. 128). Ligand exchange requires the simultaneous presence of both the dissociated and undissociated acid, and since humic acids dissociate hydrogen ions over a very wide pH range, this adsorption mechanism can also operate over the whole pH range found in soils, except possibly at the most acid end. Soils rich in hydroxides, such as those derived from basic igneous rocks and some volcanic ash, tend to have high levels of organic matter, as would be expected. Much of this humic matter will be released if the pH is raised sufficiently high for most of its acidic groups to be fully dissociated, which is usually brought about by extraction with sodium hydroxide. The adsorbed humus has an important stabilising effect on ferric hydroxide surfaces,[1] for the pure films of ferric hydroxide are unstable in soils and recrystallise as discrete particles of goethite or haematite with a much smaller surface area,[2] but the adsorbed humus stabilises them.

Adsorption of humic particles on clay surfaces by van der Waals forces is through the weak attraction that develops between atoms when brought sufficiently close together. Thus, if a humus particle comes very close to a clay surface, forces of attraction may become important, and this is particularly so if the humus particle is flexible, and can bend to touch a large area of the clay, a process which can be very relevant to the adsorption of flexible linear polymers, such as some polysaccharide polymers, by clay surfaces. These forces become stronger and the adsorption firmer as the system is dried, for this removes water from the clay surface, allowing the polymer to lie on the surface itself. The change in free energy of the system due to this adsorption is small, but because of the very large number of spots at which contact is made between the polymer and the surface, and the consequent displacement of water molecules from the surface into the free solution, the entropy of the system has been greatly reduced, making it very difficult to intercalate water molecules between the polymer and the clay surface.[3]

1 U. Schwertmann, *Nature*, 1966, **212**, 645, and *Trans. 9th Int. Congr. Soil Sci.*, 1968, **1**, 645.
2 D. J. Greenland, J. M. Oades and T. W. Sherwin, *J. Soil Sci.*, 1968, **19**, 123.
3 D. J. Greenland in *Sorption and Transport Processes in Soils*, Soc. Chem. Ind. Monog. 37, 79, 1970.

Although the adsorption involves only a very small change in free energy, it can be extremely difficult to desorb the polymer from the clay.

It is not yet possible to say if these are the only three mechanisms by which humus is held on clay surfaces, or clay held on humic surfaces. This is because the nature of the humin fraction, that is the fraction which cannot be dispersed by sodium hydroxide, is not yet understood. It is extremely unlikely that the humin fraction only consists of polymers having free flexible chains that can be held by the van der Waals forces, and it is possible that a part of it is inaccessible to the usual dispersion agents by being situated inside clay domains bonded together by iron or aluminium hydroxide bonds. If, however, this were the principal cause, treatment with suitable sesquioxide solvents, such as an acidic chloride solution to remove the easily soluble aluminium and dithionite and citrate to remove the ferric iron and any less easily soluble aluminium, should allow the removal of everything except the long-chain polysaccharides from the humic fraction.

Some properties of the clay-humus complex are different from the properties of the two individual components. In the first place, the buffer curve of an artificially made clay-humus complex is different from the sum of the curves for the clay and humus components separately, and, in general, the exchange capacity up to a pH of 7 or 8 is less than that of the sum of the two components.[1] This is similar to the effect of adsorbed polyvalent cations on the buffer curve for humic acid (see p. 307), and simply means that some of the negatively charged spots on the clay and humus are neutralised by positive charges unaffected by pH in the pH range studied. This result has the implication that the cation exchange of humus determined from the drop in the exchange capacity of a soil when treated with hydrogen peroxide may be underestimated.

Humus adsorbed on the clay usually changes the colour of the clay, but this change need not be large. In general the adsorbed humus gives clay a brown colour, but there appear to be two groups of exceptions. Some red- and orange-coloured clays may have quite an appreciable content of humic material, though the colour of the clay from which the humus has been removed has not been studied. On the other hand, calcium-saturated montmorillonite clays are black in colour, although their humus content may be quite low.

The different clay fractions in any soil may adsorb different components of the total humus, or different components of the humus adsorbed onto fine clay particles may affect their apparent size. Very little experimental work has been done on this subject,[2] but it is possible that if the clay fraction as normally dispersed from a soil is itself fractionated into size classes by

---

1 See, for example, H. E. Myers, *Soil Sci.*, 1937, **44**, 331; L. T. Evans and E. W. Russell, *J. Soil Sci.*, 1959, **10**, 119.
2 See, for example, M. A. Arshad and L. E. Lowe, *Proc. Soil Sci. Soc. Amer.*, 1966, **30**, 731; F. W. Chichester, *Soil Sci.*, 1969, **107**, 356.

sedimentation, the humus in the fraction finer than $0.1\ \mu$ will have a lower carbon content, a lower C/N ratio, and be less aromatic, and therefore probably more oxidised than the coarser fraction. It is not known if this is due to the differential adsorption of different humic fractions by different-sized clay particles, or to nearly all the clay particles being fine but that those which adsorb the more aromatic humic particles become bonded into domains that are not dispersed in the normal mechanical analysis procedures.

The turnover of organic carbon in soils

Humus formation

Humic materials are intermediates in the conversion in the soil of plant residues into carbon dioxide and simple inorganic salts, and the conversion is carried out by the soil organisms. These materials have at least two separate origins: microbial cell wall materials and extracellular gums which are predominantly polysaccharide-rich material, and products of microbial metabolism which include both polysaccharides and polyphenols. Bacteria are known to be an important source of polysaccharides and some fungi of polyphenols.

A number of soil fungi have dark-coloured hyphae, and the dark pigments have long been suspected of being either humic material or the immediate precursors of humic material. Recently, J. P. Martin and K. Haider[1] have shown that certain soil fungi, they used the Ascomycetes *Stachybotris* and *Epicoccum*, produce a number of polyphenols when growing on a glucose-asparagine medium, apparently as a byproduct in their metabolism, and some of these are excreted into the medium and some accumulate in the young mycelium. These are then converted into a range of polyphenols, they identified up to 30 including appreciable proportions of 5-methyl-pyrogallol and 2,3,5-trihydroxybenzoic acid, both of which will autoxidise at pH values over 6 and link up with other polyphenols, amino acids and peptides to give humus-like polymers both in the external medium and in their mycelium. These fungi can also decompose lignins, but they attack the aliphatic side chains and not the benzene rings, so produce additional polyphenols which also become converted to these humic-like polymers on autoxidation. These polymers have a similar chemical composition to soil humic acids, they yield the same distribution of phenols on reduction with sodium amalgam, have similar exchange capacities and have molecular weights between 5000 and 200000; though unlike the humic acid from the Urrbrae soil, the size distribution of particles throughout this range is more even than for the humic particles (see p. 286).

It is not known what proportion of soil fungi or soil bacteria can produce

1 *Proc. Soil Sci. Soc. Amer.*, 1967, **31**, 657, 766. *Soil Sci.*, 1969, **107**, 260; *Soil Biol. Biochem.*, 1970, **2**, 145.

these humic-like polymers, but it is possible all those fungi with dark mycelia may do so. If this ability is restricted to only a few groups, it could explain the relative constancy of the composition and properties of humus in different soils, although Martin and Haider showed that the distribution of the poly-phenols produced by the three fungi they used was different for the different species. They also made the observation that Basidiomycetes which can decompose lignin by splitting the benzene ring, the 'white rot' fungi, are not humus producers, presumably because they obtain their energy from the oxidation of polyphenols rather than produce them as a byproduct of their metabolism.

All three of the classical humus groups—fulvic and humic acids and humin are produced simultaneously by the micro-organisms. Thus, J. Mayaudon and P. Simonart[1] found that 30 days after glucose was added to a soil, 11·3 per cent of the carbon remained in the soil, and 20, 35 and 45 per cent of this remainder were in these three fractions. D. S. Jenkinson[2] also found that one year after he added ryegrass labelled with ^{14}C to a Rothamsted soil, the specific activities of the carbon in the humic acid and in the pyro-phosphate extract were approximately the same as for the organic carbon of the soil as a whole. The specific activity of the fulvic acid was higher than that of the humic, which, in turn, was higher than that in the humin. Unfortunately, the interpretation of these results may not be straight-forward, for the labelled material need not be an integral part of the humic fractions. This is because some soluble labelled plant material can be absorbed on to the surface, or within the pores, of these colloids and possibly be protected against decomposition.[3]

Humus decomposition

Humus is a heterogeneous mixture so would not be expected to decompose uniformly, and some humus fractions may be converted into others during the process. Thus, it is possible, though unproven, that the pyrophosphate dispersible fraction has been derived from more polyphenol-rich material by oxidation. However, there is still only limited evidence on the relative decomposabilities of the different humus fractions in field soils.

There is a rough relation between the texture of a soil and its humus content, in that sands and loamy sands tend to have a lower humus content than loams or clays. There is possibly a stronger correlation with the amount of active iron and aluminium in the soil. Thus, allophane soils, which are rich in aluminium hydroxide surfaces usually have a high humus content,[4]

1 *Pl. Soil*, 1959, **11**, 170, 181.
2 *J. Soil Sci.*, 1968, **19**, 25.
3 See, for example, D. Sauerbeck and F. Fuhr in *Isotopes and Radiation in Soil Organic Matter*, I.A.E.A., Vienna, 1968; D. S. Jenkinson, *Soil Sci.*, 1971, **111**, 64.
4 See S. Tokudome and I. Kanno, *Trans. 9th Int. Congr. Soil Sci.*, 1968, **3**, 163 for examples from Japan.

and E. G. Williams[1] found that organic carbon content of a number of Scottish soils was closely correlated to the amount of aluminium extractable with Tamm's acid oxalate and of iron extractable with dithionite. This result is in accord with work on the composting of plant residues for clay absorbs some of the products of decomposition and this absorbed material decomposes more slowly than the unabsorbed.

The decomposability of humus in soil can be increased by subjecting the soil to a series of drying and wetting treatments (see p. 237). H. F. Birch,[2] among others, has examined this point and showed that the more strongly the soil was dried, and the longer it was left dry, the greater the flush of decomposition when the soil was wetted. The full cause of this flush has not been established, but an important part of the organic matter that decomposes during the flush is humic material that has become desorbed from the soil during the drying process and which disperses into the soil solution on wetting.[3] Little is known about the composition of this fraction, but Birch found that after a Kikuyu red loam soil from Muguga, Kenya, had been dried and wetted 204 times over a period of 4 years, it had lost 63 per cent of its carbon, 46 per cent of its organic nitrogen, and all its organic phosphate, and the soil had lost its structure completely. This is a soil well supplied with active iron and aluminium, containing 8·5 per cent organic carbon, and it only loses its organic matter very slowly under cultivation.

The rate of decomposition of humus in field soils is increased by cultivating the soil, particularly by intensive cultivations during the growing season when the soil is warm and moist. Before the advent of herbicides, crops such as maize were frequently cultivated between the rows for weed control, whereas the small grain crops, such as wheat and barley, usually received little cultivation except in the cool part of the year. The effect of cultivations on the soil humus level is illustrated in Table 15.5a, which shows that at Wooster, Ohio[4] the continuous maize plots had only 35 per cent of the organic carbon present in the soil thirty years previously while the continuous wheat plots had 62 per cent.

Similarly, H. J. Haas[5] found for a number of stations in the wheat growing area of the Great Plains, that, over a period of thirty to forty years, continuous wheat gave a loss of 25 per cent, alternate wheat-fallow of 32 per cent, and a maize-sorghum rotation of nearly 50 per cent of the organic nitrogen initially present.

Birch's result implies that the various components of the soil humus should decompose at different rates in the soil in the field. This point has not received much attention, but there may be less differential effect in the field

1 With N. M. Scott and M. J. McDonald, *J. Sci. Fd. Agric.*, 1958, **9**, 551.
2 *Nature*, 1959, **183**, 1415; 1961, **191**, 731.
3 See also G. W. Skyring and J. P. Thompson, *Pl. Soil*, 1966, **24**, 289.
4 R. M. Salter and T. C. Green, *J. Amer. Soc. Agron.*, 1933, **25**, 622.
5 With C. E. Evans and E. F. Miles, *U.S. Dept. Agric.*, *Tech. Bull.* 1164, 1957.

TABLE 15.5a   Effect of crop rotation on the organic matter content of arable soils. Crops grown since 1894 without manure or fertiliser.* Weight per hectare in top 16 cm of soil

Cultivation	Carbon in 1000 kg	Nitrogen in 100 kg
Initially (1894)	22·8	24·4
After 30 years' cropping to continuous maize	8·3	9·4
After 30 years' cropping to continuous wheat	14·3	14·7
Five-course rotation:		
maize–oats–wheat–clover–timothy	17·3	17·3
Three-course rotation :†		
maize–wheat–clover	19·1	19·8

* Some plots received dressings of lime, which appeared to have no effect on the carbon or nitrogen contents of the soil.
† Since 1897.

than in the very artificial conditions of the laboratory. Thus, D. R. Keeney and J. M. Bremner[2] found that the distribution of nitrogen compounds in the humus from ten virgin prairie soils and their cultivated counterparts were very similar, in spite of cultivation reducing their nitrogen contents by nearly 40 per cent (see p. 290). This suggests that, in the field, the relative rates of the formation and decomposition of the principal humus fractions are not very dependent on the soil management.

The effect of drying and wetting, or of cultivation, on the rate of oxidation of the humic fraction is unlikely to be due solely to humus becoming soluble. It is probable that an important function of these disturbances is the gradual breaking up of stable soil crumbs, exposing humus that had been inaccessible to the soil organisms, for humic matter in pores finer than about 1 μ is inaccessible to micro-organisms, and in fine-textured soils a large proportion of the soil pores are finer than this. Thus grinding soil crumbs to pass a fine sieve,[2] or dispersing them ultrasonically[3] increases the amount of humic material that will be decomposed.

The level of organic matter in field soils

The level of organic matter in an agricultural soil is determined by the rates of addition and oxidation of plant residues and of soil humus. An uncultivated soil carrying natural vegetation, such as a forest or pasture, typically has a higher humus content than the same soil when cultivated, both because

1 *Proc. Soil Sci. Soc. Amer.*, 1964, **28**, 653. A. W. Moore and J. S. Russell (*Trans. 9th Int. Congr. Soil Sci.*, 1968, **2**, 557) confirmed this result for some Australian soils under different systems of management.
2 S. A. Waring and J. M. Bremner, *Nature*, 1964, **202**, 1141.
3 F. W. Chichester, *Soil Sci.*, 1969, **107**, 356.

it has a higher rate of addition of organic matter to the soil and because the soil is not disturbed. The humus content of farmland left to go derelict will increase for the same reason, as is shown in Table 16.8 on p. 350. Correspondingly, the higher the rate of addition of organic matter to a cultivated soil, the higher its humus content; arable rotations involving few cultivations per year have a higher content than those involving frequent alterations, particularly during the growing season when the soil is warm and moist. For any given system of farming, the humus content of the soil tends towards a value that is characteristic for that system on that soil in that climate, so it is nearly always possible to forecast the direction of change in the humus content consequent upon any change made in the system of farming. The principal exception to this statement is the effect of liming a soil on its humus content. This is a topic that has been little studied, but in general liming a very acid soil reduces its organic matter content, as would be expected, although this is based on soils from which fine root material and partially decomposed plant residues have not been removed. But on the old grassland plots at Rothamsted on Park Grass, which have been cut for hay annually for over a century, liming appears to increase the humus content, without necessarily increasing the yield of harvestable dry matter, as is shown in Table 15.6.[1]

TABLE 15.6   Effect of liming an old meadow on the organic matter content of the soil. Rothamsted: Park Grass. Fertiliser treatments since 1856. Sampled 1959. Yield: mean dry matter t/ha 1920–59

Treatment	Unlimed					Limed				
	Grass yield	pH	% C	% N	C/N	Grass yield	pH	% C	% N	C/N
Unmanured*	1·59	5·2	3·4	0·27	12·4	1·74	7·2	4·1	0·34	12·1
Minerals only†	3·66	4·9	2·8	0·23	16·5	4·42	7·0	3·4	0·30	11·4
Minerals with‡ ammonium sulphate	5·52	3·7	4·4	0·33	13·4	6·57	4·8	3·9	0·30	13·0
Minerals with§ sodium nitrate	5·27	5·7	3·3	0·25	13·2	5·05	7·2	3·7	0·32	11·8

* Plots 2, 3.   † Plot 7.   ‡ Plots 9, 11¹, 11².   § Plots 14, 16.

The rate of oxidation of humus in field soils can be estimated when a humus-rich soil, such as an old pasture or virgin prairie, is ploughed out and cropped to a succession of similar crops, such as wheat or barley, for example, for a long spell of years without any organic matter other than old roots and stubble being added to the soil. This causes an initial rapid fall in the humus content of the soil, which becomes less rapid until the humus

1 *Rothamsted Rept.* 1963, 240.

content is stabilised at a lower level. The rate of fall in the humus content, whether measured by the fall in the organic carbon or nitrogen content, follows a first order process rate, that is it is a linear function of the content. Thus, if C is the organic carbon content of the soil t years after the prairie was ploughed out, then $dC/dt = a - bC$ and

$$bC = a - (a - bC_0)e^{-bt}$$

where a is the annual addition of organic carbon to the soil, b is the fraction of the carbon that is decomposed each year, and C_0 is the initial carbon content of the soil. The assumption that a is constant implies that there is a constant and uniform return of crop residues, and that b is constant that the soil humus is all equally easily decomposable. This equation states that, after a period of years, the carbon content of the soil will approach the value a/b asymptotically, and that after a period of $0.693/b$ years the carbon content will be halfway between its initial and asymptotic value. This period is called the half-life of the humus in the soil.

W. V. Bartholomew and D. Kirkham[1] tested the validity of this equation for a number of prairie soils in North America and found half-lives varying from 10 to 45 years, P. H. Nye and D. J. Greenland[2] found values between 7 and 55 years for a number of cultivated tropical soils, and D. S. Jenkinson[3] from a study of the rate of loss of carbon from the unmanured Broadbalk plot found a half-life of 25 years, a figure also found for another Rothamsted soil by J. L. Monteith[4] from estimates of the carbon dioxide flux out of the soil. However, as already noted on p. 274, Jenkinson found that the half-life of the humus produced from fresh organic matter, such as ryegrass, is only 4 years in the Broadbalk soil.

On the other hand, a Rothamsted soil which has had its organic matter increased by annual dressings of farmyard manure, loses this additional humus very slowly once the application of the manure has ceased. Thus, on Hoosfield, which has been in barley since 1852, the organic nitrogen content in 1946 of the unmanured soil was 0·103 per cent, for the plot receiving annually 35 t/ha farmyard manure it was 0·272 per cent, but for a plot which received this amount of manure for 20 years (1852–71) and thereafter was unmanured it was 0·151 per cent. Although the nitrogen content of this soil was not determined in 1871, it was probably about 0·185 per cent, so that this plot gained 0·08 per cent nitrogen in 20 years, but only lost 0·13 per cent in the subsequent 75 years, so had an apparent half-life of about 100 years.[5] Jenkinson[6] calculated that the half-life of the organic nitrogen in the farmyard

1 *Trans. 7th Int. Congr. Soil Sci.*, 1960, **2**, 471.
2 *The soil under shifting cultivation*, Commonw. Agric. Bur. 1960.
3 *J. Soil Sci.*, 1965, **16**, 104.
4 With G. Szeicz and K. Yabuki, *J. appl. Ecol.*, 1964, **1**, 321.
5 R. G. Warren, *Fertiliser Soc., Proc.* 37, 1956.
6 In *The Use of Isotopes in Soil Organic Matter Studies*, Pergamon, 1966.

manure plot was about 25 years, a very much shorter period than the apparent half-life once additions of manure have ceased.

It is now possible to obtain an estimate of the half-life, or mean age, of the carbon in the soil humus from its $^{14}C/^{12}C$ isotope ratio, based on assumptions about the value of this ratio in the atmospheric carbon dioxide; and the age of the humus so determined may come out surprisingly high. Thus D. S. Jenkinson[1] found the half-life of the humus in the unmanured plot on Broadbalk sampled in 1881, was about 1400 years in the top 22 cm, rising to 3700 years at 45 to 67 cm deep. The corresponding figure for the humus in the top 22 cm of the old permanent unmanured grass plots sampled in 1886, was 600 years, showing that the greater the return of organic matter to the soil, the shorter its half-life.[2] A similar result was found in a Canadian chernozem,[3] where the mean half-life of the humus in the soil cropped to a wheat-sweet clover-fallow rotation was 1700 years compared with 2200 years for the same soil under a wheat-fallow rotation. R. W. Simonson[4] has quoted ages in the range 200 to 400 years for five pasture soils in the Mid-West of the USA.

The apparent age of the carbon in the various humus fractions is not, however, the same. C. A. Campbell[5] in western Canada found that, for the two soils they examined, the fulvic acid fraction was younger than the humic or humin fractions, and that, for the only soil they examined, the fraction that was hydrolysed with hot 6 N.HCl was younger than the fraction that was not hydrolysed, as shown in Table 15.7. In fact, in this chernozem soil, the hydrolysed fraction of the humic acid was too young to be dated. This result is comparable to that of Jenkinson's at Rothamsted, for he found that in the experiment in which he buried ryegrass roots labelled with ^{14}C, 70 per cent of the labelled carbon in the soil was hydrolysed in 6 N.HCl compared with only 51 per cent of the unlabelled after one year, and 64 per cent compared with 49 per cent after four years, during which time 37 per cent of the labelled carbon, but only 9 per cent of the unlabelled carbon, had been lost, presumably by oxidation.[6]

These results for the apparent age of humus in soils based on ^{14}C dating methods appear to be completely at variance with half-lives based on rates of loss of organic carbon or nitrogen from soils during cultivation.

Part of the explanation is that, in the determination of the half-life of organic carbon added to the soil during farming operations, no distinction has been made between the rates of decomposition of non-humified or

1 *Rothamsted Rept.*, 1968, Part I, 73, and see also ibid., 1969, Part I, 84.
2 Owing to the effects of A- and H-bomb testing, which much increased the $^{14}CO_2$ concentration in the atmosphere, the apparent ages of the humus in these two Rothamsted soils, sampled in 1966, are 870 and 35 years respectively.
3 C. A. Campbell, E. A. Paul *et al.*, *Trans. 8th Int. Congr. Soil Sci.*, 1964, **3**, 201.
4 *Proc. Soil Sci. Soc. Amer.*, 1959, **23**, 152.
5 With E. A. Paul *et al.*, *Soil Sci.*, 1967, **104**, 217.
6 *J. Soil Sci.*, 1968, **19**, 25.

TABLE 15.7   Half-life of humus fractions in two western Canada soils

	Chernozem			Grey-wooded podsolic	
	per cent total C	half-life years	per cent of contribution to total C oxidised annually derived from	per cent total C	half-life years
Initial acid extract	13·7	325	10	29·3	0
Fulvic acid	14·8	495			
Humic acid:				37·3	195
hydrolysable	7·1	25 ± 50	65		
not hydrolysable	33·2	1400	5		
Humin:				32·2	485
hydrolysable	7·3	465	11*		
not hydrolysable	23·5	1230	9		
Whole soil		870			250

* This includes the fulvic acid fraction contribution.

partially humified material and that of the relatively stable and long-lived humus, nor, if it takes time, between the first-formed group of humic products and the stable. There is, in fact, no adequate data to show how long it takes for added organic matter to be incorporated into this stable fraction, if it really is different from the humic substances first formed during the decomposition. The other possible explanation is that a major part of the humus is within the pores of crumbs smaller than about 0·25 mm size, which are too narrow for bacteria to enter; and that the half-life of the humus is in reality the half-life of these small crumbs in the soil. A. P. Edwards and J. M. Bremner[1] have, for quite other reasons, postulated that crumbs of this size are very stable in many soils. On this explanation one would also expect the humus of pasture soils to have a lower half-life than the corresponding arable, as is found at Rothamsted, because their crumbs have a lower bulk density and hence larger proportion of the soil volume should be accessible to the micro-organisms.

The results for the rate of oxidation of organic matter in the soil can be interpreted as showing that there are at least three different types of humic decomposition taking place in aerated soils. The first process is the conversion of added organic matter into microbial biomass and its immediate decomposition products. At Rothamsted these are very susceptible to acid hydrolysis and have a mean half-life of about four years, and are presumably converted into the typical more stable humic compounds. In so far as these are accessible to microbial attack, they have a mean half-life of about twenty-

1 *J. Soil Sci.*, 1967, **18**, 47.

five years, and for other soils this may vary between ten and forty-five years. But a portion of this fraction becomes situated in pores too fine for micro-organisms to enter, and the more stable these fine pores, the longer this portion will remain in the soil. There is still little critical information on the factors which determine what the half-life of humus will be in a given soil, although it is likely to be shorter the higher the soil temperature, provided the soil is moist, and to be longer the higher the aluminium content of the humic fractions. There is still no really good evidence if the clay content of a soil has any appreciable effect apart from any aluminium or ferric hydroxide films it may have on its surface, and its effect in increasing the number of fine pores in the soil.

The humus colloids in their natural conditions in the soil may not be very resistant to laboratory treatments commonly supposed to have little effect. Boiling a soil in water will certainly bring organic matter into solution, and G. Stanford[1] was able to bring into solution between 25 to 40 per cent of the total nitrogen in a number of soils by repeated autoclaving in 0.01 M $CaCl_2$ at $120°C$, and 66 per cent in one soil subjected to very prolonged autoclaving.[2] Initially at least a proportion of the nitrogen was set free from amino sugars which were hydrolysed and brought into solution.[3] Oven drying is also likely to cause degradation of some of the humic colloids, bringing organic matter into solution, which can then be readily decomposed by the soil micro-organisms, as H. F. Birch found for an East African soil. Grinding a soil may also affect the stability of the humic colloids, since it increases the rate of decomposition of the organic matter.[4] On the other hand, dispersing a soil in water using ultrasonics also increases the rate appreciably, presumably because it has made more of the organic matter accessible to the micro-organisms.[5] This is an aspect of humus chemistry that deserves more critical examination, as it has a very direct bearing on the significance of humic fractions with a very long half-life in a soil.

Effect of organic manures

The humus content of arable soils is increased if they receive regular dressings of farmyard manure, compost or other organic wastes. Table 15.8[6] gives the effect of annual dressings of 35 t/ha of farmyard manure or of fertilisers on the organic nitrogen content of Broadbalk soil compared with that of the unmanured plot. Unfortunately, the nitrogen content of the soil in 1843 was

1 *Soil Sci.*, 1969, **107**, 323.
2 With J. O. Legg and F. W. Chichester, *Pl. Soil*, 1970, **33**, 425.
3 With W. H. Demar, *Soil Sci.*, 1970, **109**, 190, and see D. S. Jenkinson, *J. Sci. Fd. Agric.*, 1968, **19**, 160 for similar results.
4 S. A. Waring and J. M. Bremner, *Nature*, 1964, **202**, 1141.
5 F. W. Chichester, *Soil Sci.*, 1969, **107**, 356.
6 A. E. Johnston, *Rothamsted Rept.* 1968, Part II, 98.

not determined, but it was probably close to 0·110 per cent, and the manure adds about 225 kg/ha N annually. These figures must be accepted with considerable caution, because up to 1914 they were based on only relatively few samples per plot, and the method of determining the soil nitrogen was changed in 1893. Further during the period 1914–36 a system of fallowing was introduced and from 1936–66 the land was usually fallowed once in five years and no manure was added in the fallow year.

There is still little knowledge on the amounts of plant material, as roots, stubble, weed and other debris, that is added annually to the soil in these experiments, though D. S. Jenkinson[1] estimated that the unmanured plot receives about 1000 kg/ha annually, while giving a mean yield of grain plus straw of 1400 kg/ha.

The important points that emerge from Table 15.8 are, first, that the nitrogen content of the unmanured plot soon stabilises at about 0·10 per

TABLE 15.8   Nitrogen percentage in the top 23 cm of Broadbalk soils

Plot	3	5	7	2A	2B
				Farmyard manure	
Manuring	None	PK	NPK	Since 1885	Since 1843
Year					
1865	0·105	0·106	0·117	—	0·175
1881	0·101	0·098	0·121	—	0·184
1893	0·099	0·101	0·122	0·163	0·221
1914	0·093	0·103	0·120	0·196	0·236*
1936	0·103	0·105	0·120	0·186	0·226
1945	0·106	0·105	0·121	0·191	0·236
1966	0·099	0·107	0·115	0·216	0·251

Results for 1865 and 1881 determined by the soda–lime method, which gives slightly low values compared with the Kjeldahl method used subsequently.
* Only one sample; the duplicate sample gave a nitrogen content of 0·266, probably through a bit of manure being included.

cent N and the plots receiving fertiliser at about 0·12 per cent. During the first 22 years, about 37 per cent of the nitrogen added in the manure remains in the soil and during the next 28 years just under 20 per cent does, giving a mean retention of about 25 per cent over the first 50 years. During the last 30 years, with a fallow once in 5 years, only 15 per cent is retained, which, compared with a retention of about 20 per cent for the 49 years 1865–1914. This shows that one year fallow in five on the Rothamsted clay loam does not seriously reduce the soil nitrogen content.

1 *Rothamsted Rept.* 1968, Part I, 73.

Results for the sandy loam at Woburn are similar to those at Rothamsted. Thus the nitrogen content of unmanured soil under continuous wheat or barley fell from 0·16 to 0·10 per cent over a period of 50 years, but was maintained at about 0·16 per cent if it received annual dressings of 16 t/ha of farmyard manure. This is equivalent to 25 per cent of the nitrogen in the manure being retained in the humus compared with the unmanured plots.[1] Under intensive vegetable cultivation,[2] a soil with initially 0·085 per cent N had its nitrogen content raised by 0·035 and 0·055 per cent over a period of 9 years if given 37 or 74 t/ha straw compost annually and by a further 0·027 and 0·036 during a second 9-year period. These figures are equivalent to 50 and 40 per cent and 40 and 25 per cent of the added nitrogen being retained in the soil organic matter. The efficiency of the compost in increasing the organic matter content of the soil is dropping with the heavier dressing in this example, but this is not always found. Thus M. B. Russell[3] found a linear increase of both organic carbon and nitrogen when 25, 50 or 100 t/ha were added annually for a period of 25 years to a soil.

M. Chater and J. K. R. Gasser[4] found rather lower results at Woburn when 25 t/ha of manure were added every other year over a period of 18 years for only 25 per cent of the added carbon, equal to a gain of 400 kg/ha annually, and 40 per cent of the added nitrogen remained in the soil at the end of this period. But during the next 12 years, when no further organic manure was added, one-half of this carbon and nitrogen had been lost, the annual rates of loss being 300 kg/ha and 22 kg/ha respectively.

Experiments elsewhere, using local rotations with farmyard manure being applied only once every 4 to 6 years have given results comparable to these just described. Thus K. Dorph-Petersen[5] in Denmark found that soils cropped to a six course rotation retained nearly one-third of the nitrogen added in the farmyard manure during four courses. It is probable that under normal agricultural conditions in north-western Europe, soils receiving regular additions of farmyard manure retain between one-quarter and one-third of the nitrogen in the manure in the form of soil organic matter having a C/N ratio of about 10.

Green manures can increase the humus content of the soil in the same way that farmyard manure can, but the amounts of dry matter in the green manure crop is usually small. Thus M. Chater and J. K. R. Gasser[6] found that about 15 per cent of the carbon and nitrogen added in green manure

1 E. M. Crowther in *Fifty Years at the Woburn Experimental Station* by E. J. Russell and J. A. Voelcker, Longman, 1936.
2 H. H. Mann and H. D. Patterson, *Rothamsted Rept*. 1962, 186.
3 With A. Klute and W. C. Jacob, *Proc. Soil Sci. Soc. Amer.*, 1952, **16**, 156. This result has also been found by L. A. Pinck and F. E. Allison, *Soil Sci.*, 1951, **71**, 67, and D. S. Jenkinson, *J. Soil Sci.*, 1965, **16**, 104; *Soil Sci.*, 1971, **111**, 64.
4 *J. Soil Sci.*, 1970, **21**, 127.
5 *Tidskr. Planteavl*, 1946, **50**, 555, and see G. W. Cooke, *J. Roy. Agric. Soc. Engl.*, 1957, **118**, 131.
6 *J. Soil Sci.*, 1970, **21**, 127.

crops taken either every year or every alternate year was retained in the soil. Adding straw to the soil will also increase the organic matter content, but the little data available indicates that it is surprisingly inefficient. Thus at Rothamsted[1] the addition of 6·6 t/ha of straw in alternate years for 18 years has only increased the organic carbon content by 0·1 per cent, indicating that no more than 10 per cent of the carbon added has been retained.

Effect of grass leys

Laying arable land down to grass usually increases the organic matter and the humus content of the soil, often considerably. Figure 15.5 illustrates this for the Rothamsted soil, which typically has a nitrogen content of about

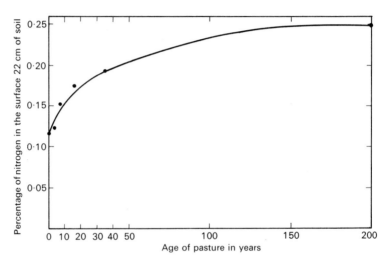

FIG. 15.5. The rate of accumulation of nitrogen in the top 22 cm of soil at Rothamsted when old arable land is laid down to grass

0·11 per cent under arable crops receiving no farmyard manure, and a content of between 0·25 and 0·35 per cent under old pasture. H. L. Richardson[2] estimated it took about 25 years under grass for the organic nitrogen content of the soil to increase half way from the old arable to the old pasture level at Rothamsted, a value also found by J. S. Russell and D. L. Harvey[3] for some Australian soils.

Short periods of grass leys, whether grazed or cut for hay or silage will usually give a definite increase in both the organic carbon and organic

1 *Rothamsted Rept.* 1958, 43.
2 *J. Agric. Sci.*, 1938, **28**, 73.
3 *Aust. J. Agric. Res.*, 1959, **10**, 637; 1960, **11**, 902. See D. J. Greenland, *Soils Fert.*, 1971, **34**, 237 for more recent Australian results.

nitrogen content of the soil, although the published figures usually include carbon or nitrogen in some of the partially decomposed plant debris, as this has not been removed before analysis. C. R. Clement and T. E. Williams,[1] for example, found that most of the increased organic matter in their soils that had been in grass for three years was in the top two centimetres of the soil, and this suggests that it probably contained a considerable proportion of non-humic material.

A more recent Rothamsted experiment illustrates the relatively small effect of short-term grass leys on the organic matter content of the soil. The experiment is in two fields, one of which had been in pasture probably for centuries and one in old arable. The treatments compared are permanent grass, continuous arable, and three years grass or lucerne followed by three years arable, and the results for the first eighteen years are given in Table 15.9.[2] The introduction of three years grass into the arable rotation has

TABLE 15.9   Effect of grass on the organic carbon content of soil. Rothamsted 1952–64. Per cent organic carbon in top 30 cm

	Old arable field			Old pasture field		
	After 6 years	*After 12 years*	*After 18 years**	*After 6 years*	*After 12 years*	*After 18 years**
Old pasture				3·22	3·82	3·75
New pasture	1·68	2·14	2·67	3·02	3·61	3·76
All arable	1·42	1·31	1·58	2·74	2·25	2·05
3-year arable after 3-year grass	1·62	1·43	1·68	2·80	2·45	2·11
3-year arable after 3-year lucerne	1·56	1·33	1·53	2·75	2·29	2·04

* The figures for 18 years were kindly supplied by A. E. Johnston.

increased the organic carbon content by up to 0·1 per cent compared with the all-arable, and, surprisingly, lucerne has behaved as the arable crops. On the other hand, leaving the old arable land down to grass for eighteen years has increased its carbon content by about 1·2 per cent—an increase of about 80 per cent above the initial value. Ploughing out the old pasture and immediately reseeding it to grass gave an initial loss of about 0·2 per cent, which has been made good after eighteen years, but the land that has been in arable has lost 1·7 per cent organic carbon, or nearly half of what the soil probably contained when it was ploughed out of grass. The Woburn ley-arable experiment also showed that this sandy loam soil under a rotation of three years grazed grass or lucerne/sainfoin cut for hay, followed by two

1 *J. Agric. Sci.*, 1964, **63**, 377.
2 *Rothamsted Rept.* 1964, 43.

years arable, had the same carbon content even after thirty years as other plots which had been in continuous arable. Thus, the Rothamsted and Woburn data give no support to the view that a three-year ley has any great effect on the organic carbon content of the soil even two years after it has been ploughed out and cropped to arable: a result that has also been found in a series of rotation experiments in which three years ley followed by three years arable was compared with six years arable on a number of experimental husbandry farms in England and Wales.[1] On only one farm did the ley-rotation increase the soil organic matter at the end of the six years; on one it maintained it, and on four it reduced it, and on these farms the continuous arable only increased the rate of loss of organic matter by about 50 per cent above the rate for the rotations with three year leys.

1 D. J. Eagle, *Min. Agric. Fish. Fd., Tech. Bull.* 20, 125, 1971.

16

The nitrogen cycle in the soil

The mineralisation of soil nitrogen

Mineralisation of soil nitrogen is the term used for the process by which nitrogen in organic compounds becomes converted into the inorganic ammonium and nitrate ions. The transformation can probably only take place through the stages

$$\text{organic N} \;\rightarrow\; \text{ammonia} \;\rightarrow\; \text{nitrite} \;\rightarrow\; \text{nitrate}$$

and as far as is known these transformations are predominantly brought about in the soil by micro-organisms.

The production of ammonia from organic matter

The soil microflora typically produce ammonia from organic compounds when they set free more nitrogen from the organic matter on which they are living than they can assimilate into their own protoplasm. Further, ammonia forms their sole important nitrogen excretion product under aerobic conditions, though if the oxygen supply is restricted amines may also be produced. Ammonia production from nitrogen-rich organic matter is thus not confined to a few groups of soil micro-organisms: it is their typical and characteristic nitrogen excretion product. Soil animals, on the other hand, probably excrete either uric acid, as do the insects, or urea as do the mammals.

This concept of ammonia production being the nitrogen waste-product in the conversion of organic matter into microbial tissue and vital energy—originally put forward by E. Marchal[1] in 1893—is fundamental for understanding the effect of adding different types of organic matter on the mineral nitrogen in the soil. Thus, when a protein, such as dried blood, is added to a soil, about 80 per cent of the added nitrogen is liberated as ammonia, the remainder of the nitrogen being retained in microbial tissue; but as increasing quantities of a carbohydrate, such as cellulose, are mixed in with the protein, so the amount of microbial tissue that can be built up is increased, with the

1 *Bull. Acad. Roy. Belg.*, 1893 (3), **25**, 727.

consequence that the proportion of nitrogen liberated as ammonia decreases until the ratio of carbohydrate to protein reaches a ratio of about 5:1 when all the nitrogen in the dried blood is needed by the micro-organisms.

There is some evidence that ammonia can be produced in soils treated with toluene, and therefore presumably not by the action of living micro-organisms. This seems to have been first noted by E. J. Russell and H. B. Hutchinson[1] in 1909, and has been re-examined in more detail by J. P. Conrad,[2] who showed that soils so treated still possess the power to hydrolyse urea and some other amides to ammonium carbonate, presumably through the action of the urease enzymes present in the soil which have been released there during the decomposition of added plant and animal tissues. Whether uric acid—the nitrogen excretion product of most soil invertebrates—is similarly decomposed has not been investigated.

Nitrification in the soil

The organisms involved

Th. Schloesing and A. Müntz[3] proved in 1877 that nitrification, used here to mean the production of nitrate from ammonia, was mainly a biological process by showing it could be stopped by antiseptics such as chloroform; and this is still the only process that is definitely known to bring about this oxidation in soils.

Various claims have been put forward that nitrate can be produced photochemically from organic compounds, but these have not yet been substantiated.[4] There seems little doubt that ammonia can be oxidised photochemically to nitrite in solution,[5] but it appears that nitrates are more liable to photochemical reduction to nitrites than the other way round. The importance of this photo-oxidation to nitrite even in tropical soils, where it can only be a surface phenomenon, is still undecided.

The first organisms proved to oxidise ammonia to nitrite and nitrite to nitrate were bacteria, but the older bacteriological techniques failed to isolate them in pure culture. Thus R. Warington at Rothamsted[6] obtained bacterial cultures, which must have been very mixed, that would convert ammonia and some organic nitrogen compounds into nitrites, but not necessarily nitrates, and others that would convert nitrites, but not ammonia,

1 *J. agric. Sci.*, 1909, **3**, 111.
2 *Soil Sci.*, 1942, **54**, 367; *J. Amer. Soc. Agron.*, 1941, **33**, 800; 1942, **34**, 1102; and *Proc. Soil Sci. Amer.*, 1944, **8**, 171.
3 *Comp. Rend.*, 1877, **84**, 301; **85**, 1018; and 1878, **86**, 982.
4 For a review, see S. A. Waksman and M. R. Madhok, *Soil Sci.*, 1937, **44**, 361.
5 G. G. Rao and N. R. Dhar, *Soil Sci.*, 1931, **31**, 379; A. S. Corbet, *Biochem. J.*, 1934, **28**, 1575; B. N. Singh and K. M. Nair, *Soil Sci.*, 1939, **47**, 285.
6 It is instructive to read his papers on this subject (*Trans. Chem. Soc.*, 1878, **33**, 44; 1879, **35**, 429; 1884, **45**, 637; 1888, **53**, 727; and 1891, **59**, 484) and to realise that in spite of the far better techniques available today many of the problems raised are still unsolved.

into nitrates; but he could not isolate these organisms as pure cultures on gelatine. It was left to S. Winogradsky,[1] to isolate the bacteria responsible. He argued that the organisms should be autotrophic, obtaining their carbon from CO_2 and their energy from the oxidation of ammonia and nitrite, and that therefore they alone of the soil organisms should grow on silica gel plates free from organic matter but containing bicarbonate and ammonium or nitrite salts. Using this technique, he isolated an ammonium-oxidising and nitrite-producing bacterium, which he called *Nitrosomonas*, and a nitrite-oxidising and nitrate-producing bacterium, which he named *Nitrobacter*. He found these bacteria grew on the calcium carbonate added to the medium to supply the bicarbonate and assumed these surfaces were essential for their growth, and that culture solutions containing glucose, gelatine or peptone were toxic to them and assumed that these substances were the cause of the toxicity. However, it is now known that the toxicity was due to toxins produced during the sterilisation of his culture media in an autoclave, for these solutions are not toxic if sterilised by ultra-filtration.[2] He also showed they were slow-growing and only formed very small colonies on his plates, about 0·1 mm in diameter. This technique, which was the only one in use for the next seventy years, requires very considerable skill[3] and only recently have simpler methods been developed capable of growing these bacteria on the large scale.[4] Later he isolated other ammonium oxidisers which he put in the genera *Nitrosococcus*, *Nitrosocystis*, *Nitrosospira* and *Nitrosogloea*.[5] All these bacteria are strict autotrophs and cannot use sugars or other organic compounds for their energy supply, although trace amounts of some amino acids such as aspartic and glutamic and some organic acids such as pyruvic and possibly acetic[6] may stimulate their growth. All isolates of *Nitrosomonas* are usually referred to as *N. europea*, although they differ appreciably among themselves. Thus, some isolates have flagellae, so are motile and some do not, and some are gram-negative and some gram-positive.[7] *N. europea* has, in the past, been considered the dominant and most widespread ammonium oxiser in the soil, but N. Walker and S. Soriano[8] have found that, at Rothamsted, it is the dominant oxidiser only on soils receiving organic manures. Two species of *Nitrobacter* have been widely recognised, *N. agilis* and *N. winogradskii*. It is likely that all these organisms occur in different strains with varying degree of adaptation to different environmental conditions.

1 *Ann. Inst. Pasteur*, 1890, **4**, 213, 257, 760; and with W. Omeliansky, *Zbl. Bakt.* II, 1899, **5**, 329, 429; and *Arch. Sci. Biol.* (St Petersburg), 1899, **7**, No. 3.
2 H. L. Jensen, *Nature*, 1950, **165**, 974; J. Meiklejohn, *Pl. Soil*, 1951, **3**, 88.
3 For a discussion of the technique see J. Meiklejohn, *J. gen. Microbiol.*, 1950, **4**, 185.
4 R. F. Lewis and D. Pramer, *J. Bact.*, 1959, **76**, 524.
5 With H. Winogradsky, *Ann. Inst. Pasteur*, 1933, **50**, 350.
6 C. Clark and E. L. Schmidt, *J. Bact.*, 1967, **93**, 1302, 1309; 1966, **91**, 367 and C. C. Delwiche and M. S. Finstein, ibid., 1965, **90**, 367.
7 For a review see J. Meiklejohn, *J. Soil Sci.*, 1953, **4**, 59.
8 *Rothamsted Rept.* 1970, Part 1, 91.

These two groups of bacteria are the only autotrophs that have been proved to carry out these oxidations, but a number of heterotrophs have been suspected of possessing this property. There are no simple methods for obtaining enrichment cultures of these heterotrophs, so it has been much more difficult to isolate organisms which can carry out these oxidations.

A number of soil organisms will produce nitrites in very small yield from ammonium and other organic nitrogen compounds; and several will produce nitrite from oximes such as pyruvic oxime $CH_2 . C(NOH) . COOH$ and from some nitro-compounds, but these are probably not oxidations. There is no evidence that any of these play an important role in the production of nitrite on the field.

There are strains of fungus *Aspergillus flavus* which, in pure culture, will produce nitrate both from organic nitrogen sources such as peptones and proteins, and from ammonium or urea, although they may only be able to oxidise ammonium ions in the presence of an abundant energy supply.[1] However, E. L. Schmidt[2] found that, while they produced nitrates from ground lucerne meal or farmyard manure when growing in these media, they did not produce any from them if the substrate was mixed with soil either when they were the only organisms present or in the natural soil environment. K. G. Doxtader and A. D. Rovira[3] also found they produced less nitrate in sterilised soil than in pure culture. They differ completely from the nitrifying autotrophs in that they only produce nitrate when growing on high nitrogen substrates, such as proteins or amino acids, or in solutions containing a relatively high concentration of ammonium ions, and they only excrete a very small proportion of the nitrogen present in the organic nitrogen or ammonium as nitrate. Thus, although they are a fairly common constituent of the zymogenous population, so are only active when growing on fresh or only partially decomposed organic matter, there is still no evidence that they contribute appreciably to the nitrate supply in the soil. However, C. T. I. Odu and K. B. Adeoye[4] isolated a fungus from a western Nigeria teak forest soil which produced nitrate not only from peptone but also from a humic acid preparation extracted with sodium hydroxide from the soil, so this fungus may also play some role in decomposing soil humus.

The biochemistry of nitrification

The nitrifying organisms obtain their energy from the oxidation of ammonium and nitrite ions, according to the equations:

$$NH_4^+ + \tfrac{3}{2}O_2 \rightarrow NO_2^- + H_2O + 2H^+ \qquad \Delta F = -84 \cdot 0 \text{ k cal}$$
$$NO_2^- + \tfrac{1}{2}O_2 \rightarrow NO_3^- \qquad\qquad\qquad\quad \Delta F = -17 \cdot 8 \text{ k cal}$$

1 E. L. Schmidt, *Science*, 1954, **119**, 187, and with O. R. Eylar, *J. gen. Microbiol.*, 1959, **20**, 473. P. Hirsch, L. Overrein and M. Alexander, *J. Bact.*, 1961, **82**, 442.
2 *Trans, 7th Int. Congr. Soil Sci.* 1960, **2**, 600.
3 *Aust. J. Soil Res.*, 1968, **6**, 141.
4 *Soil Biol. Biochem.*, 1970, **2**, 41.

The first oxidation involves the transfer of six electrons and the second of two electrons from the nitrogen compound. There is a surprising variation in the values of $-\Delta F$ for the first oxidation, for values ranging from 57 to 84 have been given by various authors, partly due to variations in the way the first oxidation has been formulated. The only proven source of carbon for their tissues is bicarbonate, and the *Nitrosomonas* group oxidises between 35 and 70 moles of ammonium per mole of carbon assimilated, and the *Nitrobacter* group between 70 and 100 moles of nitrite. This helps to explain why these organisms are such slow growers.

The oxidation of ammonium[1] to nitrite almost certainly goes through the hydroxylamine (NH_2OH) stage, and this is the only nearly-proven intermediate. The oxidation of ammonium to hydroxylamine has been little studied, but it only involves a small change in free energy $-\Delta F = 0.7$ k cal; and it is inhibited by many enzyme poisons such as allyl thiourea and other compounds containing an active $-SH$ group such as some thio- and dithio-carbamates. The oxidation of hydroxylamine to nitrite has been more intensively studied. It involves cytochromes and cytochrome oxidase systems including a copper-containing enzyme, it may go through nitrohydroxylamine $NO_2.NHOH$ and it is inhibited by hydrazine $NH_2.NH_2$.[2] But when ammonium ions are added to a suspension of *Nitrosomonas* cells, ammonium begins to disappear several days before nitrite is excreted, and no intermediates appear to be excreted, so all the stages in the oxidation of ammonium to nitrite take place inside the bacterial cell. Cell-free extracts of the enzymes will carry out the oxidation of hydroxylamine but not the oxidation of ammonium. The oxidation of nitrite to nitrate is probably a dehydrogenation reaction involving electron transfers from the hydrated nitrite ion $NO_2^-.H_2O$ to a cytochrome-c system. H. Lees and J. H. Quastel[3] showed that the growth of *Nitrobacter* was inhibited by the chlorate anion, but this did not affect the oxidation process, and this inhibition was reduced in the presence of an adequate concentration of nitrate ions.

The rate of nitrification can be reduced in field soils by various toxic substances. Ammonia itself is toxic to the two groups, but is more toxic to the *Nitrobacter* than to the *Nitrosomonas* group. Thus when high concentrations of urea or anhydrous ammonia build up in soils, nitrites may accumulate, particularly in neutral or alkaline soils of low cation exchange capacity in cool weather.[4] These soils often occur in river valleys regularly receiving silts low in clay and humus from seasonal floods. Free calcium carbonate by buffering the soil a little and so reducing the rise in pH, will reduce the nitrite concentration in the soil;[5] and medium- and fine-textured soils, with

1 For a discussion of these processes see W. Wallace and D. J. D. Nicholas, *Biol. Rev.*, 1969, **44**, 359. The values for ΔF given above are taken from this paper.
2 T. Hofman and H. Lees, *Biochem. J.*, 1953, **54**, 579.
3 *Biochem. J.*, 1946, **40**, 824.
4 H. D. Chapman and G. F. Liebig, *Proc. Soil Sci. Soc. Amer.*, 1952, **16**, 276. (They worked in Riverside, Calif.) M. K. Mahendrappa, R. L. Smith and A. T. Christiansen, ibid., 1966, **30**, 60.
5 J. H. Smith and G. R. Burns, *Proc. Soil Sci. Soc. Amer.*, 1965 **29**, 179.

an adequate exchange capacity, absorb ammonia and ammonium ions, which reduces the risk of nitrites accumulating when urea or anhydrous ammonia are used as nitrogen fertilisers.[1]

The rate of nitrification can be reduced by a number of chemicals which appear to inhibit the oxidation of ammonium to the hydroxylamine stage. Some of these, such as N-serve, 2-chloro-6-(trichloromethyl) pyridine, have been produced commercially (see p. 627). A number of herbicides and pesticides will also reduce the rate of nitrification, but usually only in concentrations well in excess of those needed for their efficient use. However, these organisms are sensitive to soil fumigants, such as methyl bromide, and nematicides.

Since the conversion of ammonium to nitrate is an oxidation, its rate of conversion must depend on the oxygen supply to the bacteria. Poorly aerated soils cannot nitrify ammonium, as is shown by an experiment of F. M. Amer and W. V. Bartholomew,[2] in which they added ammonium sulphate to a soil and drew air containing varying amounts of nitrogen through it for 21 days, and then determined the amount of nitrate-nitrogen present in it. Their results were:

percent O_2 in the air	20	11	4·5	2·1	1·0	0·4
percent of added N nitrified	46	43	38	28	21	2

These results show that nitrification is taking place in quite poorly aerated soils, though they must be interpreted with caution because the actual oxygen concentration close to the bacteria is probably lower than in the air surrounding the soil crumbs. D. J. Greenwood[3] investigated the oxygen demand of these organisms in more detail and showed that the rate of nitrification was not seriously reduced until the oxygen concentration in the water close to the bacterial surface fell to 4×10^{-6} M, which corresponds to a solution in equilibrium with air containing 0·3 per cent oxygen. This is presumably because the cytochrome-C enzyme system, which both groups of organisms possess, has a very strong affinity for molecular oxygen. Since carbon dioxide supplies all, or almost all, the carbon for the organisms' needs, growth will also be limited if this falls to too low a level, but there is no evidence that this concentration ever controls the rate of nitrification in the field.

Nitrification of ammonium salts in field soils

In general, all field soils contain micro-organisms that will oxidise ammonium to nitrate, provided the soil is not too acid, too cold or too wet, and the rate of oxidation of nitrite is more rapid than of ammonium, so that

1 For an example see J. H. Smith, *Proc. Soil Sci. Soc. Amer.*, 1964, **28**, 640.
2 *Soil Sci.*, 1951, **71**, 215.
3 *Pl. Soil*, 1962, **17**, 365, 378.

nitrite is usually only present in very low concentrations. The numbers of nitrifying bacteria found in soils is surprisingly low, so low that there is a strong suspicion that the counting technique is unreliable; and if it is, it is probably due to the difficulty of dislodging the bacteria from the soil particles on which they are adsorbed. J. Meiklejohn[1] counted ammonium and nitrite oxidisers on various Broadbalk plots on different occasions and usually found a few thousand per gram of ammonium oxidisers, varying from 360 to over 26 000, and a few hundred nitrite oxidisers, although on one occasion they rose to over 11 000. She also counted ammonium and nitrite oxidisers in a number of Rhodesian soils,[2] and usually found several thousand ammonium oxidisers in arable soils or soils under well-managed pastures, though the numbers varied from very few to about 30 000 per gram, but soils under natural poor pasture often had under 100. The number of nitrite oxidisers was usually less than the number of ammonium oxidisers, and only on occasion did they rise over 10 000. O. M. Ulianova[3] found comparable numbers of ammonium oxidisers in a range of Russian soils, while H. Barkworth and M. Bateson[4] found between 10^6 and 10^7 ammonium oxidisers in twenty-five fields sampled in the English Midlands over a twenty-seven month period, but the number of nitrite oxidisers was much more variable and only between one-tenth and one-hundredth as numerous.

Drying a soil tends to kill off the nitrifying bacteria, and in the semi-arid tropics and subtropics there may be a considerable interval between the onset of the rains after a pronounced dry season and the onset of nitrification. It is possible that, under some circumstances, the nitrite oxidisers are more sensitive than the ammonium oxidisers to desiccation, so that nitrite sometimes accumulates in the soil for a number of days after the onset of the rains.

Nitrification does not take place in very acid soils, but there is no clear relation between soil pH and rate of nitrification, probably because the effect of acidity is through the toxic effect of active aluminium ions. Thus, there are examples of soils with a pH between 4 and 5 which will nitrify slowly[5] and others with a pH over 5 which will only nitrify after they have been limed.[6] But in general over the pH range of 5 to 8 there is usually little effect of pH, though above pH 8, the ammonium oxidisers may be less sensitive to alkalinity than the nitrite. There is, however, some evidence that nitrifiers present in acid soils have become adapted to acidity and are more effective in these soils than nitrifiers taken from neutral soils.[7]

Nitrification only proceeds rapidly in warm soils, and goes on very slowly

1 *Rothamsted Rept.* 1968, Part II, 177.
2 *J. appl. Ecol.*, 1968, **5**, 291, and for similar results for some Ghanaian soils *Emp. J. exp. Agric.*, 1962, **30**, 115.
3 *Mikrobiologiya*, 1960, **29**, 813.
4 *Pl. Soil*, 1965, **22**, 220.
5 D. F. Weber and P. L. Gainey, *Soil Sci.*, 1962, **94**, 138.
6 J. W. Millbank, *Pl. Soil*, 1959, **11**, 293.
7 S. S. Brar and J. Giddens, *Proc. Soil Sci. Soc. Amer.*, 1968, **32**, 821.

when the soil temperature is below 4 or 5°C.[1] As an example of the effect of cold weather on nitrification, R. L. Fox[2] in Nebraska broadcast 90 kg/ha of ammonium nitrogen as ammonium nitrate, on a soil on 15 November, and on 1 April there was still 67 kg left, 28 kg remained on 1 May, and it was not till June that it had all been nitrified. In England grass growth begins in the spring when the soil temperature reaches 4 or 5°C, which is just about the same temperature at which nitrification begins. The rate of nitrification increases with temperature, reaching a maximum somewhere between 25 and 30°C,[3] but in the same way as nitrifying bacteria can become adapted to acid soils, they can also become adapted to warm soils, for M. K. Mahendrappa[4] found that nitrification went on most rapidly at between 20 and 25°C in soils from the northern part of the western United States, but at between 35 and 40°C in soils from the southern part. They also found that on the addition of ammonium salts, nitrites tended to accumulate when the temperatures were not favourable to nitrification. Thus they accumulated in the northern soils when they were incubated at between 35 and 40°C and in the southern at between 20 and 25°C.

The effect of the moisture content of the soil on nitrification depends on its effect on aeration at the high moisture content end, but at the dry end, the effect of suction depends on the soil. Data from the continental areas of the USA shows that the rate of nitrification is a maximum at suctions below 1 bar, provided aeration is adequate, and falls as the suction rises, but is still measurable at 15 bar suction.[5] On the other hand, in some tropical soils, the rate of nitrification may be almost independent of suction up to suctions of 15 bar, and then drop rapidly as the suction rises above this; but soils differ in the effect that high suctions have on the rate.[6] The rate of nitrite oxidation is usually greater than the rate of ammonium oxidation at high suctions, for nitrites do not normally accumulate in field soils at high suctions, but the rate of ammonium production is less sensitive to high suctions, so ammonium ions usually increase in soils kept at suctions not much above 15 to 30 bar.

The effect of soluble salts on the rate of nitrification is partly through their effect on the osmotic pressure of the soil solution and is partly specific. D. D. Johnson and W. D. Guenzi[7] found that, in the soils they were using,

1 See, for example, O. E. Anderson with R. Purvis, *Soil Sci.*, 1955, **80**, 313; with F. C. Boswell, *Proc. Soil Sci. Soc. Amer.*, 1964, **28**, 525; 1960, **24**, 286; and L. R. Frederick, *Proc. Soil Sci. Soc. Amer.*, 1956, **20**, 496.
2 With R. A. Olson and A. P. Mazurak, *Agron. J.*, 1952, **44**, 509.
3 See, for example, D. T. Parker and W. E. Larson, *Proc. Soil Sci. Soc. Amer.*, 1962, **26**, 238, B. R. Sabey, L. R. Frederick and W. V. Bartholomew, ibid., 1959, **23**, 462, and R. Ettinger-Tulczynska, *J. Soil Sci.*, 1969, **20**, 307.
4 With R. L. Smith and A. T. Christiansen, *Proc. Soil Sci. Soc. Amer.*, 1966, **30**, 60.
5 B. R. Sabey, *Proc. Soil Sci. Soc. Amer.*, 1969, **33**, 263. R. D. Miller and D. D. Johnson, ibid., 1964, **28**, 644. G. A. Reichman, D. L. Grunes and F. G. Viets, ibid., 1966, **30**, 363. J. K. Justice and R. L. Smith, ibid., 1962, **26**, 246.
6 R. Wetselaar, *Pl. Soil*, 1968, **29**, 9. H. D. Dubey, *Canad. J. Microbiol.*, 1968, **14**, 1348. J. B. D. Robinson, *J. agric. Sci.*, 1957, **49**, 100.
7 *Proc. Soil Sci. Soc. Amer.*, 1963, **27**, 663.

sodium sulphate had little effect until the osmotic pressure reached about 5 bar, and that nitrification was slow at 20 bar and ceased at 30 bar. Thus, the effect of sodium sulphate appeared to be entirely due to its effect on the free energy of the soil water. On the other hand sodium chloride depressed the rate of nitrification at a given osmotic pressure compared with the sulphate, for nitrification had almost ceased when the osmotic pressure was 10 bar. Further, at high osmotic pressures, nitrites begin to accumulate when nitrification of added ammonium starts, a result not found for the sulphate. It is not certain how far these results apply in naturally saline soils, in which the nitrifying bacteria may have become adapted to chloride.

Free ammonia in the soil typically inhibits nitrification, and *Nitrobacter* is more sensitive to ammonia toxicity than is *Nitrosomonas*. Thus, heavy dressings of urea or anhydrous ammonia in neutral or calcareous soils, and ammonium fertilisers on some high pH soils, are liable to lead to a build-up of nitrites in the soil, which may reach over 10 ppm NO_2–N and persist for an appreciable time if the soil temperature is rather low. Thus H. D. Chapman and G. F. Liebig[1] working with five citrus orchards around Riverside, Calif., found that a heavy dressing of urea given to the trees gave rise to 90 ppm NO_2–N on a soil of pH 7·7 which persisted for several months when the soil temperature was between 10 and 15°C. However, this result is not always found, for if it were the use of anhydrous ammonia on neutral soils would be very dangerous, and the reasons for nitrite build-up in some soils but not in others are not known.

The level of mineral nitrogen in field soils

The mineral nitrogen in the soil is present either as ammonium or nitrate ions. The nitrates are all dissolved in the soil solution, unless the soil dries out, but much of the ammonium is held on the exchange complex. The total quantity of mineral nitrogen in the soil is the difference between the rate it is being produced from the soil's store of organic matter by the soil population and the rate it is being removed by leaching, by growing crops and by other members of the soil population; and the proportion of nitrate to ammonium depends also on the rate of oxidation of ammonium to nitrates, the uptake of nitrates by the plant and the loss of nitrates by leaching.

Arable soils in the temperate regions, particularly if they are not too acid, have a fairly constant but low content of ammonium nitrogen, but a very variable and high nitrate content, ranging from 2 to 20 mg N as nitrate per kg of soil (i.e. 2 to 20 ppm of nitrate nitrogen) for normal soils, but rising up to 60 mg for rich garden soils. This nitrate content varies throughout the season and, under some conditions which cannot yet be specified, from hour

1 *Proc. Soil Sci. Soc. Amer.*, 1952, **16**, 276; and for an earlier example from Arizona, J. P. Martin, T. F. Buehrer and A. B. Caster, ibid., 1942, **7**, 223.

to hour.[1] Figure 16.1 shows the nitrate nitrogen content, throughout the growing season of 1915, of two parts of the Broadbalk plot at Rothamsted which receives farmyard manure annually, one part being cropped to wheat and one part fallowed.[2] Not all the details of these curves can be explained,

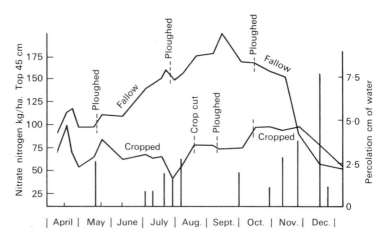

FIG. 16.1. Amount of nitrate nitrogen in cropped and fallow soils at different seasons of the year (Broadbalk Field, Rothamsted, 1915) and the amount of percolation through the 50 cm bare soil gauge

but the figure shows clearly that the fallow soil accumulates nitrate during the spring and summer, whereas the cropped soil does not, and that the fallow appears to have lost all its accumulated nitrate by early winter, presumably due to it being leached down into the deep subsoil.

Nitrification in fallow soils

Fallows can only accumulate nitrates under four conditions: there must be decomposable organic matter in the soil to provide a supply of ammonium ions, the fallow must be kept free from weeds, there must not be too much rain otherwise the nitrates are leached out of the soil, and the soil must be moist or be subjected to alternate wetting and drying. F. Crowther,[3] working in the Sudan Gezira, has given a good example of weeds reducing the accumulation of nitrates in a fallow soil. This heavy clay land used to be fallowed for several years in preparation for cotton, and Crowther showed that cutting down the weeds that grew during the short rainy season, instead of letting them die as soon as the dry season began, increased the nitrate nitrogen in

1 H. G. Thornton and P. H. H. Gray, *Proc. Roy. Soc.*, 1930, **166** B, 399.
2 E. J. Russell and A. Appleyard, *J. agric. Sci.*, 1917, **8**, 385. For a further example, see W. A. Albrecht, *Missouri Agric. Expt. Sta., Res. Bull.* 250, 1937.
3 *Emp. J. exp. Agric.*, 1943, **11**, 1.

the top 30 cm of soil from 2 to 4 ppm to 5 to 10 ppm, and also increased that in the second 30 cm. This led to a 66 per cent increase in the uptake of nitrogen by the cotton and corresponding increase in the yield of seed-cotton.

Nitrification proceeds more rapidly in soils subjected to alternate wetting and drying than in soils kept permanently moist, because the alternate wetting and drying causes a more rapid oxidation of soil humus (see p. 237). Thus, in hot regions having a pronounced dry and wet season, nitrates are produced most rapidly at the commencement of the rains, and only slowly during the rainy season itself. G. Semb and J. B. D. Robinson[1] found flushes from between 13 and 180 kg/ha of nitrate-nitrogen in thirteen soils from different parts of East Africa shortly after the onset of the rains which follow the dry season. Hence, if crops are to benefit from this early flush, they must be planted as early as possible, taking advantage of any showers that fall before the beginning of the main rains, so that they can start growing quickly when the rains break and use this nitrate before it is leached out of the soil. But the nitrates in the surface of fallow soils also increase during the dry season and may reach very high figures. Thus G. ap Griffith[2] found up to 110 ppm of nitrate-nitrogen in the top 15 cm of a fallow soil at Kawanda, Uganda, although levels of 25 ppm were more common; and W. R. Mills[3] showed it could exceed 200 ppm in the top 5 cm of soil there. This compares with 30 to 40 ppm in the top 15 cm of the Broadbalk soil receiving 35 t/ha farmyard manure annually during a summer fallow.

The cause of these high nitrates in the surface of these soils is only partly due to the nitrification that takes place in them while the soil is still moist, a part is due to the upward movement of nitrates from the subsoil, mostly from above 30 cm, due to the upward movement of water which is evaporated from the surface soil during the dry season.[4] If the surface soil is self-mulching over a more compact subsurface layer, the nitrates tend to accumulate in the surface of this more compact layer as the dry season progresses. It is possible, though unproven, that in regions receiving heavy dews, enough water may condense in the surface film of the soil each night to allow some nitrification to take place each subsequent morning. G. Drouineau[5] has suggested that this occurs on some soils in the south of France.

The direct effect of rainfall on the nitrate content of the soil is to leach it down the profile. If the profile is dry to depth, and the rainfall limited in amount, the nitrates will accumulate lower down in the soil profile, and this accumulation can be very considerable. Thus W. R. Mills[6] found over

1 *East Afr. agric. for. J.*, 1969, **34**, 350.
2 *Emp. J. exp. Agric.*, 1951, **19**, 1, and with H. L. Manning, *Trop. Agric. Trin.*, 1949, **26**, 108.
3 *East Afr. agric. J.* 1953, **19**, 53.
4 J. R. Simpson, *J. Soil Sci.*, 1960, **11**, 45, and R. Wetselaar, *Pl. Soil*, 1961, **15**, 110, 121. See also D. Stephens, *J. Soil Sci.*, 1962, **13**, 52, and J. B. D. Robinson and P. Gacoka, ibid., 1962, **13**, 133.
5 With G. Lefèvre, *Ann. Agron.*, 1951, **21**, 1; 1953, **4**, 245.
6 *East Afr. agric. J.*, 1953, **19**, 53.

400 ppm nitrate nitrogen had accumulated at 90 cm depth in a Kawanda soil under suitable climatic conditions. If the profile is fairly shallow and well drained, and the rainfall adequate, nitrates will be washed into the ground-water or the rivers.

Well-structured loams and clays can, however, hold appreciable quantities of nitrates against leaching. This is because the percolating water moves down principally through the cracks and coarse pores between the crumbs, and most of the nitrates are formed in the crumbs, so the nitrates can only get into this water by diffusion, which is a slow process. This holding of nitrates against leaching is of considerable agricultural importance in British soils, as for example the Rothamsted clay loam,[1] for it means that a part of the nitrates produced in a previous summer fallow is available for the succeeding crop, even if the autumn and winter are wet, although as E. M. Crowther[2] showed, using indirect evidence, most of this nitrate must be held in the subsoil rather than in the surface crumbs, possibly because the surface nitrates are lost by denitrification, the surface crumbs being well supplied with decomposable organic matter.

The effect of surface mulches and of shade on the nitrate content of fallow soil in the tropics and subtropics is not fully understood. In general a surface mulch or shading the soil surface reduces the nitrates in the surface soil,[3] though D. J. Greenland[4] gave an example from Ghana when the shading had little effect, and F. Leutenegger[5] an example from Mlingano, Tanzania, where a paper mulch increased it.

Nitrification in arable soils

Nitrates are always lower in cropped land than under fallow, as one would expect, because the crop will be taking up nitrates from the soil while it is growing. But, in addition, there is considerable evidence that the crop appears to depress the rate of nitrification in a soil, for the nitrogen present in the cropped soil as nitrate and in the crop is less than the nitrates in adjacent soil kept fallow. Table 16.1 illustrates the type of data from which this con-clusion is drawn, but the correct interpretation of this result is not fully established. Thus in the table only the nitrogen in the above-ground portion of the crop has been determined, that in the roots being ignored, but it is extremely unlikely that there could be enough in the roots to account for the difference. This difference occurs in the earlier part of the crop's growth, for as Table 16.1 shows, in 1915 the amount of nitrates produced in the fallow and in the cropped land was the same for the period between 26 July and

1 See, for example, R. K. Cunningham and G. W. Cooke, *J. Sci. Food Agric.*, 1958, **9**, 317.
2 *Trans. 3rd Int. Congr. Soil Sci.*, 1935, **3**, 126.
3 For an example see J. B. D. Robinson, *J. agric. Sci.*, 1961, **56**, 49.
4 *J. agric. Sci.*, 1958, **50**, 82.
5 *East Afr. agric. J.*, 1956, **22**, 81.

TABLE 16.1   Nitrogen as nitrate in cropped and fallow soil at Rothamsted

Nitrogen in kg/ha

	Hoosfield		Broadbalk		
			Un-manured	Dunged	
	June 1911	July 1912	July 1915	26 July 1915	17 Aug. 1915
In top 45 cm of fallow soil	61	51	78	139	168
In top 45 cm of cropped soil	17	15	49	41	77
In crop	26	7	23	78	78
Total	43	22	72	119	155
Deficit in cropped land	18	29	6	20	13

Expressed as mg N per kg soil

Fallow, 0–22·5 cm depth	12	8
22·5–45 cm depth	9	10
Cropped, 0–22·5 cm depth	4	2
22·5–45 cm depth	2	3

Cropping.   Hoosfield: Alternate wheat and fallow, unmanured since 1851.
Broadbalk: Continuous wheat, but fallow taken over part of this field for the third time since 1843.

17 August, the day on which the wheat was harvested. On the other hand, the deficit only starts to build up when the crop is well established.[1]

This apparent deficit could be due to four causes: the crop could excrete substances into the soil which inhibit or slow down the rate of nitrification, and although this has been postulated for grass roots, there is no evidence that annual arable crops excrete such substances. Then the crop may reduce the rate of nitrification in the soil by its effect on shading the soil, but if so one would expect the effect to be as large at the end of the growing season as earlier on. Then the crop might encourage denitrification or nitrogen losses from its leaves, but this is unlikely because, as is shown in Table 16.4, p. 346, one can get a good nitrogen balance sheet for cropped unmanured land at Rothamsted. Finally, the rhizosphere organisms on the crop roots, or other micro-organisms close to the root, may extract soluble carbohydrates or other low-nitrogen compounds from the roots, and use the soil nitrates as their source of nitrogen, or even use soluble nitrogen compounds from the

1 C. A. I. Goring and F. E. Clark, *Proc. Soil Sci. Soc. Amer.*, 1949, **13**, 261.

roots, converting these soluble compounds to insoluble soil organic soil nitrogen compounds; and there is direct evidence, using N^{15}, for this last process.[1]

Nitrification in grassland soils

Grassland appears to have a different mineral nitrogen economy from arable, for as Table 16.2 shows for unmanured grass leys and pastures at Rothamsted, a greater proportion of their mineral nitrogen is ammonium than nitrate. This ammonium content, as H. L. Richardson[2] showed, remains

TABLE 16.2   Average ammonium and nitrate contents of Rothamsted grassland soils

Years under grass	Depth sampling in cm	Nitrogen in parts per million of dry soil	
		Present as ammonium	Present as nitrate
Ley, under 1	20	1·1	0·5
Ley, under 2	20	2·2	1·1
Pasture, 59	10	4·7	1·3
Pasture, over 200	20	5·4	1·1

fairly constant between 3 and 9 ppm ammonium-nitrogen throughout the year, yet if an ammonium or nitrate fertiliser is added, this mineral nitrogen very soon disappears from the soil and has presumably been taken up by the grass. Thus C. R. Clement[3] found that during the first production year of a grazed ryegrass ley to which 22 kg/ha N as ammonium nitrate was added nine times during the growing season, the ammonium content of the top 7·5 cm varied between 2·5 and 10·6 ppm N, with a mean of 5·2, and the nitrate content between 0·0 and 6·4, with a mean of 3·1 ppm N. These figures could be higher in the second or third year for a grazed ley, but a little lower for a grass ley cut for hay or silage, even if as much as a total of 320 kg/ha N is added during the growing season. If some clover is included with the grass, the total ammonium plus nitrate nitrogen is 1 to 2 ppm N higher than for grass alone in leys cut for silage. Under grazed swards given high levels of ammonium nitrate fertiliser during the growing season, high levels of nitrate nitrogen of the order of 20 ppm have been found. Little work has been done on how many years a grass ley must be left before the nitrate levels fall to a

1 W. V. Bartholomew and F. E. Clark, *Trans. 4th Int. Congr. Soil Sci.* (Amsterdam), 1950, **2**, 112 and A. E. Hiltbold, W. V. Bartholomew and C. H. Werkman, *Proc. Soil Sci. Soc. Amer.*, 1951, **15**, 166.
2 *J. agric. Sci.*, 1938, **28**, 73.
3 I am indebted to Dr Clement for this information.

low value, but a ley that is only two or three years old may have appreciable levels of nitrate in the soil.[1]

There is still no fully satisfactory explanation of why grass cannot reduce the ammonium level below this fairly high level, nor why this ammonium is not nitrified. Richardson showed that there was nothing in the soil of the old Rothamsted pastures that inhibited nitrification, for on incubating these soils all the ammonium produced was readily nitrified, unless the soil was too acid or too low in phosphate—a result which has been confirmed by other workers. Further, if an old pasture is ploughed out, the usual British experience is that the following crop is over-well supplied with soil nitrates. Pastures in the tropics also have very low ammonium and nitrate levels,[2] though not enough data are available to establish if the ammonium is present at a fairly constant level and whether the nitrate concentration is lower than the ammonium.

The low level of nitrates in grassland soils could be due to a very low level of ammonium ions in the soil, but it could also be due to the grass roots excreting substances which inhibit nitrification, and since nitrites do not accumulate in these soils they must inhibit the oxidation of ammonium ions. J. J. Theron[3] obtained evidence that such substances can be excreted, and a number of other authors have confirmed this.[4] These exudates are not confined to grasses,[5] nor do all grasses exude them,[6] but so far nothing is known about whether there are a number of different inhibitory substances excreted or whether plants differ only in the amounts of a single substance which they excrete.

Many natural grassland soils, particularly savanna soils, have very few nitrifying organisms in them, so that if an ammonium salt is added to them, it does not nitrify. But this is at least sometimes due to the soil conditions being unsuitable for the nitrifiers. Thus J. B. Robinson[7] found some New Zealand grasslands to be almost incapable of nitrifying, but if they were limed and urea added, after a short period of time they began to nitrify normally. Other natural grasslands give soils very low in available nitrogen after they have been ploughed out, but this is not due to the absence of nitrifiers but to the absence of organisms producing ammonium from the organic matter,[8] possibly because of the very high C/N ratio of the organic residues incorporated in the soil when it is ploughed. In general, in the

1 For an example from New South Wales, see J. R. Simpson, *Aust. J. agric. Res.*, 1962, **13**, 1059.
2 For an example from the savanna region of Ghana see D. J. Greenland, *J. agric. Sci.*, 1958, **50**, 82.
3 *J. agric. Sci.*, 1951, **41**, 289.
4 For example, G. Stiven, *Nature*, 1952, **170**, 712, and P. E. Munro, *J. appl. Ecol.*, 1966, **3**, 231.
5 See, for example, D. R. E. Moore and J. S. Waid, *Soil Biol. Biochem.*, 1971, **3**, 69.
6 J. L. Neal, *Canad. J. Microbiol.*, 1969, **15**, 633.
7 *Pl. Soil*, 1963, **19**, 173.
8 For an example see P. H. Nye, *Emp. J. exp. Agric.*, 1951, **19**, 275, and J. J. Theron, *South Afric. J. agric. Sci.*, 1963, **6**, 155.

tropics, properly managed improved pasture soils probably always nitrify after the pasture has been ploughed out, provided there is an adequate supply of calcium or a sufficiently low concentration of aluminium ions in the soil.[1]

Nitrification in forest soils

The level of mineral nitrogen in forest soils, and the forms in which it is present, depends on the pH of the soil and on the type of humus in the forest floor. C. H. Bornebusch[2] found in Danish forest soils with pHs in the range 4 to 5·5 that the rate of nitrification exceeded that of ammonification under mull though it did not under mor, as is shown in Table 16.3. The reason for this is possibly that the mull is better supplied with bases and phosphates than the mor, but this may not be the whole reason, for the mull is a form of humus produced by earthworms, and N. V. Joshi[3] noted that the introduction of earthworms into some Indian black cotton soils well supplied with bases and phosphate increased the rate of nitrification though this could also be because they increase the rate of decomposition of the organic matter.

TABLE 16.3   The effect of type of humus on nitrification. Parts per million of ammonium and nitrate nitrogen in the soil

Mull soils				Raw humus soils			
				Under beech			
Layer:	pH	NH$_4$	NO$_3$	Layer:	pH	NH$_4$	NO$_3$
Old leaf litter	6·1	84	1200	Old leaf layer	5·6	252	20
Wormcasts	5·8	8	264	Upper raw humus	4·3	388	trace
Upper mull	5·4	4	48	Middle raw humus	3·7	95	trace
Lower mull	5·2	2	7·5				
				Under spruce			
Mull layer	4·3	80	26	Upper raw humus	3·6	115	0·8
Upper top soil	4·1	2	5	Middle raw humus	3·5	32	trace

Very few observations have been made on the level of ammonium and nitrate ions in tropical forest soils, but the evidence such as it is points to the similarity between tropical and temperate conditions. Thus D. J. Greenland,[4] in Ghana, found fairly high and fluctuating nitrate levels but low and fairly constant ammonium levels in the floor of a natural forest. He also found relatively high levels of readily nitrifiable nitrogen in these soils, as did H. Jenny[5] in Columbian forest soils and Y. Dommergues[6] in West African forest soils.

1   For an example see W. R. Mills, *East Afr. agric. J.*, 1953, **19**, 53.
2   *Forstl. Forsoegv. Danmark*, 1930, **11**, 1.
3   *Bull. Nat. Inst. Sci. India*, 1954, **3**, 115.
4   *J. agric. Sci.*, 1958, **50**, 82.
5   *Soil Sci.*, 1949, **68**, 419.
6   *Trans. 6th Int. Congr. Soil Sci.* (Paris), 1956, E, 605.

Losses of inorganic nitrogen from the soil

The inorganic nitrogen compounds—ammonium and nitrates—suffer several types of loss from the soil. They may be taken up by growing plants, they may be assimilated into the bodies of micro-organisms and so brought back into the organic nitrogen reservoir, they may be converted into volatile compounds and lost into the air, or they may be leached out of the soil. It is only the last two losses that will be considered here.

Ammonium can be lost to the atmosphere, in the form of ammonia, but this can only happen under alkaline conditions. Such conditions arise when ammonium sulphate is added to the surface of calcareous soils, or large dressings of urea are added to the surface of soils that are not strongly acid, particularly during periods of light showers of rain interspersed with strong drying conditions. These losses can be avoided if the fertiliser is placed sufficiently far below the soil surface for any ammonia that is released by volatilisation to be absorbed by the soil before it can reach the atmosphere (see p. 625).

Denitrification in soils

The most important source of loss of soil nitrogen as gaseous products is denitrification, that is the reduction of nitrates to nitrous oxide N_2O and free nitrogen gas N_2. This occurs in soils that are not too acid, usually above pH $5 \cdot 0$[1] under conditions of poor aeration in the presence of an active microbial population, and can therefore be important during wet periods in warm soils well supplied with decomposable organic matter. Nitrous oxide may, however, be produced by *Nitrosomonas europaea* when the temperature and phosphate supply are low but the pH is high, presumably because, under these conditions, the earlier stages in the oxidation of ammonium go quicker than the later.[2]

Denitrification can be a very active process for G. S. Cooper and R. L. Smith[3] showed that in the laboratory, soils containing 300 ppm NO_3–N could lose almost the whole amount by denitrification in between 28 and 96 hours, if there was also a source of decomposable organic matter present. These reductions are brought about by a number of groups of bacteria, some of which may only be able to carry out one stage of the reduction process from nitrate to nitrogen gas;[4] and the bacteria are predominantly facultative anaerobes in that they only use the oxygen of nitrate, nitrite or oxides of nitrogen as a hydrogen acceptor in the absence of free oxygen.[5] The reduc-

1 C. L. Valera and M. Alexander, *Pl. Soil*, 1961, **15**, 268. J. N. Carter and F. E. Allison, *Proc. Soil Sci. Soc. Amer.*, 1961, **25**, 484.
2 T. Yoshida and M. Alexander, *Proc. Soil Sci. Soc. Amer.*, 1970, **34**, 880.
3 *Proc. Soil Sci. Soc. Amer.*, 1963, **27**, 659.
4 J. H. Jordan, W. H. Patrick and W. H. Willis, *Soil Sci.*, 1967, **104**, 129.
5 C. L. Valera and M. Alexander, *Pl. Soil*, 1961, **15**, 268.

tion of nitrate goes through the nitrite stage, but nitrite normally does not accumulate in the soil, then goes direct to nitrous oxide N_2O or to nitrogen gas; under certain conditions it may go through the nitric oxide NO stage for this is sometimes found in the initial stages when nitrates are being reduced.[1] There is no good evidence that nitrates are ever reduced to ammonium ions in soils, though some nitrate may be assimilated by the bacteria, converted into their protoplasm, and later appear as ammonium on their decomposition.[2]

Nitrate reduction only takes place under conditions of a low oxygen supply. D. J. Greenwood[3] found that the concentration of dissolved oxygen in the solution bathing the bacteria must fall to about 4×10^{-6} M before reduction starts to take place actively. This corresponds to the concentration that is in equilibrium with air containing 0·3 per cent O_2. However, soils which appear to be well aerated may yet reduce nitrates to nitrous oxide, particularly if they are well supplied with readily decomposable organic matter, but this simply demonstrates that many soils, when moist, possess a range of microhabitats characterised by varying levels of oxygen supply. Thus many soils can contain clods which have no free air when at field capacity, so that the soil within the clod will be anaerobic except for a surface film 5–10 mm thick in which there is an oxygen gradient (see p. 409). Nitrification and denitrification can therefore take place simultaneously in the same soil when it is at field capacity.[4] This is the probable explanation of the many experimental results in the literature which have been interpreted as showing that denitrification can take place in aerobic soils.

Denitrification can only take place actively in soils containing an adequate supply of organic matter and when they are warm. Thus, nitrates in the deep subsoil do not suffer denitrification because of lack of bacterial activity, but, on the other hand, ploughing in decomposable organic matter, such as farmyard manure or green manures, will encourage denitrification during any short period of poor aeration; and this will be particularly marked if there is any sign of a plough pan holding up water during a rainstorm for short periods of time. Nor do surface soils denitrify during cold weather. Thus little denitrification takes place in British soils during the winter, because the temperatures are too low to allow it to go on at an appreciable rate. As the temperature rises, the process goes on more vigorously, and the proportion of nitrous oxide evolved relative to nitrogen gas decreases. It also proceeds more rapidly in neutral and calcareous soils than in acid soils. Little usually takes place in soils with a pH much below 5·0, and it probably

1 F. B. Cady and W. V. Bartholomew, *Proc. Soil Sci. Soc. Amer.*, 1963, **27**, 546; 1960, **24**, 477.
2 *Int. Rice Res. Inst.*, *Ann. Rept.* 1966, 128.
3 *Pl. Soil*, 1962, **17**, 365, 378, see also F. E. Allison, J. N. Carter and L. D. Sterling, *Proc. Soil Sci. Soc. Amer.*, 1960, **24**, 283.
4 For examples see D. J. Greenland, *J. agric. Sci.*, 1962, **58**, 227; P. W. Arnold, *J. Soil Sci.*, 1954, **5**, 116.

has a pH optimum between 8·0 and 8·6.[1] Correspondingly the proportion of nitrous oxide produced relative to nitrogen gas decreases with increasing pH, and in acid soils denitrification may stop at the nitrite stage. The roots of crops growing actively in soils well supplied with nitrates will also cause denitrification to take place under some conditions, presumably due to the high oxygen demand of their rhizosphere population.[2]

There is increasing evidence of a purely chemical pathway involving the conversion of nitrites to nitrogen gas that is operative in soils, particularly acid soils adequately supplied with organic matter. Nitrites are unstable in acid solutions, being decomposed to nitric oxide NO and nitrogen peroxide NO_2, and nitric oxide in the presence of oxygen oxidises readily to the peroxide. The peroxide reacts with water to give nitric and nitrous acids, so involves no loss of nitrogen from the system. But as noted on p. 299, nitrites react with polyphenols, lignin and humic acids under acid conditions, and a part, and often a major part of the nitrogen in the nitrite is released as nitrogen gas. D. W. Nelson and J. M. Bremner[3] consider this can be an important reaction in field soils, for they showed that if soils were incubated with added nitrite, the amount of gaseous nitrogen produced increases as the acidity of the soil increases in the pH range 5 to 7, that it is negligible if the soils are limed, but that it increases as the organic matter content of the soil increases. It is also greater in moist than in wet soils, showing it has nothing to do with a reduction, and it is as large in the fresh as in the sterilised soil. They also showed that under their conditions nearly 90 per cent of an addition of only 10 ppm nitrite-nitrogen was converted to gas in 4 days when the soil was incubated at 30°C. A number of other possible reactions in which a part of the nitrogen of the nitrite could be set free as nitrogen gas have been suggested, but there is no good evidence that they can take place in the conditions of soils in the field.[4]

There have only been a few measurements made of the rate of production of nitrous oxide in field soils because of the difficulty in the past of determining the small quantity of nitrous oxide present in the soil air, and no measurements have been made of the production of nitrogen gas because of a lack of methods available. P. W. Arnold[5] measured the rate of production of nitrous oxide from some English arable and pasture soils in the field and found a sandy loam could lose about one-tenth of its mineral nitrogen as nitrous oxide per day when moist but not wet, and it was losing nitrous oxide until the suction of water had risen to about 100 cm. Some of his soils also

1 J. M. Bremner and K. Shaw, *J. agric. Sci.*, 1958, **51**, 22, 40; R. D. Hauck and S. W. Melsted, *Proc. Soil Sci. Sci. Soc. Amer.*, 1956, **20**, 361.
2 R. C. Stefanson and D. J. Greenland, *Soil Sci.*, 1970, **109**, 203. J. K. R. Gasser, D. J. Greenland and R. A. G. Rawson, *J. Soil Sci.*, 1967, **68**, 289. J. W. Woldendorp, *Meded. Landbouwhog. Wageningen*, 1963, **63** (13); *Pl. Soil*, 1962, **17**, 267.
3 *Soil Biol. Biochem.*, 1969, **1**, 229.
4 For a review see F. E. Allison, *Adv. Agron.*, 1966, **18**, 219.
5 *J. Soil Sci.*, 1954, **5**, 116.

produced detectable amounts of nitrous oxide from ammonium but not from nitrate nitrogen when the soil was as dry as this. J. R. Burford and R. J. Millington,[1] at the Waite Institute, Adelaide, also measured the rate of production of nitrous oxide from soil under wheat, receiving 100 kg/ha N as sodium nitrate. Nitrous oxide was only produced when the soil was wet after heavy rain, and it was produced over a longer period of time in the subsoil, between 30 and 60 cm deep, than in the surface; but it was being produced in measurable quantities even though there was about 19 per cent of oxygen in the soil air. Arnold found nitrous oxide concentrations as high as 2×10^{-6} g/cm^3 in the soil air, which was about five times higher than the Australian figures. Burford and Millington calculated that their soil would lose up to 1·2 kg/ha N as nitrous oxide per day when it was wet, and Arnold's figure went up to about 3 kg/ha a day.

Nitrogen losses from soils in the field

Nitrogen losses from a soil can be due to crops removing nitrogen from the soil, nitrates being leached out of the soil, and losses due to nitrogen compounds being converted to nitrous oxide or nitrogen gas. If the mineral nitrogen status of the soil is low, crop and leaching losses appear to be the

TABLE 16.4   The nitrogen balance sheet of a soil kept free from vegetation but exposed to rain and weather.* Rothamsted drain gauge soil

Year	Per cent N in soil, top 22 cm	N in soil, top 22 cm kg/ha	Loss of N from top 22 cm kg/ha	N recovered as nitrate 1870–1915 kg/ha†
1870	0·146	3900	—	—
1905	0·102	2750	1170	1110
1915	0·098	2650	1290	1370

* Constructed from data given in N. H. J. Miller, *J. agric. Sci.*, 1906, **1**, 377; and E. J. Russell and E. H. Richards, ibid., 1920, **10**, 22.

† After deducting the amount brought down in the rain. No significance should be attached to the difference between the last two columns owing to the assumptions involved in obtaining, and the uncertainties in interpreting, the figure for the weight of nitrogen in the top 22 cm of the soil.

dominant source of loss; while if it is high, particularly under conditions of where microbiological activity is maintained at a high level by the generous use of organic manures, there can be very large losses of nitrogen not accounted for by these two sinks.[2]

1 *Trans 9th Int. Congr. Soil Sci.*, 1968, **2**, 505.
2 For a review see F. E. Allison, *Adv. Agron.*, 1966, **18**, 219.

The Rothamsted drain gauges give a good example of the losses from a bare uncropped soil being accounted for almost entirely as nitrates leaching out of the soil. These gauges, each 4 m² in area, have been kept free from vegetation since 1870, and by 1915 had lost one-third of the nitrogen originally present in the top 22·5 cm of soil. As Table 16.4 shows, practically all this nitrogen was found as nitrate.

The unmanured plot on Broadbalk, the permanent wheatfield at Rothamsted, gives an example of the full nitrogen loss from the soil being accounted for by the nitrogen in the crop removed, as shown in Table 16.5. But this

TABLE 16.5   Losses of nitrogen from a cultivated soil: Broadbalk, Rothamsted, 49 years, 1865–1914. Kg/ha in top 22·5 cm of soil

	Farmyard manure (Plot 2B)	No manure (Plot 3)	Complete artificials (96 kg N as sulphate of ammonia annually)	
			(Plot 7)	(Plot 13)*
N in soil in 1865	4860	3070	3630	3550
1914	6700	2920	3600	3640
Total change in 49 years	+ 1840	− 150	− 30	+ 90
N added in manure, seed and rain, per annum	233	8	104	104
N removed in crops per annum	56	19	52	49
N retained (+) or lost (−) by soil per annum	37·5	− 3.1	− 0·6	+ 1·8
N unaccounted for, per annum	140	(gain 8)	52	53

* These plots are almost duplicates, plot 7 receiving 112 kg each of the sulphates of soda and magnesia in addition to the 440 kg of superphosphate and 225 kg sulphate of potash given to plot 13.

table shows that when nitrogen is added either as sulphate of ammonia, or as farmyard manure, very large losses occur, no less than 70 per cent of the nitrogen in the farmyard manure being unaccounted for. Some of this loss undoubtedly occurs in the drainage water, which cannot be estimated accurately in the field experiment, but it is very unlikely it could account for any appreciable proportion of it.

Lysimeter experiments with growing crops can, however, give more detailed information about this loss, because the amount of nitrogen lost in the drainage water can be measured.[1] Table 16.6, due to J. A. Bizzell[2] at Cornell University, shows that provided the level of available nitrogen in the soil is

1 For a review of results obtained from lysimeters see H. Kohnke, F. R. Dreibelbis and J. M. Davidson, *U.S. Dept. Agric., Misc. Publ.* 372, 1940.
2 *Cornell Agric. Expt. Sta., Mem.* 252, 1943; *Mem.* 256, 1944. For an example from California, see H. D. Chapman *et al.*, *Hilgardia*, 1949, **19**, 57.

TABLE 16.6 Loss of nitrogen from soil under crops liberally supplied with inorganic nitrogen (Cornell, NY)

	Nitrogen changes in kg/ha of nitrogen			
	Timothy (9-year period)		*Market garden crops (15-year period)*	
			Nitrogen as	
	High nitrogen	*Low nitrogen*	*Sulphate of ammonia*	*Nitrate of soda*
Lost from soil	− 100	− 45	530	800
Added as fertiliser and rain	2090	790	2500	2500
Total	1990	745	3030	3300
Removed in crop	1440	720	1570	1840
Removed in drainage water	45	20	710	780
Total	1485	740	2280	2620
Unaccounted for	505	—	750	680
As per cent as fertiliser	24	—	31	28

low, there is no unaccounted loss, but if it is raised by manuring it is appreciable.

These unaccounted losses also occur when grassland is converted to arable, or when virgin prairie soil is broken up for cropping. Oxidation of the organic matter and nitrification both proceed rapidly in the initial years, but only a small part of the loss of soil nitrogen can be accounted for in the crop. Table 16.7 gives an example of this at Indian Head, Saskatchewan,[1] where the rainfall is sufficiently low for there to be very little drainage from the soil. These large losses occur in the early years after breaking up the prairie, for E. S. Hopkins and A. Leahey,[2] continuing Shutt's work at Indian Head, found that in the next sixteen years almost the whole of the loss of soil nitrogen could be accounted for by the nitrogen in the wheat. H. E. Myers and his coworkers[3] in Kansas have shown that this loss of nitrogen is proportional to the nitrogen content of the soil in excess of its equilibrium value for the rotation adopted; thus for the wheat and barley rotations used,

1 F. T. Shutt, *J. agric. Sci.*, 1910, **3**, 355.
2 *Cornell Agric. Expt. Sta., Mem.* 252, 1943; *Mem.* 256, 1944. For an example from California,
3 *Kansas Agric. Expt. Sta., Tech. Bull.* 56, 1943. For an example from the Canadian prairies, see J. D. Newton *et al.*, *Sci. Agric.*, 1945, **21**, 718; and for the Great Plains, H. J. Haas *et al.*, *U.S. Dept. Agric., Tech. Bull.* 1164, 1957.

TABLE 16.7   Losses of nitrogen consequent on breaking up of prairie land. Top 22·5 cm of soil

	Per cent	*kg/ha*
Nitrogen present in unbroken prairie	0·371	7790
Nitrogen present after 22 years' cultivation	0·254	5330
Loss from soil		2460
Recovered in crop		790
Deficit, being dead loss		1670
Annual dead loss		76

nitrogen losses only occurred if the nitrogen content of the soil exceeded 0·10 per cent.

All these results are consistent with the hypothesis that denitrification is an active process even in soils that are considered to be reasonably well aerated. The observed losses could all be proportional to the average level of mineral nitrogen in the soil, and though they are higher than the calculated mean annual loss of nitrogen for the world as a whole, this would be expected as the examples have all been chosen from soils whose nitrogen status is well above their equilibrium levels.

Gains of nitrogen by the soil

Most soils are continuously losing nitrogen by one or more of the processes so far described; they must possess, therefore, some methods for gaining it to balance these losses. Arable soils in the temperate regions, if allowed to revert either to natural forest or to pasture, rapidly gain nitrogen and organic matter. On Broadbalk, for example, a third plot adjacent to the two quoted in Table 16.5 carried a wheat crop in 1882 which was never harvested: the plot was allowed to revert to natural vegetation, and became known as the Broadbalk Wilderness. Initially the legumes *Medicago lupulina* and *Lathyrus pratensis* formed an important part of the vegetation, but by 1895 trees and shrubs had invaded the area and were growing strongly. The area was then divided into two, one half being left undisturbed and the other having all the trees and bushes cut out regularly so it became a kind of natural pasture. The legumes died out completely from the woodland and only the *L. pratensis* persisted in very small numbers in the cut-over area. The soil was sampled in 1904 and 1964. This soil contained free calcium carbonate in the top 23 cm during this whole period, for although it is not naturally calcareous, it has received heavy dressings of chalk at irregular intervals for many centuries. A neighbouring old arable field, Geescroft, went derelict in 1885 after it had been in red clover for four years. This soil was neutral at the time but it did not contain any free calcium carbonate; the pH of its surface soil had fallen to 6·1 in 1904 and to 4·5 in 1965, when it also was sampled. This field first reverted to grass, largely *Aira caespitosa*, and slowly went over to woodland, but at no time was there any appreciable proportion of leguminous plants.

Table 16.8 gives the results of the soil analysis for carbon and nitrogen, in which the figures for the different sampling depths have been adjusted for the decrease in the bulk density of the soils due to the increase in organic

TABLE 16.8   Gains in carbon and nitrogen in soils permanently covered with vegetation at Rothamsted

Depth of sample (or equivalent)* in cm	Broadbalk Wilderness (with free $CaCO_3$ in top 23 cm soil)							
	Carbon per cent				Nitrogen per cent			
	1881	1904	1964		1881	1904	1964	
			Wood-land	Meadow			Wood-land	Meadow
0–23	0·98	1·34	2·70	2·79	0·107	0·142	0·254	0·260
23–46	0·59	0·66	0·85	0·81	0·078	0·097	0·098	0·098
46–69	0·47	0·52	0·61	0·55	0·066	0·080	0·078	0·075
Rate of gain kg/ha/year		600 from 1883	620 from 1904	600		52 from 1883	33 from 1904	34

Depth of sample (or equivalent)* in cm	Geescroft (No free $CaCO_3$)					
	1883	1904	1965	1883	1904	1965
0–23	1·07	1·37	1·98	0·116	0·131	0·166
23–46	0·57	0·58	0·76	0·081	0·082	0·092
46–69	0·48	0·42	0·49	0·068	0·069	0·071
Rate of gain kg/ha/year		370 from 1885	260 from 1904		21 from 1885	15 from 1904

* The equivalent depth is the soil depth containing the same weight of mineral soil as that depth in 1881 or 1883. This increases with time as the bulk density of the soil falls as the organic matter content increases.

matter of the different layers.[1] The table shows that the two soils have been gaining carbon and nitrogen at an appreciable rate since they went derelict, with the Broadbalk Wilderness soil gaining at almost twice the rate of the Geescroft, for reasons that are not yet known. Crop yields on the two fields were comparable when they were in arable crops, but from the beginning there was a stronger growth of vegetation on the Broadbalk site, and it went

1   D. S. Jenkinson, *Rothamsted Rept.*, 1970, Part II, 113. This contains a detailed discussion of these results.

over to woodland quicker, possibly because it was closer to trees. Legumes were present in Broadbalk from the beginning, and they could have been the cause of the initial rapid gain in soil nitrogen; but the later gains must be due to other agencies.

The C/N ratio in the surface soil of both areas was initially 9·0, and it has risen to 10·7 on Broadbalk and to 11·9 on Geescroft, which had the smaller gain of organic matter. This is concordant with a high C/N ratio in the organic matter being a sign of infertility in the soil. Finally, it is interesting to note that on Broadbalk during the last sixty years, the organic matter has increased at the same rate, and with the same C/N ratio, under woodland as under the herbaceous-grassy vegetation from which the trees and bushes have been prevented from growing.

Arable soils put down to grass containing no legumes also gain nitrogen, and gains of the order of 50 kg/ha annually have been found in both temperate and tropical conditions;[1] though under savanna conditions, under which the grass is burnt annually, the gain is probably only half as large.[2] Cultivated soils or bare soils planted to forest also gain nitrogen, again at about 50 kg/ha in the temperate regions, but probably a little less in tropical.[3]

Soils can gain small amounts of nitrogen from the rain which falls on them. Thus at Rothamsted[4] rainwater carries down anually about 4 kg/ha of nitrogen as ammonia and nitrates, at Cornell[5] the figure is about 6 kg and elsewhere in the world[6] it varies from 2 to 22 kg/ha. These quantities are clearly quite inadequate to give the gains of soil nitrogen just quoted, although in some stable and mature ecological regions, such as undisturbed natural forests, it may be sufficient to balance the losses of nitrogen by leaching.

The most important natural process for increasing the nitrogen content of soils in temperate agriculture is through bacteria living in the nodules of leguminous plants, a process discussed later on in this chapter. Thus on p. 382 it is shown that a good clover or lucerne crop will increase the organic nitrogen in the soil by over 100 kg/ha annually—a quantity which can be measured without much difficulty. Symbiotic nitrogen fixation is also brought about in root nodules on certain trees and shrubs, probably by Actinomycetes though this has not been adequately proved; and this process appears to be of importance in the colonisation of land after it has been freed from the ice-sheets which covered much of the northern hemisphere, or of sandy land

1 See D. J. Greenland, *Soils Fert.*, 1971, **34**, 237 (Australian soils). C. R. Clement and T. E. Williams, *J. agric. Sci.*, 1967, **69**, 133 (cocksfoot at Hurley, England); and A. W. Moore, *Soils Fert.*, 1966, **29**, 113.
2 P. H. Nye and D. J. Greenland, *Tech. Comm. 51 Commw. Bur. Soils*, 1960.
3 A. W. Moore (temperate forests), D. J. Greenland and P. H. Nye, *J. Soil Sci.*, 1959, **10**, 284 (tropical).
4 N. H. J. Miller, *J. agric. Sci.*, 1905, **1**, 280.
5 E. W. Leland, *Agron. J.*, 1952, **44**, 172.
6 See, for example, E. Erikson, *Tellus*, 1952, **4**, 215, for a collection of published data.

left by the retreat of the sea. This process is discussed in more detail on pp. 356–7, but it has not yet been used in agriculture. It is also possible, but not yet properly proven, that nitrogen fixation may take place in nodules on the leaves of certain plants, or even by bacteria living on their leaves, processes mentioned again on p. 357.

Non-symbiotic fixation of nitrogen

Soils contain a number of free living nitrogen-fixing organisms, and though most investigations have been confined to two groups of bacteria and some blue-green algae, modern techniques based on the use of the acetylene test, or of air enriched with either the stable isotope N^{15} or the radioactive N^{13}, have demonstrated conclusively that a number of other organisms can convert atmospheric nitrogen into organic compounds. Excluding the blue-green algae and the photosynthetic bacteria, the organisms which have been proved to fix nitrogen appear to be:

1 Bacteria of the genus *Azobacter*. The common soil species are *chroococcum*, *beijerinckii* and *vinelandii*, and are relatively large organisms encased in bacterial gum. They are aerobic, and usually confined to soils well supplied with phosphate and not more acid than pH 6,[1] though species have been found in acid soils.[2] Though widely distributed, they do not occur in many soils that are apparently suitable for them.

2 Bacteria of the genus *Beijerinckia*. The first species discovered was originally described as an *Azobacter* (*A. indicum*), but H. G. Derx[3] separated it from *Azotobacter* on morphological grounds. They differ from *Azotobacter* in that they are more tolerant of acid soils, so are the common nitrogen fixer in soils with a pH below 6, and they are not found in neutral soils. They are typical inhabitants of tropical and subtropical soils, particularly in tropical red earths and latosols, and like the distribution of *Azotobacter*, they do not occur in a considerable proportion of soil samples where they would be expected.[4] They probably have a higher molybdenum demand than most strains of *Azotobacter*, which cannot be satisfied with vanadium as it can be with most *Azotobacter* strains.[5] There are probably several species of *Beijerinckia*, but there are still difficult problems of systematics in the *Azotobacter-Beijerinckia* and other related genera.

1 H. L. Jensen (*Tidsskr. Planteavl*, 1950, **53**, 622) and A. Kaila (*Maataloust. Aikak.*, 1954, **26**, 40) find none occur below pH 5·8. T. McKnight (*Queensland J. agric. Sci.*, 1949, **6**, 177) finds a few in soils between pH 5·5 and 6·0.
2 For example by Y. T. Tchan, *Proc. Linn. Soc. N.S.W.*, 1953, **78**, 83, and H. L. Jensen, *Acta agric. Scand.*, 1955, **5**, 280.
3 H. G. Derx, *Meded. Konink. Ned. Akad. Wetens.*, 1950, **53**, No. 2.
4 See, for example, J. H. Becking, *Pl. Soil*, 1961, **14**, 49; A. W. Moore, ibid., 1963, **19**, 385; J. Meiklejohn, *Emp. J. exp. Agric.*, 1962, **30**, 115; J. P. Thompson, *Trans. 9th Int. Congr. Soil Sci.*, 1968, **2**, 129.
5 J. H. Becking, *Pl. Soil*, 1961, **14**, 297; 1962, **16**, 171.

3 Some strains of the soil *Clostridia*, which are anaerobes, and of *Bacillus polymyxa* which are facultative anaerobes. These are widely distributed in soils, and both fix nitrogen under conditions of low oxygen tension.

4 A number of other bacteria, some of which are common soil inhabitants will fix nitrogen, particularly in conditions of low oxygen tension. They have been placed in the genera *Klebsiella*, *Achromobacter* and *Pseudomonas*, but their exact taxonomic position, like that of the *B. polymyxa* group, is often unclear.

5 A few soil fungi, including some *Pullularia* and yeasts.

The number of the aerobic nitrogen-fixing bacteria in most soils in which they are found is surprisingly low, usually of the order of a few hundred to a few thousand per gram. Thus H. L. Jensen[1] found that in 167 Danish soils with a pH exceeding 6·5, 49 contained no *Azotobacter*, 53 under 100, 51 between 100 and 1000, and only 19 with more than 1000. J. Meiklejohn,[2] using an improved counting technique usually only found a few hundred to a few thousand in the Broadbalk plots, and a similar number of *Beijerinckia* in some Rhodesian soils.[3] She also found up to 10^5 nitrogen-fixing Pseudomonads in these soils. *Clostridia* have usually been found in higher numbers when they have been looked for, and are usually present in numbers about 10^5 per gram of soil. It is still uncertain if these observed numbers genuinely reflect the numbers present in the soil, or if most of the bacteria stick so strongly to the surface of the soil particles that they are not dispersed in the counting solution. It is difficult to believe that if they are genuinely present in these numbers they can be of much significance in the nitrogen economy of field soils. In fact, T. Hauke-Pacewieczowa[4] found that unless there were at least 10^6 *Clostridia* per gram of soil, fixation was not appreciable.

Nitrogen fixation goes on more actively under conditions of poor than of good aeration, provided no hydrogen gas is being produced. As long ago as 1927 O. Meyerhof and D. Burk[5] showed that the nitrogen-fixing ability of *Azotobacter*, which is regarded as an aerobic organism, increased with decreasing oxygen partial pressure down to a concentration of about 10^{-6} M in the solution around the bacterial cells. C. A. Parker[6] found that his strains of *Azotobacter*, for example, fixed about 22 mg N per gram of sucrose used when the oxygen content of the air was 4 per cent, but only 8 mg when it was 20 per cent. Again *Azotobacter* will grow symbiotically with facultative anaerobes under partially anaerobic conditions on plant residues, such as cellulose, hay or straw: there must be an inadequate supply of nitrogen com-

1 *Tidsskr. Planteavl*, 1950, **53**, 622.
2 *Pl. Soil*, 1965, **23**, 227; *Rothamsted Rept.* 1968, Part II, 179.
3 *Trans. 9th Int. Congr. Soil Sci.*, 1968, **2**, 141.
4 With J. Balandreau and Y. Dommergues, *Soil Biol. Biochem.*, 1970, **2**, 47.
5 *Z. Phys. Chem.*, 1927, **139** A, 117.
6 *Nature*, 1954, **173**, 780. For earlier examples, see D. Burk, *J. Phys. Chem.*, 1930, **34**, 1195; H. L. Jensen, *Proc. Linn. Soc. N.S.W.*, 1940, **65**, 1.

pounds for the facultative anaerobes, and there must be an air supply sufficiently restricted for the fermenting bacteria to convert a considerable proportion of the cellulose into simple acids, such as lactic, butyric or acetic.[1] Thus, under waterlogged conditions, H. L. Jensen and R. J. Swaby[2] found that *Azotobacter* could fix about 14 mg of nitrogen per gram of cellulose decomposed, and most of this nitrogen was taken up by the other bacteria. C. J. Lind and P. W. Wilson[3] also found that, under these symbiotic conditions, *Azotobacter* could fix nitrogen efficiently with a much more restricted supply of available iron than in a pure culture. *Azotobacter* usually excretes little if any of the nitrogen it fixes while it is growing actively, though it may excrete most of what it fixes after active growth has ceased.[4]

Clostridia have been considered less efficient fixers of nitrogen than *Azotobacter*, fixing only 6·8 mg per gram of sugar decomposed compared with about 20 mg in pure culture, but E. D. Rosenblum and P. W. Wilson[5] showed this was principally due to the conditions in the medium not being favourable to high fixation rates; and that when these were assured, the efficiencies of the two organisms were similar. They also showed that all the soil *Clostridia* which grew actively in their media were also active nitrogen fixers. *Clostridia* appear to differ from *Azotobacter* in that they may excrete soluble nitrogen substances into the medium during their growth phase. Nitrogen fixation by *Achromobacter* and *Pseudomonas* in aerobic conditions is relatively inefficient, as they only fix between 1 and 4 mg nitrogen per gram of sugar decomposed,[6] and although Proctor and Wilson state that poor aeration increases the rate of nitrogen fixation they gave no figures for its effect on the efficiency of fixation. It is therefore possible that nitrogen fixation by all these bacteria, whether aerobic or anaerobic, takes place predominantly in soil microhabitats where the oxygen tension is low,[7] and at an appreciable rate only if there is an accessible supply of decomposable organic matter low in nitrogen. It takes place most vigorously when the level of soil organic matter is below the equilibrium value appropriate to the existing system of land use; so typically becomes important when arable land is put down to a resting crop.

Nitrogen fixation is also carried out by some blue-green algae (see p. 179), and they are of proven importance in the nitrogen economy of some rice soils (see p. 691); but although they probably fix nitrogen in most soils during periods when the soils are warm and their surface is both moist and exposed to sunlight,[8] there is little evidence as to whether the amounts they fix of any

1 E. H. Richards, *J. agric. Sci.*, 1939, **29**, 302.
2 *Proc. Linn. Soc. N.S.W.*, 1941, **66**, 89.
3 *Soil Sci.*, 1942, **54**, 102.
4 D. Burk and R. H. Burris, *Ann. Rev. Biochem.*, 1941, **10**, 587.
5 *J. Bact.*, 1949, **57**, 413, and 1950, **59**, 83.
6 M. H. Proctor and P. W. Wilson, *Nature*, 1958, **182**, 891.
7 See, for example, R. Knowles, *Proc. Soil Sci. Soc. Amer.*, 1965, **29**, 223.
8 For Broadbalk see *Rothamsted Rept.* 1969, Part I, 101.

agricultural significance. Nitrogen fixation can also be carried out in sunlight by all photosynthetic bacteria, but these normally require anaerobic conditions and certain oxidisable inorganic substrates (see p. 154). Some of these may be active in rice soils.

This discussion has assumed that *Azotobacter* and *Beijerinckia* fix nitrogen only when living free in the soil, but this assumption is sometimes invalid. Thus J. Dobereiner[1] found that the roots of the tropical grass *Paspalum notatum* may fix up to 90 kg N/ha/year when they have a species of *Azotobacter*, *Az. paspali*, growing on their roots, though the rate of fixation depends on the oxygen pressure at the root surface, being at a maximum when the pressure is 0·04 atm. The roots of sugar cane, *Panicum maximum* and *Pennisetum purpureum* may also fix appreciable amounts of nitrogen though, for these grasses, a strain of *Beijerinckia indica* was abundant on their root surfaces. This fixation can be looked upon as a type of symbiotic fixation, since the root surface of these grasses appears to be a particularly favourable environment for the bacteria. D. Harris and P. J. Dart[2] have also found that the root systems of several plant species in the Broadbalk Wilderness have appreciable nitrogen-fixing ability, although the soil in which they are growing has little.

The biochemistry of nitrogen fixation

It is possible that all the biological nitrogen-fixing systems use the same basic biochemical pathways. They all need iron, molybdenum and possibly cobalt in very small amounts, and the process is inhibited by hydrogen gas and carbon monoxide. The nitrogen gas is absorbed, the bond between the two nitrogen atoms broken, and the nitrogen is then hydrogenated to give ammonium, which combines with α-keto-glutaric acid to give glutamic acid, and it is in this form that it enters the metabolic pathways of the nitrogen-fixing organism or system. The hydrogenation enzyme system is not very specific, for in the absence of a suitable hydrogen acceptor, hydrogen gas may be produced. It will also reduce nitrous oxide to nitrogen gas and water, will hydrogenate azides and cyanides, and also acetylene to give ethylene.[3] This last reduction is believed to be specific for the nitrogen-fixing system, so that a soil or plant capable of hydrogenating acetylene to ethylene must contain nitrogen-fixing organisms. The hydrogenation enzymes are probably inhibited if the oxygen tension is appreciable, so that aerobic organisms and root nodules must have some mechanism by which they can obtain the oxygen they require for their metabolic processes but which maintains a low or very low oxygen tension close to the enzyme sites. In root nodules, this is achieved

1 With J. M. Day and P. J. Dart, *J. gen. Microbiol.*, 1972, **71**, 103.
2 *Rothamsted Rept.* 1971, Part I, 95.
3 For references, see the review papers by J. Chatt, *Proc. Roy. Soc.*, 1969, **172** B, 327, and F. J. Bergersen, ibid., 401.

through the presence of haemoglobin-like substances which transport oxygen very efficiently at low tensions.[1]

The first step in the process of fixation appears to involve the adsorption of molecular nitrogen by an enzyme system possibly consisting of an iron- and molybdenum-containing protein and of an iron-containing protein. The adsorbed nitrogen is activated by having its $N\equiv N$ bond stretched, and then hydrogenated. The hydrogen donor is not known, though in the laboratory dithionite can perform this function though pyruvate and similar substances cannot. The donor is very labile and may be a ferredoxin, ATP is directly concerned in the fixation, and 15 molecules of ATP are required for each molecule of ammonia produced. The hydrogenation system contains both a carbon monoxide sensitive and a carbon monoxide insensitive enzyme, and the function of the small amount of cobalt needed for the fixation is not known.

Symbiotic nitrogen fixation in non-leguminous plants

Symbiotic nitrogen fixation in agricultural practice is confined to certain leguminous plants, but it is not confined to this family. A number of trees and shrubs carry root nodules which have been proved to fix nitrogen.[2] They include the alder (*Alnus glutinosa*) and three genera of the family Elaeagnaceae, namely, *Hippophae*, *Shephardia* and *Elaeagnus*, which have somewhat similar nodules; and the genus *Myrica* (myrtles) and the subtropical and tropical genus *Casuarina* which have a different type of nodule. In addition, species belonging to the genera *Coriaria* and *Ceanothus* also have nodules, which may be somewhat similar to the *Myrica-Casuarina* group. These plants are characteristic of soils low in nitrogen and often of rather wet sites. The nodules are usually much larger than leguminous nodules, have irregular shapes and are perennial. The *Myrica-Casuarina* group nodules have the same structure as the *Alnus-Elaeagnus* group, but they typically have a root, which does not contain the endophyte, growing upward from the apex of the nodule or nodule lobe.

The endophyte has not been grown in pure culture, and is commonly considered to be an Actinomycete.[3] The biochemistry of nitrogen fixation in the nodule is probably very similar to that in the leguminous nodule, and the efficiency of fixation per unit weight of nodule is also similar. Thus, an alder nodule will fix about 800 mg N per gram dry matter of nodular tissue daily, and *Myrica* and *Casuarina* about 400 mg daily, during the growing season compared with about 700 mg for field beans and red clover and

---

1 For a review see F. J. Bergersen, *Bact. Rev.*, 1960, **24**, 246.
2 For reviews see W. S. Silver, *Proc. Roy. Soc.*, 1969, **172** B, 389, and D. O. Norris, *Commw. Bur. Pastures and Field Crops*, *Bull.* 46, 1962.
3 For electron microscope photographs of the endophyte in *Hippophae* see E. M. S. Gatner and I. C. Gardner, *Arch. Microbiol.*, 1970, **70**, 183.

250 mg for soyabeans.[1] Gains of about 60 kg/ha N annually have been found in soils under both alder and *Casuarina* in the field.[2]

Two other groups of plants possess symbiotic nitrogen-fixing mechanisms. Cycads have certain species of blue-green algae, belonging to the genus *Nostoc* or *Anabaena* living in the free spaces of the middle cortex of their roots, and they give rise to coralloid structures; and certain tropical plants, such as *Psychotria bacteriophila* and *Pavetta zimmermanniana* have nodules on their leaf containing the nitrogen-fixing bacteria *Klebsiella*.[3] W. S. Silver showed, however, that these leaf nodules do more than provide the plant with nitrogen, for in their absence the *Psychotria* only grew into a dwarfed plant, even if supplied with nitrogen.[4] There is also the probability that a number of plants carry a population of nitrogen-fixing bacteria on their leaves, including some *Azotobacter*, *Beijerinckia* and *Pseudomonas*, which draw their energy supplies from leaf exudates and supply the leaf with nitrogenous compounds. J. Ruinen,[5] who first studied this population which she called the phyllosphere population because it lived on the leaf surface, worked with tropical and subtropical plants including coffee and cotton, but K. Jones[6] found the phyllosphere population on the needles of Douglas fir, *Pseudotsuga douglasii*, in an English forest also fixed nitrogen at a rate comparable to the rate of gain of nitrogen by the tree. He did not, however, identify the micro-organisms responsible for this fixation.

Symbiotic nitrogen fixation in leguminous plants

Certain leguminous plants are very extensively cultivated in temperate agriculture, and correspondingly their ability to fix nitrogen has been equally extensively investigated. But D. O. Norris[7] has emphasised that any general discussion on symbiotic nitrogen fixation must be based on the fact that the great majority of leguminous genera and species are tropical, and that the ancestral ecological niche was probably on fairly strongly leached soils in tropical rain forests. The temperate and sub-tropical legumes developed from these, and some of them, particularly certain species belonging to the Trifolieae, which includes the genera *Trifolium*, *Medicago*, *Melilotus* and *Trigonella*, and the Vicieae, which includes the genera *Vicia*, *Pisum*, *Lens*, *Cicer* and *Lathyrus*, have become adapted to neutral or calcareous soils of higher nutrient status. A high proportion of the leguminous crops cultivated in temperate agriculture belong to these groups, and a very considerable pro-

1   G. Bond, *Ann. Bot.*, 1957, **21**, 373.
2   J. H. Becking, *Pl. Soil*, 1970, **32**, 611. Y. Dommergues, *Agrochimica*, 1963, **7**, 335. R. L. Crocker and J. Major, *J. Ecol.*, 1955, **43**, 427. See also W. D. P. Stewart, *J. exp. Bot.*, 1962, **13**, 250.
3   See, for example, W. S. Silver, *Proc. Roy. Soc.*, 1969, **172** B, 389, and with Y. M. Centifanto, *J. Bact.*, 1964, **88**, 776.
4   With Y. M. Centifanto and D. J. D. Nicholas, *Nature*, 1963, **199**, 396.
5   *Pl. Soil*, 1961, **15**, 81; 1965, **22**, 375.
6   *Ann. Bot.*, 1970, **34**, 239.
7   *Emp. J. exp. Agric.*, 1956, **24**, 247.

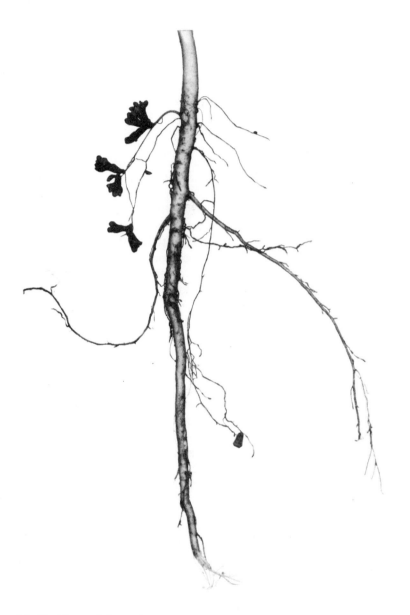

PLATE 16. Healthy nodules on lucerne rootlets (natural size).

portion of the research work done on legumes has been confined to a few species having this untypical demand for fertile soils of high calcium saturation. A great many of the statements made in this chapter may therefore be of limited validity: how limited cannot be known until far more work has been done on the tropical species.

A typical statement often heard is that leguminous crops carry nodules on their roots, which fix nitrogen, and therefore legumes raise the nitrogen status of the soil. This statement is only doubtfully true for the cultivated large-seeded crops, such as peas, beans, soyabeans and groundnuts because, even if their roots are well nodulated, which is often not so when they are grown under good farming conditions, a large proportion of the nitrogen they have fixed is removed from the land in the seed crop, and nearly all the rest is in the vines or straw which again is usually removed at harvest time. And in the tropics many legumes, grown for the ostensible purpose of raising the nitrogen status of the soil, are not even nodulated for most of the growing season. It is not known if there are genera or sub-families which never carry nodules on their roots, and the problem is complicated because a leguminous plant growing in one habitat, for example, high tropical rain forest, may rarely be nodulated, yet it may be commonly well nodulated in another habitat, for example, on the fringes of the forest.[1]

The nodules, which are found on the roots of leguminous crops, vary widely in their shape and size. They may be spherical, though they are usually elongated, or flat and grooved, or may have finger-like projections, or they may be irregular and sometimes convoluted, though they usually have a smooth surface. Their size may vary from that of a pin head to over 1 cm, but the larger nodules are never spherical but have shapes giving a high ratio of surface area to volume, possibly to ensure an adequate supply of nitrogen gas to the active nodule cells and an adequate means of disposal of the carbon dioxide produced in the nodule. Typical shapes on lucerne and clover roots are illustrated in Plates 16 and 21.

Nodules contain bacteria living symbiotically with the plant: the plant leaves supplying the carbohydrate and the bacteria the amino acids for the combined organism. But the bacteria become parasitic if for any reason the carbohydrate supply is restricted, as, for example, by keeping the plant in the dark,[2] or by restricting the boron supply[3] so that the vascular strands supplying carbohydrates to the nodule fail to develop, for the nodules form close to the vascular strands, as can be seen in Plate 17. This parasitism also occurs normally in old nodules, resulting in their decay, and in nodules of some annuals such as peas and vetches at the time of flowering or setting of seed.[4]

1 For an example from the Yangambi forests (Zaire), see C. Bonnier, *I.N.E.A.C., Sci. Ser., Publ.* 72, 1957.
2 H. G. Thornton, *Proc. Roy. Soc.*, 1930, **106** B, 110.
3 W. E. Brenchley and H. G. Thornton, *Proc. Roy. Soc.*, 1925, **98** B, 373.
4 J. S. Pate, *Aust. J. biol. Sci.*, 1958, **11**, 366, 496.

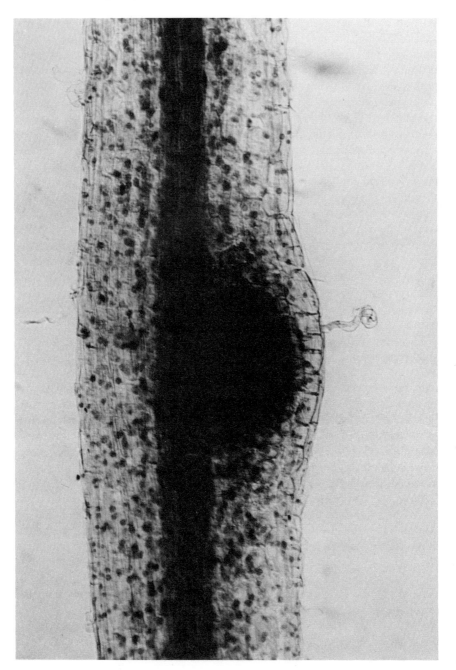

PLATE 17. A very young nodule on a lucerne rootlet seen by transmitted light. The dark lines across the centre of the rootlet are the vascular bundles. ($\times 250$).

Conditions necessary for nitrogen fixation

Nodules can only fix nitrogen actively if the plant is adequately supplied with all the mineral elements essential for active growth. But, in addition, all nodulated plants require small quantities of cobalt,[1] which is probably a constituent of leghaemoglobin, and have a higher molybdenum requirement than un-nodulated; some species have a higher calcium or higher pH requirement, and a few a higher boron and sulphur requirement. Thus W. E. Brenchley and H. G. Thornton[2] in 1926 showed that the symbiosis between the nodule of the plant could only be complete with broad beans (*Vicia faba*) if it was adequately supplied with boron, as otherwise the vascular tissue between the nodule and the root did not develop properly; and it is probable that this need is higher with broad beans than many other legumes, though this point does not seem to have received detailed investigation. Again lucerne and groundnuts may have a higher sulphur demand than many other legumes. Magnesium is probably essential for nitrogen fixation, but although many legumes have an appreciable calcium demand,[3] there is no evidence that calcium plays any role in the fixation process. The pH tolerances of legumes probably depend partly on their tolerance to aluminium ions and partly on their calcium requirements. Lucerne and red clover, for example, only grow well on soils of fairly high pH or calcium status, and in general they need soils of pH 6 or above for optimum growth. On the other hand, some white clovers and subterranean clovers can fix nitrogen in soils of pH 4·2 to 4·5[4] although they usually respond to liming if the soil is as acid as this; while other plants, and particularly many tropical legumes, can fix nitrogen actively at these low soil pHs, provided they can extract calcium from the soil. Thus, W. A. Albrecht[5] found that soyabeans could nodulate well and fix nitrogen actively in a soil of pH 4·2 provided the calcium supply was adequate. So far no quantitative study seems to have been made on the ability of acid-tolerant temperate legumes, for example, lupins or gorse (*Ulex europaeus*) to fix nitrogen in acid to very acid soils.

The reason for this difference in pH or calcium requirements of the various legumes is not completely known. It is partly due to the toxic effect of aluminium ions,[6] for legumes vary in their sensitivity to aluminium ions very considerably, and different strains of Rhizobia that can form a nodule differ in their response to acidity, though it is not known how far this is a pure

1 See, for example, C. C. Delwiche, C. M. Johnson and H. M. Reisenauer, *Plant Physiol.*, 1961, **36**, 73. E. G. Hallsworth *Nature*, 1960, **187**, 79.
2 *Proc. Roy. Soc.*, 1925, **98** B, 373.
3 D. O. Norris, *Aust. J. agr. Res.*, 1959, **10**, 651. J. M. Vincent and J. R. Colburn, *Aust. J. Sci.*, 1961, **23**, 269.
4 See, for example, H. L. Jensen, *Proc. Linn. Soc. N.S.W.*, 1947, **72**, 265.
5 *J. Amer. Soc. Agron.*, 1933, **25**, 512, and with C. B. Harston, *Proc. Soil Sci. Soc. Amer.*, 1942, **7**, 247.
6 T. M. McCalla, *Missouri Agric. Exp. Sta.*, *Res. Bull.* 256, 1937, and D. O. Norris, *Emp. J. exp. Agric.*, 1956, **24**, 247, and *Nature*, 1958, **182**, 734.

calcium response. Some nodulated legumes have only a very low calcium demand and some appear to have a much higher demand, though this may only be a reflection of the differing abilities of legumes to take up calcium from soils poorly supplied with this element. Thus D. O. Norris[1] has shown that on some strongly leached acid soils in Queensland the typical tropical legumes could take up adequate amounts of calcium for active growth and nodulation, whereas temperate species of *Trifolium* and *Medicago* failed to grow properly, yet a dressing as low as 250 kg/ha of calcium carbonate was sufficient to give reasonable, and 1250 kg/ha good, growth and nodulation.

The molybdenum requirements of leguminous plants are entirely, or almost entirely, for the nitrogen-fixing mechanisms; for the plants will grow perfectly well, and their roots will be well nodulated, in soils very low in molybdenum, but the nodules will fix no nitrogen.[2] Again, if the molybdenum is in short supply, the molybdenum in the plant will be concentrated in the nodules rather than in the roots or tops. H. L. Jensen and R. C. Betty,[3] for example, found the nodules of lucerne contained 6 to 20 ppm of molybdenum in the dry matter, which was 5 to 15 times as much as the rest of the root system, which, in turn, had a higher content than the tops. It is only when the supply is adequate that the molybdenum content of the tops begins to rise. In Australia and New Zealand there are some soils on which subterranean clover will not grow unless it is given a dressing of a molybdenum salt. Dressings of a few hundred grams per hectare of molybdenum as sodium molybdate[4] are common, but examples of appreciable responses to 3 and 20 g have been reported.[5] If the concentration of molybdenum in the soil is too high, it accumulates in the leaves of the leguminous plants and causes the soil to become 'teart'[6] (see p. 47).

The bacteria in the nodules must be supplied with energy if they are to fix nitrogen, and hence with oxygen for the oxidation of the carbohydrate. G. Bond[7] has estimated that, in some of his experiments with soyabeans, 16 per cent of the total carbohydrates synthesised by the plant were respired by the nodules, and of the total respiration from the plant, 57 per cent was from the tops, 18 per cent from the roots and 25 per cent from the nodules. Hence a well-nodulated leguminous crop in the field needs a better oxygen supply to its roots, which respire more carbon dioxide than does a cereal or

1 *Nutrition of Legumes*, Ed. E. G. Hallsworth, London, 1958.
2 A. J. Anderson, *Aust. Counc. Sci. Indust. Res., Bull.* 198, 1946; and H. L. Jensen, *Proc. Linn. Soc. N.S.W.*, 1945, **70**, 203.
3 *Proc. Linn. Soc. N.S.W.*, 1943, **68**, 1; see also K. G. Vinogradova, *C.R. Acad. Sci. (U.S.S.R.)* 1943, **40**, 26.
4 H. C. Trumble and H. M. Ferres, *J. Aust. Inst. agric. Sci.*, 1946, **12**, 32.
5 See, for example, A. J. Anderson, *Aust. J. Counc. Sci. Indust. Res.*, 1946, **19**, 1.
6 For the margin between optimum molybdenum contents for nitrogen fixation and the minimum for teartness, see H. L. Jensen, *Proc. Linn. Soc. N.S.W.*, 1946, **70**, 203.
7 *Ann. Bot.*, 1941, **5**, 313.

root crop.[1] P. W. Wilson[2] calculates that for a crop fixing 1 kg nitrogen per hectare per day in summer, nodular respiration should amount to 3·3 g/m^2 per day of carbon dioxide, a figure that should be compared with the normal rates of soil respiration given in Table 18.1. Bond[3] also showed that if soil aeration is reduced, the growth of soyabeans given nitrates or ammonia is not much affected, but it is definitely reduced if they are nodulated plants receiving no nitrogen fertiliser.

Nitrogen fixation, however, is a hydrogenation reaction, and must therefore take place in an environment low in oxygen. Nodules that are fixing nitrogen actively contain leghaemoglobin—which is related to the haemoglobins present in animal blood—the function of which appears to be the active transport of oxygen to the bacteria while maintaining a low oxygen tension in the bathing solution. The pink colour due to this leghaemoglobin is characteristic of nodules fixing nitrogen; and if it is absent it is usually a sure sign that nitrogen fixation is not taking place within the nodules.

Nodules are more efficient fixers of nitrogen than the non-symbiotic bacteria, for nodules fix 1 g of nitrogen for every 15 to 20 g of carbohydrate oxidised, whereas the non-symbiotic bacteria use about $2\frac{1}{2}$ times as much carbohydrate. A healthy lucerne nodule can fix up to 100 mg of nitrogen daily per gram of dry matter, which corresponds to 1·36 times its own nitrogen content, and nodules of subterranean clover can fix up to 50 mg daily, though for each crop the average fixation per unit of dry matter in the nodules is usually about half of these rates.[4] For these rates to be possible, the lucerne nodules must contain 10 to 25 ppm and the clover 4 to 8 ppm of molybdenum in the dry nodule tissue.

The nodule bacteria

The nodule bacteria are classified in the genus *Rhizobium*, though they were formerly called *Bacillus radicicola*. The bacteria are typically rod-shaped when grown on suitable media and when actively growing in healthy nodules, but they may have a shape like an X, Y, T or club if growing in unfavourable conditions in media or in the nodule,[5] and they have characteristic banded and branched shapes in the older cells of the nodule. They seem to have affinities with *Agrobacterium* (*Achromobacter*) *radiobacter*, a soil bacterium found near the roots of many plants and particularly legumes, and with *Agrobacterium* (*Phytomonas*) *tumifaciens*, a soil bacterium which is often

1 See, for example, P. Hasse and F. Kirchmeyer, *Ztschr. Pflanz. Düng.*, 1927, **A10**, 257.
2 *The Biochemistry of Symbiotic Nitrogen Fixation*, Madison, 1940.
3 *Ann. Bot.*, 1950, **15**, 95. For the corresponding work with red clover, see T. P. Ferguson and G. Bond, *Ann. Bot.*, 1954, **18**, 385.
4 H. L. Jensen, *Proc. Linn. Soc. N.S.W.*, 1947, **72**, 265
5 C. Bonnier, *I.N.E.A.C. Sci. Ser., Publ.* 72, 1957.

parasitic on plant roots, producing swellings and tumours on them,[1] and *Bacillus polymyxa*.[2]

Nodule bacteria can be grown fairly easily in simple culture media, though some species need some of the vitamin B group for active growth.[3] They fall into two fairly well-defined groups; those that grow slowly on culture media and make the medium more alkaline and those that grow rapidly and make it more acid.[4] The first group are characteristic of tropical legumes adapted to acid soils and the second of temperate legumes adapted to neutral soils, though some tropical legumes, such as *Leucaena leucocephala* (formerly *L. glauca*), are nodulated by fast-growing strains of Rhizobia[5] and some temperate legumes adapted to acid soils, such as gorse (*Ulex*), are nodulated by slow-growing strains. These bacteria exist in a large number of species or strains which differ in the host plants they can infect, the longevity and size of the nodules they produce on a given host, their serological reactions, their resistance to different phages, their morphology and particularly in the number of flagellae they possess, and their production of gum; but there is little detailed correlation between most of these properties, and there are still no generally accepted criteria for their classification.

The nodule bacteria of temperate cultivated legumes have been divided into species according to the host plants they infect. E. B. Fred and his Wisconsin coworkers[6] recognised six so-called cross-inoculation groups based on the assumption that all bacteria in one group will infect all plant species in that group and none outside it. Fred considered the bacteria in each group belonged to a different species, and gave them the following names:

1 *R. meliloti* which infect lucerne and sweet clover (*Medicago*, *Melilotus* and *Trigonella*).
2 *R. trifolii* which infects clovers.
3 *R. leguminosarum* which infects peas and vetch (*Pisum*, *Lathyrus*, *Vicia* and *Lens*).
4 *R. lupini* which infect lupins and serradella (*Lupinus* and *Ornithopus*).
5 *R. japonica* which infects soyabeans.
6 *R. phaseoli* which infect a few species of *Phaseolus*, e.g. *P. vulgaris*, *P. coccineus* and *P. angustifolia*.

It is doubtful if there are any theoretical grounds for accepting this classification as it stands, though it has a certain validity for the first three

---

1 For a discussion of these affinities, see H. J. Conn, G. E. Wolfe and M. Ford, *J. Bact.*, 1940, **39**, 207, and 1942, **44**, 353. They put these two groups into a new genus *Agrobacterium*.
2 K. A. Bisset, *J. gen. Microbiol.*, 1959, **20**, 89. S. Hino and P. W. Wilson, *J. Bact.*, 1958, **75**, 403. D. O. Norris, *Aust. Inst. agri. Sci.*, 1959, **25**, 202.
3 P. H. Graham, *J. gen. Microbiol.*, 1963, **30**, 245. For a review, see E. K. and O. N. Allen, *Bact. Rev.*, 1950, **14**, 273.
4 See, for example, D. O. Norris, *Pl. Soil*, 1965, **22**, 143.
5 M. J. Trinick, *Exp. Agric.*, 1968, **4**, 243.
6 *Root Nodule Bacteria and Leguminous Plants*, Madison, 1932.

groups, but it is of little help in practice because, if a bacterial strain is to have any agricultural value, it must form nodules capable of fixing appreciable amounts of nitrogen; and no single strain can form effective nodules on all plant species in any of the first three cross-inoculation groups. Thus, Australian experience has shown that the *R. trifolii* group must certainly be divided into three,[1] if the bacterium is to form effective nodules, which infect the following *Trifolium* species:

1 *T. repens, pratense, hybridum, procumbens,* and *fragiferum.*
2 *T. subterraneum, incarnatum, glomeratum* and *alexandrinum.*
3 *T. ambiguum.*

D. O. Norris[2] also found that the Rhizobia nodulating the equatorial high-altitude indigenous clovers differ from the temperate groups and themselves probably fall into four groups, each nodulating its own specific group of *Trifolium* species.

A much clearer understanding of the relationship between bacterial strain and plant species can be obtained by studying it in the tropics, where there is a very wide range of indigenous leguminous plants. There is no good evidence for the existence of any cross-inoculation groups among this vast collection of species. What has been found is that among many of the cultivated and semi-cultivated tropical and subtropical legumes, the ability of different bacterial strains to form effective nodules, or any nodules at all varies very widely. Some strains can nodulate a wide range of species and genera, whereas others can only form nodules on a few species of a single genus.[3] The pattern of effectiveness can also vary considerably. Cowpeas, for example, can be effectively nodulated by a wide range of bacterial strains, while others, such as *Leucaena leucocephala* (*L. glauca*) have become specialised and can only be effectively nodulated by a few strains, which themselves may be specialised to a restricted range of plants.

The formation and morphology of nodules

This is a subject on which our knowledge is still limited, in spite of a vast amount of research which has gone into it, and only a brief account, possibly only applicable to the legumes of temperate agriculture, can be given here.[4]

Actively growing roots of leguminous plants secrete substances into the soil which stimulate the multiplication of nodule bacteria, and related species

1 See, for example, J. M. Vincent, *Proc. Linn. Soc. N.S.W.*, 1954, **79**, 4, who also breaks down the *Medicago-Melilotus* group into three subgroups, species from each genus occurring in two of them.
2 With L't Mannetji, *E. Afr. agric. for. J.*, 1964, **29**, 214.
3 See, for example, O. N. and E. K. Allen, *Soil Sci.*, 1939, **47**, 63, and *Bot. Gaz.*, 1940, **102**, 121, and J. C. Burton, *Proc. Soil Sci. Soc. Amer.*, 1952, **16**, 356.
4 For a detailed discussion see E. K. and O. N. Allen, *Hanbd. Pflanzenphys.*, 1958, **8**, 48, and P. S. Nutman, *Biol. Rev.*, 1956, **31**, 109.

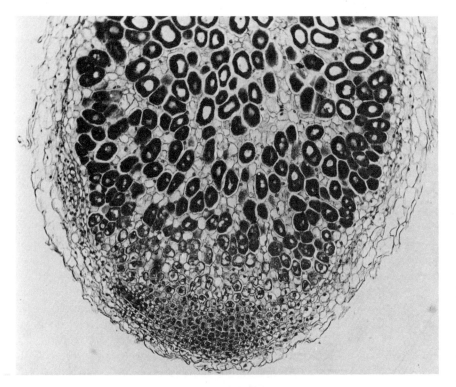

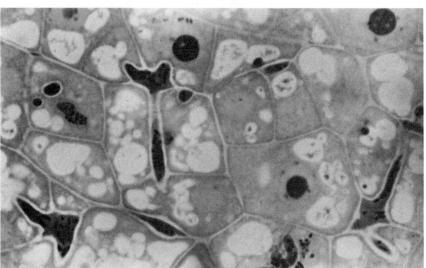

PLATE 18. (*a*) Transverse section of the meristematic end of a lucerne nodule. Note the cells filled with bacteria. (× 180).
(*b*) Infection threads in a root of subterranean clover with some rhizobia released into the cortical cells. The densely stained round bodies in some of the cells are nucleoli. (× 1600).

such as *Agrobacterium radiobacter*, in their rhizosphere.[1] The nodule bacteria. in their turn, produce a substance, probably β indole-acetic acid,[2] possibly from tryptophan excreted by the legume roots.[3] This causes a proportion of the root hairs of leguminous, but not other plants to curl, and if the nodule bacteria belong to a suitable strain, they will enter a proportion of these curled root hairs.

The bacteria grow down a proportion of these root hairs in the form of a fine infection thread made up of a thin line of bacterial cells[4] until they penetrate the inner cortical cells. They will then cause a proportion of the cells they penetrate to start proliferating, and they themselves will proliferate in these cells until they have almost filled them. In the case of red clover (*T. pratense*) only a tetraploid cell can begin proliferation and the proliferating cells are predominantly tetraploid, but it is not known how general this feature is. These proliferating cortical cells remain within the root endodermis and form the nodules as is shown in Plate 17. These changes are illustrated in Plates 18 and 19. The cross section of a lucerne nodule is illustrated in Plate 18a, and it shows the enlarged cortical cells forming the body of a nodule, which contain the bacteria; and Plate 18b, taken at a higher magnification, shows parts of the infection threads between the enlarged cells in which the bacteria are multiplying. Plate 19 shows the cells nearly filled with bacteria.

Nodules themselves differ in shape and size, partly as a response to soil conditions and partly as a characteristic of the particular bacterial strain–plant variety interaction. They may be roughly spherical, cylindrical, flattened and often bidentate or with coralloid branching, or they may have an entirely irregular shape. In general, a nodule will only contain a single bacterial strain.[5]

Once the bacteria have filled a proliferating cell, they change their form into a bacteroid. Plate 19b shows that in the nodules of subterranean clover, the bacteria become enlarged during this change. Only the bacteroid possesses the nitrogen fixing enzymes, but it has lost the ability to multiply and it also appears to have lost most of the ribosomes the cell originally contained. It may therefore have lost its ability to synthesise proteins. It is usually larger than the actively multiplying forms, and in some nodules appreciably larger; and it usually contains granules of poly-β-hydroxybutyric acid, which are probably food reserves for the cell.

The bacteroids in the cortical cells are themselves surrounded by mem-

1 H. Nicol and H. G. Thornton, *Proc. Roy. Soc.*, 1941, **130** B, 32.
2 K. V. Thiman, *Proc. Natl. Acad. Sci. Wash.*, 1936, **22**, 511; *Trans. 3rd Comm. Int. Soc. Soil Sci.*, New Brunswick, A 1939, 24, and H. K. Chen, *Nature*, 1938, **142**, 753.
3 A. D. Rovira, *Pl. Soil*, 1956, **7**, 178.
4 For electron microscope photographs of an infection thread in soyabeans see D. J. Goodchild and F. J. Bergersen, *J. Bact.*, 1966, **92**, 204.
5 This paragraph is based on P. S. Nutman's chapter 10 in *Soil Nitrogen*, eds. W. V. Bartholomew and F. E. Clark. *Agronomy*, 10, 1965.

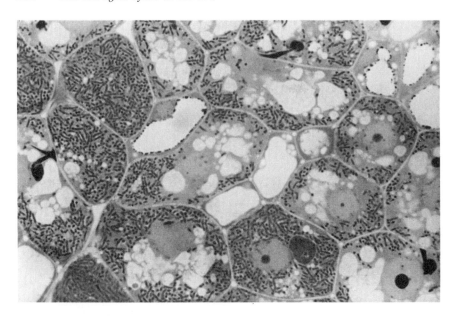

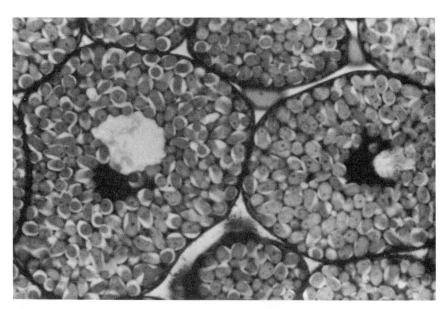

PLATE 19. (*a*) Multiplication of rhizobia in the cortical cells of subterranean clover after release from infection threads. Some of these threads show vesicles at their tip from which rhizobia escape into the cell. (× 1000).
(*b*) The bacteroid zone in subterranean clover. (× 1600).

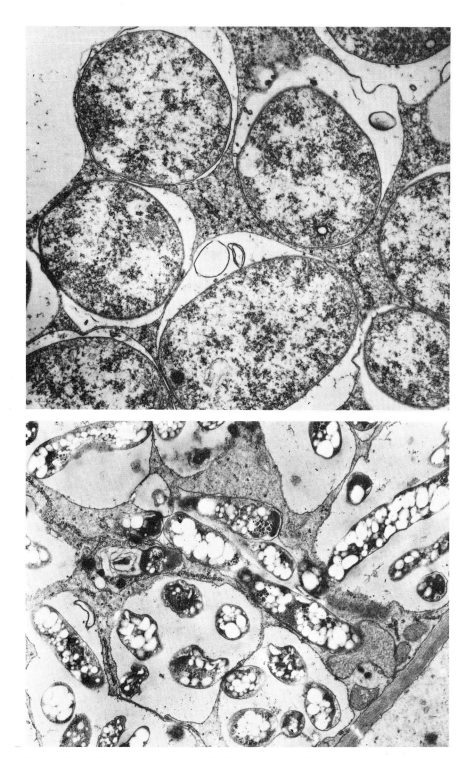

PLATE 20. (*a*) Bacteroids enclosed singly in their membrane; subterranean clover. (× 25 000).
(*b*) Groups of bacteroids enclosed within each membrane envelope; *Phaseolus vulgaris*. (× 34 000).

branes which are part of the host cell. The number of bacteroids enclosed within a membrane depends on the plant species. There is only one bacteroid within each membrane in the clovers, pea, medics and groundnuts, as is shown in Plate 20a; while in the cowpea, soyabean and acacia there are several, usually much smaller bacteroids, within each membrane,[1] as is shown in Plate 20b. Thus in the soyabean, there are four to six within each membrane,[2] and there are about 10^5 such groups within a single cell. If the nodule is a few millimetres in size, it may contain 35 000 cells, giving something over 10^{10} bacteroids per nodule. The bacteroids within their envelopes are bathed in a solution containing leghaemoglobin, which they synthesise: this allows them to function with oxygen concentration of only 10^{-8} M in the bathing solution.[3] It is presumed that there is an interchange of nutrients between the liquid within the envelope and the membrane, the membrane excreting energy substances, possibly fumarate or succinate and absorbing glutamic acid. As the nodule ages the cells lose their leghaemoglobin, become brown in colour and a large vacuole appears in each, crowding the bacteroids and the cell contents into a dense mass against the periphery of the cell wall; and finally the bacteroids break up into bacteria which digest the remaining cell contents and attack the cell wall. The nodule now becomes necrotic, and the bacteria are released into the soil. The period of maximum development and leghaemoglobin content in the nodules of annual legumes is probably just before flowering.[4]

This description of infection and nodule formation only applies if the bacterial strain can form an effective or fully healthy nodule. But this process can fail to be completed for a number of reasons, and since these have been studied in more detail with red clover than with other legumes, the various causes of failure will be described for this species.[5]

The establishment of an effective symbiosis in clover can be blocked by incompatibility at any of the following stages of development:

1 At the primary infection of the root hair.
2 In the growth of the infection thread.
3 In the release of the bacteria from the infection thread.
4 In their multiplication within the cytoplasm of the host cell.
5 In bacteroid formation.
6 In bacteroid persistence.
7 In the functioning of the bacteroids.

In some of these, the cause has been traced to a bacterial defect, in others to

1 P. J. Dart with F. V. Mercer, *J. Bact.*, 1966, **91**, 1314; with M. Chandler, *Rothamsted Rept.* 1971, Part I, 99.
2 F. J. Bergersen and M. J. Briggs, *J. gen. Microbiol.*, 1958, **19**, 482.
3 F. J. Bergersen, *Trans. 9th Int. Congr. Soil Sci.*, 1968, **2**, 49.
4 D. C. Jordan and E. H. Garrard, *Canad. J. Bot.*, 1951, **29**, 360.
5 F. J. Bergersen, *Aust. J. biol. Sci.*, 1957, **10**, 233, and P. S. Nutman, *13th Symp. Soc. gen. Microbiol.*, 1963, 51.

a hereditary defect in the host plant, and in still others it involves a specific interaction between the bacteria and the host plant.

The lack of bacteroid persistence appears to be the commonest cause of the nodules being ineffective. Nodules which are ineffective for this reason differ from effective ones in that they are very much smaller, often short lived and, although they are typically far more numerous than effective ones on the roots, yet the total volume of bacteroid, or red nodular, tissue per plant is very much smaller. However, the rate of nitrogen fixation per unit volume of red tissue appears to be about the same for this type of ineffective nodule as for an effective one.[1] Plate 21 illustrates white clover roots carrying these two types.

Other types of ineffective nodules which can occur in the field have not been fully studied. In equatorial Africa, lucerne at Muguga (Kenya) and groundnuts at Yangambi (Zaire) sometimes have swellings on their roots which look like nodules, but their colour is white or pale green, and they appear to be a proliferation of cortical parenchyma without the cellular differentiation of the normal nodule.[2]

The process of infection and nodule formation can be disturbed if the nitrate or ammonium concentration around the plant roots is too high. A high nitrate concentration reduces the proportion of root hairs that can be infected, though this reduction can in part be counteracted by adding a suitable quantity of glucose to the solution.[3] Nitrate not only reduces the number of nodules, it also decreases their volume. The volume can, in fact, be halved by a concentration of nitrate insufficient to affect the number of nodules.[4]

A legume root can carry nodules formed by several bacterial strains though each nodule usually only contains one bacterial strain, at least for the temperate legumes. The root can normally only carry a limited number of nodules per unit length; hence, if root growth ceases fairly early in the season, as it may do with peas, for example, the root system can become saturated with nodules, and once this has happened no further bacteria of any other strain can produce additional nodules on the root. Root saturation will not be shown by plants such as clovers whose root system continues to develop through much of the growing season.

The number of nodules produced on unit length of root depends on the bacterial strain, the genetic constitution of the plant and the density of plant roots in the soil. P. S. Nutman[5] has shown that clovers possess several genes which determine the number of nodules a given bacterial strain can produce and the period in the growth of the plant when they are produced; and he

1 H. K. Chen and H. G. Thornton, *Proc. Roy. Soc.*, 1940, **129** B, 208.
2 C. Bonnier, *I.N.E.A.C. Sci. Ser.*, *Publ.* 72, 1957.
3 H. G. Thornton, *Proc. Roy. Soc.*, 1936, **119** B, 474.
4 H. G. Thornton and H. Nicol, *J. agric. Sci.*, 1936, **26**, 173.
5 P. S. Nutman, *Ann. Bot.*, 1948, **12**, 81, and 1949, **13**, 261; *Heredity*, 1949, **3**, 263; *Proc. Roy. Soc.*, 1952, **170** B, 176.

also found that clovers selected for abundant nodulation produce a larger number of lateral roots than those selected for sparse nodulation, whether or not the plants were nodulated. He also showed that the number of nodules produced on a root of clover or lucerne is greater on plants grown singly than in pairs or larger groups,[1] and that this depressing effect of high root density could be reduced if charcoal were added to the rooting medium. When this charcoal was extracted with suitable organic solvents it was found that it had absorbed substances secreted by these roots which when added to the rooting medium sometimes depressed but sometimes stimulated nodulation.

The full significance of the role of charcoal around the roots on nodulation is not yet known. J. T. Vantsis and G. Bond[2] earlier noted that $\frac{1}{2}$–2 per cent of activated charcoal added to sand cropped to peas increased the efficiency of nitrogen fixation, though in this example the number of nodules was diminished though their size was increased. They also obtained evidence that several substances were being excreted, for if animal charcoal was used the growth of the peas was depressed. Nor is this effect confined to the laboratory, for A. J. Anderson and D. Spencer[3] in Australia, and C. Bonnier[4] in Zaire (Belgian Congo) have both noted that clovers in one case, and *Puereria javanica* in the other, were much better nodulated, and the plants were stronger with a darker green colour where wood charcoal was present in the soil; and Bonnier also noted that charcoal was not necessary, for the roots were also well nodulated when they grew through rotten wood.

Under some conditions, some of the soil micro-organisms appear to excrete substances into the soil which prevent both the rhizobia and the rhizosphere bacteria from multiplying on the surface of legume roots, and this toxic principle can be de-activated by heat and is probably absorbed on the charcoal.[5] If this observation is correct, it may well supply part of the explanation of this beneficial effect of charcoal. Bentonite or fuller's earth in the rooting medium also can have the same stimulating effect on nodulation and growth, but the clay close to the root changes colour, going blue-green close to a red clover root, faint orange near a lucerne, and orange or brown near a vetch. This colour change is not confined to leguminous roots, for flax roots coloured the bentonite saffron yellow; nor is it confined to nodulated red clover plants, for un-nodulated plants also produce it; nor do all inoculated red clover plants produce it. The colour developed depends on the pH, but it is not known what the compounds responsible for these colours are.[6]

1 *Ann. Bot.*, 1953, **17**, 95, and 1952, **16**, 79, and E. R. Turner, *Ann. Bot.*, 1955, **19**, 149. For a possible similar result for soyabeans see G. H. Elkan, *Canad. J. Microbiol.*, 1961, **7**, 851.
2 *Ann. appl. Biol.*, 1950, **37**, 159.
3 *J. Aust. Inst. agric. Sci.*, 1948, **14**, 39.
4 *I.N.E.A.C. Sci. Ser., Publ.* 72, 1957.
5 F. W. Hely, F. J. Bergersen and J. Brockwell, *Aust. J. agric. Res.*, 1957, **8**, 24.
6 P. S. Nutman, *Nature*, 1951, **167**, 288, and E. R. Turner, *J. Soil Sci.*, 1955, **6**, 319.

Nodules already present on the roots of clover can affect the number of new nodules produced. Roots infected by ineffective strains of bacteria carry far more but much smaller nodules than those infected by effective strains. This is illustrated in Plate 21, which shows the root system of red clover inoculated with an effective and an ineffective strain of bacteria. Initially nodulation is as rapid with either strain, but the effective nodules inhibit further nodulation as soon as nodule growth is properly started.[1]

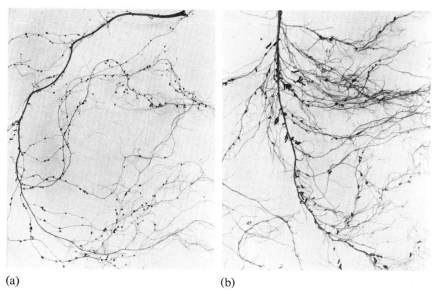

(a)   (b)

PLATE 21. (*a*) Nodules formed by an ineffective strain (Coryn Strain) of the nodule bacteria on red clover roots.
(*b*) Nodules formed by an effective strain (Strain A) of the nodule bacteria on red clover roots.

The longevity of nodules

The amount of nitrogen fixed by a leguminous crop depends very largely on the longevity of the nodules on its roots. Four factors affect longevity: the physiological condition of the plant, the moisture content of the soil, parasites in the nodule, and the strain of bacteria forming the nodule.

The longevity of a nodule depends on the physiological condition of the plant. The nodules of annual plants tend to die at flowering and seed set, presumably because at this time the flowers and developing seeds are drawing on the carbohydrate reserves of the plant very heavily, and the young seeds

1 P. S. Nutman, *Ann. Bot.*, 1949, **13**, 261, and 1952, **16**, 79.

may also be drawing on the nitrogen compounds in the nodules.[1] Again the cutting or hard grazing of clovers, for example, may cause the death of the nodules, presumably because the carbohydrate supply to the nodules is interrupted. Perennial legumes differ appreciably in the longevity of their nodules. Perennial clovers, for example, have nodules that are normally only short-lived and they will be shed by winter. Legumes that are shrubs or trees may, however, carry nodules for several years, and this group may include some herbaceous legumes such as lupins.[2]

Nodules only seem to remain on the roots of many leguminous crops if the soil is kept moist, and the first effect of the onset of drought is for the crop to shed its nodules; though unfortunately no systematic work has been done on the moisture deficit in the plant, or the suction of the water in the soil, at which shedding is severe. In the laboratory J. I. Sprent[3] found that soyabean nodules cease to fix nitrogen when their fresh weight falls below 80 per cent of their fully turgid value. This effect is very noticeable on exotic legumes in parts of Africa, for even quite moderate breaks in the rains can cause the nodules of exotic legumes to be shed.

Nodules can also be short-lived through being parasitised by the larvae of insects. Thus E. G. Mulder[4] in Holland, and G. B. Masefield[5] in Great Britain, found that the nodules of pea and field bean crops can be heavily attacked by the larvae of the pea weevil *Sitona lineata*, and Masefield[6] showed that the severity of this attack could be reduced by irrigation in dry seasons, so this practice prolongs the life of the nodule.

The number of *Rhizobia* in the soil

P. S. Nutman at Rothamsted has found no clear relation between the number of leguminous plants of any particular species growing in a soil and the number of *Rhizobia* that will form nodules on their roots; but the bacterial numbers are affected by soil conditions particularly by the pH of the soil if in the acid range. Thus, on the Park Grass plots at Rothamsted,[7] which have been under permanent grass for several centuries, there are very few *Rhizobia* in soils more acid than pH 4·2, the numbers tend to increase up to a pH of about 7, and are independent of the pH in the range 7·0 to 7·8. However, the numbers are extremely variable from plot to plot. Thus the numbers of *R. trifoli* in the rhizosphere of white clover, in the top 2·5 cm of the soil, and in the 7·5 to 10 cm depth varied in the range 10^8 to 10^{11}, 10^4 to 10^8, and

1 See J. S. Pate (*Aust. J. biol. Sci.*, 1958, **11**, 366, 496) for a description of this on the roots of field peas and field vetch.
2 J. O. Harris, *Amer. J. Bot.*, 1949, **36**, 650.
3 *New Phytol.*, 1971, **70**, 9.
4 *Pl. Soil*, 1948, **1**, 179.
5 *Emp. J. exp. Agric.*, 1952, **20**, 175.
6 *Emp. J. exp. Agric.*, 1961, **29**, 51.
7 With G. J. S. Ross, *Rothamsted Rept.*, 1969, Part II, 148.

10^5 to 10^6 per gram soil for soils with a pH of 7 or over, and of *R. leguminosarum* 10^4 to 10^{10} in the rhizosphere and 10^4 to 10^6 in the soil at each depth. *R. lupini* differed from the other two in that it was more common in the soils with a pH below 7, and its numbers were between 10^3 and 10^9 in the rhizosphere and 10^2 and 10^7 in the soil at each depth. These bacteria only occurred in plots where the appropriate host legume also occurred, and, in particular, there are no medics in this field and no *R. meliloti*.

The picture is not so clear for arable soils, but in general the rhizobial numbers are very much smaller. Land which had been in root crops since 1843, except for three years in barley during the years 1853 to 1855, contained between 10 and 10^3 *R. leguminosarum* in February 1967. It was then sown to beans (*Vicia faba*) and though the number of *Rhizobia* rose to 10^5 per gram soil in the rhizosphere by May, the number in the bulk of the soil had not altered. This field also contained about the same number of *R. trifoli*, but only small numbers of *R. lupini* and *R. meliloti*. But Broadbalk, which also contained, on a geometric mean average, about 3×10^4 and 3×10^2 *R. leguminosarum* and *R. trifoli*, in spite of a very low number of suitable hosts, only contained about 13 *R. meliloti* per gram of soil in spite of *Medicago lupulina* being a very common weed in two plots.[1]

There can also be varieties in the strains of a given species in a soil sample, particularly if the soil is acid. Thus on the Park Grass, in plots more acid than pH 5·5 only a proportion of the bacteria would form effective nodules, and the rest gave relatively ineffective nodules. This result has also been found by C. L. Masterson[2] in Ireland, for his *R. trifoli* isolates from mineral soils of varying pH tended to be relatively ineffective from the more acid soils. This result is in accord with the common experience in Great Britain that the bacteria nodulating the few clover plants on acid upland pastures appear to be ineffective, but that if the land is drained and given a good dressing of basic slag the clovers multiply and the bacteria in their nodules are effective.[3] The unimproved soils must contain a small proportion of effective strains which can only multiply when the soil conditions are improved.

The inoculation of plants with *Rhizobia*

A soil may contain no suitable *Rhizobium* for an introduced legume which is otherwise suited to the area. Thus, in Great Britain *R. meliloti* is a relatively rare organism, and most of those present in the soil are not effective on lucerne, yet lucerne is otherwise well adapted to many neutral and calcareous soils in the south and east of England. Similarly, soyabeans are well adapted to large areas of the Mid-West of the United States, except that the soils do

1 *Rothamsted Rept.* 1968, Part II, 179.
2 *Trans. 9th Int. Congr. Soil Sci.*, 1968, **2**, 95.
3 M. P. Read, *J. gen. Microbiol.*, 1953, **9**, 1. See also A. J. Holding and J. King, *Pl. Soil*, 1963, **18**, 191 for Scottish hill pastures.

not naturally contain a suitable nodule bacteria. In many parts of Australia, also, subterranean clover has been found very well suited as a pasture legume over large areas of the Continent, except again that the soils do not contain suitable nodule bacteria. It has therefore been necessary to develop methods for inoculating the soil with suitable strains of these organisms.

In practice the appropriate rhizobial strain is inoculated onto the seed usually shortly before it is sown. It is now common practice either to mix the seed with a peat or a compost carrying a very high population of the bacteria, or to incorporate the bacterial suspension with finally divided calcium carbonate if the *Rhizobium* is a fast-growing acid-producer, but otherwise with a finely ground rock phosphate[1] and pellet the seed with this mixture using gum arabic as an adhesive. The seed-coats of some legumes such as some varieties of subterranean clover and *Centrosema pubescens* contain a diffusible substance that inhibits the growth of a wide variety of bacteria, including *Rhizobia*,[2] which complicates the problem of inoculation, but it is possible that either pelleting or peat or compost inoculum may absorb these substances.[3] However, these introduced *Rhizobia* often do not last in the soil for long after the crop has been removed from the land, so in general these introduced leguminous crop must be re-inoculated each time they are sown.

The choice of rhizobial strain for inoculation is not quite straightfoward, for several strains may be equally effective when tested under laboratory conditions, but differ greatly in performance in the field. Part of the reason is that strains can differ greatly in their ability to infect roots in the field, which is a property difficult to evaluate in pure culture work in the laboratory. As an example of the problem J. M. Vincent[4] compared the effect of inoculating subterrancan and crimson clover (*T. incarnatum*) in nine field trials with five strains which were all equally effective in the laboratory and found one strain gave nodules on 42 per cent of the young plants and another only 9 per cent, with corresponding effects on the dry matter yields. The relative order of the strains, however, depended somewhat on the soil type. J. Brockwell and his colleagues,[5] for example, have shown that for subterranean clover on the south-eastern side of Australia, there is no one strain of inoculum that is best for the whole region and for the different varieties of the clover. For maximum response to the inoculation, the strain must be matched to the variety, the climate and the soil; and there is no reason to doubt that this finding is of wide validity if a new leguminous species is to be introduced on the continental scale.

1 For a review see J. M. Vincent, Chap 11 in *Soil Nitrogen*, Eds, W. V. Bartholomew and F. E. Clark, *Agronomy*, 10, 1965.
2 G. D. Bowen, *Pl. Soil*, 1961, **15**, 155.
3 See, for example, J. A. Thompson, *Aust. J. agric. Res.*, 1961, **12**, 578, and K. P. John, *J. Rubb. Res. Inst. Malaya*, 1966, **19**, 173.
4 *7th Int. Grassland Conf.*, 1956.
5 With W. F. Dudman *et al.*, *Trans. 9th Int. Congr. Soil Sci.*, 1968, **2**, 103.

The conditions needed for successful inoculation are well understood for most soils. The most important condition is that the surface soil should remain moist from the time of sowing until nodules begin to form, for if the soil becomes dry in this period, all the introduced bacteria are likely to be killed. Pelleting the seed has been found of great value in areas of erratic rainfall. High soil temperatures are also unfavourable, so the seed should be sown well before the hot weather arrives. In some soils, particularly where there is a high population of bacteria capable of forming ineffective nodules, high bacterial numbers may be needed for each seed: J. A. Ireland and J. M. Vincent[1] have given an example where 10^6 bacteria were needed per seed of subterranean clover to ensure that its nodules were all effective on some red basaltic soils in New South Wales, whereas only 100 per seed are needed under good conditions.[2]

Inoculation may become of importance in some tropical legumes. C. Bonnier,[3] working at Yangambi, Zaire, found that groundnuts carried a rather sparse and erratic population of nodules early in the season, which only became abundant towards the end when it was too late for them to make a full contribution to the harvest yield. However, he found a strain that would give much more vigorous nodulation early on, and this gave an appreciably better plant with a higher yield. This late nodulation of several tropical legumes may be the reason why they often respond to a dressing of nitrogen fertiliser given in the seedbed.

The amount of nitrogen fixed by leguminous crops in the field

The amount of nitrogen fixed per hectare by a leguminous crop depends on the number of nodules per hectare, their size and longevity, and the bacterial strains in them. In turn, it also depends on the conditions of growth and management of the crop, and in particular on the availability of water and the nutrient status of the soil. Little is known about the effect of drought on the rate of fixation of nitrogen by a leguminous crop, except that the nodules are shed from the roots of many legumes shortly after the onset of a drought, and this effect is particularly marked for a number of leguminous crops grown in tropical and subtropical regions. This effect is probably important even in southern England, for A. J. Low and E. R. Armitage[4] considered that, in the area in which they worked, a grass-clover ley needed about 22 cm of irrigation water to give a dry matter yield of 10 tons per hectare as hay, whilst a grass ley given 50 kg/ha N gave this yield without any additional water.

1 *Trans. 9th Int. Soc. Congr. Soil Sci.*, 1968, **2**, 85.
2 R. A. Date, *Trans. 9th Int. Congr. Soil Sci.*, 1968, **2**, 75.
3 *I.N.E.A.C., Sci. Ser. Publ.* 72, 1957; 76, 1958.
4 *J. agric. Sci.*, 1959, **52**, 256.

The rate of nitrogen fixation should be dependent on the total volume of active nodule tissue the crop is carrying, but very few field observations have been made on this. G. B. Masefield[1] made some studies on the weights of nodules carried by some crops both in temperate and tropical regions, though he worked on the weight of nodules per plant rather than per acre. On the whole British crops tend to have a higher weight of nodules per plant than tropical crops, possibly because of the lower soil temperatures and higher soil moistures. Thus broad beans (*Vicia faba*) and dwarf beans (*Phaseolus vulgaris*) in England were found to carry 4·5 g nodules per plant, while groundnuts in West Africa had between 1·5 and 3 g, soyabeans in Malaya up to 3·3, and cowpeas up to 4 g; and plants having these weights of nodules typically have between 100 and 1000 nodules, so the individual nodules vary between 1 and 40 mg in weight. Some plants such as lupins in Kenya can have nodules weighing very much more than this, but they will have very few per plant and probably they are relatively inefficient per gram of nodule, as they usually show a large volume that is only pale pink in colour.

Nitrogen fixation can only go on actively if the crop is healthy and the nutrient supply adequate. A good supply of calcium or magnesium, potassium, phosphate, sulphate, borate, molybdate and cobalt is essential, though different leguminous crops have very different requirements. As an example among the temperate legumes white clover and lupins can be grown on acid soils while red clover and lucerne need neutral or calcareous soils for vigorous growth; and among tropical and subtropical legumes soyabeans and *Stylosanthes gracilis* grow well on acid soils while *Desmodium uncinatum* and *Centrosema pubescens* have appreciable calcium demands.[2] Again, there is a great variation in the levels of phosphate required for maximum growth. Thus J. E. Begg[3] found some of the indigenous legumes on the eastern side of Australia did not respond to phosphorus or sulphur added to the soils in which they were growing, but introduced legumes, such as white clover (*T. repens*), gave a very appreciable response, and much higher yields than the indigenous on fertilised soils, a result also found for introduced legumes that had become naturalised in these areas. In addition, an adequate ammonium or nitrate supply is necessary for rapid establishment of the crop from seed, particularly if the seed is small; and if a leguminous crop is to be sown on worn-out land, it is nearly always desirable to give an initial dressing of a nitrogen fertiliser to help it get established.

While a high level of soil nitrates may depress the amount of nitrogen fixed by crops such as clovers and lucerne, it may be essential for high yields of crops such as soyabeans and groundnuts. Some results obtained by A. H. Bunting on the black cotton soil of the Rainlands Research Station, Sudan,

1 *Emp. J. exp. Agric.*, 1952, **20**, 175; 1955, **23**, 17; and 1957, **25**, 139.
2 See, for example, C. S. Andrew and D. O. Norris, *Aust. J. agric. Res.*, 1961, **12**, 40, and C. T. I. Odu, A. A. Fayemi *et al.*, *J. Sci. Fd. Agric.*, 1971, **22**, 57.
3 *Aust. J. exp. Agric. anim. Husb.*, 1963, **3**, 17.

illustrate this very clearly. Table 16.9 shows that groundnuts responded to 50 kg/ha of N, in this particular year very profitably, but sorghum only responded to 25 kg/ha N and sesame did not respond at all. P. H. Le Mare[1] has also shown that groundnuts responded to nitrogen at Kongwa and Nachingwea in Tanzania, although oddly enough not on the nitrogen-deficient soil of Urambo. Unfortunately, no observations were taken on the nodulation, or of the effectiveness of the nodules, on these crops. However, J. R. Seeger[2] found that at Yangambi, Zaire, a nitrate fertiliser would increase both the nodulation and the yield of groundnuts, and F. O. C. Ezedinma,[3] at Ibadan, W. Nigeria, also found the same result for cowpeas (*Vigna sinensis*).

TABLE 16.9   Response of groundnuts, sorghum and sesame to nitrogen. Rainlands Research Station, Sudan, 1954–55. Yields: grain or seed in 100 kg/ha. Nitrogen given as ammonium nitrate

	kg/ha N				
	0	*25*	*50*	*75*	*100*
Groundnuts	8·9	10·4	12·2	12·7	13·1
Sorghum	25·0	29·8	29·4	29·4	26·8
Sesame	5·4	5·3	5·3	4·9	5·4

The reasons for this response to nitrogen have not been worked out, but in the examples quoted for groundnuts, it may have been due to poor nodule development due to the strain of Rhizobium present in the soil. But it could also be due to an effect of bright sunlight, for E. B. Fred[4] at Wisconsin has shown that the amount of nitrogen fixed by soyabean plants growing in pots and exposed to bright sunlight increased as the amount of nitrogen fertiliser added to the pots was increased, yet if the plants started in the shade until well established, and then transferred to the sun, even in the absence of added nitrogen, the plants were fixing nitrogen very actively. This is illustrated in Table 16.10. Alternatively, it could be due to the depressing effect of a high soil temperature on nodule fixation. Thus D. R. Meyer and A. J. Anderson[5] showed that the nodules of subterranean clover fixed nitrogen actively at 20°C and at its maximum rate at 25°C, but hardly at all at 30°C, and they give reasons for assuming this effect is not due to carbohydrate starvation in the nodule. J. C. Lyons and E. B. Earley[6] also found

1 *Emp. J. exp. Agric.*, 1959, **27**, 197.
2 *Bull. Inst. agron. Gembloux*, 1961, **29**, 197.
3 *Trop. Agric. Trin.*, 1964, **41**, 243.
4 *Proc. Natl. Acad. Sci. Wash.*, 1938, **24**, 46.
5 *Nature*, 1959, **183**, 61. See also A. H. Gibson, *Aust. J. biol. Sci.*, 1963, **16**, 28 for confirmatory experiments.
6 *Proc. Soil Sci. Soc. Amer.*, 1952, **16**, 259.

TABLE 16.10   Effect of nitrogen supply and shade on nitrogen fixation by soyabeans at Madison, Wisconsin. 5–8 plants per pot. Sown, 1/7/37. Harvested, 30/8/37. Nitrogen added or pots put in shade, 28/7/37

Nitrogen added per pot as $Ca(NO_3)_2$ mg	Where pot kept	Dry matter in plants, g	Per cent nitrogen in dry matter	Nitrogen fixed, mg
0	in sun	4·2	1·14	2·6
5		20·7	1·26	35·6
10		28·5	1·28	43·7
25		30·8	1·45	63·3
50		37·7	1·72	67·7
0	in shade	8·9	2·20	25·6
0	in shade till 9–8–37 then in sun	18·7	1·98	54·7

that in a hot dry season in Illinois, nitrogen fertiliser increased the yields of soyabeans, but it had little effect in a cool moist one, though this was probably due to the much poorer nodulation in the former than the latter seasons.

The actual amounts of nitrogen fixed by leguminous crops in the field is difficult to estimate, because of the difficulty of determining accurately the nitrogen content of a soil on the one hand, and the amount of denitrification taking place during the growing season. Some typical figures have been given by T. L. Lyon and J. A. Bizzell[1] at Cornell. They found that two courses of a rotation consisting of one-year clover cut for hay, and four years' grain crops added 220 to 340 kg/ha of nitrogen to the soil and 340 to 450 kg to the harvested crops, compared with similar rotation in which timothy grass replaced the clovers, indicating that each clover crop had fixed between 280 and 390 kg/ha of nitrogen, and this is comparable with the amount of nitrogen other workers have found a good crop of lucerne can fix in a year.[2] Soyabeans, field beans (*Vicia faba*) and peas harvested for grain on the other hand, depleted the soil of nitrogen as much as ordinary cereal crops: they fixed between 110 and 220 lb. of nitrogen, but it all appeared in the harvested material. Table 16.11 gives some of their results for five courses of a two-year rotation of legume-cereal,[3] the cereal being either rye or barley, and it shows that the yield of the following cereal crop largely depends on the amount of nitrogen the legume adds to the soil.

These results show in the first place that leguminous crops fix very varying amounts of nitrogen, in this particular experiment from about 80 to 500

1 *J. Amer. Soc. Agron.*, 1933, **25**, 266; 1934, **26**, 651.
2 See, for example, R. C. Collison, H. G. Beattie and J. D. Harlan, *New York Agric. Expt. Sta., Tech. Bull.* 212, 1933; E. S. Hopkins and A. Leahey, *Canada Dept. Agric., Publ.* 761, 1944.
3 T. L. Lyon, *Cornell Agric. Expt. Sta., Bull.* 645, 1936.

TABLE 16.11   Amount of nitrogen fixed by leguminous crops, and their influence on a following cereal crop. Mean of two experiments, each five courses of legume-cereal (rye or barley) rotation

| | Nitrogen harvested in | | Gain or loss of nitrogen in the soil per rotation kg/ha | Total nitrogen fixed by legume kg/ha | Yield of cereal grain 100 kg/ha |
	Leguminous crop kg/ha	Cereal crop kg/ha			
Lucerne	335	74	136	505	29·2
Clover	140	57	129	290	24·4
Sweet clover	190	57	94	300	23·7
Soyabeans	197	32	−9	180	14·8
Field beans	115	28	−22	80	13·3
Cereal every year	—	25	−11	—	10·9

kg/ha, and even the figure of 80 kg/ha is probably high for some crops, for A. G. Norman[1] found that soyabeans growing on Iowan prairie soils and giving good yields, often fixed no more than 20 to 25 kg of nitrogen, though if this crop is grown under conditions of low available nitrogen in Mississippi it may fix between 100 and 150 kg/ha.[2] The second result established by these experiments is that because a leguminous crop fixes nitrogen it need not enrich the soil in nitrogen. There is a general tendency for leguminous crops grown for their seed—peas, field beans, soyabeans and groundnuts—to reduce the nitrogen content of the soil, and legumes grown for their leaf—clovers, sweet clovers and lucerne—to increase the nitrogen content, though they do not necessarily.[3] Naturally if the large-seeded legumes are grown as a green manure and ploughed in, all the nitrogen they fix is returned to the soil. However, there may be exception to this rule, for the statement is continually being made in farming journals and textbooks that field beans grown for seed enrich the soil in nitrogen.

Tropical legumes that are either not harvested or used for hay or grazing can also give high rates of fixation. G. H. Gethin Jones[4] found that land rested under *Glycine javanica*, a creeping legume, which was not harvested increased the nitrogen content of a Kikuyu red loam outside Nairobi, Kenya, at the rate of 180 kg/ha nitrogen annually for the first five years of rest and at 110 kg/ha annually for the second five year period. E. R. Orchard and G. D. Darby[5] found the soil under a wattle plantation (*Acacia mollissima*)

1 *Proc. Soil Sci. Soc. Amer.*, 1947, **11**, 9.
2 J. L. Cartter and E. H. Hartwig, *Adv. Agron.*, 1962, **14**, 391.
3 For an example from Utah where the nitrogen content of land under lucerne for 16 years remained about the same, see J. E. Greaves and L. W. Jones, *Soil Sci.*, 1950, **69**, 71.
4 *E. Afr. agric. J.*, 1942, **8**, 48.
5 *Trans. 6th Int. Congr. Soil Sci.*, 1956, D, 305.

in Natal increased the nitrogen content at the average rate of 200 kg/ha over a thirty year period, and this was on a strongly weathered leached acid (pH 4·8) soil initially of low nitrogen content. A. W. Moore[1] at Ibadan, Western Nigeria found *Centrosema pubescens* growing with a *Cynodon* grass in a grazed pasture increased the nitrogen content of the top 30 cm of the soil by over 600 kg/ha compared with the soil under the pure grass during the two-year period the pasture was down for.

The processes by which leguminous crops add organic nitrogen compounds to the soil are difficult to study in detail. On the whole crops which have a very extensive root system, particularly a system continually sending out new roots which can later carry nodules, appear to leave more nitrogen behind in the soil than those with a restricted root system, particularly when nearly all the nodules are formed during a relatively short part of the growing season. Clovers and lucerne in the temperate regions, for example, may add between 150 and 200 kg/ha annually to the soil, and they are making new roots which become nodulated throughout the growing season. As a consequence they can add between 5 and 8 tons of root fibre per hectare to the soil annually and this can easily contain the 150 to 200 kg of nitrogen they add to the soil. Peas and beans, on the other hand, have a far smaller root system, so the weight of root fibre is very much less, and the nodules are nearly all formed within a few weeks. Further, very few experiments seem to have been made on the proportion of the nitrogen fixed by a leguminous crop that remains in its root system, but in some pot experiments by J. S. Pate[2] in Northern Ireland, up to 90 per cent of the nitrogen in the nodules of peas and vetches had been transferred to the tops before the plant died, a result confirmed by I. E. Miles[3] in Mississippi, who further found that the subtropical summer forage legumes—velvet beans (*Stizolobium deeringianum*), lespedeza and kudzu (*Pueraria phaseoloides*)—retained 20 to 30 per cent of the nitrogen they fixed in their root system.

Another factor which affects the amount of nitrogen a legume adds to the soil is the rate of mineralisation of the nitrogen in its root residues. The roots of annual legumes, such as Korean lespedeza grown as a summer crop in areas with moist warm winters, may decompose so rapidly that all the nitrogen in their roots has been converted into nitrates and leached out of the soil by the time the next summer crop is planted,[4] whereas the roots of a perennial crop, such as lucerne, may leave a higher proportion of the nitrogen they contain as soil organic nitrogen.

There is little doubt, however, that leguminous crops can add nitrogen to the soils by processes other than the death of their root systems. One process is the sloughing-off of dead nodules from the roots; this occurs towards the

1 *Emp. J. exp. Agric.*, 1962, **30**, 239.
2 *Aust. J. biol. Sci.*, 1958, **11**, 366, 496.
3 *Miss. Agric. Expt. Sta., Circ.* 126, 1946.
4 P. E. Karrakar, C. E. Bortner and E. N. Fergus, *Kentucky Agric. Expt. Sta., Bull.* 557, 1950.

end of the growing season, and can be encouraged by cutting a grass-legume mixture for hay or silage.[1]

A second process that may sometimes occur is the excretion by nodules of soluble organic compounds, such as the amino acids, aspartic acid and β-alanine.[2] This excretion has been very difficult to prove both in the laboratory and in the field, but it seems to occur when nitrogen fixation is taking place more rapidly than carbohydrate synthesis.[3]

Pot experiments, in which the legume and non-legume are grown in a nitrogen-poor sand, however, can give clear examples of the non-legume using nitrogen fixed by the legume, and although the experiments cannot prove that any of the nitrogen used by the non-legume was excreted by the legume, this is the simplest explanation of the results. Thus H. G. Thornton and H. Nicol[4] at Rothamsted grew lucerne and Italian ryegrass together in sand, and found at the end of July, four months after sowing, that the lucerne, which contained 1000 mg of nitrogen, had apparently transferred 80 mg to the ryegrass, and a month later, when it contained 1300 mg it had transferred 250 mg to the grass. They considered it was unlikely any appreciable amount of nitrogen had been transferred as a result of the death and decomposition of nodules. This result, however, probably only holds if the combined crop is grown in a sand or a sand-soil mixture low in available nitrogen, and even then one may have to choose the leguminous crop carefully.

It is very difficult to determine if and when any transfer of nitrogen from a legume to a non-legume occurs in the field, when arable cropping is used. Table 16.12, which gives the result of a field experiment at Rothamsted[5]

TABLE 16.12   Yields and nitrogen contents of different oats-vetch mixtures. Field experiment at Rothamsted

Weight of seed sown kg/ha		Yield of dry matter tons/ha			Per cent nitrogen in dry matter		Yield of nitrogen kg/ha		
Oats	Vetches	Oats	Vetches	Total	Oats	Vetches	Oats	Vetches	Total
220	—	5·37	—	5·37	1·14	—	61·5	—	61·5
165	55	4·60	1·38	5·98	1·33	2·70	61·0	37·3	98·3
110	110	3·83	2·31	6·14	1·29	2·72	49·3	63·0	112·3
55	165	2·72	3·00	5·72	1·40	2·86	38·0	86·0	114·0
—	220	—	3·58	3·58	–	3·06	—	111·0	111·0

1 For an example see J. R. Simpson, *Aust. J. agric. Res.*, 1965, **16**, 915.
2 A. I. Virtanen. For an account of his work on this subject, see his book *Cattle Fodder and Human Nutrition*, Cambridge, 1938.
3 For a discussion, see P. W. Wilson, *The Biochemistry of Symbiotic Nitrogen Fixation*, Madison, 1940.
4 *J. agric. Sci.*, 1934, **24**, 269, and see also J. Nowotnowna, ibid., 1937, **27**, 503.
5 *Rothamsted Ann. Rept.* 1932, 148.

using a mixture of oats and vetches, shows the typical results that have been found in temperate regions. Mixed cropping causes the yield per acre of both the oats and the vetches to be reduced, though the total yield per acre is increased for certain mixtures. It reduces the total uptake of nitrogen by the oats but increases the nitrogen content of its dry matter, while reducing that in the vetches. This result does not prove that no transfer of nitrogen from the legume to the cereal takes place, but the simplest explanation is that the legume, by fixing its own nitrogen, does not need to draw on the soil nitrates, so that as the number of cereal plants is reduced, the nitrate supply per plant is increased. Further, the higher the number of cereal plants, the more the legume is shaded and hence the amount of nitrogen fixed per plant is reduced, although if any excretion of nitrogen did take place, shading should encourage it.

The principal process of transfer in pastures of nitrogen from legumes to grasses growing in association with them is through the urine and dung of the grazing animal, and this transfer begins in the first grazing after the leguminous nodules have started fixing nitrogen at the beginning of the grazing season. The magnitude of this transfer is illustrated in Table 16.13,

TABLE 16.13   The distribution of clover nitrogen between clover and grass. Palmerston North. Yield of nitrogen in kg/ha

| Age of sward | *Mown* | | | | *Grazed, urine and dung returned* | | | |
	N in grass	N in clover	Total N	Per cent harvested N in grass	N in grass	N in clover	Total N	Per cent harvested N in grass
First year	123	462	585	21	285	258	543	51
Second year	152	434	586	26	309	350	659	47
Third year	211	388	599	35	425	304	729	58
Grass alone without clover	About 60 kg per year				About 90 kg per year			

taken from some work of J. Melville and P. D. Sears[1] with a ryegrass–white clover pasture at Grasslands, Palmerston North, New Zealand. This table shows how as the sward on the mown area ages, so a greater proportion of the nitrogen fixed by the clover appears in the grass, although the total amount of nitrogen harvested annually remains constant at about 590 kg/ha. This result, therefore, does not support the theory that the clover roots are excreting soluble nitrogen compounds into the soil, though it does not dis-

1 *N.Z. J. Sci. Tech. A*, 1953, **35**, *Suppl.* 1, 30.

prove it.[1] The results do, however, show that the clover is fixing 530 kg/ha of nitrogen per year, which is harvested in the grass and clover leaf, and if, as T. W. Walker[2] suggests, as much as 50 per cent of what appears in the tops is left behind in the soil as grass and clover roots, the white clover must be fixing about 800 kg/ha of nitrogen annually. In the grazed series, with the return of urine, a greater proportion of the nitrogen fixed by the clover appears in the grass, as the urine is an effective nitrogen fertiliser.

These rates of nitrogen fixation by white clover are much higher than are usually considered operative in the United Kingdom, possibly because white clover grows for nearly every month of the year in New Zealand. But some experiments made by W. Holmes and D. S. MacLusky[3] at the Hannah Dairy Research Institute, Ayr, on a very well-fertilised ryegrass and white clover sward, receiving no nitrogen fertiliser but repeatedly mown for high-protein dried grass and a corresponding pure grass sward receiving varying levels of a nitrogen fertiliser, also repeatedly mown for dried grass, showed that the yield of dry matter and of protein in the grass of the grass-clover sward was only equalled on the grass-alone plots if it received 180 kg and 320 kg/ha respectively of fertiliser nitrogen, given as a number of dressings throughout the growing season. These experiments indicate, therefore, that even in the United Kingdom very large transfers of nitrogen from clover to grass can occur, and R. E. Hodgson[4] at Beltsville has found very similar results there.

This discussion shows, therefore, that whilst there is very little evidence that crops such as peas and vetches can transfer any nitrogen to a cereal crop growing with them, there is very firm evidence that clover, in a well-managed grass-clover sward, can transfer over 100 kg/ha of nitrogen to the grass annually, probably through decomposition of its nodules and ephemeral roots. This figure is naturally only found under very good conditions, and transfers of the order of 20 to 50 kg is probably more common in English leys, though in many cases the transfer is less and may be negligible.[5] In normal pastures, which are grazed and not mown, the transfer is higher, through the return of the urine of the grazing animal.

The residual value of a leguminous crop

A leguminous crop, particularly if it is a pasture legume, will increase the level of available nitrogen in the soil, and the following crop will respond to this nitrogen. This is illustrated in Table 16.14 which gives the yield of winter wheat after a one-year clover, ryegrass, or ryegrass-clover ley, which were

1 For a discussion and analysis of the nitrogen transfer in this experiment, see T. W. Walker, H. D. Orchiston and A. F. R. Adams, *J. Brit. Grassland Soc.*, 1954, **9**, 249.
2 *J. Sci. Food Agric.*, 1956, **7**, 66.
3 *J. agric. Sci.*, 1955, **46**, 267.
4 *Better Crops and Plant Food*, Nov. 1953.
5 T. W. Walker, H. D. Orchiston and A. F. R. Adams, *J. Brit. Grassland Soc.*, 1954, **9**, 249.

cut three times during the grazing season.[1] Wheat after ryegrass needed about 62 kg/ha N as fertiliser to give the same yield as wheat after clover, and wheat after ryegrass plus clover needed somewhat less than this. The amount of readily-nitrifiable-nitrogen in the soil just before the wheat was sown is also given, and it is roughly inversely proportional to the response of the wheat to nitrogen, or proportional to the yield of wheat without any additional fertiliser.

TABLE 16.14   Effect of the composition of a 1-year ley on the yield of the suceeding wheat crop at Rothamsted. Mean yields for plots given K fertiliser

Ley in 1961	Clover	Ryegrass			Ryegrass–Clover	
N given to ley kg/ha	0	0	125	250	0	125
Dry matter in 3 cuts t/ha	10·1	6·80	11·8	12·9	10·8	12·1
N in dry matter kg/ha	290	64	158	223	166	178
	Wheat yield in 1962. Grain in t/ha					
N given to wheat kg/ha						
0	6·05	3·98	4·32	4·29	4·96	4·53
62	7·01	5·91	5·84	5·75	6·44	6·08
125	7·14	7·25	6·75	6·89	6·95	6·95
	Mineralisable N in soil. October 1961 in p.p.m.					
	34	21	25	28	32	30

The residual effect of different legumes depends on the legume—a point which has already been noted on p. 380, and illustrated in Table 16.11. These results showed that lucerne and clovers enriched the soil in nitrogen and benefited the succeeding crop markedly, while soyabeans and beans taken for seed had little effect on the soil nitrogen and much less effect on the succeeding crop. And this result that annual leguminous crops taken for seed, such as beans, peas, soyabeans and groundnuts, do not enrich the soil in nitrogen and do not have any appreciable beneficial residual effect on succeeding crops appears to be generally true under many conditions. Thus, A. E. V. Richardson and H. C. Gurney[2] found at the Waite Institute, Adelaide, that 125 kg/ha of sulphate of ammonia increased the yield of wheat following a corn crop by 6 hl/ha, following a pea crop, 5 hl/ha, and it had no effect on wheat following a fallow.

Leguminous or grass crops that are left down for several years can increase the soil's nitrogen status for several years after they have been ploughed out compared with land under continuous arable and receiving no farmyard manure. But the interpretation of many older experimental results is uncertain

1 F. V. Widdowson, A. Penny and R. J. B. Williams, *J. agric. Sci.*, 1963, **61**, 397.
2 *Emp. J. expt. Agric.*, 1933, **1**, 193, 325.

because inadequate care was taken to ensure that the amount of plant nutrients removed from the soil by the various cropping sequences were comparable. A recent Rothamsted experiment has been designed to overcome the criticisms that have been raised against earlier experiments, and some results of the most recent design are given in Table 16.15.[1] In this

TABLE 16.15   Residual effects of three-year grass and legume leys. Rothamsted: mean of experiments in 2 fields for 2 years (1969–70). Yield in tons per hectare

Crop after ley	*First: potatoes*				*Second: wheat*			*Third: barley*	
Nitrogen added kg/ha N	0	75	150	225	0	50	100	0	50*
Lucerne	40·0	45·5	45·9	47·9	6·4	7·5	7·8	4·7	5·7
Grass–clover	40·0	45·5	47·4	46·4	6·4	7·4	7·7	4·7	5·5
Grass	37·5	43·5	45·5	46·2	5·9	6·7	7·7	4·4	5·3
Arable	30·3	40·0	42·0	45·5	5·3	6·7	7·5	3·9	5·3

* 40 kg N given in 1969.

experiment the yields of three test crops, potatoes followed by wheat followed by barley are compared when grown in rotation with a three-year ley or with a one-year grass–clover ley followed by two years of arable crops. Three leys are compared, lucerne, a grass–clover, and a pure grass given high nitrogen manuring, and the leys are cut regularly and the produce removed. The subsequent manuring of the plots is designed to compensate for the different amounts of potash and phosphate removed by the various crops. The table shows that the two leguminous leys apparently increase the nitrogen status of the soil for at least three years, but in the second and third year their effect is less than that given by 50 kg/ha N as fertiliser.

1 This table is derived from data given in the *Rothamsted Reports* for 1969 and 1970. For a recent summary of other Rothamsted and Woburn experiments, see D. A. Boyd, *Rothamsted Rept.* 1967, 316.

17

The temperature of the soil

The soil derives its heat almost entirely directly from the sun and loses much of it by radiation back into the sky. The opacity of the atmosphere to solar radiation and to longwave back radiation is therefore an important factor in quantitative studies on soil temperature. The sun has an effective radiation surface temperature of $6000°K$, so the incoming solar radiation has a fairly broad wavelength band centred at about $0·5$ μ; while the effective radiation surface temperature of the earth is of the order of 280 to $300°K$, so it radiates energy in the broad wavelength band 3 to 100 μ but centred around 10 μ. The clear atmosphere is fairly transparent to incoming solar radiation and it has a transparency window around 10 μ. This is bounded on each side by water vapour and carbon dioxide absorption bands centred on 6 μ and 20 μ.

Water droplets absorb and scatter both the short and the long wavelength radiant energy, and quite small amounts of cloud will reduce the input of solar energy to the soil, but reduce the loss of heat from the soil by absorbing and reradiating back to the soil surface longwave energy.

The mean daily and monthly amounts of solar energy received at Rothamsted are given in Table 17.1, and Fig. 17.1 shows the difference between the

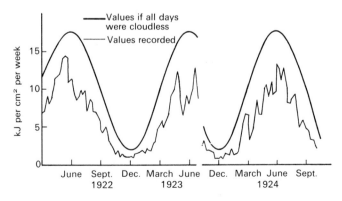

FIG. 17.1. Solar radiation received at Rothamsted

TABLE 17.1 Mean value of the daily and monthly radiation at Rothamsted. Period 1956–69

	$kJ/cm^2/day$	$kJ/cm^2/month$	
January	0·23	7·1	
February	0·40	11·1	
March	0·77	24·0	
			42·2
April	1·13	33·9	
May	1·55	48·0	
June	1·76	52·7	
			134·6
July	1·53	47·4	
August	1·25	38·9	
September	0·96	28·8	
			115·1
October	0·54	16·8	
November	0·25	7·5	
December	0·15	4·6	28·9
Total for year			320·8

actual energy received and the amount that would be received if there were no cloud. The amount of energy lost by back radiation is discussed on p. 449.

The temperature of the surface layer of a bare soil on a clear sunny day is controlled by the rate it is absorbing solar energy, as is shown in Fig. 17.2 for a Rothamsted soil in July. The surface temperature is seen to vary in phase with the incoming radiation during the day; but during the night it continues to fall, though much more slowly than during the day. The air temperature curve at 1·2 m above the soil surface lags behind the soil temperature curve during the day and reaches a lower maximum value, but during the evening and night it is almost the same as the soil surface. The soil temperature below the soil surface follows the changes in the surface temperature though it lags behind the surface and the diurnal variation is reduced, and both these changes become more marked with depth. This behaviour is shown in Fig. 17.3 for the bare Rothamsted soil in early August. It shows that at 20 cm depth, the maximum temperature occurs at about 1900 hours and the diurnal variation is already small, and it becomes negligible at about 40 cm: a result found in many other places for clear warm or hot days.[1]

The mean monthly soil temperatures also show a clear seasonal trend, that for the surface soil being approximately in phase with the incoming solar radiation, and the seasonal change becoming smaller, and the time at which

1 For Muguga, Kenya, see J. S. G. McCulloch, *Quart. J. Roy. Met. Soc.*, 1959, **85**, 51; for Illinois, J. E. Carson and H. Moses, *J. appl. Met.*, 1963, **2**, 397; for Arizona desert, W. R. Guild, *J. Meteor.*, 1950, **7**, 140.

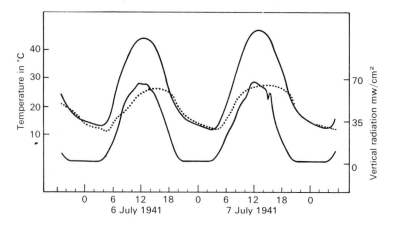

FIG. 17.2. Relation between incoming solar radiation and the air and soil surface temperatures on two cloud-free days at Rothamsted. Upper full curve: Temperature of the surface of the soil. Lower full curve: Solar radiation received by soil surface. Middle dotted curve: Air temperature 1·2 m above the ground (in the screen)

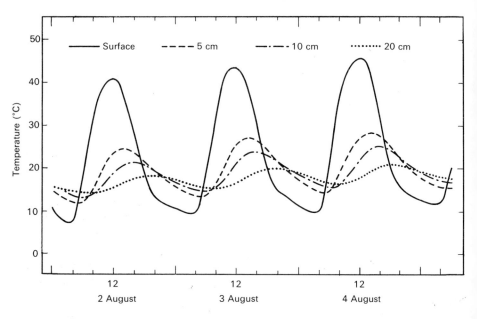

FIG. 17.3. The damping of the daily temperature wave with depth in a bare Rothamstead soil

the maximum temperature is reached becoming later in the year with depth. Figure 17.4 shows the seasonal change of temperature at three depths for an Oxford soil,[1] in this example under short grass. Whereas at a depth of 30 cm the maximum temperature is reached in mid July and the range between mid winter and mid summer is 14°C, at 3 m it is not reached till September, and the range is only 5·5°C. In temperate regions the seasonal variation does not become small until the depth is 6 to 8 m, but in equatorial regions, as for example at Muguga, Kenya (lat. 1° 13′ S) it is small at 1 m.

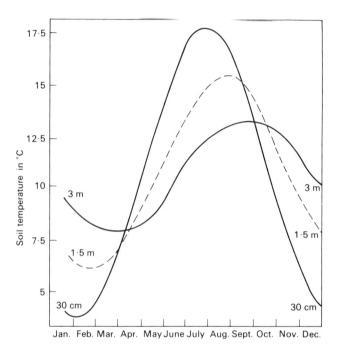

FIG. 17.4. Annual variation of soil temperature with depth under short grass. Oxford, Radcliffe Observatory

The temperature of any given volume of soil is dependent on the difference between the rates of flow of heat, the heat fluxes, into and out of that volume. In Chapter 20 the fate of the incoming solar radiation that falls on the soil will be discussed in some detail, so only a brief outline of the heat budget of a soil will be given here. Only a part of the incoming radiation R received by the soil is absorbed, for a fraction s, the albedo of the surface, is reflected or scattered, so the amount absorbed is $R(1 - s)$. The absorbed radiation is

1 Original data by A. A. Rambaut, Racliffe Observations 1898–1910, and the data for Fig. 17.4 computed from B. A. Keen, *The Physical Properties of the Soil*, Longman, 1931.

dissipated in four different ways: part is reradiated back to the sky as long-wave radiation, B; part is used for evaporating water from the soil and is dissipated as latent heat; part may raise the temperature of the surface soil and be dissipated as sensible heat to the air, since it will raise the temperature of the air in contact with it; and part will be conducted into the body of the soil.

When the soil surface is moist, most of the net radiation $R(1 - s) - B$ absorbed is used to evaporate water, but as the soil becomes drier an increasing quantity is dissipated either as sensible heat to the air or as a heat flux into the soil. Table 17.2 gives an example of this distribution of sinks for the absorbed radiation in a wet and a moist soil in Arizona.[1] In this

TABLE 17.2   Dissipation of absorbed energy by a bare soil. Fluxes in $J/cm^2/day$

Soil	Net radiation	Latent heat term	Sensible heat to air	Heat to soil
Wet	1690	1730	-4	-33
Moist	1370	940	289	142
		incoming radiation 3060		

example the wet soil is dissipating rather more energy as latent heat than it is absorbing as net radiation, so it is drawing heat from the soil and cooling its surface; whereas the moist soil is dissipating only about 70 per cent of the net radiation it is absorbing as latent heat, and 10 per cent is entering the soil and warming its surface. The reason less heat is absorbed by the drier soil is partly because its albedo is greater, so more of the incoming radiation is reflected, and partly that its surface temperature is higher so it is losing more heat by back radiation. This transference of a part of the absorbed heat to sensible heat is the cause of the air temperature following the soil temperature, as shown in Fig. 17.2.

The heat flux dQ/dt flowing into or out of the soil depends on the temperature gradient dT/dz in the soil and the thermal conductivity k, for in any given soil volume the heat flux is given by

$$dQ/dt = -k \, dT/dz$$

where z is measured along the gradient. The thermal conductivity of the soil is very dependent on its air content, for the value of k for still air is $5 \cdot 7 \times 10^{-5}$, while that for water is $1 \cdot 4 \times 10^{-3}$ and for soil solids about 4×10^{-3}, all in cgs units, so that the conductivity of air is only about one-hundredth that of water and soil solids. The more compact the soil, and the wetter the soil, the greater its conductivity.

1 C. H. M. van Bavel, *UNESCO Arid Zones Res.*, 1965, **25**, 99.

The effect of a heat flux on the rate of change of temperature at a given depth depends on the difference between the heat flux into and out of a small volume of soil at that depth, and is given by

$$\frac{dT}{dt} = \frac{k}{\rho c}\frac{d^2 T}{dz^2}$$

where ρ is the density of the soil and c is its specific heat. The quantity $k/\rho c$ is called the diffusivity of the soil, or sometimes its temperature conductivity. The diffusivity may either increase or decrease with increasing moisture content since k, ρ and c all increase with it, but, in general, starting with a dry soil, it first increases, then, after reaching a flat maximum, may decrease

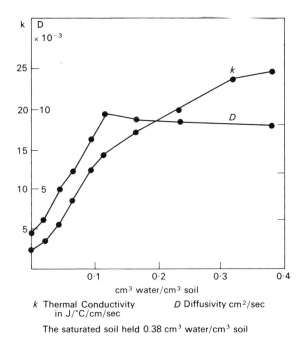

k Thermal Conductivity in J/°C/cm/sec *D* Diffusivity cm²/sec

The saturated soil held 0.38 cm³ water/cm³ soil

FIG. 17.5. Variation of thermal conductivity, and thermal diffusivity of a sandy loam soil with moisture content

slightly. Figure 17.5 shows the relation between the thermal conductivity, and the diffusivity of a sandy loam soil, where the moisture content has been expressed on a volume basis.[1]

A temperature gradient in an unsaturated soil, that is, a soil that contains air, allows water to move from warmer to cooler parts of the soil, for water will evaporate in the warmer region, diffuse as vapour through the air spaces and condense in the cooler region, so this transfer of vapour also transfers

1 A. F. Moench and D. D. Evans, *Proc. Soil Sci. Soc. Amer.*, 1970, **34**, 377.

heat which increases the apparent thermal conductivity of the soil.[1] In soils in which the matric potential of the subsoil does not fall below 15 bar, the vapour transfer contribution to the conductivity is not usually an important factor, for it is unlikely to increase the apparent conductivity by more than 10 or 20 per cent.

There are few published data on the seasonal or diurnal heat fluxes into and out of soils, but Table 17.3 gives some values for the changes in the heat content of three American soils[2] and some early data for a sandy soil at Dresden in central Europe.[3] The soils gain between 7 and 10 kJ/cm^2 in the six months from March to August, which represents up to about 4 per cent

TABLE 17.3   Seasonal heat flux into and out of a soil. Changes in heat content in top 3 m in kJ/cm^2

	Jan.–Feb.	March–April	May–June	July–Aug.	Sept.–Oct.	Nov.–Dec.	Summer Total
Kentucky, USA	−2·07	2·38	4·43	2·07	−2·36	−4·43	9·88
Texas, USA	−1·50	3·12	4·73	1·50	−3·13	−4·74	9·35
Ottawa, Canada	−1·93	2·66	4·60	1·93	−2·70	−4·60	9·19
Dresden, Germany	−1·53	1·93	4·17	1·08	−2·32	−3·49	7·18

of the incoming radiation; and for the months of May and June the net daily heat flux into the soil is of the order of 70 to 90 J/cm^2. The heat flux is into the soil during a sunny summer day during the middle of the day, but as the sun gets low in the sky the flux reverses, and it is from the soil to the air in the evening, night and early morning. An example for a sunny day in July at Argonne, Illinois, is given in Fig. 17.6.[4] During the period that the flux is from the soil to the air, the air temperature is equal to the soil temperature, as can be seen in Fig. 17.2. The actual values of the inflow and outflow of heat on any day depends on local conditions, but the net inflow can represent a considerable proportion of the incoming solar energy under some conditions. Thus J. L. Monteith[5] at Rothamsted measured a downward flux of about 200 J/cm^2 and an upward flux of about 80, giving a net downward flux of 120 J/cm^2 on a sunny day in August, which represented 10 per cent of the incoming radiation or nearly 20 per cent of the net radiation for that day. Net daily fluxes, upwards or downwards, are much smaller for equatorial soils, for their seasonal temperature wave has almost ceased at 1 m, compared with the 6 to 8 m in higher latitudes.

1 D. A. de Vries, *Soil Sci.*, 1952, **75**, 83; *Nether. J. agric. Sci.*, 1953, **1**, 115.
2 E. B. Penrod and O. W. Stewart, *Soil Sci.*, 1967, **104**, 86.
3 Data from P. Schrieber quoted from O. G. Sutton, *Micrometeorology*, McGraw-Hill, 1953.
4 Taken from J. E. Carson and H. Moses, *J. appl. Met.*, 1963, **2**, 397.
5 *UNESCO Arid Zones Res.*, 1956, **11**, 123. For other Rothamsted figures see H. L. Penman and I. F. Long, *Quart. J. Roy. Met. Soc.*, 1960, **86**, 16.

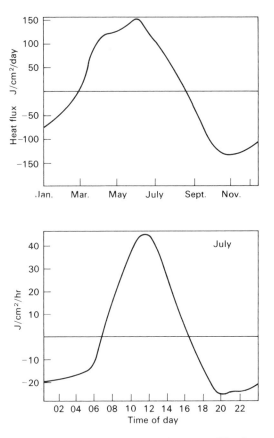

FIG. 17.6. Heat flux into and out of the soil at Argonne, Illinois, over the year, and during a sunny day in summer

The control of bare soil temperatures in practice

The factors which affect the temperature of a given soil are the amount of radiation it receives, its albedo, and its moisture and air contents, and any factor that affects the rate of evaporation of water from the moist soil. The amount of radiation a soil receives depends on its aspect, and in the northern hemisphere a south-facing slope is warmer than a horizontal surface, which, in turn, is warmer than a north-facing slope. Correspondingly a west-facing slope is normally somewhat warmer than an east-facing slope. This effect of aspect can be very important in the spring when soils are warming up. Thus J. W. Ludwig[1] at Oxford showed that maize germinated appreciably quicker if sown on a south than on a north slope if sown in late March or April, but

1 With E. S. Bunting and J. L. Harper, *J. Ecol.*, 1957, **45**, 205.

if sown in May there was little difference because the soil temperature on the north slope was by that time high enough for good germination.

Moist soils have albedos of about 0·10 to 0·15, and this increases as the soil dries, and light coloured soils have higher albedos than dark. The greater the albedo of a soil, the cooler it tends to be. Thus, covering the soil with a thin film of white powder will reduce the surface temperature of the soil, and with a film of black powder will increase it. J. W. Ludwig and J. L. Harper[1] at Oxford found that dusting the surface of their soil with either chalk or a light-coloured soil reduced the maximum daily soil temperature at a depth of 5 cm by 5°C compared with dusting it with a black or grey soil, although the daily minimum temperature was only reduced by 1°C. These figures refer to the mean temperatures during three weeks in April, and maize sown at the beginning of April took 20 days for 50 per cent of the seeds that finally germinated to germinate under the dark-coloured film compared with 32 days under the light, and the final germinations were 70 and 44 per cent of the seed sown. Similarly, G. Stanhill[2] working in the Negev in Israel showed that under the hot desert conditions prevailing there a film of white magnesium carbonate 0·05 mm thick increased the albedo from 0·31 to 0·64 and reduced the maximum daily temperature at 2 cm depth by about 7 to 10°C over a period of a month, though it increased the minimum temperature by about 2°C.

Mulches applied to the surface of a soil affect the amount of heat received and the way it is dissipated. Mulches of dead vegetation, of straw, stover or dead grass, for example, immobilise to some extent the air within the mulch; and because still air has a very low thermal conductivity, heat is only slowly transmitted from the surface of the mulch to the soil surface. Thus soils under these mulches remain cold in spring in regions which have a cold winter, and germination of seed sown under such mulches is delayed and usually poor. This can be very damaging in regions with cold winters and hot summers where crops requiring warm soils in which to germinate, such as maize or sorghum, are grown; for these mulches will delay sowing dates and so shorten the growing season.[3] As will be discussed on p. 803, this type of mulch can be very valuable for soil conservation if this effect on shortening the growing season can be overcome. In addition the soil remains moist and often rather poorly aerated.

A film of transparent polythene laid on the soil surface will increase the soil temperature. This is because, although it is transparent both to solar radiation and to back radiation during clear nights, its under-surface becomes covered with a film of water droplets which act as a barrier to radiation losses from the soil surface into outer space. Table 17.4 gives the effect of

1 *J. Ecol.*, 1958, **46**, 381.
2 *Agric. Met.*, 1965, **2**, 197.
3 For an example with maize in the U.S. see W. R. van Wijk, W. E. Larson and W. C. Burrows, *Proc. Soil Sci. Soc. Amer.*, 1959, **23**, 428; *J. Agron.*, 1962, **54**, 19.

different types of mulch or surface covering on the soil temperature at Manhattan, Kansas, for the summer months[1], and it shows very clearly the higher mean soil temperatures under transparent polythene, and the lower under mulches. This additional warmth in the soil under polythene is of increasing commercial importance for the early production of high-priced crops such as strawberries and cantaloupe melons.

TABLE 17.4   Effect of different mulches on the soil temperature. Manhattan, Kansas

Depth	Polythene film	Bare soil	2·5 cm layer gravel painted black	aluminium	Wheat straw 10 t/ha
1 cm	34·4	29·1	27·4	25·9	23·8
4 cm	31·8	28·0	27·1	25·4	23·6
16 cm	29·0	26·8	26·0	24·2	23·0

The effect of the moisture content of a soil on its temperature is complex. On the one hand, a moist soil conducts heat upwards or downwards much better than a dry soil, but during a sunny day the surface of a dry soil warms up much quicker, and during a clear night cools much quicker than a wet surface. This greater temperature fluctuation at the dry surface is rapidly damped with depth, and at 7 to 10 cm there need be little difference in temperature due to the soil's moisture content. In the same way a sandy soil and a clay soil in the same region will have very different surface temperatures, that for the sand having a much greater diurnal variation, but at depths of 5 to 10 cm the temperature difference between the soils can be small. Sandy soils are however commonly said to warm up quicker in the spring than clay soils, and in fact sands are often referred to as warm soils and clays cold soils. This is because seeds germinate sooner in spring and the young plants grow quicker on the sandier soils. The reasons for the differences in the rate of germination and early growth, however, may not be due to differences in soil temperature but to differences in soil aeration and pore space distribution, as will be discussed on p. 418. Heavy-textured soils in spring are typically wet after the winter and possess few air-filled pores; seeds tend to germinate poorly in them and they contain few pores into which the young roots can grow. On the other hand, in spring, lighter-textured soils contain considerably more air-filled pores and it may be that fact that accounts for the earlier spring crops on these soils. This will result in the surface of the sandier soils warming up much quicker in the middle of the day than the clay soils, but because their thermal conductivity will be lower, due to

1 Taken from R. J. Hanks, S. A. Bowers and L. D. Bark, *Soil Sci.*, 1961, **91**, 233. For other examples see P. E. Waggoner, *Trans. 7th Int. Cong. Soil Sci.*, 1960, **1**, 164; J. E. Adams, *Agron. J.*, 1962, **54**, 257; D. E. Miller and W. C. Bunger, ibid., 1963, **55**, 417.

their higher air content their temperature at, say, 5 to 10 cm deep need not be very different. This appears to be what is found in practice; for instance, the sandy loam soil at Woburn is definitely earlier than the clay loam at Rothamsted, yet in spring its temperature is only about 0·5°C higher at a depth of 10 cm, while in autumn its temperature is about the same and in winter it may freeze earlier. There is still very little published data on the temperature differences between soils of different texture in the same locality.

Irrigating a soil in summer will always lower its surface temperature, but this cooling effect need not go very deep. G. E. P. Smith[1] in Arizona has given an example where irrigation cooled the soil by 5°C at 2·5 cm, 2°C at 5 cm and 1°C at 7·5 cm depth. Conversely, if the soil is loose before irrigation, irrigation may raise the temperature at 5 cm, because of its higher thermal conductivity. Thus F. A. Brooks[2] found that the soil temperature at 10 cm was between 3 and 4°C warmer when irrigated than on a neighbouring plot which had a loose dry tilth, although the maximum surface temperature of the dry soil was appreciably higher.

This effect of irrigation in increasing the thermal conductivity of a soil can be important in practice at night time during periods of clear skies and strong temperature inversion when radiation frosts are liable to occur, particularly if the surface soil is bare and dry because of its low thermal conductivity. If the soil is made moist, or if it is rolled to be made more compact,[3] its conductivity is increased, more heat can be conducted from the body of the soil to its surface, and its surface temperature will drop much less; for although the total radiation loss is the same, a greater proportion of the heat radiated will come from the soil and a smaller proportion from the air. These effects are large enough to prevent a light frost from harming crops at nights.

The influence of vegetation on soil temperature

Vegetation has the same general effect as a mulch, for it reduces both the diurnal and seasonal fluctuations, because it intercepts a part or all of the incoming radiation and of the back-radiation from the soil. Its effect depends on the degree with which its leaves shade the soil surface from the sky, and in general its effect is roughly proportional to the proportion of soil shaded. Under a complete canopy, the leaves will absorb all the incoming solar radiation and will be the direct source of all back radiation from the surface into space. The soil surface will only receive longwave radiation from the canopy and its back-radiation will all be absorbed and scattered by the canopy. In general the air temperature above a crop will be lower than the sur-

1 With A. F. Kinnison and A. G. Carns, *Ariz. Agr. Exp. Sta. Tech. Bull.* 37, 1931.
2 Quoted by S. J. Richards *et al.*, in *Soil Physical Conditions and Plant Growth*, Agronomy Monograph No. 2, 1952, 332.
3 For an example from Victoria, Australia, see S. F. Bridley, R. J. Taylor and R. T. J. Webber, *Agric. Met.*, 1965, **2**, 373.

face soil temperature on a clear night, whereas it will be the same as the surface temperature of a bare soil. The difference between the air temperature above a crop and the surface soil temperature on a clear day depends on the rate of transpiration by the crop, but if the crop dissipates any of its net radiation as sensible heat the air temperature will typically be higher than the surface soil temperature.

Vegetation affects the seasonal changes in the soil surface temperature. Soil under vegetation warms up slower in the spring and cools down slower in the autumn than bare soil. But little is known about the magnitude of this effect or how it varies from year to year. At Rothamsted at 10 cm below the surface, bare soil is always cooler than turf, except in June and July, and the difference between these two is largest in autumn. On a ten-year average (1930–39) soil under turf is 1·2°C warmer in October and November than bare, but is only 0·6°C warmer during the winter and spring, whilst from May to August it is within 0·3°C of the bare soil temperature.

The effect of vegetation on the heat flux into and out of a soil depends on the thickness and height of its canopy. A complete canopy of thick grass differs from that of a forest in that the thick grass maintains an almost stagnant body of air among its leaf blades, which insulates the soil surface against appreciable heat fluxes, whereas in a forest there is a deep column of fairly still air in which convection can take place easily. This allows heat to be transferred from the soil to the air whenever the air becomes cooler than the soil surface, although in many forests the flux from the soil to the air may be low because of the large proportion of air pores in the forest floor.

This difference in type of canopy can be of great importance in affecting the incidence of frost during clear nights in spring. Thus C. E. Cornford[1] found the minimum air temperatures 90 cm above the ground in some flat land in Kent after a clear still night at the end of May were: above bare soil 9·7°C, in a wood 9·4°C, above a short grass meadow 7·6°C and above a long grass meadow 6·1°C. The small difference between the wood and the bare soil is due to the circulation of the air mass in the wood brought about by it being cooled at canopy level and warmed at the soil surface. This cooling effect of vegetation only affects the air near the canopy, for the air a few metres above the canopy will be appreciably warmer. As already noted, straw mulches have the same effect as grass and as another example from Kent, W. S. Rogers[2] found that in some strawberry beds there could be 7·5°C temperature difference on either side of the straw mulch on a clear spring night, and the minimum air temperature might be 4°C lower over a straw mulch than a bare soil.

Since frost damage in spring can be so serious on fruit and horticultural crops, it is worth while summarising the methods the grower has at his

1 *Quart. J. Roy. Met. Soc.*, 1938, **64**, 553.
2 *Imp. Bur. Hortic., Tech. Comm.* 15, 1945.

disposal for minimising this damage. These fall into two groups—ensuring the maximum transfer of heat from the soil to the air around the frost sensitive parts of the crop, and preventing cold air reaching these sensitive parts. The former is achieved by keeping the soil surface as free from weeds and mulch as possible, as well as compact and damp. This not only ensures the maximum transfer of heat from the subsoil to the air, but also the maximum transfer of water from the soil to condense as dew on the plant. The latter can be achieved either by ensuring that air draining into the area is as warm as possible, which can be achieved by treating the higher lying land in this way, or by managing the crop so its sensitive parts are as high as possible above the ground. Careful management can often raise the temperature by 5°C, or even more, compared with poor management.

This blanketing effect of vegetation or mulches can also be important in winter, for it reduces the penetration of frost into the soil. Thus, E. J. Salisbury[1] quotes an example where frost penetrated a sandy loam to a depth of 5·5 to 8.5 cm if bare, to a depth of 2·5 to 3·5 cm under rough grass, to less than 2 cm under some bushes, to 1·5 cm in an open hazel copse where there

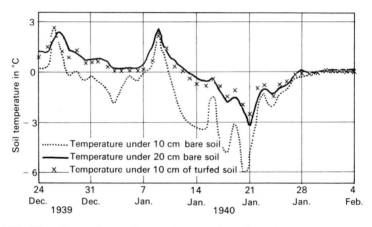

FIG. 17.7. The effect of a turf covering on the sub-surface temperature of the Rothamsted soil during a hard frost

was no litter and did not enter at all where there was litter. R. K. Schofield[2] found at Rothamsted and Woburn that a grass cover was even more effective, being equivalent to 10 cm of soil as is shown in Fig. 17.7, that is, if frost penetrated 20 cm in bare soil, it only penetrated 10 cm under short grass. F. A. Post and F. R. Dreibelbis,[3] in Ohio, have given another example; they found that soil became frozen to a depth of 5 cm or more many

1 *Quart. J. Roy. Met. Soc.*, 1939, **65**, 337.
2 *Quart. J. Roy. Met. Soc.*, 1940, **66**, 167.
3 *Proc. Soil Sci. Soc. Amer.*, 1942, **7**, 95.

times during the winter when under wheat, only a few times under grass and not at all under forests. Snow also protects the soil against penetration of frost; in an intensely cold winter, during which there is a great deal of snow, the soil may be warmer and frozen to a lesser depth than in a milder winter without snow.

This effect of vegetation in reducing fluctuations in the soil surface temperature, and in particular the maximum temperature in the middle of the day, is of particular importance in the tropics. Tree crops such as coffee and cocoa, which do not give complete canopy cover of the soil, either need to be planted under shade trees or to have the soil surface covered with a mulch, for only in this way can they maintain an active root system in the surface soil, which is required for the uptake of nutrients. As will be noted on p. 379, however, shade trees have other effects on the crop besides that on the soil surface temperature.

The effect of soil temperature on crop growth

The rate of germination of seeds, of seedling growth, and of root growth depends on the soil temperature, being negligible below a certain temperature, rising to a maximum and then falling off as the temperature rises. The temperature at which appreciable growth starts, at which the maximum rate is achieved, and at which root damage becomes appreciable differ both for the stage of growth reached, and for the species of crop considered: the temperatures typically being lower for temperate than for tropical crops, though the temperature range at which some temperate crops can grow is probably very wide.[1] Thus the optimum soil temperature for many temperate crops is about 20°C, while for crops such as sorghum and cotton it is probably over 30°C.

Soil temperature affects the type of root growth a plant makes: low temperatures encourage white succulent roots that suberise slowly and show little branching, while high temperatures encourage a browner, finer and much more freely branching root system which suberises fairly rapidly. Many temperate grass species growing in continental regions with a hot summer make root growth only in spring and autumn, while crops adapted to hot regions, such as maize, sorghum or cotton, will make active root growth in mid summer. However, the soil temperature at which the roots grow fastest is usually higher than the temperature which encourages the most extensive root system.

Low soil temperatures can be very harmful at seeding time to crops, such as maize or sorghum, which need a high soil temperature for active seedling growth, for the young seedling will grow very slowly while the soil will normally contain many weakly parasitic fungi which will grow actively and

1 For an excellent review on this subject, though now rather old, see R. M. Hagan in *Soil Physical Conditions and Plant Growth*, Agronomy Monograph No. 2, 1952, 367.

rapidly at these low temperatures and so will weaken or kill the seedlings. On the other hand, seedlings of the temperate zone cereals, which are adapted to grow actively at these lower temperatures are relatively resistant to their attack.

The early growth of seedlings requiring a fairly high temperature for growth are therefore encouraged by any practice that warms the soil and discouraged by any that cool it. This can raise difficulties in regions liable to serious erosion in spring, for one of the easiest ways of controlling this is by anchoring the previous year's crop residues in the surface of the soil, to act as a mulch, and this mulch will tend to lower the soil temperature. Thus the optimum soil temperature for the early growth of maize is about 25°C, and in regions where the soil temperature at seeding time is appreciably below this, as it is in most of the American Corn States, stubble mulch farming will tend to lower yields.[1] On the other hand, the optimum soil temperatures for wheat is very much lower than this, so stubble mulches will have little effect on early growth. The agricultural value of these practices is discussed again on p. 798. Normal vegetation mulches have their greatest benefit on root growth when the soil is too warm for good root growth, for then the cooling effect of the mulch will be beneficial. Only certain transparent mulches will raise soil temperatures, as discussed on p. 397.

The rate at which roots can take up water and nutrients from the soil also increases with soil temperature, and, for some crops at least, this rate may continue to increase until the temperature gets sufficiently high to start harming the root.[2] The effect of low soil temperature in reducing the rate of water uptake may be important in springtime, when the soil can be cool but the sun bright and the wind dry, giving conditions encouraging strong transpiration, and its effect on nutrient uptake has been studied for phosphate by several workers. Since the rate of phosphate uptake increases with increasing temperatures, it means that phosphate fertilisers tend to increase uptake more noticeably when the soil is cool, in autumn or spring, than in summer; and crops well supplied with phosphate tend to have a much wider temperature range at which growth is optimum than those poorly supplied.[3] The rate of nutrient uptake, however, is almost certainly different for each crop, but little appears to be known about the relative effect of temperature on their uptake.

1 See, for example, W. R. van Wijk, L. E. Larsen and W. C. Burrows, *Proc. Soil Sci. Soc. Amer.*, 1959, **23**, 428, and W. C. Burrows and L. E. Larsen, *Agron. J.* 1962, **54**, 19.
2 For an example on absorption of water by cotton, see M. E. Bloodworth, *Trans. 7th Int. Congr. Soil Sci.*, 1960, **1**, 153.
3 For an example with barley see J. F. Power, D. L. Grunes *et al.*, *Agron. J.*, 1963, **55**, 389, and with *Phaseolus* beans S. B. Apple and J. S. Butts, *Proc. Amer. Soc. Hort. Sci.*, 1953, **61**, 325.

18

The soil atmosphere

The soil pores that are not filled with water contain gases, and these gases constitute the soil atmosphere. Its composition differs from that of the free atmosphere because the plant roots and organisms living in the soil remove oxygen from it and respire carbon dioxide into it, so that it is richer in carbon dioxide and poorer in oxygen than the free atmosphere. Since the roots of most crops can only function actively if they have an adequate oxygen supply, there must be present in soils mechanisms or processes which allow the transfer of oxygen from the atmosphere to the soil organisms and plant roots, and of carbon dioxide from there to the atmosphere at rates adequate to meet the needs of the crop. The two factors affecting the magnitude of these transfers are the rate at which the soil organisms and the plant roots are converting oxygen into carbon dioxide, and the rate that oxygen can move from the atmosphere to the sites of active oxygen demand.

The rates of oxygen consumption and carbon dioxide production in normal agricultural soils adequately supplied with oxygen depend on the soil moisture content and temperature, as discussed on pp. 222–5, on the ease of decomposition of the organic matter in the soil, and on the activity of the crop's roots and their associated rhizosphere organisms. The respiratory quotient, RQ, of a well-aerated soil, that is the ratio of the volume of carbon dioxide produced to the volume of oxygen consumed, is close to unity;[1] it only rises above unity when there are anaerobic pockets present in the soil. A determination of this quotient in a soil is therefore a sensitive method for checking if the whole body of the soil is aerobic.

There is a very wide range in the published figures for oxygen consumption and carbon dioxide production in soils, because this depends on so many factors. J. L. Monteith[2] found, for a bare Rothamsted clay loam soil, a carbon dioxide flux of 1·5 $g/m^2/day$ in winter and 6·7 $g/m^2/day$ in summer, and the temperature dependence of the flux followed approximately:

$$R = R_0 Q^{T/10},$$

1 For an example see A. J. Dixon and B. J. Bridge, *Nature*, 1968, **218**, 961.
2 With G. Szeicz and K. Yabuki, *J. appl. Ecol.*, 1964, **1**, 321. N. J. Brown, E. R. Fountaine and M. R. Holden (*J. agric. Sci.*, 1965, **64**, 195), working with a sandy clay loam at Silsoe obtained similar results with potatoes and kale.

where R_0 is the flux at 0°C and R at T°C. He found Q had a value of about 3, and the results of these determinations are given in Fig. 18.1. The mean daily flux was 4 g/m² and the annual flux 1·44 kg/m² corresponding to a loss of 0·4 kg/m² of carbon. In this example the carbon content of the top 46 cm of soil was 12·2 kg/m², so this annual loss represents 3·1 per cent of the soil carbon. He also found that beans, kale and short grass increased the summer flux by 2 to 3 g/m²/day, but barley gave fluxes up to 10 g/m²/day in excess of the bare soil. These figures should be compared with the rates of carbon dioxide assimilation of the crops of 14 to 22 g/m²/day.

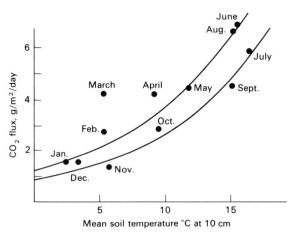

FIG. 18.1. The relation between the daily soil respiration and mean soil temperature at Rothamsted for a bare soil. (October 1960–September 1961.) The two curves are the plot of $R_0 Q^{T/10}$ for $Q = 3$ and $R_0 = 1·2$ upper curve and $R_0 = 0·9$ lower curve

These figures of Monteith may be rather low, because J. A. Currie,[1] also at Rothamsted, but using a more direct method of measuring fluxes, obtained much higher summer figures than Monteith. His results are given in Table 18.1, and since he measured both the oxygen consumed and carbon dioxide produced, was able to demonstrate that the respiratory quotient was almost exactly 1·00, except for the land under kale in July when it was 1·05. In another experiment he showed that the soil under a kale crop, which gave a dry matter yield of about 12 t/ha, absorbed about 17·5 t/ha of oxygen during the eighteen week period from July to November, when it was making most of its growth, of which probably about half was due to the oxygen absorbed by the roots and their associated organisms.[2] The results of many other workers tend to fall within the ranges given above. F. E. Clark and W. D. Kemper,[3] quote mean figures for the oxygen consumption of bare soil as

1 In *Sorption and Transport Processes in Soil*, Soc. Chem. Ind. Monog. 37, 152, 1970.
2 *Rothamsted Rept.*, 1967, 33.
3 In *Irrigation of Agricultural Lands*, Agronomy Monog. 11, 472, 1967.

TABLE 18.1   Oxygen consumption and carbon dioxide production from a bare soil and a soil under kale at Rothamsted (g/m^2/day)

Soil	July		January	
	cropped	*bare*	*cropped*	*bare*
Oxygen consumption	24	12	2·0	0·7
Carbon dioxide production	35	16	3·0	1·2
Soil temperature at 10 cm	17°C		3°C	

between 2·5 and 5 g/m^2/day, which are doubled if the soil is cropped. L. G. Romell[1] working in Sweden, found consumptions of between 13 and 20 g/m^2/day for forest soils and an *RQ* of about unity. F. Hilger[2] working in Zaire (Belgian Congo) found respirations of 12 to 15 g/m^2/day of carbon dioxide under forest, but if this was cleared and cultivated it rose up to 35 g, and E. D. Schulze[3] working in Puerto Rico found respirations up to 10 g under deciduous forest, which would be a fairly open type of forest, and 25 to 50 g/m^2/day under gallery or high-rainfall forest.

The rate of oxygen uptake by soils is usually large compared with the amount of oxygen present in a soil. If a soil is using 7 g/m^2/day of oxygen, and if this use is all assumed to take place in the top 25 cm of soil, this amounts to $2·8 \times 10^{-5}$ g/cm^3 soil/day or $3·25 \times 10^{-10}$ g/cm^3/sec. If the soil contains 20 per cent by volume of air, and if this air contains 20 per cent of oxygen, there will be $5·8 \times 10^{-5}$ g/cm^3 soil of oxygen, which is only about twice the daily use. Thus, if the surface of the soil was sealed completely against the entry of oxygen into the soil, the soil's oxygen supply would only last for about two days.

The mechanisms of gaseous transfer

The principal paths along which oxygen and carbon dioxide move in the soil are those pores containing air which form a continuous system stretching from the surface into the deeper layers of the soil. Problems of soil aeration are normally only of importance when this system of pores is absent, either due to the soil being waterlogged or to the coarsest pores being sufficiently fine that they are not emptied of water by drainage. There are a number of processes operative which help in these gaseous transfers. Changes in soil temperature and in atmospheric pressure cause the soil gases either to expand or contract; rain carries dissolved oxygen into the soil as well as pushing a body of air in front of its wetting front as it penetrates into the soil; and even

1 *Medd. Skogsförsöksanst.*, 1922, **19**, 125.
2 *Bull. Inst. agron. Gembloux*, 1963, **31**, 154.
3 *Ecology*, 1967, **48**, 652.

gusts of wind blowing over the soil surface will cause some atmospheric air to be sucked into the top 1 or 2 cm of a loose soil.[1] But L. G. Romell[2] in a classic paper showed that the only process which is normally of any significance is gaseous diffusion along the concentration gradients in the air-filled pores. This allows oxygen to diffuse to the root from the atmosphere and carbon dioxide from the root to the atmosphere, for this is the direction of the two gradients.

The rate of gaseous diffusion from, say, the neighbourhood of the plant root to the atmosphere depends on the difference in gas concentration between these two regions, the length of the diffusion path and the diffusion coefficient of the gas; and it is much less dependent on the shape of the air-filled pores than is the hydraulic conductivity of the soil. For many soils under normal field conditions the rate of diffusion is approximately proportional to the proportion of the soil volume that contains air.

Early work by H. L. Penman[3] and C. H. M. van Bavel[4] suggested that, if D is the rate of diffusion of a gas in the soil and D_0 that in free air, $D/D_0 = 0.6\,S$ to a first approximation, where S is the proportion of the soil volume occupied by air. Later work by J. A. Currie[5] showed that if this was written $D/D_0 = \alpha S$, the value of α depends both on the soil structure and on the moisture content. Some of his results are illustrated in Fig. 18.2.[6] The relation between α and S for a dry sand and a wet sand are quite different because compaction of a dry sand does not fundamentally alter the shape of the pore space, while wetting a sand introduces water wedges around all the points of contact of the sand particles, and this alters the shape of the air pores very considerably and so has a much greater effect on the diffusion coefficient. Columns of soil crumbs behave like sand grains at the wet end, but once a certain moisture content, or air space, has been reached, the value for α becomes about 0.6. The point at which this happens is when all the pore space between the crumbs contain air, but the crumbs themselves contain little if any. This shows that in structured soil, the diffusion coefficient is primarily determined by the intercrumb air spaces, and the air spaces within the crumbs only make a minor contribution. The shape of the curve for any particular soil depends on the soil structure, and the actual value of α in the wet end tends to be higher for a well-structured than a puddled soil.[7]

Factor α is composed of two parts, one part being the true tortuosity and the other a shape factor. Since the air pores are not straight tubes, but follow tortuous paths around the soil particles, the actual distance between two

1 See, for example, B. A. Kimball and E. R. Lemon, *Proc. Soil Sci. Soc. Amer.*, 1971, **35**, 16.
2 *Medd. Skogsförsöksanst.*, 1922, **19**, 125. B. A. Keen has given a detailed summary of Rommell's work in his book *The Physical Properties of Soils*, Longman, 1931.
3 *J. agric. Sci.*, 1940, **30**, 437, 570.
4 *Soil Sci.*, 1952, **73**, 91.
5 *Brit. J. appl. Phys.*, 1960, **11**, 318; 1961, **12**, 275.
6 Taken from *Sorption and Transport Processes in Soil*, Soc. Chem. Ind. Monog. 37, 152, 1970.
7 J. W. Bakker and A. P. Hidding, *Neth. J. agric. Sci.*, 1970, **18**, 37.

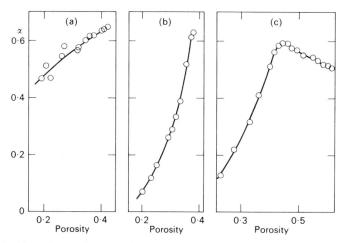

FIG. 18.2. Plot of α against the air content of sand and soil: (a) dry sand; (b) wet sand; (c) wet soil crumbs

points in the path l_e is greater than the straight line distance l between them, and the effect of this is to decrease the apparent diffusion coefficient in the soil by a factor $(l/l_e)^2$, which is known as the tortuosity factor. The ratio l_e/l is defined as the tortuosity of the path, and is greater the more tortuous the path, while the tortuosity factor, as here defined, becomes smaller as the path becomes more tortuous. Further, the pore is not a cylindrical tube but has a very varying cross-sectional area, and this still further reduces the coefficient by a shape factor f, so that $\alpha = f(l/l_e)^2$. Figure 18.2 shows that at high moisture content when there are only few air filled pores, α can be as low as 0·1 or 0·2, while it is 0·6 for a dry soil. It is not possible in practice to determine f and l/l_e separately, and α is often taken as equal to the tortuosity factor.

Currie[1] also compared the diffusion coefficient of a gas in a dry soil in its natural structure with the diffusion coefficient within the individual soil crumbs, and found that the within-crumb diffusion coefficient was only about one-fifth that of the between-crumb coefficient for a given air space. He used a theoretical equation developed by H. C. Burger[2] which can be written

$$D_c/D_0 = S_c/[1 + (k_c - 1)(1 - S_c)],$$

where D_c and S_c are the diffusion coefficient and volumetric air space within the crumb and k_c is a constant. He showed that k_c was a useful measure of soil structure, being high for poorly structured soils, and is, in fact, a measure of the tortuosity of the larger pores within the crumb. The value of k_c varied from 4·2 for crumbs from an old Rothamsted pasture soil to 7·9 for a Rothamsted arable soil that had received no farmyard manure for over 100 years and

1 *J. Soil Sci.*, 1965, **16**, 279.
2 *Phys. Ztschr.*, 1919, **20**, 73.

was in poor physical condition to 11·0 for crumbs of a silty clay soil from a restored open cast coal site, in which the soil had completely lost its structure. This was not connected with the air space, for the air spaces of the three groups of crumbs were 0·36, 0·25 and 0·30.

Diffusion of oxygen through a water film is about 10^{-4} times that in air, the diffusion coefficients being approximately $2·4 \times 10^{-5}$ compared with $2·1 \times 10^{-1}$ cm^2/sec. It will therefore diffuse quicker through a column of air 1 m deep than through a water film 1 mm thick. But transport of oxygen through a water film in the soil is affected by the presence of soil organisms, for they live in the water and will themselves be taking up oxygen from it. This means that oxygen can only diffuse through thin films of water in the soil, the thickness depending on the level of microbial activity. J. A. Currie[1] has shown that if oxygen is diffusing into a spherical waterlogged crumb, the maximum radius r of the crumb for which oxygen can just reach the centre, so the whole of the crumb contains some oxygen, is given by

$$r^2 = 6DC/M,$$

where D is the diffusion coefficient of oxygen in the crumb, C is the concentration of oxygen in the water at the outer surface of the crumb, and M is the rate at which the micro-organisms and plant roots in the crumb are using up oxygen; and the assumption that must be made for this simple equation to be valid is that M is a constant independent of the oxygen concentration around the organisms, an assumption which D. J. Greenwood[2] has shown to be reasonable.

The value of D depends on the size and tortuosity of the pore space within the crumb. If $D = \alpha D_0$, where D_0 is the diffusion coefficient of oxygen in free water, the value of α is greater the more open the structure within the crumb. Currie showed that the value of α for dry crumbs of an old Rothamsted pasture soil, which had a pore space of 0·34 cm^3/cm^3, was between 0·08 and 0·13; for an arable soil which had received regular annual dressings of farmyard manure, with a pore space of 0·27 cm^3/cm^3, was between 0·034 and 0·063; and for an arable soil which had no manure or fertiliser for over a century, with a pore space of 0·21 cm^3/cm^3, was between 0·015 and 0·044. Thus, D for wet crumbs probably falls within the range 0·5 to $2·0 \times 10^{-6}$ cm^2/sec. The value of M depends on the soil temperature, the amount of decomposable organic matter and the activity of the plant roots. At Rothamsted for a bare soil in winter it is about 50×10^{-12} g/cm^3/sec of oxygen corresponding to about 1 g/m^2/day, going up to over 500×10^{-12} for

1 *Soil Sci.*, 1961, **92**, 40. D. J. Greenwood and G. Berry. (*Nature*, 1962, **195**, 161) also obtained the same result but gave a more general solution of the differential equation by allowing M to depend on the oxygen concentration. For a computer simulation solution to this problem see P. J. Radford and D. J. Greenwood, *J. Soil Sci.*, 1970, **21**, 304. For experimental confirmation of the equation see D. J. Greenwood and D. Goodman, *J. Sci. Fd. Agric.*, 1964, **15**, 578, 781.
2 With D. Goodman, *J. Sci. Fd. Agric.*, 1965, **16**, 152.

cropped soil in summer. C is the concentration of oxygen in the soil water at the air-water interface, and if the oxygen concentration in the air just outside the wet crumb is assumed to be the same as in the free atmosphere, it is equal to about 10^{-5} g oxygen per cm^3 water, a figure somewhat dependent on the temperature, as the solubility of oxygen in water is temperature dependent. Putting these figures into the equation gives values of r ranging from about 0·4 to about 1 cm, so that, under summer conditions, waterlogged crumbs larger than about 1 cm in size are likely to have an anaerobic core, whereas in winter the minimum size of crumb that is aerated throughout is over 2 cm.

Wet well-drained soils of rather coarse structure can therefore consist of a mosaic of anaerobic volumes embedded in an aerobic matrix. This result explains an old result of E. J. Russell and A. Appleyard[1] that the composition of the gases dissolved in the soil water is quite different from that in the soil pores in that it is almost oxygen-free and consists primarily of nitrogen and carbon dioxide. It also explains why denitrification can proceed in soils in which the oxygen content of the soil air has not been appreciably reduced below its value in the atmosphere (see p. 344).

The application of this diffusion exchange process to the transfer of carbon dioxide predicts that the carbon dioxide content of the soil air should not rise above 1 per cent if there are no anaerobic pockets.[2] This is primarily due to the solubility of carbon dioxide in water being thirty-one times greater than that of oxygen, for the ratio of its diffusion coefficient in water to that of oxygen is 0·74. This has the consequence that if the oxygen and carbon dioxide fluxes in a waterlogged crumb are equal,[3] as they usually are in the absence of anaerobic pockets, for each 1 per cent reduction in the oxygen gradient between the inside and the outside of the crumb, there is only a 0·05 per cent increase in the carbon dioxide concentration, so that if the centre of the crumb is just free from oxygen, so the difference in oxygen concentration is 20 per cent, the difference in carbon dioxide concentration is 1 per cent.

The theory that the principal mechanism of gas exchange between the soil and the atmosphere is by diffusion through air pores would predict that a soil in which the surface structure has collapsed into a thin crust should rapidly become anaerobic whenever the crust becomes wet, for its pores are too fine to be emptied of water by drainage. Surface crusts, as distinct from the waterlogging of the soil surface, are not found to be as damaging to the aeration as would be expected.[4] This is due to the effect of wind,[5] for provided there are a certain number of cracks and holes through the crust, as there usually are, wind blowing across the surface of the soil will suck air through these holes, and through the air spaces just under the crust, and so maintain

1 *J. agric. Sci.*, 1915, **7**, 1.
2 D. J. Greenwood, *J. Soil Sci.*, 1970, **21**, 314.
3 See, for example, J. S. Bunt and A. D. Rovira, *J. Soil Sci.*, 1955, **6**, 119, 129.
4 N. J. Brown, E. R. Fountaine and M. R. Holden, *J. agric. Sci.*, 1965, **64**, 195.
5 D. A. Farrell, E. L. Graecen and C. G. Gurr, *Soil Sci.*, 1966, **102**, 305; and see D. R. Scotter and P. A. C. Raats, ibid., 1969, **108**, 170.

an oxygen concentration there not very different from that of the atmosphere.

A further consequence of the diffusion mechanism for gas exchange is that if the soil layers are absorbing oxygen at a uniform rate down the profile, a depth will be reached at which the atmosphere will become oxygen-free. This depth need not be very great if the air porosity of the soil is low and the biological activity of the subsoil is high, but in general the principal zone of biological activity is the surface soil. However, if crops are taking their water from depth, the roots at depth will have an appreciable oxygen demand so the oxygen diffusion path will be appreciable. As an example R. C. Lipps and R. L. Fox[1] found that when lucerne was growing actively and taking water from a depth of 2 m, the oxygen content at that depth was 16 per cent, but in the autumn when active growth ceased, so the oxygen demand of the roots fell, it rose to 20 per cent, and it maintained this concentration down to a depth of 3·6 m.

There has been a great deal of interest taken in the possibility of measuring electrometrically the rate that oxygen can diffuse through the water films in a moist soil to a zero oxygen sink. L. H. Stolzy and J. Letey[2] have reviewed the results obtained by this method and have concluded that most plants need an oxygen flux of about 3 or 4×10^{-9} g/cm^2/sec of oxygen if they are to function normally. Unfortunately, it has not proved possible to relate this flux, which is a flux across a unit area of water surface, to the diffusion rates as already defined. But more recent work has indicated that results obtained by this method are subject to a number of errors which seriously affect their interpretation,[3] and it is now uncertain how much reliance can be placed on the quantitative results that have been obtained.

The composition of the soil atmosphere

The composition of the soil air is determined by the rate of removal of oxygen and its rate of replenishment on the one hand, and the rate of production of carbon dioxide and its removal on the other. The rate of production and removal of carbon dioxide and of replenishment of oxygen fluctuate continuously with every change in structure, moisture content and temperature of the soil, so that both the carbon dioxide and the oxygen concentrations in the soil air are continuously fluctuating, and their fluctuations do not necessarily follow each other closely. Figure 18.3 shows these fluctuations in the air drawn from a 15 cm depth out of the dunged wheat plot on Broadbalk over a period of twenty-one months.[4]

It can be seen that, during periods of good aeration, the decrease in the

1 *Soil Sci.*, 1964, **97**, 4.
2 For a review of both their work and of others see *Adv. Agron.*, 1964, **16**, 249.
3 For a review see D. S. McIntyre, *Adv. Agron.*, 1970, **22**, 235.
4 E. J. Russell and A. Appleyard, *J. agric. Sci.*, 1915, **7**, 1.

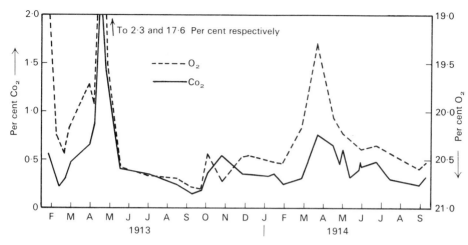

FIG. 18.3. The oxygen and carbon dioxide content of the soil air in the dunged plot of Broadbalk under wheat

oxygen concentration is numerically equal to the decrease in the carbon dioxide, but at times of poor aeration it is greater. J. T. Wood and D. J. Greenwood[1] have shown theoretically that the first result would be expected if the soil contained no volumes that are anaerobic, and the second if it did. The soil to which Fig. 18.3 refers receives 35 t/ha annually of farmyard manure, so it would not be surprising if it contained anaerobic volumes during wet periods; and it should now be possible to test the validity of Wood and Greenwood's conclusion, for anaerobic decomposition produces substances such as nitrous oxide, methane and ethylene, all of which can be measured relatively easily chromatographically.

The composition of the air in the surface layers of an arable soil, as determined by the older methods of withdrawing an appreciable volume of air from the soil for analysis, does not usually differ widely from that of the atmosphere when the main fluctuations in composition are smoothed out. The oxygen content is usually over 20·3 per cent, the nitrogen about 79 per cent, and the carbon dioxide between 0·15 and 0·65 per cent, compared with 20·96, 79·01 and 0·03 per cent respectively in the atmosphere. Table 18.2 gives some typical figures for the composition of the air in the surface soils of north-western Europe. In general, during periods of poor aeration, the fall in the oxygen content of soil air is greater than the rise in the carbon dioxide, a result explained by D. J. Greenwood[2] as being due to the more rapid diffusion of carbon dioxide than of oxygen through water films, because of its greater solubility in water. Modern methods based on the analysis of very small volumes of gas extracted from a soil however, usually give

1 *J. Soil Sci.*, 1971, **22**, 281.
2 *J. Soil Sci.*, 1970, **21**, 314.

TABLE 18.2   Composition of the air in soils, per cent by volume

Soil	Usual composition		Extreme limits observed		Analyst
	Oxygen	*Carbon dioxide*	*Oxygen*	*Carbon dioxide*	
Arable, no dung for 12 months	19–20	0·9	—	—	J. B. Boussingault and Léwy*
Pasture land	18–20	0·5–1·5	10–20	0·5–11·5	Th. Schloesing *fils*†
Arable, uncropped, no manure:					E. Lau,‡ mean of determinations made
sandy soil	20·6	0·16	20·4–20·8	0·05–0·30	frequently during a
loam soil	20·6	0·23	20·0–20·9	0·07–0·55	period of 12 months.
moor soil	20·0	0·65	19·2–20·5	0·28–1·40	Values at depths of
Sandy soil,					15 cm, 30 cm and
dunged and cropped:					60 cm, not widely
potatoes, 15 cm	20·3	0·61	19·8–21·0	0·09–0·94	different. (30 cm
serradella, 15 cm	20·7	0·18	20·4–20·9	0·12–0·38	values given here.)
Arable land, fallow	20·7	0·1	20·4–21·1	0·02–0·38	
unmanured	20·4	0·2	18·0–22·3	0·01–1·4	E. J. Russell and A.
dunged	20·3	0·4	15·7–21·2	0·03–3·2	Appleyard§
Grassland	18·4	1·6	16·7–20·5	0·3–3·3	

* *Ann. Chim., Phys.*, 1853, **37**, 5. † *C.R.*, 1889, **109**, 618, 673. ‡ Inaug. Diss., Rostock, 1906.
§ *J. agric. Sci.*, 1915, **7**, 1.

somewhat lower oxygen and higher carbon dioxide concentrations than were given by the older methods.[1]

Under tropical conditions the CO_2 content of the soil air may rise much higher, and the oxygen content fall lower, than these figures during the warm rainy seasons,[2] presumably because of the very rapid evolution of carbon dioxide by the soil organisms on the one hand, and the heavily restricted air space in the soil on the other.

The carbon dioxide content of the soil air increases, and the oxygen content decreases, with depth; and this can be very marked during wet periods in heavy or badly drained soils. Figure 18.4 shows this variation for a well-drained sandy soil and a badly drained heavy soil in the apple orchards of Cornell,[3] and Table 18.2 shows it for a cacao soil in north-west Trinidad during the rainy and the dry seasons.[4] These show that the carbon dioxide content can rise to very dangerous heights in a subsoil where the crop roots ought to be active. Hardy also showed, in more detail than is given in Table 18.3, that the CO_2 gradient was nearly constant down the profile during the dry season, but was much higher in the surface than subsurface layers during the wet.

1 See, for example, H. R. B. Hack, *Soil Sci.*, 1956, **82**, 217.
2 For some results at Pusa see J. W. Leateher, *India Dept. agric. Mem. Ser.*, 1915, **4**, 85.
3 D. Boynton, *Cornell Agric. Exp. Sta. Bull.* 763, 1941.
4 H. Vine, H. A. Thompson and F. Hardy, *Trop. Agric. Trin.*, 1942, **19**, 175, 215; 1943, **20**, 13.

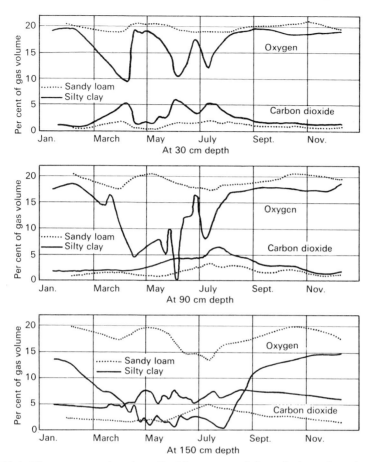

FIG. 18.4. The oxygen and carbon dioxide content of the soil air at three depths in a sandy loam and a silty clay apple orchard (Cornell)

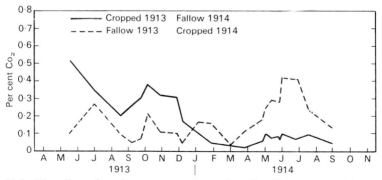

FIG. 18.5. The effect of a wheat crop on the carbon dioxide content of the soil air (Rothamsted, Hoosfield, wheat and fallow)

TABLE 18.3   The oxygen and carbon dioxide content of soil air under cacao (Rivers Estate, Trinidad)

Depth of sampling in cm	Oxygen content		Carbon dioxide content			Carbon dioxide gradient per cent per cm	
	Wet Oct.– Jan.	Dry Feb.– May	Wet Oct.– Jan.	Early dry Feb.	Late dry April– May	Wet Oct.– Jan.	Dry April– May
10	13·7	20·6	6·5	1·0	0·5	0·65	0·05
25	12·7	19·8	8·5	2·1	1·2	0·13	0·06
45	12·2	18·8	9·7	4·3	2·1	0·04	0·07
90	7·6	17·3	10·0	6·7	3·7	0·01	0·06
120	7·8	16·4	9·6	8·5	5·1	−0·01	0·06
Observed CO_2 diffusion rate from the soil in g/m²/day			13·4	14·8	35·0		

The carbon dioxide content of the soil air is higher in cropped than in fallow land (Fig. 18.5), and this effect is most noticeable when the crop is growing actively. The effect of the type of crop on the composition of the soil air is not well established, as the short period fluctuations in composition and changes in the tilth of the surface soil consequent upon growing the crops usually swamp any differences between them. There is, however, little

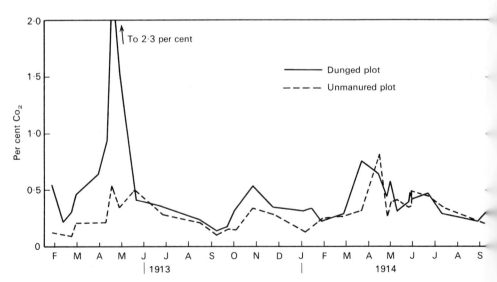

FIG. 18.6. The effect of annual dressings of farmyard manure on the carbon dioxide content of the soil air (Broadbalk, wheat)

doubt that the air in soils carrying lucerne or permanent pastures is richer in carbon dioxide than that in arable soils.

The effect of dressings of farmyard manure on the carbon dioxide content of the soil air is less than one might expect, presumably because although the rate of production of carbon dioxide is increased, the air space in the soil, and hence the rate of diffusion of the carbon dioxide into the atmosphere, is also increased. Figure 18.6 shows the carbon dioxide contents of two neighbouring plots on Broadbalk, one of which has been unmanured since 1843, and the other which has received 35 t/ha of farmyard manure practically every year since then. The main difference between them lies in the carbon dioxide contents when the soil is wet and diffusion slowed down. The soil was wet throughout the period April–May 1913, whereas it was dry throughout this period in 1914, and this is the probable cause of the much greater difference between the carbon dioxide contents of the two plots in the former spring.

Green manuring, particularly if the crop is fairly succulent, would be expected to put up the carbon dioxide of the soil air, though no measurements appear to have been made on the magnitude of this increase. It is just possible that it may be sufficient to account for the poor germination and early growth which is sometimes observed if seeds are sown too soon after a green manure crop has been ploughed in.

Soil aeration and microbial activity

Most aerobic soil micro-organisms have a very efficient enzyme system for absorbing free oxygen based on cytochrome oxidase as their terminal oxidase system, which only falls to half the rate for a fully aerobic system when the oxygen concentration in the solution at the site of the enzyme has fallen to about 2.5×10^{-8} M.[1] For most organisms, however, the site of enzyme activity is not at the outer surface of the organisms in contact with the solution, but within the bacterial cytoplasm which is separated from the solution by a cell wall, and sometimes by extra-cellular polysaccharide gums. Thus, the actual oxygen concentration in the bathing solution necessary to reduce the activity of the terminal oxidase by half is considerably greater than this, and in the complex environment of the soil, lies between 3 and 6×10^{-6} M.[2] This should be compared with the concentration of oxygen in solution when in equilibrium with air at 20 to 25°C of 2.7×10^{-4} M, which is between fifty- and a hundredfold larger. There are, however, a number of organisms incapable of anaerobic metabolism which have their growth affected by oxygen concentrations considerably higher than this, presumably due either to the organisms not using a cytochrome-oxidase

1 Quoted by D. M. Griffin, *New Phytol.*, 1968, **67**, 561.
2 D. J. Greenwood, *Pl. Soil*, 1961, **14**, 360; 1962, **17**, 365, 378. Also with D. Goodman, *J. Sci. Fd. Agric.*, 1964, **15**, 579.

system, or using this system but having a high resistance to oxygen diffusion to the sites of enzyme activity.

This result has an important consequence for it implies that a soil is fully aerobic as long as the oxygen concentration in the solution does not fall below the figure corresponding to a partial pressure of 1 per cent of the pressure of oxygen in the free atmosphere. Until the oxygen concentration has fallen to about this, the microbial population will use a fully aerobic metabolism, but, once it has fallen below, the metabolism will rapidly become characteristic of an anaerobic system. Products of reduction such as nitrous oxide from nitrates, or hydrocarbons such as methane and ethylene from organic matter decomposition, will occur only if there are soil volumes containing active micro-organisms and an oxygen concentration of less than between 2 to 4×10^{-6} M. Now that the presence of these gases or vapours can be easily detected and their concentration measured by the use of gas chromatography, this technique offers one of the most reliable methods available for determining if the whole body of the soil is aerobic or if there are anaerobic pockets within the soil crumbs or clods.

The effect of carbon dioxide concentration in the soil air, as distinct from the oxygen concentration, on microbial activity has been little studied. A. Macfadyan[1] found that, on an acid sandy heath, a concentration of 0·25 per cent carbon dioxide markedly reduced its oxygen consumption, and a concentration of 0·8 per cent suppressed it. He also found that, for the soils he worked with, the finer the texture the higher the carbon dioxide concentration needed to reduce the oxygen consumption of the soil by a given proportion; and a grass pasture on clay was the most tolerant of his soils. This implies that the sensitivity of the soil organisms to carbon dioxide increases the better aerated the soil in its normal range of moisture conditions.

Poor aeration affects root growth and functioning principally through the products of reduction metabolised by bacteria when using an anaerobic metabolism, as will be discussed in more detail on p. 675; for the root itself contains efficient cytochrome-oxidase systems. It may be necessary to maintain a rather higher oxygen concentration just outside the root epidermis than outside a bacterial cell, for oxygen may have to diffuse through the living protoplasm of several cells, or in the air spaces around several cells if the whole root is to be adequately aerated.

Soil aeration and plant growth

Plant roots need oxygen to function, and, for most agricultural crops once they have passed the seedling stage, the source of oxygen is the soil air. The roots of marsh plants, such as rice, for example, and many seedlings, have a second source, which is an internal pathway in which oxygen from the atmosphere diffuses from the leaves through the stem to the root. The root

1 *Soil Biol. Biochem.*, 1972, **4**.

systems of crops will also need an aerobic soil if they are sensitive to substances produced by soil bacteria when growing in anaerobic conditions, for some of these substances are very toxic to root cells, and some may upset the balance of growth-factors, such as auxins and abscissic acid within the plant, resulting in epinasty of the leaves or abscission of the reproductive organs. These aspects of aeration and oxygen demand of roots are discussed in more detail in Chapter 25. The plant may also be weakened sufficiently by these substances to become more susceptible to pests and diseases.

It is possible to determine when the root cells of a number of plants begin to be affected by lack of oxygen because they start to convert glucose into ethyl alcohol which appears in the xylem exudate.[1] This alcohol can upset plant development badly, particularly at flowering time for some plants, such as peas and many cereals,[2] causing very serious loss of yield. However the root cells of plants adapted to poor aeration do not behave in this way, for instead of sugars going over to ethyl alcohol, malic acid accumulates in the cells instead, and this is not toxic to the cells.[3] The roots of a number of plants also begin to accumulate γ-amino butyric acid when subjected to waterlogging, and this compound may be a specific indicator of anaerobic conditions. It does not, however, appear to be toxic to the plant.[4]

The oxygen concentration in the water outside the root may need to be much higher than the minimum concentration needed within the cell for its metabolism to remain aerobic, because the oxygen must diffuse from outside the root to all the active cells in the root, and this requires an oxygen gradient. Little is known about the actual gradients necessary, and, in any case, the relevant oxygen diffusion coefficient is likely to depend on the past history of the root, for it affects such properties as the thickness of the cell walls and the air spaces within the roots. It is probable that the root system of most crops functions normally if the oxygen partial pressure outside the root is between 0·06 and 0·02 bar,[5] with the minimum pressure being higher, the higher the temperature. For British conditions, where the soil temperatures rarely rise above 15°C, most roots probably do not have their behaviour affected by lack of oxygen until the partial pressure has fallen below 0·02 bar.

A root subjected to oxygen shortage has its growth or rate of elongation decreased,[6] and the rate of transfer of water[7] and nutrients[8] from the soil to

1 J. M. Fulton and A. E. Erickson, *Proc. Soil Sci. Soc. Amer.*, 1964, **28**, 610. *Trans. 8th Int. Congr. Soil Sci.*, 1964, **2**, 171. E. F. Bolton and A. E. Erickson, *Agron. J.*, 1970, **62**, 220.
2 A. E. Erickson with D. M. van Doren, *Trans. 7th Int. Congr. Soil Sci.*, 1960, **4**, 428; with R. A. Cline, *Proc. Soil Sci. Soc. Amer.*, 1959, **23**, 333.
3 R. M. M. Crawford, *J. exp. Bot.*, 1967, **18**, 458; with P. D. Tyler, *J. Ecol.*, 1969, **57**, 235.
4 J. M. Fulton, A. E. Erickson and N. E. Tolbert, *Agron. J.*, 1964, **56**, 527.
5 D. J. Greenwood, *Trans. 9th Int. Congr. Soil Sci.*, 1968, **1**, 823.
6 C. L. Baneth and N. H. Monteith, *Pl. Soil*, 1966, **25**, 143 (sugar cane). A. R. Grable and E. G. Siemer, *Proc. Soil Sci. Soc. Amer.*, 1968, **32**, 180.
7 C. R. Clement, *Pl. Soil*, 1964, **20**, 265. J. Letey, O. R. Hunt *et al.*, *Proc. Soil Sci. Soc. Amer.*, 1961, **25**, 183. R. M. Hagan, *Pl. Physiol.*, 1950, **25**, 748.
8 For a short review see M. Fried, K. Tensho and F. Zsoldos in *Isotopes and Radiation in Soil—Plant Nutrition Studies*, Proc. Symp. FAO/IAEA, Ankara, 1965, 233.

the xylem tissues is also decreased, possibly both because of reduction in the rate of transfer from the soil into the root cells and from these cells into the xylem stream.[1] Increasing the carbon dioxide concentration around the root has much the same effect as decreasing the oxygen concentration, though this effect is usually small until its partial pressure is about equal to that of oxygen.[2] Plants may sometimes respond to carbon dioxide partial pressures around their roots that are higher than are found in well-aerated soils,[3] for there is some evidence that moderate pressures play a part in causing lime-induced chlorosis of sensitive plants on calcareous or over-limed soils.[4]

In the field, plant roots tend to grow into the coarser pores or channels, which are the regions in the soil profile that only become poorly aerated when the soil is waterlogged, or when water is standing on the soil surface. In so far as roots grow through clods or peds that have an anaerobic core, oxygen can diffuse within the root from regions where it is in contact with oxygenated water to these regions, provided they are not too extensive. Further, root growth takes place faster in regions where the aeration is good than where it is poor. It is, in fact, possible to increase the power of roots to penetrate tight soils by supplying oxygen to the root tips, which can be done experimentally by incorporating calcium peroxide.[5]

Attempts have been made to specify the minimum air content of a soil, at field capacity, needed for crop roots to be well aerated, but these figures have no general validity although they can be very useful for a particular crop on a particular group of soils.[6] Thus, a very wide range of crops can be grown on well-structured clays, although their air content at field capacity is low and the root systems will be much more restricted than if growing on a well-drained sandy soil. As already noted on p. 397, the crop with the well-developed root system may begin growth earlier in the spring than that with a poorly developed system, and that with the good root system may be economic to grow while that with the poor system uneconomic.

A striking feature of many crops growing on soils that are poorly aerated is their pale colour, due, in part at least, to nitrogen deficiency. This can be very striking in a wet spring, and it is often possible to reduce the damage done by wetness by applying additional amounts of nitrogen fertiliser as a top dressing. This may have to be done from the air if the soil is too wet to carry a tractor and fertiliser drill.

1 R. E. Shapiro, G. S. Taylor and G. W. Volk, *Proc. Soil Sci. Soc. Amer.*, 1956, **20**, 193.
2 For an early review see M. B. Russell in *Soil Physical Conditions and Plant Growth*, Agronomy Monog. 2, 1952: L. C. Hammond, W. H. Allaway and W. E. Loomis, *Pl. Physiol.*, 1955, **30**, 155; D. G. Harris and C. H. M. Van Bavel, *Agron. J.*, 1957, **49**, 11.
3 For an example with peas see G. Geisler, *Pl. Physiol.*, 1967, **42**, 305.
4 J. C. Brown, O. R. Hunt *et al.*, *Soil Sci.*, 1959, **88**, 260. For the effect of low oxygen content see E. F. Wallihan, M. J. Garber *et al.*, *Pl. Physiol.*, 1961, **36**, 425.
5 L. D. Baver and R. B. Farnsworth, *Proc. Soil Sci. Soc. Amer.*, 1941, **5**, 45; T. W. Scott and A. E. Erickson, *Agron. J.*, 1964, **56**, 575.
6 For an example with citrus see J. Patt, D. Carmeli and I. Zafrir, *Soil Sci.*, 1966, **102**, 82.

Crops probably differ very considerably in the effect of periods of poor aeration on their yield, though these effects have been little studied. A well-known example is the difference between maize and sorghum, for sorghum yields are much less affected than maize yields in wet seasons, for reasons that are not yet known.

19

The water in soils

Where and how water is held

Soil holds water in two ways: in the interstices or pores or capillaries between the solid particles, and by adsorption on the solid surfaces of the clay and organic matter particles.[1] The mechanism of the adsorption of water on the clay particles has already been discussed in Chapter 8; the mechanism of capillary condensation will be considered here. But at the outset it is important to realise that the water held in even a fairly dry soil cannot be sharply separated into capillary and adsorbed water.

The soil capillaries are not straight uniform tubes, and for that reason it may be best to drop this word and use the words interstices or pores to describe the spaces between soil particles. Soil particles have irregular shapes and so leave irregular spaces between them; they aggregate in crumbs and clods which are separated from each other by spaces; and the soil is filled with channels or tunnels made either by the roots of plants which have since decayed, or by the larger soil fauna, e.g. earthworms and insects. These spaces form an interconnected system filling the whole soil—the pore space of the soil—whose geometry in most soils is characterised by very rapid changes in the width of the space: relatively wide spaces may only connect with other wide spaces through necks which are much narrower.

The water in the soil thus exists in thin films or sheets of very irregular shape and thickness, sometimes bounded by solid particles and sometimes by air menisci. The properties of such thin films, or sheets of water, are dealt with under the heading of capillary phenomena in many books of physics, physical chemistry or colloid science. For the present purpose the following properties of these films are relevant.

Consider a liquid in a capillary tube having a boundary with air. The boundary layer between the liquid and the air is called the meniscus. The meniscus is usually curved; it may make a definite angle—the angle of con-

---

1 For two recent books on this subject see E. C. Childs, *Introduction to the Physical Basis of Soil Water Phenomena*, Wiley, 1969; D. Hillel, *Soil and Water*, Academic Press, 1971.

tact—with the walls of the tube; and it puts the liquid column under a tension T, given by

$$T = \frac{2\sigma}{r} \cos \alpha$$

if the liquid is in a circular tube of radius r, where σ is the surface tension of the liquid and α is the angle of content or angle of wetting, which is usually zero for the system soil mineral particles-water-air; but may be appreciable if the soil contains much organic matter.

A study of the detailed processes of adsorption and capillary condensation shows that it is not even theoretically possible to divide up the water held by a soil into adsorbed and capillary water for three reasons. Water, unlike alcohols and in particular the long-chain alcohols, does not form a definite monomolecular layer on the clay particles; it has a zero angle of wetting with clay particles and with the adsorbed water films on them, and it can exist in capillaries only a few molecules in thickness—so narrow that, for most soils, the water would be in adsorbed multimolecular films.[1] Hence, concepts such as that of hygroscopic water, which is meant to represent the total amount of adsorbed water a soil can hold, are without theoretical foundation.

Suction and pF curves for soils

The intensity with which water is held by a soil can be measured in units based on the concept of suction. If water rises a height h cm in a capillary tube standing in water, one can speak of the air-water meniscus exerting a suction of h cm on the water; alternatively, a suction of h cm of water must be applied to the base of the tube to suck the water down to the same level as the water outside. Thus, water, held by a soil at a suction of h cm, can only be removed from that soil if a suction greater than this is applied to it. Suction can therefore be measured as a pressure, which can either be expressed in absolute units such as dynes per square centimetre or as bars or, in conventional units, as the pressure exerted by a column of water h cm high. If T is measured in dyn/cm^2, $T = g\rho h$ where g is the acceleration due to gravity, about 980 cm/sec^2, and ρ is the density of water in g/cm^3, which is about unity. One can also express the suction in terms of the diameter of a circular capillary tube in which the air-water interface exerts the same suction on the water in the tube as is exerted by the air-water interface in the soil. Since the surface tension of the air-water interface is about 75 dyn/cm the approximate conversion formula is $d = 3/h$, where d is the equivalent diameter of the pore in millimetres and h is the suction in millibars. A suction of 1 bar thus corresponds to an interface of 3 μ equivalent diameter. Since this suction is exerted by the air-water menisci on the soil particles, it pulls the soil particles

1 R. K. Schofield, *Trans. 1st Comm. Int. Soc. Soil Sci.* (Bangor), 1938, A, 38.

together and is the mechanism which causes a soil to shrink on drying, and it is equivalent to putting a soil under an isotropic compressive stress of this magnitude.

The intensity with which water is held in a capillary tube bounded by an air-water meniscus can also be defined in terms of the reduction of its free energy below that of water in bulk, and is measured in units of energy per unit of mass, such as joules per kilogram (J/kg). It is a matter of convenience whether this reduction of free energy is expressed in these units, or in units of energy per unit volume, that is as a pressure or in head of water or suction.[1] The units of conversion are 10^6 dyn/cm^2 = 1 bar = 100 J/kg = 1020 cm water.[2] As already noted the last unit, suction, differs from the other two in that it involves both the density of water and the acceleration due to gravity. In normal research practice little distinction is made between pressure and head, for the difference between them is normally too small to be significant. Since the range of free energies of the water in a soil between its dry and its wet state is very large, R. K. Schofield[3] suggested that the logarithm of the free energy, measured as a head in centimetres and denoted by pF, would be a more convenient unit than the free energy itself; and such a unit is almost essential if one wishes to graph the moisture content of a soil against its suction over a wide moisture content range.

The free energy of the water in a soil depends on the soil temperature, for the surface tension of the air-water interface decreases as the temperature rises. This means that for a soil at constant moisture content, where the air-water menisci remain in pores of about the same diameter, as the temperature rises, so the free energy reduction of the water becomes less. This is illustrated in Fig. 19.1,[4] which shows that the temperature effect becomes greater, the drier the soil.

Water can, however, have its free energy reduced for other reasons than being put under a suction. Thus, it is reduced if a salt is dissolved in it, for this increases its osmotic pressure. The salts usually present in the soil solution do not affect its surface tension appreciably, so that the lowering of free energy of water containing dissolved salts in the soil is approximately the sum of the lowering due to the air-water menisci and the dissolved salts. These free energy reductions are commonly called potentials, that due to the menisci being called capillary or matrix potential, and that due to the dissolved salts the osmotic potential, and the combined effect being the total water potential in the soil.

It is important to note that as a soil becomes drier, or as the water contains

1 For a full discussion see A. T. Corey, R. O. Slatyer and W. D. Kemper in *Irrigation of Agricultural Lands*, Agron. Monogr. 11, 1967.
2 1 atmosphere = 1·013 bar = 1033 cm water at 4°C.
3 *Trans. 3rd Int. Congr. Soil Sci.*, 1935, **2**, 37.
4 S. A. Taylor, *UNESCO Arid Zones Res.*, 1965, **25**, 149. The data for the first figure are given in S. A. Taylor and G. L. Stewart, *Proc. Soil Sci. Soc. Amer.*, 1961, **24**, 171; see also G. S. Campbell and W. H. Gardner, *Proc. Soil. Sci. Soc. Amer.*, 1971, **35**, 8.

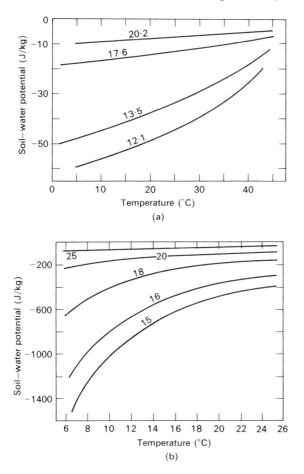

FIG. 19.1. The effect of temperature on the water potential in a soil at constant moisture content: (a) at small potentials (Millville silt loam); (b) at large potentials (Benjamin silty clay loam)

more dissolved salts, its free energy is reduced below that of free water, so the quantity is negative; and the greater the reduction of free energy, the larger is this negative quantity. However, if it is measured as a suction or as a head, the greater the reduction in free energy, the greater is the suction; so this has a positive sign. It is essential therefore to be quite clear what definition of potential or free energy one is using, otherwise there can be a very severe muddle in the use of the words *increasing* and *decreasing* in connection with this quantity.

There are a number of methods available for measuring this reduction of free energy.[1] The total water potential can be measured as the reduction in

1 For a review see F. Cope and E. S. Trickett, *Soils Fert.*, 1965, **28**, 201.

the vapour pressure of the soil water. If e and e_s are the actual vapour pressure of the water in the soil at temperature $T°K$ and the saturation vapour pressure of water at that temperature, then the reduction in free energy $\Delta F = RT/M.ln(e/e_s)$, where M is the molecular weight of water and R is the gas constant. This reduces to $3.17 \times 10^6 log(e/e_s)$ mbar at 20°C, so if the relative humidity of the soil water is 99 per cent, its suction is about 13 bar. In practice it is difficult to determine free energy reductions less than 1 bar when expressed as a suction.[1] The total water potential can also be determined directly by measuring the osmotic pressure of a solution in equilibrium with the soil water when separated from the soil by a membrane permeable to water but not to the ions constituting the dissolved salts, but these techniques are really only applicable to moist soils because of the difficulty of ensuring contact between the soil water and the membrane when the soil begins to become dry.[2]

The matric potential is measured directly as a suction using a tensiometer or ceramic plate, but these are only suitable for suctions less than about 0.8 bar. Above this value it can be measured by a pressure plate technique,[3] which is suitable for tensions up to the level at which the soil water films cease to be continuous, but is of only limited value when the capillary conductivity of the water becomes so low that it takes a long time for the soil sample to reach equilibrium. In practice its useful range is from 1 to 15 bar, but it is of very limited value for poorly structured clays. It can also be measured as an osmotic pressure by determining the osmotic pressure of a solution of a high molecular weight compound, such as a polyethylene glycol, that is in equilibrium with the soil when separated from it by a membrane that is permeable to water and to simple ions but not to these large molecules.[4] It can also be determined indirectly by allowing a plaster-of-paris block, which has pores requiring suctions of between 1 and 10 bar to empty them of water to come into equilibrium with the soil, and measuring its electrical conductivity; but the method is unsuited to soils containing much soluble salts.

The soil moisture-characteristic curve

This is the name given to the curve obtained when the moisture content of a soil is plotted against its suction, or more usually its pF. Since the suction is inversely proportional to the effective radius of the pores containing the air-water menisci, the slope of the moisture-characteristic curve plotted against the suction gives a picture of the size distribution of the pores in the soil; and

1 J. L. Monteith and P. C. Owen, *J. Sci. Instr.*, 1958, **35**, 433.
2 L. A. Richards and G. Ogata, *Proc. Soil Sci. Soc. Amer.*, 1961, **25**, 456.
3 See, for example, L. A. Richards *Soil Sci.*, 1941, **51**, 377; 1949, **68**, 95; *Agric. Engng.*, 1947, **28**, 451.
4 B. Zur, *Soil Sci.*, 1966, **102**, 394. L. J. Waldron, L. T. Manbeian, ibid., 1970, **110**, 401.

in particular it will show if some pore sizes are much more common than others. Fig. 19.2 gives these two curves for the suction range 0–30 mbar, for a soil composed of crumbs of a Gault clay soil 1–2 mm in size.[1] In this example there is a concentration of pores that are emptied by a suction of 7 mbar, corresponding to pores about 0·4 mm in size. But, in fact, the moisture-characteristic curve must be interpreted with great caution, for most soils do not possess a unique curve, because over the range of suctions of interest in agriculture the moisture content at a particular suction is likely to depend on its past history of wetting and drying.

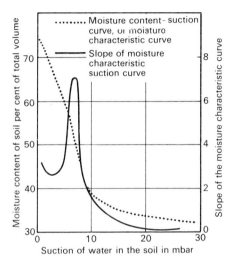

FIG. 19.2. The moisture content-suction curve and its differential for a Gault clay. (Soil initially in crumbs 1–2 mm size)

The pore space of sandy soils is nearly always cellular, that is it consists of volumes having a relatively large effective radius bounded by necks having a smaller radius. When the wet soil dries, that is when water is being sucked out of the soil, the effective diameter of a pore is that of its largest bounding neck in contact with air, for the pore cannot be emptied until the suction is sufficient to break the meniscus in this neck. But when the dry soil is wetting, the pore cannot fill until the suction has fallen low enough for the meniscus of maximum surface area to have been formed, and then the rest of the pore fills immediately. Such a soil holds more water at a given suction when water is being withdrawn than when it is being added. Figure 19·3 illustrates this dependence of the moisture-characteristic curve on past history, and such a curve is called a hysteresis curve, and the loop a hysteresis loop.[2]

1 Taken from E. C. Childs, *Soil Sci.*, 1940, **50**, 246.
2 For a full discussion of the hysteresis loops in non-swelling soils, see E. C. Childs, *The Physical Basis of Soil Water Phenomena*, Wiley, 1969.

This cellular pore space can produce a hysteresis loop for a second reason, for as the dry soil wets, a certain proportion of the coarser pores will have the air they contain sealed within them, because the finer necks will become filled and the air enclosed as a bubble. In so far as these hysteresis curves are obtained fairly rapidly these bubbles often form in the same pores, and so their presence need not prevent the hysteresis curve being reproduced. But the air in the bubble slowly dissolves in the water, and diffuses out into the atmosphere. The full effects of entrapped air on the hysteresis curves has not been fully worked out, as there are a number of observations for which there is no satisfactory quantitative explanation.[1]

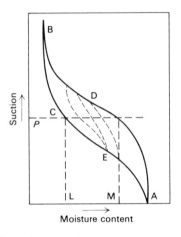

Fig. 19.3. Diagrammatic suction—moisture content curve showing hysteresis loops. At a suction of *P*, the soil may have any moisture content between L and M

A soil can be in equilibrium with moisture at a suction of P in Fig. 19·3 for any moisture content between L and M, for if the soil begins to dry from the wet end, from point A for example, and by the time it has reached point D it begins to wet up, the new wetting curve will lie inside the hysteresis loop. In the same way if the dry soil is wetting up and it begins to dry when it has reached point E the new drying curve will lie inside the loop. Thus the pores that hold water at point L are always full when the suction is P, and the pores that hold air at point M always hold air at this suction; but there is a group of pores, whose volume is represented by LM which may or may not hold water at this suction depending on the past wetting and drying regimes to which the soil has been subjected.

Most soils differ from sands in that they swell on wetting and shrink on drying, because of the changes in volume of the clay fraction with moisture

1 See, for example, J. W. Carey, *Soil Sci.*, 1967, **104**, 174; A. J. Peck, *Aust. J. Soil Res.*, 1969, **7**, 79.

content. Thus, for these soils, the size distribution of pores depends on the soil moisture content. But as a moist soil dries, the forces causing it to shrink can be very considerable, for the greater the suction of the water, the greater the forces pulling the soil particles together. On the other hand as such a soil wets, the forces pulling the soil particles apart become weaker as it wets, so it may take a very long time for the dry crumbs to have swollen to their initial size on rewetting. W. W. Emerson,[1] for example, found that subsoil crumbs of the Rothamsted clay-loam soil took over three months to swell to their fully saturated volume after they had been dried to 15 bar suction. This resistance to swelling gives, therefore, a time-dependent hysteresis curve, being more marked the higher the clay content.

The forces set up between soil particles, pulling them together as the soil dries, are likely to cause repacking of the soil particles if they have been disturbed. Figure 19.4, taken from some work of L. A. Richards,[2] shows this

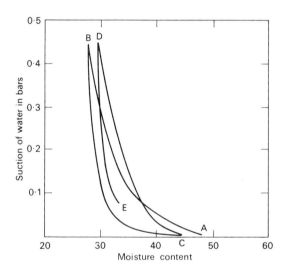

FIG. 19.4. Wetting and drying curve for a Tama silt loam soil, showing hysteresis. The course taken is A–B–C–D–E

for a disturbed silt loam soil. Drying the saturated soil to a suction of about 450 mbar water has caused a repacking of the particles so that on rewetting, the moisture content of the saturated soil has fallen from about 50 to 45 per cent; but on redrying to 450 mbar, the moisture content has risen from about 27 to 30 per cent, due to some of the coarser pores having become finer. This effect is usually not seen if an undisturbed soil is taken through a wetting and drying cycle repeatedly. Thus, undisturbed subsoils, in so far as they show

1 *J. Soil Sci.*, 1955, **6**, 147.
2 *J. Amer. Soc. Agron.*, 1941, **33**, 778.

hysteresis curves, give closed curves if taken between saturation and wilting point at 15 bar suction, for they will have been subjected to these cycles over a long period of time, so there is no irreversible repacking of the soil particles; but if they are dried to a higher suction, the particles may undergo repacking. This effect is usually most noticeable with clay subsoils, which typically show a very narrow hysteresis loop.[1]

There are three other points of importance about the moisture characteristic curve of a soil. Firstly, well-structured soils typically have a point of inflection on the drying curve corresponding to the emptying of the coarser pores between the crumbs, which is usually complete at suctions less than 50 mbar. Secondly there is a tendency for the pF-moisture content curve to be approximately linear over much of the moisture range of importance for

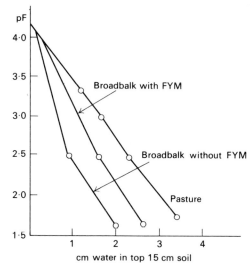

Fig. 19.5. Effect of level of organic matter on the pF-moisture content curve for Rothamsted soil

crop production, particularly for soils well-supplied with humus. This is illustrated in Fig. 19.5 for three Rothamsted soils,[2] and shows that the moisture characteristic curve based on the pF scale is almost linear for the pasture soil but is concave for the arable soil that has received no farmyard manure. Thirdly poorly-structured clay soils tend to have few coarse pores, so that suctions, sometimes in excess of 1 bar, must be applied before the clay begins to lose water, or begins to have any air-filled pores. Such a soil will have a pF-moisture content curve that is strongly convex. Loss of

1 See, for example, D. Croney, *Geotechn.*, 1952, **3**, 1.
2 P. J. Salter, and J. B. Williams, *J. agric. Sci.*, 1969, **73**, 155.

structure due to compaction,[1] either brought about by traffic, such as tractors and machinery running over the land, or by weather, reduces the coarse pores and increases the fine pores, so increases the amount of water held at moderate suctions.

The effect of temperature on the moisture characteristic curve is complicated. A rise of temperature reduces the surface tension of the air-water meniscus, which reduces the suction of the water. But it also causes an increase in volume of any entrapped air and this will cause an expansion of the soil and hence a widening of the pores.[2] This will also cause a fall in the suction. Thus, both effects work in the same direction, but their relative magnitude depends on the amount of entrapped air and the ease with which a rise in pressure of this air will result in an expansion of volume.

The movement of water in soils

Water can move by liquid flow through a soil under a pressure gradient, under gravity (drainage), or under a suction gradient (capillarity), and the rate of movement is controlled by the size and continuity of the pores containing the water and by its viscosity. Under a given pressure gradient water moves fastest through a soil when it is saturated, that is when all the pores are full of water, but as the soil becomes unsaturated, that is as the proportion of pores containing air increases, the rate of flow decreases, usually very rapidly. This is because the pores which get emptied of water first are the widest, so, as the soil dries out, water flow takes place in increasingly thinner films along paths becoming increasingly more tortuous; and as the films become thinner, so the effect of viscosity rapidly becomes of more importance. This can be readily understood if it is borne in mind that the rate of flow of water through a capillary tube depends on the fourth power of its diameter, so that if the pressure gradient remains constant, halving the diameter reduces the flow sixteenfold.

The rate of flow at constant temperature under a suction gradient can be expressed formally by the equation:

$$Q = -K \, ds/dl$$

where Q is the water flux across a unit area perpendicular to the direction of flow, ds/dl is the suction gradient, and K is the hydraulic or capillary conductivity of the soil under the particular circumstances prevailing. These two words, hydraulic and capillary, are often used interchangeably, but some authors restrict the term hydraulic conductivity to saturated soils and keep capillary conductivity for unsaturated soils. K is, in fact, dependent on the suction of the water in the volume, not on the suction gradient, so if the soil

---

1 J. E. Box and S. A. Taylor, *Proc. Soil Sci. Soc. Amer.*, 1962, **26**, 119. J. N. S. Hill and M. E. Sumner, *Soil Sci.*, 1967, **103**, 234.
2 R. Gardner, *Soil Sci.*, 1955, **79**, 257. R. S. Chahal, ibid., 1965, **100**, 262.

moisture is maintained at constant suction, K is a constant, provided the temperature also remains constant. But K depends on properties of the liquid as well as of the soil, and it is possible to define a soil permeability k which is dependent only on the soil properties and not on the liquid, for $K = k\rho g/\eta$, where ρ is the density of water, g the acceleration due to gravity and η is the viscosity of the water. A third unit that is sometimes used is the diffusivity D, which is defined by $Q = -D \, dm/dl$, where m is the moisture content of the soil, so that dm/dl is the moisture content gradient in the soil. The relation between diffusivity and conductivity is:

$$D = K \, ds/dm$$

where ds/dm is the slope of the moisture-characteristic curve at suction s.

The units that K is measured in depend on the units used to express the reduction of free energy of the water in the soil. If they are expressed heads or suction, ds/dl is formally dimensionless and K is expressed as a velocity, in cm/sec, but if they are expressed as an energy per unit mass, e.g. joules per kilogram, K is expressed as a time, usually seconds.

Figure 19.6 shows the relation between the capillary conductivity and suction, between the diffusivity and moisture content, and between suction and moisture content for a Californian clay soil and a sandy loam,[1] and they show how rapidly the conductivity and diffusivity can drop with increasing suction or decreasing moisture content, even for a relatively small increase in suction at the wet end. The diffusivity curves, however, must be interpreted carefully, because there is no unique relationship between suction and moisture content over the moisture range where hysteresis is appreciable, so that diffusivity depends on the past moisture history of the soil.[2] The capillary conductivity, on the other hand, is very much more strongly dependent on moisture content and much less affected by the hysteresis effect.[3]

This strong dependence of conductivity on suction is due, as already noted, to the rapidly increasing effect of viscosity and tortuosity in slowing down the rate of flow under a given suction gradient. W. D. Kemper and J. B. Rollins[4] calculated the effective thickness of the water films at different suctions in a soil to give the experimentally determined conductivities based on the moisture-characteristic curve for that soil, allowing for the effects of viscosity and tortuosity, and their results are given in Fig. 19.7. It shows, as one would expect, that the effective film thickness is a great deal less than the pore diameter that contains the air-water meniscus. Thus at 0·3 bar the air-water meniscus should be in pores about 10^{-2} mm diameter, but the effective

1 These figures are taken from W. R. Gardner, *Soil Sci.*, 1960, **89**, 63, and *UNESCO Arid Zone Res.*, 1960, **15**, 37.
2 A. Klute, F. D. Whisler and E. J. Scott, *Proc. Soil Sci. Soc. Amer.*, 1964, **28**, 160.
3 G. C. Topp and E. E. Miller, *Proc. Soil Sci. Soc. Amer.*, 1966, **30**, 156; D. A. Rose, *J. Soil Sci.*, 1971, **22**, 490.
4 *Proc. Soil Sci. Soc. Amer.*, 1966, **30**, 529.

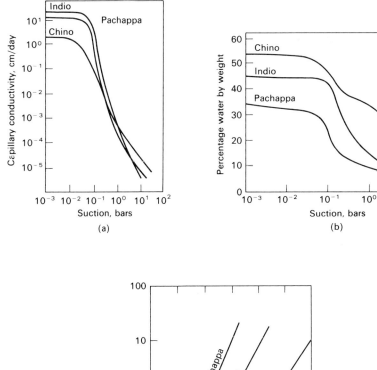

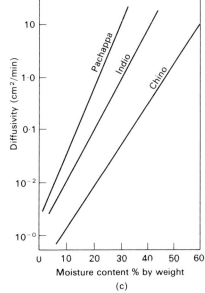

FIG. 19.6. Relation between the capillary conductivity and suction, and between diffusivity and moisture content for Pachappa sandy loam, Indio loam and Chino clay

film thickness for conductivity is fifty times smaller. It is in fact possible to calculate the capillary conductivity-moisture content curve from the mois-ture-characteristic curve for soils, making some assumptions about the distribution of pore space, which gives a curve reasonably in agreement with the experimentally determined curves.[1]

Capillary conductivity is dependent on the continuity of the water films in the soil, and for loam and clay soils these remain continuous to relatively high suctions, often exceeding 50 bar. But the films may become discon-tinuous for sandy soils at suctions as low as 10 bar.[2] Continuity can be proved by dissolving a salt in the water, for if the water is moving through the soil as a liquid, it will carry the dissolved salt with it.

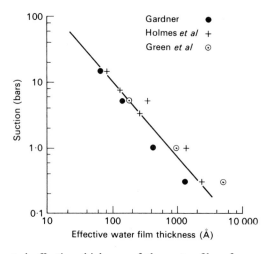

FIG. 19.7. Estimated effective thickness of the water films from measurements of hydraulic conductivity of a montmorillonite paste. Data: W. R. Gardner, *Proc. Soil Sci. Soc. Amer.*, 1956, **20**, 317; J. W. Holmes *et al.*, *Trans. 7th Int. Cong. Soil Sci.*, 1960, **1**, 188; R. E. Green *et al.*, *Proc. Soil Sci. Soc. Amer.*, 1964, **28**, 15

Water, however, can move down a suction gradient by vapour diffusion as well as by liquid flow, for the vapour pressure of the water decreases as its suction increases. When this type of movement takes place, any salt dis-solved in the water will not move with the vapour, so that if initially the salt concentration in the water is uniform, vapour transfer will cause the salt to become more concentrated from the volumes losing vapour and more dilute in those gaining vapour. Vapour transfer can take place only through the pores containing air (see p. 405), so that the smaller the air space, the slower the rate of diffusion of the vapour. Further, it takes place more rapidly the

1 R. J. Kunze, G. Uehara and K. Graham, *Proc. Soil Sci. Soc. Amer.*, 1968, **32**, 760.
2 R. D. Jackson, *Proc. Soil Sci. Soc. Amer.*, 1965, **29**, 144.

greater the vapour pressure gradient in the soil, so can only be important when the suction has risen to an appreciable value. It is, however, very slow in a really dry soil, for the soil will then absorb the water vapour strongly, but it becomes much more rapid once water films have begun to form in suitable soil interstices. It is the major process operative in soils of moderate dryness, but by the time the suction has fallen to 15 bar, liquid flow is usually more important than vapour flow.[1]

The discussion so far has assumed that the soil temperature remains constant, although this cannot be strictly true when vapour transport is taking place, for the evaporation of water cools the moister soil and condensation warms the drier. But in normal field conditions there are always temperature gradients in the soil profile, and if the suction gradients are not too large, water vapour will diffuse from the warmer to the cooler regions. Again this will go most rapidly in moderately dry soil, when there is a reasonably large volume of air-filled pores for the vapour to move through. This vapour transfer from warm to cool volumes of the soil will tend to increase the suction of the water in the warm areas, because they are losing water, and decrease the suction in the cooler areas, so this sets up a suction gradient and liquid water will tend to move from the cool to the warm regions along this gradient. This liquid flow can also be followed by dissolving a salt in the water, for it will be found that the salt moves from the cool to the warm region.[2] However, under conditions where vapour transfer is minor or negligible compared with liquid flow, water will tend to flow from warmer to cooler regions because, for a given moisture content, the suction of a soil rises as its temperature falls. This can be of practical consequence in moist soils during periods of frost, for water will move from the warmer subsoil to the surface soil, freeze there, and will sometimes cause frost heaving.[3] This is a separate cause of frost heaving from that due to the freezing of a waterlogged soil, for then the heaving is due to the water already in the zone that is being frozen.

Entry of water into a soil

The rate that free water will enter a moist soil when run over its surface as irrigation or when it falls as rain, depends on the distribution of pores and cracks in the soil profile; for it will move easily only in pores that are emptied of water by suctions of about 10 mbar, it will move slowly through pores requiring a suction as high as 40 mbar, and very slowly through finer pores.[4]

1 D. A. Rose, *Brit. J. appl. Phys.*, 1963, **14**, 256, 491, and see also H. E. Jones and H. Kohnke, *Proc. Soil Sci. Soc. Amer.*, 1952, **16**, 245.
2 C. G. Gurr, T. J. Marshall and J. T. Hutton, *Soil Sci.*, 1952, **74**, 335; R. D. Jackson, D. A. Rose and H. L. Penman, *Nature*, 1965, **205**, 314.
3 For a discussion see S. Henin and O. Robichet, *Comp. Rend.*, 1951, **232**, 2358, and E. J. Kinbacher and H. M. Laude, *Agron. J.*, 1955, **47**, 415.
4 See, for example, R. M. Smith, D. R. Browning and C. G. Pohlman, *Soil Sci.*, 1944, **57**, 197; G. R. Free, G. M. Browning and G. W. Musgrave, *U.S. Dept. Agric.*, *Tech. Bull.* 729, 1940.

The rate of movement through these pores is dependent on the soil temperature through its effect on the viscosity of water. The higher the soil temperature, the finer the pore through which water can flow under gravity at a predetermined rate. This restriction of water movement to the coarser pores has the consequence that, in general, water does not move uniformly through the soil, but is restricted to certain preferred channels. If the soil was initially rather dry, the wetting front may move fairly uniformly through the soil, but once it is wet, much of the body of the soil will not have any water percolating through it, unless it is a fairly coarse sandy soil.

The rate of entry of water into a soil is called its infiltration rate, and this rate is initially high for all soils if they are dry. But once they are wet, this rate is dependent on the distribution, continuity and stability of the coarse pores. Very permeable soils will have infiltration rates as high as 10^{-2} cm/sec, or about 35 cm/h, while soils of low permeability will have rates of 10^{-5} cm/sec or 0·03 cm/h, or less.

The infiltration rate of most medium- or fine-textured soils kept flooded becomes less with time for a number of reasons. If the soil is initially dry, wetting the soil will entrap air in the coarser pores; and even if the soil was initially moist some of the air present in the wider pores may become entrapped. These air bubbles will block these pores and slow up the passage of water through them, although with time the air in these bubbles will dissolve in the water and diffuse into the atmosphere. Again most soil crumbs, if they contain much clay or organic matter, swell on wetting, and this swelling will also make the pores more narrow. But since this swelling may be a slow process for the heavier-textured soils, such as clay loams and clays, cracks produced by the crop drying the soil in summer often persist through the autumn and early winter, allowing water to drain through the soil, but they may become almost sealed by late winter and spring so the soil will have become almost impermeable.[1]

The infiltration rate of a soil can only be maintained if the system of coarser pores is maintained. The zone where this system is most likely to collapse is in the surface of the soil, for wet soil crumbs are weak and can easily be broken by fast-falling rain drops, which cause soil particles to become detached from the crumbs and block the coarser surface pores. The crumbs and walls of the coarser pores in some soils may collapse spontaneously on wetting, and this is particularly liable to happen if a dry soil is suddenly flooded; or they may collapse slowly if the soil becomes waterlogged for any reason. The maintenance of permeability in the surface layer of the soil is one of the major problems of good soil management in soils of rather low structural stability, and failure to maintain permeability can lead to loss of crop through poor aeration, and to loss of water or soil by run-off and soil erosion.

1 For an example from Rothamsted see W. W. Emerson, *J. Soil Sci.*, 1955, **6**, 147.

When a limited amount of water is added to a deep dry soil in good structure, the water tends to move down with a fairly sharply defined wetting front, but once the free water on the soil surface has all percolated into the soil, the wetting front soon becomes almost stationary. This line of demarcation between the wet and the dry soil may remain fairly sharp for a matter of days or sometimes weeks; and it is likely to remain sharper the steeper the slope of the moisture-characteristic curve at the pF of the water in the front, and to become diffused quicker the more gentle the slope of this curve at this pF.[1] Figure 19.8 illustrates this difference of behaviour between soils;[2]

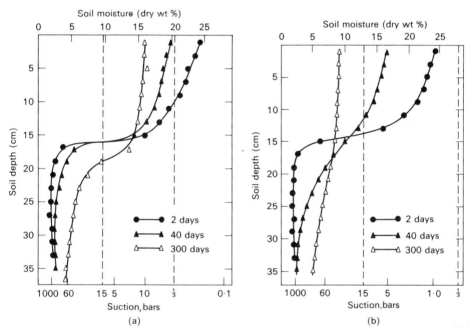

FIG. 19.8. Rate of movement of a limited volume of water into a dry soil column. (a) Clarion loam. (b) Mumford silty clay loam

it shows how the boundary between the wet and the dry soil remains fairly sharp for the Clarion loam, but disappears after a time for the Mumford silty clay loam. The reason for this is the very great resistance to flow of the water from the thin films at the wetting front to the still thinner films in the dry soil. With time the front will always become diffuse because of vapour transfer, particularly if there is a temperature gradient from the warmer moist soil to the cooler drier soil.

1 F. D. Whisler and A. Klute, *Proc. Soil Sci. Soc. Amer.*, 1965, **29**, 489.
2 From D. J. Sykes and W. E. Loomis, *Soil Sci.*, 1967, **104**, 163.

This slow movement of water from a moist to a dry soil, once the suction of the moist soil has risen above a certain value, can be illustrated by an experiment made by F. J. Veihmeyer.[1] He packed a cylinder 60 cm long with soil containing 22 per cent moisture, at a suction of about 200 mbar water and placed it between two other cylinders containing the same soil, but with about 14 per cent moisture at a suction of about 15 bar. After 139 days the moisture content of the wetter soil had fallen to about 17 per cent and the drier soil had its moisture content increased to a distance of about 30 cm from the original wet front. L. A. Richards[2] has also given a good example of this effect. He found it took over three months for equilibrium to be reached when 7·5 cm of air-dry soil was wetted with water under a suction of 1·5 bar, though if the soil was previously damp, equilibrium was attained in two to three days.

This same effect is seen if a water-table rises in a dry soil and then stays at a constant level.[3] Water will be sucked up by capillarity into the dry soil giving a fairly uniform wetting front, but after a time the front will cease to rise and the relatively sharp boundary will only slowly become diffuse for the same reason.

In the field, however, temperature gradients can be an important cause of water movement in a soil profile. An experiment by N. F. Edlefson and G. B. Bodman[4] at Davis, California, illustrates this point very clearly. They irrigated heavily a deep silt loam, and then covered its surface to prevent any evaporation of water and determined the amount of water in each 15 cm layer down to 2·7 m at increasing intervals of time after the soil surface had

TABLE 19.1   Water draining out of a silt loam profile. (Davis, Calif.)

Sampling dates	3/10/34– 6/10	6/10– 17/10	17/10– 30/11	30/11/34– 8/5/35	8/5– 31/8	31/8/35– 22/1/37
No. of days of drainage	3	11	44	159	115	510
Cm water draining out of						
top 1·2 m soil	24·2	3·8	4·0	0·5	6·8	1·0
top 2·7 m soil	44·5	10·4	10·1	2·0	13·4	3·3

ceased to be waterlogged, which was on 3 October 1934. Some of their results are given in Table 19.1, which shows that from the fourteenth to the fifty-eighth day after the commencement of the experiment 10 cm of water had drained out of the profile, of which 4 cm had come from the first 1·2 m, that

1 *Hilgardia*, 1927, **2**, 125.
2 *J. Amer. Soc. Agron.*, 1941, **33**, 778.
3 For an example see R. E. Moore, *Hilgardia*, 1939, **12**, 383.
4 *J. Amer. Soc. Agron.*, 1941, **33**, 713.

drainage almost ceased throughout the winter and spring, but as the soil warmed up in the summer another 13·4 cm left the profile, while during the next summer a further 3·3 cm left. This soil was still losing water during summertime by drainage two and a half years after this irrigation. But water movements in this profile are complicated by temperature gradients, for during the spring of 1936 the top 1·2 m and the bottom 45 cm were both losing water, but the middle section from 1·2 to 2·2 m below the surface was gaining water.

Field capacity

The moisture content of a number of deep well-drained soils, in which the profiles have been wetted by rain or irrigation, fall to a fairly definite value, and the suction of the water rises to a fairly definite value, a few days after all the surface water has seeped into the soil, and these values only change slowly thereafter if evaporation is prevented. A soil in this condition is said to be a field capacity, and the suction of the water in the soil in this condition lies in the range 50 to 350 mbar, depending on the soil and the soil temperature.

As already noted the soil temperature affects the rate that water can move through the soil under a given suction gradient due to its effect on its viscosity, so that raising the soil temperature increases the speed with which water can leak out of the profile, and this leak is usually large enough to cause the suction of the residual water to rise, in spite of the effect of a rise of temperature in lowering the suction of the water if the water content had remained constant. This effect of temperature on moisture content can be quite noticeable in the United Kingdom, for example, as between winter and summer.[1] The size distribution of the pores in field soils which can be emptied of water in the field capacity range of suctions can be altered by management and by alterations in soil structure; and in particular any practice which encourages collapse of soil structure and a rise in bulk density of the soil typically converts pores which could be emptied of water by drainage with pores which cannot, so increases its moisture content at field capacity.

A soil at field capacity is as dry as it can be under drainage alone—it can be described as wet but well drained, so its air content depends entirely on the proportion of coarse pores. The Rothamsted clay loam, for example, may contain only 1 to 2 per cent of air in this condition. Most British soils drain to a suction of about 50 mbar in winter, which corresponds to emptying pores wider than 60 μ equivalent diameter, but in the subtropics in the hot season drainage can give suctions as high as 350 mbar corresponding to emptying pores about 10 μ diameter. Thus, at Rothamsted, drainage of

1 See, for example, P. J. Salter and J. B. Williams, *J. Soil Sci.*, 1965, **16**, 1; 1967, **18**, 318.

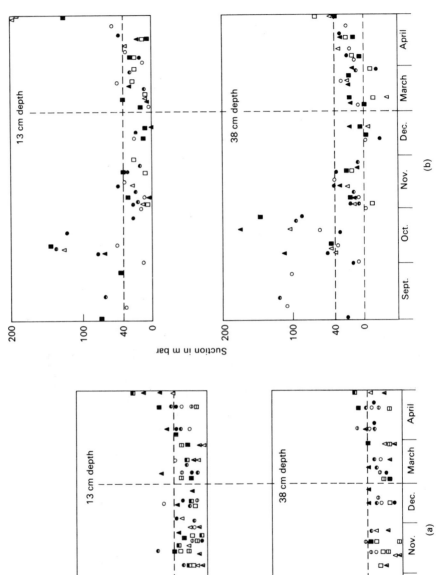

(b)

(a)

water out of a block of undisturbed soil 1·5 m deep ceases 2 to 3 days after rain, although there will be a trickle of a few hundredths of a millimetre a day for a further 2 to 3 days in summer and 10 to 12 days in winter. This soil loses 11 mm of water during this time so that it contains only 0·7 per cent by volume of air, showing that a well-drained soil need not have an appreciable air content.

There have been relatively few determinations of the actual suctions of water in wet well-drained soils in the field, under conditions of low evaporation; but drainage will probably only raise the suction of water to about 50 mbar in many temperate soils in winter time. Figure 19.9a, due to R. Webster and P. H. T. Beckett,[1] illustrates this for a number of sites on two soil series—the Frilford and the Fyfield—in the Oxford district from October 1961 to April 1962, excluding two months when the ground was frozen. These soils are well-drained loamy sands to sandy loams overlying sandy loams to sandy clay loams, derived in part from Jurassic limestone; and Fig. 19.9a shows that the soil suction, at both 13 and 38 cm depth, rises to a fairly definite upper limit of 50 mbar during the winter period. Fig. 19.9b gives the suctions in a number of soils on the Sherborne series, which is also apparently well drained and derived from the same limestone, but with loam to clay loam textures. The suction does not rise to any very definite suction, though it approaches 40 mbar rather than 50, but this could be due to drainage being sufficiently slow at these low soil temperatures for rain to have occurred before equilibrium has been reached. It is interesting to note that although the profile shows no signs of impeded drainage, on several occasions water was at a positive pressure at both depths. Webster and Beckett also showed that soils belonging to two other series, with textures between sandy loam and clay loam but overlying limestone rubble, both behaved as the Frilford-Fyfield soils; and soils from three other series whose profiles showed signs of impeded drainage, showed even less signs of a definite field capacity, and positive water pressures were found more frequently.

1 *J. agric. Sci.*, 1972, **78**, 379.

FIG. 19.9. Actual suctions of water in field soils during periods of low evaporation. (a) Loamy sand to sandy loam soils (Frilford and Fyfield Series). Period 1 November 1962 to 30 April 1963.

depth	mean	66% confidence limit		field capacity
		lower	upper	
13 cm	35 mbar	26	48	50 cm
38 cm	43 mbar	22	62	50 cm

(b) Calcareous loam over clay loam (Sherborne Series). Period 1 November 1962 to 30 April 1963.

depth	mean	66% confidence limit		
		lower	upper	
13 cm	29 mbar	17	31	possible field
38 cm	13 mbar	8	27	capacity about 40 cm

It is not yet certain under what conditions soils possess a definite field capacity. They must obviously be well drained, but it is probable that only soils whose capillary conductivity falls rapidly with increasing suction in the suction range 50–300 mbar will show this property clearly. These are presumably soils with two fairly distinct pore-space distributions—a continuous system of coarse pores and cracks, extending around the principal crumbs or peds—the structural pores—and a rather more discontinous system of finer pores within the crumbs. Thus rather structureless fine sand and silt soils do not possess a definite field capacity, but show the phenomenon of delayed drainage, that is when drainage out of the profile continues at a measurable rate for weeks after water has ceased to be added to the soil; and the silt loam soil illustrated in Table 19.1 is an example of such a soil.

The moisture content of a given soil layer when wet but well drained can be affected by changes of texture in the profile below it. Thus a finer-textured soil layer overlying a coarser-textured layer, particularly if the boundary between the two layers is fairly sharp, will hold more water against drainage than if it is underlain by material of its own texture or of a finer texture.[1] This is very noticeable if the coarser-textured layer is a sand, for there will be a sudden change of capillary conductivity at this junction when the sand has lost its interstitial water, so water will only be able to leak out from the finer-textured soil very slowly because of the low capillary conductivity of the sand. This effect is naturally not seen in a coarse-textured soil when it overlies a finer. Again, if only a limited amount of water is added to a dry soil, so that a wetting front is formed, the suction of the water in the moist soil close to the front is very much higher than corresponds to field capacity, and the effect of the front on the suction may be noticeable for a considerable distance above it, the actual distance depending on the texture and structure of the soil.

The apparent field capacity is sometimes dependent on the rate of wetting. If a dry soil is flooded, there is likely to be a considerable volume of air entrapped in the coarse pores and this may slow up drainage so much that the soil will reach its apparent field capacity at a higher moisture content than if it had been wetted slowly. This effect is likely to be most noticeable either in soils high in organic matter, which wet slowly, or in clay or clay loam soils, which will absorb more air and have a finer set of pores than the more coarse-textured sands or sandy loams.[2]

Field capacity has been found to be a concept of very considerable utility for many purposes, but it is not a well-defined soil property. In so far as it is defined as the moisture content of a deep soil 2 days after all free water

1 For examples see J. R. Eagleman and V. C. Jamison, *Proc. Soil Sci. Soc. Amer.*, 1962, **26**, 519; D. E. Miller and W. C. Bunger, ibid., 1963, **27**, 586: and for the mathematical basis R. J. Hanks and S. A. Bowers, ibid., 1963, **27**, 530.
2 For an example see E. Bresler, W. D. Kemper and R. J. Hanks, *Proc. Soil Sci. Amer.*, 1969, **33**, 832.

has left the soil surface and in the absence of evaporation, it is a measurable property of the soil, though not necessarily a reproducible one.

Drainage to a water-table

A water-table affects the suction of water in a wet soil which possesses a fairly definite field capacity to a height above the table equal to the suction at field capacity when drainage has ceased. The suction at the top of the water-table is by definition zero, and it then increases uniformly with height above it, by one centimetre for each centimetre rise, until it has reached its value for the soil at field capacity, and it then remains constant up to the soil surface. The zone above the water-table at which the suction is increasing is called the capillary fringe, though this term is not restricted to this condition: it is also used for the zone above the water table when water is rising by capillarity into a dry soil and in this case the fringe is thicker than in the former, for the suction at the top of the fringe will be much higher than corresponds to field capacity. The height of the top of the capillary fringe above the water table is therefore about equal to the soil suction at field capacity, which in England is typically about 50 mbar.

Evaporation of water from a bare soil

The rate of loss of water from a damp soil is controlled by the evaporating power of the air and the heat energy falling on it, and so long as the soil can supply water to the surface fast enough to keep it moist, this rate of loss persists. The rate falls sharply, however, as soon as the rate of loss exceeds the rate of supply sufficiently for the surface to become dry; for then instead of the water vapour being produced at the soil surface, and diffusing immediately into the air to be rapidly removed by the air currents, it is produced below the soil surface and must diffuse through the soil pore space under a small concentration gradient before it reaches the atmosphere. Even a dry layer 1 to 2 mm thick can appreciably reduce the rate of evaporation.[1]

The magnitude and rate of evaporation losses from deep soils without a water-table can be illustrated with results obtained during the last seventy years from the fallow drain-gauges at Rothamsted. The rate of loss of the first 12 mm of water is controlled by the evaporating power of the air, and in summer this lasts for about five days, and the rate then drops rapidly from 2·5 mm per day to about 1 to 2 mm per week, so that it is only after a six-week drought that 23 mm has been lost by evaporation from the bare soil; and the three-month drought of 1921, which lasted from June to August, only caused a total loss of 32 mm,[2] and it is unlikely that more than 6 mm

---

1 H. L. Penman, *J. agric. Sci.*, 1941, **31**, 454, and for a mathematical analysis see J. Philip, *J. Met.*, 1957, **14**, 354.
2 H. L. Penman and R. K. Schofield, *J. agric. Sci.*, 1941, **31**, 74.

of water moved up from the soil layer below 22 cm in depth into the layer 0 to 22 cm deep. Figure 19.10 illustrates this result for the moisture content of a bare soil at Rothamsted on 27 June 1870, after a prolonged drought. Within the accuracy of the determination no water had moved up from below 45 cm and only about 6 mm from below 22 cm. The figure also gives

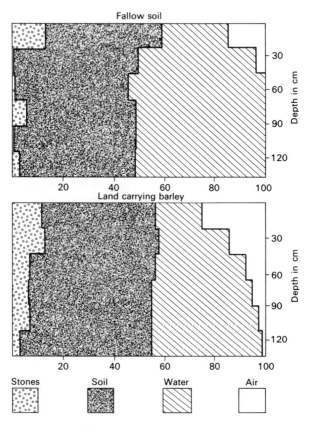

Fig. 19.10. Volumes of soil, water and air at different depths of a fallow and of a cropped soil at Rothamsted on 27 June 1870, after a prolonged drought (R. K. Schofield)

the moisture distribution in a neighbouring plot carrying barley, and shows by way of contrast how the crop reduced the water content of the layer 110 to 135 cm deep. The figure further illustrates the importance of specifying the moisture contents on a volume and not a weight basis—a point that will be stressed later on—for the fallow plot happens to have a larger average pore space than the cropped, and hence can hold more water; so that if the comparison was made between moisture contents calculated on a weight basis, the barley would be judged to have removed much more water from the subsoil than it in fact had done.

This result, that a bare Rothamsted soil can only lose about 12 mm of water rapidly by evaporation is not necessarily true in the laboratory, for there H. L. Penman[1] showed that it could lose about 30 mm at the rate of 2·5 mm per day, that is, at the same initial rate as the drain-gauge on typical June days but for over twice as long. But after it has lost about 30 mm its rate of loss falls very rapidly to the 2 mm or less per week, just as with the drain-gauge. The difference in behaviour between the drain-gauge and the laboratory soil is due to the high evaporation rate drying out the soil surface, and in consequence of preventing water moving to the surface at an appreciable rate; for the laboratory results were obtained by keeping the column of soil continuously in a constant-temperature room. But if the surface of this column was heated by shining an electric fire over it for twelve hours a day, to simulate the effect of a sunny June day, the soil then behaved like the drain-gauge soil—it only lost about 12 mm of water at the rate of 2·5 mm a day before the rate of loss dropped rapidly to the 2 mm or less per week. Fig. 19.11 illustrates this effect of potential evaporative power of the air on

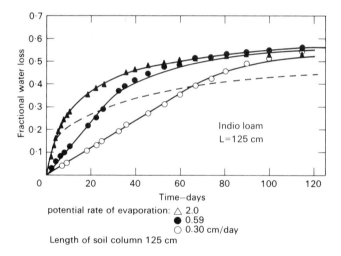

FIG. 19.11. Cumulative water loss by evaporation from a wet column of an Indio loam soil

the rate of loss of water from a cylinder of Indio loam soil when the temperature is kept constant.[2] It shows as Penman found, that initially the rate of loss is equal to the evaporative demand, but that once a certain amount of water

1 *J. agric. Sci.*, 1941, **31**, 454.
2 Taken from W. R. Gardner and D. I. Hillel, *J. Geophys. Res.*, 1962, **67**, 4319. For a further example see T. A. Black, W. R. Gardner *et al.*, *Proc. Soil Sci. Soc. Amer.*, 1969, **33**, 655.

has been lost, which is as soon as the rate of upward movement of water to the surface is less than the potential demand, the rate drops to that controlled by diffusion. The dotted curve in the graph gives the locus of points at which the liquid supply becomes limiting; and all the drying curves above this line are parts of the same diffusion-controlled curve. They also showed that once diffusion controls the rate of loss from the column, the total amount of water lost t days after diffusion control had started is proportional to $\sqrt{t}$, a result which would be expected on theoretical grounds if the water loss is solely due to diffusion.[1]

A further consequence of these results is that during the winter half-year at Rothamsted, from October to April, the evaporation from the soil is controlled by the evaporative power of the air, or by the amount of radiation received from the sun; for this is sufficiently low, and the rainfall sufficiently frequent, for the soil surface never to be properly dried out. But during the warmer months, from May to September—and even in some years in March, April and October—the evaporation is normally controlled by the number of days the soil surface remains wet, which, in turn, depends on the frequency of the rain showers and on their intensity.[2]

This rainfall distribution effect is of great practical importance in arid countries having summer rainfall, as in the dry farming wheat belt of the Canadian prairies where summer fallows are taken to store water for a succeeding crop. Frequent light falls of rain on such fallows are evaporated soon after they fall and so cannot contribute to the soil moisture supply. Thus, J. W. Hopkins[3] has computed that at Swift Current, Saskatchewan, if 25 mm of rain falls on one day, 16 mm remains in the soil ten days afterwards, whereas if it falls as five separate showers, each of 5 mm, on five consecutive days only 8 mm remained ten days afterwards. The problems of water conservation in these areas is discussed in more detail in Chapter 28.

In subtropical and tropical regions having a prolonged hot dry season, soils will lose water from considerably greater depths than they do in temperate regions. There is still some uncertainty about the depth that water will move up to the surface because a reduction in the moisture content or of the suction in the subsoil need not be due to the upward movement of water—it could equally well be due to the slow downward movement of water, and Table 19.1, p. 436, shows that this can go on for over two years after a deep soil is wetted.

The depth from which a bare soil loses water in the tropics depends on a number of factors. One important factor is the amount of soil cracking that takes place during drying, for water vapour diffuses very much faster from

1 See, for example, D. A. Rose, *Int. Soc. Sci. Hydrol. (Wageningen)*, 1966, 171; and *Br. J. appl. Phys. (J. Phys. D)*, Ser. 2, 1968, **1**, 1779.
2 R. K. Schofield and H. L. Penman, *J. agric. Sci.*, 1941, **31**, 74; H. L. Penman, *Quart. J. Roy. Met. Soc.*, 1940, **66**, 401.
3 *Canad. J. Res.*, 1940, **18** C, 388.

the surface of even deep cracks into the atmosphere than through the equivalent depth of soil. The loss is much more important in deep-cracking
montmorillonitic clays than from kaolinitic soils which show little of it.
Further, the greater the average windspeed over the soil surface, the more
rapid the transport of water vapour from the sides of the cracks into the
atmosphere;[1] and gusts of wind can cause air currents in the surface layers
of loose or open soils, which also increase considerably the water vapour
flux through a dry surface crust.[2] A second factor is the magnitude of the
temperature gradient in the soil profile and its diurnal variation, for this
affects the rate and the direction of diffusion of water vapour in the profile.
This is usually more important in the surface layers of the soil than in the
deep subsoil, both because the temperature and suction gradients are
greater there and also both the air space is greater. Thus, C. W. Rose,[3]
working at Alice Springs, Central Australia, showed that, under the very
hot strong drying conditions prevailing there, the downward flux of water
vapour during the day in the surface soil was of the same order of magnitude
as the upward flux of liquid water once the soil suction exceeded 0·3 bar.

There is still little direct field evidence on the depth from which soils lose
water in subtropical and tropical regions having a long hot dry season. Part
of the difficulty of interpreting some of the earlier work is due to the neglect
of loss of water from the subsoil by slow drainage, as is illustrated in Table
19.1. H. C. Pereira[4] studied both the loss of water and the rise in suction of
the soil water of a bare soil at Kongwa in central Tanzania after the rainy
season in two successive years, and in each it took about one month for the
soil at 15 cm and 3 or 4 months for the soil at 30 cm to be dried to a suction
of 15 bar. At the end of a seven-month dry season the soil had been dried to a
suction of about 7 bar at a depth of 45 cm, and to about 1 bar at a depth of
60 cm. The mean monthly soil temperature at 7·5 cm depth towards the end
of the dry season exceeded 33°C.

The amount of water a soil can lose by evaporation after it has been
wetted depends not only on the evaporative power of the air, but also on the
rate of downward movement of the water into the subsoil, and on the amount
of water the surface soil can hold against drainage. On this basis a dry soil
flooded by irrigation may hold more water in its surface than if wetted
slowly, due to the trapped air reducing the rate of infiltration into the subsoil,
with the consequence that an irrigated soil is likely to lose more water by
evaporation if it is irrigated by flooding than by sprinkler.[5]

1 J. E. Adams, J. T. Ritchie *et al.*, *Proc. Soil Sci. Soc. Amer.*, 1969, **33**, 609; with R. J. Hanks,
  ibid., 1964, **28**, 281.
2 D. A. Farrell, E. L. Graecen and C. G. Gurr, *Soil Sci.*, 1966, **102**, 305.
3 *Aust. J. Soil Res.*, 1968, **6**, 31, 45. For another example see D. D. Fritton, D. Kirkham and
  R. H. Shaw, *Proc. Soil Sci. Soc. Amer.*, 1967, **31**, 599.
4 With R. A. Wood *et al.*, *Emp. J. exp. Agric.*, 1958, **26**, 213.
5 For an example see E. Bresler, W. D. Kemper and R. J. Hanks, *Proc. Soil Sci. Soc. Amer.*,
  1969, **33**, 832.

Soils with a water-table near the surface

Evaporation from a bare soil with a water-table fairly close to the surface, and by this is usually meant within 2 m of the surface, is controlled both by the evaporative power of the air and by the rate at which water will move up from the water-table to the soil surface by liquid flow through the capillary pores and water films in the soil under the suction gradient caused by the soil surface drying out. This, in turn, depends on the relation between the capillary conductivity and the suction in the soil profile. Soils whose capillary conductivity falls relatively uniformly with increase in suction, such as many alluvial fine sandy or silty loams, will lose water by evaporation at a given rate from a deeper water-table compared with a soil possessing a well-marked field capacity at a fairly low suction. It is possible that losses of the order of 0·5 mm/day can take place under fairly strong drying conditions from a water-table 2 m below the surface in certain soils.[1]

The rate of loss of water by evaporation can be very considerably reduced in soils with a fairly high water-table either by applying a mulch of dry sand or stones, or by hoeing the surface to let it dry out, so ensuring that water transfer through the mulch to the atmosphere only takes place through the vapour phase. This reduction in evaporation can be very marked for soils with a water-table between 1 and 2 m below the soil surface.

Evaporation of water from saline soils

Evaporation of water from a saline soil differs from that of a non-saline soil in that, under comparable conditions, the saline soil is likely to lose more water by evaporation during a drought than the non-saline, because it will maintain a moist surface for a longer time. There are two principal reasons for this. In the first place, as water evaporates from the surface of the saline soil, the salt concentration in the soil surface increases, which lowers the vapour pressure of the solution and increases its osmotic pressure. This reduces its rate of evaporation and so allows solution to move up from the subsoil for a longer time than if no salts were present. Once the soil surface becomes dry, and covered with a salt crust or efflorescence, the rate of evaporation drops, just as in a salt-free soil. In consequence of these effects, salts may increase the depth from which water can move up from a water-table to be evaporated at the surface, though the effect is most noticeable at moderate concentrations, for at high concentrations a salt crust soon forms.[2]

Vapour pressure gradients will also move salts to the surface indirectly, for during daytime when evaporative conditions are high, there is a marked

1 For examples see R. E. Moore, *Hilgardia*, 1939, **12**, 383; L. A. Richards, W. R. Gardner, G. Ogata, *Proc. Soil Sci. Soc. Amer.*, 1956, **20**, 310; W. R. Gardner and M. Fireman, *Soil Sci.*, 1957, **85**, 244.
2 See, for example, M. A. Quayyum and W. D. Kemper, *Soil Sci.*, 1962, **93**, 333.

downward temperature gradient, which will give an appreciable downward vapour flux. This will result in an increased upward flux of water carrying salts with it, so the daytime temperature gradient acts as a pump, pumping salts from the subsoil into the surface as long as the soil remains moist enough for the upward movement of liquid water to remain appreciable.[1]

These effects have the consequence that once a salt patch starts to form on the soil surface, it grows at the expense of the neighbouring soil; so instead of the soil being covered with a uniform thin film of salt, it tends to have a large number of salt patches, as is illustrated in Plate 46 it is seen to a much less extent if water is moving up from a water-table, or if the soil surface is almost dead level as happens sometimes in an old lake bed. It can also have very important results in irrigated fields where much salt is present, for these salt patches can be prevented from appearing only by maintaining a level surface and a uniform permeability in the surface soil, and it is essential to avoid water standing in pools. It is also essential for the water-table to be sufficiently deep so that it cannot influence the moisture regime of the surface soil.

1 R. D. Jackson, D. A. Rose and H. L. Penman, *Nature*, 1965, **205**, 314.

20

Water and crop growth

The water requirements of crops

Water is essential for plant growth, and it is needed in much larger quantities than are the plant nutrients, for whereas a large proportion of any nutrient absorbed by the plant is retained, the outstanding characteristic of the water is its continuous one-way flow from the soil through the roots up the stems into the leaf surface, where it is evaporated mainly inside the stomata, through which it diffuses into the air. The evaporation of water from the leaf stomata is referred to as transpiration, and the rate of transpiration is controlled by the microclimate outside the leaves, if the plant is growing in a moist soil.

Transpiration can be considered as the process by which water is moved through a series of resistances, namely, to the movement through the soil to the root surface, across the root membranes into the root vascular system, up the vascular system from the root to the leaf cells, from the inside of the leaf cells to their outside, and to the diffusion of water vapour from the outside of the leaf cell through the stoma into the body of the atmosphere. The resistance to the upward flow in the vascular system and transfer across of walls of the leaf cells only appears to limit the rate of transpiration when plants are taking their water from a deep subsoil. J. B. Passioura,[1] for example, has pointed out that the diameter of the principal conducting channel in the fine roots of many grasses and cereals is only about 50 μ, and that viscous resistance is an important rate limiting factor to water flow through it. Thus, the rate that water can be conducted to leaves from a deep subsoil will be controlled by the number of fine roots growing down into this subsoil. This is in accord with experience that crops can rarely take all their water from depths of over 3 m and maintain full turgor in their leaves during the middle of the day. It is the last resistance that controls the rate of transpiration from crops growing in a moist soil with a well-developed root system.

The rate that water vapour diffuses into the atmosphere, that is the rate of transpiration, is controlled partly by the energy falling on the leaf, for this

1 *Aust. J. agric. Res.*, 1972, **23**, 745.

controls the amount of water that can be evaporated, and partly by the rate the water vapour diffusing out of the stomata can be dispersed into the free atmosphere. The energy available for evaporation can be derived from the energy budget for the plant, which can be written in the form

$$R(1 - s) - B = H = P + G + A + E$$

where R is the incoming energy falling on the crop, s is the reflection coefficient or albedo of the absorbing surface, so $R(1 - s)$ is the amount of energy absorbed, B is the flux of back radiation from the absorbing surface into space, so $R(1 - s) - B$ is the amount of incoming radiant energy that is converted into forms other than radiation. This quantity is called the net radiation, H, absorbed by the surface. During the summer at Rothamsted about half the incoming radiation is converted into net radiation, about 30 per cent of the remainder being re-radiated as longwave radiation and about 20 per cent reflected back.[1]

The net radiation is dissipated partly as energy in chemical compounds formed during the process of photosynthesis P, partly as heat energy warming up the crop and the ground below the crop G, partly it goes to warm up the air in contact with the crop and the soil A, called sensible heat, and partly to evaporate water from the crop and soil E, called latent heat. The magnitudes of these quantities for a sunny summer day over a barley field at Rothamsted in kJ/cm^2 are: incoming solar radiation 2·05, reflected radiation 0·49, back radiation 0·34, net radiation 1·22. Of the net radiation 1·22 was dissipated as water vapour on this day, 0·11 was transferred from the air to the crop, that is A was negative, and 0·12 was transferred to the soil. On this particular day, the net radiation and latent heat flux were equal, a result that is not uncommon during English summer weather when water is not limiting. It is often convenient to express these energy fluxes in terms of the amount of water this amount of energy would evaporate, so that

$$E \text{ mm water evaporated} = E \text{ in J/cm}^2 \text{ divided by } L$$

where L is the latent heat for evaporating water and equals 246 at 10°C, 244 at 20°C and 242 at 30°C approximately.

The maximum proportion of the photosynthetically active component of the net energy that can be used in photosynthesis is 13 per cent, and some crops can reach this level for short periods provided they are well manured and well supplied with water.[2] But taken over periods of weeks, the efficiency is very much lower. Thus H. L. Penman[3] found that a well-manured irrigated grass pasture giving 14 t/ha dry matter over the growing season at Woburn,

1 H. L. Penman, *Brit. J. appl. Phys.*, 1951, **2**, 145. J. L. Monteith and G. Szeicz, *Quart. J. Roy. Met. Soc.*, 1961, **87**, 159.
2 For an example with maize see E. R. Lemon, *Agron. J.*, 1960, **52**, 697; *Crop Sci.*, 1961, **1**, 83.
3 *J. agric. Sci.*, 1962, **58**, 349.

in England, only converts about 2 per cent of the net radiation into chemical energy of the dry matter, and most crops have a lower efficiency than this.

The heat content of the standing crop changes very little during the day, but on clear, sunny days there can be an appreciable heat flux into the soil from sunrise till about 3 p.m. in the afternoon and a corresponding heat flux out from 3 p.m. till sunrise, as discussed on p. 394. This flux into the soil can amount to over 10 per cent of the net radiation at Rothamsted, but is unlikely to exceed 5 per cent over the 24 hours. It is into the ground during spring and summer, till August in the northern hemisphere, and out of the ground thereafter. Taken over periods of several days, no great error is made if it is assumed that the only two sinks for net radiation are conversion into latent heat by evaporating water and into sensible heat by warming the air, and neglecting the other terms in the energy balance equation, so that $H = E + A$.

Net radiation can be measured using net radiometers, but such measurements are usually only made for research purposes. The incoming solar radiation is increasingly being measured at meteorological stations, but in the past the usual observation was hours of bright sunshine. Back radiation is only measured when net radiation and incoming radiation are measured, so must nearly always be calculated from other meteorological data. The reflection coefficient or albedo of the reflecting surface is usually an assumed figure. H. L. Penman[1] showed in his classical 1948 paper that R and B could be computed accurately enough for useful estimates of E to be made using the relations

$$R = R_A(a_1 + 0.62 \, n/N)$$

$$B = \sigma T^4(0.47 - 0.065\sqrt{e_d})(0.17 + 0.83 \, n/N),$$

where R_A is the incoming solar radiation if the atmosphere were completely transparent, so can be computed from the solar constant, the latitude of the station and the time of year and is given in suitable tables, n/N is the hours of bright sunshine as a proportion of the maximum possible if there were no cloud, σ is the Stefan–Boltzmann constant (5.67×10^{-12} watts/cm^2), T the temperature of the absorbing surface in °K, e_d the vapour pressure in millibars of the air at its dew point, and a_1 depends somewhat on latitude and is about 0.16 at Rothamsted and 0.25 in tropical regions.[2]

The albedo of a surface depends on the sun's elevation, being lower the higher the sun is in the sky. It also depends on the wavelength of the incident light,[3] being low for light in the visible band (0.3–0.7 μ) but much larger for light in the near infrared (0.7–3.0 μ). Thus, it does not depend entirely upon the visible colour of the leaves, being about the same for pale green as for dark green leaves. It is, however, dependent on the structure of the leaf

1 *Proc. Roy. Soc.*, 1948, **193** A, 120. The numerical values given here are not those given in this paper, but revised figures given in *J. agric. Sci.*, 1970, **75**, 69.
2 J. Glover and J. S. G. McCulloch, *Quart. J. Roy. Met. Soc.*, 1958, **84**, 56.
3 For some figures see C. S. Yocum, L. H. Allen and E. R. Lemon, *Agron. J.* 1964, **56**, 249.

canopy, both because this includes the angle the leaves make with the in-coming light, and the proportion of energy reflected by leaves low down in the canopy that is absorbed by leaves higher up.[1] To a good approximation, during the period of the day when evaporation is high, that is when the sun's elevation is appreciable, the albedo for clear water is about 0·05, for short grass completely covering the surface 0·27–0·28, for evergreen moist tropical high forest 0·05, for normal forests where a proportion of light reaches the ground 0·15–0·20, for dry soil about 0·20 and for wet soil 0·10, with consider-able variations between soils (see p. 396), and for short annual crops 0·20–0·25, and for taller crops, such as cotton and maize, 0·15–0·20, depending on leaf area and type of crop.[2]

Sensible and latent heat fluxes from the absorbing surface involve the transfer of warm air and of humid air from the surface into the atmosphere, and cause a temperature and vapour pressure gradient above it. The fluxes can be related to the gradients by the relations:

$$A = -aK_H\, \partial T/\partial z \qquad\qquad E = -bK_W\, \partial e/\partial z$$

where the gradients are averaged over appreciable areas of land surface and over appreciable intervals of time because they are constantly varying at any one point due to the atmospheric turbulence. The quantities K_H and K_W are defined as the eddy coefficients for heat and vapour diffusion, and since they both rely on the same process of eddy diffusion, they are usually considered to be equal.[3] Under these conditions the ratio of the two quantities a and b is the psychometric constant γ which equals $Pc_p/L\varepsilon$, where P is the atmo-spheric pressure, c_p is the specific heat of the air at constant pressure, ε is the ratio of the density of water vapour to dry air at equal temperature and pressure and L is the heat of vaporisation of water at the prevailing tempera-ture ($\gamma = 0.66$ at $P = 1000$ mbar and $T = 20°C$).

The ratio of the sensible to latent heat transfer A/E is known as the Bowen ratio, and the gradients are often replaced by the mean air temperature and humidity at the absorbing surface and in the air at some selected height above the surface. If T_1 and e_1 are the temperature and humidity of the air at the absorbing surface and T_2 and e_2 at the selected height, then

$$A/E = \gamma(T_1 - T_2)/(e_1 - e_2),$$

so that if these temperatures and humidities are known, Bowen's ratio can be calculated, and this value for A inserted into the heat budget equation and E calculated. But the temperature and the humidity at the absorbing surfaces

1 For an example for cotton see D. A. Rijks, *J. Appl. Ecol.*, 1967, **4**, 561.
2 J. L. Monteith, *Quart. J. Roy. Met. Soc.*, 1959, **85**, 386; 1961, **87**, 159. D. A. Rijks, *J. Appl. Ecol.*, 1967, **4**, 561, and D. M. Gates and R. J. Hanks in *Irrigation of Agricultural Lands*, Agron. Monog. 11, 1967.
3 For some measurements over pasture see N. R. Rider, *Phil. Trans.*, 1954, **246** A, 481.

are not normally known, and cannot easily be directly measured. This equation can also be written in the form

$$A/E = (\gamma/\Delta).(e_s - e_a)/(e_s - e_d),$$

where $\Delta = (e_s - e_a)/(T_1 - T_2)$, and e_s and e_a are the saturation vapour pressures at temperatures T_1 and T_2, so that Δ is the mean slope of the vapour pressure-temperature curve in this temperature range. In this equation e_1 has been replaced by e_s, that is the air is assumed saturated at the evaporating surface, and e_d is the vapour pressure of the air at the selected height at its dew point.

The Bowen equation cannot be determined from simple meteorological measurements because the temperature of the evaporating surface is not known, and for a complex surface, such as a crop, probably has no unique meaning. Penman, in his 1948 paper, overcame this difficulty by introducing a second method for computing the rate of evaporation of water, based on the evaporative power of unsaturated air blowing over a wet surface. The simplest relation controlling this flux is that it is proportional to the product of the saturation deficit of the air and a function dependent on the velocity of the air flow at some selected height above the surface. He showed that, to a reasonable approximation, for evaporation from an open water surface

$$E = 0.26(1.0 \times 6.2 \times 10^{-3} u_2)(e_s - e_d),$$

where u_2 is the wind velocity at a height of 2 m above the water surface in km/day, $(e_s - e_d)$ is the saturation deficit of the air in mbar, and E is in mm water per day.[1] The Penman equation for E is obtained by eliminating $(e_s - e_d)$ between this and the Bowen equation to give

$$E = \frac{\Delta}{\Delta + \gamma} H + \frac{\gamma}{\Delta + \gamma} E_a = \frac{qH + E_a}{q + 1}$$

where $q = \Delta/\gamma$ and $E_a = 0.26 (1.0 + 6.2 \times 10^{-3} u_2)(e_a - e_d)$ and $(e_a - e_d)$ is the saturation deficit of the air at the height of 2 m. The term E_a is often referred to as the aerodynamic term in the equation, in contrast to H which is a radiation term, and the expression defining it is based on experimental data. A correction term must however be added to this formula for computing E_a if the atmospheric pressure is appreciably less than 1013 mb, due to the height of the land surface above sea level. Under these conditions $E_a = E_a$ (sea level)$(1 + 5 \times 10^{-5} h)$ where h is the altitude of the land surface in metres. The quantity γ also decreases with increasing altitude, since it is proportional to the atmospheric pressure, which is approximately equal to $(1013 - 10^{-2} h)$ mb.

This equation relates the rate of evaporation of water from an open water surface to the net radiation and the evaporative power of the air each with

1 The first term in the bracket is sometimes taken as 0.5 in place of 1.0.

its own weighting factor. The factor Δ/γ depends on the temperature and it increases from 0·9 at 5°C to 1·6 at 15°C and 2·8 at 25°C, so that as the temperature rises, the weighting factor for the net radiation term increases and for the aerodynamic term decreases, showing that in cool conditions the aerodynamic term is likely to be more important than the net radiation, and in hot the net radiation is likely to be the more important.

This basic form of the Penman equation is rarely used in practice because net radiation figures are not normally available for the site, so these must be computed. The exact form of the equation used relies on whether the incoming solar radiation R is measured or computed from the hours of bright sunshine. The equation as normally used can be tedious to calculate, but tables[1] and computer programmes are now available for the routine calculation of the net radiation, the aerodynamic and the psychometric terms.

The effect of a crop, compared with an open water surface, is to modify the aerodynamic term in the equation and, again empirically, for a short grass turf well supplied with water:

$$E_a = 0\cdot26(1\cdot0 + 6\cdot2 \times 10^{-3}u_2)(e_a - e_d);$$

so the expression for the potential evaporation rate E_T is

$$E_T = (qH_T + E_a)/(q + 1),$$

where H_T is the net radiation received at the crop surface. This equation should apply to other short agricultural crops, but it cannot be generalised in its existing derivation for other crops. It is possible to derive from a semi-theoretical basis a formal expression for the transport equation for water vapour through a crop, which in energy units is

$$E = \frac{\rho c_p}{\gamma} \frac{(k^2 - \phi)(e_1 - e_2)(u_2 - u_1)}{[\ln\{(z_2 - d)/(z_1 - d)\}]^2}$$

where e_1 and e_2 are the water vapour pressure and u_1 and u_2 the wind speed at heights z_1 and z_2 above the ground, d is a constant known as the zero plane displacement and allows for the fact that the effective height of evaporation of water is above ground level; k is van Karman's constant (about 0·42), ρ is the density of dry air, c_p is the specific heat of air at constant pressure and ϕ is a term to take account of any serious deviation of the lapse rate above the crop from the adiabatic rate. It is usually ignored because it is small if the observations are averaged over a reasonable interval of time.

This equation can be written as:

$$E = \frac{\rho c_p}{\gamma} \frac{(e_1 - e_2)}{r_a},$$

where r_a is a resistance to the transport of the water vapour from height z_1 to height z_2. For crop studies, it is convenient to take z_2 as the height of the

1 J. S. G. McCulloch, *E. Afr. agric. for J.*, 1965, **30**, 286.

meteorological screen, and to identify z_1 as the level within the crop where the wind velocity is zero. Then

$$r_a = [\ln(z_2 - d)/z_0]^2/k^2 u_2$$

where $z_0 = z_1 - d$.

To reach the atmosphere from this level, where the wind velocity is zero, the vapour must move through a canopy resistance r_c, which is determined by the stomatal resistance to diffusion, the population of stomata on the leaf surface, the leaf area per unit of ground area, or the leaf area index of the crop. Formal analysis leads to a modified form of Penman's equation,[1] namely,

$$E = \frac{qH_T + \rho c_p/\gamma.(e_s - e_a)/(r_a)}{q + 1 + r_c/r_a}$$

In the numerator, the second term corresponds to, but is greater than E_a in the simple Penman equation, and in the denominator there is an extra term r_c/r_a. For short grass well supplied with water, these two amounts compensate for each other, so that integrated over a day, E is approximately equal to E_T the potential evaporation rate. For other crops the adjustment may not be so complete, and both r_a and r_c need to be considered as crop factors in evaporation.

The length z_0 is known as the roughness length of the crop, and is the main factor determining the value of r_a. It depends on the height of the crop, and for crops that do not bend too strongly in the wind, the roughness length is about one-tenth the crop height,[2] it is therefore small for short grass but can be very large for a forest. The larger the value of z_0, the smaller is r_a which means that the rougher aerodynamically is the crop the less is the resistance of transfer of water vapour from the surface of a leaf to the atmosphere; and typical values of r_a are about 1 sec/cm for short grass, 0·35 sec/cm for tall agricultural crops and 0·025 sec/cm for a forest. The transpiration is likely to be greater from a tall crop, because it is well ventilated, than a short crop, and as crops such as grass and lucerne grow taller so their transpiration demands increase,[3] and correspondingly because a tall crop transpires more water than a short one, it will have a lower effective temperature.[4]

These values for z_0 or r_a are of only limited value for prediction purposes, however, because the value of z_0 depends on the windspeed. As the wind velocity through a crop increases, the leaves will begin to flutter, which increases the roughness. But if the leaves are small, they will tend to set themselves in the wind, which will make the surface smoother. Thus in general

1 J. L. Monteith, *Soc. Exp. Biol. Symp.*, 1965, **19**, 205.
2 C. B. Tanner and W. L. Pelton, *J. Geophys. Res.*, 1960, **65**, 3391, and G. Szeicz, G. Endrodi and S. Tjachman, *Water Resources Res.*, 1969, **5**, 380.
3 See C. B. Tanner and W. L. Pelton, *J. Geophys. Res.*, 1960, **65**, 3391, and K. J. Mitchell and J. R. Kerr, *Agron. J.*, 1966, **58**, 5.
4 For an example see J. L. Monteith and G. Szeicz, *Quart. J. Roy. Met. Soc.*, 1962, **88**, 496.

z_0 begins to decrease as the wind velocity increases, but at a certain velocity it begins to increase. The zero plane displacement, d, also depends upon the wind velocity for crops such as long grass and the cereals that bend in the wind. At Rothamsted, for windspeeds between 1 and 5 m/sec, r_a tends to be relatively constant over a moderate range of windspeeds.[1]

The canopy resistance depends on the degree of stomatal opening, except when the leaves are wet, with dew, rain or irrigation water, when r_c and so r_c/r_a will be zero. Many agricultural crops have a value of r_c between 0·3 and 0·5 sec/cm when well supplied with water and when their leaves are dry,[2] rising to 1·0 to 1·5 for forests;[3] but if the stomata begin to close due to water shortage r_c will rise to quite high values, up to 3·5 to 7·5 sec/cm for orange orchards in Arizona[4] and up to 15 for a droughted prairie in Nebraska.

Typical values of r_c/r_a are therefore between 1 and 4 for agricultural crops, but up to 50 for forests. The larger the value of r_c/r_a the smaller the value of E for a given net radiation, and for values of r_c/r_a equal to 10, E/E_0 is about 0·25 at 20°C.

This effect of stomatal resistance decreasing the rate of evaporation of water has the consequence that water on the outside of a leaf evaporates more easily than water inside a leaf. Thus under comparable evaporation conditions, the rate of evaporation from the wet leaves of an agricultural crop, for which r_s/r_a is between 1–2 will be 1·3–1·6 times as fast as transpiration from the dry canopy; and for a forest where r_c/r_a may be as large as 15, the rate of evaporation from a wet canopy will be about 5 times as high as from the dry. This result emphasises that spray irrigation, by maintaining a wet leaf canopy, will increase evaporation during and just after irrigation. It also shows that dew is a relatively inefficient mechanism for providing a crop with water.[5]

The use of the Penman equation in practice

Penman's approach to transpiration and evaporation has made possible a much clearer understanding of the factors that control the water use of crops. Initially, however, it was technically difficult to test the validity of his approach in practice because of the difficulty of determining rates of evaporation from large areas of land; but with the introduction of weighing lysimeters holding many tons of soil on the one hand, and of the use of neutron moisture meters on the other, it is now possible to test the equations with considerable precision. However, the more appropriate the equation to a given crop, the more

1 H. L. Penman and I. F. Long, *Quart. J. Roy. Met. Soc.*, 1960, **86**, 16. J. L. Monteith, *Symp. Soc. Exp. Biol.*, 1965, **19**, 205.
2 J. L. Monteith, *Symp. Soc. Expt. Biol.*, 1965, **19**, 205.
3 G. Szeicz, G. Endrodi and S. Tjachman, *Water Resources Res.* 1969, **5**, 380.
4 C. H. M. van Bavel, J. E. Newman and R. H. Hilgeman, *Agric. Met.*, 1967, **4**, 27.
5 P. E. Waggoner, J. E. Begg and N. C. Turner, *Agric. Met.*, 1969, **6**, 227; T. A. Black, C. B. Tanner and W. R. Garner, *Agron. J.*, 1970, **62**, 66.

quantities it contains, such as roughness factors and canopy resistances, that must be determined from the experimental data.

The original Penman equation referred to evaporation from an open water surface E_0 and Penman himself used it to predict evaporation over periods of weeks or seasons rather than over hours or days. He compared his estimate E_0 with the amount of water transpired by short grass kept moist, E_T, and showed that E_T/E_0 was independent of E_0 provided the grass was kept short and remained green and there was no shortage of water; though he found the value of this ratio depended on the season of the year, being 0·8 in summer and 0·7 in spring and autumn at Rothamsted.[1] Many other workers have also determined E_T/E_0 for different canopies. Thus, J. Glover and J. Forsgate[2] at Muguga in Kenya almost on the Equator found that the transpiration from short grass kept moist followed the Penman equation, even on a daily basis with $E_T/E_0 = 0·75$, and it did not depend on the time of the year. H. C. Pereira and his colleagues[3] also at Muguga found that transpiration from forests also followed the Penman equation with $E_T/E_0 = 0·9$ for high altitude moist evergreen tropical high forest to $E_T/E_0 = 0·8$ for a wide range of pine, cypress and bamboo forests. An important part of the reason that E_T/E_0 is less than unity is that E_0 is calculated for a surface of albedo 0·05, whereas the albedo of vegetation is always greater, up to 0·25 for short grass, though approaching 0·05 for high forest.

Table 20.1 illustrates one use that can be made of the Penman equation. It gives the computed average evaporation from well-watered short grass at Rothamsted and the average rainfall; and it shows that on average a soil moisture deficit of about 75 mm will be built up by the end of July, which will not have been made good until early November. In summers and autumns that are drier than average, much higher deficits will be built up: in 1970 the potential deficit was about 150 mm at the end of July, and it was not until early October that it began to be reduced. In years of below-average winter rains, the deficit may not be made good till the following spring. Table 20.1 also gives the total amount of incoming radiation, expressed as the depth of water that it would evaporate if it was dissipated entirely as latent heat. Maps have been published for Great Britain in which the average maximum soil moisture deficit for any area can be read off, from which an estimate of the likely value of irrigation for any cropping sequence can be made.

The second important aspect of the Penman equation is that the net radiation is divided into a latent heat and a sensible heat term; and this approach emphasises that sensible heat is an additional method of dissipating absorbed solar energy. Sensible heat transfer depends on the temperature difference

1 H. L. Penman, *Quart. J. Roy. Met. Soc.*, 1950, **76**, 372; *J. Soil Sci.*, 1949, **1**, 74.
2 *Quart. J. Roy. Met. Soc.*, 1964, **90**, 320. They gave $E_T/E_0 = 0·86$ due to using an incorrect value for the psychometric constant in the Penman equation.
3 *Int. Symp. Forest Hydrology*, Pergamon, 1966, 435; and with P. H. Hosegood, *J. Soil Sci.*, 1962, **13**, 299; *E. Afr. agr. for J.*, 1962, **28**, special issue.

TABLE 20.1   Mean incoming radiation, rainfall and potential evaporation at Rothamsted*

	Incoming radiation (1956–69) mm	Potential evaporation E_T 1967–70 mm	Rainfall R 1948–69 mm	$R - E_T$ mm
January	29	5	59	54
February	45	10	51	41
March	97	30	43	13
April	137	50	47	− 3
May	194	70	54	− 16
June	213	90	56	− 34
July	193	80	60	− 20
August	157	65	66	1
September	117	45	60	15
October	68	20	63	43
November	30	5	71	66
December	19	0	68	68
Total	1300	470	698	

* I am indebted to Dr H. L. Penman for these figures. The values for E_T have been rounded to the nearest 5 mm.

between the leaves of the crop and the air, and leaf temperature is controlled by the heating of the leaf due to it absorbing solar energy and its cooling due to transpiring water. Thus, sensible heat transfers are intimately bound up with transpiration rates; and it may happen that transpiration rates are sufficiently high for the leaf to be cooler than the air, and so there will be a transfer of sensible heat from the air to latent heat. This effect will be particularly noticeable when warm dry air is blowing over a rough crop, for the incoming air will encourage transpiration, which will cool the leaves, so cool the air as well as adding water vapour to it.

Grass pastures well supplied with water, will dissipate nearly all the net radiation as latent heat during summer in Great Britain, so the Bowen ratio over such a cover will be small, but if the water supply to the leaves begins to limit the rate of evaporation, an increasing proportion of the net radiation will be converted into sensible heat, and the ratio will rise. This is illustrated in Table 20.2 for short grass at Rothamsted during two sunny spells in August,[1] the first after a period of dry weather and the second after the soil had been moistened by rain. The incoming radiation was about the same for the two periods and there was only a small difference in the net radiation, but whereas

1 I am indebted to Dr J. L. Monteith for this table.

sensible heat transfer to the air during the warmer part of the day was small when water was not limiting, it was actually larger than the latent heat transfer when water was in short supply. If the surface is rough, the energy dissipated as latent heat will become increasingly greater than the net radiation. Thus O. T. Denmead[1] showed that at Canberra, Australia, forests will cool the air more than wheat fields, because of their much greater aerodynamic roughness.

TABLE 20.2   Mean daily energy balance over short grass at Rothamsted after drought (3–6 Aug. 1960) and after rain (16–19 Aug. 1960). $J/cm^2/12$ hours

| | At end of drought | | | | After rain | | | |
	H	G	A	E	H	G	A	E
0800–2000 hr	1010	69	488	455	950	39	33	880
2000–0800 hr	− 33	− 28	− 60	56	− 71	− 69	− 98	84
Daily total in mm	3·95	0·17	1·73	2·06	3·55	− 0·12	− 0·26	3·90
Incoming radia-tion		6·75 mm				6·75 mm		
Bowen ratio 0800–1600 hr		1·3				0·13		

H, net radiation; G, heat flux into the ground; A, sensible heat to the air; E, latent heat used for evaporating water.

Penman's equation can be used to show the relation between E_T, the evaporation from a field in an annual crop with E_0, the open water surface evaporation. Initially the soil is bare, so E_T is the evaporative flux from bare soil, which is very dependent on the moisture content of the surface soil, being small when the surface is dry. As the crop grows both its leaf area index and the aerodynamic roughness of the surface increase, and since the crop takes its water from below the soil surface, E_T becomes increasingly independent of the wetness of the surface soil. Figure 20.1 gives a typical picture of the way E_T/E_0 increases during the season for a maize crop.[2] It reaches a plateau when the leaf expansion is complete and silking has just begun, and later it drops more rapidly than one would expect from loss of leaf, suggesting that the physiological activity of the leaves is declining. If Penman's estimate for E_0 had been used, rather than the evaporation from a class A open water-pan, the ratio of E_T/E_0 at maximum would have been about unity. Under some conditions, the ratio will rise well above unity due to the roughness of the canopy.

Annual crops, even when they have a leaf area index well over unity, rarely completely shade the soil, so that evaporation takes place from the soil sur-

1 *Agric. Met.*, 1969, **6**, 357.
2 From O. T. Denmead and R. H. Shaw, *Agron. J.*, 1959, **51**, 725.

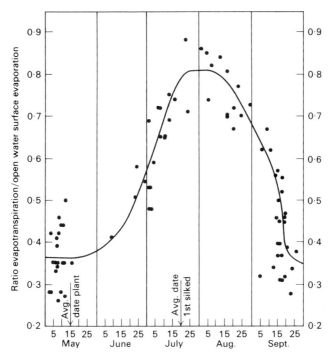

Fig. 20.1. The ratio of evaporation from maize to evaporation from an open-water surface throughout the growing season

face whenever it is moist, even when the crop has made its maximum growth. Thus at Rothamsted H. L. Penman and I. F. Long[1] found that for spring wheat in June, with a leaf area index of 2, about half the evaporation from the field was evaporation from the soil surface, when the surface was moist; and American experience with maize in July and August, when its leaf area index is at a maximum, is that up to half the evaporation from a maize field is evaporation from the soil surface.[2] In these maize fields, between 15 to 25 per cent of the net radiation is absorbed at the soil surface. In so far as the soil surface is dry, this net radiation will be dissipated as sensible heat, so will raise the Bowen ratio for the canopy. But the closer the maize is planted, the less energy reaches the soil surface and the more is absorbed by the leaves, which increases the transpiration of the crop and reduces the Bowen ratio. This has the consequence that the amount of water lost by evaporation will increase as the plant population increases, and if water is in short supply, the closer-spaced crop will wilt before the wider-spaced.[3]

1 *Quart. J. Roy. Met. Soc.*, 1960, **86**, 16.
2 C. B. Tanner, A. E. Peterson and J. R. Love, *Agron. J.*, 1960, **52**, 373. O. T. Denmead, L. J. Fritschen and R. H. Shaw, ibid., 1962, **54**, 505. K. W. Brown and W. Covey, *Agric. Met.*, 1966, **3**, 73.
3 G. M. Aubertin and D. B. Peters, *Agron. J.*, 1961, **53**, 269.

The evaporative power of the air depends on the moisture conditions of the earth's surface over which it has been travelling. Wind blowing over a hot desert and striking an irrigated crop will increase the evaporative demand in that crop considerably above the level due to net radiation. The effect is naturally greatest at the leading edge of the crop, that is, where the wind first hits it, and it rapidly falls to a fairly constant level within 60 to 80 m from the leading edge;[1] and the effect then decreases slowly over distances measured in tens of kilometres.[2] This means that the crop at the leading edge should be irrigated more frequently than the crop further in, though there seem to be no examples of where this is done in practice. The effect can even be seen to a small extent in England, for D. C. Davenport and J. P. Hudson[3] have given an example of wind blowing over a dry barley stubble which, when it hit a field of sugar-beet, also caused increased transpiration at the leading edge of the crop and neither the air temperature nor the saturation deficit had reached their stable values 70 m behind the leading edge. This effect can also be seen on the large scale in north-western Europe in winter,[4] for the south-westerly winds that have crossed the Atlantic are warmer than the cool land, and comprise an important part of the heat input to the vegetation during the winter months.

There is not a great deal of information on the maximum rates of transpiration crops can maintain before the stomatal resistance of the leaf limits the rate, but it is probable most crops, when their foliage covers the land surface reasonably, can maintain rates of 5 mm day provided the soil is well supplied with water. O. T. Denmead and R. H. Shaw[5] found that, both for grass and maize, stomatal resistance began to be limiting once this rate was exceeded. It is possible that crops adapted to strong drying conditions may be able to maintain higher rates, for maize and sorghum can sustain rates of 8 to 10 mm per day in the Great Plains areas of the central US,[6] but it is not known what the importance of stomatal resistance is for these varieties.

Effect of water shortage on transpiration

Crop leaves cease to transpire at their full potential once their stomata begin to close, and this occurs as soon as the water content of the leaf falls below its fully turgid value due to the transpiration rate having exceeded the rate of supply of water to the leaf. Stomatal closing increases the resistance of

1 For examples from the Gezirah in the Sudan see J. P. Hudson, *Expt. Agric.*, 1965, **1**, 23; D. W. Head, *Emp. J. expl. Agric.*, 1964, **32**, 263.
2 For an example from Texas see E. R. Lemon, A. H. Glasser and L. E. Satterwhite, *Proc. Soil Sci. Soc. Amer.*, 1957, **21**, 464.
3 *Agric. Met.*, 1967, **4**, 339.
4 For heat budgets for Denmark see H. C. Aslyng and S. E. Jensen, *Roy. Vet. Agric. Coll. Copenhagen*, Yearb. 1965, 22.
5 *J. Agron.*, 1962, **45**, 385.
6 H. F. Rhoades, J. A. Bondurant *et al.*, *Nebraska Agric. Exp. Sta. Bull.* 424, 1954. J. T. Musick and D. W. Grimes, *Kansas Agric. Exp. Sta. Tech. Bull.* 113, 1961.

diffusion of water vapour from the outside of the mesophyll cells through the stomata to the atmosphere, and an increase in r_c results in an increase in the proportion of net radiation that is dissipated as sensible heat, and hence in an increase of leaf temperature. The stronger the evaporative conditions around the leaf, whether due to incoming solar radiation or to advection, the sooner will any check to the leaf water supply result in stomatal closing. Correspondingly, the drier the soil the lower is the maximum rate that the roots can supply water to the leaves, so the less strong the evaporative conditions need to be for the canopy resistance to begin to control transpiration. G. Szeicz and I. F. Long[1] found that the surface resistance of a grass-clover pasture at Rothamsted remained at a low value for moderate transpiration rates (2 to 3 mm/day), until the soil suction in the top 25 cm soil had risen to 3 to 4 bar, and then the resistance increased linearly with increase of suction up to 12 bar, when it began to increase more rapidly. Figure 20.2 also illustrates the effect of reduction in the matric potential of a soil on the distribu-

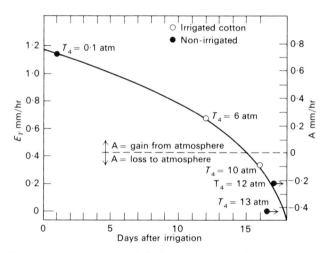

FIG. 20.2. Relation between E_T and the heat exchange with air, A, for an irrigated clay soil in Texas.
$\bar{T}_4$ is the mean value of the suction of water in the top 4 feet (1·2 m) of the profile. The value for $\bar{T}_4 = 12$ and 13 bar are for unirrigated cotton

tion of net energy between evapotranspiration and sensible heat for a cotton crop in Texas during two weeks following an irrigation.[2] The crop is removing sensible heat from the air just after it has been irrigated, and two-thirds of the energy used for evaporation is derived from this source. But as the soil dries, so both the amount of heat being removed from the air and the evaporation rate fall, and by the time the mean soil suction in the top 1 to 2 m has

1 *Water Resources Res.*, 1969, **5**, 622.
2 E. R. Lemon, A. H. Glaser and L. E. Satterwhite, *Proc. Soil Sci. Soc. Amer.*, 1957, **21**, 644.

reached 10 bar, the crop is warming the air instead of cooling it, and its evaporation rate has fallen to one-third of its initial value.

When the water content of a leaf falls, the free energy of the water in the leaf does likewise, due partly to an increase in the osmotic pressure of the cell cytoplasm and partly to a decrease in the turgor pressure within the cell; and after the leaf has lost a certain proportion of its water, the turgor pressure has either fallen to zero or to a relatively low value, and the leaf wilts. For the leaves of many plant species, the lowering of the free energy of the water in the leaf when it has wilted is approximately equal to the lowering caused by the osmotic pressure, and it occurs at a water potential of about −12 to −15 bar.[1] In the past it has not been easy to measure the free energy of the leaf water either in the laboratory or in the field, so it has been common to measure the water deficit in the leaf, that is the amount of water the leaf will absorb if floated on water. Typically, the turgor pressure drops rapidly and linearly from a value of about 10 bar to about 2 bar as the relative water content of the leaf falls from 100 to about 85 per cent, and then more slowly, and more dependent on the crop, to about zero as the leaf water content falls to 50 to 70 per cent, the exact figure depending on the crop.[2]

A wilted leaf does not necessarily cease to lose water, for it will lose some through diffusion through the epidermal cells and the cuticle layer that is present on the outer surface of these cells. The better-developed and the thicker this layer, the less water will be lost from a leaf in which the stomata are closed; and xerophytic plants, that is, plants well adapted to water stress, have leaves with a well-developed layer, and hence with a very high resistance to water vapour diffusion except through their stomata. Plants adapted to moist shady conditions have leaves with thin cell walls and a thin and poorly developed cuticle layer, so will lose water relatively easily through the epidermal cells. The stomata in the leaves of some xerophytic plants are sparse and on the under-surface, and may close when the water deficit in the leaf is small, so these plants will have a high canopy resistance, particularly in the middle of the day, as found by C. H. M. van Bavel[3] for some citrus species in Arizona and southern California.

When a crop is growing in an initially moist soil during a period of drought, a time comes when the leaves wilt at periods of maximum potential evaporative demand, because the roots cannot supply the leaves with water sufficiently fast to maintain turgor, but as the demand falls, the water deficit will have been reduced sufficiently for the leaf to regain turgor later in the afternoon. As the drought continues, so the leaves regain turgor later and later in the day, until a time comes when the leaves are still wilted at sunrise, and they will not regain turgor until after rain or irrigation has supplied some

1 W. R. Gardner and R. H. Nieman, *Science*, 1964, **143**, 1460. R. O. Slatyer and W. R. Gardner, *Symp. Soc. Exp. Biol.*, 1965, **19**, 113.
2 W. R. Gardner and C. F. Ehlig, *Pl. Physiol.*, 1965, **40**, 705.
3 With J. E. Newman and R. H. Hilgeman, *Agric. Met.*, 1967, **4**, 27.

water. At this time, the suction of the soil in the root zone is commonly found to be about 15 bar, which corresponds approximately to the free energy of the water in the wilted leaf. The plant in this condition is said to be permanently wilted, and the soil at this moisture content is said to be at its permanent wilting point. On the picture presented here, the permanent wilting point of a soil is a leaf-determined and not a soil-determined property; it is only soil-determined if the soil has a sufficiently coarse texture for the water films to become discontinuous before the matric potential has fallen to -15 bar, and then permanent wilting will take place at a potential corresponding to the onset of this discontinuity.

Crops differ in the reaction of their leaves to further desiccation. The leaves of many species adapted to semi-arid conditions continue to lose water, but at a slower rate, and the roots continue to dry the soil, and even after a long period of wilt the leaves will regain turgor and the stomata function normally once water has been supplied. But in course of time all leaves will become dried and killed; those of species not adapted to drought will be killed or drop off the plant after a short wilt, and those well-adapted after a much longer wilt; and the leaves of some species that have been moderately severely wilted may regain turgor, but the wilt may have so damaged the guard cells of their stomata that they cease to function properly. Sorghum and many semi-arid grasses are examples of crops whose leaves can survive a relatively long wilt without damage, maize leaves only survive a shorter wilt and have their guard cells damaged if the wilt is moderately severe,[1] and tomato leaves can be easily killed. Crops that have previously been growing in moist soil under no water stress, particularly if they have received a high dressing of nitrogen fertiliser, have leaves that are much more easily killed by wilt than corresponding crops that have previously been under stress and that have only had moderate levels of nitrogen fertiliser.

The rate that roots can take water up from the soil depends on the length of absorbing roots in the soil, the volume of soil they permeate, and the rate that water moves from the body of the soil to the root; while the rate water can be moved from the root to the leaf depends on the difference between the water potential of the leaf and the soil in contact with the root, the permeability of the root membranes, and possibly the distance within the plant between the absorbing part of the root and the transpiring leaf. As will be described on p. 533 when a plant is drawing water from deep subsoil through long very fine roots, it is quite possible that these long lengths of roots may be interposing a considerable resistance to water movement, particularly at times of rapid transpiration. The permeability of the root membranes to water depends on the soil temperature, and at low soil temperatures can be very small, so small that plants may wilt on a sunny day in early spring when the soil is cool but moist.[2]

1 J. Glover, *J. agric. Sci.*, 1959, **53**, 412.
2 R. O. Slatyer and J. F. Bierhuizen, *Aust. J. biol. Sci.*, 1964, **17**, 115.

The rate water can move through the soil to the root decreases as the soil dries due to the rapid drop in the capillary conductivity of the soil. If the soil is sandy, the water films will become discontinuous before a soil suction of 15 bar has been reached, so that plant roots can only dry the soil to a suction as low as this if the water can move as a vapour to the root and be absorbed in this form. But in loams and clays there is continuity of water films to suctions much in excess of 15 bar. Figure 20.3, taken from some work by

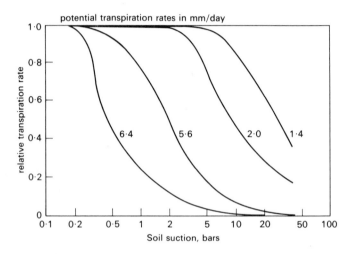

Fig. 20.3. Relative transpiration rate of maize as a function of the soil suction for different potential transpiration rates on a silty clay loam in Iowa

O. T. Denmead and R. H. Shaw,[1] shows the limitation that capillary conductivity imposes on the transpiration rate of maize growing on a silty clay loam soil in Iowa, characterised by a pF-moisture content curve that is almost linear over the moisture range between field capacity and permanent wilting point. If the transpiration rate is only 1·4 mm/day, the capillary conductivity of the soil can almost maintain this rate until the suction of the water exceeds 12 bar, whereas if the rate is 6·4 mm/day, capillary conductivity is unable to supply water at this rate once the suction exceeds 0·3 bar. Figure 20.3 also shows that there is no discontinuity in the curves at 15 bar suction, and, in fact, if the evaporating conditions are mild, the crop will be transpiring at a rate of about 0·8 mm/day, even though the leaves are wilted. Some drought-tolerant crops can, however, dry the soil to 15 bar suction before their rate of transpiration falls below the potential rate. Kikuyu grass, *Pennisetum clandestinum*, growing in a lysimeter with a soil depth of 1·2 m at Muguga, Kenya, for example,[2] maintained transpiration at its potential rate

1 *Agron. J.*, 1962, **54**, 385.
2 J. Glover and J. Forsgate, *Quart. J. Roy. Met. Soc.*, 1964, **90**, 320.

of about 4·8 mm/day until the soil suctions in the root zone had risen to 15 bar, when the grass wilted and its maximum rate of transpiration fell to 2·8 mm/day. J. H. Palmer,[1] at Griffith, NSW, found that cotton would maintain transpiration at its potential rate of 10 mm/day, until the soil suction rose to 10 bar, when it wilted and its rate fell; but it was still transpiring at 3 mm/day when the suction in its root zone had risen to 30 bar.

W. R. Gardner[2] has given numerical examples of the calculated rate of movement of water to a root for soils with given capillary conductivity–soil water suction curves, and has shown that water can move at appreciable rates as liquid water in relatively dry soils. Using the Pachappa sandy loam, whose capillary conductivity curve is given in Fig. 19.6, page 431, and a daily water uptake of 0·1 cm^3/cm length of root, which is a common figure for root uptake, he finds that the suction at the root surface must be 7 bar if the soil suction is 5 bar, and must be 28 bar if it is 15 bar, for this rate of uptake; and in this example the root will be reducing the soil water content 4 cm away from the root after ten days. Thus the soil suction at which roots cease to take water from a soil is determined by the suction developed at this surface which, in turn, is dependent on the potential that can be developed in the leaves of the plant, provided continuity of water films is maintained in the soil.

The results that have been given so far are for plants growing in a restricted volume of soil, and they do not apply in detail to field crops growing on deep soils. Under these conditions the roots tend to dry the superficial layer of the soil to about 15 bar, but before this point has been reached they will have grown into a deeper layer of moist soil. The soil is dried to about 15 bar from the surface down to the bottom of the root zone, and it is only when the roots can go no deeper that they begin to raise the suction above 15 bar. Thus, H. C. Pereira[3] found that during a drought the roots of *Arabica* coffee growing on a deep red loam soil in Kenya had removed water from more than 3 m depth of soil but had only dried the top 1·5 m to 15 bar suction. The leaves of the bushes were wilted all day but otherwise showed little other sign of stress. Neighbouring coffee, on a site where the soil was only 2 m deep, had removed 6 cm more water from the profile than was held at 15 bar suction, and the bushes had shed most of their leaves, and those that were left showed severe signs of drought.

Similar results have been found at Muguga, Kenya, on a similar deep soil. Table 20.3 shows the moisture content under two forest plantations in October 1965, after 138 days without any effective rain, and also under a star grass (*Cynodon dactylon*) pasture in October 1956, after a still more severe dry season.[4] The eucalypt forest had dried the top 6 m of soil to a

1 With E. S. Trickett and E. T. Linacre, *Agric. Met.*, 1964, **1**, 282.
2 *Soil Sci.*, 1960, **89**, 63. See also *Agron. J.*, 1964, **56**, 41, and I. R. Cowan, *J. appl. Ecol.*, 1965, **2**, 221; E. I. Newman, ibid., 1969, **6**, 1, 261.
3 *J. agric. Sci.*, 1957, **49**, 459.
4 I am indebted to the Physics Department, E.A.A.F.R.O., for the data in this Table.

TABLE 20.3   Extraction of water from soil profiles by forests and pastures. Muguga, Kenya. Soil: deep Kikuyu red loam. Forests: sampled October 1965. Grass: sampled October 1956

Depth in cm	Water content of soil in cm				
	0–150	150–300	300–450	450–600	0–600
Eucalyptus saligna					
in field	36·4	40·2	41·7	46·5	164·8
at 15 bar suction	37·9	41·2	42·7	49·0	170·8
deficit below 15 bar suction	1·5	1·0	1·0	2·5	6·0
Pinus patula					
in field	39·3	47·8	51·8	48·9	187·8
at 15 bar suction	39·3	48·0	50·8	48·2	186·3
deficit below 15 bar suction	0·0	0·2	−1·0	−0·7	−1·5

Depth in cm	Water content of soil in cm				
	15–76	76–137	137–198	198–258	258–320	15–320
Cynodon dactylon						
in field	13·8	16·1	18·8	19·0	20·0	87·7
at 15 bar suction	15·8	18·6	21·8	20·7	22·2	99·1
deficit below 15 bar suction	2·0	2·5	3·0	1·7	2·2	11·4

suction a little greater than 15 bar and the pine to a little less, and both had taken water from below this depth. The trees were not showing visible signs of wilt at the time of sampling, but the difference between the water use by the eucalypt and the pine probably reflects the greater ability of eucalypt to remove water from this soil, and so to survive a longer drought than the pine on this site.

These results cannot be used to estimate the amount of water the crops have transpired compared with the amounts they would have transpired if water had not been limiting. But M. Dagg[1] showed that the grass *Cenchrus ciliaris* could build up a deficit of 32 cm water in the top 3 m of the profile without its transpiration rate being appreciably affected, though by the time it had built up a deficit of 45 cm, and taking water from 6 m, the transpiration had fallen well below its potential value; but it continued to dry the soil slowly, and before the end of the drought it had built up a deficit of 54 cm and dried the top 6 m to below its permanent wilting point. Tea at Mlanje, Malawi, will dry the top 4 m of soil to 15 bar suction and take water from 5 m depth during the dry season, but its transpiration rate falls below the potential rate when

1   *J. Hydrol.*, 1969, **9**, 438.

it has taken 20 to 25 cm water, or only 40 per cent of that held between 0·33 and 15 bar in the top 4 m; and it is transpiring very slowly by the time it has dried this depth of soil to 15 bar.[1]

These results show how the amount of water a crop can take from a soil before its transpiration rate begins to drop below its potential rate depends on its depth of rooting. Some factors affecting this will be discussed on p. 532, but an important factor for short season crops is their pattern of root growth. The root system of most annual crops ceases to grow after the crop has come into flower, so the crop does not have much time to develop a deep root system. At Muguga, where perennial crops are very deep-rooting, maize sends few roots below 150 cm, and in Great Britain in dry summers, the roots of annual crops and also of grass pastures, rarely extend to depths much greater than 120 cm, and they probably do not dry the soil to suctions of 15 bar to depths much in excess of 50 cm, except on coarse sands. Figure 20·4 illustrates the depth from which a ryegrass ley takes its water

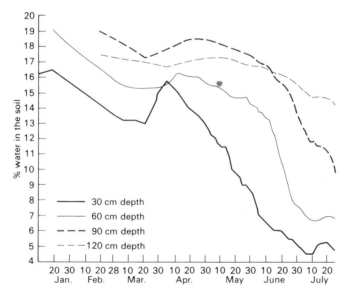

FIG. 20.4. The soil moisture content at different depths in a soil under a ryegrass ley at Hurley

during a dry summer from a well-structured loam soil at Hurley.[2] In wet summers in England many crops will send few if any roots to depths of 50 cm. The depth of rooting of a crop in a given soil, however, depends on the water regime, and frequent irrigations encourage shallow rooting. Thus lucerne, which is potentially a deep rooting crop, and citrus which is shallow rooting,

1 D. H. Laycock, *Trop. Agric. Trin.*, 1964, **41**, 277. S. T. Willatt, *Agric. Met.*, 1971, **8**, 341.
2 E. A. Garwood and T. E. Williams, *J. agric. Sci.*, 1967, **68**, 281.

both take their water from the same depth under the standard irrigation schedule used in Arizona and Southern California, as is shown in Table 20.4.

A crop will take some water from below its depth of rooting if this deeper soil is moist because, as the roots dry the soil in their neighbourhood, water will move up from below by capillary flow under the suction gradient that has been set up. The amount of water that can move up is about the same as can move up to the surface of the bare soil under drying conditions. This rate of upward movement can be of the same order as the rate of transpiration for a few days, but once 1·5 to 2·5 cm of water has moved up, the upward flux becomes slow and is inadequate to maintain the turgidity of the crop if the transpiration rate is appreciable, though it will continue to supply water slowly to grasses such as *Cenchrus ciliaris* that are very drought-tolerant.

TABLE 20.4   Depth from which lucerne and citrus trees take their water under standard irrigation practice. Soil—Sandy loam in California

Depth	Proportion of water taken from these depths	
	Lucerne	*Citrus*
0–60 cm	62	66
60–90 cm	15	16
90–120 cm	12	10
Below 120 cm	11	8

The management of grass may, however, affect its depth of rooting. Thus J. E. Goode[1] at East Malling, Kent, found a ryegrass-white clover sward dried the soil to a depth of 45 cm and took some water from the 45 to 90 cm layer if kept short, but took appreciably more water from this layer if allowed to grow tall.

Some annual crops are less sensitive to drought than others because their root systems grow downwards faster, for any factor that encourages rapid deep rooting into moist soil reduces the harmful effect of drought during the growing season. R. M. Hagan and Y. Vaadia[2] have given an example of this by comparing the irrigation requirements of water melons and cantaloupe melons at Davis, California. If both are planted on soils that have been irrigated water melons will need no additional irrigation because they have a very fast growing and deep root system, whereas cantaloupes need several subsequent irrigations because of their weak slow-growing roots.

1 *East Malling Ann. Rept.*, 1955, 64.
2 UNESCO, *Arid Zones Res.*, 1960, **15**, 215.

The effect of water shortage on crop growth

The yield of many potentially high-yielding varieties of crops, grown under conditions of a high level of available nutrients, can be reduced by quite minor water deficits during the growing season;[1] for they can only give their maximum yield if given a high level of nitrogen fertiliser and if their stomata remain open during the whole day, and in particular during the middle of the day when the potential rate of photosynthesis is at a maximum. However, if the crop is being grown with irrigation, maximum yield may involve a very inefficient use of water. Thus O. W. Howe and H. F. Rhoades,[2] working with maize on a very fine sandy loam soil in Nebraska, obtained a yield of 9·6 t/ha grain if the moisture tension in the upper part of the root zone never exceeded 400 mbar, and this required 55 cm water of which 36 cm was applied as irrigation. But if the amount of irrigation water was halved, so that the moisture tension rose to appreciably higher values before irrigation was given, the yield only dropped to 9·0 t/ha, while if no irrigation was given the yield was 4·1 t/ha. The effect of a moderate water deficit on crop yield can be quite small if the crop is being grown under conditions of low nutrient supply, for under these conditions it is the nutrient supply and not the potential rate of photosynthesis that is limiting yield. Thus C. O. Stanberry and M. Lowrey[3] at Tucson, Arizona, found that winter barley receiving no nitrogen fertiliser gave the same yield if it was irrigated when the suction in the root zone reached 150 mbar or 8 bar, but at a high level of nitrogen fertiliser, the more frequent irrigation gave the higher yield. Their actual yields of grain in t/ha were:

Nitrogen added kg/ha	0	67	135	270
Infrequent irrigation	0·87	3·05	2·95	3·57
Frequent irrigation	0·82	3·49	4·54	4·81

As another example kikuyu grass, *Pennisetum clandestinum*, at Muguga, Kenya, did not have its daily rate of dry matter production, which was between 20 and 45 kg/ha, reduced until all the soil in the root zone, in this example only 1·2 m deep, was dried to a suction of 15 bar, but then growth almost ceased.

Water deficits can, however, reduce yields seriously if they occur at certain critical periods in the growth of the crop; and the most usual critical period is during flower formation and fertilisation, for pollen production, and the viability of the pollen produced can be seriously reduced by a water deficit at this time.[4] Some results by J. S. Robins and C. E. Domingo[5] for maize on

1 For two recent reviews on crop responses to irrigation see P. J. Salter and J. E. Goode, Crop Responses to Water at Different Stages of Growth, *Commonw. Bur. Hort. Plant. Crops. Res. Rev.*, **2**, 1967; *Irrigation of Agricultural Lands*, Amer. Soc. Agron. Monogr. 11, 1967.
2 *Proc. Soil Sci. Soc. Amer.*, 1955, **19**, 94.
3 *Agron. J.*, 1965, **57**, 36.
4 For reviews see R. A. Fischer and R. M. Hagan, *Expt. Agric.*, 1965, **1**, 161.
5 *Agron. J.*, 1953, **45**, 618.

a fine sandy soil in Washington will illustrate this. They found the yield of maize when under no moisture stress was 8·7 t/ha, which was only reduced to 8·6 t/ha if the maize was allowed to wilt just before tasselling; but it was reduced to 5·0 t/ha if it wilted for 6 to 8 days during tasselling, and if it was only irrigated at tasselling, so suffered water shortage at other times, it yielded 6·0 t/ha.

When an annual crop produces its flowers over a relatively long season and is being grown for its seed or fruit, water shortage may result in a much more serious reduction in plant growth than in the yield of the marketable part of the plant. Cotton, for example, can have its total dry matter yield and its height seriously reduced by water shortage, yet its yield of seed-cotton need be little affected, as is illustrated in Table 20.5.[1] Here water shortage halved

TABLE 20.5   Effect of water shortage on the growth of cotton. Davis, California

Water applied cm	Relative growth of cotton		
	Total dry matter minus seed-cotton	Height	Yield of seed-cotton
33	51	73	90
61	81	91	96
109	100	100	100

the dry matter production of leaf and frame, reduced the crop height by 25 per cent, but only reduced the yield of seed-cotton by 10 per cent. Similarly, water shortage, if it does not occur seriously at flowering time, increases the grain-straw ratio in cereals, for it reduces the yield of straw more than that of the grain. This is illustrated in Table 20.6 taken from some early work of J. A. Widtsoe[2] in Utah.

TABLE 20.6   Yield of wheat with increasing quantities of irrigation water. Greenville Farm, Utah

Irrigation water: cm	12·5	19	38	63	125
Yield of grain, t/ha	2·47	2·68	2·86	2·89	2·95
Yield of straw, t/ha	3·1	3·5	4·1	4·6	6·0
Ratio grain/straw	0·80	0·76	0·69	0·63	0·49

Average rainfall during growing period, 12·5 cm.

1 From F. Adams, F. J. Veihmeyer and L. N. Brown, *Calif. Agric. Expt. Sta., Bull* 668, 1942.
2 *Utah Agric. Exp. Sta., Bull.* 116, 1912.

Irrigation, however, does more than simply supply water to a crop, for most of the plant nutrients are in the surface layer of the soil, which is the first to dry out on the onset of drought. The crop may then still be well supplied with water, but its growth is limited by lack of nutrients because the roots cannot remove them from the dry surface zone. Irrigation, by moistening the soil surface, allows the crop to take up its nutrients again; and a part of the reason why frequent small irrigations are often better than infrequent large ones lies in the ability of the crop to make much better use of the nutrients in the surface soil.

This effect is most noticeable with intensively managed grass, because it is harvested several times during the summer, so the grass foliage remaining after harvest cannot contain any large reserves of nutrients, even if it showed considerable luxury uptake. Thus nearly all the nutrients in the new growth must be taken up from the soil after harvest. The effects of irrigation on grass production illustrate the role of water on crop growth very clearly for this reason.

Grass growth on fertile soils kept well watered is typically limited by the incoming solar energy, while transpiration is controlled by net radiation and under these conditions most of the net radiation is dissipated as transpiration. Since net radiation is closely correlated with incoming radiation, rates of photosynthesis and of transpiration should be closely correlated. H. L. Penman[1] has shown that this correlation is very close for grass at Woburn, as is shown in Fig. 20.5. Thus he finds that a ryegrass ley produced 270, 230 and 190 kg/ha dry matter per 1 cm water transpired at a high, medium and low level of nitrogen fertiliser; and at the high nitrogen level, the energy content of the dry matter in the grass equals about 1 per cent of the incoming solar radiation. He also found that the rate of dry matter production fell below these figures if the water deficit in the soil exceeded 5 cm for the high nitrogen or 2·5 cm for the low nitrogen plots. This correlation is only valid when the rate of transpiration is sufficiently high that little net radiation is converted to sensible heat, and when photosynthesis remains proportional to the intensity of the incoming radiation.

Penman himself did not investigate the reason why the grass on the high-nitrogen plots could tolerate a higher water deficit than the low-nitrogen plots, but an experiment by E. A. Garwood and T. E. Williams[2] at the Grassland Research Institute, Hurley, gives the most probable explanation. They showed that a ryegrass ley, whose growth had been checked in early summer by drought, started growing again if a nitrogen fertiliser was placed in the moist subsoil. Thus during a seven-week period from late May to early July unirrigated ryegrass produced 540 kg/ha dry matter, but if some nitrogen was injected into the moist subsoil at a depth of 45 cm in late May, it gave a yield

---

1 *J. agric. Sci.*, 1962, **58**, 349, and for later results ibid., 1970, **75**, 75. See W. Stiles and T. E. Williams, *J. agric. Sci.*, 1965, **65**, 351 for corresponding figures for Hurley.
2 *J. agric. Sci.*, 1967, **69**, 125 for yields, and ibid., 1967, **68**, 281 for water use.

of 1450 kg/ha and had taken up 27 per cent of the applied nitrogen, while in both plots the grass transpired about 9 cm water and the water deficit in the soil increased from 5·5 to 12·2 cm. However, once the subsoil at this depth was dry, the grass could not respond to this nitrogen, so the grass gave negligible growth between mid July and mid August, yet transpired 12 cm of water and increased the soil deficit to 14·8 cm.[1] This need to maintain a moist surface soil for nutrient uptake is one of the reasons that a number of high yielding crops on the eastern side of England respond to irrigation at quite low moisture deficits, when their roots are capable of building up very much larger deficits before transpiration is seriously affected.

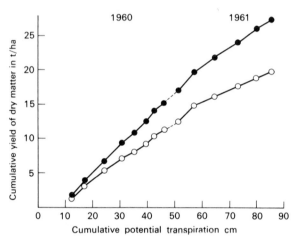

FIG. 20.5. Cumulative dry matter production of irrigated Italian ryegrass plotted against its cumulative transpiration (Woburn).

●—● 75 kg/ha N applied after each cut
○—○ 37·5 kg/ha N applied after each cut

Crops which take up much of their nutrients early on in the season and only manufacture carbohydrates later on may not have their final yield much affected, even if they build up quite considerable soil water deficits in late summer or early autumn. Thus A. P. Draycott and D. J. Webb,[2] working at Brooms Barn in Suffolk, on a sandy loam soil, showed that sugar-beet could build up a deficit of 15 cm water without its yield, which was 6 t/ha sugar, being affected; but in another year a deficit of 19 cm reduced the yield of sugar from 8·5 t/ha with irrigation to 7·5 t/ha without.

1 T. E. Williams, *Grassland Res. Inst.*, *Ann. Rept* 1969, 136.
2 *J. agric. Sci.*, 1971, **76**, 261. H. L. Penman found at Woburn that deficits in excess of 10 cm in September began to limit sugar-beet yields compared with fully irrigated plots.

The available water content of a soil

The available water content of a field soil is defined as the amount of water a crop can remove from the soil before its yield is seriously affected by drought. This concept was put forward by F. J. Veihmeyer,[1] who defined it as the amount of water a soil held between its condition at field capacity and at its permanent wilting point. He found that the yields of many irrigated fruit orchards in central California were little affected by drought if the top 1·8 m of soil was allowed to dry out to its permanent wilting point before the next irrigation was given.[2] This concept has only proved useful in practice under very limited conditions,[3] and then only for soils which hold most of their available water, as so defined, at suctions close to field capacity. These typically are coarse textured soils, not finer than fine sandy or silt loams, but soils with this texture commonly occur on alluvial river terraces, which constitute the major proportion of all irrigated lands of the world. This concept has never been of any value on irrigated clays.[4]

The effect of soil texture on the available water content of a soil is very dependent on the actual structure of the soil. Figure 20.6, taken from a study by A. J. Thomasson and P. D. Smith,[5] gives the scatter of values of field capacity, permanent wilting point and available water content for over 100 samples of surface soil collected by the Soil Survey of England and Wales. It shows that field capacities, on a volume basis, are relatively independent of texture in the loam to clay range, that the permanent wilting percentages tend to increase as the texture becomes finer, and that the soils with the highest volume of available water are the loams.[6] However, the rate that roots can extract water from a soil depends on the soil capillary conductivity, which decreases very rapidly as the soil suction rises, as the shape of the moisture content-suction curve is an important factor in controlling this rate. This results in practice in the water held at low suctions being more readily available to a fast growing crop than that held at high suction. Figure 20.7 gives typical curves for four contrasting textures of the proportion of the total available water held at different suctions,[7] and shows that a much greater proportion of this water is held at low suctions for loamy sands and sandy

1 *Hilgardia*, 1927, **2**, 125.
2 With A. H. Hendrickson, *Calif. Agric. Exp. Stn.*, *Bull.* 479, 1929; *Bull.* 573, 1934, for peach and prune orchards; *Bull.* 667, 1942, for apples and pears; *Bull.* 668, 1942, for cotton; *Proc. Hort. Soc. Amer.*, 1931, **28**, 151, for grapes; ibid., 1937, **35**, 289, for walnuts; and for growth rates of fruit on peaches, pears and prunes, ibid., 1942, **40**, 13.
3 For a relevant review see R. M. Hagan, Y. Vaadia and M. B. Russell, *Adv. Agron.*, 1959, **11**, 77; and F. J. Veihmeyer, *Trans. Amer. Geophys. Un.*, 1955, **36**, 425.
4 M. R. Lewis, *US Dept. Agric., Tech. Bull.* 432, 1934; *Plant Physiol.*, 1935, **10**, 309, for pears; J. R. Furr and C. A. Taylor, *US Dept. Agric. Tech. Bull.* 640, 1939.
5 I am indebted to Mr Thomasson for permission to quote from this unpublished work.
6 See, for example, P. J. Salter and G. Berry, *J. Soil Sci.*, 1966, **17**, 93; G. W. Petersen, R. L. Cunningham and R. P. Metelski, *Proc. Soil Sci. Soc. Amer.*, 1968, **32**, 866.
7 S. J. Richards and A. W. Marsh, *Proc. Soil Sci. Soc. Amer.*, 1961, **25**, 65.

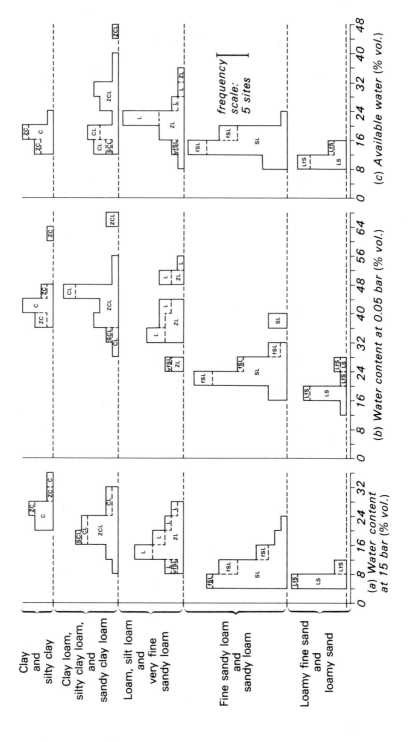

Fig. 20.6. Histograms of water content in mineral surface (A) horizons of five texture groups: (a) 15 bar (b) 0·05 bar and (c) available water (b–a)

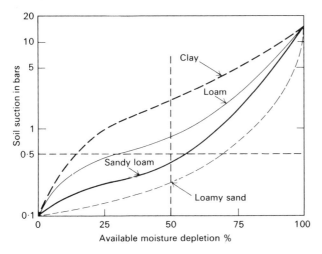

Fig. 20.7. Relation between relative moisture depletion and suction for soils of different textures

loams than for loams or clays; and this greater proportional availability at low suctions—in the sandy loams—counterbalances, to some extent, the advantage of the higher available water content of the finer-textured loams.

The available water content of a soil can be raised somewhat by increasing its organic matter content, though it is doubtful if this is of any significance in farming practice. On the clay loam soil of Broadbalk field at Rothamsted, 93 dressings of 35 t/ha farmyard manure given most years between 1843 and 1943 increased the available water in the plough layer (0–22·5 cm) from 5·0 cm, the value for the unmanured plot, to 6·8 cm, and had little or no effect below this depth.[1] P. J. Salter and J. B. Williams[2] found that 50 t/ha of farmyard manure, given twice a year for five years to the sandy loam soil at Wellesbourne, in Warwickshire, had probably only increased the available water in the plough layer (0–15 cm) by 0·5 cm, and 15 t/ha annually for nine years on a fine sandy loam at Efford in Hampshire had no measurable effect, although it increased both the field capacity and the permanent wilting point a little.[3] The effect of salts in the soil solution, which raises its osmotic pressure, on the available water in a soil is discussed on p. 750.

It is worth noting that a crop with a well-established root system growing under a high transpiration demand, that has just been irrigated, will use a certain amount of water from its root zone that would have drained out of that zone if the transpiration demand was low; so that the amount of available water held in the profile by an irrigated crop will depend, to some

1 E. W. Russell and W. Balcerek, *J. agric. Sci.*, 1944, **34**, 123. P. J. Salter and J. B. Williams (*J. agric. Sci.*, 1969, **73**, 155) found rather lower figures than this for samples taken in 1967.
2 *J. hort. Sci.*, 1968, **43**, 263.
3 P. J. Salter, J. B. Williams and D. J. Harrison, *Expl. Hort.*, 1965, **13**, 69.

extent, on the rate of water use and of downward drainage. It is not unusual for crops to use up to 5 cm of water that is potentially capable of draining out of the profile.[1] Further, C. H. M. van Bavel[2] has shown that appreciable amounts of water that have drained through the root zone of the crop will move back from the subsoil into the root zone as the soil dries out and transpiration proceeds, and this may amount to 4 mm/day for several days, although the total upward flux only amounted to 25 mm.

Effect of moisture stress on the quality of the crops

Water stress will affect the composition of the dry matter of the crop in various ways. It tends to make leaves more xerophytic, thickening their cell walls and increasing the deposits of cutin outside their epidermal cells. Thus, the fresh green weight of herbage produced by crops under water stress will often drop more rapidly than the dry weight, since the percentage dry matter in the green weight increases with moisture stress.[3] Possibly also because water stress increases the osmotic pressure of the cell sap, it increases the percentage of sugar in sugar-cane and often in sugar-beet, although the yield per acre may be reduced. A certain amount of water stress at the end of the growing season helps to mature many fruits, making them firmer and less liable to bruising.[4] On the other hand, too severe a stress will prevent the fruit from swelling properly, giving smaller or shrivelled fruit. This can be noticeable if the stress is severe both for cereal grain and for coffee berries,[5] as well as for fruits having a much higher moisture content.

Moisture stress can affect the chemical composition of leaves in other ways. A moderate stress improves the quality of some tobacco leaves,[6] improving the body and aroma of the tobacco. It will also raise the hydrocyanic acid content of sorghum forage, particularly if nitrogen manuring has also been given.[7] As a matter of field experience, the nitrogen content of cereal grains tends to be higher in dry than in wet years, but this may simply be a reflection of the nitrogen supply to the crop; for in wet years nitrates will be washed out of the soil, and in dry years the total plant growth will be reduced and no nitrates washed out, so that there will be an ample supply when the grain is forming. But fertiliser experiments indicate that this is not the full explanation, and it is possible that water stress genuinely encourages a higher

1 See, for example, D. R. Nielsen, D. Kirkham and W. R. van Wijk, *Proc. Soil Sci. Soc. Amer.*, 1959, **23**, 408; J. C. Wilcox, *Canad. J. Soil Sci.*, 1962, **42**, 122.
2 With K. J. Brust and G. B. Stirk, *Proc. Soil Sci. Soc. Amer.*, 1968, **32**, 317, and see ibid., 310, 322.
3 For an example with ladino clover see R. M. Hagan, M. L. Peterson *et al.*, *Proc. Soil Sci. Soc. Amer.*, 1957, **21**, 360.
4 F. L. Overley, E. L. Overholsen *et al.*, *Washington Agric. Exp. Sta. Bull.* 268, 1932, and A. L. Ryall and W. W. Aldrich, *J. agric. Res.*, 1944, **68**, 121.
5 J. A. N. Wallis, *J. agric. Sci.*, 1963, **60**, 381.
6 C. H. M. van Bavel, *Agron. J.*, 1953, **45**, 611.
7 C. E. Nelson, *Agron. J.*, 1953, **45**, 615.

protein grain. C. H. Bailey[1] examined the nitrogen content of spring wheat gathered from sixteen countries in Minnesota and found a close connection between seasonal rainfall and protein content, which is unlikely to be affected by the leaching of nitrates at the drier end of the range. His results were:

Rainfall: 1 Apr.–1 Sept. (cm)	30–35	35–40	40–45	45–50	50–55	55–60
Nitrogen, % in grain	2·62	2·41	2·14	2·35	2·26	2·04

The efficient use of water

The first major point to stress in any discussion on the efficient use of water is that a weedy crop, or a crop receiving inadequate levels of fertiliser will use as much water as a clean, well-manured crop. In fact, well manured crops often come to maturity earlier than an inadequately fertilised crop. And, as will be discussed on p. 789, these remarks are as applicable to semi-arid regions as to irrigated crops. Thus, any agricultural practice that increases crop yields automatically increases crop production per unit of water used.

The second major point is that the more frequently a crop is irrigated, the higher the water use, even if there is no through drainage, because evaporation of water from the soil surface is added to transpiration by the crop. Infrequent heavy irrigations thus often give a more economical use of water than frequent light ones. S. A. Taylor,[2] for example, working in Utah where evaporation rates are high, finds that crops use 10 to 20 per cent more water if frequently irrigated than if they are given less frequent but heavy irrigations.

The third major point is that if the growth of the crop is set back badly, by, say, water shortage at a critical period, the overall efficiency of water use will also be depressed, because the effect of this set back on transpiration is nearly always less than on growth or yield. On the other hand, water shortage at a non-critical time may reduce total transpiration while having only a small effect on yield, and this may increase the efficiency of water use, as judged by the yield of the harvested product. Further, a crop needs a certain amount of water to give any yield at all, so that additional water supplies above this level raises the overall efficiency of water use from zero to a potentially useful level. M. E. Jensen and J. T. Musick[3] give an example of the yield of sorghum being raised from 3·4 t/ha, if supplied with 38 cm of water to 9 t/ha if given an additional 20 cm, so the efficiency of water use has gone up from 88 to 150 kg/ha of grain per cm of water used. They also find that for winter wheat in Kansas, the yield increases from about 780 kg/ha if 20 cm water are available to 3·4 t/ha if 58 cm are, so the efficiency rises from 38 to 57 kg/ha per cm of water.

1 *Minnesota Agric. Expt. Sta., Bull.* 131, 1913.
2 *Trans. Amer. Geophys. Un.,* 1955, **36**, 425.
3 *Trans. 7th Int. Congr. Soil Sci.* (*Madison*), 1961, **1**, 386.

O. J. Kelley gives a similar example for lucerne grown in Yuma, Arizona.[1] On the plots that were irrigated whenever the soil suction at 30 cm depth reached 175 mbar, the yield of hay was 130 kg/ha per cm of water applied; those which were irrigated when the suction reached 600 mbar at 45 cm depth yielded 120 kg/ha, and those allowed to wilt before irrigation 100 kg/ha. These figures are for plots receiving adequate amounts of phosphate, and the corresponding figures for those receiving inadequate phosphate are 84, 75 and 70 kg/ha per cm of water respectively.

1 *Adv. Agron.*, 1954, **6**, 67. The data are from Stanberry, Converse and Haise.

21

Soil structure and soil tilth

Soils in their natural condition have at least some of their individual particles clustered into clods and crumbs, or peds as they are often called, through them being bonded together by the clay particles present in the soil; and the size distribution of these peds, or its converse—the size distribution of the pore spaces between them—determines the soil structure and, in part, the soil tilth.

There is not yet any agreed definition of the phrase 'soil structure', and in particular whether a collection of particles in uniform or close packing can be said to have a structure or not. Authors who consider every material must possess a structure would say that a heap of sand grains would have a single grain structure, and a well-puddled block of clay would have a massive structure. But if a soil is thought of as a medium for plant growth, some word is wanted for the characteristic of a soil that distinguishes it from these uniform blocks of material, and the phrase 'soil structure' is often used to describe this characteristic. On this use of the phrase, soil structure implies the existence of cracks or pores between volumes of soil whose particles may be in close packings, so that a characteristic of a soil possessing a structure is that its apparent bulk density is lower than if it was without structure, and the magnitude of this reduction throughout the year would be a useful measure of the development and stability of its structure. On this concept of structure, the size, spatial distribution and longevity of the spaces between soil particles that are large enough to be filled by other soil particles, but are not so filled, constitute the properties that distinguish a soil with structure from a structureless soil.

The size distribution of pores and cracks can be put into several different size classes. Water only moves freely through pores wider than about 0·3 mm, which need suctions of less than 10 mbar to empty them of water, and the young roots of many plants also need pores of about this size for easy entry. Water only drains out of soil under gravity if it can move through pores larger than 60 to 30 μ, that is pores which need suctions of about 50 to 100 mbar to empty them of water. Root hairs and the larger members of the soil micro-organisms such as protozoa and fungi with coarse mycelia need pores larger than about 10 μ to grow into or move in, and the smaller micro-organisms need pores larger than 1 μ for movement, which require a suction of 3 bar to empty them of water. Water also has a relatively high capillary conductivity in pores as wide as this, so plant roots can take water relatively easily from them. Thus, very roughly, the pores can be divided into coarse, above about 200 μ in size, medium from this size down to about 20 μ, fine, from this size down to about 2 μ and very fine, which are finer than this.

Pores finer than 2 μ in size affect a group of properties of great importance in soil cultivations known as consistency or tilth properties. These include such properties as plasticity, mouldability and stickiness of the soil when wet, and hardness and crumbliness of the soil when dry. The pore sizes which contribute to these properties are not accurately known, but are probably less than a few hundred Ångstroms, that is pores which need suctions of over about 100 bar to empty them of water. If a clay soil has a large number of pores finer than about 100 Å, it is likely to have a poor tilth: one that a farmer would probably describe as raw, while if most of the pores are larger than this the tilth would be friable or mellow.

The other aspect of soil structure is concerned with the size, shape and appearance of the individual lumps of soils, or peds, which form the body of the soil. A soil surveyor, when describing a soil profile, will give a description of both the cracking pattern and the ped structure, though the relative importance of the two depends on the soil. Some soils have so little cohesion between the soil particles that they do not possess any noticeable peds, and the visible pore space pattern is one of channels rather than cracks. Other soils have cracks that are so narrow when the soil is moist that a soil lump must be broken carefully to display the ped structure, though this may be obvious when the soil is dry.

A number of systems of description and classification of peds have been devised, and that developed in the USA is a good example of an adequate system.[1] The soil surveyor has to note three main groups of properties: the type of structure or the shape of the peds, the class of the structure or their size, and the grade of the structure or their distinctiveness and durability. There are four primary types of structure, some of which are illustrated in Plates 22 to 25, laminar, in which the natural cracking is mainly horizontal, such as is often found in the A_2 horizon of leached soils or in soils rich in kaolinite; prism-like, in which vertical cracking is better displayed than horizontal and the faces of the peds are fairly smooth and flat, and which is often found in the B horizon of clay loams and clay soils; blocky, in which vertical and horizontal cracking are about equally strongly developed, so the peds have roughly equal axes, and the peds may have either flat or rounded surfaces, but when wet they swell and the surfaces of contiguous peds fit exactly one with the other leaving no gaps between them; and spheroidal

1 *U.S. Dept. Agric.*, *Soil Survey Manual*, 1951; and see G. R. Clarke, *The Study of the Soil in the Field*, 5th Edn, Oxford, 1971.

PLATE 22. Profile of the Rothamsted Clay-with-flints soil (0–1·5 m). Note the characteristic rectangular cracking in the subsurface soil and the flat inclined surfaces in the subsoil.

PLATE 23. Fine blocky soil structure (natural size).

PLATE 24. Fine prismatic soil structure (natural size).

which differs from blocky by the surfaces of the peds being rounded and not fitting the surfaces of their neighbours so, when wet, there are appreciable pores between them. These types can be further subdivided, as, for example, the prism-like is divided into columnar, when the upper end of the prism is rounded, and prismatic when it is not; the blocky into angular blocky when the surfaces intersect at relatively sharp angles, and subangular when the edges are rounded; and spheroidal into granular, when the ped is relatively non-porous, and crumb when porous. There are three grades of structure, if one ignores the structureless class, namely, weak, when the peds are barely discernible in the undisturbed soil; strong, when the soil consists of well-defined peds which only adhere weakly to one another, and medium for structures intermediate in development between these two.

Soil peds may have a well-developed structure of their own, particularly if the soil is a clay or loam under natural vegetation which is little disturbed by man or his grazing animals. Under these conditions, any large aggregate or clod is seen to be built out of smaller clods or crumbs, and these out of still smaller granules. This can often be seen by scratching the surface of a clod with a sharp needle, or putting a dry crumb in water, for both will loosen granules from the ped or crumb surface. Crumbs that are built out of definite granules are said to possess a well-developed micro-structure or marked granularity; and in fine-textured soils, granularity and mellowness of tilth

PLATE 25. Rounded and flat tops of prismatic aggregates (half natural size).

are closely associated. A wet deflocculated or puddled clay or loam soil usually dries out into hard clods devoid of microstructure, but adding a soluble salt, such as gypsum or an adequate amount of sodium chloride to flocculate the soil, will increase the granularity of the clods.

The actual distribution of these granules, and of their distribution relative to the sand particles, can be seen by examining thin sections of soil crumbs under a petrological microscope. This technique was introduced by W. Kubiena,[1] who named some of the types of microstructure which can be seen

1 See his book, *Micropedology*, Ames UP, 1938.

in different soils, and the classification of these structures has been greatly expanded by R. Brewer.[1] In typical English arable soils, these granules lie between sand grains whose surface appears to be clean, and they prevent the sand grains falling into close packing because they are larger than the pores present in such a system. But, in some soils, the sand grains can be seen to have a film of oriented clay particles on their surface, and these films adhere to each other sufficiently strongly to hold the sand particles apart in open packing against moderate mechanical stress.

The individual clay particles present in the granules cannot be seen under a petrological microscope, and in thin sections they usually appear as a plasma whose opacity depends on the thickness of the section and on the amount of iron and manganese in or on the clay. The plasma may be composed of films of clay particles, all of which have the same orientation, in which case a thin section of the film will be birefringent; and it sometimes contains birefringent spots, up to about 1 μ in size, but usually it appears to be structureless. The last condition is probably typical of an intimate mixture of clay and humic colloids bonded together. These soils, in which the plasma shows birefringent spots, probably contain small granules or domains of clay particles, in which all the particles have about the same orientation and are held together by di- or polyvalent cations between the layers of adjacent particles.

The concept of the domain structure of clay particles in soils was first put forward by W. W. Emerson,[2] and has been developed by J. P. Quirk. Quirk[3] studied the fine microstructure of clays and clay soils using their adsorption isotherms for nitrogen at 78°K and tetrachloroethane at room temperature: the characteristic of each of these liquids being that they cause no swelling of the clay. A plot of the slope of the adsorption isotherm against the relative vapour pressure of the liquid gives the proportion of the total pore volume that is present in pores of different sizes. They found that for the clays and clay soils they used, the pore volumes present were usually concentrated in three zones: one of about 20 to 40 Å due to spaces between overlapping lattices at the broken edge of the clay particle, one at around 200 Å, probably due to spaces between contiguous clay particles held together in at least partially random orientation, and one at 1000 Å or above, probably due to spaces between small packets of clay particles and, again, held together in more or less random arrangement. The word domain is used for the group of clay particles containing pores of about 200 Å but separated from their neighbours by pores around 1000 Å. Plate 26 gives an electron microscope photograph of the broken edge of a micaceous clay, showing the parallel channels between groups of overlapping layers, and Fig. 8.4 on p. 145 gives a diagrammatic representation of a domain.

---

1 See his book, *Fabric and Mineral Analysis of Soils*, New York, 1964.
2 *J. Soil Sci.*, 1959, **10**, 235.
3 With L. A. G. Aylmore, *J. Soil Sci.*, 1968, **18**, 1, and C. R. Panabokke, ibid., 1962, **13**, 71.

PLATE 26. Electron micrograph of a replica of a fracture surface of a clay core of Willalooka illite. The arrows show the direction of shadowing. ($\times 16000$).

The principal structural units present in soils can be put into various size groups, each with a definite name.[1] These are:

Domains: groups of clay particles, probably up to 5 μ in size, with a concentration of pores around 1000 Å in size, which themselves may consist of smaller packets with a concentration of pores around 200 Å.

Granules: groups of domains and silt and fine sand particles aggregated together in compound particles up to 0·5 mm size. The aggregates in this group are sometimes called microaggregates or microcrumbs, and the upper size is about the largest granules that does not break up into finer granules when wetted with water. This figure is not much larger than the size of clay granules A. P. Edwards and J. M. Bremner[2] found in soils whose clay did not disperse easily, which was up to 0·25 mm.

Crumbs: a collection of granules having a size of several millimetres. There is no particular property which defines an upper limit to a crumb, but in general aggregates larger than about 5 to 10 mm are too large for a good seedbed.

Clods: soil aggregates larger than about 1 cm in size. They have normally to be broken down into crumbs when preparing a seedbed.

1 R. G. Williams, D. J. Greenland and J. P. Quirk, *Aust. J. Soil Res.*, 1967, **5**, 7, suggested a classification like this, but they used different names and size limits.
2 *J. Soil Sci.*, 1967, **18**, 64.

In a well-structured soil, each type of aggregate is built up out of smaller units in a semi-random packing so that there are appreciable volumes of structural pores between each group of units.

The stability of soil structure

Soil structure can be destroyed or seriously reduced by wetting a soil when it is dry, and it can be even more seriously reduced if it is subjected to compression or shearing forces when wet. A soil with a stable structure will be much more resistant to wetting after drying, or to cultivation during wet weather, than a soil with a weak structure. A large number of methods have been devised to measure the stability of structure, or its resistance to destruction, under a variety of conditions using a wide range of measurable structural properties,[1] but since so many factors affect structural stability, there is not, and probably there cannot be a single universal method for doing so.

A soil can lose structure on wetting because the clay deflocculates spontaneously if put in water of a low salt content. If a crumb of such a soil is put in water, it swells and its surface becomes very diffuse and clay particles become dispersed in the water and form a stable suspension. A soil showing this property is typically characterised by having sodium forming an appreciable proportion of the exchangeable cations on the clay. For many soils, this happens if the percentage exceeds 15; and if it is between 7 and 15 per cent, the crumb will swell and its surface become rather diffuse, but the clay particles will not spontaneously disperse in the water. These soils can occur in irrigated areas where the soil and the water contain too much sodium in relation to calcium (see p. 758).

A dry soil crumb or clod usually slakes to a greater or lesser extent on wetting. By slaking is meant the falling off from the crumb of fine crumbs or granules, whose surface remains clear and sharp in the water. A soil with a very weak structure will slump on wetting, that is the surface tension forces will be stronger than the forces bonding the particles or granules together so they will be pulled together into closest packing. But the forces holding particles or granules together are usually much stronger than these surface tension forces.

Wetting a moist or dry clod, particularly if the soil is a loam or clay, will often cause a break-up of the clod into smaller clods, or will cause cracks to form in the clod due to entrapped air. If a clod is immersed in water, or if it is wetted rapidly, all the outer pores will be filled with water, and surface tension forces will pull water into many of the coarser pores. This will compress the air in the clod, which can only escape by bursting the clod either into small clods or large crumbs, or by causing cracks to develop in it. If the clod is dry,

1 For a description of the methods that have been used in W. Europe see *West European Methods for Soil Structure Determination*, Ghent, 1967.

it will also have adsorbed air, which will be displaced from the clay surface on wetting. A dry montmorillonite clay will, for example, adsorb about 15 cm^3 air per 100 g of clay or about 0·3 mg/g clay.[1] This entrapment of air has the consequence that the faster a dry clod is wetted, the greater the disruption the entrapped air will cause, but wetting under vacuum[2] or slow wetting by rainfall at rates of 2·5 mm/hour, which is typical of gentle rain,[3] will give minimum break-up due to this cause as is illustrated in Fig. 21.1. Heavy rain causes an additional slaking of these crumbs, due to the mechanical impact of the fast-falling large raindrops (see p. 780).

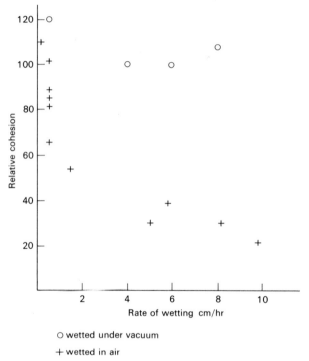

○ wetted under vacuum

+ wetted in air

FIG. 21.1. Effect of rate of wetting of a grassland soil on the break up of soil crumbs 2–5 mm in size

C. R. Panabokke and J. P. Quirk[4] showed that the uneven swelling of clods on wetting, for those soils which swell, also gives rise to forces that break up a clod, but for the soils they worked with this was not important if the wetting occurred under suctions exceeding 30 mbar.

1 V. C. Jamison, *Proc. Soil Sci. Soc. Amer.*, 1953, **17**, 17; J. R. Runkles, A. D. Scott and F. S. Nakayama, ibid., **22**, 15.
2 E. W. Russell and R. V. Tamhane, *J. agric. Sci.*, 1940, **30**, 210.
3 W. W. Emerson and G. M. F. Grundy, *J. agric. Sci.*, 1954, **44**, 249. For a critical discussion of these effects see W. D. Kemper and E. J. Koch, *US Dept. Agric. Tech. Bull.* 1355, 1966.
4 *Soil Sci.*, 1957, **83**, 185.

This instability of structure due to slaking, for a given soil, is usually more noticeable the lower its humus content, possibly due to clods with the higher humus content wetting more slowly than those with a lower content, for the humus may make the soil surfaces more hydrophobic, in which case the water films make a positive angle of contact with the soil surface. However, D. O. Robinson and J. B. Page[1] found that the humic colloids present in a normal mineral soil did not appear to affect the angle of wetting appreciably. But Quirk and Panabokke showed that a more likely reason is that some of the humus strengthens the coarser pores which, in turn, reduces the tendency for cracks to form as the clod wets. Crack formation reduces the rate of wetting of a clod, for cracks allow water to flow through them more rapidly than through the finer channels constituting the original pore space.

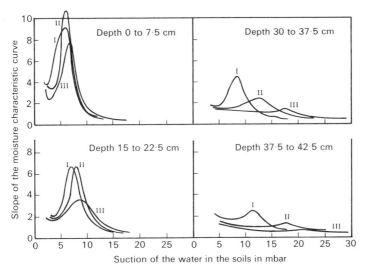

FIG. 21.2 The variation in the stability of the pore space distribution down the profile of an Upper Lias clay. Curve I is for the first, Curve II for the second and Curve III for the third wetting. (Soil initially in crumbs 1–2 mm size)

An important group of methods for measuring the stability of soil structure is based on measuring the size distribution of the crumbs that small dry clods break down into on wetting. The methods differ in the initial moisture of the clods that are used, their rate of wetting and the mechanical forces the crumbs are subjected to during the measurement of their sizes. The size distribution can be determined by such methods as separation of the crumbs on a bank of sieves under water—a group of methods that have been very widely used, or the size distribution of the pores between the crumbs can be measured. This is done by determining the moisture content-suction curve for a shallow column of the soil crumbs, after each cycle of drying the column

1 *Proc. Soil Sci. Soc. Amer.*, 1950, **15**, 25.

and rewetting it. The slope of this curve, plotted against suction, gives the pore-size distribution between the crumbs. Figure 21.2 taken from some work of E. C. Childs[1] and designed to determine whether a clay soil is suitable for mole draining or not, illustrates the results obtained on an old pasture soil, and show clearly that the surface crumbs are very stable to successive wetting and drying treatments, but the subsoil crumbs are very unstable. This is, in fact, an example of a clay quite unsuited to mole-draining because of the instability by slaking of the clay when wetted.

These methods give little indication of the strength of the soil crumbs that are water-stable by these tests. This can be measured by subjecting the crumbs to a known internal swelling pressure. S. Henin[2] introduced this method originally, but under conditions where the swelling pressure could not be accurately controlled. His method was first to saturate the clods or crumbs with a non-polar liquid immiscible with water—he used benzene—and then to put them into water. As the crumb wets and swells, it will transmit a swelling pressure to the benzene in the soil pores, and because the benzene is effectively incompressible, it will transmit this pressure to the rest of the crumb. W. W. Emerson[3] used an alternative and more controllable method of increasing the swelling pressure tending to disrupt a crumb, by leaching a solution of 1N sodium chloride through a bed of soil crumbs to saturate the soil with exchangeable sodium ions; and then to reduce the concentration of sodium chloride percolating through the column, and determine the concentration at which the crumbs collapsed and the column lost permeability. Soils with very stable structure maintain permeability even if distilled water is percolated through it, while soils with a weak structure will lose permeability at a concentration of between 10^{-2} and 10^{-1} N.

This test is done under conditions when the crumbs are put under no mechanical force, but in the field this condition often does not apply, for soils must often be cultivated when they are wet. Emerson and M. G. Dettman[4] found that crumbs stabilised with organic matter would lose most of their stability if the soil was puddled when wet, showing that the organic matter bonds holding the soil particles could be broken fairly easily; and on the basis of this observation, Emerson[5] suggested a quantitative test based on a more gentle working of the soil at a moisture content corresponding to field capacity and determining the stability of small wet lumps of this reworked material. Soils with strong crumbs, particularly some with calcium carbonate in the clay fraction, gave lumps that remained stable; but if these lumps were then shaken up with water, those from some dispersed and from others remained flocculated. Many calcium-saturated soils can be dispersed

1 *Soil Sci.*, 1940, **50**, 246; 1942, **53**, 84.
2 *Pl. Soil*, 1948, **1**, 167.
3 *J. Soil Sci.*, 1954, **5**, 233, and with M. G. Dettman, ibid., 1959, **10**, 215.
4 *J. Soil Sci.*, 1959, **10**, 227.
5 *Aust. J. Soil Res.*, 1967, **5**, 47.

by this process, for the wet working will pull some of the clay particles apart through distances sufficient to overcome the energy barrier to deflocculation.

H. C. Pereira[1] also introduced an empirical test to measure the wet strength of crumbs, with the object of determining the ability of a surface soil structure to withstand a heavy rainstorm, rather than for stability against cultivation when wet. To do this it is necessary to discover the amount of water that runs off the surface of an undisturbed moist surface soil when it is subjected to an artificial rain storm of 2·5 cm falling in 10 min, and which is under a suction of 10 mbar.

The creation of structural pores in undisturbed soils

The most important group of agents that create structural pores in undisturbed soils are plant roots. The roots of many plants grow by forcing a root tip, only a few tenths of a millimetre in diameter, into soil pores of about their size, and the young root then swells making the pore larger. This swelling involves soil particles, domains or granules being moved relative to their neighbours, and this relative movement sets up shears which will cause these structural units to be twisted relative to each other; and since they are not spheres, this will enlarge some of the pores between them, if they were in fairly close packing. Further, one of the functions of a plant root is to extract water from the soil, and if the soil shrinks as it loses water, this drying of the soil will cause it to crack. Thus the more intimately the roots permeate through such a soil, the more intimate will be the system of cracks and pores that will be developed. Cracks produced by shrinkage can be long-lived, for once a loam or clay clod has been dried it may take months before the soil has swollen again to its full volume.[2] In these cracking soils, roots help to create their own spatial environment for growth, as the root dries the soil, cracks develop around the root into which a new young root can grow.

The other group of agents that can create structural pores are some of the larger members of the soil fauna. Earthworms are the most important group in many temperate soils, for they make their burrows through a considerable depth of soil, and they make wormcasts, which are fine ribbons of soil excreted by the worm and left in a little heap. The larvae of a number of insects, particularly beetles, make channels in the soil as they move about in search of food; and ants make numerous channels in the upper part of the soil. R. D. Green and G. P. Askew[3] found on some Romney Marsh pastures that the ants *Lasius flavus* and *L. niger* made numerous channels about 1 mm in diameter which penetrated to depths of 1·5 m, and that a proportion of the soil in the anthills came from depths of 70 to 100 cm. In the tropics, ants and

1 *J. Soil Sci.*, 1956, **7**, 68.
2 W. W. Emerson, *J. Soil Sci.*, 1955, **6**, 147.
3 *J. Soil Sci.*, 1965, **16**, 342.

termites are the principal animals that create pores, channels and burrows in the soil (see p. 207).

The stabilisation of structural pores

The role of the clay fraction

The structural pores in a soil can only be stable against wetting if there is some agent holding the soil particles apart against the surface tension forces set up as it is wetted, for these pull the soil particles together. The most important soil component resisting the collapse of structure is the clay. A pure fine sand is a most unfavourable environment for root growth, for any channels or cracks made in such a soil are filled up by collapse of the sand grains on wetting. Clay particles, by being in domains and granules, help to keep sand particles apart on wetting, provided they retain their individuality, that is, provided they are flocculated.

All clays are flocculated if the concentration of salts in the soil solution is high enough. This must be very high for many clays containing a large proportion of exchangeable sodium ions, and in normal circumstances such clays usually deflocculate if they are wetted by rain (see p. 137). Clays with a high proportion of exchangeable calcium ions form metastable domains and granules, for they are stable in water if not subjected to a force sufficiently large to pull the individual clay particles apart; but once they have been separated from each other, many types of clay form a deflocculated paste which needs a calcium ion concentration of over 10^{-3} M for flocculation.[1] Soils containing calcium carbonate, particularly if finely divided so it remains in equilibrium with the soil solution, are usually flocculated because, in these soils, the concentration of the anions present—bicarbonate, nitrate and sulphate, which are neutralised by calcium, is usually higher than this. If a soil contains free gypsum, or calcium sulphate, this is sufficiently soluble to maintain flocculation.

Clays differ in the ease with which they can be deflocculated, depending on the number of positive charges they carry. Soils and clays containing appreciable areas of aluminium hydroxide give very stable granules,[2] for they are difficult to disperse even if they contain an appreciable proportion of exchangeable sodium ions.[3] Ferric hydrous oxides are probably much less important than aluminium hydroxide, because of their much smaller surface area in most soils, though ferric hydroxides may be as efficient. Most workers[4] who have studied the effects of the ferric oxides in soils, however, have not

1 For the Rothamsted soil see W. W. Emerson, *J. Soil Sci.*, 1954, **5**, 233.
2 T. L. Deshpande, D. J. Greenland and J. P. Quirk, *J. Soil Sci.*, 1968, **19**, 108; G. R. Saini *et al.*, *Canad. J. Soil Sci.*, 1966, **46**, 155.
3 B. L. McNeal *et al.*, *Proc. Soil Sci. Soc. Amer.*, 1968, **32**, 187; 1966, **30**, 308.
4 See, for example, J. F. Lutz, *Proc. Soil Sci. Soc. Amer.*, 1936, **1**, 43; D. S. McIntyre, *J. Soil Sci.*, 1956, 7, 302; M. N. Arca and S. B. Weed, *Soil Sci.*, 1966, **101**, 164.

distinguished between their effect and that of the associated aluminium hydroxides on structure stability. Soils derived from basalts and other basic igneous rocks tend to have an extremely stable structure for this reason. Less is known about the stability due to broken edge positive charges and exchangeable aluminium ions, but field experience is consistent with the stability of clay domains and granules containing such charges and ions being similar to the stability of calcium-saturated clays that do not contain calcium carbonate.

Films of calcium carbonate also can be a stabilising agent, and these may be responsible for the structure stability to rendzina soils, which are soils formed from soft limestone where much of the limestone is present in the clay fraction.[1]

Stabilisation due to humus

The polysaccharides in the humus are the most active fraction in stabilising soil structure. The uncharged polymers are adsorbed on the surface of soil particles by van der Waal forces, as described on p. 148, and the charged polymers can absorb clay particles along their broken edges due to the positive charges on these edges, and along their surface due to exchangeable polyvalent cations sharing their valency bonds with the carboxylate groups on the polymer and with the negative charges on the clay surface. The other components of the humic colloids are probably bonded to clay particles in the same way as the negatively charged polysaccharide polymers, but would be expected to be more active since they carry a higher density of carboxyl groups.

Most of the soil polysaccharide gums are believed to be of bacterial origin, and much of the work on their role as structure stabilisers has been done with gums derived from specific groups of bacteria. The effectiveness of the different gums varies widely[2] for reasons that are not fully known, but are probably concerned with the length of the polymer; for effectiveness usually increases with molecular weight, and gums with molecular weight of the order of 1 million appear to be more effective than those of appreciable lower molecular weight.[3] An effective gum can have a marked stabilising effect if present in concentrations of the order of 0·02 per cent in the soil: an amount that could well be present in a soil containing 1 per cent of organic carbon, or 2 per cent organic matter.[4]

1 N. C. Mehta, H. Streuli *et al.*, *J. Sci. Fd. Agric.*, 1960, **11**, 40; D. J. Greenland, G. R. Lindstrom and J. P. Quirk, *Proc. Soil Sci. Soc. Amer.*, 1962, **26**, 366.
2 M. J. Geoghegan and R. C. Brian, *Biochem. J.*, 1948, **43**, 5; C. E. Clapp, R. J. Davies and S. H. Waugaman, *Proc. Soil Sci. Soc. Amer.*, 1962, **26**, 466. For a review see R. F. Harris, G. Chesters and O. N. Allen, *Adv. Agron.*, 1966, **18**, 107.
3 See also G. Dell' Agnola and G. Ferrari, *J. Soil Sci.*, 1971, **22**, 342.
4 J. P. Martin and D. G. Aldrich, *Proc. Soil Sci. Soc. Amer.*, 1955, **19**, 52, and see D. A. Rennie, E. Truog and O. N. Allen, ibid., 1954, **18**, 399.

Much of the evidence for the importance of these bonds comes from experiments on the effect of removing polysaccharides from soils by oxidation with sodium periodate and extraction of the oxidation products with tetraborate.[1] While this treatment destroys the material that stabilises structure in many arable soils, it does not destroy the structure in many grassland or forest mull soils, but sequential treatment with periodate followed by sodium pyrophosphate removes the stabilising agent.[2] However, treatment with the pyrophosphate alone does not destroy the stability of these soils. These observations are interpreted as showing that in pasture and forest mull soils there are two groups of stabilising agents—polysaccharides and true humic colloids; but the interpretation is uncertain because pyrophosphate removes not only an active humus fraction low in polysaccharides, but iron, aluminium and calcium ions, all of which are active in structure stabilisation. The result that polysaccharide gums are the principal constituent of humus that stabilises the structure in agricultural soils has also been found by G. Chesters,[3] who showed that crumb stability, in the soils he worked with, was much better correlated with the polysaccharide content of the soil than with the total humus content. This conclusion implies that the polysaccharides are active primarily through the large number of weak van der Waal bonds between the uncharged polymer and the surfaces of the mineral soil particles.

The reason why only the polysaccharide fraction of the humus appears to be responsible for stabilising the structure of arable soils may be that in these soils there is no adequate mechanism for the other fractions of the humus to become sufficiently intimately mixed with the soil particles. This intimate mixing may be a very slow process brought about in undisturbed pasture and forest mull soils through the action of earthworms. Humic colloids do, however, have a small effect in stabilising the structure of arable soils, as determined from the effect of extracting the soil with pyrophosphate.[4]

The polysaccharide polymers certainly, and probably the other humic colloids, are present as ropes or nets of polymers, not as individual molecules.[5] Thus a rope or net can have a large number of clay particles attached to it, or it can cover the surfaces of a number of contiguous clay domains and hold them apart against surface tension forces. These ropes do not enter the clay domains but coat the outside of domains and granules sealing a number of pores by covering them over as shown in Plate 29a.[6] Thus they only reduce the interdomain or intergranule swelling, they do not affect the

1 D. J. Greenland, G. R. Lindstrom and J. P. Quirk, *Proc. Soil Sci. Soc. Amer.*, 1962, **26**, 366.
2 C. E. Clapp and W. W. Emerson, *Proc. Soil Sci. Soc. Amer.*, 1965, **29**, 127, 130. R. C. Stefanson, *Aust. J. Soil Res.*, 1971, **9**, 33.
3 With O. J. Attoe and O. N. Allen, *Proc. Soil Sci. Soc. Amer.*, 1957, **21**, 272.
4 R. C. Stefanson, *Aust. J. Soil Res.*, 1971, **9**, 33.
5 For electron micrographs see S. L. Rawlins, J. A. Kittrick and W. H. Gardner, *Proc. Soil Sci. Soc. Amer.*, 1963, **27**, 354, for the kind of structure likely to be present in soils.
6 J. R. Burford, T. L. Deshpande *et al.*, *J. Soil Sci.*, 1964, **15**, 192, but much of the evidence comes from work with soil conditioners, for which see B. G. Williams, D. J. Greenland and J. P. Quirk, *Aust. J. Soil Res.*, 1966, **4**, 131; 1967, **5**, 77.

intradomain swelling.[1] But they cannot cause the domains or clay particles to increase their separation, that is they can stabilise but not create structure. The structure produced by these polymers is stronger, the greater the area of sesquioxide surfaces, for these bind carboxyl groups strongly by ligand exchange, so soils high in sesquioxide surfaces can have a very stable structure due to the very stable and strong bonds between the humic colloids and these surfaces.

Soil conditioners

Much of our knowledge of the role of organic polymers in stabilising structure has been derived from a study of the interactions between some synthetic linear organic polymers, known as soil conditioners, and the soil. They were first introduced into soil science research by Montsanto, an American chemical company, who produced a product under the trade name of Krilium which was a vinyl acetate-maleic anhydride polymer. Later a further polymer that has been widely used was introduced—a hydrolysed poly-acrylonitrile. Both of these contain carboxylic groups spaced along the polymer chain, so are long chain poly-anions in the pH range found in soils. They stabilise clay domains through electrostatic attraction with positive charges on clay particles and, as W. W. Emerson[2] showed, they were much more efficient in acid than neutral soils because of the strength of the polymer-aluminium-clay bond. They can probably also be bonded to uncharged silica surfaces by hydrogen bonding through the silanol groups on these surfaces.

Much interest has also been taken in the uncharged polymer polyvinyl alcohol, which can be manufactured with a range of molecular weights, and hence presumably polymer lengths. Uncharged polymers differ from charged in that they can diffuse into soil pores more easily, as they are only adsorbed by weak hydrogen bonds of van der Waal forces; and the lower the molecular weight, the further they can penetrate into soil crumbs.[3] Adsorption, and hence stabilisation, is greatly helped by drying the soil, and once the polymer is adsorbed, it is very difficult to desorb because of the large entropy gain due to the adsorption (see p. 148). It cannot, for example, be desorbed by treating the soil with pyrophosphate to remove the active aluminium ions[4]— a treatment which desorbs much of an anionic polymer.

Stabilisation by the decomposition of organic matter

If decomposable organic matter is added to a soil, there is a rapid improvement in the stability of the soil structure. This is brought about partly by the

1 B. K. G. Theng, D. J. Greenland and J. P. Quirk, *Aust. J. Soil Res.*, 1967, **5**, 69.
2 *J. agric. Sci.*, 1956, **47**, 117, and with M. G. Dettman, *J. Soil Sci.*, 1960, **11**, 149. This result was not found by R. M. Hedrick and D. T. Mowry, *Chem. Ind.*, 1952, 652.
3 B. G. Williams, D. J. Greenland and J. P. Quirk, *Aust. J. Soil Res.*, 1966, **4**, 131.
4 W. W. Emerson, *J. Soil Sci.*, 1963, **14**, 52.

production of polysaccharide gums by the soil bacteria, and partly by the growth of fungal and actinomycete hyphae growing over soil particles, for some of these hyphae are adsorbed on to or stick on to the surfaces over which they grow. Many of the soil bacteria produce gums, either as extra-cellular or capsular gums or as cell-wall gums, and it is probable that the bulk of the humic polysaccharides are of bacterial origin. M. L. Jackson,[1] using an electron microscope, showed that a proportion of soil crumbs in the soils he examined appeared to be composed of clay particles clustered around a bacterium, as is illustrated in Plate 7c, or a group of bacterial cells. Little is known about the factors that affect the amount of high mole-cular weight gums produced by bacteria when fresh organic matter is added to a soil, but a low level of available nitrogen in the immediate neighbour-hood of the bacteria may be favourable.[2]

Filamentous fungi, which multiply when decomposable organic matter is added to a soil, contain species whose mycelia bind clay or soil particles to their surface and will stabilise soil pore surfaces over which they grow. Fungi differ markedly in the laboratory in their power to do this. Thus, R. J. Swaby[3] found that fungi with fast-growing woolly hyphae were better aggregators of loam and clay soils than those with slow-growing or smooth hyphae. R. B. Aspiras,[4] using ultrasonic vibrations to disperse soil crumbs, found great differences between the strength of the fungal bonds between fungi, but the relative strength of bond as between different fungi depended to some extent on the soil. He also found that the fungal bond did not appear to involve polysaccharides, and in fact fungi are poor producers of high molecular weight gums,[5] and that those fungi whose mycelia bound soil particles strongly to their surface produced more stable crumbs than those stabilised by bacterial gums. The importance of fungi in the stabilisation of soil struc-ture in the field has been difficult to establish, a problem not helped by the difficulty of estimating the number and length of the individual bits of hyphae in a soil. In so far as they hold soil particles on their surface and run between a number of soil particles, they will help stabilise structure in the same way as polysaccharide gums, but will be more effective stabilisers, as they actively grow through the wider pore spaces touching a number of soil particles and clay along their length. R. D. Bond[6] in Australia has shown that some fungi stabilise sandy soils by forming a weft of hyphae covering sand grains,

1 With W. Z. Mackie and R. P. Pennington, *Proc. Soil Sci. Soc. Amer.*, 1947, **11**, 57.
2 C. J. Acton, D. A. Rennie and E. A. Paul, *Canad. J. Soil Sci.*, 1963, **43**, 206; S. M. Schwartz, P. G. Freeman and C. R. Russell, *Proc. Soil Sci. Soc. Amer.*, 1958, **22**, 409.
3 *J. gen. Microbiol.*, 1949, **3**, 236, and see E. Griffiths, *Biol. Rev.*, 1965, **40**, 129.
4 With O. N. Allen *et al.*, *Proc. Soil Sci. Soc. Amer.*, 1971, **35**, 283.
5 M. Stacey, *J. Chem. Soc.*, 1947, 853.
6 *Nature*, 1959, **184**, 744; see also R. H. Thornton, J. D. Cowie and D. C. McDonald (ibid., 1956, **177**, 231) for an example from New Zealand, B. C. Barrett (ibid., 1962, **196**, 835) from England, and S. M. Savage, J. P. Martin and J. Letey (*Proc. Soil Sci. Soc. Amer.*, 1969, **33**, 149, 405) from the USA.

66 Bacteria
22 Fungi

33 Actinomycetes
All together

Sterile
Soil inoculum

(*a*) The water-stable soil structure built up by mixed populations of different types of micro-organisms (Rothamsted soil and glucose).

Initial
7 Bacteria

10 Fungi
15 Bacteria

10 Actinomycetes
30 Bacteria

(*b*) The break-up of the water-stable built up by the mycelia of the fungus *Aspergillus nidulans* (initial), after it had been heat-killed and the soil inoculated with a number of species of other organisms.

PLATE 27. The effect of microbial action on soil structure.

PLATE 28(*a*). The furrow slice of a Saxmundham chalky boulder clay soil after a three year lucerne ley.

which holds them together and by waterproofing the weak sand structure, which makes it difficult to wet.

There is some evidence that the decomposition of fungal hyphae yields products that are more effective in stabilising structure than the hyphae themselves. This possibility was first suggested by F. Y. Geltser,[1] but it has been little investigated,[2] and the conditions under which it occurs and the reasons for it are not known. Plate 27 illustrates some experimental results of R. J. Swaby[3] for structure build-up by different groups of micro-organisms, and the effect of different micro-organisms on structure stabilised by fungal hyphae. In these experiments fungi definitely appear to be the most effective

1 See her book *The Significance of Micro-organisms in the Formation of Humus*, Moscow, 1940, and for an English summary, *Soils Fert.*, 1944, **7**, 119.
2 J. P. Martin, J. O. Erwin and R. A. Shepherd, *Proc. Soil Sci. Soc. Amer.*, 1959, **23**, 217, E. Griffiths and D. Jones, *Pl. Soil* 1965, **23**, 17.
3 *J. gen. Microbiol.* 1949, **3**, 236.

PLATE 28(*b*). The furrow slice of a Saxmundham chalky boulder clay soil on old arable land.

structure stabilisers of the Rothamsted clay-loam soil, and they give no support to Geltser's hypothesis that stabilisation is due to substances produced by the bacteria decomposing fungal tissue.

Plant roots growing in a soil may stabilise the soil structure in their neighbourhood, for some exude a gum or mucigel around, and for some distance behind, the growing tip. In addition they carry a microbial population on their surface and in the soil close to the surface, which may be producing microbial gums, and there may also be fungal hyphae growing from within the root into the soil itself. The roots of some grasses seem to be particularly strong stabilisers and improvers of structure, for reasons that have not been fully explored. On the one hand, D. M. Webley[1] found that the rhizosphere of the grasses he examined was particularly rich in gum-producing bacteria, and, on the other, fine dead roots appear to hold fine crumbs around them,

1. With R. B. Duff *et al.*, *J. Soil Sci.*, 1965, **16**, 149.

as is shown in Plate 31. This could be due to the roots growing through pre-existing crumbs, or to fungal hyphae and bacterial gums holding fine granules together into fine crumbs.

The effect of the decomposition of roots on the stability of the channels they have made has not received much study. In a number of cases, perhaps in general, the inside of the root is eaten out by mites or decomposed by micro-organisms while the outer cells, perhaps those that are suberised, still form a coherent cylinder. These hollow tubes allow air to penetrate into the subsoil at all times, unless the land is flooded with water, for water will move through them rapidly under drainage. As the outer cells decompose, any gum-producing bacteria living between the soil and the root will help stabilise the wall of the soil pore which the root occupied or created. Thus, the more extensive the root system, the greater is likely to be the number of soil volumes which were close to root channels that have been stabilised. Roots differ considerably in their ability to stabilise their channels on their death, but this is a topic that has received very little attention. G. W. Cooke and R. J. B. Williams[1] found that when a 3-year lucerne ley, on a poorly structured boulder clay soil at Saxmundham in Suffolk, was ploughed out in autumn, the soil had an excellent structure compared with old arable as is shown in Plate 28, and it could be worked down to a seedbed very easily, even in wet weather. Yet in the following year, the whole benefit had disappeared. They also noted that the lucerne roots, at the time of ploughing out, contained 25 per cent of water-soluble carbohydrates, so the excellent structure on ploughing out may have been produced from these by the soil population earlier on; but all these stabilising substances were decomposed after 12 months in winter wheat. This soil contains between 35–40 percent of coarse sand and only about 20 percent of clay, and forms very compact clods which do not crack on drying, and into which the roots of most crops will not penetrate. Normal additions of farmyard manure do not affect their micro-structure appreciably, probably because microbial activity is at a very low level within the clods due to the lack of any mechanism for transferring decomposable organic matter into their interior.

Undisturbed soils are also a favourable habitat for earthworms in temperate regions, if the soil is not acid or too sandy. These animals are important agents in structure stabilisation, partly because they exude a mucus from their skin as they move through their burrows, which may stabilise the burrow walls,[2] though its stabilising power has not been studied in detail, and partly because they produce wormcasts (see p. 202). These consist of a heaped coil of a thin ribbon of soil which has passed through the earthworm gut and has had the undigested portion of any plant remains eaten by the worm thoroughly macerated and mixed with it. As the cast ages after it has been formed,

1 *Rothamsted Rept.*, 1971, Part II, 122.
2 C. Jeanson, *Geoderma*, 1967, **1**, 325. *Comp. Rend.*, 1971, **272**, 422.

the stability of its structure increases,[1] presumably due to microbial activity in these fine ribbons; and these stabilised large granules or small crumbs can last a very long time in the soil.

R. J. Swaby[2] found that the increase in the stability of soil structure in a wormcast depended on the soil it was ingesting. Thus, the crumb stability of casts of *Allobophora nocturna*—the most common casting species at Rothamsted, was appreciably greater than the soil from which it had been derived in pasture soils, but little different on arable soils with a much less stable structure. He also showed that there were very high numbers of gum-producing bacteria in the casts produced from pasture soils. A. C. Evans[3] also showed that these casts had a higher proportion of structural pores finer than 50 μ, which increased the volume of readily available water the soil could hold.

The cause of the longevity of soil structure stabilisers in soils is probably associated with the longevity of some of the soil humus (see p. 319), though no critical study has been made of the relative ages of polysaccharides and non-polysaccharides in soils. It was suggested that old humus is old because it is inaccessible to micro-organisms, not because it is very resistant to decomposition; and that it is inaccessible either because it is in pores finer than about 1 μ, the size of a bacterium or actinomycete hyphae, or is in volumes bounded by pores finer than this. Bacterial gums differ in the rapidity with which they can be decomposed if added to a soil,[4] but all gums are normally readily degradable on a time-scale of one year, so their longevity does not lie in their inherent resistance to decomposition. However E. Griffiths and R. G. Burns[5] have also found that the extra-cellular polysaccharide films produced by the yeast *Lipomyces starkeyi* when mixed with soil becomes more resistant to decomposition if water containing either tannic acid or soluble products of decomposing vegetation is percolated through the soil. If the behaviour of this gum is typical of other fungal and bacterial gums, this could be an important mechanism for the long-term stabilisation of soil structure under forest, for example.

The gums are produced on the walls of pores larger than 1 μ, so if the gums are to be long-lived, the pores in which they are produced must be made more narrow. This may happen when the soil dries, if it shrinks on drying; but any process that moves soil granules relative to one another will tend to entrap gums in fine pores, and allow them to bond the granules or domains together. It is possible also that some of the gum diffuses into pores finer than 1 μ close to where it has been produced, and becomes inaccessible to any exoenzymes produced by a micro-organism subsequently growing in its neighbourhood.

1 J. N. Parle, *J. gen. Microbiol.*, 1963, **31**, 13, and for a general discussion see J. E. Satchell in *Soil Biology*, Eds. A. Burges and F. Raw, Academic Press, 1967.
2 *J. Soil Sci.*, 1950, **1**, 195, and for a similar example see A. K. Dutt, *J. Amer. Soc. Agron.*, 1948, **40**, 407.
3 *Ann. appl. Biol.*, 1948, **35**, 1.
4 J. P. Martin, J. O. Erwin and R. A. Shepherd, *Proc. Soil Sci. Soc. Amer.*, 1965, **29**, 397.
5 *Pl. Soil*, 1972, **36**, 599.

It is also possible that any of these gum molecules which become bonded to aluminium ions through ligand bonds are more resistant to hydrolysis than the free molecules.[1]

The discussion so far has been concerned with the processes involved in the improvement of the structural stability of soils brought about by the addition of decomposable organic matter to mineral soils low in active humic colloids. This improvement due to one or a few additions of organic matter is typically short-lived and most of it will have disappeared over a few years after the additions have ceased. But if large additions are given to a soil annually over periods of decades, or arable land is laid down to well-managed grass for similar periods, the structural stability will continue to improve, though the rate of improvement drops, but the length of time the improvement is maintained after these large annual additions of organic matter cease, or after land is ploughed out from grass, also increases.

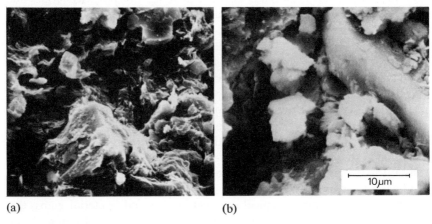

(a)   (b)

PLATE 29. Scanning electron microscope photographs of crumb surfaces: (*a*) old grassland; (*b*) old arable.

The immediate cause of this long-lived improvement is not really known. It is probably due in part to the accumulation in the soil of material other than polysaccharides, and in part to the association between the organic colloids and the mineral particles slowly becoming more intimate as the content of the polysaccharides slowly increases. As the rate of production of these gums increases, they will become with time more evenly distributed around individual soil particles or micro-aggregates, and also an increasing proportion will become situated in long-lived pores finer than 1 μ. P. R. Stuart and A. J. Low[2] have examined the surface of soil micro-aggregates using the stereoscan electron microscope and shown that the individual soil mineral particles and micro-aggregates of an old pasture soil have become

1 J. P. Martin, J. O. Erwin and R. A. Shepherd, *Proc. Soil Sci. Soc. Amer.*, 1966, **30**, 196.
2 I am indebted to these authors for showing me these unpublished photographs.

coated with a film of organic colloids, which is lacking in the corresponding old arable soil, as is illustrated in Plate 29. This film presumably is responsible for the much greater wet strength of the pasture crumbs compared with the arable, and for increasing the proportion of soil pores coarser than $0.02–0.05\ \mu$ at the expense of the finer pores.

Figure 21.3, due to W. W. Emerson,[1] summarises in pictorial form our present ideas on the way clay particles help to build up structure. The clay particles are shown as domains consisting of several clay particles held together face-to-face, and the domains may be held edge-to-face through aluminium bonds, or edge-to-face, edge-to-edge or face-to-face through organic polymers. These polymers can also bond clay particles to the surface of siliceous silt or sand particles.

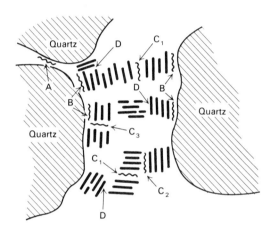

FIG. 21.3. Possible arrangements of domains, organic matter and quartz in a soil crumb.

Type of bond: A Quartz–organic matter–quartz
B Quartz–organic matter–domain

C Domain–organic matter–domain $\begin{cases} C_1 \text{ face–face} \\ C_2 \text{ edge–face} \\ C_3 \text{ edge–edge} \end{cases}$

D Domain–domain, edge–face

The creation and protection of soil structure in arable soils

The problems of managing arable soils, from the point of view of the maintenance or creation of a suitable soil structure primarily concern the production of a seedbed on the one hand, and of maintenance of coarse pores that allow rapid, but not too rapid, downward movement of rain or irrigation

1 *J. Soil Sci.*, 1959, **10**, 235.

water on the other. They are therefore more concerned with the creation and maintenance of suitable clod and crumb structures rather than with granule or domain structure. On a typical arable field, the volume of the coarser pores decreases from seedbed to harvest, and the bulk density increases. Part of the increase of bulk density is due to natural causes—the slaking of clods on wetting filling up coarser pores, and part due to the passage of cultivation and harvesting machinery, including the tractor pulling these implements. This traffic can be a very important cause of loss of structure or increase in bulk density, as G. W. Cooke and R. J. B. Williams[1] have shown on a boulder clay at Saxmundham in Suffolk. Small plots that were cultivated and sown to barley, and which had no traffic on them at all had a bulk density after harvest of 1·3 compared with 1·63 on the remainder of the field, which was cultivated in the normal way with tractors and implements. It is interesting to note that the bulk density of the soil under a three-year grass ley, which gave a marked improvement in soil structure, was also 1·32, compared with a bulk density of 1·13 for soil under old pasture.

The traditional method of converting a stubble to a seedbed is by cultivating the soil usually beginning with a plough, which cuts and inverts a furrow, followed by passes over the land with cultivators and harrows to break the furrow slice down into crumbs of a size suitable for a seedbed. But the ease with which these processes produce a good seedbed depends on the moisture content of the soil. If the soil is too dry, the clods produced by the plough will be too hard to be broken down by cultivators or harrows. If it is too wet, the crumbs or clods will be stuck together under the shearing stresses set up by the implement, and in addition the wheels of the tractor and the cultivator implements will press the soil crumbs together and make the soil particles flow, filling up all the coarse pores and leaving a wide ribbon of structureless soil. Thus, the range of moisture contents at which a soil can be cultivated is a character of prime importance for ease of cultivation. In general, if a soil has a moisture content above its lower plastic limit, which is the moisture content at which the soil particles begin to stick to each other when worked under moderate pressure, cultivation is liable to create larger clods devoid of coarse pores, and the more rapidly the soil crumbs become really sticky and begin to flow under pressure as the moisture content rises above this point, the more harm can be done by cultivation if carried out at these moisture contents. On the other hand, for sandy soils, compressing the soil when wet will tend to give clods more stable to weathering, if the moisture content is correct. D. G. Vilensky,[2] for example, found that if cultivation only set up gentle pressure on the soil particles, a moisture content just before all the soil pores were full of water gave maximum stability, and the stability

1 *Rothamsted Rept.*, 1971, Part II, 122.
2 *Trans. 1st Comm. Int. Soc. Soil Sci.*, 1934, 17; and with V. N. Germanova, *Pedology*, 1934, **34**. For further examples see W. T. McGeorge, *Arizona Agric. Exp. Sta.*, *Tech. Bull.* 67, 1937, and T. F. Buehrer and M. S. Rose, ibid., *Tech. Bull.* 100, 1943.

fell off rapidly on either side of this moisture content. As the pressure of working increases, so the optimum moisture content decreases,[1] and the range of moisture contents giving a structure almost as stable as the maximum increases.

There is no definite suction at which the lower plastic limit occurs, but it is nearly always drier than field capacity in temperate soils, so that a wet loam or clay soil, just after drainage has ceased, is too wet for cultivation. The rate of increase of stickiness or mouldability of the soil as the moisture content increases depends on many factors such as the silt and clay content, the degree to which the clay particles are bound together into stable granules, and the organic matter content of the soil. The moisture content at which cultivations can be profitably carried out at the drier end depends on the rate the strength of the clods increase with decrease in moisture content, and on the number of planes of weakness within them, due to their content of structural pores, crumbs and cracks. Poorly structured soils with more than a certain proportion of clay can have a very narrow range of moisture contents within which cultivation implements can be used, and it is literally true that a soil can be too wet one afternoon, in a suitable condition the following morning, but too dry in the afternoon.

The moisture content of a soil, as it affects its compressibility, can have another very important consequence in cultivation. Most ploughing in temperate countries is now done with wheel tractors with two wheels running on top of the furrow bottom, so that this is subjected not merely to compaction but to surface smear also, if the surface is moist and the tractor driving wheel slips at all seriously. This causes a discontinuity in the coarse pores between the surface and the subsoil, for not only have all the root and worm channels been broken, but their ends in the furrow bottom have been closed. This can easily lead to water ponding on the furrow bottom in the cultivated layer, and causing anaerobic conditions to develop. Even short periods of anaerobic conditions can rapidly lead to loss of structural pores,[2] for reasons that have not been fully investigated but possibly because these conditions are conducive to the bacterial degradation of polysaccharide gums into smaller molecules which have no ability to bond soil particles together, or possibly due to some consequence of ferric ions being reduced to ferrous.

These compaction problems may be even more serious if cultivations are being done on ploughed land, for the full depth of loosened soil under the wheels of the tractor, and to a much less extent of the cultivation implement, will be compacted and may well lose all its coarser structural pores;[3] this

1 S. Henin, *Ann. Agron.*, 1936, **6**, 455.
2 N. H. Pizer (*J. Sci. Fd. Agric.*, 1962, **13**, 391, and *J. Roy. Agric. Soc. Engl.*, 1961, **122**, 7) has given examples from eastern England; F. Hunter and J. A. Currie, (*J. Soil Sci.*, 1956, **7**, 75,) for spoil heaps of surface soil.
3 B. D. Soane (*J. Inst. Agric. Eng.*, 1970, **25**, 115) has published measurements of compaction under tractor wheels.

may be hidden from view, for the harrow or shallow cultivator will pull loose soil over the depression made by the tractor wheel. If there are bits of readily decomposable organic matter in the soil, such as bits of sugar-beet leaf or even farmyard manure, in the compacted zone, these will be centres of anaerobic decomposition, which will weaken or destroy any structural pores that may have been left.

Zones or volumes of compaction in the soil typically reduce the volume of soil tapped by the root system of the crop, making it more stunted and less branched. This is very noticeable for crops growing on soils where the structure above, or just below, the plough furrow has collapsed. This has the consequence that even if the root system is developed enough to take up all the water the crop needs from the soil, it will be much less efficient in tapping the soil for nutrients which only move slowly by diffusion through the soil solution. J. K. Coulter,[1] for example, showed that the amount of potassium ryegrass could take up from some subsoils depended on their structure, whereas the amount of sodium taken up was nearly independent of structure because nearly all of it moved to the root by mass flow (see p. 546).

The ideal conditions for a seedbed are that it should consist of soil crumbs, none much finer than 0·5 to 1 mm, and none much coarser than 5 to 6 mm, in a fairly firm packing.[2] Fine crumbs will block the coarser pores needed for drainage and coarser crumbs are likely to have anaerobic centres during wet weather. Such a seedbed, if its structure is stable and so lasts while the plants are young, gives maximum early growth of the crop, and allows maximum rate of nitrification of the soil organic matter. The problem of seedbed preparation is therefore to get the field soil as close to this condition as possible by the time the seed should be sown, and in circumstances that will allow the maintenance of this condition for as long as possible.

Traditionally the principal agent which weakens the furrow slice of soil and allows cultivation implements to break it down to crumbs is the action of the weather. Allowing a clod to dry and wet not only causes some slaking of crumbs from its outer surface but also causes cracks to develop internally due to uneven shrinking of the clod during drying and perhaps to some compression of entrapped air during wetting. This weakens the clod, so allowing it to be more easily shattered by mechanical impact from a cultivator tine when it has become drier again.

The other important climatic agent causing the break-up of clods is frost. The effect of frost on a soil depends on its moisture content, the rate of freezing, and the amount of water left unfrozen. Slow freezing tends to cause

1 *Rothamsted Rept.* 1966, 57.
2 A. G. Doyarenko, *Russian J. agric. Sci.*, 1924, **1**, 451. for podsolised soils near Moscow and V. U. Kvasnikov, ibid., 1928, **5**, 459, for semi-arid soils near Samara. See M. Krause, *Landw. Jahrb.*, 1931, **73**, 603, for a German summary of this work. For later work see R. E. Yoder, *Proc. Soil Sci. Soc. Amer.*, 1938, **2**, 21; H. Dittrich, *Bodenk. PflErnähr.*, 1939, **16**, 16; and I. S. Cornforth, *J. agric. Sci.*, 1968, **70**, 83.

much of the ice to be formed in relatively thick wedges or films,[1] and if the moisture content is not too high, considerable volumes of soil will be left almost ice-free and be subjected to the high pressure produced during the expansion of water on freezing. If these soil volumes have a fairly high content of unfrozen water and so are still plastic, that is, if they are plastic at suctions of 30 to 100 bar on the drying curve, this pressure will give crumbs of great stability and of a desirable size[2]—between about 3 and 30 mm for the Rothamsted soil. But they have a smaller internal pore space than the original clod, owing to the pressure they have been subjected to, so that when the soil is thawed, much of the ice becomes free water and drains away. Hence, the effect of slow freezing and then thawing of wet clods of a heavy soil is to leave the soil drier than it was before the frost and with a more stable crumb structure of a desirable size distribution.

This desirable frost tilth is only produced under fairly restricted conditions, which happen to be commonly fulfilled on loam and clay soils of temperate climates. The soil must be sufficiently moist at the time of freezing, and the rate of freezing must be sufficiently slow, to allow the formation of continuous ice films separating volumes of unfrozen soil.[3] Freezing sandy soils tend to destroy the structure because the volume of water that does not freeze at, say, $-5°C$ is small, so nearly all the water forms a continuous block of ice in which the sand particles are embedded. The same result would also be expected if the wet soil consisted of fine crumbs separated by water films which would freeze at temperatures not far below the freezing point of pure water, for again the frozen soil is in essence a block of ice with small crumbs dispersed in it. Further, if frozen soil is exposed to very dry air, so the surface soil suffers freeze-drying, the dry soil is typically powdery, though this result may be more characteristic of soils high in fine sand rather than of loams and clays.[4]

It is worth stressing, in conclusion, that presentday methods of cultivation are only really appropriate on many soils if the weather conditions are suitable. Thus, if the land lies wet continuously from after harvest till after seed-time, the farmer cannot produce a good seedbed on medium- and fine-textured soils. Again, in difficult years and with skilful management, it may be possible to produce a shallow seedbed, but below this there may remain wet structureless clods, often with bits of decomposable organic matter occluded in them, and they can have such a poor structure that no crop roots penetrate them; the clods may still be there intact at the end of the growing season. In such conditions, the soil may need quite heavy dressings of ferti-

1 For the results of Swedish experiments bearing out this point, see S. Eriksson, *Lantbr. Hogsk. Ann.*, 1941, **9**, 80. For photographs of the structures produced by frost in arable soils see W. Czeratzki, *Z. PflErnähr.*, 1956, **72**, 15.
2 G. Torstensson and S. Eriksson, *Kgl. Lantbr. Akad. Tidskr.*, 1942, **81**, 127.
3 E. Jung, *Z. PflErnähr.*, 1931, A **19**, 326.
4 F. Bisal and K. F. Nielsen, *Soil Sci.*, 1964, **98**, 345.

lisers to compensate for the very small proportion of the surface soil the roots can exploit.

The stability of seedbed structure

It is often possible to prepare a suitable seedbed, particularly in light-textured soils, but the whole structure may collapse due to slaking or under heavy rain. Slaking involves the breaking of fine granules off a crumb or the dispersion of sand or silt particles from a crumb into the soil solution. These will then fill up the coarser pores in the soil surface giving a crust devoid of

PLATE 30. Crust formation on a weakly-structured light sandy loam subjected to 25 mm of simulated rainfall in drops 2mm in diameter.

coarse pores,[1] in which the individual particles may be so firmly interlocked that the crust becomes extremely firm and dries to a very hard layer that is difficult to break up again into crumbs. Plate 30 illustrates such a crust, which was produced by 25 mm of rain, falling as droplets 2 mm in size on natural crumbs between about 3 and 5 mm in size, taken from a well-structured sandy loam soil.[2] If the structure is weak and the seedbed rather loose, the slaking may continue to the bottom of the plough layer. The crust often prevents the shoots of germinating seeds penetrating it, even if it is only a few

1 For a micromorphological study of surface crests see D. D. Evans and S. W. Buol, *Proc. Soil Sci. Soc. Amer.*, 1968, **32**, 19.
2 This is described by A. G. M. Bean and D. A. Wells, *Nat. Inst. Agric. Eng.*, *Rept.* 23, 1953.

millimetres thick, and this can lead to a complete crop failure. Slaking is very difficult to prevent in many soils, if heavy rain falls between sowing and complete emergence of the seedling plants. Crusting will also result in water standing on the soil surface after rain, which may cause anaerobic conditions to build up in the soil, which, in turn, may damage the crop so it is unable to make full normal growth subsequently.

The collapse of seedbed structure is very common on soils containing more than 70 per cent of sand, particularly if much of this is in the fine sand fraction; and it is usually more serious on soils low in organic matter. There is no one level of organic carbon a soil must contain if it is to have its practicable maximum stability, but general experience in eastern England is that once the organic carbon content has fallen below 1·5 per cent, structure deterioration occurs during the growing season; if it falls below 1 per cent, this deterioration is likely to cause serious difficulties in soil management. Levels as high as 1 per cent can be very difficult to maintain on many well-drained sandy soils in all-arable systems of farming, so that instability of structure during cultivation and during the growing season is a normal feature for these soils.

This trouble also occurs on some non-calcareous soils in which crumb structure is unstable to wetting, but in which the clods do not weather down easily to crumbs, possibly because they are so compact that they wet very slowly. Seedbeds on these soils may easily have water standing on their surface during wet weather. Further, if they are cultivated in the autumn and sown to winter wheat, for example, the finer crumbs in the plough layer may collapse during a wet winter, giving conditions that are very unsuitable for growth of the young crop in spring.[1] The Saxmundham soil, illustrated in Plate 28 is of this type.

A seedbed can only remain aerobic if water can penetrate easily from the cultivated layer to the subsoil across the layer just below the furrow bottom which is subjected to compression and smear by the tractor wheels, or by compaction alone if the tractor is running on the soil surface. Two methods are available for maintaining the permeability of this layer—subsoiling or regular deep ploughing. Subsoiling is probably most effective if done when this layer is dry, which is usually after harvest in those years with a dry summer in England. This practice will be discussed in more detail on p. 801, and effective subsoiling is illustrated in Plate 51. Deep ploughing brings the top of the subsoil layer to the surface so it can be weathered to crumbs, and puts better-structured soil at the bottom of the plough layer. It must, however, be used with discretion if the subsoil has an unstable structure—if it is very high in fine sand, for example. Marked easing of the problem of preparing a seedbed in spring, and marked reduction in the winter-killing of wheat are found where such a compacted layer had existed and is broken up.

1 For an example of such a soil see G. W. Cooke and R. J. D. Williams, *Rothamsted Rept.*, 1971, Part II.

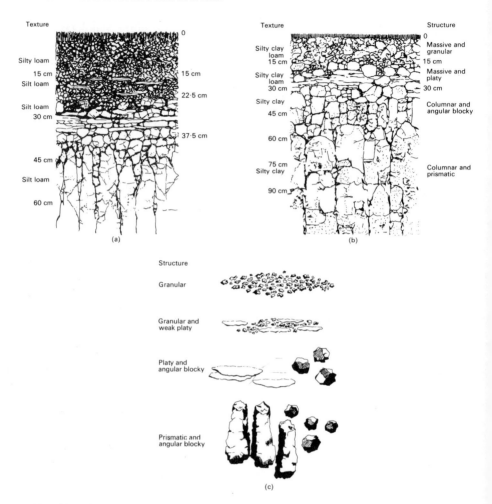

Fig. 21.4. (a) Soil structure of a well structured profile of pasture soil; (b) of a poorly drained, weakly structured arable soil; and (c) the constituent parts of the various horizons in the first profile. (From 'Modern Farming and the Soil', HMSO Crown Copyright 1971)

By way of summary Fig. 21.4 shows diagrammatically the structural units in two soil profiles in eastern England.[1] The first profile is for a basically well-structured silt loam under old pasture and shows the very well-developed stable granular structure in the surface layer but typical inherent weakness

1 This figure is taken from a report entitled Modern Farming and the Soil by the Agricultural Advisory Council on Soil Structure and Soil Fertility, particularly of English arable soils. It contains a wealth of detail on the problems of maintaining the condition of these soils under intensive systems of cropping.

of the structure of the clods and crumbs in the subsoil due to the high fine sand and silt content. This gives a platy structure at a depth of 30–35 cm which will thicken and become impervious when the land comes under arable crops unless great care is taken with the cultivations. The second profile is for a typical non-calcareous boulder clay which is poorly drained because the fissures below 30 cm are blocked by mud which has dispersed from the clods higher up. In this profile the platy layer initially at 30 cm depth is thickening upwards due to water ponding on it causing the collapse of the crumbs and clods that become waterlogged. Plate 22 shows the good natural system of stable cracks in the Rothamsted soil profile, which are the cause of its natural reasonably good drainage in spite of its texture.

Stabilising the structure of arable soils

It has long been known that the structure or tilth of a soil can be modified by adding suitable improvers to it, or growing suitable crops on it. Claying or marling sands, particularly blowing sands, chalking heavy soils, applying dressings of farmyard manure or composts, ploughing in green manures, and laying land down to long leys or pastures, are all methods that have been used by farmers for improving the condition of their soils.

Soils composed of almost pure sands are extremely difficult to cultivate because of their almost complete lack of structure. In some regions they have been reclaimed by spreading heavy dressings of clay or marl, at the rate of 250 to 500 t/ha, on the soil surface, letting it weather, and mixing it with the top 15 to 20 cm of the soil. This will reduce the tendency of the seedbed to slump and of the sand to blow when the surface is bare and dry. However, the practice is expensive even if clay pits are close to the sands, and it has only been used in England during the last few decades with the help of a Government subsidy.

Spreading chalk on some loam and clay soils at rates in excess of 50 t/ha was traditional in many parts of England, and this was considered, among other things, to make these soils more easy to cultivate, but there was and still is no real knowledge about the conditions under which this happens. Liming an acid soil could affect the stability of its crumbs in two ways. It will convert exchangeable aluminium ions into polymerised hydroxy-aluminium ions or aluminium hydroxide films on the surface of clay particles and exchangeable calcium ions will take their place. There is no experimental evidence that this affects the stability of the structure or the ease of working the soil.[1]

1 See W. W. Emerson and M. G. Dettman, *J. Soil Sci.*, 1960, **11**, 149, for some experiments on this.

Liming an acid soil will also increase the concentration of calcium ions in the soil solution, particularly if sufficient lime is applied for some free calcium carbonate to be left in the soil, and this will reduce the tendency for clay particles to deflocculate. W. W. Emerson[1] showed that Rothamsted soil which had been in arable cultivation for a long time and which was low in organic matter began to deflocculate if the calcium ion concentration fell below 10^{-3} M, and noted that it was only necessary for the soil solution to be in equilibrium with air containing 0·1 per cent CO_2 for the calcium ion concentration in a soil containing free calcium carbonate to be above this. E. C. Childs[2] also noted that drainage water from a limed clay soil had less suspended silt and clay particles than that from an unlimed soil. Regular annual additions of calcium carbonate to some plots of a brick earth soil at Versailles, which are very low in organic matter, have also given a remarkable good surface structure compared with neighbouring plots not receiving these dressings.[3]

L. de Leenheer[4] noted that heavy dressings of calcium carbonate, of the order of 50 t/ha, sometimes gave a marked visible improvement in the structure of soils low in organic matter, even if they were neutral, and there is evidence that excess calcium carbonate may reduce the cohesion between clay particles in moist clods[5] and reduce the size of the water stable crumbs in a soil.[6] On the other hand, the traditional view could be based on the slow stabilisation of structure brought about by these heavy dressings if the calcium carbonate dissolves and is reprecipitated as films around the clay particles. The Rothamsted soil is naturally acid but the old fields may still contain over 2 per cent $CaCO_3$ in the plough layer, although no chalk has been given for over a century, due to these old heavy dressings.

Some interest has been taken for many years in eastern England in the possibility of improving the structure, or easing the problem of cultivation, of some of the more difficult clay soils in this region, by adding gypsum to them. There has been very little field evidence published for its effectiveness, though recently G. W. Cooke and R. J. B. Williams[7] have shown that 25 t/ha of gypsum to the Saxmundham soil reduced its bulk density and the force needed to push a penetrometer probe into the soil one year after its application. Presumably its effect will be short-lived as it is slowly soluble in the soil solution, though once it has diffused into the body of the clods and weakened them, it will only slowly diffuse out again into the percolating water.

1 *J. Soil Sci.*, 1954, **5**, 233.
2 *J. agric. Sci.*, 1943, **33**, 136.
3 S. Henin *et al.*, *Le Profile cultural*, Paris, 1960, p. 293.
4 *Trans. 8th Int. Congr. Soil Sci.*, 1964, **2**, 561.
5 E. W. Russell and J. J. Basinski, *Trans. 5th Inst. Congr. Soil Sci.*, 1954, **2**, 166.
6 W. T. H. Williamson, J. Pringle and J. R. H. Coutts, *Int. Symp. Soil Structure*, Ghent, 1958, 176.
7 *Rothamsted Rept.*, 1971, Part II, 122.

Value of farmyard manure

Crumb stability in arable soils can usually be increased if regular dressings of farmyard manure are given, though the amounts required may be very large. Annual applications of 35 t/ha for a century have made a measurable increase in the crumb stability of the Rothamsted soil, increasing the proportion of water stable crumbs over 0·5 mm in size from 28 to 55 per cent on Broadbalk, which is in continuous wheat, and from 54 to 70 per cent on Barnfield, in continuous root crops.[1] Yet seventy annual dressings of 15 t/ha on the Saxmundham boulder clay had no measurable effect on structure stability,[2] and eighteen annual dressings of 75 t/ha on the Woburn sandy loam cropped to an intensive vegetable rotation,[3] and twenty-eight annual dressings of 40 t/ha on a loam soil in Ohio, only gave small improvements.[4] A. J. Low[5] made the interesting observation that, in the very unstable fine sandy loam soil at Fernhurst, 75 t/ha annually did not affect the stability of the structure appreciably, but it gave a very large earthworm population which maintained aeration and drainage by the number of their burrows.

Surface mulches of organic matter, such as compost, straw or mature grass, for example, improve the soil tilth under them very considerably. Part of the action is undoubtedly that they protect the surface tilth from raindrops, but in part they cause a build-up of structure underneath them due to increased biological activity in the surface layers of the soil.

Value of grass and lucerne leys

The structure of arable soils depends on the system of farming practised. Cropping systems requiring much cultivation, and crops leaving only a small weight of roots and residues behind, give soils with a less stable structure than systems requiring the minimum of cultivations and using crops that leave a large weight of root residues. Thus systems based on continuous wheat and barley tend to give more stable soils than those with many root crops. In the past, before the use of selective herbicides, root crops needed much intertillage during the growing season for weed control—a practice that is much less used at the present day—and they also leave behind much less root residues than the cereals. But the major cause of structure deterioration with root crops occurs at harvest, for they are commonly harvested in autumn, when the soil may be wet and evaporation low. Cereals, on the other hand,

1 E. W. Russell unpublished. M. G. Dettman and W. W. Emerson (*J. Soil Sci.*, 1959, **10**, 215) established the same result using a different technique. M. B. Russell *et al.*, *Proc. Soil Sci. Soc. Amer.*, 1952, **16**, 156, found 25 annual applications of 25 t/ha to a Sassafras silt loam gave a small and 100 t/ha a considerable improvement of structure.
2 G. W. Cooke and R. J. B. Williams, *Rothamsted Rept.*, 1971, Part II, 122.
3 R. J. B. Williams and G. W. Cooke, *Soil Sci.*, 1961, **92**, 30.
4 L. Havis, *Ohio Agric. Expt. Sta.*, Bull. 640, 1943.
5 I am indebted to Mr Low for this unpublished information.

may leave two to three times the weight of root residues in the soil, they need no cultivations after they have been sown, and they are harvested earlier on a firmer land, so there is much less likelihood of structure deterioration.

Laying land down to grass for three to four years has often been claimed as an efficient method of improving soil structure. Figure 21.5 due to A. J. Low[1] illustrates the increase of water-stable crumbs on a clay-loam soil in Northamptonshire under a grass-clover ley; and shows that there is a rapid

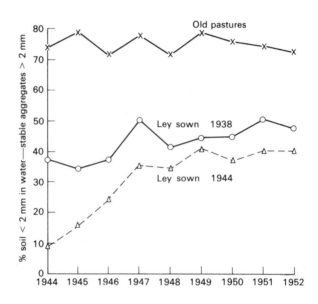

Fig. 21.5. Annual changes in water stability of a clay–loam under grass pastures of different ages. Each point represents the mean of a spring and autumn sampling

improvement in stability during the first three or four years, and then any further improvement takes place slowly; but even after twelve years the stability of the structure is much less than for the soil under permanent pasture. This benefit from a ley is not always found, and Low has given examples where the benefit was very much less. It is not known what factors are important, but Low concluded that the better the soil structure initially, the more rapidly it improves, that grazing tends to give a slower improvement than taking hay crops due to the treading of the animals; and that improvement takes place more slowly in soils high in fine sand and silt, which are soils in which it is most difficult to maintain good structure.

1 *J. Soil Sci.*, 1955, **6**, 179. For American data showing the rapid build-up of structure in arable soils due to grass or grass-clover leys, see C. H. M. van Bavel and F. W. Schaller, *Proc. Soil Sci. Soc. Amer.*, 1950, **15**, 399, and A. J. Wisniewski *et al.*, ibid., 1958, **22**, 320.

This general result has been found by other workers. Thus M. G. Dettman and W. W. Emerson at Rothamsted,[1] assessing structure by the technique of saturating the soil with sodium ions and measuring the loss of permeability of a thin column of soil in 0·05 N NaCl, obtained the following values of the structure index for soils at Rothamsted:

Over 80 years' continuous root crops, unmanured	5
With 35 t/ha annually farmyard manure	18
Normal arable soil, after 4 years' continuous fallow	18
After 4 years' wheat	43
After 4 years' grass	50

Lucerne can also be a crop that improves soil structure, as already noted. E. T. Sykes[2] has given an example of this on a boulder clay near Cambridge, and Plate 28 gives an excellent visual example of this on the boulder clay at Saxmundham, and Emerson found that at Woburn it was as effective as a three-year grazed ryegrass ley. His results for the top 5 cm of soil were

3 year grass (grazed)	64
3 year lucerne (cut for hay)	64
3 year arable	35
Clean cultivated headland	14

The type of grass, and perhaps its management, sometimes influences the soil structure very strongly. Thus, A. L. Clarke[3] found ryegrass, with an extensive fine root system gave better improvement than a phalaris grass with fewer coarser roots, and T. Pavlychenko[4] found crested wheat grass (*Agropyron cristatum*) an excellent structure improver on the Canadian prairies, while slender wheat grass (*A. pauciflorum*) had little value. Similarly, H. C. Pereira[5] showed that a three-year elephant grass (*Pennisetum purpureum*) pasture gave a better structure than either *Paspalum notatum* or Rhodes grass (*Chloris gayana*) on an earthy red latosol at Kawanda, Uganda. On the other hand, the common British pasture and hay grasses show no consistent difference in their ability to improve soil structure, although perennial ryegrass appears to be the most reliable improver.[6] The reasons why grasses

1 *J. Soil Sci.*, 1959, **10**, 215.
2 *Agric.*, 1956, **63**, 421.
3 With D. J. Greenland and J. P. Quirk, *Aust. J. Soil Res.*, 1967, **5**, 59.
4 *Natl. Res. Counc. Canada, Publ.* 1088, 1942.
5 With E. M. Chenery and W. R. Mills, *Emp. J. expt. Agric.*, 1954, **22**, 148.
6 W. W. Emerson, *Trans. 5th Int. Congr. Soil Sci.*, 1954, **2**, 64, for some Rothamsted results, and C. R. Clement and T. E. Williams, *J. Soil Sci.*, 1958, **9**, 252, for some from a clay soil at Drayton and a loam at Hurley. However, J. Pringle and J. R. H. Coutts (*J. Brit. Grassland Soc.*, 1956, **11**, 185) found timothy (*Phleum pratense*) better than ryegrass on some Aberdeen soils.

sometimes differ in their ability to improve structure have not been quantitatively investigated, but a property which may be of importance for some grasses is the tensile strength of their roots.[1] The roots of some grasses grow into soil crumbs, holding them like beads on a chain, and this continues even after the root is dead as illustrated in Plate 31. Thus the greater the tensile strength of the root, and the longer the dead root lasts until it is decomposed, the more noticeable will be the build-up of structural units.

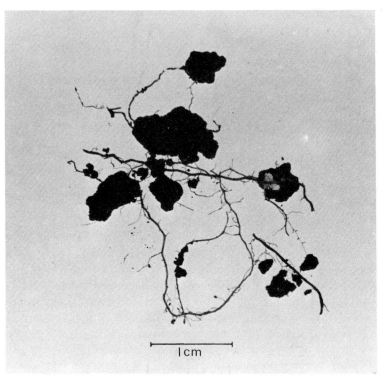

PLATE 31. Soil crumbs held together by grass roots. (Ordinary chernozem on clay, 10–20 cm, Khrenovaia Station.)

The depth of soil in which the structure is improved by the grass has not been studied in detail. Emerson found that four-year leys at Rothamsted could improve the stability of the structure at 10 to 20 cm depth, a result also found by A. Troughton[2] on two Welsh fine sandy loam soils, while C. R. Clement and T. E. Williams[3] found that three-year leys at Hurley gave a

1 T. M. Stevenson and W. J. White, *Sci. Agric.*, 1941, **22**, 108.
2 *Emp. J. expt. Agric.*, 1961, **29**, 165.
3 *Int. Symp. Soil Structure*, Ghent, 1959, 166.

marked improvement in the top 2 cm, but little below. Thus, when the ley was ploughed out using a 20 cm ploughing depth, the structure of the ploughed surface was no better than that of an older arable soil. This result is also found when the subsurface soil is so tight that grass roots cannot penetrate it, and it is then usually desirable to plough out the ley, loosen the soil and reseed; this will often allow deeper rooting of the grass and a much better build-up of the earthworm population.

The visual effects of a three-year grass or lucerne ley on the soil condition can be striking, as already noted. After the ley has been ploughed out, the land may be in magnificent tilth that allows a seedbed to be made over a very wide moisture range, whereas old arable soil may either be so sticky and plastic or so hard that seedbed preparation is impossible; and Low has given an example of this at Jealotts Hill and Cooke and Williams at Saxmundham. Low has described the various changes that occurred in the soil referred to in Fig. 21.5. Initially it was very sticky when wet, but after two years the stickiness became less noticeable and more of the fine particles were in water-stable granules larger than 50 μ. The soil remained compact during this period with few visible pores and when dry tended to break up into angular aggregates about 2 or 3 cm in size. Subsequently, these aggregates became smaller and less angular, until after eight years they were about 5 mm in size and rather more rounded, although they were still fairly closely packed in the soil. Low noted it was difficult to distinguish between the soil that had been under the ley for three years and for eight years: initially there were only few earthworms, but after eight years there were over $1 \cdot 5 \times 10^6$ per hectare. This build-up of earthworms may affect the land after the ley has been ploughed out, for Low has also given an example from Derbyshire, on a fine sandy loam, where the ley gave no improvement in structure, but after it was ploughed out the high earthworm population that built up under it kept the soil just below the surface permeable to water during the following wheat crop. It is important to note that it takes many decades under pasture before the full benefit of the pasture structure is obtained. Low finds, for example, that on the soil referred to in Fig. 21.5, a fifty-six-year-old pasture had on the average only 56 per cent of soil in water-stable crumbs over 2 mm in size compared with 73 per cent for the pasture illustrated, which was about 110 years old.

The cause of the improvement in structure is only partially understood. There is no evidence that the type of humus produced from grass differs from the bulk of the soil humus, and it has approximately the same ratio of polysaccharide to non-polysaccharide material, and the same distribution of fractions in the different molecular weight ranges.[1] Grass roots themselves do not appear to be responsible for the change in structure, but it is probable

1 J. H. A. Butler and J. N. Ladd, *Aust. J. Soil Res.*, 1969, **7**, 219; G. D. Swincer, J. M. Oades and D. J. Greenland, ibid., 1968, **6**, 225; J. M. Oades, ibid., 1967, **5**, 103.

that the characteristic structure of old pasture is the direct consequence of prolonged earthworm activity.

The structure built up by three to four years of a grass crop is lost fairly rapidly after the ley is ploughed out. A. J. Low[1] found that on the sandy loam soil at Jealott's Hill, a three-year ley gave a marked improvement in structure, much of which was still present after a year's cropping to kale, but almost the whole effect had disappeared after the following wheat crop; and this is probably a fairly representative result for many temperate soils. This result has also been found at Saxmundham where, as shown in Plate 28, the very noticeable effect of a three-year lucerne or grass ley disappeared within twelve months of the ley being ploughed out.

These results do not apply to land ploughed out from old pasture on loam and clay soils, for on some of these soils farmers consider it may take several decades of arable cropping for the full effects of the pasture to be exhausted. There is still only limited experimental evidence on this point. Low notes that the percentage water-stable aggregates on the fields he investigated dropped rapidly during the first few years of arable cultivation, and then more slowly; and he considered this drop was more rapid than the drop in the organic nitrogen content of the soil. On the soil type referred to in Fig. 21.5 he noted that after four years' cultivation, the crumbs from a pasture which was over a hundred years old were still not sticky when wet and they crumbled readily, although four years after a pasture that was only twenty-two years old they were beginning to be sticky and were losing their crumbliness.

Grasses may improve one aspect of structure that could be fairly long lasting if suitable methods of cultivation are used. One important aspect of structure in some soils is the presence of channels down which rain or irrigation water can move from the surface to the deeper subsoil, and in pasture soils dead roots produce such channels when they have been decomposed.[2] Ploughing-out a grass ley will cut all the channels at the depth of ploughing and will often seal their tops by smearing soil over them. If the grass is killed by a herbicide, and a shallow cultivation only is used to produce a seedbed for the following crop these channels can be maintained, which will often give a marked improvement in the ease with which the surface soil will drain.

There has been much discussion on how far the improved yield of arable crops after a grass ley is due to the improved structure it has brought about, and how far to the increased level of organic nitrogen, and therefore of nitrogen supply. This is a very difficult question to answer because the consequences of poor structure can to some extent be ameliorated by additional

1 With F. J. Piper and P. Roberts, *J. agric. Sci.*, 1963, **60**, 229. For similar results at Woburn see R. J. B. Williams and G. W. Cooke, *Soil Sci.*, 1961, **92**, 30.
2 W. W. Emerson, *J. agric. Sci.*, 1954, **45**, 241; K. P. Barley and R. H. Sedgley, *Soils Fert.*, 1959, **2**, 155.

fertilisers.[1] The fact that, given a high enough level of fertiliser, it is often possible to get high yields of crops with very restricted root systems on soils of poor structure complicates the interpretation of experimental results. A further complication is that leys may only improve the structure of the top few centimetres of soil during a two or three year period; when ploughed out, a soil with unimproved structure, and so liable to slump or cap badly, is likely to be turned up on to the surface.[2]

1 For an example see C. R. Clement, *J. Brit. Grassland Soc.*, 1961, **16**, 194, and D. A. Boyd, *Rothamsted Rept.*, 1965, 216.
2 For an example see C. R. Clement, *Pl. Soil*, 1964, **20**, 265.

22

The development and functioning of plant roots in soil

The soil affects the plant primarily through its effect on the root system, and these effects should form the basis of this book. Unfortunately, investigations on the root systems of crops growing in the field are difficult to carry out, and in consequence our knowledge of the interaction between the soil, the roots and the aerial parts of the plant, and, as important, between neighbouring root systems, is scanty.

Plant roots have a number of separate functions; they absorb nutrients and water from the soil, and transport these from where they were absorbed to the stems; they are the site of synthesis of a number of plant hormones or plant growth regulators; they may act as food storage organs, as, for example, in the root vegetables; and they may anchor the plant into the soil.

Only recently has it been possible to begin studying the effects of soil conditions on the production of growth regulators in the root system, and the consequent effect of variation in supply of these to shoot growth. A number of gibberellins and cytokinins are produced in the root, of which the quantitative assay is still difficult, and these have marked effect on shoot growth. In general, if the root is under stress for any reason, such as drought, salinity or low temperature, the production of these regulators is probably reduced.[1]

The longevity of roots depends on the plant. Perennial dicotyledonous plants and gymnosperms typically have a system of perennial roots, which may be very thick, and which serve partly as conduction channels of nutrients taken up by more ephemeral roots, and partly as anchors to hold the plant in the ground.

Some plants have a well-developed taproot, which, if it is not impeded, grows more or less vertically downwards. Many of the other roots are ephemeral, and are adapted for the uptake of water and nutrients. Typically root systems are branched, with a primary root coming from the seed or from the base of the plant, and branches grow out of these, the first order laterals, and branches out of these—the second order laterals, and sometimes third

---

1 See, for example, Y. Vaadia and C. Itai in *Root Growth*, Ed. W. J. Whittington, Butterworth, 1969; H. Kende and D. Sitton, *Ann. N.Y. Acad. Sci.*, 1967, **144**, 235.

order out of these. In addition the younger roots of many plants—but not all—have root hairs growing out of their surface.

A typical young root consists of a growing tip, behind which are the elongating meristem cells, which differentiate into the cells forming the epidermis, the cortex and the stele, which itself differentiates into the endodermal and phloem cells and the xylem vessels.

Some of the epidermal cells behind the meristem produce root hairs, which are thin filamentous extensions from the cell, and which grow into the soil. As a given part of the root ages, it changes colour from white to brown due to suberisation in the endodermal cells, caused by a band of a waxy substance suberin being laid down within their radial and transverse walls, and root hairs become less common. Later still the roots of many plants, excluding the grasses and cereals, may start to have secondary thickening due to the development of a cambium behind the endodermis and the laying down of an outer corky layer over the root surface. The cell arrangement in the epidermal and cortical layer is such that there is an appreciable proportion of intercellular space between the individual cells, so this layer is like a sponge, and the spaces are filled with a solution of much the same composition as that outside the root surface. These intercellular spaces form the water free-space in the root.

The cell walls of the epidermal and cortical cells contain polysaccharide and polyuronide gums, composed of hexose, pentose and amino-sugars, and containing carboxylic acid groups which are dissociated at the pH prevailing outside the cells. Thus the cell walls are negatively charged and have a diffuse double layer outside them in which cations are concentrated and anions excluded. The water volume affected by this double layer is known as the Donnan free space. These gums may be excreted in sufficient quantity by the epidermal cells to coat the young root with a mucilagenous layer, known as a mucigel[1] and may attain a thickness of 1 μ. Obvious mucigel development is only found on some roots, but there has been no systematic investigations which have found what conditions encourage its formation.

Absorption of water and nutrients can take place across the cell walls bounding the water free-space into the cell cytoplasm, and so into the xylem; and under these conditions the absorbing power of a root is dependent on the surface area of its water free-space, which is dependent on the root volume for fine roots, since the volume of the stele is much less than that of the cortex. This is probably the reason why the root volume is often a more useful parameter than root surface area in laboratory studies of nutrient uptake.[2] In the field, however, a problem arises which has not received much attention. Few roots are much finer than 60 μ, or root hairs than 10 μ, and pores of these sizes are emptied of water by suctions of 50 mbar and 300 mbar respectively. It is

1 For electron microscope photographs, see H. Jenny and K. Grossenbacher, *Proc. Soil Sci. Soc. Amer.*, 1963, **27**, 273.
2 I am indebted to Dr R. Scott Russell for this observation.

PLATE 32. A root of Italian ryegrass (*Lolium multiflorum*) growing in a sandy loam soil, showing development of root hairs.

not therefore obvious how the root cells maintain contact with water in finer than about 10 μ pores. It is possible that an important function of root hairs is to maintain this contact, and if so they will be the major location of water and nutrient adsorption when the water suction exceeds about 100 mbar.[1] It is very difficult to prove this directly, but Plate 32 gives a photograph of the root hairs on a ryegrass root growing in a sandy loam soil. If they exude mucigel, this would allow them to maintain liquid contact as the suction rises appreciably above say 500 mbar. G. C. Head[2] has concluded, from a study of photographs of fruit tree roots growing in the East Malling root laboratory, that root hair development is more likely to take place when a root is in contact with a soil crumb than when in an air space, though this conclusion can only be partially true, as can be seen for the ryegrass root shown in Plate 32. Roots producing an abundant mucigel sheath will also be able to maintain liquid contact as the soil dries; but there appear to be no critical investigations on the role of mucigel as a replacement for root hairs. The hyphae of mycorrhizal fungi that are growing in the soil can probably function in the same way as root hairs. This is proven for phosphate uptake where, because the mobility of phosphate is so low, its uptake is very dependent on the area of absorbing surface in liquid contact with the soil; but little is known about how far the hyphae will also transmit water from the soil to the root.

The soil just outside the root surface contains the rhizosphere population and is a zone of great microbial activity. Its primary energy source is derived from the plant, possibly in part from the mucigel and in part from soluble organic compounds excreted by the roots. Some of the effects of this population on plant growth have already been discussed in Chapter 13, including the bringing into solution of ferric iron and manganese ions through the production and excretion into the soil of organic acids that form soluble chelation compounds with them. However, this population competes with the roots for nutrients, competition which can be particularly severe if any nutrient is in short supply. This is illustrated in Table 22.1 for the uptake of phosphate by sterile or with non-sterile roots of barley from a culture solution kept at pH6, to which phosphoric acid labelled with ^{32}P was added.[3] It shows that the sterile root took up more phosphate than the non-sterile root, that a greater proportion of the uptake was translocated to the tops, and that a greater proportion of the phosphate remaining in the root was present as soluble phosphate esters or inorganic phosphate; whereas a greater proportion of the phosphate held by the non-sterile roots was present as nucleic acids and phospho-proteins and -lipids, which are the typical forms of phosphate in microbial tissue. This effect is most noticeable in the most dilute

1 See K. P. Barley, *Adv. Agron.*, 1970, **22**, 159.
2 *Trans. 9th Int. Congr. Soil Sci.*, 1968, **1**, 751. This paper gives a photograph showing this.
3 D. A. Barber and U. C. Frankenburg, *A.R.C. Radiobiology Lab.*, *Ann. Rept.* 1967, 26.

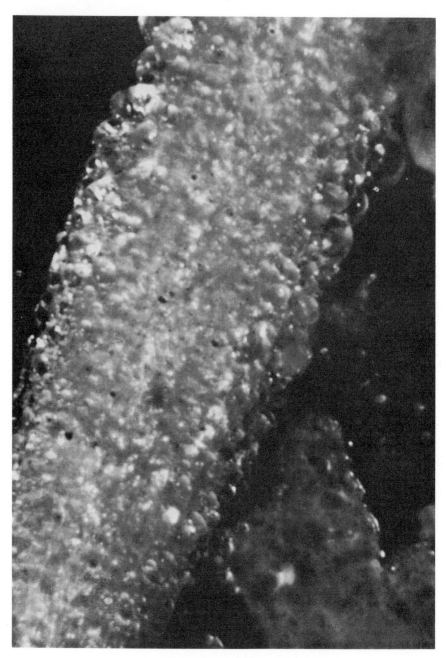

PLATE 33. Liquid droplets exuded from an apple root. ($\times$ 65).

TABLE 22.1 Uptake of phosphate by sterile and non-sterile barley roots from a culture solution at pH6

Concentration of P in solution	Uptake of P µg per plant		% P translocated to tops			% root P present as soluble esters or inorganic P		
	Sterile	Non-sterile	Sterile	Non-sterile	Ratio	Sterile	Non-sterile	Ratio
3.2×10^{-8} M	0·32	0·15	6·5	0·7	9·3	80·0	37·2	2·1
3.2×10^{-7} M	3·40	1·53	15·3	3·3	4·6	78·6	46·6	1·7
3.2×10^{-6} M	26·9	22·5	36·0	24·1	1·5	81·2	58·3	1·4
3.2×10^{-5} M	79·5	77·6	50·7	44·8	1·1	88·0	71·3	1·2

solution, but can be detected at 3×10^{-6} M and possibly also at 3×10^{-5} M. Plate 34 reproduces a radio-autograph of a section of a sterile and non-sterile root growing in a solution labelled with ^{32}P, and it shows clearly the concentrations of ^{32}P labelled material on the outside of the non-sterile root, a feature that is absent from the sterile.[1] It is not possible to be certain how important this effect is in natural soils, for roots of crops growing in soils low in phosphate typically have mycorrhizal roots, and the hyphae of the fungal component in the soil may compete effectively with the rhizosphere population for phosphate (see p. 259).

Head has also published photographs of roots growing in a soil which show drops of liquid apparently being exuded from roots and root hairs (Plate 34), but the significance of this observation for root or rhizosphere activity is not known. As the root ages, the rhizosphere population in conjunction with small soil arthropods hastens the disappearance of most of the cortical cells, which, in turn, reduces nutrient uptake by these older parts of the root system. Many plant roots are, in fact, ephemeral organs, as is illustrated in Plate 35 for apple tree roots growing in the East Malling root laboratory.[2]

The principal factors determining the rate that the root system of a plant will grow in a soil are: the amount of nutrients available to the root both from the aerial parts of the plant and from the soil, its oxygen supply, the soil moisture content and its osmotic pressure, the soil temperature, the level of toxins and pathogens in the soil, and the system of pores into which the roots can grow and the shear strength and compressibility of the soil. To a limited extent some of these factors interact with each other, in that the harmful effect of an unfavourable factor can be diminished by improving some of the others. Since many soils are very heterogenous, restricted root growth in a pocket of soil unfavourable for root penetration may be compensated for

1 D. A. Barber, J. Sanderson and R. S. Russell, *Nature*, 1968, **217**, 644.
2 W. S. Rogers, *J. Pomol.*, 1939, **17**, 99, and see his chapter in *Root Growth*, Ed. W. J. Whittington, Butterworth, 1969.

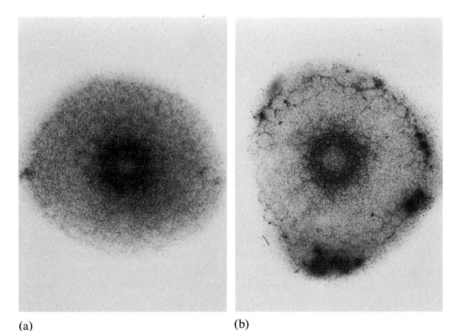

(a)   (b)

PLATE 34. Distribution of ^{32}P in (*a*) a sterile root; (*b*) a non-sterile root. Three-week-old barley plant growing in a culture solution containing 3×10^{-6} M P.

by improved root growth in more favourable pockets of soil accessible to the root system. Again, if a soil has an uneven distribution of plant nutrients, root growth, and particularly increased root branching is likely to occur in those volumes where nutrients are concentrated. This can be seen in the way roots will ramify through small volumes of farmyard manure, or in the vicinity of bands of nitrogen and phosphate fertiliser placed close to seed at drilling time.[1]

Roots must have an adequate energy and nutrient supply for growth, which initially comes from the seed, but later on the energy supply comes from the leaves. The distribution of the carbohydrates and amino-acids or proteins produced in the leaves between the tops and the roots depends on many factors, but the more fertile the soil and the more favourable its moisture content, the smaller is the proportion of the carbohydrates available for the root system; and it is in soils of moderately low fertility or during periods of drought that the root system may draw more heavily on the tops for the carbohydrates synthesised in the leaves. For annual crops the rate of growth of the root system reaches its maximum value earlier in the season than that of

1 A. J. Ohlrogge with M. H. Miller, *Agron. J.*, 1958, **50**, 95, with W. G. Duncan, ibid., 1958, **50**, 605.

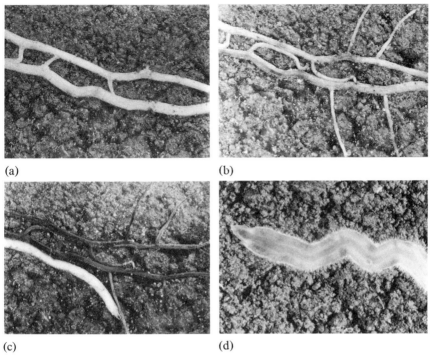

(a)   (b)

(c)   (d)

PLATE 35. Apple tree roots growing in a silt loam, photographed on: (*a*) 6th; (*b*) 25th; (*c*) 35th day after the two thick roots had first entered the area; (*d*) Pea root growing in a moderately compacted loam soil.

the tops, and also falls off earlier, as illustrated in Fig. 22.1, for the depth of penetration of the roots and the dry weight of the tops of groundnuts at Kongwa in Tanzania.[1]

Root growth typically ceases when the plant is filling its fruit or seed, and many roots die during this period. Thus about one-quarter of the dry weight of spring wheat at Woburn is in the roots at ear emergence, but only about one-tenth at harvest—the root weights being 160 and 85 g/m^2 to a depth of 60 cm with high nitrogen and 70 and 30 g/m^2 for low.[2] This period is also of great importance for young fruit trees, for if they are allowed to carry too large a crop, the fruit will so impoverish the root system that all the starch reserves in the roots will be used by the fruit, and most of the roots will be killed. Similarly, overgrazing of pasture plants will reduce the amount of carbohydrates available for the root system, which, in turn, will reduce new growth. Any serious shortage in the soil of an essential plant nutrient will also restrict the root system.

1 J. D. Lea, *Trop. Agric., Trin.*, 1961, **38**, 93.
2 *Rothamsted Rept.* 1965, 93.

Roots require an oxygen supply to their growing points and absorbing regions, and for most dryland crops this must come principally from the soil. To an extent which varies widely between crops, oxygen can diffuse from the leaves down the stem to the roots through intercellular spaces; and this pathway is better developed in young than in mature plants.[1] In some plants, such as rice and other swamp plants, the diffusion rate through this internal pathway is so high that oxygen diffuses out of the root into the soil (see p. 685).

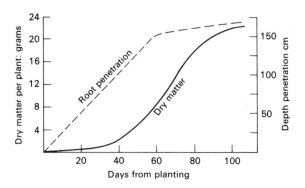

FIG. 22.1. Depth of rooting and dry matter production of Natal Common groundnuts at Kongwa, Tanzania

The oxygen concentration needed at the root surface, or the oxygen diffusion rate through the soil needed by roots, differs very considerably between crops[2] though the roots of many agricultural crops may need rates faster than 30×10^{-10} g/cm^2/sec, and in particular a good oxygen supply sometimes helps a root system grow in an unfavourable environment. Thus roots will penetrate into a more compact soil layer if the oxygen supply is good than if it is poor.[3]

Roots normally only grow into moist soils, but the roots of some semi-arid grasses, such as *Bouteloua curtipendula* and *Eragrostis trichodes*[4] will grow into soils as dry as or drier than the permanent wilting point. W. S. Rogers[5] found that apple roots would not grow into soils drier than 0·5 bar, and J. F. Bierhuizen and C. Ploegman[6] found the rate of elongation of roots of lettuce, spinach and broad beans (*Vicia faba*) was much slower when the soil water was at a suction of 1 bar compared with 0·1 bar, yet D. R. Peters[7]

1 D. J. Greenwood, *New Phytol.*, 1967, **66**, 337, 597.
2 L. H. Stolzy, J. Letey *et al.*, *Proc. Soil Sci. Soc. Amer.*, 1961, **25**, 463.
3 R. M. Hopkins and W. H. Patrick, *Soil Sci.*, 1969, **108**, 408; J. L. Tackett and R. W. Pearson, *Proc. Soil Sci. Soc. Amer.*, 1964, **28**, 600; T. W. Scott and A. E. Erickson, *Agron. J.*, 1964, **56**, 575.
4 M. H. Salim, G. W. Todd and A. M. Schlehuber, *Agron. J.*, 1965, **57**, 603.
5 *J. Pomol.*, 1939, **17**, 99.
6 *Meded. Dir. Tuinb.*, 1958, **21**, 484.
7 *Proc. Soil Sci. Soc. Amer.*, 1957, **21**, 481.

found the growth rate of maize roots was only halved as the soil suction rose from a low value to 8 bar. Wheat, oats and barley on the other hand probably will not grow into soil as dry as this.[1] However, the effect of moisture content on root growth is often confounded with its effect on the shear strength of the soil, for the drier the soil, the greater its resistance to shear.

The effect of soil temperature on root growth depends on the crop. Roots of crops adapted to temperate regions will grow and function in cooler soils than will those of tropical crops, though it is difficult to distinguish, in the field, between the air and the soil temperature on the growth of roots. The roots of at least some cool temperate crops, such as some grasses and cereals, will grow slowly at 3°C, and at an appreciable rate of 7°C, while tropical plants need much higher temperatures. Thus, cotton roots, and the roots of some maize and sorghum varieties make little growth below 16°C. The optimum soil temperature for root growth is also higher for tropical than temperate crops: for cotton it is about 30°C, perhaps a little lower for maize and a little higher for some strains of star or Bermuda grass (*Cynodon dactylon*), whereas for Canadian and Kentucky bluegrass (*Poa pratensis*) it is about 10°C and 15°C respectively.[2]

The soil temperature around plant roots also affects the growth of tops, for reasons that are only now being investigated. Low temperatures appear to reduce the uptake of water and nutrients,[3] and one can see plants wilting in spring when the soil is cold and moist but the evaporative power of the air is high. Low soil temperature also appears to reduce the uptake of phosphate by crops, for J. R. Gingrich[4] has given examples of wheat not responding to phosphate fertilisers in spring until the root temperature exceeds 10°C. The interpretation of these observations may not be straightforward, however, for the effect of low soil temperature on shoot growth may be through its effect on the production of gibberellins and cytokinins, which are produced in the root and translocated to the shoots, where they have marked effect on the growth and appearance of the crop.[5]

The principal root poisons are too high a concentration of carbon dioxide, the products of reduction of soil organic matter such as ethylene and of inorganic substances such as hydrogen sulphide (see p. 680), and some inorganic ions such as aluminium and manganese in acid soils, and heavy metal ions such as nickel, chromium and cobalt, which occur naturally in some soils and also as a result of contamination by industrial effluents in others. In addition, there are a number of organic substances secreted by the

1 M. H. Salim, G. W. Todd, A. M. Schlehuber, *Agron. J.*, 1965, **57**, 603.
2 For cotton, C. H. Arndt, *Pl. Physiol.*, 1945, **20**, 200; for maize, R. R. Allmares, W. C. Burrows and L. E. Larson, *Proc. Soil Sci. Soc. Amer.*, 1964, **28**, 271; for the grasses, E. M. Brown,
3 *Missouri Agr. Exp. Sta., Res. Bull.* 299, 1939.
4 For a review see K. F. Nielsen and E. C. Humphries, *Soils Fert.*, 1966, **29**, 1.
5 *Agron. J.*, 1965, **57**, 41.
  For some preliminary results with maize see R. K. Atkin and G. E. Barton, *Grassld. Res. Inst., Ann. Rept.* 1970, 59.

roots of some plants which inhibit the growth of roots of other plants from growing near them (see p. 553). Roots can also be attacked by a number of pathogens, some of which are of most importance in well-aerated soils, as, for example, some free-living nematodes which attack the growing points of otherwise healthy root systems (see p. 194). However, many of the weak un-specialised fungal root parasites are of most importance when a root is already damaged or weakened (see p. 246).

The effect of pore sizes and soil strength

The ability of the plant root to find space in which to grow, or to force its way into the soil, is often the most important factor limiting plant growth. There is not a great deal of systematic information on the minimum size of pore into which a root can grow without having to enlarge it, but it is probable that only a few plants have primary roots which are initially finer than 0·1 mm, and the roots of temperate cereals need spaces of at least 0·2 mm,[1] and the young growing roots of most trees and herbaceous plants and young taproots are considerably larger than this. J. Finney, at Jealotts Hill, Bracknell, found that winter wheat had primary seminal roots between 390 and 450 μ in diameter, with first laterals between 320 and 370 μ, second laterals between 300 and 350 μ, and root hairs between 8 and 12 μ.[2] Roots are therefore larger, and usually very much larger than the pores present in a compact structureless soil even if it is a sand, so that rooting is usually only possible in a structured soil. It is worth bearing in mind that soils at field capacity in Great Britain, hold water with a suction of about 50 cm corresponding to pores with a diameter of 60 μ, and in warmer countries the suction may be up to six times higher, so the maximum size of a pore holding water will be about six times smaller (see p. 437). Thus, soils can be well drained yet contain no pores wide enough for the primary roots of plants to enter. On the other hand plants growing in a coarse sand, typically have a very well branched and extensive root system if other conditions are favourable.

Roots must, in consequence, force their way into a soil, so they can only grow in soils which are compressible. Many attempts have been made to measure the effect of the compaction or shear strength of a soil on root elonga-tion.[3] Roots will rarely enter a light-textured soil if its bulk density exceeds 1·7 to 1·8, or a heavy-textured soil if it exceeds 1·5 to 1·6. It is possible that the roots of large-seeded plants, such as peas, which have a well-developed tap root, can exert a maximum pressure as high as 5 or 10 bar, but cereal roots appear to have their growth affected by very much smaller soil pressures. M. J.

1 L. K. Wiersum, *Pl. Soil*, 1957, **9**, 75; T. Kubota and R. J. B. Williams, *J. Agric. Sci.*, 1967, **68**, 227 (barley needs pores >0·25 mm): G. M. Aubertin and L. T. Kardos, *Proc. Soil Sci. Soc. Amer.*, 1965, **29**, 290 (maize).
2 Quoted by A. J. Low, *Chem. Ind.*, 1972, 373.
3 For a review see K. P. Barley and E. L. Graecen, *Adv. Agron.*, 1967, **19**, 1 and M. J. Goss and M. C. Drew, *A.R.C. Letcombe Lab.*, *Ann. Rept.* 1971, 9.

Goss[1] grew barley seedlings in a bed of glass beads 1 mm in diameter, so the minimum size of pore in the bed was 0·16 mm, and found that if the bed was put under an external pressure of 0·2 bar, root elongation was reduced and under a pressure of 0·5 bar was reduced by 80 per cent. The roots growing under an external pressure were much thicker and the number of laterals per unit length of seminal root was much increased, so that the root volume was not greatly affected by pressures in this range. Figure 22.2 illustrates these effects for pressures of 0·15 and 0·25 bar.

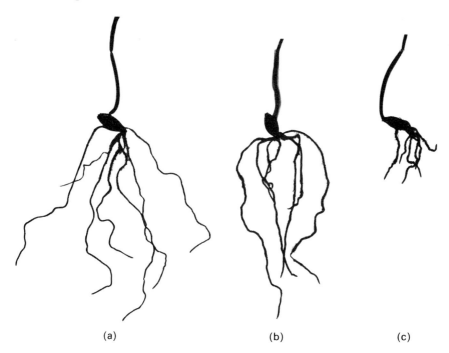

(a)   (b)   (c)

FIG. 22.2. The effect of external pressure on the growth of barley rootlets in a medium of Ballotini beads 1 mm diameter (pore neck diameter 0·15 mm). External pressure applied in bars: (a) 0; (b) 0·15; (c) 0·25

   The mechanism of root growth is that the meristem cells behind the root tip elongate longitudinally pushing the root tip forward. The effect of increasing resistance to the forward movement of the root tip is a development of root hairs from the epidermal cells just behind the meristem, which anchor these cells so allowing the full pressure producing the elongation of the meristem cells to be transmitted to the root tip. It is also possible that mucigel production by the meristematic cells reduces the friction between the elongating cells and the soil particles.[2] The anchoring effect of root hairs is well illustrated by

1 *A.R.C. Letcombe Lab., Ann. Rept.* 1969, 43; 1970, 6.
2 D. D. D. Jones, J. Morre and H. H. Mollenhauer, *Amer. J. Bot.*, 1966, **53**, 621.

an experiment made by L. H. Stolzy and K. P. Barley.[1] They found a pea root devoid of root hairs only needed a force of 4 g/mm length of root to pull it out of the soil, while a similar root with root hairs needed 20 g. This anchorage also allows the tip to move forward along the line of least resistance, but if it comes to a more compact zone the root behind the tip may buckle or kink, as is shown in Plate 35d,[2] for a pea root growing through moderately compacted loam. Such a compact zone can be produced by a cultivation implement smearing the surface of soil clods or by the furrow wheel of a tractor, when ploughing, smearing the bottom of the furrow.[3]

The force a root can exert appears also to depend on its diameter, for K. P. Barley[4] showed that the fairly thick roots of the two grasses *Phalaris tuberosa* and *Paspalum dilatatum* could penetrate clods that the thin roots of perennial ryegrass and cocksfoot were unable to. The pressure also depends on the turgor of the root cells, which, in turn, depends on the leaf turgor. During periods when the leaf is wilting, as may happen in the middle of the day in hot semi-arid areas, root growth ceases. When roots force their way through the soil, they compact the soil in their neighbourhood and this may reduce the infiltration rate in the soil.[5] Once the roots have died and started to decompose, however, the infiltration rate may rise since water will run down these old root channels easily,[6] provided the channel walls are stable.

Roots growing in poorly-structured clay soils are sometimes concentrated on the surfaces of cracks separating larger blocks of structureless clay. Although the main roots may not penetrate the clay, some fine secondary or tertiary roots, or root hairs may be able to do so.[7] The roots on the clod surfaces are often flattened by the pressure exerted by the adjacent clay clods when they swell on wetting in the autumn or winter, after they have shrunk by drying in the summer. This same effect can be seen when roots grow between a clay and a stone, as in the Rothamsted clay-with-flints soil. J. E. Weaver and R. W. Darland[8] have given examples of prairie grasses in Nebraska growing on soils with a compact B horizon overlying a looser C having an unbranched root system lying on the face of the clods or peds in the B horizon, but being profusely branched in the C.

Depth of rooting in the field

If a crop is growing on a deep well-drained soil of good pore space distribution, the depth of rooting is very dependent on the water supply. In general,

1 *Soil Sci.*, 1968, **105**, 297.
2 K. P. Barley, *Trans. 9th Int. Congr. Soil Sci.*, 1968, **1**, 759.
3 R. E. Prebble, *Expl. Agric.*, 1970, **6**, 303; J. K. Coulter, *Rothamsted Rept.* 1964, 49.
4 *Aust. J. agric. Sci.*, 1953, **4**, 283.
5 For an example see K. P. Barley, *Soil Sci.*, 1954, **78**, 205.
6 W. W. Emerson, *J. agric. Sci.*, 1954, **45**, 241.
7 R. A. Champion and K. P. Barley, *Soil Sci.*, 1969, **108**, 402.
8 *Ecol. Monogr.*, 1949, **19**, 309.

most of the available plant nutrients are present in the surface soil, so this is where most of the roots of a crop are concentrated. Under conditions when the surface soil is not disturbed and it is protected from strong evaporation by a mulch of dead vegetation lying on the surface, the plant roots may develop right up to the soil surface; and this can be a method for getting phosphates into a crop when growing on a soil that converts water-soluble phosphates into very insoluble forms.

Frequent light showers of rain, or frequent irrigation, by keeping the surface soil always moist, also encourage shallow rooting. Most crops only display their ability to root deep if they grow in conditions of drought, where the only way for the crop to obtain water for transpiration is for the root system to tap the water in the deeper subsoil. Thus C. A. Thompson and E. L. Burrows[1] found that lucerne roots only penetrated to about 1 m when irrigation water was supplied in 5 cm irrigations, but to 1·5 m if irrigated with 12 cm irrigations, while on deep silts it will penetrate to 6·9 m.[2] This result is not always found, for E. A. Garwood[3] found that the depth of rooting of ryegrass on the loam soil at the Grassland Research Institute at Hurley was not affected by supplementary irrigation. Annual crops, such as the cereals, maize and groundnuts, do not grow actively in the soil long enough for them to develop a really deep root system, and even where the soil conditions are suitable for really deep rooting, they rarely attain depths in excess of 2 m, for the roots cease growing on the onset of flowering. Depths such as these are, however, unusual in temperate countries, and normally they do not exceed 1 to 1·5 m.

Perennial crops growing on deep porous soils in climates with a pronounced wet and dry season can be very deep-rooting, and may continue transpiration all through the dry season if there is available water within their root zone. Again, on the deeply weathered basaltic soils of the Kenya highlands, some perennial grasses and trees will root to 5 or 7 m, drying the soil to the wilting point at this depth, whenever there is sufficient rain in the rainy season to wet it as deep as this (see p. 466). Since this depth of soil may hold 50 cm of available water, this only happens when the seasonal rainfall is over 80 or 100 cm. It is interesting to note that at Muguga, near Nairobi, the grass *Cenchrus ciliaris* will send its roots down to 3·7 m within the first six months after planting, and in the following dry season the roots will grow below 6·1 m even if grazed.[4] M. Dagg and P. H. Hosegood[5] determined the volume of star grass (*Cynodon dactylon*) roots down the soil profile at Muguga to a depth of 3 m, and found the following volumes:

1 *New Mexico Agric. Exp. Sta., Bull.* 123, 1920.
2 F. L. Duley, *J. Amer. Soc. Agron.*, 1929, **21**, 224; T. A. Kiesselbach *et al.*, ibid., 1929, **21**, 241.
3 *J. Brit. Grassland Soc.*, 1967, **22**, 176.
4 M. Dagg, *J. Hydrol.*, 1969, **9**, 438; H. C. Pereira, P. H. Hosegood and M. Dagg, *Expl. Agric.*, 1967, **3**, 89.
5 *East Afr. Agr. For. Res. Org., Ann. Rept.* 1964, 58. See P. H. Hosegood (*East Afr. Agr. For. J.*, 1963, **29**, 60) for similar results for Kikuyu grass (*Pennisetum clandestinum*) and wattle (*Acacia mollissima*), and with P. Howland, *East Afr. Agr. For. J.*, 1966, **32**, 16, for four other trees (to 4·5 m depth).

	Volume of stargrass roots $cm^3/10^4$ cm^3 soil					
Depth in m	0·0–0·3	0·3–0·6	0·9–1·2	1·5–1·8	2·1–2·4	2·7–3·0
Volume	218	48	22	8·1	5·8	4·2

so that although grass rooted to well below 3 m, two thirds of the total root volume was in the top 30 cm of soil. Unfortunately the length of the roots was not measured, but if the roots at 3 m had a diameter of 0·25 mm, and grew vertically downwards and were evenly spaced, which visually appears to be a reasonable assumption, the average spacing between the roots would be a little over 1 cm at this depth. Such a system should be quite capable of drying most soils uniformly to the wilting point during a dry spell (see p. 466), as is observed at Muguga.

On the other hand, as the rainfall becomes less, so the depth of rooting of the vegetation becomes more shallow, because it is restricted, in the main, to the volume of soil that is regularly wetted. This is illustrated in Plate 36 for a range of prairie species at Hays, Kansas,[1] after a period of moist years, and in Plate 37 after a series of dry years. In the wetter years 65 per cent of the species send their roots down to below 1·5 m, many reach 2·7 m and a few reach 6 m, whereas in the dry years most of the grass roots are in the top 60 cm and few herbaceous roots penetrate deeper than 1·5 m. Figure 22.3 illustrates this same effect of low rainfall in making the root system more shallow for wheat in the Great Plains.[2] This dependence of depth of rooting on water supply only holds if the plant is producing sufficient carbohydrates to allow good root development; and this limitation becomes of great importance in the correct management of pastures. Hard grazing of pastures reduces surplus carbohydrate production, for then all carbohydrates produced are needed to make good the loss of leaf, and hence such pastures are typically shallow-rooted. Hard grazing is therefore a practice only suited to moist climates; as the climate becomes more arid, the herbage must be allowed to grow longer and the grazing must be lighter. W. R. Hanson and L. A. Stoddart[3] found that, on the semi-arid ranges of Utah, the roots of bunch wheat grass (*Agropyron inerme*) only penetrated to about 39 cm on overgrazed range and to about 60 cm on protected range, and E. L. Florey and D. F. Thrussell, working on New Mexico ranges, found the roots of black and blue grama grasses (*Bouteloua eripoda* and *gracilis*) penetrated to about 120 cm on properly grazed range, about 60 cm on overgrazed, and to under 30 cm on badly overgrazed.

1 F. W. Albertson, *Ecol. Monogr.*, 1937, **7**, 481; with J. E. Weaver, ibid., 1943, **13**, 1.
2 J. E. Weaver, *Root Development of Field Crops*, New York, 1926.
3 *J. Amer. Soc. Agron.*, 1940, **32**, 278. For further examples see L. A. Stoddart and A. D. Smith, *Range Management*, New York, 1943, and J. L. Schuster, *Ecol.*, 1964, **45**, 63. For the effect of different grazing methods on the depth of the root zone of pastures in W. European conditions see E. Klapp, *Pflanzenbau*, 1943, **19**, 221.

The closeness of spacing of plants in a field affects rooting depth. An isolated plant in a large weed-free field, and where water is short, will tend to have a shallow root system with a very great lateral spread. On the other hand, if plants are too close together they tend to be small, and also to have a shallow root system. There is, therefore, usually a spacing that encourages a

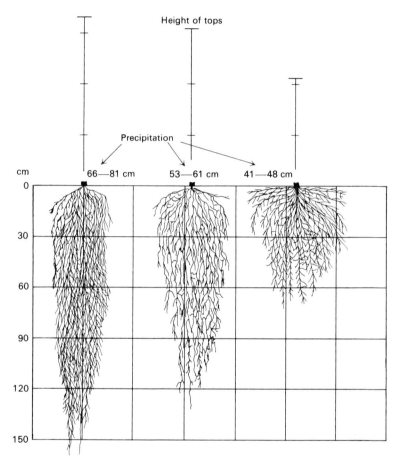

Fig. 22.3. The effect of the amount of rainfall on the depth of rooting of winter wheat (very fine sandy loam—silty loam in the Great Plains)

maximum depth of rooting, and which also encourages the roots to penetrate a layer that is too compact for the roots to penetrate if at an appreciably different spacing.[1] Very close spacing of a rapidly growing grass can, in theory, be used to conserve water under some conditions, for this will encourage a shallow root system. The grass will then only be able to dry the soil to a

1 J. K. Coulter, *Rothamsted Rept.*, 1965, 40.

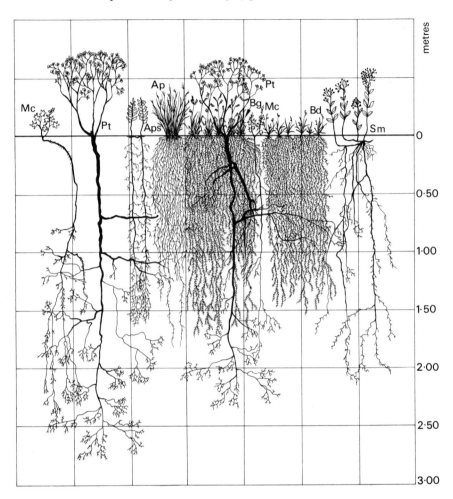

PLATE 36. The root systems of plants in a typical short-grass prairie after a run of years with average rainfall (Hays, Kansas). Al *Allionia linearis*, Ap *Aristida purpurea*, Aps *Ambrosia psilostachya*, Bd *Buchloe dactyloides*, Bg *Bouteloua gracilis*, Kg *Kuhnia glutinosa*, Lj *Lygodesmia juncea*, Mc *Malvastrum coccineum*, Pt *Psoralia tenuiflora*, Sm *Solidago mollis*, Ss *Sideranthus spinulosus*.

shallow depth, but will maintain surface permeability, so allowing most of the water falling in heavy rainstorms to penetrate into the soil. Unfortunately, in practice, it is difficult to ensure a sufficiently uniform dense stand; for any areas where the stand is not thick, the grass will root deeply and so dry the soil to depth.[1]

1 H. C. Pereira, R. A. Wood, H. Brzostowski *et al.*, *Emp. J. exp. Agric.*, 1958, **26**, 213.

To 3 metres

PLATE 37. The root systems in the same area as Plate 36 after a run of drought years. (For key, see Pl. 36.)

The effect of fertilisers on the root system of a crop depends on the fertility of the soil; but if the soil is low in a nutrient, suitable manures or fertiliser will give a stronger, better-developed root system, which has the potential to go deeper, and in particular it will have a greater ability to penetrate a rather compacted layer.[1] P. Newbould[2] measured the effect of incorporating a complete fertiliser into the surface of a Woburn soil on the weight of roots produced, and the uptake by barley of phosphate and calcium from different depths: his results are given in Table 22.2. They show that the fertiliser increased the weight of roots down to 40 cm, which was his lowest sampling depth, increased the uptake of calcium ions by much more than the increase in root weight, and increased the uptake of phosphate by less than this increase. The experiment does not allow any detailed interpretation of this result, but the phosphate supply in the surface soil of the fertilised plot was probably sufficient to reduce the demand of phosphate from the deeper roots. Similarly,

1 J. B. Fehrenbacher, P. R. Johnson *et al.*, *Trans. 7th Int. Congr. Soil Sci.*, 1960, **3**, 243.
2 In *Ecological aspects of the mineral nutrition of crops*, Ed. I. H. Rorison, Blackwell 1969.

S. T. Willett[1] showed that maize growing on a sandy loam of low nutrient status could only take water from the top 120 cm of soil if unmanured, but from 150 cm if adequately fertilised; and J. B. Lawes and J. H. Gilbert,[2] over 100 years ago, showed that, in a drought, unmanured grass at Rothamsted could only take water from the top 67 cm, but if suitably manured from 90 cm.

TABLE 22.2   Effect of fertilisers on the rooting of barley. Sandy loam soil: Fertiliser incorporated in top 13 cm soil. Results to harvest. Weights and absorptions relative to 5–10 cm unfertilised results

Soil depth	Root weight		Ca—uptake		P—uptake	
	none	NPK	none	NPK	none	NPK
5–10 cm	100	202	100	526	100	141
15–20	59	157	160	567	68	101
25–30	8	77	85	437	22	58
35–40	10	52	75	395	15	32

Deep-rooting does allow the crop to take up additional nutrients if the nutrient supply in the surface soil is inadequate. Thus, P. H. Nye and W. N. M. Foster[3] showed that savanna grasses and their associated herbs in Ghana brought up 30 per cent of their phosphate uptake, about 6 to 12 kg/ha P annually, from below 30 cm depth; whereas annual crops, such as maize and millet, only took between 7 to 15 per cent from below this depth, probably because of their more restricted root range.

This effect of fertilisers on the root system is also illustrated in Plate 38, which shows the stunted root system of an eight-year-old gooseberry bush growing on a poor sandy soil without added nutrients, and the much deeper and more extensive root system of a similar bush given a good dressing of farmyard manure before planting. But a soil that is uniform and permeable but of low nutrient status often carries crops with a sparse but well-distributed root system which is often well supplied with mycorrhiza. Under these conditions the roots will ramify in any pocket of soil well supplied with nutrients, such as rotting fragments of basaltic or granite rock or any pockets of farmyard manure provided they are in anaerobic volume.

The root systems of trees often illustrate this dependence of type on soil conditions. Many trees growing in natural forest with an undisturbed floor of litter tend to have an extensive much-branched root system filling the first few centimetres of the soil and a few deep tap roots. But if these trees are planted in open land where litter cannot accumulate, and particularly if this is done in rather dry regions, many of these trees will develop a root system

1 *Rhod. J. agric. Sci.*, 1969, **7**, 51.
2 *J. Roy. Agric. Soc. Engl.*, 1871, **7**, 1.
3 *J. agric. Sci.*, 1961, **56**, 299.

filling several metres of the soil. F. Hardy[1] has given a good illustration of this with cacao. In humid conditions, or badly drained soils, it has a very superficial root system containing many mycorrhizas which, with the layer of leaf litter that accumulates, forms a mat on the surface. In drier areas on well-drained soils, on the other hand, it has a deep root system, carrying no

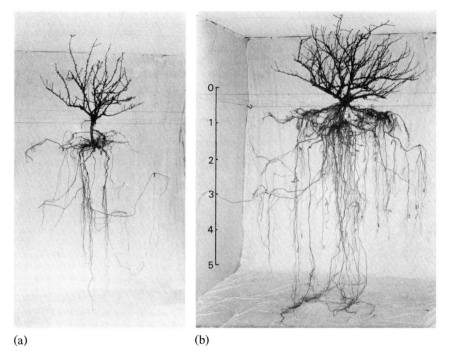

(a)   (b)

PLATE 38. The effect of nutrients on the root development of eight-year-old goose-berry bushes in a sandy soil. (Wrotham Heath, Kent.)
(*a*) No manure; (*b*) With 110 tons per hectare of farmyard manure in the year of planting. The roots of the manured bush, resting on the ground, should continue downwards. The measure is marked in feet.

visible mycorrhiza and forming no surface mat. Similarly the method of soil cultivation used in orchards and perennial crops affects the root systems of the plants, for the regular hoeing or cultivation of the soil to kill weeds will also kill all the surface-feeding roots of the crop. Weed control using herbicides, particularly if it leaves the dead weed cover as a surface mulch, will, therefore, encourage surface rooting, as is illustrated in Plate 52 and discussed in more detail on p. 805.

1 *Trop. Agric., Trin.*, 1944, **21**, 184; 1943, **20**, 207.

It is often necessary to improve the conditions in the subsoil if deeper root-ing is required. The most common subsoil condition inhibiting deep-rooting is poor drainage, for roots rarely penetrate into soil containing reduced sub-stances. Poor drainage is often a consequence of a subsoil pan, in which case subsoiling is necessary. If the subsoil is acid, deep placement of lime will sometimes encourage deeper rooting as, in some circumstances, will deep placement of phosphate. These circumstances may occur when the surface soil is dry at periods when the crop has a high phosphate demand, for phos-phate in the subsoil will remain in moist soil for a longer period than in the surface.

The mean daily growth rate of a root in a soil naturally depends on very many factors, but daily rates of 2 to 5 cm for annual crops are not unusual in the early period of growth, dropping off to lower figures as the plant ages. J. D. Lea[1] working with groundnuts found primary root growth rates of 3 cm/day for about the first month, falling to about half this during the second month, and to very little during the third, on a fairly permeable soil in Tanzania; on a cracking black montmorillinitic clay in central Sudan, the growth rate was only about 1 cm/day.

By way of summary, Table 22.3 illustrates the relative lengths and surface areas of roots and root hairs for several crops.[2] Taking oats as an example

TABLE 22.3   Length and surface area of roots and root hairs of various crops. Roots and root hairs in a soil cylinder 1 cm^2 area, 15 cm deep

	Roots			Root hairs			% soil volume occupied by roots and root hairs
	Number	Length m	Surface area cm^2	Number in thousands	Length km	Surface area cm^2	
Soyabeans	—	6·3	8·80	133	0·13	6·03	0·91
Oats	102	9·8	6·85	137	1·74	74·0	0·55
Rye	142	13·9	11·3	272	3·66	167	0·85
Poa pratensis	1840	83·5	25·0	1120	11·2	344	2·80

from this table, if the roots were evenly spaced in the soil, and growing verti-cally downwards, they would be 1·2 mm apart, and if the root hairs were at right-angles to the roots, they would be 0·07 mm apart. Table 22.4 illustrates the size and weight of the root systems of some grasses and crops growing in the central prairie states of the USA and Canada, when the grasses were

1 *Trop. Agric., Trin.*, 1961, **38**, 93, and see W. L. Balls, *Emp. Cott. Grow. Rev.*, 1951, **28**, 81 (cotton in Egypt); and F. S. Nakayama and C. H. M. Van Bavel, *Agron. J.*, 1963, **55**, 271 (sorghum in Arizona).
2 From H. J. Dittmer, *Science*, 1938, **88**, 482; *Soil Conserv.*, 1940, **6**, 33.

TABLE 22.4   Quantity of roots found in upper 10 cm of soil under different crops. Numbers and lengths of roots in metres or kilometres per half square metre. Weight of roots in tons per hectare

	Prairie grasses			Crops	
	Agropyron cristatum	*Agropyron pauciflorum*	*Poa pratensis*	*Lucerne*	*Wheat near harvest*
Underground stems:					
Length in m	23·0	4·5	28·0	47·0	15·0
Weight per ha	3·09	0·84	2·19	2·07	0·53
Main roots:					
Numbers in thousands	11·7	8·0	16·3	—	1·5
Length in m	935	563	1140	—	179
Average dia in mm	0·4	0·5	0·19	—	0·32
Weight per ha	1·17	0·64	1·33	—	0·34
Branches, first order:					
Number in millions	0·65	0·53	1·94	0·0054	0·082
Length in km	39·2	21·2	75·5	0·25	4·30
Average dia in mm	0·09	0·1	0·04	—	0·09
Branches, second order:					
Number in millions	7·85	2·28	10·57	0·152	1·10
Length in km	55·0	16·0	95·1	4·85	15·4
Average dia in mm	0·04	0·03	0·018	—	0·03
Branches, third order:					
Number in millions	3·13	1·15	—	0·738	0·83
Length in km	9·5	3·1	—	5·16	6·67
Average dia in mm	0·01	0·007	—	—	0·008
All branches:					
Number in millions	11·6	3·96	12·6	0·895	2·02
Length in km	103·4	40·2	170·6	10·3	26·4
Weight per ha	1·76	0·94	2·00	—	0·53
All roots:					
Weight per ha	6·00	2·44	5·52	2·38	1·39

growing in fairly pure stands.[1] The table gives some idea of the variation between the amounts of roots formed by different grasses in the top 10 cm of the soil, which contained a large proportion of the total weight of the roots. As noted on p. 515, the soil structure forming powers of crested and slender wheat grass (*Agropyron cristatum* and *A. pauciflora*) reflects the difference in the amounts of coarser root material produced by these two species.

It is also worth stressing a final point. A vigorous crop normally requires a soil of a certain minimum depth to ensure that it is under no severe nutrient or moisture stress at critical times in its growing season. Tree crops, which are long-lasting and accumulate considerable quantities of plant nutrients in their above-ground trunk and branches, can be very sensitive to shallowness of soil, and it is not unusual to see trees planted on shallow soils low in

1   T. K. Pavlychenko, *Natl. Res. Counc. Canada, Publ.* 1088, 1942. For a summary of root distribution of European agricultural crops see J. Köhnlein and H. Vetter, *Ernterückstande und Wurzelbild*, Paul Parey, 1953, and S. Gericke, *Bodenk. PflErnähr*, 1945, **35**, 229.

nutrients grow rapidly when young, but at a certain age cease growth, because the main plant nutrients in the soil have been exhausted, and they then become weakened and either succumb to disease or are blown over by high winds in wet weather.[1]

The uptake of nutrients by roots

The sources from which plant roots extract nutrients

There are three possible sources from which roots can extract their nutrients: the soil solution, the exchangeable ions, and the readily decomposable minerals; and it is very difficult to separate out the relative importance of these three sources for any particular plant. C. G. B. Daubeny, in a classical investigation made over a hundred years ago,[2] showed that only a part of the total supply of plant nutrients was available to the plant. But it was soon found that there was no sharp division between available and unavailable plant foods in the soil, for different species of plants had differing powers of extracting the nutrients. The older investigators attributed the different 'feeding powers' of plants to differences in the total acidity of their sap; thus they assumed the reason that leguminous plants could often extract more nutrients from a soil than could the grasses was because they had a more acid root sap,[3] although later work showed that, if anything, they have a less acid sap.[4]

J. von Liebig first introduced the idea that the exchangeable cations in the soil are the primary source of nutrients for the plant, which he expressed in the following words: 'The power of the soil to nourish cultivated plants is, therefore, in exact proportion to the quantity of nutritive substances which it contains in a state of physical saturation. The quantity of other elements in a state of chemical combination distributed through the ground is also highly important, as serving to restore the state of saturation, when the nutritive substances in physical combination have been withdrawn from the soil by a series of crops reaped from it.'[5] This hypothesis that the exchangeable cations are the principal source of these nutrients for the plant was extended by W. Knop[6] in a remarkable investigation of considerable historical interest, and also by O. Kellner.[7] M. Whitney in 1892,[8] and later in collaboration with F. K. Cameron,[9] developed the hypothesis that the plant obtains its nutrients from the soil solution, and hence that a study of the nutrients in the soil solution should lead to an accurate assessment of the fertility of the soil.

1 For examples see W. R. Day, *Forestry*, 1953, **26**, 81.
2 *Phil. Trans.*, 1845, 179.
3 B. Dyer, *Trans. Chem. Soc.*, 1894, **65**, 115; *Phil. Trans.*, 1901, **194** B, 235.
4 H. Kappen, *Landw. Vers.-Stat.*, 1918, **91**, 1; W. Thomas, *Plant Physiol.*, 1930, **5**, 443.
5 *Natural Laws of Husbandry*, London, 1863.
6 *Die Bonitierung der Ackererde*, Leipzig, 1871.
7 *Landw. Vers.-Stat.*, 1886, **33**, 349.
8 *U.S. Weather Bur. Bull.* 4, 1892; *U.S. Dept. Agric., Farmers' Bull.* 257, 1906.
9 *U.S. Dept. Agric. Bur. Soils, Bull.* 22, 1903.

Thus, the early investigators had shown that at least some of the nutrients in the soil solution, the exchangeable ions, and readily decomposable minerals are available to the plant roots. Yet these are not three independent sources, for the soil solution is in equilibrium with the exchangeable cations and the phosphate compounds in the soil; and if any nutrient except nitrate is removed from the solution, at least a part of this loss will be made good from the non-soluble nutrient reserves in the soil. The solid material of the soil, in fact, keeps the soil solution well buffered both for pH and for all nutrients except nitrates. But since the power of different plants to extract nutrients from a soil differs, the roots must presumably be able to use at least some proportion of the nutrients contained in the more readily decomposable minerals in the soil.

The soil solution

The water in the soil contains soluble salts, and hence whenever this aspect of the soil water is relevant, it is usually known as the soil solution. As has often happened in the history of agricultural science, the first investigations on the chemical composition of these dissolved salts was made in France. Th. Schloesing[1] in 1866 devised a method of collecting the soil solution based on displacement by water. He placed 30 to 35 kg of freshly taken soil containing 19·1 per cent of water in a large inverted tubulated bell jar and poured on it water, coloured with carmine, in such a way as to simulate the action of rain. The added water at once displaced the soil water and caused it to descend so that it could be collected: a sharp horizontal line of demarcation between the added and the original water persisted throughout the experiment, even when eight days were occupied in the descent. A typical analysis of the displaced liquid in milligrams per litre was:

CaO	MgO	K_2O	Na_2O	Nitric acid	Carbonic acid	Sulphuric acid	Chlorine	SiO_2	Organic matter
264	13·5	6·9	7·8	305	118	57·9	7·4	29·1	37·5
Concentration in m eq/litre									
9·44	0·66	0·15	0·25	4·84	1·90	1·18	0·21	0·48	

The solution, therefore, contained about 850 mg of material per litre, or about 160 parts per million of the soil was present in the solution. Its concentration was about 10^{-2} N, calcium was the principal cation, and nitrate the principal anion though bicarbonate and sulphate were also appreciable.

These results are typical of the relatively few determinations of the composition of the soil solution of non-saline soils that have been published. Table 22.5 gives further examples of the ionic concentrations in some American soils, and it shows that in neutral soils the calcium ion concentration is

1 *Comp. Rend.*, 1866, **63**, 1007; 1870, **70**, 98.

between 3–10 mM, and that nitrate is usually present in higher concentration than are the other anions. The calcium ion concentration is in general determined by the concentration of the anions, as both the nitrate and bicarbonate concentrations are dependent on microbial and root activity. On the other hand, the calcium ion concentration in an acid soil can be low, and magnesium may be present in a higher concentration. The magnesium, potassium and sodium ion concentrations in soils not well supplied with these ions are

TABLE 22.5   Composition of soil solution in some American soils. Concentrations in m eq/litre

	pH	Ca	Mg	K	Na	NH$_4$	NO$_3$	Cl	HCO$_3$	SO$_4$	cation/anion	Total concentration
Sandy loam*	5·82	8·1	1·7	1·1	0·8	0·4	8·0	0·3	1·2	1·0	1·15	11·3
	6·11	10·6	2·5	2·7	0·6	0·4	9·6	1·6	1·7	3·6	1·02	16·7
	7·19	21·0	1·2	0·7	1·8	0·4	15·6	2·2	1·1	7·0	0·97	25·5
Loam	7·65	6·4	1·7	0·3	1·4	0·4	4·4	1·9	3·4	1·7	0·90	10·8
Acid†	3·9	1·0	1·4	0·4	0·4	0·1	3·8	0·2	n.d.	0·8	0·96	4·8
			Al 0·63			Mn 0·58						

* F. M. Eaton, R. B. Harding and T. J. Ganje, *Soil Sci.*, 1960, **90**, 253.
† J. Vlamis, *Soil Sci.* 1953, **75**, 382.

between 10^{-4} and 10^{-3} M, and in the acid soil both aluminium and manganese are of the order of 2×10^{-4} M. These results were obtained from solutions derived from the soils when they were wet. These authors did not determine the phosphate ion concentration in their soil solutions, but it is typically about 10^{-5} to 10^{-6} M.

The composition of the soil solution depends on the moisture content of the soil, the rate of growth of the crop and the activity of the microbial population. These factors are difficult to separate out in the field, though they can be in the laboratory. Table 22.6 gives an example of this separation for an English clay soil that was air-dried, then wetted to one of two levels and incubated for periods up to 53 days.[1] The results are consistent with the following generalisations:

1 The concentration of nitrate and chloride varies inversely with the moisture content, since all of these ions are in solution.
2 The concentration of potassium is little affected by change in moisture content, in contrast to the sodium ion concentration which behaves as the nitrate and chloride. Thus the sodium ions behave as if they are nearly all in solution.
3 The bicarbonate is relatively unaffected by dilution, implying that it is controlled by other factors, and the sulphate varies less than expected, probably because some is held as an adsorbed phase on soil particles.

1 S. Larsen, *J. Sci. Fd. Agric.*, 1968, **19**, 693.

TABLE 22.6 Effect of soil moisture content and period of incubation on composition of the soil solution. Clay soil: pH 5·7, incubated at 20°C. Concentrations in m eq/litre

	Ca + Mg	K	Na	NO$_3$	Cl	HCO$_3$	SO$_4$	Si(OH)$_4$	cation/anion
After 2 days									
Moisture content									
18·3%	47·8	1·7	5·6	30·4	3·8	1·8	17·0	0·6	1·04
24·0%	37·9	1·5	4·3	22·1	2·6	1·6	14·4	0·6	1·07
Ratio 1·31	1·26	1·1	1·30	1·38	1·4	1·1	1·18	1·0	
After 53 days									
18·3	68·9	1·5	6·6	53·1	3·5	1·9	14·3	0·8	1·06
24·0	53·2	1·6	5·2	41·6	2·5	1·9	13·0	0 7	1 02
Ratio 1·31	1·29	0·9	1·27	1·28	1·4	1·0	1·10	1·1	

The phosphate ion concentration was not determined by Larsen, but the general result found by other workers is that it also is relatively independent of the concentration of the soil solution. Micro-organic activity results in an increase in the nitrate content of the soil, with little change in the concentration of the other anions, and in neutral or weakly acid soils the calcium ion concentration increases to neutralise the nitrate ions produced, a point which is also brought out by Table 22.6.

The soil solution of many leached agricultural soils under normal conditions of moisture content is found to contain up to about 0·05 per cent of soluble matter, and the osmotic pressure of the solution varies from 0·2 to 1 bar, being higher the nearer the moisture content is to the wilting point. Thus, even at the wilting point the soil solution is dilute, and has an osmotic pressure appreciably lower than the 10 to 20 bar of the root sap.

The movement of nutrients to the roots

The rate of uptake of nutrients by the root system of a crop depends in part upon the rate that they are brought from the body of the soil to the surface of the root, which is predominantly dependent on soil factors, but to some extent on the rate they can be transferred from the root surface into the main xylem stream in the plant stem, and this is predominantly dependent on plant factors.[1] These plant factors include the size of the root system and its rate of growth on the one hand, and a number of other factors which will not be discussed in detail here, but include the demand of the tops for the various nutrients, the mechanisms regulating the transport from any particular segment of the root to the xylem, and the physiological condition of the cortical and epidermal cells of the root. The rate of uptake of water may also be controlled by the viscous resistance to flow through the central conducting channel in the root. This is normally considered to be only a minor component in the total resistance to the transfer of water from the outside of a root to

1 For a recent review see K. P. Barley, *Adv. Agron.*, 1970, **22**, 159.

the outside of a leaf stoma; but J. B. Passioura[1] has shown that this resistance may be a rate-limiting factor when grasses, whose roots may only have channels 50 μ in diameter, are drawing water from depths of 3 to 6 m.

There are, in theory, two quite distinct processes through which nutrients are transferred from the body of the soil to the root surface, though, in practice, it is not possible to make a sharp distinction between them. These processes are known as mass flow and diffusion. Mass flow is brought about by transpiration, for the water taken up by a root to meet this demand contains ions dissolved in it, which are therefore carried to the root surface. Diffusion occurs when ions are being taken up faster than they are carried to the surface by mass flow, for this sets up a concentration gradient between the root surface and the body of the soil down which the ion will diffuse.

The water demands of arable crops growing on reasonably fertile soil will usually bring more calcium and often more sodium to the root surface than is translocated to the tops, about as much magnesium but much less phosphate; while the amount of potassium carried to the root is sometimes in balance and sometimes short of uptake, depending on the concentration in the soil solution and the potassium demand of the crop. Typically in many fairly fertile British soils the concentration of calcium in the soil solution is about 10^{-2} M, of potassium 10^{-3}–10^{-4} M, and of phosphorus 10^{-5}–10^{-6} M, so that if a crop uses 30 cm of water during its growth season, this amount of water would carry about 400 kg/ha calcium, 4 to 40 kg potassium and 0·3 to 0·03 kg phosphate, whereas the crop will take up and translocate to the above ground part about 10 to 30 kg/ha calcium, 50 to 250 kg potassium and about 20 to 30 kg phosphorus. J. L. Brewster and P. B. Tinker[2] have given an example of the conditions around a leek root (*Allium porrum*) growing in a clay loam at the Oxford University Field Station at Wytham. Mass flow brought seven or eight times as much calcium to the root surface as was taken up, so the calcium ion concentration just outside the root was about 1·2 times that in the soil solution, it brought 1·5 to 2 times as much magnesium, but only 7 to 10 per cent of the amount of potassium that was taken up, so the potassium ion concentration just outside the root was 2·5 to 6 times smaller than in the soil solution. The actual orders of magnitude found for uptake fluxes into roots growing in solutions of the order of concentration found in normal soils are of the order of 10^{-12} to 10^{-13} moles of cations per cm length of root per second for onions and leeks,[3] and 10^{-14} to 10^{-15} moles of phosphate for barley.[4]

The role of diffusion

Recently there have been a number of quantitative studies on the soil factors which control the rate ions can move from the soil to the root surface

1 *Aust. J. agric. Res.*, 1972, **23**, 745 gives his results for wheat.
2 *Proc. Soil Sci. Soc. Amer.*, 1970, **34**, 421.
3 C. J. B. Mott and P. H. Nye, *Soil Sci.*, 1968, **106**, 18.
4 D. T. Clarkson and J. Sanderson *A.R.C. Letcombe Lab. Ann. Rept.*, 1970, 22.

by diffusion. This work involves the solution of differential equations under specified boundary conditions, for which only approximate numerical values can be obtained with a computer. The basic assumptions are that diffusion only takes place through the water films, that the diffusive flux at any point is proportional to the concentration gradient at that point, that the root is a uniform cylinder and the flux to the root is entirely radial. The assumption that diffusion only takes place through the water films is justified for, in general, the migration of ions along the clay surfaces to the root is very much slower than the diffusive flux.[1]

The basis of the theoretical work has been to calculate the diffusion coefficient or diffusive conductivity of the ion in the soil water, which is defined as the ratio of the flux to the concentration gradient, knowing its value in a solution of the same salt concentration as the soil solution. If D is the diffusion coefficient in the soil and D_0 in the solution, then if the ion is not absorbed by the soil, such as the nitrate ion for example, D can be written as

$$D = vfD_0$$

where v is the proportion of the soil volume that contains water and f is defined as an impedance factor, corresponding to the tortuosity factor for gaseous diffusion. W. D. Kemper[2] has suggested that it involves three terms, the true tortuosity of the water films, a term dependent on the viscosity of the water close to the clay surfaces, and a term dependent on the diffuseness of the electrical double layer, for if it occupies a significant proportion of the water film this will affect the volume of water through which anions can diffuse. D. L. Rowell[3] working with a Lower Greensand sandy clay loam, found that when the water in the soil was at a suction of 100 mbar, $v = 0.4$, $f = 0.5$ so $vf = 0.2$; while at 15 bar suction, $v = 0.07$, $f = 0.01$, so $vf = 7 \times 10^{-4}$, or 300 times smaller than at field capacity.

But if the ion is adsorbed by the soil, and the ionic concentration in the soil water is in equilibrium with the amount of that ion adsorbed on the soil particles, then, as the root absorbs some of the ions from the solution, the ionic concentration in the solution close to the root will drop, but some of the ions absorbed on the soil surface will go into solution so will counteract this reduction to some extent. This reduces the diffusion by a factor dC/dE, where C is the concentration of the ion in the solution, and E is the total amount of the ion per unit volume of soil, that is the amount in solution and the amount adsorbed on the soil surfaces; dC/dE is, in fact the buffering power of the soil for that ion (see p. 99) as determined from the desorption curve, where it is assumed that ionic desorption is a rapid process relative to the rate of change of its concentration. The more strongly buffered the soil, or the more strongly the ion is adsorbed by the soil, the smaller is dC/dE and the smaller the effec-

1 D. L. Rowell, M. W. Martin and P. H. Nye, *J. Soil Sci.,* 1967, **18**, 204.
2 With D. E. L. Maasland and L. K. Porter, *Proc. Soil Sci. Soc. Amer.,* 1964, **28**, 164, and with L. K. Porter *et al., Proc. Soil Sci. Soc. Amer.,* 1960, **24**, 460.
3 With M. W. Martin and P. H. Nye, *J. Soil Sci.,* 1967, **18**, 204.

tive diffusion coefficient.[1] The buffering factor can be about 10^{-2} for potassium and 10^{-3} for phosphate, though the exact value depends on the soil and the level of exchangeable potassium or of labile phosphate. The result of these mathematical studies has been to show that if an ion is being supplied entirely by diffusion,[2] the total amount taken up from the commencement of uptake to time t is a function of $\sqrt{(Dt)}$.

The diffusion coefficients for ions in dilute solution are of the order of 10^{-5} cm^2/sec for ions that are not adsorbed, such as nitrates; of 10^{-7} cm^2/sec for cations, such as calcium, magnesium and potassium in moist soils and of 10^{-9} cm^2/sec for strongly adsorbed anions, such as phosphate. In the absence of transpiration, a root with a diameter of 1 mm will scarcely have affected the potassium and phosphate concentrations in the solution at distances from it of about 1 cm and 1 mm after 10 days uptake.[3] These calculations also show the very important role of root hairs in the uptake of phosphate. Thus, if a root of diameter 0·5 mm has 100 root hairs per mm length, and each root hair is 0·02 mm diameter and about 1 mm long, which are reasonable figures for a number of crops,[4] during the first ten days the root hairs will take up four to five times as much phosphate as the root without root hairs would have alone,[5] and it will be about sixteen days before the contribution of the soil from outside the root hair zone will exceed that from the inside of the zone; with potassium this will happen after a few hours.[6]

Further, nearly all the phosphate will be taken from very close to the root hair. Thus, D. G. Lewis and J. P. Quirk calculated that after four days 94 per cent of the phosphate taken up by a root hair had come from the soil within 0·1 mm of its surface. This factor can also be important in the uptake of potassium by a young root in conditions of a high potassium demand, growing in a soil where the potassium is strongly held on clay particles. M. C. Drew and P. H. Nye[7] have given an example of this using ryegrass, where the root hairs increased potassium uptake by 70 per cent.

The rate of uptake of ions

The rate of uptake of nutrients by plant roots growing in reasonably fertile

1 See, for example, L. V. Vaidyanathan, M. C. Drew and P. H. Nye, *J. Soil Sci.*, 1968, **19**, 94; and P. H. Nye, *Trans. 9th Int. Congr. Soil Sci.*, 1968, **1**, 117.
2 For a discussion see S. R. Olsen and W. D. Kemper, *Adv. Agron.*, 1968, **20**, 91, and for a numerical example P. H. Nye and F. H. C. Marriott, *Pl. Soil*, 1969, **30**, 459.
3 For examples see E. Farr, L. V. Vaidyanathan and P. H. Nye, *Soil Sci.*, 1969, **107**, 385 (onion root, uptake of K and Ca); M. C. Drew and P. H. Nye, *Pl. Soil*, 1970, **33**, 545 (ryegrass, uptake of P); S. R. Olsen, W. D. Kemper *et al.*, *Proc. Soil Sci. Soc. Amer.*, 1962, **26**, 222 (maize uptake of P).
4 M. C. Drew and P. H. Nye, *Pl. Soil*, 1969, **31**, 407 (for ryegrass); D. G. Lewis and J. P. Quirk, *Pl. Soil*, 1969, **26**, 445, 454 (for wheat); and D. R. Bouldin, *Proc. Soil Sci. Soc. Amer.*, 1961, **25**, 476 (for maize).
5 D. G. Lewis and J. P. Quirk, *Pl. Soil*, 1967, **26**, 99, 119, 445, 454 (using wheat); M. C. Drew and P. H. Nye (*Pl. Soil*, 1970, **33**, 545) found a rather lower value of two- to threefold for ryegrass.
6 P. H. Nye, *Pl. Soil*, 1966, **25**, 81, with M. C. Drew, ibid., 1969, **31**, 407.
7 *Pl. Soil*, 1969, **31**, 407.

soils is, however, not usually limited by the rate that the ions can be transported from the soil to the root surface, but by the rate the ion can move from the soil solution into the root tissues. S. R. Olsen and W. D. Kemper[1] illustrated this by showing that if an oat crop has as large a root system as found by Dittmer in Table 22.3, and if the whole of the system of roots and root hairs took up phosphorus as fast as short lengths of active maize roots, namely, 4×10^{-12} g/cm^2/sec, the roots would take up ten times more phosphate than was present in the crop.

Lewis and Quirk also worked out that for the wheat plants they were studying, root systems of 5 m and 1·5 m would take up phosphate at the observed rate from a soil somewhat low and somewhat high in soluble phosphate, yet the plants had between 50 and 90 m of root. However, if crops are taking up all their water from the deep subsoil, where the root system is sparse and the soluble phosphate in the soil solution very low, the roots will only be tapping a very small proportion of the soil for phosphate, while tapping a larger volume for potassium and a much larger volume for magnesium, for example, so phosphate uptake will fall to a very low level compared with other nutrients.

The factors that determine the rate of ion uptake, such as the effect of its external concentration, are not fully known. P. H. Nye and his colleagues[2] have studied the validity of the assumption that the rate of uptake is dependent on the ionic concentration just outside the root surface, by defining α as

$$F = \alpha c$$

where F is the uptake flux of the ion into the root. Nye terms α the coefficient of root absorbing power, and he has studied the dependence of α on concentration under conditions in which all other plant factors have been standardised as far as possible. He finds that α has its highest value when the ionic concentration is low, it falls slowly with concentration over a fairly wide range, then more rapidly, and at high ionic concentration the uptake flux may become independent of concentration. Thus, Nye and P. B. Tinker,[3] using some water culture results of J. F. Loneragan and C. J. Asher,[4] found the value of α fell from $1·7 \times 10^{-3}$ cm/sec to $1·2 \times 10^{-3}$ cm/sec when the phosphate concentration increases twenty-five-fold from 4×10^{-8} M to 100×10^{-8} M, but it fell fifteenfold when the phosphate concentration increased a further twenty-fivefold to $2·5 \times 10^{-5}$ M.

It is possible that α only decreases appreciably with increasing concentration when there is an element of luxury consumption of that element taking place. The value of α for potassium appears to be about ten times smaller than for phosphate at low concentrations; that for sodium and magnesium about 100 times and for calcium about 1000 times smaller than for potassium.

1 *Adv. Agron.*, 1968, **20**, 91.
2 *Pl. Soil*, 1966, **25**. 81. *J. appl. Ecol.*, 1969, **6**, 293; with R. Bagshaw and L. V. Vaidyanathan, *J. agric. Sci.*, 1969, **73**, 1. J. L. Brewster and P. B. Tinker, *Proc. Soil Sci. Soc. Amer.*, 1970, **34**, 421.
3 *J. appl. Ecol.*, 1969, **6**, 293.
4 *Soil Sci.*, 1967, **103**, 311.

The concentration around the plant root is controlled by the relative rates of ion uptake and supply. The more rapidly the root takes up an ion, or the slower the replenishment, the faster does its concentration fall; and one of the factors of most importance for replenishment is the buffering capacity of the soil for that ion. The greater the release of an adsorbed ion for a given fall in concentration, the smaller will be the fall in concentration. There is little quantitative data on the concentration of different nutrient ions at the root surface, but it is probable that if a crop responds to a fertiliser the concentration of that nutrient ion around the root has fallen sufficiently to limit the rate of uptake appreciably.

Selectivity in uptake

Roots possess a definite selectivity for the uptake of certain ions compared with others, in that the ionic composition of the aerial part of a plant is different from the ionic composition of the soil solution. Typically, the plant contains much more phosphate and definitely more potassium relative to calcium than does the soil solution. But this concept of selectivity involves two separate processes—selectivity of uptake and selectivity of translocation. Only a proportion of the ions taken up by a root cell need be translocated: some may remain in the root system and either be used in the growth of the root itself or be concentrated within or around the root cells; and others may

TABLE 22.7   Uptake and translocation by root segments. Barley root segments 3·5 mm long: uptake in 10^{-12} moles/segment/day

Distance of segment behind tip	1 cm	44 cm
P uptake:	395	335
translocated	53	241
% translocated	12	42
Ca uptake:	89	58
translocated	51	2
% translocated	36	3

be excreted by the cells after uptake, such as sodium, by the so-called sodium pump process. D. T. Clarkson[1] has given a good example of this distinction using the uptake and translocation of calcium and phosphate by segments of barley roots close to the growing tip and 44 cm behind the tip. Their results are given in Table 22.7, which shows that most of the phosphate taken up by the young root was used for root growth, but most of the calcium taken up by the older root accumulated in the root and very little was translocated. This Table shows also that the root segment 44 cm behind the growing point is

1 With J. Sanderson and R. S. Russell, *Nature*, 1968, **220**, 805.

taking up both phosphate and calcium although its cortical cells are already disorganised and the thickened layer around the endodermis is intact.

The exact biochemical processes involved in ionic uptake are not yet agreed, but there is a definite element of competition between ions for uptake. Thus, the uptake of magnesium by a root depends not only on the magnesium concentration in the solution around the root, but also on the potassium concentration; and the higher this is the smaller is the magnesium uptake. On the other hand, the uptake of magnesium is much less dependent on the calcium or the sodium ion concentration around the root, and root cells show so little selectivity as between calcium and strontium, that the ^{89}Sr isotope is commonly used in nutrient uptake studies as a tracer for calcium. This has the consequence that adding a potassium fertiliser to a soil low in magnesium may induce magnesium deficiency in the crop. These effects are illustrated in Table 22.8, which gives the effect of adding calcium, magnesium or potassium salts to two soils cropped to Kentucky bluegrass (*Poa pratensis*) or sweet clover (*Melilotus alba*).

TABLE 22.8   The effect of a given increase of a cation in the soil on the composition of the plant.* Milli-equivalents of cation per 100 g dry matter in the plant

Added to soil	Bluegrass				Sweet clover			
	Ca	Mg	K	Total	Ca	Mg	K	Total
Basal	35	42	54	131	102	62	47	211
+ Ca	41	37	53	131	124	57	45	226
+ Mg	30	63	50	143	100	97	42	239
+ K	23	24	95	142	72	38	102	212

* C. E. Marshall, *Missouri Agric. Exp. Sta., Res. Bull.* 385, 1944. The results given are for the means of the crops on two soils and at two levels of calcium supply.

These selectivity processes depend on the plant species: the ionic composition of the growing leaves of a crop, for example, depends not only on the ionic composition of the soil solution, but also on the plant species. Thus some species, such as lucerne for example, have a higher calcium-potassium ratio in their leaves than others such as most grasses; and even within one plant, not only does the composition of a leaf change as it ages, but also the composition of a younger leaf may be different from that of an earlier formed leaf at the same stage in their development. These differences between plant species become very noticeable if an ion is in excess of its normal value in the soil solution, for then plants can often be classified into accumulators or rejectors of that ion. Some plants growing in acid soils, high in either aluminium or manganese, will accumulate concentrations of aluminium or manganese in their leaves far in excess of that needed without their growth being affected;

and some plants growing in salt marshes will accumulate very high concentrations of sodium chloride in their leaves, yet other plant species, growing in these same soils, will have compositions little affected by these high concentrations.

Ionic absorption must maintain overall electric charge neutrality and this is achieved by an excess of either hydrogen or bicarbonate ions being excreted if there is an excess adsorption of cations or of anions; the sourse of the hydrogen ions probably being from malic or fumaric acids whose rate of synthesis in the absorbing cells is conditioned by the hydrogen ion demand, and the bicarbonate anions are a normal byproduct of cell metabolism. The pH of the solution outside the root will accordingly rise if more anions than cations are being absorbed, and fall if more cations are. Which of these will happen usually depends on the source of nitrogen the roots are using. If all the nitrogen is being taken up as nitrate, anion uptake is usually in excess, and the pH outside the roots may rise by over one pH unit,[1] whilst if it is being taken up as ammonium it will fall. As an example, R. K. Cunningham[2] analysed Italian ryegrass samples collected from a number of British sites, and showed that the roots took up between 1·1 and 3·0 times as many anions as cations, if all the nitrogen was taken up as nitrate. He also found, both for these samples of ryegrass and for a number of other crops growing near Rothamsted, that the higher the nitrogen content in the dry matter of the tops, the larger this ratio, as one would expect. Thus, the soil pH in the neighbourhood of many roots, particularly if they are taking up their nitrogen as nitrate, is higher than in the body of the soil.

Root exudates

Roots excrete a number of soluble compounds into the soil.[3] Some of these form the energy source for members of the rhizosphere population, and many of these compounds may be excreted by a very wide range of plant species. But many root exudates are specific either to most members of the same genus, or to a group of related genera, or to a few species of a single genus; and these may affect very specific groups of root pathogens or root symbionts. Many of the pathogens will exist in the soil in a resting phase, either as a spore, an egg or a seed, which is activated only after it has absorbed a certain amount of this exudate. These pathogens may be present as seeds, such as plant parasites belonging to the genera *Cuscuta*, *Striga* and *Orobanche*, or as eggs, such as some gall-forming nematodes, or as spores of root-invading fungi, or the organisms may be symbionts such as Rhizobia or mycorrhizal fungal spores.

Root exudates can be toxic to some root pathogens or pests. Thus the roots

1 D. Riley and S. A. Barber, *Proc. Soil Sci. Soc. Amer.*, 1969, **33**, 905.
2 *J. agric. Sci.*, 1964, **63**, 97, 103, 109.
3 For a recent summary see *Biochemical Interactions among Plants*, Nat. Acad. Sci. (Washington DC), 1971.

of asparagus excrete a soluble carbohydrate into the soil that is toxic to the stubby root nematode, *Trichodorus christiei*,[1] and some species of *Tagetes*, the African marigold, excrete a polythienyl that is toxic to a wide range of plant parastic nematodes;[2] and beans, *Vicia faba*, reduce populations of wireworms, the larvae of elaterid beetles belonging to the genus *Agriotes*.[3] Some plants excrete inorganic toxins into the soil. Thus some species of flax excrete prussic acid HCN, which protects the roots from attack by some pathogenic fungi.[4]

Excretions from the roots of some plant species affect the growth of other roots, usually of other species. This can be very noticeable at the seedling stage, and may be the explanation of the observation so frequently made in the field, that even small weeds can interfere with a germinating crop more than one would expect from their power of competing for light and nutrients. The roots of some couch grasses appear to excrete substances which interfere with the uptake of nitrogen and potassium by other plants growing in association with them. Thus, J. D. Bandeen and K. P. Buchholtz[5] showed that the couch grass *Agropyron repens* growing with maize very seriously affected its growth, and that even a high dressing of nitrogen and potassium did not allow the maize to give a good yield, although the grass could only have taken up a small proportion of the added fertiliser. Table 22.9 gives some of their results. In the same way another couch grass *Digitaria scalarum*, growing in

TABLE 22.9   Effect of *Agropyron repens* on the yield of maize.

Fertiliser in hg/ha		Grain yield in t/ha	
N	K	*No grass*	*With grass*
0	0	5·90	0·65
220	0	6·27	0·57
0	185	6·92	0·84
220	185	7·57	2·52

tea and coffee plantations, will affect the bushes so seriously that production falls to an uneconomic level. Again, the bushes appear to be suffering from acute nutrient deficiency, which cannot be rectified by normal dressings of fertiliser.

Root excretions from a number of plant species will inhibit nitrification, and those from some grasses are particularly effective[6] (see p. 341).

1 R. A. Rohde and W. R. Jenkins, *Maryland Agric. Exp. Sta., Bull.* A-97, 1958.
2 A. F. G. Slootweg, *Nematol.*, 1956, **1**, 192; J. H. Uhlenbroek and J. D. Bijloo, *Recl. Trav. chim. Pays-Bas*, 1958, **77**, 1004; 1959, **78**, 382; 1960, **79**, 1181.
3 A. C. Evans, *Ann. appl. Biol.*, 1944, **31**, 235.
4 See, for example, M. I. Timonin, *Soil Sci.*, 1941, **52**, 395.
5 *Weeds*, 1967, **15**, 220.
6 D. R. E. Moore and J. S. Waid, *Soil Biol. Biochem.*, 1971, **3**, 69.

Crops can, however, have a harmful effect on a succeeding crop if they leave residues behind that lock up soil nitrates on decomposition. A well-known example of this is the depressing effect a crop of sorghum can have on its successor. Thus, H. E. Myers and A. L. Hallsted[1] found that in a two-year rotation in Kansas, winter wheat yielded on the average 1000 kg/ha after maize, but only 800 kg/ha after kafir during the period 1918–41. J. P. Conrad[2] showed that this effect is due to the high sugar content in the crown roots of the sorghum; it was equivalent to nearly 2000 ppm of sucrose in the top foot of soil just under the plant and fell off to 15 to 20 ppm 22 cm away. On the other hand, maize roots only returned the equivalent of 2 to 3 ppm of sucrose to the soil. This depressing effect of the sorghum is therefore due both to the locking up of nitrates by the soil micro-organisms as they decompose this sugar, and also to the unfavourable soil structure that may ensue when these sugars are decomposing.[3] It can be rectified by adding nitrates to the soil, or, as Myers and Hallsted also showed, by taking a leguminous crop just before the sorghum. Thus they found that, if oats followed sorghum grown after lucerne or sweet clover, there was enough nitrogen left in the soil from the lucerne or clover to provide for both the sorghum residues and the oats, but if soyabeans were used as the leguminous crop, the sorghum residues depressed both the nitrate nitrogen content of the soil and the yield of oats to about one-third of their normal values, for soyabeans do not enrich the soil appreciably with nitrogen.[4]

1 *Proc. Soil Sci. Soc. Amer.*, 1943, **7**, 316.
2 *J. Amer. Soc. Agron.*, 1927, **19**, 1091; 1937, **29**, 1014; 1938, **30**, 475.
3 J. F. Breazeale, *J. Amer. Soc. Agron.*, 1924, **16**, 689.
4 For further examples which show that soyabeans taken as a grain crop do not enrich the soil in nitrogen, see R. E. Uhland, *Missouri Agric. Expt. Sta.*, *Bull.* 279, 1930. For further references, see H. B. Brown, *Louisiana Agric. Expt. Sta.*, *Bull.* 265, 1935; F. S. Wilkins and H. D. Hughes, *J. Amer. Soc. Agron.*, 1934, **26**, 901.

23

The sources of plant nutrients in the soil: Phosphate

The phosphates present in the soil

These can be divided into three groups:

1 Phosphates present in the soil solution. This is always negligible compared with the other forms.
2 Phosphates present in the soil organic matter.
3 Inorganic phosphates including both definite phosphate compounds and surface films of phosphate held on inorganic particles.

In general more, and often considerably more, phosphate is present in the inorganic than the organic form, though there are soils, particularly soils on ancient peneplains and some strongly leached tropical soils, where the organic phosphate is the principal reserve of plant available phosphate, although it itself is not available.

Organic phosphates

There is still considerable doubt on the phosphate compounds present in organic matter (p. 291), but it is almost certain none of them are directly available to the plant. The importance of the organic phosphates is twofold. On soils low in phosphate, put down to grass, for example, the rate of build-up of humus may be limited by lack of phosphate, for the C/P ratio in humus is usually between 100 and 200, and inorganic or available phosphate can be converted into unavailable forms by the process. Again, in soils low in inorganic phosphate, the phosphate supply to the crop can be very dependent on the relation between the periods when the soil humus is decomposing and releasing phosphate and those when the crop is making demands on the soil phosphates.

More work has been done on the factors controlling the liberation of phosphate by the decomposition of organic matter than its lock-up due to humus accumulation. As described on p. 237, one of the important factors which controls the rate of mineralisation of this phosphate is the number of times the soil becomes really dry between rewettings, and another factor is temperature, for the warmer the soil the more rapid the rate of decomposition can be.[1]

This source of phosphate is naturally the more important the lower the level of available inorganic phosphate, but in fact in the great majority of arable soils it is of minor importance. There are a number of tropical soils, however, where the major part of the potentially plant available phosphate is in the organic matter in the surface soil, so that if the surface gets washed away by rain, or removed by bulldozing or any other cause, the subsurface soil which now becomes the surface is so low in phosphate, and usually in many other elements as well, that crop growth becomes exceedingly slow unless heavy dressings of phosphate are given. Work on organic phosphates in soils is limited by the difficulty of determining this fraction, and none of the methods[2] in use are really satisfactory.

The soil solution also contains some organic phosphates, and the amount of phosphate present in this form may be several times greater than the soluble inorganic phosphate.[3] There is some doubt, however, about its value as a source of phosphate for plants, for Pierre and Parker found it was a poor source for maize, soyabean and buckwheat; but it is not a homogeneous fraction and A. Wild and O. L. Oke,[4] who fractionated it, showed that some of the fractions contained myo-inositol monophosphate and the fractions that hydrolysed easily were readily available to clover seedlings grown aseptically, while those resistant to hydrolysis were of little value.

Calcium phosphates

Calcium phosphates exist in several forms, the most important of which for our purposes are:

$Ca(H_2PO_4)_2 . H_2O$, monocalcium phosphate, which is water soluble, and the dominant phosphate of superphosphates.

$CaHPO_4 . 2H_2O$ dicalcium phosphate (hydrated) and $CaHPO_4$, the dehydrated form. The dihydrate is metastable and goes over to the dehydrated form relatively easily, and both are only slightly soluble in water. Dicalcium phosphate is present in many phosphate fertilisers made by treating monocalcium phosphate with ammonia or calcium hydroxide.

$Ca_8H_2(PO_4)_6 . 5H_2O$, octacalcium phosphate.[5]

$Ca_3(PO_4)_2$ tricalcium phosphate, which is certainly formed in high-temperature slags, but there is still doubt if it can be formed by precipitation from aqueous solutions.[6]

1 See, for example, M. T. Eid, C. A. Blake *et al.*, *Iowa Agric. Exp. Sta., Res. Bull.* 406, 1954.
2 For these methods, see for example M. L. Jackson, *Soil Chemical Analysis.* Englewood Cliffs, N.J., 1958.
3 W. H. Pierre and F. W. Parker, *Soil Sci.*, 1927, **24**, 119; A. Wild, *Z. PflErnähr.*, 1959, **84**, 220.
4 *J. Soil Sci.*, 1966, **17**, 356.
5 W. E. Brown, J. P. Smith, *et al.*, *Nature*, 1962, **196**, 1048, 1050.
6 U. Schoen, G. Barbier and S. Henin, *Ann. Agron.*, 1954, **4**, 441, claim that a precipitated tricalcium phosphate exists, giving an apatite-like X-ray diagram but on ignition to 900°C giving β–$Ca_3(PO_4)_2$.

$Ca_{10}(PO_4)_6(OH)_2$ hydroxyapatite, which is the phosphatic constituent of bones and teeth.

$Ca_{10}(PO_4)_6F_2$, fluorapatite, the principal phosphatic constituent of mineral phosphates in the great deposits in North Africa, Florida and elsewhere, and which form the principal commercial sources of phosphates.

The mono- and dicalcium phosphates do not dissolve congruently in weakly acid, neutral or alkaline solutions, that is, the ratio of phosphorus to calcium dissolved differs from that in the solid phase: relatively more phosphate than calcium comes into solution, so the solid phase becomes more calcium rich. Thus if a pellet of monocalcium phosphate is placed in water or a dilute solution, only a proportion of the pellet dissolves, the residue is a porous pellet of about the same size from which 60 to 80 per cent of the phosphate has gone, leaving an open framework of dicalcium phosphate, which is hydrated if the water is cool and its reaction mildly acid, and anhydrous if hot or alkaline.[1] W. L. Lindsay and H. F. Stephenson[2] showed that, as a consequence of the pellet losing more phosphate than calcium, the saturated solution diffusing out of it is acid, with a pH between 1 and 1·5, and had a composition between 4 and 4·5 M in phosphorus and about 1·3 M in calcium, and it was in equilibrium with the residual dicalcium phosphate, being more acid if the dicalcium phosphate was anhydrous rather than hydrated. If a granule of dicalcium phosphate is shaken up with water, or with a dilute calcium solution, the ratio of dissolved phosphate to calcium is higher for the anhydrous than for the hydrated form, and the residue is probably hydroxy-apatite if the solution is weakly acid and octacalcium phosphate if neutral or alkaline. But equilibrium between the solution and the solid phase is reached very slowly, particularly if the solution is neutral or alkaline.

An important part of the chemistry of calcium phosphates in soils is that of apatite, and particularly hydroxy-apatite. When a calcium hydroxide solution is added to a monocalcium phosphate solution, or sufficient is added to a phosphoric acid solution to give a precipitate, the precipitate has the crystal structure of hydroxyapatite; but the crystals are so small that their chemical composition may differ appreciably from a calcium-phosphorus ratio of 1·67:1 because the composition of the monolayer on the crystal surface is different from that of the unit crystal cell.[3] Thus apatite crystals in bone are hexagonal tablets about 300 Å across and 25 to 30 Å thick, that is, three unit cells thick, so only about one-third of the unit cells have the composition of apatite, the other two-thirds have only one face attached to

1 W. E. Brown and J. R. Lehr, *Proc. Soil Sci. Soc. Amer.*, 1958, **22**, 29 and 1959, **23**, 7.
2 *Proc. Soil Sci. Soc. Amer.*, 1959, **23**, 12, 18.
3 For the most important papers on this subject see: H. Bassett, *Trans. Chem. Soc.*, 1917, **111**, 620; J. A. Naftel, *J. Amer. Chem. Soc.*, 1936, **28**, 740; P. W. Arnold, *Trans. Faraday Soc.*, 1950, **46**, 1061; and S. B. Hendricks and W. L. Hill, *Proc. Nat. Acad. Sci.* (Washington), 1950, **36**, 731; J. G. Arvieu, *Bull. Ass. fr. Étude Sol*, 1969, No. 4, 5. For a review on phosphates of bone see W. F. and M. W. Neuman, *Chem. Rev.*, 1953, **53**, 1, as well as Hendricks and Hill.

an apatite cell, and the other face is free. Such crystals have a surface area of about $100 \, m^2/g$. Precipitates formed by adding pure lime water to phosphoric acid solutions have the hydroxy-apatite structure with a surface area between 10 and 20 m^2 but impurities such as citrate[1] and possibly carbonate ions present in the solution reduce the crystal size.

A consequence of this extremely small crystal size is that these precipitates are not true hydroxy-apatite, for the composition of the surface layer can vary appreciably from that of true apatite, due to the exchange of the hydrion $(H_3O)^+$ for calcium, and possibly of the hydroxyl for phosphate. Thus precipitates giving the hydroxy-apatite structure can have calcium-phosphorus ratios between 1·4 and 1·8.

There is a marked reaction between the apatite surface and carbonate ions in solution. The surface can absorb a monolayer of carbonate ions relatively easily, and, in fact, the fine crevices between the apatite crystals in dentine of teeth are filled with a calcium carbonate film forming on them. Correspondingly, the surface of calcite will absorb a film of phosphate, initially thought to be apatite,[2] but it is uncertain if it is apatite or octacalcium phosphate, for films of these two phosphates have laminar structures sufficiently similar for them not to be readily distinguished by X-ray analyses.[3] The presence of this carbonate layer also slows down the rate at which apatite crystals in a solution of calcium, phosphate and carbonate ions reaches equilibrium. Thus J. S. Clark[4] was able to measure the solubility product of hydroxy-apatite quite easily if he worked in CO_2-free solutions, particularly if the precipitate was aged for a few hours at 90°C, but he failed to get an equilibrium value in the presence of CO_2, even after long periods of ageing at high temperatures. This has the consequence that equilibrium between the solid apatites in the soil and calcium and phosphate ions in solution is reached exceedingly slowly; and in such a labile system as the soil solution, the evidence is that a true equilibrium is never in fact reached.

Hydroxy-apatite is not stable in the presence of fluoride ions, for it will absorb fluoride from very dilute fluoride solutions, but this absorption goes exceedingly slowly. Thus bones buried in the soil under conditions in which the phosphate does not dissolve pick up fluoride from the soil solution, but the reaction may take thousands of years to come to completion. It is for this reason that the fluorine content of a buried bone can sometimes give a good indication of its age, and that most rock phosphate deposits are fluor-apatites, for they have been formed from phosphates being precipitated in the sea and picking up fluoride ions from it.

Natural fluor-apatites, which constitute the normal rock phosphates rarely

1 D. Patterson, *Nature*, 1954, **173**, 75.
2 G. Nagelschmidt and H. L. Nixon, *Nature*, 1944, **154**, 428.
3 F. Amer and A. Ramy, *J. Soil Sci.*, 1971, **22**, 267; W. E. Brown, J. P. Smith *et al.*, *Nature*, 1962, **196**, 1048.
4 *Canad. J. Chem.*, 1955, **33**, 1696.

ever have the ideal formula, but there is usually some replacement of PO_4 by CO_3 with a corresponding replacement of calcium by sodium to maintain electrical neutrality, but in addition some PO_4 is replaced by FCO_3, and about 40 per cent of the CO_3 in fluor-apatite is in this form, which has the consequence that there are between 4·6 and 5·9 moles of PO_4 instead of 6 in the formula weight, and between 1·2 and 2·86 moles of F instead of 2, with hydroxyls substituting for fluoride in the low fluoride apatites.[1] The solubility of these apatites depends very considerably on their composition, so the fertiliser value of different rock phosphates differs widely.

There is also considerable evidence that a small proportion of the phosphate ions on the surface of the fluor-apatite crystals can be replaced by carbonate $CO_3{}^{2-}$, and that this replacement reduces the small tendency that the fluor-apatite crystals have to grow. Certainly the rock phosphates that are easiest to make superphosphate from, and which are of most use as phosphate fertilisers when simply ground fine, are those which contain the higher amounts of non-calcite carbonate ions, and their presence on the lattices seems to counteract the reduction in the solubility of phosphate one would expect from the replacement of hydroxyl by fluoride.[2]

Many soils contain apatites in their sand fraction, in a form that is very insoluble,[3] and it is probable that much of this is present as inclusions within the minerals composing this fraction.[4] Until the minerals have weathered, this phosphate has no agricultural importance.

Iron and aluminium phosphates

In well drained soils one of the principal aluminium phosphates that have been recognised belongs to the plumbogummite group of minerals having the general formula $M.Al_3(PO_4)_2(OH)_5H_2O$ where M is usually barium, strontium, yttrium or cerium. They have been found in some Australian,[5] Indian[6] and Welsh[7] soils by means of the electron probe, but they are probably very resistant to weathering. The other principal phosphates found in some well-drained acid soils belong to the isomorphous variscite-barrandite-strengite group; variscite having the formula $AlPO_4.2H_2O$, strengite $FePO_4.2H_2O$, and barrandite being a mixture of the two in almost any proportions. The only iron phosphate mineral that has been found in crystals which are recognisable in the petrological microscope is vivianite, $Fe_3(PO_4)_2$.

1 G. H. McClellan and J. R. Lehr, *Amer. Mineral.*, 1969, **54**, 1374.
2 See J. H. Caro and W. L. Hill, *J. Agric. Fd. Chem.*, 1956, **4**, 684, for a discussion of this.
3 See E. G. Williams and W. M. H. Saunders, *J. Soil Sci.*, 1956, **7**, 90, 189, for some Scottish soils.
4 See M. P. Cescas, E. H. Tyner and J. K. Syers, *J. Soil Sci.*, 1970, **21**, 78, for examples from New Zealand.
5 K. Norrish, *Trans. 9th Int. Cong. Soil. Sci.*, 1968, **2**, 713.
6 V. A. K. Sarma and G. S. R. Krishna Murti, *Geoderma*, 1970, **3**, 321.
7 D. A. Jenkins in *Micromorphological Techniques and Applications*, Soil Survey Monog. 2, 1970, Harpenden.

$8H_2O$, a ferrous phosphate that can often be found in waterlogged or badly drained soils.

There are also a few minerals found in nature which may occur in soils as crystals too small to have been detected yet, such as sturrettite $3Al(OH).2PO_4.3H_2O$. A number of crystalline phosphates of iron and aluminium have been prepared in the laboratory,[1] and W. L. Lindsay and H. F. Stephenson[2] have shown that when granulated superphosphate is added to a soil, crystals of two of these, a taranakite $H_6K_3Al_5(PO_4)_8.18H_2O$ and one with a composition $H_8K(Al.Fe)_3.(PO_4)_6.6H_2O$ may be formed. Both of these contain potassium and the second has an aluminium-iron ratio of 2 or less.

It is likely that iron and aluminium phosphates occur in many soils in the form of films, a few molecules thick at the most. These films are probably held on the surface of hydrated ferric and aluminium oxide films, or on ferric and aluminium ions forming part of the surface layer of clay crystals. The properties of these phosphate-hydrous oxide films have been discussed on p. 131, for the adsorption of the phosphate is probably a ligand exchange reaction with water or hydroxyl ions coordinated with the iron or aluminium ion.

Soils containing iron or manganese concretions may also contain appreciable amounts of phosphate, probably as ferric phosphates, as a constituent of the concretion,[3] and this phosphate is in a form unusable by plants as long as the soil remains aerobic. A number of tropical soils containing iron concretions may have nearly all their phosphate locked up in this way, as do a number of soils on old land surfaces in New Zealand.[4] But in general most of the mineral phosphates present in non-calcareous soils is associated with the clay fraction.

The solubility products of soil phosphates

It should be possible to prove if the phosphate present in a soil is present as a definite chemical compound or not by determining the apparent solubility product of the soil phosphate. Thus, if the concentration of phosphate in the soil solution is controlled by the solubility of hydroxy-apatite, the activities of the phosphate, calcium and hydrogen ions in the solution must be consistent with its solubility product. If the formula of hydroxy-apatite is

1 For a discussion on the conditions under which the various crystalline forms of ferric and aluminium phosphates are produced, see J. E. Salmon, *J. Chem. Soc.*, 1952, 2316, and with R. F. Jameson, ibid., 1954, 4013, C. V. Cole and M. L. Jackson, *J. Phys. Coll. Chem.*, 1950, **54**, 128, J. F. Haseman, J. R. Lehr, and J. P. Smith, *Proc. Soil Sci. Soc. Amer.*, 1951, **15**, 76.
2 *Proc. Soil Sci. Soc. Amer.*, 1959, **23**, 440.
3 See D. A. Jenkins in *Micromorphological Techniques and Applications*, Soil Survey of England and Wales, 1970.
4 J. D. H. Williams and T. W. Walker, *Soil Sci.*, 1969, **107**, 22, 213.

written as $Ca_5(PO_4)_3.OH$, then when apatite is in equilibrium with its saturated solution,

$$a_{Ca}^5 \times a_{PO_4}^3 \times a_{OH} = \text{constant}$$

where a_{Ca} is the activity of the calcium ions in solution. If the logarithms of the activities are used, and pCa is written for $-\log a_{Ca}$, this equation can be rewritten in the form[1]

$$10(\text{pH} - \tfrac{1}{2}\text{pCa}) - 3(\text{pH} + \text{pH}_2\text{PO}_4) = \text{constant}$$

where the constant is 14·7 at 25°C for a hydroxy-apatite made in the absence of all carbonates and aged at 90°C, presumably to obtain a fairly stable surface.[2] R. K. Schofield called the first term (pH $- \tfrac{1}{2}$pCa) the lime potential in the solution, and the second term (pH $+$ pH$_2$PO$_4$) the phosphoric acid potential. It is in fact the negative logarithm of the activity of phosphoric acid in the solution, and is proportional to its negative thermodynamic potential. The larger this term, the lower is the activity of the phosphoric acid, so, in general, the lower the phosphate concentration. This term differs from the lime potential in that the higher the lime potential, as defined above, the greater is the activity of calcium hydroxide in the solution because the lime potential is defined as pH $- \tfrac{1}{2}$pCa instead of pOH $+ \tfrac{1}{2}$pCa. The pH of a soil in 0·01 M CaCl$_2$ solution is 1·14 units higher than its lime potential.

The determination of the phosphoric acid potential (pH $+$ pH$_2$PO$_4$) in a soil is not always straightforward, because phosphate is sufficiently immobile in field soils for its potential to vary from site to site; so a representative sample of the soil taken for analysis takes time to come into equilibrium when shaken up in a suitable salt solution.[3] It can be further complicated in neutral soils because, at least under some conditions, the neutral molecule $[CaHPO_4]^0$ exists in the soil solution, and must be allowed for if a correct figure for the phosphate potential is to be obtained; and the error involved in ignoring it increases both with the pH from pH 5 upwards, and with the concentration of the calcium chloride solution in which the soil is shaken.[4]

It has not been found easy to prove under what conditions hydroxy-apatite controls the solubility of phosphate in the soil solution. This is partly because, under conditions prevailing in the soil, the solubility constant depends on whether hydroxy-apatite is being precipitated or dissolved; and partly because the surface of the crystal may possess an isoelectric point, which implies that the electric double layer outside the surface is either

1 For p(H$_2$PO$_4$) = constant + 2pH + p(PO$_4$) and p(OH) = constant $-$ pH.
2 J. S. Clark, *Canad. J. Chem.*, 1955, **33**, 1696; see also D. R. Wier, S. H. Chien and C. A. Black, *Soil Sci.*, 1971, **111**, 107.
3 R. E. White, *Pl. Soil*, 1964, **20**, 1, 184. *Aust. J. Soil Res.*, 1966, **4**, 77. H. C. Aslyng (*Act. Agric. Scand.*, 1964, **14**, 260) discusses in detail the precautions that must be taken to get reproducible and reliable results.
4 S. Larsen, *J. Soil Sci.*, 1965, **16**, 275; H. E. Jensen, *Pl. Soil*, 1970, **33**, 17; M. D. Webber and G. J. Racz, *Canad. J. Soil Sci.*, 1970, **50**, 243.

enriched in calcium ions or in the anion, depending on whether the pH of the solution is above or below this point.[1] This has the consequence that the solution appears to be supersaturated with respect to hydroxy-apatite if the pH is above the isoelectric point, and undersaturated if below.

The relation between the lime and phosphoric acid potentials for a saturated solution of dicalcium phosphate is:

$$2(pH - \tfrac{1}{2}pCa) - (pH + pH_2PO_4) = 0\cdot66 \text{ at } 25°C,$$

and for octacalcium phosphate:

$$8(pH - \tfrac{1}{2}pCa) - 3(pH + pH_2PO_4) = 11\cdot7,$$

under conditions when the solid phase dissolves congruently.[2] The corresponding relations for a saturated solution of variscite is:

$$3(pH - \tfrac{1}{3}pAl) - (pH + pH_2PO_4) = -2\cdot48 \text{ at } 25°C,$$

and for strengite:

$$3(pH - \tfrac{1}{3}pFe) - (pH + pH_2PO_4) = -6\cdot3 \text{ at } 25°C.$$

But these values are only valid if all the aluminium or iron in solution is present as the simple ion, that there is no gibbsite or aluminium hydroxide present in the case of variscite, and the pH is sufficiently low to ensure the phosphates dissolve congruently.[3] Although in some soils the phosphate and aluminium hydroxide or ferric hydroxide potentials are consistent with variscite or strengite controlling the solubility of the phosphate, this may not be the correct interpretation, for, if the soils are made rather more acid, more aluminium than phosphate comes into solution due to the dissolution of aluminium hydroxide, and if phosphate is added to the acidified soil it is precipitated in a form less soluble or more basic than variscite.[4] A characteristic of the solubilities of these microcrystalline phosphates is that they come to equilibrium in their bathing solutions very slowly, in periods measured by months rather than days;[5] and minor changes in the solution, such as in the concentration of carbon dioxide dissolved in the water can affect both the rate and the apparent equilibrium concentration.

The consequence of the solubility of aluminium ions controlling that of the phosphate is that liming a soil increases the solubility of aluminium phosphates but reduces that of the calcium phosphates, as is shown in Fig. 23.1.[6] Thus increasing the pH of variscite in the presence of gibbsite from 5 to 6

1 S. Larsen, *Nature*, 1966, **212**, 605.
2 E. C. Moreno, W. E. Brown and G. Osborn, *Proc. Soil Sci. Soc. Amer.*, 1960, **24**, 94, 99.
3 B. W. Bache, *J. Soil Sci.*, 1963, **14**, 113.
4 A. W. Taylor and E. L. Gurney, *J. Soil Sci.*, 1962, **13**, 187; *Soil Sci.*, 1962, **93**, 241; 1964, **98**, 9.
5 See, for example, W. L. Lindsay *et al.*, *Proc. Soil Sci. Soc. Amer.*, 1959, **23**, 357; 1960, **24**, 177.
6 W. L. Lindsay and E. C. Moreno, *Proc. Soil Sci. Soc. Amer.*, 1960, **24**, 177 and H. C. Aslyng, *Roy. Vet. Agric. Coll. Copenhagen, Yearbook*, 1954, 1.

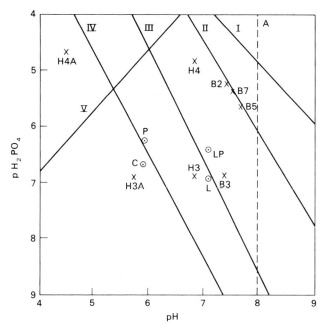

FIG. 23.1. The relation between pH and the phosphoric acid potential in one Danish and two Rothamsted field experiments. I, II, III, IV, V are solubility curves for dicalcium phosphate dihydrate, calcium octaphosphate, apatite, fluorapatite and variscite respectively.

A Calcium carbonate in equilibrium with air containing 0.03% CO_2.

H Hoosfield soils; 3 and 3A no superphosphate, 4 and 4A with superphosphate; A plots receive sulphate of ammonia.

B Broadbalk soils; 2 with farmyard manure, 3 unmanured, 5 with phosphate without nitrogen, 7 with phosphate with sulphate of ammonia.

Trystofte, Denmark, sampled 1949

C Control. P Superphosphate, 1943. L Limed, 1943

increases the concentration of phosphate in the solution from 2.2×10^{-6} to 2.4×10^{-3} M in the presence of 5×10^{-3} M Ca, while increasing the pH of hydroxy-apatite from 6 to 7 in the presence of 5×10^{-3} M Ca reduces the phosphate concentration from 3.1×10^{-3} to 2.5×10^{-7} M; and even for octacalcium phosphate, increasing the pH from 7 to 8 reduces the phosphate concentration from 7.7 to 1.0×10^{-5} M in a calcium solution of this concentration.

A number of authors have, in fact, plotted the lime or aluminium hydroxide potential against the phosphate potential for a range of soils, and Fig. 23.1 and Fig. 23.2[1] give examples of the kind of results that are obtained. Figure 23.2 shows that for soils with a pH above about 6.5, the $H_2PO_4^-$

1 B. Ulrich and P. K. Khanna, *Geoderma*, 1968, **2**, 65.

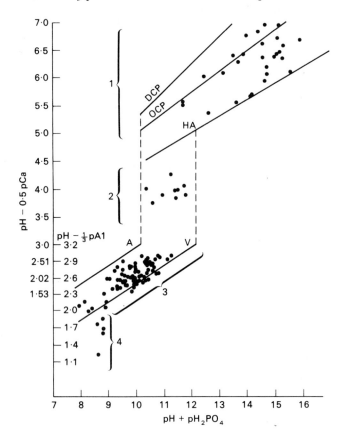

FIG. 23.2. The relation between the $H_2PO_4^-$ potential and the lime or alumina potentials for 230 German soils
DCP, OCP and HA are solubility lines for dicalcium phosphate, octacalcium phosphate and hydroxyapatite respectively
A, V are solubility lines for amorphous aluminium phosphate and variscite
1 Range of calcium phosphate
2 Transition from calcium to aluminium phosphate
3 Range of aluminium phosphate
4 Transition from aluminium to ferric phosphates

concentration in the soil solution is controlled by the solubility of calcium phosphates more soluble than hydroxy-apatite and less soluble than octacalcium phosphate, unless the soil contains more than 2 per cent $CaCO_3$, when it lies on or above the octaphosphate line. Correspondingly, if the soil is more acid than about pH 4 in 0·01 M $CaCl_2$ solution, the $H_2PO_4^-$ concentration is controlled by aluminium phosphates more soluble than variscite, but less soluble than amorphous aluminium phosphate.[1] Figure 23.1 shows that adding superphosphate to a field soil over a period of years decreases the phosphate potential without appreciably affecting the lime potential,[2] and liming an acid soil or making a neutral soil more acid affects the lime potential but has little effect on the phosphate potential, and it will take a soil from a point less soluble than hydroxy-apatite to almost the octacalcium phosphate line. However, it does affect the phosphate in that liming increases the amount of phosphate a resin can extract from the soil[3] probably because, as noted on p. 570, the bonding energy of adsorbed phosphate is about five times greater on acid than on neutral soils.

Both Figs. 23.1 and 23.2 show that the solubility of phosphate in neutral or calcareous soils which have received little, if any, phosphate fertiliser corresponds either to hydroxy-apatite[4] or a rather more soluble phosphate; and for calcareous soils which have received much superphosphate, the solubility corresponds to that for octacalcium phosphate, and this remains the stable form of the added phosphate.[5] There is also some evidence that the presence of magnesium ions in the soil solution stabilises the octa-calcium phosphate, and reduces the rate it is converted to apatite,[6] in soils not containing free carbonate. However, if very large amounts of mono calcium phosphate are added, dicalcium phosphate may be formed.[7] The solubility of phosphate from fluorapatite is less than from hydroxy-apatite, and as the fluoride ion concentration rises above 10^{-8} M, characteristic for a saturated fluorapatite solution, the solubility drops still further.[8] Fluoride ion concentration in soils probably vary between 5×10^{-6} to 5×10^{-5} M,[9] and this variation can give quite appreciable variations in the relation between the phosphate and the lime potentials in soils in which fluorapatite is the source of soluble phosphate; but little work has been done on the consequences of these relatively high fluoride concentrations on the solubility of different calcium or other phosphates.

The sorption processes operative in soils

The reason that the solubility diagram is often of little help in interpreting the form of the soil phosphate which controls the phosphate concentration in

1 See also E. O. Huffman, *Fertiliser Soc., Proc.* 71, 1962.
2 For another example see H. E. Jensen, *J. Soil Sci.,* 1971, **22**, 261.
3 A. W. Taylor and E. L. Gurney, *Proc. Soil Sci. Soc. Amer.,* 1965, **29**, 482.
4 For another example see S. Larsen, *J. Soil Sci.,* 1966, **17**, 121.
5 M. D. Webber and G. E. G. Mattingly, *J. Soil Sci.,* 1970, **21**, 121. L. C. Bell and C. A. Black, *Proc. Soil Sci. Soc. Amer.,* 1970, **34**, 583. C. C. Weir and R. J. Soper, *J. Soil Sci.,* 1963, **14**, 256.
6 C. S. Martens and R. C. Hariss, *Geochim. Cosmochim. Acta.,* 1970, **34**, 621.
7 S. Larsen and A. E. Widdowson, *J. Soil Sci.,* 1970, **21**, 364, but see F. Amer and A. Ramy, ibid., 1971, **22**, 267 for an alternate interpretation.
8 R. P. Murrmann and M. Peech, *Proc. Soil Sci. Soc. Amer.,* 1968, **32**, 493.
9 S. Larsen and A. E. Widdowson, *J. Soil Sci.,* 1971, **22**, 210.

the soil solution in near-neutral or acid soils is that this phosphate is sorbed on the surface of clay particles, as can be proved by the use of ^{32}P labelled phosphate solutions. This technique measures the amount of soil phosphate that is exchangeable with the labelled phosphate, and it shows first of all that only a small proportion of the total soil phosphate is accessible for exchange in soils low in phosphate, or that have not received phosphate fertiliser for a long time.[1] But it also shows that the accessible phosphate has an apparent wide scatter of accessibility, for while much of the exchange takes place rapidly, exchange is continuing to take place over a long period of time, measured in weeks or months, at a decreasing rate.[2] The full cause of this continuing slow exchange is not known: it could be due in part to some of the phosphate being present in cracks of molecular dimensions, so the slow rate of exchange would be a diffusion controlled process. It could also be due to a slow absorption of phosphate from the solution, or to a slow diffusion of labelled phosphate into the absorbing film; or it could be due to some of the surface phosphate having a very high activation energy for exchange, that is, it is bound so strongly to the surface that there is only a very low probability that it will exchange with a phosphate ion in the bathing solution.

This technique also shows that when phosphate is first absorbed by the soil, it is all isotopically exchangeable, but with time this amount drops until, after about one month for the Rothamsted soil, only about 70 per cent remains exchangeable, and this then may remain constant over a period of years, although the proportion that is rapidly exchangeable may continue dropping for several years.[3]

The surfaces that are active in phosphate sorption contain di- or trivalent cations: calcium, aluminium and ferric ions. They may be active by having a part of their positive charge neutralised by the negative charge on the clay or humic surfaces and part by the phosphate anion, or the aluminium may be at the broken edge of the clay sheet, or the active aluminium and iron may be present as surface films of their hydroxides or oxides. It has not been simple to determine the relative importance of these various processes for holding phosphate in any particular soil, because of lack of methods for distinguishing between them.

One group of methods, developed by S. C. Chang and M. C. Jackson,[4] is based on the assumption that neutral 0·5 N sodium fluoride will only displace phosphate held by aluminium ions and will displace all the phosphate so

1 See O. Talibudeen, *J. Soil Sci.*, 1958, **9**, 120 for examples from Rothamsted.
2 C. D. McAuliffe, N. S. Hall *et al.*, *Proc. Soil Sci. Soc. Amer.*, 1948, **12**, 119. R. S. Russell, J. B. Rickson and S. N. Adams, *J. Soil Sci.*, 1954, **5**, 85. O. Talibudeen, *J. Soil Sci.*, 1957, **8**, 86.
3 O. Talibudeen, *J. Soil Sci.*, 1958, **9**, 120. See also S. Larsen and A. E. Widdowson, ibid., 1971, **22**, 5.
4 *Soil Sci.*, 1957, **84**, 133. *J. Soil Sci.*, 1958, **9**, 109. For the effect of the pH of the fluoride solution see C. V. Fife, *Soil Sci*, 1959, **87**, 13, 83; 1962, **93**, 113; 1963, **96**, 112.

held, that a subsequent treatment with 0·1 N sodium hydroxide will displace all the phosphate held by ferric ions, and that 0·5 N sulphuric acid will dissolve all the phosphate held by calcium ions, but none held by the iron or aluminium. These assumptions are certainly not generally valid. The fluoride dissolves very considerably more aluminium than phosphate from many soils,[1] so it presumably brings into solution all phosphate associated with this readily extractable aluminium; but it will also displace some phosphate held by calcium and by iron,[2] though most of the phosphate held by iron hydroxides and oxides in soil is held with a strength sufficient to ensure that it is not displaced by the neutral fluoride.

The second group of methods seeks correlations between the amount of phosphate a group of soils hold or will absorb from a standard solution and the amount of aluminium or ferric iron soluble in solutions such as Tamm's acid oxalate or in neutral pyrophosphate. A strong reducing solution, such as dithionite, is not suitable for ferric iron since these absorption reactions are primarily surface reactions, so only solvents removing surface films should be used.

The results of a great deal of work, using both methods, have shown that in well-drained mildly acid temperate soils aluminium that is removable by acid potassium chloride or Tamm's acid oxalate is the principal agent holding phosphate on the soil surface.[3] It is probably not possible, even in theory, to separate out the various positions of the aluminium, whether on the broken edge of a clay lattice, or an aluminium ion either partially or fully neutralised with hydroxyls,[4] or aluminium associated with organic matter.[5] However, in many poorly drained temperate soils and possibly some tropical soils[6] ferric oxide or hydrated oxide surfaces are important, though little critical work has been done on a range of tropical soils. Unfortunately, there is little evidence for the relative surface areas of iron and aluminium hydroxides in soils, so it is not known how far these experimental results merely reflect this ratio. When phosphate fertilisers are added to a soil, the phosphate becomes adsorbed initially by aluminium ions, but slowly a proportion migrates to ferric ions. A considerable proportion of the phosphate adsorbed either by aluminium or iron, as high as two thirds in some soils, rapidly loses isotopic exchangeability, the proportion probably usually being higher for the iron than the aluminium bound phosphate.[7] Some of the phosphate that loses isotopic exchangeability also ceases to be extracted by these solvents

1 H. L. S. Tandon, H. L. Motto and L. T. Kurtz, *Proc. Soil Sci. Soc. Amer.*, 1967, **31**, 168.
2 S. M. Bromfield, *Aust. J. Soil Res.*, 1967, **5**, 225. *Soil Sci.*, 1970, **109**, 388.
3 S. M. Bromfield, *Nature*, 1964, **201**, 321.
4 W. T. Franklin and H. M. Reisenauer, *Soil Sci.*, 1960, **90**, 192.
5 E. G. Williams, N. M. Scott and M. J. McDonald, *J. Soil Fd. Agric.*, 1958, **9**, 551. R. D. Harter, *Proc. Soil Sci. Soc. Amer.*, 1969, **33**, 630.
6 S. C. Chang and W. K. Chu, *J. Soil Sci.*, 1961, **12**, 286 (Taiwan soils). J. A. R. Bates and T. C. N. Baker, ibid., 1960, **11**, 257 (Nigerian forest soils).
7 H. L. S. Tandon and L. T. Kurtz, *Proc. Soil Sci. Soc. Amer.*, 1968, **32**, 799.

so has presumably either become incorporated in stable crystals or has diffused below the active surface film.[1]

There is also evidence, from thermodynamic considerations, that surface films of phosphate on ferric hydroxide surfaces are stable in the soil, while surface films of phosphate on aluminium hydroxide are not, for they crystallise to form aluminium phosphate exposing a new aluminium hydroxide surface to continue the adsorption.[2] This is because the formation of aluminium phosphate from aluminium hydroxide and the phosphate ion decreases the free energy of the system by about 120 kJ/moles, whereas the formation of ferric phosphate from ferric hydroxide and the phosphate ion increases it by about 60 kJ/mole.[3]

Unfortunately, in most of this work, the role of calcium has not been examined in detail. However, J. P. Leaver and E. W. Russell[4] found that for a soil with a very high capacity for absorbing phosphate, due to its high content of active iron and aluminium, calcium ions would absorb phosphate appreciably even when the pH was as low as 5.

The adsorption isotherm

The relation between the amount of phosphate a soil will absorb from a phosphate solution and the concentration of the phosphate left in the solution is known as the adsorption isotherm. It depends on the pH of the soil, and for a given soil, the higher the pH in the range 5 to 8, the lower the amount of phosphate sorbed at a given concentration in the bathing solution. As already noted there is rarely any genuine equilibrium reached, for the soil continues to absorb phosphate from the solution slowly to very slowly over very long periods of time. It is common practice, therefore, to determine the adsorption isotherm either after a fairly short period, usually of the order of a few hours after the rapid adsorption has ceased, or over a period of several days when the rate has fallen to a very low level.

The adsorption isotherm falls into two parts, a curved part in which the phosphate concentration in the soil solution does not exceed 5×10^{-4} M, so this covers the normal range of concentrations found in the soil, and a linear portion at higher concentrations. In the first region, the soil adsorbs phosphate very strongly at low concentrations, but as the concentration rises the gradient of the curve falls. This strong adsorption is due to an increase in the entropy of the system rather than a decrease in its free energy.[5]

1 For examples from Rothamsted see O. Talibudeen, *J. Soil Sci.*, 1958, **9**, 120, and with P. Arrambarri, *J. agric. Sci.*, 1964, **62**, 93 (neutral and calcareous soils), and P. B. Manning and M. Salomon, *Proc. Soil Sci. Soc. Amer.*, 1965, **29**, 421, for an acid soil (pH about 5) where most of the fertiliser phosphate accumulated as 'Chang and Jackson' aluminium phosphate.
2 A. W. Taylor *et al.*, *Proc. Soil Sci. Soc. Amer.*, 1964, **28**, 49; 1965, **29**, 317.
3 M. C. Gastuche, J. J. Fripiat and S. Sokolski, *Pedologie*, 1963, **13**, 155.
4 *J. Soil Sci.*, 1957, **8**, 113; but see A. Wild, ibid., 1953, **4**, 72, for an alternative explanation of the results.
5 D. Muljadi, A. M. Posner and J. P. Quirk, *J. Soil Sci.*, 1966, **17**, 238.

This isotherm follows the Langmuir equation approximately, namely

$$c/y = c/s + 1/ks$$

where c is the concentration of phosphate in the bathing solution, y is the amount adsorbed per gram of soil, and k and s are constants, k representing a bond energy and s the phosphate maximum adsorption, assumed to be as a monolayer.

For a number of soils the plot of c/y against c gives a slightly curved line rather than the straight line required by the equation, as is shown in Fig. 23.3, which could be due to the bond energy decreasing with the amount of phosphate sorbed; and D. Gunary[1] showed that for twenty-four British soils the

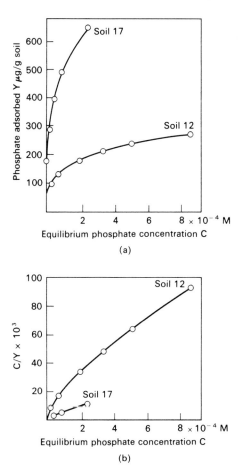

FIG. 23.3. (a) Phosphate adsorption isotherm. (b) Langmuir plot of the isotherm for two soils

1 *J. Soil Sci.*, 1970, **21**, 72.

addition of a term $a\sqrt{c}$ to the righthand side of the equation, where a is a constant for each soil, gave isotherms that fitted the experimental data very closely, though he could give no theoretical interpretation of the additional term. The importance of this work is that it defines a maximum phosphate adsorption for a soil, which is assumed to be on a monolayer, and if Gunary's equation is accepted, the simple Langmuir equation would underestimate this maximum by 30 to 50 per cent.

S. R. Olsen and F. S. Watanabe[1] had previously shown that, for the group of soils they worked with, the adsorption isotherm followed the Langmuir equation approximately when the concentration of phosphate in the bathing solution did not exceed about 5×10^{-4} M, and so were able to determine a Langmuir maximum phosphate adsorption, which increased the longer the soil was in contact with the solution. They also measured the surface area of their soils, using ethylene glycol, and found that the effective area occupied by a phosphate ion was 2200 Å^2 and 5200 Å^2 for an acid and a calcareous soil respectively, and the bonding energy of the phosphate was five times greater for the acid than the calcareous soil. This shows that the phosphate is very unevenly distributed over the soil surface. But soils will absorb much more phosphate than is needed to form a monolayer. Moreover, they found that a Davidson clay sorbed 0·8 m mole/100 g of soil as a monolayer, but 19 m mole from a concentrated solution. J. R. Woodruff and E. J. Kamprath[2] showed that the Langmuir maximum phosphate adsorption of three out the five acid soils they studied was halved if they were limed sufficiently to neutralise the exchangeable aluminium, while liming hardly affected the maximum adsorption of the other two.

The adsorption isotherm for phosphate can exhibit a very important peculiarity. If a kaolinitic clay is brought into equilibrium with phosphate at a certain concentration, and the concentration is then reduced while the pH is kept constant, less phosphate is desorbed than would be expected from the adsorption isotherm, but the amount of phosphate that is isotopically exchangeable follows the adsorption isotherm.[3] This is illustrated in Fig. 23.4. Thus, washing out of phosphate results in some of the absorbed phosphate being transferred to a non-isotopically exchangeable fraction. Yet this phosphate remains on the surface of the film for it can be displaced by other anions that can displace hydroxyls or water molecules coordinated with an aluminium or ferric ion, such as bicarbonate, citrate, or silicate.[4] This phosphate behaves as if it were held by ligand adsorption except that some of it is not isotopically exchangeable.

The phosphate isotherm can also be plotted as the amount of phosphate Q that a soil will give up to or take up from a phosphate solution over a short

1 *Proc. Soil Sci. Soc. Amer.*, 1957, **21**, 144.
2 *Proc. Soil Sci. Soc. Amer.*, 1965, **29**, 148.
3 U. Kafkafi, A. M. Posner and J. P. Quirk, *Proc. Soil Sci. Soc. Amer.*, 1967, **31**, 348.
4 S. Nagarajah, A. M. Posner and J. P. Quirk, *Proc. Soil Sci. Soc. Amer.*, 1968, **32**, 507.

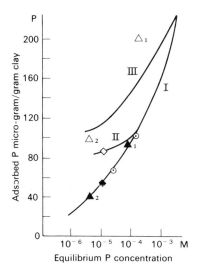

FIG. 23.4. Adsorption and desorption of phosphate by a kaolinite clay. Solution: pH 5·7, 0·01 M KCl. Curve I is the P adsorption isotherm. Curves II and III are desorption curves produced by washing out adsorbed phosphate with water. ◆ is the amount of isotopically exchangeable phosphate corresponding to the point ◇ on the desorption curve. ▲ are the amounts of isotopically exchangeable phosphate corresponding to the points △ produced by re-adsorption of P after washing. ⊙ are two points on the adsorption isotherm showing that all the adsorbed P is isotopically exchangeable

period of time, such as one or two hours, against the activity I either of monocalcium phosphate or of phosphoric acid. P. H. T. Beckett and R. E. White[1] found that for one of the soils they used, derived from the Upper Greensand, the Q/I curve was approximately linear, as is shown in Fig. 23.5, so they could determine, by extrapolation, a quantity Q_0, the amount of phosphate that could be released if the phosphate concentration was reduced to zero. The particular samples used for the results illustrated in the diagram were taken straight from the field, and the sample as a whole was clearly not in equilibrium. This is due to the very slow equilibration of phosphate over distances measured in centimetres in field soils. The slopes of these curves give the phosphate buffer capacity of the soil, which is a parameter needed when calculating the diffusion coefficient for the phosphate ion in the soil. The buffer capacity differs widely for different soils,[2] and Webber and Mattingly found that, for the soils they examined, this capacity increases with clay content and the amount of extractable phosphate the soil contains, but decreases with increasing calcium carbonate content. Beckett and White also showed

1 *Pl. Soil*, 1964, **21**, 253.
2 See, for example, M. D. Webber and G. E. G. Mattingly, *J. Soil Sci.*, 1970, **21**, 111 for soils being depleted of phosphate, and H. E. Jensen, *Pl. Soil*, 1970, **33**, 17, for soils being enriched.

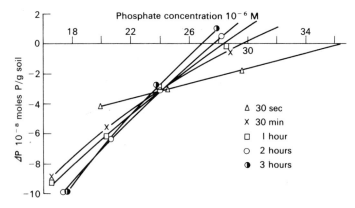

FIG. 23.5. The relation between phosphate desorbed and the phosphate concentration for a field sample of an Upper Greensand soil, after different times of contact between the soil and the solution

that if an anion exchange resin is used to extract a fraction of the phosphate from the soil, and the soil is then stored moist for three or four months, the buffer capacity and the value of Q_0 were the same as for the original soil if one-third of the original isotopically exchangeable phosphate was removed, as is shown in Table 23.1. Thus, an appreciable proportion of the isotopically exchangeable phosphate can be transferred to the readily extractable pool, without the phosphate potential of the soil being appreciably affected.

TABLE 23.1   The effect of removing phosphate on the Q/I curve. Soil: Upper Greensand. Initial isotopically exchangeable P in soil about 80×10^{-8} moles/g

Quantity of P removed moles/10^8 g soil	Period of storage after depletion days at 25°C	Activity of H_2PO_4 M × 10^{-6}	$\frac{\Delta Q}{\Delta I}$	Q_0 moles/10^8 g soil
0	120	42·5	4·3	19·0
8	126	30·0	6·0	18·5
16	117	32·7	5·0	17·0
38	94	38·5	5·5	21·5
95	95	10·0	4·5	5·0

Summary of the types of phosphate in soils

It can now be appreciated that phosphate is held in soils by a number of different mechanisms, each of which affects the concentration of phosphate in the soil solution in a different way. Soil phosphates can be put into the following groups, based on their accessibility to the soil solution and their ease of desorption:

1 The phosphate that rapidly equilibrates with the soil solution when its phosphate concentration is altered, as determined by Q/I curves with equilibration times of a few hours.
2 The phosphate that rapidly equilibrates with phosphate labelled with ^{32}P.
3 The phosphate that can be displaced by anions that themselves are capable of ligand exchange with aluminium or ferric ions.

For as far as is known all the first group is included in the second, but not all the second may be included in the third, for isotopically exchangeable phosphate sorbed by calcium ions may not be all displaceable by the anions in group 3. Strong anion exchange resins will extract all the isotopically exchangeable phosphate from a soil, and for some soils will extract more than this, but how much more, and from what other sources, is not known. But in each of these groups there is some phosphate that is not readily accessible to the soil solution because it is situated in micropores through which diffusion is a slow process.

4 Phosphate that can be brought into solution by dilute acid, dilute alkali or neutral ammonium fluoride. It is not known how much phosphate in this group is not included in the first three.
5 Phosphate that comes into solution very slowly from insoluble microcrystalline calcium, aluminium and iron phosphates present in the soil.

There are also phosphates inside mineral grains and concretions that are inaccessible to the soil solution and make no contribution to its phosphate concentration.

The source of phosphates in the soil used by crops

Phosphate dissolved in the soil solution

The immediate source of phosphates for crops growing in a soil is probably that of the inorganic phosphate ions in the soil solution. This concentration varies widely for different soils. It can be below 10^{-8} M in some very poor tropical soils,[1] it is in the order of 10^{-6} M in temperate soils known to be deficient in phosphate, and concentrations of 10^{-5} M occur in many soils of moderate phosphate status and it can exceed 10^{-4} M in some soils known to be well supplied.[2] (10^{-5} M corresponds to 0·3 ppm of P in the soil solution.) Now a soil which has a soil solution 10^{-5} M in P and which, for example, holds 6 cm of available water in the top 30 cm, will have less than 0·04 kg/ha of P in solution to that depth; and if a crop uses 37 cm of water during its growth, there will only be about 0·2 kg/ha of P dissolved in it, yet it may take

1 *Rubber Res. Inst. Malaya, Ann. Rept.* 1967, 70.
2 See, for example, J. S. Burd and J. C. Martin, *J. agric. Sci.*, 1923, **13**, 265. W. H. Pierre and F. W. Parker, *Soil Sci.*, 1927, **24**, 119. L. J. H. Teakle, *Soil Sci.*, 1928, **25**, 143. W. L. Nelson *et al., Agronomy Monog.* 4, 1953, 170.

up 10 to 20 kg/ha of P during the growing season. The soil solution can, therefore, only be an adequate source of phosphate if soil phosphate goes from the solid to the solution phase at least as quickly as the crop roots can extract it from the soil solution, and if crop roots can extract phosphate readily from these dilute solutions. It has been found that the phosphate concentration in the soil solution of many soils does, in fact, remain approximately constant when a crop is growing in it; and M. Fried and his coworkers [1] have shown that in a normal soil, moderately well supplied with phosphate, soil phosphate can go into solution at a very much higher rate than roots can take it up.

The soil solution, however, can only be an adequate source of phosphate if the crop can removè phosphate from these dilute solutions at an adequate rate. Experiments to measure the minimum concentration of phosphate needed for good plant growth are technically difficult to carry out, but it is now well established that plants differ considerably in the minimum concentration that must be maintained around their roots if they are to make satisfactory growth. Perennial ryegrass needs a higher phosphate concentration than grasses adapted to poor habitats, such as *Festuca ovina* and *Nardus stricta*,[2] although a given species can become adapted to low phosphate concentrations. Thus wild white clover (*T. repens*) native to soils low in phosphate may not respond to phosphate on that type of soil, while plants native to high phosphate soils will make poor growth but will respond strongly to added phosphate.[3] Plants adapted to poor soils can make reasonable growth if the concentration around their roots is as low as 10^{-7} M, but most agricultural crops need at least 10^{-6} M and normally need about 10^{-5} M or even higher for maximum yields.

The concentration needed for good crop growth probably also depends on the degree of dissociation of the phosphate anion. Most, possibly all, crops take up $H_2PO_4^-$ more readily than the HPO_4^{2-} ions[4] and above pH 7 the relative concentration of the divalent is greater than that of the monovalent ion. Taking the second dissociation constant of phosphoric acid as $pK_2 = 7.2$ at $18°C$,[5] the proportion of the phosphate ions in the solution present as HPO_4^{2-} are approximately pH 5, 0·6; pH 6, 6; pH 7, 39; pH 8, 86; pH 9, 98·4; and the remainder are almost entirely $H_2PO_4^-$, the proportion of H_3PO_4 and PO_4^{3-} being negligible in this pH range.

The minimum phosphate concentration needed for reasonable growth depends in part on the microbiological environment of the root. Many rhizosphere organisms, for example, will compete with the root for phosphate, when the phosphate concentration is low, reducing considerably the amount

1 *Soil Sci.*, 1957, **84**, 427.
2 A. D. Bradshaw, M. J. Chadwick *et al.*, *J. Ecol.*, 1960, **48**, 631. For other examples see I. H. Rorison, *New Phytol.*, 1968, **67**, 913; C. J. Asher and J. F. Loneragan, *Soil Sci.*, 1967, **103**, 225.
3 R. W. Snaydon and A. D. Bradshaw, *J. expt. Bot.*, 1962, **13**, 422.
4 C. A. Hagen and H. T. Hopkins, *Pl. Physiol.*, 1955, **30**, 193.
5 R. G. Bates and S. F. Acree, *J. Res. Nat. Bur. Stand.*, 1943, **30**, 129.

of phosphate available to the plant[1] (see p. 526). On the other hand, if the phosphate level is low, but not too low, mycorrhizal roots are likely to form which are more efficient phosphate scavengers than uninfected roots (see p. 259).[2] Also, under some conditions, the rhizosphere micro-organisms will contain species that will solubilise phosphate from compounds of very low solubility and so increase the phosphate supply to the plant (see p. 244). They do this by excreting acids such as 2-keto gluconic, citric, or other hydroxy-acids which chelate calcium, aluminium and ferric ions, thus increasing the solubility of phosphates associated with them. It is also possible that the mucigel gums secreted near the growing tips of some plant roots will take up phosphates adsorbed on iron and aluminium hydroxide surfaces.[3]

Phosphate uptake, particularly from solutions of low concentration, is affected by the presence of other ions. Thus, ammonium ions can assist the uptake of phosphate by maize roots, or at least its translocation from the soil to the leaves,[4] and aluminium ions can cause the precipitation of aluminium phosphate within the free space of roots.[5]

The rate of uptake of phosphate by a root depends on the phosphate concentration close to its surface, but as the root takes up phosphate it lowers the concentration there, so the rate of uptake over a period of time depends on the rate of diffusion of phosphate from the body of the soil through the water films to the root surface. The factors controlling this rate have already been discussed on p. 547. Since the rate of diffusion of phosphate in most soils is very slow, most of the phosphate taken up comes within a fraction of a millimetre of the absorbing surface. Further, the drier the soil, the more tortuous is the diffusion path for the phosphate, so the slower its rate of diffusion to the root surface, so the phosphate supply to a plant depends much more on the size of the root system, the density of its root hairs, and the intensity of its ramifications through the soil than does the supply of most other nutrients.

The rate of uptake also depends on the soil temperature, being low in cold weather. For instance, the leaves of young barley at Rothamsted have a purplish colour, due to phosphate deficiency, in a dry cool spring, but have green and healthy leaves in a warm, moist spring, although in both types of year the crop gives a good response to phosphate. The same observations have been made with potassium, for its deficiency symptom of a bright yellow leaf is also much more noticeable in cool, dry springs.[6]

The texture of the soil affects the concentration of phosphate in the soil

1 D. A. Barber with B. C. Loughman, *J. expt. Bot.*, 1967, **18**, 170; and with J. Sanderson and R. S. Russell, *Nature*, 1968, **217**, 644.
2 For a review see J. W. Gerdemann, *Ann. Rev. Phytopath.*, 1968, **6**, 397; and for Rothamsted results D. S. Hayman and B. Mosse, *New Phytol.*, 1971, **70**, 19.
3 S. Nagarajah, A. M. Posner and J. P. Quirk, *Nature*, 1970, **228**, 83.
4 F. C. Leonce and M. H. Miller, *J. Agron.*, 1966, **58**, 245.
5 For an example with barley see D. T. Clarkson, *Pl. Physiol.*, 1966, **41**, 165; with rice see P. R. Hesse, *Pl. Soil*, 1963, **19**, 205.
6 I am indebted to Dr Mattingly for this information.

solution that will allow a root to take up phosphate at a given rate, as an example given by S. R. Olsen and F. S. Watanabe[1] shows. They solved the diffusion equation giving the uptake of phosphate Q by a plant root during a time t in the form

$$Q = aB(C_0 - C_r)f(Dt/Ba^2),$$

where a is the radius of the root, B the buffer capacity of the soil, D the diffusion coefficient of phosphate in the soil, C_0 and C_r the phosphate concentrations in the body of the soil and at the root surface and f is a function of Dt/Ba^2. Table 23.2 gives the results of their calculations for three soils, a Pierre clay, a Apishapa silty clay loam and a Tripp fine sandy loam, for the value of C_0 needed by a maize root if it is to take up 31 mg P per g root per day, which is a typical figure for a fast-growing maize crop. They also give the value of C_0 for the soil, the amount of water-soluble phosphate that must be added to the soil, and the amount of phosphate extractable by sodium bicarbonate (available phosphate) the soil must contain to give the necessary value of C_0. It is interesting to note that the necessary level of available phosphate is almost independent of texture, while the amount of phosphate that must be added and the necessary concentration of phosphate at the root surface are very dependent on texture. Table 23.2 indicates that clay soils may need a higher level of phosphate manuring, so appear to be more phosphate deficient than lighter-textured soils, largely because they are more strongly buffered; yet when adequately supplied with phosphate they do not need to have such a high concentration of phosphate in the solution.

TABLE 23.2   Effect of soil texture on phosphate concentration C_0 in the soil solution for a given rate of uptake by a maize root.

Soil	Diffusion coefficient 10^{-7} cm^2/sec	Buffer capacity	C_0 10^{-6} M. Actual	C_0 10^{-6} M. Necessary	P to be added mg/100 g soil	Available P necessary mg/100 g soil
Clay	5·40	261	1·0	3·5	4·05	2·20
Silty clay loam	3·23	86·9	2·9	9·2	2·90	2·18
Fine sandy loam	1·12	28·0	7·1	27·2	2·16	1·96

Plant-available phosphate in the soil

Soil chemists for a long time have been concerned with advising farmers how much fertiliser it is economically justifiable to give a crop on a particular field and have devoted a great deal of time, for upwards of a century, devising simple and reliable methods to help them.

Initially they looked for a chemical that would dissolve the same amount

1 *Soil Sci.*, 1970, **110**, 318.

of phosphate from the soil as would the plant roots, ignoring the fact that different plants extract different amounts of phosphate from the same soil. But they soon realised that any standardised chemical extraction technique which placed the soils in the order of crop responsiveness to phosphate was all that was needed, so recent work has been concerned with determining the correlation coefficient either between the amount, or the logarithm of the amount, of phosphate extracted using various techniques and either the responsiveness of a crop to a phosphate fertiliser or the yield or phosphate uptake of a crop grown on a range of soils, usually representative of a region or a country. The crops may either be grown in field trials or in pots in a greenhouse, and the chemist chooses the method that gives the highest correlation coefficient. If the experiment is done with crops in the field the correlation coefficients rarely exceed 0·7.

It is now clear that there cannot be a universal simple and reliable method of soil analysis that will allow an accurate forecast of the amount of phosphate a crop can take up from a soil, for this depends, as already noted, not only on the phosphate concentration in the soil solution and its rate of diffusion to the root surface, but also on the extensiveness of the root system and the amount of root hairs it carries; and this depends on soil and climatic factors unrelated to its phosphate status. Thus, potentially high yielding crops growing in favourable seasons on shallow soils will make quite different demands on the soil phosphate resources than a lower yielding crop in an unfavourable season on a deeper soil.

Methods of analysis for the level of available phosphate in soils are in most demand for annual crops, many of which need a good supply early on in their growth period, so they can only draw on phosphate that is readily accessible and desorbable. The principle underlying the methods in common use for temperate soils involves determining the amount of phosphate a soil sample, usually ground to pass a 2 mm sieve, will release to a mild extractant using short extraction times and a small volume of extractant per unit weight of soil. The extractants used include 0·5 M sodium bicarbonate,[1] probably the most generally useful for soils that are not too acid, dilute mineral acids such as hydrochloric, dilute hydroxy-acids or their acid salts, such as lactic,[2] dilute citric acid, acid fluoride solutions,[3] and anion exchange resins that are not too strong an absorbant for phosphate.[4] In addition, the isotopically-exchangeable phosphate is used, or a seedling plant is grown in a soil to which a known amount of labelled phosphate has been added, and the ratio of ^{32}P to ^{31}P in the plant is determined.[5]

1 S. R. Olsen, C. V. Cole *et al.*, *U.S. Dept. Agric.*, *Circ.* 939, 1954. R. A. Olson, M. B. Rhodes *et al.*, *Agron. J.*, 1954, **46**, 175.
2 H. Egner, H. Riehm, and W. R. Domingo, *Kungl. Lantbr. Hogsk. Ann.*, 1960, **26**, 204.
3 R. H. Bray and L. T. Kurtz, *Soil Sci.*, 1945, **59**, 39.
4 F. Amer, D. R. Bouldin *et al.*, *Pl. Soil*, 1955, **6**, 391. I. J. Cooke and J. Hislop, *Soil Sci.*, 1963, **96**, 308. B. W. Bache and N. E. Rogers, *J. agric. Sci.*, 1970, **74**, 383.
5 S. Larsen, *Pl. Soil*, 1952, **4**, 1; 1961, **14**, 43; 1963, **18**, 77; 1964, **20**, 135.

Finally, the phosphate potential of the soil solution when the soil is equilibrated with a standard salt solution, such as 0·01 M calcium chloride, has been used. The soil solution contains much less phosphate than a crop will take up from the soil, whereas the amounts extracted by a number of the other extractants, or estimated by the two radioisotope methods, are of the same order of magnitude as crop uptake. However, if the soil phosphate is equilibrated with some carrier-free labelled phosphate, the $^{32}P/^{31}P$ ratio in the calcium chloride solution is the same as in the bicarbonate extract, although this extracted ten times as much phosphate as the chloride.[1]

All these methods have inherent limitations for as far as the adviser or the farmer is concerned. This is shown in Table 23.3 for the response of sugar-beet to phosphate in eastern England.[2] The crop is only grown on soils that

TABLE 23.3   The frequency of response of sugar-beet to phosphate fertilisers in groups of experiments arranged by soil analysis. Responses to 55 kg/ha P in kg/ha sugar.

P soluble in 0·3 N HCl	Number of experiments	Mean response	Number of centres with response			
			Over 340	330–180	170–0	Depression
0·2–2·5	53	490	27	14	6	6
2·7–4·7	54	190	13	11	15	15
4·8–8·0	54	90	6	15	18	15
Over 8·0	55	90	7	14	11	23

are not far from neutral and usually of light texture, but the table shows that the crop may not respond to phosphate in soils low in available phosphate and may respond on soils that are high. It is probable that much of this scatter in response is due to factors other than the availability of phosphate, so it would not be appreciably reduced by any other single determination. Thus, D. A. Boyd and W. Dermott[3] found that the depth of free-draining soil can have a large influence on the responsiveness of potatoes to superphosphate, as is shown in Table 23.4. The actual value of the available phosphate, as determined by some methods such as anion resin extraction, also depends on the time of year the soil sample is taken; at Rothamsted and Woburn it may vary by a factor as large as two or three, being at a maximum in late winter.[4]

1 D. A. Nethsinghe quoted by G. E. G. Mattingly and A. Pinkerton, *J. Sci. Fd. Agric.*, 1961, **12**, 772. See also E. G. Williams and A. H. Knight, ibid., 1963, **14**, 555.
2 R. G. Warren and G. W. Cooke, *J. agric. Sci.*, 1962, **59**, 269. The dilute HCl was the most efficient of the extractants used; bicarbonate soluble P was not determined in this series of experiments.
3 *J. agric. Sci.*, 1964, **63**, 249.
4 I. P. Garbouchev, *J. agric. Sci.*, 1966, **66**, 399. M. Blakemore, ibid., 139.

TABLE 23.4   Response of potatoes to 55 kg/ha P as superphosphate. Response in t/ha

Free draining to (in cm)	Lowland England		Wales and N.W. England	
	Response	No. of fields	Response	No. of fields
Below 45	1·6	34	5·2	19
30–45	2·9	33		
Less than 30	4·4	29	7·8	9

The general experience in Great Britain is that the bicarbonate method is the most appropriate for a wide range of crops and soils.[1] Responses to phosphate fertilisers are likely to be small when the solution extracts contain more than about 20, 30 or 45 ppm P in the soil for grass, potatoes or swedes,[2] and 40 ppm for sugar-beet.[3] G. W. Cooke,[4] who has recently re-examined much British field data, concluded that appreciably lower figures are usually adequate, such as 10, 15, and 20 ppm for grass swards without legumes, cereals and sugar beet, and potatoes and outdoor vegetables respectively. An appropriate fertiliser policy would therefore be to apply sufficient phosphate at regular intervals to maintain the level of available phosphate in the soil, as determined by some standard method, at a level known to be adequate for the system of cropping that is to be used. In some parts of Great Britain the acid lactate is marginally better on acid soils, but on a number of soil types no method has any practical predictive value.

Attempts have been made to estimate the amount of phosphate that should be added to a soil to give near maximum response, and these have been based on determining the amount of phosphate a soil sample must absorb to become in equilibrium with a solution 10^{-5} M in P, after a fixed period of time such as one day.[5] This measurement picks out soils that absorb phosphate strongly, so need high levels of phosphate manuring,[6] and often shows that acid soils need more phosphate than neutral or calcareous soils to raise their equilibrium concentration to 10^{-5} M, which is at least sometimes in accord with field experience. This measurement is, in fact, closely correlated with the phosphate buffer capacity of the soil as determined from the Q/I curve. However, crops differ in the concentration of phosphate they need for a high

1 R. J. B. Williams and G. W. Cooke, *J. agric. Sci.*, 1962, **59**, 275. D. A. Boyd, *Min. Agric. Fish. Fd.*, *Tech. Bull.* 13, 1965. S. McConaghy and J. W. B. Stewart, *J. Sci. Fd. Agric.*, 1963, **14**, 329. For a recent review see L. J. Hooper, *Fertiliser Soc.*, *Proc.* 118, 1970.
2 R. J. B. Williams and G. W. Cooke, *J. agric. Sci.*, 1962, **59**, 275.
3 A. P. Draycott, M. J. Durrant and D. A. Boyd, *J. agric. Sci.*, 1971, **77**, 117.
4 *Fertilizing for Maximum Yields*, Crosby Lockwood, 1972.
5 P. G. Ozanne and T. C. Shaw, *Trans. 9th Int. Congr. Soil Sci.*, 1968, **2**, 273. R. L. Fox and E. J. Kamprath, *Proc. Soil Sci. Soc. Amer.*, 1970, **34**, 902
6 R. L. Fox, D. L. Plucknett and A. S. Whitney, *Trans. 9th Int. Congr. Soil Sci.*, 1968, **2**, 301.

yield, and their ability to take up phosphate depends on the amount of active aluminium in the soil, so if this method is used, these two factors must be considered when deciding on the appropriate equilibrium concentration.

The development of methods of general applicability to tropical soils has been more difficult than in temperate, because although in some regions methods developed in the temperate regions such as dilute acid extracts, phosphate potential, and moderately strong anion exchange resins are useful, in other areas they have no value whatever for predicting phosphate response. In some areas part of the reason is that the phosphate released by the mineralisation of organic matter at the beginning of the rains can make a much larger contribution to the plant available phosphate than in most temperate soils.[1] Some types of soil hold their phosphate more strongly than temperate soils, but the crops grown can extract this phosphate, so extractants such as 0·1 N sodium hydroxide, or very strongly absorbing anion exchange resins[2] must be used in place of the less strongly absorbing resin normally used. For some soils no phosphate extraction method appears to have any value, but there may be an inverse linear relation with pH, phosphate responses ceasing when the pH is above about 6·5.[3] In other soils the level of water-soluble silica appears to be the most useful determination, the phosphate response falling as the silica concentration rises.[4] As a result of these difficulties, it is often more reliable to use field experiments to determine phosphate response and to divide the region up into subregions where, on the whole, an economic response is likely or is not likely.[5]

Phosphate fertilisers

The principal primary sources of phosphate for fertilisers are certain mineral rock phosphates derived from the phosphate of organisms living in past geological eras, and containing up to 80 per cent of apatite, usually in the form of fluor-apatite. Important commercial sources of supply are the deposits in North Africa, America, certain Pacific islands such as Nauru and Christmas Island, and the Kola Peninsula in USSR. These mineral phosphates have only a limited use as fertiliser, as the apatite crystal has a very low solubility—it will maintain a concentration of about 10^{-7} M in

1 M. T. Friend and H. F. Birch, *J. agric. Sci.*, 1960, **54**, 341. P. H. Nye and M. H. Bertheux, *J. agric. Sci.*, 1957, **49**, 141.
2 D. H. Saunder, *Soil Sci.*, 1956, **82**, 457; with H. R. Metelerkamp, *Int. Soc. Soil Sci. Trans. Comm. IV & V*, 1962, 847.
3 D. Stephens, *E. Afr. Agric. For. J.*, 1969, **34**, 401. J. B. D. Robinson, ibid., 436, P. K. Garberg, ibid., 1970, **35**, 396.
4 H. F. Birch, *J. agric. Sci.*, 1953, **43**, 229, 329; P. K. Garberg, *E. Afr. Agric. For. J.*, 1970, **35**, 396.
5 See, for example, P. R. Goldsworthy and R. Heathcote, *Emp. J. exp. Agric.*, 1963, **31**, 351 (groundnuts). *Expl. Agric.*, 1967, **3**, 29 (sorghum), 263 (maize) for N. Nigeria. M. A. Scaife, *J. agric. Sci.*, 1968, **70**, 209 (maize) for W. Tanzania.

phosphate if shaken up with water,[1] so it is necessary to break up this crystal lattice before they can be used as a general fertiliser. This is done either by treatment with a mineral acid or by high temperature sintering processes. Far the most widely used process in the past was to treat the rock with enough sulphuric acid to convert the apatite into the water-soluble monocalcium orthophosphate monohydrate, the excess calcium reacting with the sulphuric acid to give gypsum. This is the superphosphate of commerce, which has been manufactured since the 1840s and contains between 28 and 32 per cent of the monophosphate and 50 to 60 per cent gypsum. This fertiliser is sold in the United Kingdom on a guaranteed content of water-soluble P_2O_5, which is usually between 18 and 22 per cent, corresponding to 8 to 10 per cent water-soluble P.

Since about 1930 increasing amounts of other water-soluble phosphate fertilisers have been coming on to the market, based on treating the rock phosphate with sufficient sulphuric acid to convert the phosphate to phosphoric acid. This phosphoric acid may then be added to more rock phosphate to give a much more concentrated superphosphate, containing up to 85 per cent of monocalcium phosphate or 21 per cent water-soluble P; or it may be neutralised with ammonia to give either mono- or diammonium phosphates, or a mixture of the two, which in practice contain about 21 per cent water-soluble P; or it may be concentrated to give superphosphoric acid, which when neutralised with ammonia gives a mixture of ammonium pyrophosphate, tripolyphosphate and higher polyphosphates. These may contain up to 26 per cent water-soluble P and are coming into use as concentrated liquid fertilisers, as they are more soluble than the ammonium orthophosphates.

A further group of phosphates are based on the water-insoluble dicalcium phosphate which will maintain a phosphate concentration of about 10^{-3} M in solution. The normal reason for their manufacture is to save sulphuric acid, for if rock phosphate is treated with nitric or hydrochloric acid, the calcium nitrate or chloride admixed with the monocalcium phosphate is too hygroscopic to allow the mixture being used as a fertiliser, and the processes used to remove these involve the conversion of the monocalcium to the anhydrous dicalcium phosphate.[2]

High-temperature phosphates, all of which are water insoluble, are also manufactured to a limited extent. Probably the commonest of these is made by fusing rock phosphate with silica, using soda as a flux, to give a calcium silicophosphate and calcium silicate, known as silicophosphate, sodaphosphate or Rhenania phosphate. Rock phosphate is sometimes fused with serpentine or magnesium silicate, and the molten slag rapidly quenched to give a glass, which is then ground. Other high temperature phosphates are

1 K. D. Jacob and W. L. Hill, *Agronomy Monog.*, **4**, 1953, 301.
2 M. H. R. J. Plusje, *Fertiliser Soc.*, *Proc.* 13, 1951, W. D'Leny, ibid., *Proc.* 24, 1953, and F. T. Nielsson, L. D. Yates *et al.*, *J. Agric. Food Chem.*, 1953, **1**, 1050.

tricalcium phosphate and calcium metaphosphate $Ca(PO_3)_2$, which if pure contains over 31 per cent P, and potassium metaphosphate,[1] which contains about 25 per cent P and 35 per cent K, which is in limited commercial use in areas where transport costs of the fertiliser are high, and valuable horticultural and fruit crops are grown; for it is one of the most concentrated phosphate fertilisers manufactured. Both these metaphosphates hydrolyse to orthophosphate in the soil. Most of these will maintain a phosphate concentration of between 10^{-5} and 10^{-4} M in solution.[2]

The commonest high-temperature phosphates of commerce are basic slags, which are byproducts of the steel industry. They have a variety of compositions, and correspondingly of agricultural value,[3] but the most useful again contain calcium silicophosphate and calcium silicate, and the less useful crystalline fluor-apatites. These are liming materials as well as phosphate fertilisers, for the calcium silicate hydrolyses readily to calcium hydroxide and silica, and the silicophosphate is probably a solid solution of very variable composition.

The reaction between water-soluble phosphate fertilisers and the soil

When a granule of a fertiliser containing a water-soluble phosphate is added to a soil, a complex series of chemical reactions may take place in the soil in its neighbourhood. The granule will absorb water from the soil into which the phosphate will dissolve to give a saturated or nearly saturated solution, and it will then diffuse out from this granule into the soil solution. If the phosphate is monocalcium phosphate, it dissolves incongruently giving a solution with a pH between 0·6 and 1·5, which is between 3 and 4·5 M in phosphate and approximately 1 M in calcium, leaving a residue of dicalcium phosphate.[4] If the phosphate is ammonium phosphate, the saturated solution has a pH between 4 and 8, depending on the ratio of ammonium to phosphate.[5] The solution diffusing from a granule of monocalcium phosphate is very reactive, and, if the soil is non-calcareous, will dissolve iron and aluminium compounds in the soil which react with the phosphoric acid to give precipitates of potassium-containing phosphates, such as the aluminium-containing taranakite $(H_6K_3Al_5(PO_4)_8 . 18H_2O$ and the iron-aluminium phosphate $H_8K(Al, Fe)_3(PO_4)_6 . 6H_2O$, as well as amorphous aluminium phosphate and variscite. It is probable that in many temperate soils taranakite is the first compound precipitated, and only after all the exchangeable potassium has been used up are the simple aluminium phosphates formed;[6] it is also

1 See F. J. Harris, *Fertiliser Soc., Proc.* 76, 1963.
2 Quoted from K. D. Jacob and W. L. Hill, *Agronomy Monog.*, 1953, **4**, 301.
3 See, for example, G. G. Brown and K. F. J. Thatcher, *Fertiliser Soc., Proc.* 96, 1967.
4 W. L. Lindsay and H. F. Stephenson, *Proc. Soil Sci. Soc. Amer.*, 1959, **23**, 12, 18. J. R. Lehr, W. E. and E. N. Brown, ibid., 3.
5 E. O. Huffman, *Fertiliser Soc., Proc.* 71, 1962; *J. Agric. Fd. Chem.*, 1963, **11**, 182.
6 A. W. Taylor and E. L. Gurney, *Proc. Soil Sci. Soc. Amer.*, 1965, **29**, 18.

probable that the taranakite and other aluminium phosphates are precipitated in preference to the ferric phosphates, unless the soil contains much amorphous ferric hydroxide.

The distance the phosphoric acid diffuses from a fertiliser granule depends on the amount of water-soluble phosphate in the granule and on its size, but E. C. Sample and A. W. Taylor[1] found that for 6 mm granules containing 70 per cent of water-soluble monocalcium phosphate in a silt loam soil, the diffusion zone had a radius of about 17 mm after three weeks. This diffusion zone can be traced quite easily by autoradiography if some ^{32}P is incorporated into the phosphate, as has been shown by D. R. Bouldin and C. A. Black,[2] although they also found that the phosphate was not always precipitated uniformly in a spherical shell, but was sometimes concentrated more in rings, similar to Liesegang rings or in a series of spots.

The consequence of this movement of phosphate from fertiliser granules into the soil by diffusion is that, for a period after the phosphate has been added, there is a great variability in the phosphate potential from point to point in the soil, being much lower near each granule than in the bulk of the soil. Thus, the phosphate potential of a fertilised soil can be pictured as a plateau of fairly constant potential with troughs of lower potential around each fertiliser particle; it is from these troughs that the plant roots can most easily take up their phosphate. This extremely patchy distribution would be expected to last a fairly long time in an undisturbed soil because of the low solubility of the phosphate in the soil water and the low diffusion coefficient due to the strong phosphate buffering of the soil.

There has been much discussion in the past about whether adding a soluble phosphate fertiliser, and particularly superphosphate, to a soil makes it more acid. Its long-continued use on plots at Rothamsted and Woburn has not had any measurable effect on the soil pH, though a small acidifying effect might have been anticipated as the fertiliser may contain some free acid. In so far as the monocalcium phosphate reacts with iron or aluminium hydroxide surfaces, this should set free the calcium so increasing the calcium status of the soil, but this amount is only about 4·5 kg/100 kg single superphosphate and about 12 kg/100 kg concentrated superphosphate. On the other hand, in neutral soils, if the monocalcium phosphate is converted to hydroxyapatite, this will remove about 15 kg of exchangeable calcium for every 100 kg of single superphosphate that is so converted, so one would expect to find the pH of the soil reduced, particularly if it is rather poorly buffered. E. O. Alban,[3] for example, found that ten annual applications of 80 kg/ha P as triple superphosphate reduced the pH of a loam soil from 7·04 to 6·62 in the top 15 cm, with smaller reductions down to 45 cm. Similarly, S. Larsen[4]

1 *Proc. Soil Sci. Soc. Amer.*, 1964, **28**, 196. G. L. Terman, *Fertiliser Soc., Proc.* 123, 1971.
2 *Proc. Soil Sci. Soc. Amer.*, 1964, **18**, 255.
3 *Soil Sci.*, 1961, **92**, 212.
4 *Pl. Soil*, 1964, **21**, 37.

found that on neutral and calcareous soils, heavy dressings of triple super-phosphates increased the manganese uptake by oats, probably due to the reduction in the pH of the soil that they brought about.

Superphosphate, however, can increase soil acidity in the special case when its use allows the level of soil organic matter to increase. On some very phosphate-deficient soils in Australia, the use of superphosphate on the pastures allows more carbon to be converted into humus, with the concurrent locking up of fertiliser phosphate in organic forms and the creation of humic carboxylic acids. C. M. Donald and C. H. Williams,[1] in New South Wales, found that in some soils every 100 kg/ha of superphosphate added, containing about 7·5 kg P, increased the humus content by about 2 t/ha which would need about 90 kg/ha calcium to bring its pH to neutrality, yet the super-phosphate only contained about 22 kg/ha Ca, most of which was present as gypsum. As a consequence the pH of the soils fell by about 0·045 units for every 100 kg/ha superphosphate added.

Comparative value of different phosphate fertilisers

The relative manurial value of different phosphate fertilisers is usually compared with that of monocalcium phosphate present in superphosphate, because this is still the most widely used phosphate fertiliser. The earlier experiments used powdered single superphosphate and the more recent have used granular concentrated superphosphate as the standard. These two, how-ever, need not be equivalent because of the different chemical interactions between a powdered and a granular superphosphate, but general experience has been that the granular is more effective than the powder, particularly when placed in bands close to the seed and early on in the growing season. This is presumably because the phosphate in the volume of soil close to the granule, which is in the form of taranakite or amorphous aluminium phos-phate, is as available to crops as is the sorbed phosphate derived from the powder, a result established by A. W. Taylor[2] for taranakite, though the aluminium phosphates were only as effective on neutral and calcareous soils, being less effective on acid soils.

There is some evidence that ammonium phosphates are more valuable than calcium monophosphate on some soils. Part of the reason may be that the phosphate solution diffusing from an ammonium phosphate granule is much less acid than from the calcium phosphate, so will dissolve little, if any, aluminium. On calcareous soils, the ammonium may make the soil in the immediate neighbourhood a little more acid as it becomes nitrified,[3] and there is some evidence that the presence of ammonium ions around the root surface increases the uptake of phosphate anions by the root cells, perhaps

1 *Aust. J. agric. Res.*, 1954, **5**, 664; 1957, **8**, 179.
2 With W. L. Lindsay *et al.*, *Proc. Soil Sci. Soc. Amer.*, 1963, **27**, 148.
3 For an example with maize see G. L. Terman, *Fertiliser Soc., Proc.* 123, 1971.

because the uptake of ammonium ions by the root involves its exchange for hydrogen ions, so makes the soil more acid and raises the concentration of phosphate ions in its neighbourhood.[1]

The effect of granule size on the effectiveness of a phosphate fertiliser depends on its solubility; the less soluble the phosphate the larger must be its surface area in contact with the soil solution, if it is to be quick acting. Thus, a granular dicalcium phosphate is much less effective than a granular water-soluble phosphate. But the two may be equally effective if used as a powder intimately mixed with the soil,[2] though, as shown in Table 23.5, the dicalcium phosphate takes longer to exert its full effect than the water-soluble monoammonium phosphate; and, in fact, the finest granules of dicalcium phosphate may be more effective than those of the water-soluble phosphate for the second crop.

TABLE 23.5 Effect of size of granule on rate of uptake of phosphate by a crop. Mean of 2 soils: Hartsell fine sandy loam and Mountview silt loam. Two successive crops of oats. Relative uptake in arbitrary units

Granule size in mm	2·0–1·2		0·59–0·42		0·30–0·25	
crop	First	Second	First	Second	First	Second
Monoammonium phosphate	608	211	423	246	331	217
Dicalcium phosphate	30	57	90	253	136	376

The fertiliser value of so-called water-insoluble phosphates depends on their solubility in water. A saturated solution of dicalcium phosphate has a phosphate concentration of about 10^{-3} M, which is between 10 and 100 times more concentrated than the phosphate concentration in the soil solution in soils from which crops can take all the phosphate they need for their growth. One would expect, therefore, that this fertiliser, if added as a powder, should be as effective as powdered superphosphate, at least a few weeks after it has been added to a soil; and the results of field experiments are reasonably consistent with this. High temperature silicophosphates, such as basic slags, phosphates made by fusing rock phosphate with silica sand or serpentine, and some other high temperature phosphates, such as calcium metaphosphates, will maintain a phosphate concentration of the order of 10^{-5} M in their neighbourhood, while rock phosphates maintain concentrations of about 10^{-6}–10^{-7} M in neutral or mildly acid soils. These phosphates are all more soluble in acid than neutral conditions, so tend to be of more value in acid than neutral or calcareous soils; and they are of more value to crops adapted to taking their phosphate from relatively dilute rather than relatively

1 D. Riley and S. A. Barber, *Proc. Soil Sci. Soc. Amer.*, 1971, **35**, 301.
2 Taken from D. R. Bouldin, J. D. De Ment and E. C. Sample, *J. agric. fd. Chem.*, 1960, **8**, 470.

concentrated solutions, and to taking their phosphate throughout the growing season rather than having a high demand when young. Table 23.6 illustrates these points [1] and shows that swedes will use the more insoluble phosphates more efficiently than potatoes, and that for both crops the insoluble fertilisers become of decreasing value as the soil pH rises towards neutrality. It also shows that dicalcium phosphate has approximately the same value as superphosphate even on neutral soils.

TABLE 23.6   The effect of soil and crop on the relative values of phosphate fertiliser. Field experiments in the United Kingdom 1951–53. Kg P from superphosphate to give the same response as 100 kg P from the fertiliser

Soils	Swedes			Potatoes		
	Very acid $<pH$ 5·5	Acid pH 5·5– 6·5	Neutral $>pH$ 6·5	Very acid $<pH$ 5·5	Acid pH 5·5– 6·5	Neutral $>pH$ 6·5
No. of expts.	10	22	3	10	15	9
Dicalcium phosphate	97	85	95	122	62	84
Silico- phosphate	90	84	52	92	56	30
Gafsa rock phosphate	91	86	12	34	37	4

The fertiliser value of mixtures presumed to contain dicalcium phosphate have, however, raised a number of problems, for some have a smaller value than would be expected if they were pure dicalcium phosphate. A part of the reason lies in the granular size of the mixture, for dicalcium phosphate is of low value if present in a compact granule; and a part sometimes lies in the chemical process producing the dicalcium phosphate fertiliser, for some of the phosphate may be precipitated as the much less soluble hydroxy-apatite. Table 23.7 illustrates the phosphate fertiliser value of three nitrophosphate fertilisers in which the water-insoluble moiety was intended to be dicalcium phosphate. [2] It shows that barley was only responding to the water-soluble moiety though it responded to dicalcium phosphate, while ryegrass gave the same response to the insoluble part as it did to the dicalcium phosphate.

Table 23.7 also raises a problem of very great practical importance, namely, the warranty farmers should be given by the manufacturer about the composition of phosphate fertilisers. In the United Kingdom all phosphate fertilisers sold must be accompanied by a declaration of their total and their

1 From G. W. Cooke, *J. agric. Sci.*, 1956, **48**, 74.
2 From G. E. G. Mattingly and A. Penny, *J. agric. Sci.*, 1968, **70**, 131.

water-soluble content of phosphate, a requirement based on the assumption that only the water-soluble part is of value for many crops on many soils. Other countries, however, have other requirements, based on the assumption that water-insoluble phosphates, such as dicalcium phosphate, are just as valuable for many purposes, and that these should be available to the farmer if he can obtain an economic benefit from their use; and these countries have assumed that the value of these elements could be assessed either by their solubility in neutral or in alkaline ammonium citrate,[1] but, as can be seen from Table 23.7, neither test is of any value for barley though, with the exception of rock phosphate, the neutral ammonium citrate is useful for ryegrass.

TABLE 23.7   Fertiliser value of citrate soluble phosphate. Rothamsted Soils (acid)

Fertiliser	% total P soluble in			Kg P as superphosphate to give same response as 100 kg P in fertiliser	
	Water	ammonium citrate			
		Neutral	Alkaline	Barley	Ryegrass
Dicalcium phosphate	1	96	96	60	118
Nitrophosphate (a)	5	93	77	1	107
(b)	26	100	91	25	111
(c)	50	100	95	53	93
Basic slag	—	93	48	22	93
Gafsa rock	—	18	3	7	96

Table 23.7 also illustrates another point of great importance, which is that no one single test can assess the fertiliser value of a phosphate fertiliser, for it depends on the crop: a test that is satisfactory for ryegrass cannot be satisfactory for barley. It is probable that, eventually, the manufacturer will have to guarantee the chemical composition of the phosphate the fertiliser contains, leaving it to the farmer to choose the actual compound used on the farm. This problem will also arise if polyphosphates come into use, for neither water-solubility nor citrate-solubility measure their fertiliser value adequately.

Rock phosphates differ considerably in their fertiliser value ranging from samples that are ineffective on all soils and for all crops to others that can be as good as superphosphate for some crops on soils with a pH below 6. The essential characteristic that determines their value is probably their content of phosphate-bound carbonate: the higher this is, the more available or

1 For reviews of the value of these tests see P. F. J. van Burg and G. E. G. Mattingly, *Fertiliser Soc., Proc.* 75, 1963, and G. L. Terman, W. M. Hoffman and B. C. Wright, *Adv. Agron.*, 1964, **16**, 59.

soluble is the phosphate.[1] To some extent this is correlated with the solubility of the phosphate in neutral ammonium citrate.[2] But the fertiliser value of rock phosphates is erratic; it is sometimes ineffective on acid soils with crops where one would have expected it to be suitable.[3] It is a fertiliser of strictly limited value for intensive temperate agriculture, since it cannot maintain a sufficiently high phosphate concentration in the soil solution for high yields of crops with a high initial phosphate demand; and it is usually inert in neutral and calcareous soils.[4] It can, however, be suitable for many acid soils in tropical agricultural systems, where very high yields are not sought and sulphur is not deficient, because it is the cheapest source of phosphate. Basic slags also differ greatly in the value of the phosphate they contain, only that soluble in 2 per cent citric acid or possibly neutral ammonium citrate having appreciable value as a phosphate fertiliser.

Increasing interest is being taken in polyphosphates both because they are very concentrated phosphate fertilisers and because they are very water-soluble, and the phosphate is not easily precipitated by polyvalent cations,[5] as these are chelated by the polyphosphates. Polyphosphates more condensed than pyrophosphate hydrolyse fairly rapidly in soils. Thus, trimeta-phosphate hydrolyses to tripolyphosphate within one to two days, and tripolyphosphate to pyro- and ortho- within about one week.[6] Pyrophosphate hydrolyses more slowly, particularly in cold or cool soils, its rate of hydrolysis increases with temperature up to about $35°C$,[7] and is higher in acid than calcareous soils.[8] The hydrolysis is enzymatic, and is faster the higher the biological activity in the soil,[9] and it can also occur on or within the roots of some crops, for example, wheat, barley and maize,[10] though little is yet known about the rates different crops can take up pyrophosphate compared with orthophosphate from solution. Trimetaphosphate is not absorbed by soils, so can move fairly readily in the soil until it is hydrolysed, and tripoly- and pyrophosphate are adsorbed more strongly than orthophosphate by most, but not by all, soils.[11] Thus, in general orthophosphates penetrate deeper into soil crumbs than does pyrophosphate. However, not enough studies have yet been made to know how generally valid the above statements are.

1 J. H. Caro and W. L. Hill, *J. agric. fd. Chem.*, 1956, **4**, 684. R. H. Howeler and C. M. Woodruff, *Proc. Soil Sci. Soc. Amer.*, 1968, **32**, 79.
2 G. L. Terman, S. E. Allen and O. P. Engelsted, *Agron. J.*, 1970, **62**, 390.
3 For a review see G. W. Cooke, *Fertiliser Soc.*, *Proc.* 92, 1966.
4 G. W. Cooke, *J. agric. Sci.*, 1956, **48**, 74; with F. V. Widdowson, ibid., 1959, **53**, 46.
5 A. V. Slack, J. M. Potts and H. B. Shaffer, *J. agric. fd. Chem.*, 1965, **13**, 165.
6 R. W. Blanchar and L. R. Hossner, *Proc. Soil Sci. Soc. Amer.*, 1969, **33**, 141, 622.
7 C. D. Sutton, D. Gunary and S. Larsen, *Soil Sci.*, 1966, **101**, 199.
8 L. R. Hossner and J. R. Melton, *Proc. Soil Sci. Soc. Amer.*, 1970, **34**, 801.
9 C. D. Sutton, and S. Larsen, *Soil Sci.*, 1964, **97**, 196; 1966, **101**, 199. I. Hashimoto, J. D. Hughes and O. D. Philen, *Proc. Soil Sci. Amer.*, 1969, **33**, 401.
10 J. W. Gilliam, *Proc. Soil Sci. Soc. Amer.*, 1970, **34**, 83.
11 R. W. Blanchar and L. R. Hossner, *Proc. Soil Sci. Soc. Amer.*, 1969, **33**, 141, 622. D. Gunary, *Nature*, 1966, **210**, 1297.

The movement of fertiliser phosphate in the soil

Phosphates from a phosphate fertiliser only move very slowly from where the fertiliser has been placed, unless the soil is cultivated or disturbed, because of the very low solubility of the soil phosphates. The soil solution of moderately fertile soils in S. E. England is about 5×10^{-6} M in P, so that if 30 cm of water drains through the soil annually, it will only move about 0·5 kg/ha P. This has the consequence that little phosphate will be transferred from the surface to the subsoil by leaching. However, the soil solution may contain more organic phosphorus compounds in solution than inorganic, and field evidence indicates that in soils receiving regular additions of organic matter, whether from crop residues as in grassland or from farmyard manure in arable, phosphate becomes somewhat more mobile.

These points can be illustrated from some Rothamsted results. Table 23.8 gives the total phosphate content of four groups of plots on Barnfield[1] which

TABLE 23.8 The movement of phosphate down the soil profile under arable cropping. Rothamsted: Barnfield: permanent mangolds. Superphosphate and manure added annually since 1845. Total P in soil (mg/100 g) in 1959

Depth in cm	*No P*	*35 t/ha farmyard manure*	*33 kg/ha P*	*35 t/ha manure 33 kg/ha P*
0–22·5	67	126	121	188
22·5–30	45	60	51	75
30–37·5	43	50	47	59
37·5–45	41	43	40	47
	P soluble in 0·01 M CaCl$_2$ in 10^{-6} M			
0–22·5	0·5	12·3	2·6	21·9

has been in root crops for most years since 1845, two of which have received 33 kg/ha P as single superphosphate for 115 years and two farmyard manure containing about 40 kg/ha P for this period. There is no evidence that any phosphate has moved below 37·5 cm on the plots receiving all their phosphate as superphosphate, and there was no great movement below 30 cm. In the plots receiving all their phosphate as farmyard manure there was appreciably more downward movement, and in the plots with both superphosphate and farmyard manure there was clear evidence of movement below 37·5 cm. Table 23.8 also gives the equilibrium concentration of phosphate in a 0·01 M CaCl$_2$ solution, and this is appreciably higher on plots receiving farmyard manure, yet the amount of phosphate that can be extracted with Olsen's bicarbonate method is about the same for the farmyard manure as for the superphosphate plots.

1 G. W. Cooke and R. J. B. Williams, *Water Treatm. Exam.*, 1970, **19**, 253. *Rothamsted Rep.* 1963, 56. See also G. W. Cooke *et al.*, *J. Soil Sci.*, 1958, **9**, 298, and *Rothamsted Rept.* 1969, Part II, 91; 1970, Part II, 68, for some Saxmundham data.

Table 23.9 shows that under grassland some phosphate from super-phosphate moves down below 37·5 cm, and it also shows that liming an acid soil does not appear to affect this downward movement.[1]

TABLE 23.9   Movement of phosphate down the soil profile under permanent grass. Rothamsted: Park Grass. Phosphate added annually since between 1856 and 1859. About 33 kg/ha P as superphosphate. Sampled in 1959

Manuring	No lime				Limed every 4th year since 1903			
	No P	P	P+NH$_4$	P+NO$_3$	*No* P	P	P+NH$_4$	P+NO$_3$
Depth in cm	Total P in soil (mg/100 g)							
0–22·5	50	136	140	125	57	149	122	142
22·5–30	45	99	87	72	56	93	78	74
30–37·5	41	73	70	59	51	80	65	67
37·5–45	40	57	57	48	49	58	57	52
	pH							
0–22·5	5·0	5·1	3·7	5·7	7·2	7·0	5·1	7·2
22·5–45	5·4	5·2	4·2	5·9	6·8	6·5	5·1	6·9

These results demonstrate two important points. The first is that phosphate fertilisers applied to a soil can hardly affect the phosphate content of water draining out of the land into springs and rivers, although some fertiliser could be washed off the surface by heavy rain just after it has been applied. The second is that, unless the soil is very sandy and low in iron and aluminium hydroxides, the level of phosphate in the subsoil can only be increased by ploughing-in a phosphate fertiliser deep, or by injecting it either as a liquid or a solid behind a cultivator or subsoil tine. This very slow movement of phosphate through a soil profile may only apply to well-drained soils. Phosphate may move down seasonally water-logged soils much more freely,[2] possibly being carried down by ferrous ions.

The uptake of fertiliser phosphate by crops

In general, if a phosphate fertiliser is added to a soil, an annual crop usually takes up only about 5 to 10 per cent of the phosphate added, even if it responds well to the phosphate, though phosphate-demanding crops on phosphate-poor soils may give higher recoveries. Thus, in a number of Scottish experiments,[3] swedes took up about 20 per cent of the added phosphate for dressings of 18 kg/ha P and about 15 per cent for dressings of 35 kg/ha. Table 23.10 gives an example, using swedes for phosphate-deficient

1 R. G. Warren and A. E. Johnston, *Rothamsted Rept.* 1963, 240.
2 R. Glentworth and H. G. Dion, *J. Soil Sci.*, 1949, **1**, 35. A. J. McGregor, ibid., 1953, **4**, 86.
   J. K. R. Gasser with C. Bloomfield, ibid., 1955, **6**, 219. With G. W. Cooke, ibid., 248.
3 A. M. Smith and K. Simpson, *J. Sci. Food Agric.*, 1950, **1**, 208; 1956, **7**, 754.

English soils.[1] Higher recoveries, however, can be found. Thus G. W. Cooke quotes recoveries of 23 per cent by short-term leys and kale on low phosphate

TABLE 23.10   Uptake of added phosphate by swedes on phosphate-deficient soils. Mean of 27 experiments, 1942. Phosphorus added as superphosphate

P added in fertiliser kg/ha	Yield of roots in t/ha	Increase in yield due to each increment of phosphate	P in roots kg/ha	Uptake of P from each increment	
				kg/ha	As % of P added
0	15·0	—	3·1	—	—
28	38·0	23·0	7·5	4·4	16
55	42·5	4·5	10·0	2·5	9

soils at Rothamsted.[2] Again, if a moderate dressing of phosphate is suitably placed relative to the seed of swedes on a phosphate-poor soil, the additional phosphate taken up by the much better crop may equal 50 per cent of the fertiliser applied. But in these examples it is not legitimate to argue that all the additional phosphate came from the fertiliser, much has certainly come from the soil due to the much more vigorous root system of the manured crop. However, an example from Kericho, Kenya, of the effect of super-phosphate on the response of elephant grass (*Pennisetum purpureum*) to superphosphate shows a recovery as high as 67 per cent, most of which must have come from the fertiliser.[3] The grass was given the equivalent of 15 kg/ha P four years running, eight cuts of the grass were taken, and the total yield of dry matter was increased from 28 to 54 t/ha, and the uptake of phosphorus from 30 to 69 kg/ha.

The use of techniques based on incorporating ^{32}P in phosphate fertilisers has led to a clearer understanding of the relative value of the added fertiliser phosphate compared to the soil phosphate as the source of the phosphate taken up by the crop.[4] Thus, it can be shown that the young plant will take up nearly all its phosphate from a band of soluble fertiliser suitably placed with respect to the seed, and that crops will take up nearly all their phosphate from the fertiliser if a heavy dressing of a water-soluble fertiliser is used.[5] Correspondingly, if a fertiliser containing a mixture of water-soluble and water-insoluble phosphates is used, the crop will take up a higher proportion of phosphate from the soluble than from the insoluble, and the presence of

1   I am indebted to Drs Crowther and Cooke for this table.
2   *Fertiliser Soc., Proc.* 92, 1966.
3   R. Child, N. A. Goodchild and J. R. Todd, *Emp. J. exp. Agric.*, 1955, **23**, 220.
4   S. Larsen, *Pl. Soil*, 1952, **4**, 1. O. Gunnersson and L. Fredrikson, *Bull. Document.* 1952, No. 42. M. Fried, *Proc. Soil Sci. Soc. Amer.*, 1953, **17**, 357.
5   For example, see J. Mitchell, *J. Soil Sci.*, 1957, **8**, 73. But for an exception to this generalisation, using barley, see G. E. G. Mattingly and F. V. Widdowson, *Pl. Soil*, 1958, **10**, 161.

the soluble phosphate depresses the uptake of phosphate from the insoluble, if it had been used by itself.[1] These results show that the plant roots tend to take up their phosphate from those volumes of the soil where the phosphate potential is low; and the lower the potential in these spots, and the more of them there are, the less phosphate in proportion will be taken up from the high potential volumes.

The rainfall during the growing season also affects the ratio of soil to fertiliser phosphorus taken up by a crop. In wet seasons the crop will take nearly all its phosphate up from the surface soil, which is where the phosphate fertiliser is, so will tend to take a relatively large amount of its phosphate from the fertiliser. But in dry seasons, this phosphate becomes unavailable when the surface soil becomes dry, so the crop must take much of its phosphate from soil layers that have not been enriched with the fertiliser phosphate.[2]

The conditions which control the uptake of phosphate by a crop from a moderate dressing of a water-soluble phosphate fertiliser added to a soil can now be understood. Low uptake can be a consequence of adding the fertiliser to a soil of high phosphate status, the most usual cause in well-farmed land, or of a poor crop in which growth is limited by some factor other than the phosphate supply, or of an inefficient method of applying the phosphate, so that the roots of the crop cannot take it up when they require it because it is distributed through too large a volume of soil, and, finally, because the soil is so low in phosphate that after it has absorbed the added phosphate, the phosphate concentration in the soil solution is still so low that the crop can take up little of the added phosphate. This is found in some very strongly phosphate-fixing tropical soils.[3]

Uptake of fertiliser phosphate by a young crop can be increased by placing a water-soluble phosphate fertiliser close to the seed, as is often done with cereals, for example, by combine-drilling. This is very effective for soils low in phosphate, for, in general, only about half as much phosphate need be given with combine-drilling as is necessary with broadcasting;[4] the benefit is much less if the soil is well supplied.[5] This practice can be very helpful if the soil has an appreciable phosphate fixing power. Thus, J. W. S. Reith[6] found in Scotland that placement of phosphate for swedes and turnips increased the yield 34 per cent on soils derived from granites, 100 per cent from those derived from slates and 165 per cent on those from basic igneous rocks, and this order followed the phosphate sorbing power of the soils. However, some

1 J. T. Murdock and W. A. Seay, *Proc. Soil Sci. Soc. Amer.*, 1955, **19**, 199.
2 For examples from Scotland see K. Simpson, *J. Sci. Fd. Agric.*, 1956, **7**, 745.
3 For some examples from East Africa see P. H. Le Mare, *J. agric. Sci.*, 1968, **70**, 271. *Cotton Res. Corp. Rept. W. Region Tanzania*, 1964–65, 33.
4 See, for example, A. H. Lewis, *J. agric. Sci.*, 1941, **31**, 295; with A. G. Strickland, ibid., 1944, **34**, 73. E. M. Crowther, *J. Min. Agric.*, 1945, **52**, 170. F. Hanley, ibid., 1947, **54**, 354.
5 For examples from England see J. R. Devine and M. R. J. Holmes, *Exp. Husb.*, 1964, 11.
6 *Emp. J. exp. Agric.*, 1959, **27**, 300.

crops, such as potatoes and lucerne,[1] for example, benefit from starter doses of water-soluble phosphate, that is from phosphate placed close to the seed so the young rootlets can take up phosphate easily from volumes of soil, even where the phosphate concentration in the soil solution is high. Starter doses or combine-drilling can also be helpful in cold soils,[2] for low soil temperatures restrict root uptake; so increasing the phosphate concentration in the neighbourhood of the roots may allow earlier growth in the spring, as already noted on p. 402.

There has been much discussion on whether liming an acid soil affects the availability of soil phosphates to a crop. One would expect that raising the pH of a soil by liming it would reduce the availability of the calcium phosphates and increase that of the aluminium and iron phosphates, but the data, such as exist, suggest that liming most soils has little effect on the phosphate potential (see Fig. 23.1, p. 563). Yet the experimental evidence is often that liming does increase the phosphate uptake by a crop and reduce its response to a phosphate fertiliser, just as if it were a phosphate fertiliser in itself, implying that the aluminium phosphates have become more readily available to the crop.

The full explanations of the experimental results on the effect of lime on phosphate uptake from acid soils are not known. Part of the effect is often that liming the soil encourages the crop to make better root growth, so it can tap a larger volume of soil for phosphates; part may sometimes be due to the consequent increase in the rate of decomposition of organic matter in the soil, resulting in a greater liberation of phosphate; and part may also be due to a consequent decrease in the concentration of iron or aluminium ions or complex ions in the soil solution which may be interfering with the translocation of phosphate from root to tops (see p. 659). However, in so far as liming raises the pH of the soil, it may cause the desorption of some adsorbed phosphate (see p. 131) and also the slow hydrolysis of some aluminium phosphates, for M. D. Webber and G. E. G. Mattingly[3] found that the phosphate concentration in the soil solution slowly rose from 1·9 to 5·3 × 10^{-6} M P in a limed acid Woburn sandy loam without the pH changing by more than 0·1 units. But there is probably still no incontrovertible direct field evidence that liming really increases the availability of the soil phosphates.

Uptake of phosphates from fertiliser also seems to be encouraged if ammonium ions are intimately associated with the phosphate, such as if an ammonium phosphate is used.[4] This effect may only be noticeable when the crop is young, and it may explain the very satisfactory results many farmers

1 F. V. Widdowson, A. Penny and R. J. B. Williams, *Exp. Husb.*, 1964, 10.
2 C. D. Sutton, *J. Sci. Fd. Agric.*, 1969, **20**, 1.
3 *Rothamsted Rept.* 1965, 61; and see *J. Soil Sci.*, 1970, **21**, 121.
4 R. A. Olson and A. F. Dreier, *Proc. Soil Sci. Soc. Amer.*, 1956, **20**, 509. D. A. Rennie and R. J. Soper, *J. Soil Sci.*, 1958, **9**, 155. A. C. Caldwell, *Trans. 7th Int. Congr. Soil Sci.*, 1960, **3**, 517. For a review, see D. L. Grunes, *Adv. Agron.*, 1959, **11**, 369.

have had when they have combine-drilled ammonium phosphate with the seed.

The residual value of phosphatic fertilisers

A crop usually only takes up a small fraction of the phosphate added as fertiliser to the soil, and in many examples the uptake is less in the succeeding year, and it may be still less in the years following. Thus, often only between 20 and 30 per cent has been taken up after 4 or 5 years, on the assumption that all the extra phosphate taken up from the fertilised soil came from the fertiliser. This is illustrated in Table 23.11,[1] using the results of seven English grassland hay experiments made between 1930 and 1935, and it also shows that the recovery is similar on acid as on neutral soils.

TABLE 23.11   Uptake of phosphate by hay and grass in successive years after phosphate manuring. Per cent of added P recovered in hay or mowings. 55 kg/ha P added as superphosphate

Year	2 neutral soils	3 acid soils	2 soils grass mown*
Application	7·7	8·3	13·7
1 year after	5·7	5·0	9·4
2 years after	1·8	3·3	5·8
3 years after	1·5	2·3	2·1
Total recovery	16·7	18·9	31·0

* One acid and one neutral soil.

The old explanation of this type of result was that the phosphate fertiliser added to the soil slowly reverted to unavailable forms, or was fixed by the soil in such forms; and the phenomenon was known as phosphate reversion or fixation. However, if any water-soluble phosphate is added to a normal soil, it is rapidly converted to an insoluble form, that is, it becomes fixed, but just because it has lost much of its solubility, it has not necessarily lost its fertiliser value. This can be demonstrated very easily by showing that the response of crops to repeated small dressings of phosphate to a soil gradually drops. Thus, in Australia many wheat soils which used to be very phosphate responsive have ceased to require phosphate after about 2 to 2·5 t/ha of super-phosphate containing about 150 kg/ha P have been applied since the land was opened up.[2] Current farming experience in countries using large amounts

1 Figures taken from E. J. Russell, *Min. Agric. Bull.* 28, 3rd edn, 1939, and from *Basic Slag Reports to the Ministry of Agriculture*, 1933 and 1934.
2 A. J. Anderson and K. D. Maclachlen, *Aust. J. Agric. Res.*, 1951, **2**, 377, K. Woodroffe and C. H. Williams, ibid., 1953, **4**, 1927. C. S. Piper and M. P. C. de Vries, ibid., 1964, **15**, 234.

of phosphate fertiliser is that it is becoming increasingly difficult to find soils on which responses can be obtained to a phosphate fertiliser, other than to starter doses of water-soluble phosphate to some phosphate-demanding crops. It should be noted that some soils, such as soils very high in pure quartz sand, have only a limited power of converting a water-soluble phosphate to an insoluble form, so that if fairly heavy dressings of superphosphate, for example, are given to such soils, much of this will be leached out.[1]

Our general understanding of the conditions under which phosphate fertilisers are likely to have a well-marked residual effect has been helped by the concept of the soil's pool of labile phosphate at a definite thermodynamic potential. If enough phosphate is added to lower this potential appreciably when it has become incorporated with the soil, that is, it has increased the concentration of phosphate in the soil solution appreciably, it will have a long-continued residual effect, for the potential rapidly settles down to a reasonably constant value, except in so far as phosphate is removed from the pool by cropping. Thus, all phosphate fertilisers from which the phosphate can migrate from the fertiliser to the labile pool will have the same residual effect at equivalent dressings of phosphate. The reason that the residual effect of a water-soluble phosphate fertiliser is less than its direct effect is that, when first applied, it gives a large number of soil volumes with very low potentials, but these potentials increase slowly as the phosphate equilibrates with the labile pool over a period usually of months or sometimes years. Phosphate added to soils whose phosphate potential is already low will have little effect on the potential, so will show little residual effect on a following crop, while soils with a very high phosphate potential may have their potential only little reduced by a normal dressing of a phosphate fertiliser, so again the fertiliser will have only a small residual value for the following crop. The residual value of a phosphate fertiliser, once it has entered the labile pool, can be estimated using any chemical method that gives a useful measure of the available phosphate in the soil.[2]

The rate of loss of effectiveness of water-insoluble phosphate fertilisers differs from that of a water-soluble phosphate because of the time taken for the phosphate present in the insoluble form to enter the labile phosphate pool in the soil; but once in the pool it will lose availability at the same rate as will an initially water-soluble one. It may happen that because of the slowness in entering the pool, a water-insoluble fertiliser may be slightly more effective than one that is water-soluble two or three years after each has been applied, but it will usually have been less effective in the first one to two years. Thus, dicalcium phosphate or a metaphosphate sometimes has a better residual value than a superphosphate, but the superphosphate normally gives a better initial response. This residual response also depends on the form of

1 For an example see G. E. G. Mattingly, *J. agric. Sci.*, 1970, **75**, 413; P. G. Ozanne, D. J. Kirton and T. C. Shaw, *Aust. J. agric. Res.*, 1961, **12**, 409.
2 See, for example, K. Elk, J. R. Webb *et al.*, *Proc. Soil Sci. Soc. Amer.*, 1961, **25**, 21.

the fertiliser, and for dicalcium phosphate and the moderately low-soluble metaphosphate and magnesium ammonium phosphate the residual effect of the granular fertiliser may be larger than of the corresponding powder and even larger than from superphosphate itself, though in the year of application the powder is likely to be more effective.[1]

Some of these points can be clearly demonstrated from recent Rothamsted results. Since it takes time for a phosphate fertiliser to equilibrate with a soil, superphosphate broadcast and mixed in the seedbed is more available to a young plant than the residue from the previous season's application, but by harvest the difference between the fresh and the residue is less because the roots have had time to tap a large volume of soil for their phosphate. This is illustrated in Table 23.12,[2] which also shows that the additional phosphate

TABLE 23.12   Percentage fresh superphosphate equivalent of residues from super-phosphate and rock phosphate. Rothamsted: Sawyer's Field (pH 5·2 in 0·01 M $CaCl_2$): Barley, 1955. Kg P as fresh superphosphate to give the same yield or uptake as 100 kg P given in 1954

Phosphate given in 1954	At 6 weeks old from			At Harvest from		
	Yield	P uptake	$^{32}P/^{31}P$	Yield	P uptake	$^{32}P/^{31}P$
Superphosphate						
41 kg/ha P	21	21	16	25	45	51
82	20	23	17	27	40	50
164	29	27	31	25	43	47
Rock phosphate						
164 kg/ha P	2	3	2	7	12	12

taken up by the barley in the latter part of the season from the residues does not increase the response of barley to the phosphate as measured by the yield of dry matter.

The rate at which phosphate fertilisers lose their availability depends on the crop and the soil, if this is measured by yield response, but not necessarily if measured by the decrease in fresh superphosphate equivalent. This is defined as the amount of fresh superphosphate that must be added to the untreated soil for the crop to give the same yield or phosphate uptake as on the soil containing the residual fertiliser, and is expressed as a percentage of the added phosphate whose residual value is being measured. The reason why these two measures of loss of availability may differ appreciably is because the phosphate response curves for different crops in the same soil may differ appreciably. This can be illustrated by two experiments, one on two acid soils with a pH of about 5, and one on a slightly calcareous soil.

1 G. E. G. Mattingly, A. Penny and M. Blakemore, *J. agric. Sci.*, 1971, **76**, 131. J. R. Devine, D. Gunary and S. Larsen, ibid., 1968, **71**, 359.
2 G. E. G. Mattingly and F. V. Widdowson, *J. agric. Sci.*, 1963, **60**, 399.

The experiment on the acid soils compared the run-down in response to a heavy dressing of phosphate—165 kg/ha P—given in various forms and applied in the autumn of 1959, with annual dressings of 13·7 and 27·4 kg/ha P given as superphosphate, as measured by potatoes, barley and swedes. Some of the results are given in Table 12.13,[1] which shows that, for yield response,

TABLE 23.13   Residual effect of phosphate fertilisers on two acid Rothamsted soils. Sawyer's Field pH 5·0–5·5 (in 0·01 M $CaCl_2$); Great Field pH 4·8–5·2. Annual applications of P as granular superphosphate (8% P). Yields and responses in t/ha

Fertiliser	Potatoes (tubers)*			Swedes (roots)		
	1960	1963–65	1967–69	1960	1963–65	1967–69
Yield in absence of P	36·0	26·4	31·2	26·8	19·8	12·7
Responses to:						
13·7 kg/ha P annually	− 0·02	3·51	2·32	17·1	23·4	17·4
27·4 kg/ha P annually	3·06	5·97	3·64	19·8	29·4	24·2
165 kg/ha P applied in 1959 as:						
superphosphate	10·5	3·66	1·46	34·4	21·9	9·0
K-metaphosphate	10·4	3·64		35·2	19·2	
basic slag	9·75	2·16		34·8	20·5	
gafsa rock	6·40	2·75	− 0·30	32·7	19·8	11·0

* The 1965 basic slag figures for potatoes were abnormally low and have been excluded.

the residual value of the phosphate fertilisers decreases faster for potatoes than for swedes; but this is due to the different shape of the response curves for the two crops, for in 1965 the fresh superphosphate equivalent of the residual effects of the heavy dressings were between 11 and 13 per cent for potatoes and 12 and 16 per cent for swedes as measured by yield, though between 16 and 22 per cent for potatoes and 11 and 17 per cent for swedes as measured by phosphate uptake. This shows that, in fact, the potatoes were more efficient than the swedes in taking up phosphate from the soil, but they have a higher phosphate demand early on in the season if they are to give a high yield. Mattingly also showed that the solubility of the residual phosphate was either less than or equal to that of fluorapatite in the two acid soils.

The experiment on the slightly calcareous soil measured the rate of run-down of a dressing of 60 kg/ha P given as superphosphate over a period of five years when the land was farmed in a four-course rotation. Table 23.14 illustrates some of the results of this experiment for the twenty-one years ending in 1954.[2] It shows that potatoes, barley and ryegrass respond best to fresh phosphate, but the response in the following four years is almost constant for potatoes and barley, though it falls off somewhat for ryegrass. It

1 G. E. G. Mattingly, *J. agric. Sci.*, 1968, **70**, 139. *Rothamsted Rept.*, 1969, Part I, 53.
2 *Rothamsted Rept.* 1954, 154.

TABLE 23.14   Residual effect of superphosphate on Hoosfield (21 seasons). 60 kg/ha P given as superphosphate once every five years

	Yield No P	Response in year of application	Response to phosphate added years previously			
			1	*2*	*3*	*4*
Potatoes t/ha						
Low nitrogen	11·45	14·5	7·5	8·3	7·5	6·3
High nitrogen	12·10	18·3	12·1	10·8	9·5	9·5
Barley grain kg/ha	3080	390	160	240	240	160
Ryegrass hay kg/ha	2120	320	300	240	140	140
Wheat grain kg/ha	3550	50	− 60	10	− 20	40
Soluble P in the plots manured as above						
Soluble in NaHCO$_3$ (ppm)	8	24	19	17	15	15
Soluble in 0·01 M CaCl$_2$(10^{-6}M)	0·11	0·70	0·55	0·45	0·35	0·30

also shows that the superphosphate dressings have increased both the bicarbonate-soluble phosphate and the phosphate concentration in the soil solution five years after an application.

Other experiments at Rothamsted and Woburn demonstrate the very long-lasting effect of regular dressings of phosphate, whether given as farmyard manure or superphosphate. The Hoosfield Exhaustion Plots at Rothamsted on a slightly calcareous soil, can serve as an example. Some of the plots received about 30 kg/ha P as superphosphate annually from 1856 to 1901 and some others about 35 t/ha farmyard manure from 1875 to 1901, since when they all have only received nitrogen fertilisers. In the five years 1949 to 1953 barley took up an extra 4·5 kg/ha P each year from the superphosphate plots and 5 kg/ha from the manure plots, compared with the uptake from the plots which had not received any phosphate, as can be seen from Table 23.15. The phosphate plots received about 1100 kg/ha P between 1856 and 1901, and in 1953 they contained about 400 kg/ha P more than the unmanured, so the land still contained 35 per cent of the phosphate that had been added, and the barley is taking up about 1·2 per cent of this phosphate each year.[1]

The residual value of this phosphate can also be measured by comparing the response of different crops to fresh superphosphate on the no-phosphate and on the with-phosphate plots. Table 23.16[2] shows that the residual value of the phosphate was almost enough to give a full crop of barley and sugar-beet,

1 R. G. Warren, *Fertiliser Soc., Proc.* 37, 1956.
2 A. E. Johnston, R. G. Warren and A. Penny, *Rothamsted Rept.* 1969, Part II, 39.

TABLE 23.15   Mean yield and composition of barley on Hoosfield Exhaustion Plots (1949–53)

	Treatment 1856–1901			
	No P, No K	*P*	*PK**	*Farmyard manure†*
	Mean yield in t/ha			
Grain	1·35	2·32	2·48	2·58
Straw	1·59	2·40	2·59	2·57
	Nutrients in total crop kg/ha			
P	4·6	9·1	9·6	10·5
K	18·9	29·6	42·5	38·0

* PK 1856–75, P alone 1876–1901.
† 1876–1901.

being equivalent to about 30 kg/ha P as fresh superphosphate, but the residuals were only equivalent to about 6 kg/ha fresh P for potatoes, though this figure may be exceptionally low. In this field, the three crops gave effectively the same yield on the plots receiving 56 kg/ha P as fresh superphosphate whether or not they had received phosphate between 1856 and 1901. This result is not always found at Rothamsted, for on other fields crop yields on

TABLE 23.16   Response of crops to superphosphate on Hoosfield Exhaustion Plots. 56 kg/ha P as superphosphate given. Yields in t/ha

Previous history	*Potatoes (tubers)*		*Sugar-beet (sugar)*		*Barley (grain)*	
	No P	*P*	*No P*	*P*	*No P*	*P*
No phosphate given	12·8	32·7	3·84	5·74	2·04	3·50
Phosphate given 1856–1901	21·2	32·7	5·69	6·02	3·12	3·48
Fresh superphosphate equivalent of residues kg/ha P	6		27		34	

plots receiving 56 kg/ha P, or even 85 kg/ha P, as fresh superphosphate are lower on those plots which have not received any phosphate fertiliser in the past than on those which have received annual dressings, and Fig. 23.6 illustrates this for potatoes on Agdell field at Rothamsted. Two plots on this field have received no phosphate fertiliser since 1848 and two plots received 7 kg/ha P once every four years. There is no indication from the phosphate

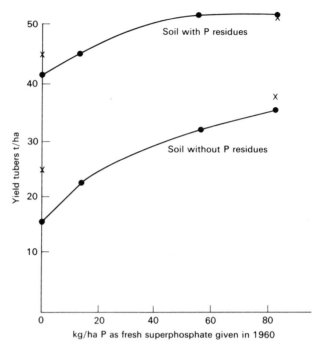

FIG. 23.6. Response of potatoes to fresh phosphate and residual phosphate. Agdell field, Rothamsted 1960. × 56 kg/ha P given as superphosphate in 1959

response curves that potatoes could give the same yield on the two groups of plots, whatever the phosphate dressing given.[1]

The long-continued fertiliser plots at Rothamsted also give some evidence of the fate of the residual phosphate in the soil. G. E. G. Mattingly[2] showed that on the permanent wheat and barley fields—Broadbalk and Hoosfield—between 35 and 45 per cent of the total phosphate left in the soil from long-continued fertilising with superphosphate is present in the labile phosphate pool, and he found this same proportion, namely, 38 per cent, in the Hoos-field Exhaustion Plots, although on these plots only 35 per cent of the total phosphate added to the soil still remains in the soil. This can only mean that as the crop removes phosphate from the pool, a proportion of the phosphate that has gone into a non-labile form goes over into the labile form to maintain this proportion, even when two-thirds of the added phosphate has been removed by crops.

1 G. Barbier and S. Trocme (Z. *PflErnähr*, 1965, **109**, 113) have found the same result.
2 *Rothamsted Rept.*, 1957, 61. *J. agric. Sci.*, 1957, **49**, 160. For other examples see S. R. Olsen, F. S. Watanabe *et al.*, *Soil Sci.*, 1954, **78**, 141. G. Barbier, C. Lesaint and E. Tyszkiewicz, *Ann. Agron.*, 1954, 923. W. W. Moschler, R. D. Krebs *et al.*, *Proc. Soil Sci. Soc. Amer.*, 1957, **21**, 293.

Most of these Rothamsted results are typical of those found for many other soils. Thus S. Larsen[1] found that a single dressing of 250 or 500 kg/ha P as granular triple superphosphate mixed in with twenty-three different soils lost its liability according to a first order reaction with a half-life between 1 and 6 years, of which one-half were between two or three years. There was a tendency for the half life to be longer on acid than neutral soils; they also used a peat and for this the half life was 56 years. For the purpose of compensation in Great Britain, a water-soluble phosphate fertiliser is assumed to have lost half its value after one year, three-quarters after two, and lost all its value after three years. The results given here, and a number of others from several Experimental Husbandry Farms in England and Wales,[2] suggest that this convention favours the incoming tenant on well-farmed land because of the long-lasting effect of regular phosphate additions, although the figures for the first two years often appear to be reasonable.

There are, however, soils in which even heavy dressings of phosphate rapidly lose their fertilising effect, and these are probably all high in active aluminium hydroxide and perhaps ferric hydroxide also. Thus, E. G. Williams[3] found two Aberdeenshire soils, one derived from a basic igneous and one from a granitic till, in which the fertiliser effect of a dressing of 250 kg/ha P as superphosphate had a fertiliser equivalent of only 60 kg/ha for turnips one year after application and too small an effect to be measured three years after. O. R. Younge and D. L. Plucknett[4] have given an even more extreme example from Hawaii where a dressing of 650 kg/ha P as superphosphate was needed for the grass *Digitaria decumbens* and the pasture legume *Desmodium intortum* to give their maximum yield, and its effectiveness decreased rapidly during the subsequent four years. A dressing of 1300 kg/ha P, however, maintained maximum yields for at least six years. On soils such as these, it may be preferable to use relatively insoluble phosphates such as fused serpentine phosphate or a high temperature calcium magnesium phosphate rather than a water soluble one, if only moderate dressings are to be given.[5]

It is worth noting that most tropical soils behave as typical temperate soils in that dressings of superphosphate supplying 10–20 kg/ha P have residual effects lasting a number of years. This is illustrated in Table 23.17 for an experiment made on a soil derived from basement complex rocks at Makaveti in the Machakos district of Kenya,[6] and in this experiment the effect of

1 With D. Gunary and C. D. Sutton, *J. Soil Sci.*, 1965, **16**, 141.
2 J. H. Williams, D. A. Boyd and K. Farrar, *Min. Agric. Fish. Food, Tech. Bull.* 20, 1971, 42.
3 *J. Sci. Fd. Agric.*, 1950, **1**, 244.
4 *Proc. Soil Sci. Soc. Amer.*, 1966, **30**, 653. For another example of a high fixing soil see E. J. Kamprath, *Agron. J.*, 1967, **59**, 25.
5 For an example see K. D. McLachlan, and B. W. Norman, *Aust. J. exp. Agric. anim. Husb.*, 1969, **9**, 38, 341.
6 E. Boswinkle, *Emp. J. exp. Agric.*, 1961, **29**, 136.

40 kg/ha P is still appreciable five years after application. Two crops were planted in this experiment each year, except in 1952, the first crop being maize but, except in 1954, the maize failed to give a significant yield of grain due to drought. On the loess soils of northern Nigeria, dressings as low as 5 kg/ha P as superphosphate have appreciable residual effect three years after application.[1]

TABLE 23.17   Residual effect of superphosphate on a tropical red earth. Makaveti, Kenya: Soil from basement complex. Yield and responses to superphosphate (18% P) given in 1952 in kg/ha. Millet: *Pennisetum typhoideum*; beans: *Phaseolus vulgaris*

Year crop	1952 Millet	1953 Beans	1954 Maize	1954 Millet	1955 Millet	1956 Millet	1957 Beans
Control yield. No P	405	760	1920	410	1250	785	785
Response to:							
20 kg/ha P	525	160	590	120	110	85	40
40 kg/ha P	720	250	785	175	180	125	125

There is one other process which may make a phosphate fertiliser change to a relatively unavailable form, but which nevertheless will let it have a slow long-lasting effect. If a soil low in organic matter and in phosphate is put down to grass for a period of years and is manured with a phosphate fertiliser, the organic matter content of the soil may rise very considerably and fairly rapidly, and this organic matter will be locking up phosphate, which in these conditions has been derived from the fertiliser. Thus in New South Wales, C. M. Donald and C. H. Williams[2] found that adding superphosphate to a subterranean clover pasture, about one half of the added phosphate was converted into organic phosphate with a consequent increase of about 76 lb of soil organic nitrogen per 100 lb of superphosphate added; and in these examples the ratio of C:N:S:P in the organic matter remained at 155:10:1·4:0·7, the same as in the soil of the unimproved pastures.

Responsiveness of crops to phosphate fertilisers

Three separate types of relation between crop and phosphate fertiliser can be recognised:

1 The sensitiveness of a crop to phosphate shortage. Crops such as swedes, potatoes and barley have their yields very seriously reduced by a shortage

1 M. Greenwood, *Emp. J. exp. Agric.*, 1951, **19**, 225. For other examples from Africa see E. W. Russell, *Fertiliser Soc., Proc.* 101, 1968.
2 *Aust. J. Agric. Res.*, 1954, **5**, 664, and 1957, **8**, 179. For an example from New Zealand, see R. H. Jackman, *Soil Sci.*, 1955, **79**, 207.

of phosphate when growing in soils in which wheat and oats may make fair growth and some grasses good growth. This can be seen in part from Table 23.14. It is also probable that sorghum is more sensitive to low phosphate in a soil than is maize.[1]

2 The responsiveness of a crop to added phosphate. On the whole this is the same property as the first, for crops which are sensitive to phosphate shortage respond best to a dressing of phosphate. But it includes in addition the ability to continue to respond to phosphate as the fertiliser dressing is increased. Thus maize will respond to a considerably higher dressing of phosphate than will groundnuts.[2]

3 The ability to use phosphate from relatively insoluble phosphate fertilisers. Crops such as swedes, turnips and mustard among the brassica crops, lupins, lucerne and sweet clover among the leguminous, and buckwheat, all can use phosphate from relatively insoluble forms, such as ground North African rock phosphate, much more easily than can crops such as potatoes, the temperate cereals and cotton. Sorghum can however respond to relatively insoluble phosphates, and is markedly superior to maize in this respect.[2]

1 P. H. Nye, *Emp. J. exp. Agric.*, 1954, **22**, 101, and L. R. Doughty, *E. Afr. Agric. J.*, 1953, **19**, 30.
2 P. H. Le Mare, *Emp. J. exp. Agric.*, 1959, **27**, 197.

24

The sources of plant nutrients in soils

Sodium, potassium, magnesium, calcium

These ions occur in soils as constituents of the silicates composing the mineral fraction of the soil, as exchangeable cations, and as simple inorganic salts; but they probably only occur in the humic matter as exchangeable cations, though they occur in non-humified plant material. Most silicates high in sodium, calcium and magnesium are relatively easily weatherable, while some of the potassium silicates are more resistant. Thus, the non-clay fraction of weathered and leached soils usually only contain a few parts per thousand of the first three elements but up to 1 or 2 per cent of potassium. Table 24.1,[1] which refers to six English non-calcareous soils after acid washing to remove these exchangeable cations, illustrates these points.

TABLE 24.1 Mineral composition of different fractions of six English soils. Content of element in m eq/100 g of fraction

Element	K	Na	Ca	Mg
Coarse sand	2–15	1–4	0·5–6	0·5–3
Fine sand	20–30	15–25	4–10	0·5–6
Silt	35–55	15–35	3–10	10–30
Clay	35–45	5–10	0–5	45–65

These cations are also present in soils as exchangeable ions. Thus, over half the total calcium content of a non-calcareous soil may be in the exchangeable form, though calcium also occurs as calcium carbonate in calcareous soils or in soils recently limed, and as gypsum in some arid soils. Exchangeable magnesium usually constitutes a smaller proportion of the total magnesium than is the case for calcium, and usually forms a smaller proportion of the total exchangeable ions than does calcium, particularly in soils of moderate or high pH, though it often exceeds calcium in acid soils and subsoils. It is also present as magnesium carbonate in soils containing

1 *Rothamsted Ann. Rept.* 1965, 61.

dolomite, and as a soluble salt in some saline soils. Sodium forms an even smaller proportion of the exchangeable ions than does magnesium, except in alkali and saline soils, and in saline soils it is present as the chloride, sulphate or bicarbonate. Exchangeable potassium in a soil is usually lower than the exchangeable magnesium, and potassium salts do not form any appreciable proportion of the salts present in a saline soil.

The level of exchangeable sodium in at least some British soils appears to be controlled by the level of sodium in the rainwater. Thus, J. Bolton[1] finds that the activity ratio of sodium to calcium plus magnesium in the rainwater at Rothamsted is about the same as in the soil solution for the British soils he examined. Using C. M. Stevenson's[2] figures for the composition of rainwater at Rothamsted, its sodium activity ratio varied from 8 to 29 $\times$ 10^{-3} M$^{1/2}$ in winter and from 4 to 8 $\times$ 10^{-3} in summer; and the sodium activity ratios in the British soils varied from 4 to 19 $\times$ 10^{-3}. Further, the rain brought down about 13 kg/ha Na annually, and the unmanured Rothamsted soil may lose about 11 kg/ha Na annually, so the amount of sodium entering and leaving the soil, when allowance is made for removal in crops, is about in balance, which is again consistent with the composition of the rainwater being the factor controlling the amount of exchangeable sodium that is present in unmanured soils in eastern England.

The exchangeable calcium, magnesium and sodium are the major immediate reserve of these ions available to the plant roots in non-calcareous non-saline soils, but the exchangeable potassium often forms only a part of the reservoir of these ions. P. Newbould and R. S. Russell,[3] for example, working with some British soils, and using calcium salts containing ^{45}Ca, showed the primary source of calcium for crops to be that which was isotopically exchangeable with the added calcium, and this, in turn, was about equal to the exchangeable calcium.

It is not possible to do the corresponding experiment with magnesium, but R. C. Salmon and P. W. Arnold[4] have shown that for a range of British soils exhaustively cropped with ryegrass in the glasshouse, the amount of magnesium removed by the crop was very closely correlated with the drop in the exchangeable magnesium: about 10 per cent may have been non-exchangeable, but the analytical method used for determining the exchangeable magnesium may have underestimated it by this amount. Under normal arable-farming conditions, however, the annual losses of magnesium do not appear to be large, being perhaps about 10 kg/ha for arable crops at Rothamsted, although heavily manured grass crops removed from the land will contain much more than this; and between one-third and one-half of this loss is

1 *J. Soil Sci.*, 1971, **22**, 419.
2 *Quart. J. Roy. Met. Soc.*, 1968, **94**, 56.
3 *Pl. Soil*, 1963, **18**, 239. P. Newbould, *J. Sci. Fd.* Agric., 1963, **14**, 311. See also D. E. Davies, W. H. MacIntire *et al.*, *Soil Sci.*, 1953, **76**, 153 for similar results with some American soils.
4 *J. agric. Sci.*, 1963, **61**, 421. See also: G. Michael and G. Schilling. *Z. PflErnähr.*, 1957, **79**, 31, for similar results with some German soils.

made good by the magnesium in the rainfall. On the other hand, soils commonly contain between 0·2 and 0·8 per cent of magnesium, which corresponds to between 10 and 40 t/ha Mg in the top 30 cm of soil, of which about two-thirds may be in the clay fraction;[1] and the weathering of these silicates may be sufficiently rapid to make good any net loss of magnesium either through leaching or through removal in the crop. This is probably the reason why magnesium deficiency is still uncommon.

Potassium, like magnesium, is an important constituent of soil silicates, particularly of soil clays—many soils containing between 0·5 and 2·0 per cent, and their clays between 1·5 and 3·0 per cent K. But if a soil is cropped exhaustively, the amount of potassium removed in the crop is usually larger, and may be much larger, than the loss of exchangeable potassium, as is shown in Table 24.2[2] for twenty British soils. Crops can, in fact, only take up large amounts of potassium from a soil if the soil can release a large amount of non-exchangeable potassium.

TABLE 24.2   Release of non-exchangeable soil potassium to ryegrass. 400 g soil; ryegrass manured with nitrogen and phosphate. Up to 13 harvests taken. Quantities of K in m eq/100 g soil

Number of soils	Uptake by ryegrass		Initial exchangeable K	Fall in exchangeable K	Release of non-exchangeable K
	Range	Mean			
5	0·15–0·33	0·22	0·21	0·12	0·10
4	0·45–0·50	0·48	0·25	0·15	0·33
6	1·4–2·1	1·64	0·56	0·33	1·31
5	2·7–4·1	3·19	1·01	0·67	2·51

The uptake of sodium, calcium and magnesium by crops

Most soils contain enough sodium for crop growth and responses to sodium fertilisers are confined to crops with a definite sodium requirement such as sugar-beet, mangolds and some brassica crops. Some of these crops, sugar-beet, for example, have a definite sodium demand if they are to give their maximum yield, and this demand is independent of, and possibly greater than, the potassium demand. This is illustrated in Table 24.3 which summarises the results of forty-two fertiliser experiments with sugar-beet[3] in the United Kingdom. The soil exchangeable sodium is present in two forms, that which is held loosely on the clay sheets and that held tightly on specific spots,

1 R. C. Salmon, *J. Sci. Fd. Agric.*, 1963, **14**, 605.
2 P. W. Arnold and B. M. Close, *J. agric. Sci.*, 1961, **57**, 295. See also G. S. Fraps, *Texas Agric. Exp. Stn. Bull.* 391, 1929.
3 P. B. H. Tinker, *J. agric. Sci.*, 1965, **65**, 207. See also A. P. Draycott, J. A. P. Marsh and P. B. H. Tinker, ibid., 1970, **74**, 568, and J. C. Holmes, W. D. Gill, *et. al.*, *Exp. Husbandry*, 1961, **6**, 1.

probably on the broken edges of the sheets, which may be the principal component in humid soils (see p. 101). P. B. Tinker[1] found that it is the loosely-held sodium that is the principal source for annual crops, and that beet only responds to sodium fertilisers if the soil contains less than 0.05 m eq/100 g of readily exchangeable sodium.

TABLE 24.3 Effect of sodium and potassium fertilisers on the yield of sugar-beet. 42 experiments 1959–62. Na: 0, 100 and 200 kg/ha Na as agricultural salt. K: 0, 125 and 250 kg/ha K as potassium chloride. Yield of sugar in t/ha

	0	Na_1	Na_2
0	6·43	7·01	7·21
K_1	6·82	7·21	7·24
K_2	6·97	7·20	7·31

Most soils also contain enough calcium for crop growth if the crop is not affected by acidity. But most crops with a high calcium demand are also sensitive to the relatively high concentrations of aluminium or manganese ions typically present in acid soils, so these soils must have their pH raised, by adding calcium carbonate or lime, if the crop is to grow well. But there are some crops which are tolerant of acid soils, whose yield may be limited by the calcium supply. Thus, potatoes may have their yield limited by lack of calcium on some acid soils in the United Kingdom, as may cotton and groundnuts on some tropical acid soils.

The roots of some legumes have a higher calcium demand during the process of root infection by Rhizobia than for growth; this demand can be met by pelleting the seed with calcium carbonate much more cheaply than by applying the carbonate to the soil.

In certain exceptional soils, such as those derived from serpentine, the magnesium to calcium ratio in the exchange complex may be so high that agricultural crops suffer from a true calcium deficiency; but these soils tend to be very infertile as they often also contain high levels of available nickel and other heavy metals and may also suffer from molybdenum deficiency.[2]

Most soils also contain enough magnesium for maximum crop yields, so yield responses to magnesium fertilisers are still uncommon. Responses are normally found on three types of soil: sandy soils low in exchangeable magnesium, a few acid soils low in exchangeable magnesium which have only received dressings of a calcitic limestone,[3] and some soils very high in potassium due to the interaction between potassium and magnesium mentioned on p. 551. European experience indicates that crop response to a magnesium

1 *Soil Chemistry and Fertility*, Int. Soc. Soil Sci., Aberdeen, 1967, 223.
2 For a short symposium see *Ecol.*, 1954, **35**, 259, and also R. B. Walker, H. M. Walker and R. Ashworth, *Pl. Physiol.*, 1955, **30**, 214.
3 W. E. Chambers and H. W. Gardner, *J. Soil Sci.*, 1951, **2**, 246, and W. H. Bender and W. C. Eisenmenger, *Soil Sci.*, 1941, **52**, 297, for American examples.

fertiliser—which is usually a dolomitic limestone on acid soils or a calcined kieserite (magnesium sulphate) on neutral soils—is unlikely if the soil contains more than 0·4 m eq/100 g Mg, and many crops do not respond even if it is only half of this amount.[1] Thus, P. B. H. Tinker,[2] working with sugar-beet in East Anglia, found responses to magnesium on soils with exchange capacities between 5 and 10 m eq/100 g, if they contain less than 0·2 m eq Mg, or alternatively if the equilibrium concentration of magnesium when the soil is shaken with 0·01 M calcium chloride is less than 3×10^{-4} M. Below this concentration he found a fair correlation between response and concentration. He used dressings of 110 kg/ha Mg but considered that about 60 kg/ha would probably be as effective; and the mean responses he obtained, on responsive soils, was about 7 per cent—about the same as for potassium or sodium.

The effect of the exchangeable magnesium content of the soil on the magnesium content of the crop is particularly important for grass, because of the tendency of grasses with a low magnesium content to induce hypomagnesaemia or milk fever in dairy cattle. R. C. Salmon[3] investigated the relation between the magnesium content of ryegrass grown in pots and the activity of magnesium ions in the soil solution, and found that it was proportional to the square root of the ratio of the magnesium ion activity to the sum of the magnesium and calcium ion activities. Thus, if the magnesium content of the herbage is low, it is necessary to increase the magnesium ion activity, or the exchangeable magnesium, fourfold to double the content, which means that very heavy dressings of magnesium fertiliser, of the order of 350–1700 kg/ha Mg, are needed on low-magnesium pastures.[4]

The uptake by crops of some other cations, such as rubidium, caesium, strontium, barium and radium has received some attention because although they are not plant nutrients ^{137}Cs and ^{90}Sr can enter the soil from fall-out from nuclear explosions and so get taken up by crops growing on the soil. The general experience is that crop roots take up potassium and rubidium or calcium and strontium in much the same ratio as their activities in the soil solution, although there may be differences in the translocations of these ions after uptake; but that caesium and barium are taken up less strongly. There are exceptions to this because some crops are known to be strontium and others barium and radium[5] accumulators compared with calcium; but these are usually restricted to plants adapted to high strontium and barium soils.

1 J. W. S. Reith, *J. Sci. Fd. Agric.*, 1963, **14**, 417. M. R. J. Holmes, *J. agric. Sci.*, 1961, **58**, 281, J. A. Birch, J. R. Devine and M. R. J. Holmes, *J. Sci. Fd. Agric.*, 1966, **17**, 76. For Woburn results see J. Bolton with A. Penny, *J. agric. Sci.*, 1968, **70**, 303, with D. B. Slope, ibid., 1971, **77**, 253.
2 *J. agric. Sci.*, 1967, **68**, 205. For more recent data see A. P. Draycott and M. J. Durrant, ibid., 1970, **75**, 137; 1971, **77**, 61.
3 *Soil Sci.*, 1964, **98**, 213.
4 K. M. Wolton, *N.A.A.S. Quart. Rev.*, 1963, **14**, 122.
5 K. A. Smith, *Pl. Soil*, 1971, **34**, 369, 643 (radium and barium); W. O. Robinson, R. R. Whetstone and G. Edgington, *U.S. Dept. Agric.*, Tech. Bull. 1013, 1950.

Since soils hold exchangeable barium and caesium much more tightly than calcium and potassium, and strontium somewhat more strongly than calcium, the ratio of calcium to strontium and to barium in the crop is about one to two and fifty times greater than on the exchange complex; and the ratio of potassium to rubidium or caesium in the plant about equal to unity and to twenty-five times greater.[1] The ratio of exchangeable calcium to exchangeable strontium is usually in the range 100 to 800, and to exchangeable barium 100 to 10 000.[2]

The availability of soil potassium to crops

Potassium differs from sodium and magnesium in that crop responses to potassic fertilisers are common in productive agriculture, and in fact a high level of use of potassic fertilisers is one of the surest signs of intensive cropping systems. The immediate source of potassium for crops is that in the soil solution, so uptake of potassium by a crop is dependent on the potassium ion concentration close to the root surface, on the rate potassium ions can diffuse from the exchange sites on the soil surfaces to the root surface, and on the intimacy with which the roots ramify through the soil, which, in turn, affects the length of the diffusion paths of potassium in the soil solution. These effects have been discussed on p. 546.

The exchangeable ions are the immediate reservoir which replenishes the soil solution, so the ease with which roots can use these ions will be determined in part from the curve relating a change in the exchangeable potassium to the activity ratio of the potassium ion, defined as the ratio of the potassium ion activity to the square root of the calcium and magnesium ion activities at equilibrium, as described on p. 98. Potassium differs from the other cations in that, for many soils, loss of exchangeable potassium will be slowly made good, in whole or in part, by some of the non-exchangeable potassium being released to the exchangeable pool.

Water and sand culture experiments have shown that the rate of potassium uptake by roots depends on the potassium concentration just outside their absorbing surfaces and that it is not very sensitive to quite large changes in the concentration of other cations, such as calcium, magnesium or sodium in the solution.[3] The minimum potassium concentration for active growth of many crops, including most of the valuable grasses, is about 5×10^{-5} M, but some crops, such as clovers and perhaps barley, need concentrations of 10^{-4} M or above.[4] The rate of uptake of potassium by crop roots in the soil is therefore determined by the ability of the soil to maintain potassium

1 R. G. Menzel, *Soil Sci.*, 1954, **77**, 419. R. S. Russell, *Ann. Rev. Pl. Physiol.*, 1963, **14**, 271.
2 H. J. M. Bowen and J. A. Dymond, *Proc. Roy. Soc.*, 1955, **144**B, 355.
3 For an example with ryegrass and flax see A. Wild, D. L. Rowell and M. A. Ogunfowora, *Soil Sci.*, 1969, **108**, 432.
4 C. J. Asher and P. G. Ozanne, *Soil Sci.*, 1967, **103**, 155.

concentrations at least as high as those in the solution just outside the absorbing root surfaces. In soils where the concentration at the root surface is limited by diffusion, a concentration of 5×10^{-5} M in the soil solution is usually inadequate for good growth of all crops, a concentration of 5×10^{-4} M is adequate for many crops, but 5×10^{-3} M or above are needed by some crops;[1] but at these high levels mass flow will probably be a more important process than diffusion.

The rate of supply of potassium to the roots of a plant in a soil depends both on the rate of transpiration of the crop, for this flow of water through the soil carries potassium ions along with it, and on the rate of diffusion of potassium ions from the soil to the root surface, when the roots are taking up potassium faster than the transpiration stream carries it to their surface. In general for soils well supplied with potassium and having a relatively high concentration of potassium in the soil solution, mass flow is adequate or more than adequate to meet the crop's requirements; but once the concentration falls below a certain level, diffusion increasingly is the process by which the potassium reaches the roots. The actual concentration range at which mass flow becomes inadequate depends on the rate of transpiration, and the potassium demands of the roots, but once the concentration has fallen below 10^{-3} M, diffusion is usually becoming the dominant process.

The rate at which diffusion supplies potassium to the root surface depends on the reciprocal of the buffer capacity of the soil for potassium, that is, the reciprocal of the slope of the curve relating the amount of exchangeable and soluble potassium per unit volume of soil to the concentration of potassium in the soil solution (see p. 546); and the total amount of potassium that diffusion can supply over the period of growth of the crop is governed by the amount of potassium the soil will release as the potassium ion concentration in the solution decreases. This amount will depend on the shape of the curve relating the change in the exchangeable potassium of the soil to the change in the potassium ion activity ratio in the solution, usually defined as the ratio of the potassium ion activity to the square root of the activities of the calcium plus magnesium ions (see p. 98). Thus potassium release from the exchangeable to the soluble state depends on the potassium activity ratio in the solution, and it is for this reason that so many soil studies have been concerned with the effect of the activity ratio on the uptake of potassium by a crop.

The effect of growing a crop on the potassium activity ratio in the soil solution is, in general, for the ratio and the level of exchangeable potassium to fall. If the activity ratio was initially high, it is probable that the uptake of potassium will be nearly equal to the drop in exchangeable potassium; but if the activity ratio is only at a moderate level, both it and the exchangeable potassium may drop to a certain level, and then remain approximately con-

1 F. Haworth and T. J. Cleaver, *J. Sci. Fd. Agric.*, 1963, **14**, 264; 1965, **16**, 600.

stant as the crop continues to take up potassium. The rate of supply to the crop is now limited by the rate that this non-exchangeable potassium can enter the exchangeable pool. In temperate regions, where winter temperatures are too low to allow crop growth to take place, but not so low that the soil is frozen, the exchangeable potassium and the activity ratio will increase during winter, as the non-exchangeable potassium slowly comes back into equilibrium with the exchangeable.

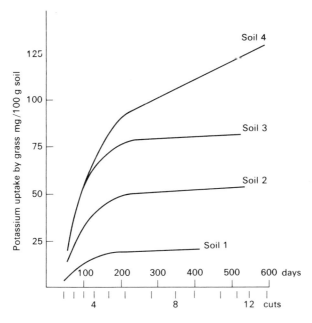

FIG. 24.1. Potassium released to ryegrass by four English soils each with about 20% clay.

	Soil 1	Soil 2	Soil 3	Soil 4
% K in clay fraction	1·96	2·66	1·69	1·78
Exchangeable K mg/100 g soil	9·0	14·0	33·5	72·0

The rate of release of non-exchangeable potassium depends either on the potassium ion concentration or on its activity ratio in the solution outside the clay particles. The amount of this type of potassium that a soil can release is usually determined by taking a soil sample, putting it through a 2 mm sieve, if necessary mixing it with a pure quartz sand or crushed flint, and growing a series of crops on it, and giving the crop adequate amounts of nitrogen and phosphate, and any other nutrient necessary for good growth except potassium. The crop is harvested at intervals and the amount of potassium taken up is determined. A grass is usually chosen as the test crop, since its tops can be repeatedly harvested without the need for replanting.

Figure 24.1 gives some typical curves for the cumulative uptake of potassium by ryegrass from four British soils, chosen because they all have about the same clay content.[1] Three of these soils lose a certain amount of potassium readily, equal to about three times the drop in exchangeable potassium, and then lose potassium much more slowly, at a rate of between 3 to 10 mg/100 g soil per year.

Some soils will however continue to lose potassium at an appreciable rate for a very long time under intensive cropping. Soil 4 in Fig. 24.1 is an example from a temperate soil: it releases potassium at the rate of 35 mg/100 g annually after the exchangeable potassium has been reduced to its floor level. This soil is derived from the Upper Greensand and contains the potassium zeolite clinoptilolite in its coarse clay and fine silt fraction,[2] which weathers readily to release potassium, if the potassium activity ratio falls too low. Some boulder clays in eastern England behave in the same way, though the silicate supplying the potassium has not been identified. Thus, the Saxmundham soil in Suffolk still supplies over 100 kg/ha K annually to crops, though it has been in continuous arable for nearly sixty years without receiving any potassium fertiliser.[3] In the same way weatherable potassium silicates, such as biotites and andesine, will weather under humid tropical conditions to release large amounts of potassium. H. W. van der Marel[4] found in Sumatra soils containing these minerals which would liberate 200 to 250 kg/ha K annually to sisal and sugar-cane over long periods of time, and A. S. Ayres[5] in Hawaii found soils that would liberate over 4 t/ha K over a four-year period to elephant grass (*Pennisetum purpureum*) without the exchangeable potassium in the top metre of soil being affected.

The results for the first three soils in Fig. 24.1 show that there are two types of non-exchangeable potassium, one that can be extracted and used by some roots relatively easily, and one that is only very slowly available. There is naturally no sharp distinction between these two groups, one merges into the other, but there are no accepted names for them. It is probable that the second, slowly available fraction, is derived from the weathering of the micaceous core in this type of clay mineral, and this potassium is sometimes called core or native potassium. The more readily available fraction is presumably situated nearer the edges of the clay sheets and has been adsorbed from the soil solution; and on adsorption the ions have caused adjacent sheets to collapse around them. This collapse will have entrapped some other hydrated cations which distort the sheet so that the rate of diffusion of potassium from between these sheets is much higher than from between the unweathered sheets. Once native potassium has been removed from between these sheets,

1 Taken from P. W. Arnold and B. M. Close, *J. agric. Sci.*, 1961, **57**, 295.
2 *Rothamsted Rept.* 1965, 72; 1966, 65.
3 P. J. O. Trist and G. W. Cooke, *J. agric. Sci.*, 1966, **66**, 327. *Rothamsted Rept.* 1965, 232.
4 *Soil Sci.*, 1947, **64**, 445.
5 *Proc. Soil Sci. Soc. Amer.*, 1947, **11**, 175.

they never collapse as perfectly afterwards when in the soil. There are unfortunately no accepted chemical methods for determining this readily extractable potassium, though extraction with hot 1N HNO_3 is often employed.[1]

When a soil receives more potassium fertiliser than the crop takes up, a part of the absorbed potassium enters the non-exchangeable form. W. E. Chambers[2] found that on Broadbalk field about ten units of potassium had to be added to the soil to increase the exchangeable potassium by one unit, though he was not able to determine what proportion of the remaining nine units had been converted to a non-exchangeable form, and what had been leached below the plough layer. Some of his results are given in Table 24.4.

TABLE 24.4   Effect of potassium fertilisers on the potassium status of Broadbalk soil, Rothamsted

Plot nos.	Manurial treatment			Farmyard manure 2
	N, P, Na, Mg 12, 14	N, P, K, Na, Mg 7, 13	P, K, Na, Mg 5	
Period 1852–1921		*In kg/ha per year**		
K added as fertiliser or manure	0	95	95	190
K removed by crop	27	47	18	58
K left in soil	− 27	48	77	132
Exchangeable Potassium				
mg/100 g 1944	8·6	26·8	33·4	51·0
1966	8·5	33·2	36·9	65·5
Potassium Activity Ratio at commencement of season†				
$\times\ 10^{-4}M^{\frac{1}{2}}$	12	19	129	212

* W. E. Chambers, *J. agric. Sci.*, 1953, **43**, 473.
† *Rothamsted Ann. Rept.* 1964, 69. These figures are only approximate as they refer to the mean value of several classical plots.

A proportion of the potassium which has been fixed has entered the readily available pool as can be seen from Table 24.5, which gives the rate of release of potassium to ryegrass from soil samples taken from the classical four-course rotation experiment on Agdell field at Rothamsted.[3] One-half of the field had three crops and a fallow and the other three crops with either beans

1 P. Moss and J. K. Coulter, *J. Soil Sci.*, 1964, **15**, 284, O. F. Haylock, *Trans. 6th Int. Congr. Soil Sci.*, 1956 B, 403.
2 *J. agric. Sci.*, 1953, **43**, 473. For more recent figures A. E. Johnston, *Rothamsted Rept.* 1969, Part II, 93.
3 P. W. Arnold and B. M. Close, *J. agric. Sci.*, 1961, **57**, 381.

or clover as the fourth; and one-third of the area received no fertiliser, one-third had potassium and phosphate only once in four years, at the rate of 230 kg/ha K, and one-third had nitrogen in addition. The table shows that the plots that had received no potassium fertilisers for over a century released 22 mg K/100 g soil that was not exchangeable, which is equivalent to over 550 kg K/ha. However, the continued use of potassium had increased the release by 18 and 6 mg/100 g of surface soil for the land that was fallowed and the land that had beans or clover; but that during the last 250 days of the experiment, all the soils were releasing about 2 mg K/100 g soil, although the exchangeable potassium remained at 7 to 9 mg/100 g. This rate of release is equivalent to about 50 kg/ha during a growing season. Other field results at Rothamsted bear out the principal results given in this table, namely, that a potassium-depleted soil will have a fairly definite amount of exchangeable potassium (8 mg/100 g), a fairly definite minimum potassium concentration of 0.5–1.5×10^{-4} M when in equilibrium with 0.01 M $CaCl_2$, and will release about 50 kg/ha K annually.[1] These figures appear to be typical for many temperate loams and clays.

TABLE 24.5   Rate of depletion of potassium from an unfertilised and a fertilised soil. Agdell Rotation Experiment 1848–1951. Extraction of potassium using 10 cuts of ryegrass, mg K/100 g soil

Treatment	K removed in			Exchangeable K		Release non-exchangeable K
	First 4 cuts	*Last 4 cuts*	*All 10 cuts*	*Initially*	*After 10 cuts*	
			Surface soil			
No K	16·6	1·6	25·7	12·0	9·0	22·7
With K fallow	38·8	2·5	51·6	19·0	8·0	40·6
clover	27·0	2·1	36·6	15·0	6·9	28·6
			Subsoil			
No K fallow	19·6	1·4	26·4	15·5	9·9	20·7
With K fallow	25·4	1·4	33·5	17·0	9·4	25·9

Total K in the clay fraction of the soil 380 mg/100 g soil
To convert to kg K/ha in 22 cm depth of soil multiply by 29.
Duration of experiment about 470 days, of which the 4th cut was taken on about day 130 and the 6th on day 220.

It is not yet possible to account for all the potassium added to a soil as fertiliser over a long period of years. In the Agdell experiment the difference between the amounts of potassium added to the soil and removed by the crops is about 1950 and 950 kg/ha K for the fallow and the beans/clover

1 O. Talibudeen and S. K. Dey, *J. agric. Sci.*, 1968, **71**, 95, 405. I. P. Garbouchev, ibid., 1966, **66**, 399. R. G. Heddle and K. Simpson, ibid., 1969, **73**, 49.

rotations, but the increase in the exchangeable and readily-extractable potassium in the top 45 cm is only 950 and 500 kg/ha K, so only about half of the residual potassium is accounted for. This is equivalent to an annual loss of 9·5 and 5·0 kg/ha K, which could be lost by drainage, for A. Voelcker[1] found that the Broadbalk plot receiving potassium fertiliser but no nitrogen lost 8 kg/ha more potassium in the drainage water than the unmanured plot. These experiments, unfortunately, can give no definite evidence on what proportion of the potassium not extracted by ryegrass has been converted into a more firmly held form within the clay particle and what proportion has been leached from the soil.[2]

These results cannot be converted directly to results in the field, for the amount of potassium a crop takes from a soil when diffusion is the principal process supplying it to the root, depends on the extensiveness of the root system, that is on the soil structure. Thus J. K. Coulter[3] showed that ryegrass would take up nine or ten times as much potassium from a poorly-structured soil if it had been put through a 2 mm sieve than if used in its natural condition, compared to a factor of only about two for a well-structured soil.

Potassium manuring of soils

Much attention has been paid to the development of chemical methods for determining the available potassium in a soil, where availability is defined as the amount of potassium a crop will take up from the soil, or alternatively where it is used as a measure of the ability of the soil to supply the potassium needed for maximum crop yield under the system of agriculture being practised. The general principle of all the methods is to extract a representative sample of the soil with a suitable reagent using a suitable technique, the success of the method is measured by the correlation coefficient between the amount of potassium extracted in the laboratory and the response of the crop to a potassium fertiliser in the field. The discussion has already shown that there can be no perfect method because the response of the crop to potassium depends on many factors, such as the effect of the rate at which the crop is removing potassium from the soil, the potassium concentration in the soil, and the effect of moisture content on the rate of diffusion of potassium from the clay to the root surfaces, on the one hand; and on the other, factors such as the effective volume of soil tapped by the crop roots. It is likely that over a short period of time, particularly if the soil is moist, the rate of uptake will be directly dependent on the potassium ion concentration in the soil solution; over intermediate periods it will depend on the amount of exchangeable potassium which the soil releases for a given drop in activity ratio, and, over longer periods, on the rate at which non-exchangeable potas-

1 *J. Chem. Soc.*, 1871, **24**, 276.
2 See, for example, T. M. Addiscott and A. G. Johnston, *J. agric. Sci.*, 1971, **76**, 553.
3 *Rothamsted Rept.* 1966, 57.

sium is released to the readily exchangeable form at a given potassium ion concentration in the soil solution. Further, crops differ in the minimum potassium concentration they require for optimum growth, and this again may vary with the stage of development of the crop. Thus, kale, for example, will extract more potassium from a soil than will ryegrass, and this more than barley or potatoes.[1]

No really critical studies have been made on the relative importance of these various factors in controlling crop response to a given level of potassium fertiliser. The potassium activity ratio by itself has some value, but it need give no indication of the amount of potassium that can be removed from the soil before it falls to a low value. It is possible that the most efficient method of prediction of fertiliser response will be based on some combination of the activity ratio at the beginning of the cropping season and on the amount of readily exchangeable potassium held by the soil, or on the potassium buffering power (see p. 99) of the soil; but both of these are usually fairly closely correlated with the exchangeable potassium, which, in turn, is also fairly closely correlated with the amount of non-exchangeable potassium that is released for a given drop in activity ratio. Since the concept of exchangeable potassium is not precise, T. M. Addiscott[2] suggested it should be defined in relation to the crop being considered. He showed that the potassium activity ratio of a soil that has been leached with N ammonium acetate for the determination of exchangeable potassium is about 1.9×10^{-4} M$^{\frac{1}{2}}$, and of one that has been leached with sodium bicarbonate in Olsen's method about 5.9×10^{-4} M$^{\frac{1}{2}}$, and of one that has been extracted with an H-resin (he used Zeocarb 225) about 3.3×10^{-5} M$^{\frac{1}{2}}$. He also showed that potatoes can take up little potassium once the activity ratio falls below about 3.2×10^{-4} M$^{\frac{1}{2}}$, while ryegrass would be taking up potassium down to 6.7×10^{-5} M$^{\frac{1}{2}}$. Thus, Olsen's bicarbonate method should be more appropriate for potatoes and ammonium acetate for ryegrass.

Table 24.6, due to R. G. Warren and G. W. Cooke,[3] illustrates the value and limitations of these methods, and the extractant chosen for the table is 1 per cent citric acid, because it gave the best correlation with response of the range of extractants chosen. This table should be compared with Table 23.3, page 578, which gives the corresponding results for phosphate responses. It shows that beet responded to potassium on some soils high in potassium, and did not respond on some soils low in potassium, for reasons that are not known.

A single dressing of a potassium fertiliser cannot, in practice, always supply all the potassium needed for maximum yield of the crop if the soil was initially low in potassium, a fact that has already been noticed for young

1 R. J. B. Williams, G. W. Cooke and F. V. Widdowson, *J. agric. Sci.*, 1963, **60**, 353.
2 *J. agric. Sci.*, 1970, **74**, 119, 123, with J. D. D. Mitchell, ibid., 495.
3 *J. agric. Sci.*, 1962, **59**, 269. For a comparison of different extracts for different crops see L. J. Hooper, *Fertiliser Soc. Proc.* 118, 1970.

TABLE 24.6   Response of sugar-beet to potassic fertiliser on soils of different potassium status. English sugar-beet experiments 1936–46. Potassic fertiliser— 125 kg/ha K in presence of 100 kg/ha N. Response as extra sugar in kg/ha

K soluble in 1% citric acid mg K/100 g soil	Number of centres	Mean response	Number of centres with response			
			over 690	650 to 300	290 to 130	0 to −1030
2·6–5·7	94	650	43	31	12	8
5·8–9·5	92	280	14	29	28	21
over 9·6	62	90	6	5	25	26

vegetable crops, and which also applies to potatoes, for example. This can be illustrated by some results obtained on the Exhaustion Plots at Rotham-sted.[1] Some of these plots received about 100 kg/ha K annually from 1857 to 1874 and then 150 kg from 1876 to 1901, and others received no potassium. After 1901 no further potassium has been given and the land cropped to cereals most years. In 1957 and 1958 a number of crops were grown on these plots with different levels of fresh potassium fertiliser. Table 24.7, gives some of the results of this experiment, and it shows that the potassium residues

TABLE 24.7   Crop responses to potassium fertiliser residues and fresh additions of potassium fertiliser. Rothamsted, Hoosfield Exhaustion Plots. Rate of K fertiliser, barley and swedes 63 kg/ha. Potatoes and sugar-beet 125 kg/ha. Figures in brackets = uptake of K by crop in kg/ha

Yields in t/ha	Yield without residues or fertiliser	Response to		
		Residue	Fertiliser without residue	Fertiliser with residues
Barley: Grain	3·24 (43)	0·20 (16)	0·23 (13)	0·23 (20)
Swedes: Roots	37·6 (88)	9·1 (47)	2·5 (20)	10·0 (54)
Potatoes: Tubers	17·1 (38)	10·5 (34)	14·2 (58)	19·6 (81)
Sugar Beet: Roots	31·2 (121)	6·0 (83)	4·8 (32)	2·5 (82)

left in the soil since manuring ceased in 1901 were adequate to supply the potassium needs of swedes and sugar-beet, and were more effective than the 63 kg given as fertiliser to the swedes or the 125 kg to the beet on the plots which had not received any potassium since 1857. Potatoes, on the other hand, gave a larger response to 125 kg of fresh potassium than to the residues, but a still larger response to both the residues and the fresh fertiliser; and the evidence from the response curves suggests that this difference would persist if still higher levels of fresh potassium fertiliser had been given. This result is

1 R. G. Warren and A. E. Johnston *Fertiliser Soc. Proc.* 72, 1962, and see *Rothamsted Rept.*, 1954, 59 for a similar result with timothy (*Phleum pratense*).

thus similar to the response of potatoes to residual phosphate fertiliser which has been illustrated in Fig. 23.6 on p. 600, and shows that it takes time to build up the potassium status of the soil, presumably because it is difficult to distribute a single dressing of fertiliser evenly through the depth of cultivation. A single heavy dressing of a potassium fertiliser can, however, have the undesirable effect of increasing the osmotic pressure of the soil solution, due to the high concentration of the associated anion, usually chloride, which can harm the growth of young plants during spells of dry weather. It is for this reason that large dressings of farmyard manure are more valuable than fertilisers for building up the relatively high concentration of potassium ions in the solution needed by many young vegetable crops, for example, if they are to make rapid early growth.[1]

Crops differ in the proportion of the potassium added as fertiliser that they take up, when all other nutrients are present in ample supply. Grass will take up about 80 per cent of added potassium, whether the dressing is large or small, as will be shown in the next section. At Rothamsted kale and potatoes will take up to 50 per cent of the added potassium, as does wheat on the Broadbalk plots receiving a balanced fertiliser, but in general the cereals take up less, as measured by their content at harvest time, though if the behaviour of wheat is at all typical they may take up over 50 per cent at heading time.[2]

The effect of leys on the potassium economy of soils

The grass crop can have a very great influence on the potassium status of the soil, whether it is poorly or well supplied, for it is a very strong extractor of potassium. Grass therefore competes very strongly with clover in grass-clover mixtures for a limited potassium supply, perhaps because it has a much more extensive root system in the surface soil; so it is only possible to have a vigorous clover component in a sward if the soil's potassium status is adequate. But grass can take up very large amounts of potassium from soils receiving a high level of potassium fertiliser. Table 24.8 illustrates this for a four-year ryegrass ley at the Grassland Research Institute, Hurley,[3] for as the level of potassium fertiliser added to the ley increases, not only does the yield of the grass increase, but so also does its potassium content, so that an annual dressing of 250 kg/ha K is inadequate to keep the soil in potassium balance. The results for the last three years of the ley, that is excluding the year of establishment, were that for every 100 kg of potassium added as fertiliser, 80 kg was harvested in the crop, over the range of fertiliser used, and since the grass with no potassium fertiliser removed about 165 kg per year, the 250 kg potassium added as fertiliser was quite inadequate to make good

1 F. Haworth and T. J. Cleaver, *J. Hort. Sci.*, 1961, **36**, 202; 1963, **38**, 40; 1966, **41**, 299.
2 F. Knowles and J. E. Watkins, *J. agric. Sci.*, 1931, **21**, 612, W. E. Chambers, ibid., 1953, **43**, 479.
3 M. J. Hopper and C. R. Clement, *Trans. Comm. 2nd and 4th Int. Soc. Soil Sci.*, 1966, 237.

this deficiency. Table 24.8 also shows that the grass had reduced the exchange-able potassium in the soil to a threshold value of 6·8 mg/100 g soil and the potassium ion concentration, when the soil was in equilibrium with 0·01 M $CaCl_2$, to 0·8 × 10^{-4} M compared with 1·4 × 10^{-4} M for the high potash plots, unless more than 125 kg K was added annually.[2] There were also plots receiving this same heavy dressing of nitrogen and 125 kg/ha K annually which were grazed by sheep, and after 4 years the exchangeable potassium in the surface 15 cm of the soil was 25 mg/100 g soil. The following year these plots received the same level of nitrogen, but were cut for silage, and in the absence of potassium fertiliser the exchangeable potassium fell to 7·9 mg, almost to the threshold value, and if given 125 kg/ha of potassium to 8·8 mg.

TABLE 24.8   Uptake of soil potassium by ryegrass when manured with potassium fertilisers. Grassland Research Station, Hurley. Soil. Nitrogen: as Nitrochalk (ammonium nitrate/calcium carbonate). Potassium: as potassium chloride, half applied in March, half in June. Totals for 4 years of the ley

Potassium added as fertiliser kg/ha	Potassium removed in grass kg/ha	Decrease in soil potassium kg/ha	Exchangeable potassium at end of experiment mg/100 g	Yield of grass dry matter t/ha	% potassium in dry matter
0	660	660	6·8	36	1·9
250	840	590	6·8	38	2·2
500	1030	530	6·6	39	2·6
1000	1290	290	9·9	40	3·2

The foregoing results have considerable importance when fertiliser pro-grammes are being discussed on the farm. It is obviously uneconomic to try to maintain the potassium status of a soil under intensively managed grass cut for silage, but this has the corollary that the crop following a ley, even a one-year grass ley if heavily manured with nitrogen, may need a substantial dressing of a potassium fertiliser if its yield is not to be limited by a shortage of available potassium. This point is of especial importance if winter wheat follows the ley, for it has a substantial potassium requirement, particularly up to heading time, after which the total potassium in the crop may decrease to about one-half at harvest time.

Nitrogen

By far the largest part of the combined nitrogen in a soil is in the organic matter fraction, and in the humus fraction in normal arable soils. This nitro-gen is not available to crops, except possibly to crops with well developed mycorrhiza. The soil also contains a small amount of nitrogen as nitrate

1 *Grassland Res. Inst. Ann. Rept.*, 1965, 21.

dissolved in the soil water, and as ammonium either as an exchangeable ion or fixed within the edges of clay particles in the same way that potassium is. The nitrate and readily exchangeable ammonium rarely compose more than 1 or 2 per cent of the nitrogen present in the soil. It has been found difficult to determine the amount of fixed ammonium in the soil, since the methods in use have been based on treating the soil with a hydrofluoric acid-hydrochloric acid mixture to decompose much of the clay, and determining the amount of ammonium so released; but some of the ammonium produced by this treatment may have been derived from the organic matter itself.[1] The fixed ammonium probably only forms a small proportion of the nitrogen in most surface soils, but may constitute up to half the nitrogen present in the deeper subsoil, particularly in deep tropical soils.[2] Whereas all the nitrate and the readily exchangeable ammonium is available to plants growing in the soil, and the exchangeable ammonium is potentially readily oxidisable to nitrate, much of the fixed ammonium is held in a form that is not accessible to the bacteria or plant roots.[3]

The principal source of nitrogen for crops growing on land not receiving any nitrogen fertiliser is that which is released as ammonium by the soil population decomposing the soil organic matter; for the other source, rainfall, only adds a few kg/ha annually.[4] Many attempts have been made to estimate the amount of mineral nitrogen so released during the growing season based on incubating the moist soil aerobically for a standard time though this depends on a number of factors not all of which are within the control of the farmer. Many workers have tried to divide the organic nitrogen in the soil into two groups, the readily mineralisable and the remainder, but although this is often a convenient concept, it can have no absolute validity, for the amount of organic nitrogen that can be mineralised depends, among other factors, on the frequency and intensity of the drying of the soil.

The concept of readily mineralisable nitrogen is, however, useful in practice, and provided the technique for determining it is properly standardised there is often a good correlation between this nitrogen fraction in the soil and the yield or nitrogen uptake by a cereal crop for groups of soils in a region of fairly similar climates and soil types. The correlation coefficient between the readily mineralisable nitrogen and either the yield or nitrogen uptake by the crop, or the response to a moderate level of nitrogen fertiliser, can be as high as 0·7 to 0·8.[5] But since there can be no exact relationship between a routine

1 See for example J. R. Freney, *J. agric. Sci.*, 1964, **63**, 297, and for a method to minimise this error J. A. Silva and J. M. Bremner, *Proc. Soil Sci. Soc. Amer.*, 1966, **30**, 587.
2 G. Rodrigues, *J. Soil Sci.*, 1954, **5**, 264. A. W. Moore and C. A. Ayeke, *Soil Sci.*, 1965, **99**, 335.
3 See, for example, C. A. Bower, *Proc. Soil Sci. Soc. Amer.*, 1050, **15**, 119, and F. E. Allison *et al.*, *Proc. Soil Sci. Soc. Amer.*, 1953, **17**, 107; *Soil Sci.*, 1953, **75**, 373.
4 See E. Erikson, *Tellus*, 1952, **4**, 215 for world figures and C. E. Jung, *Trans. Amer. Geophys. Un.*, 1958, **39**, 241 for America. Rothamsted receives 4–6 kg/ha annually.
5 For a review of work before 1955 see G. W. Harmsen and D. A. van Schreven. *Adv. Agron.*, 1955, **7**, 299. For later examples J. K. R. Gasser *J. Sci. Fd. Agric.*, 1961, **12**, 562; 1963, **14**, 269;

determination of available nitrogen based on an incubation technique and the amount of mineral nitrogen produced in the soil during the period when it is most needed by the crop, purely chemical methods for measuring some property assumed to be related to the readily mineralisable nitrogen have been developed which are quick and more adapted to routine use. Thus, the ammonium produced in anaerobic incubation of the soil,[1] or by gentle oxidation of the soil with alkaline permanganate[2] or by gentle hydrolysis with cold sodium hydroxide,[3] or barium hydroxide,[4] or boiling or autoclaving the soil with water and determining the total nitrogen brought into solution,[5] all give estimates of the nitrogen supplying power of the soil that are comparable in reliability with the mineralisable nitrogen. All these methods are, however, based on the assumption that the principal source of nitrogen for the crop is that produced by decomposition of organic matter in the surface soil, and they become inapplicable in so far as the crop is taking up nitrogen from the deeper subsoil. They are also less applicable to crops that receive much intertillage, such as potatoes, root crops or maize, because the intertillage itself increases the oxidation of organic matter and therefore the production of nitrate. Further, if crops are being grown without any soil tillage, based on a suitable herbicide regime, the amount of mineralisable nitrogen produced in the soil will be decreased, so that higher levels of nitrogen fertiliser will have to be used (see p. 798).

Fallowing land increases the nitrate content of the soil, and it was an old belief that every additional ploughing given to a bare fallow in summertime increased the yield of the following winter wheat crop. This is because the greater the volume of soil that is dried between each rainstorm, the greater release of nitrate nitrogen when the soil is wetted. Table 24.9 illustrates this effect of fallowing the land on the subsequent winter wheat crop on Broadbalk: since 1935 one-fifth of the field is fallowed annually, so it is possible to determine the effect of fallow on the subsequent four wheat crops. The table shows that the fallow was equivalent to a dressing of about 160 kg/ha N in the first crop after the fallow, but the effect is small for subsequent crops. Figure 16.1, p. 336, shows that this is approximately the extra amount of nitrate in the fallow soil at the end of the summer. Yet as this diagram shows, and as has been confirmed by J. K. R. Gasser,[6] the amount of nitrate in the surface soil of the fallow and the non-fallow plots is about the same in the late autumn and winter. Some results of H. L. Richardson, quoted by Gasser

and D. J. Eagle, *J. Sci. Fd. Agric.*, 1961, **12**, 712; 1963, **14**, 391. See C. R. Clement and T. E. Williams, *J. Soil Sci.*, 1961, **13**, 82, for the effect of leys.

1 S. A. Waring and J. M. Bremner, *Nature*, 1964, **201**, 1951.
2 E. Truog, *Canad. J. Soil Sci.*, 1959, **39**, 120.
3 A. H. Cornfield, *Nature*, 1960, **187**, 260.
4 D. S. Jenkinson, *J. Sci. Fd. Agric.*, 1968, **19**, 160.
5 D. R. Keeney and J. M. Bremner, *Agron. J.*, 1966, **58**, 498. G. Stanford, *Soil Sci.*, 1970, **109**, 190.
6 *Pl. Soil*, 1962, **17**, 209.

TABLE 24.9   Response of wheat to a one-year fallow. Broadbalk 1935–1964.*
Yield of grain in t/ha

Nitrogen manuring in kg/ha N as sulphate of ammonia	Years after fallow			Increase in yield over third and fourth years	
	First	*Second*	*Third and fourth*	*In first year*	*In second year*
0	2·32	1·35	1·35	0·97	0·00
48	2·71	1·85	1·71	1·00	0·14
96	2·94	2·45	2·24	0·70	0·21
144	3·07	2·84	2·57	0·50	0·27
37 t/ha FYM	3·21	2·92	2·58	0·63	0·34

* From *Rothamsted Rept. 1968*, Part II 46.

are summarised in Table 24.10, which indicate that at least some of this nitrate was held in the subsoil between 25 and 70 cm deep, which was as deep as he sampled. F. J. Sievers and H. F. Holtz[1] found a similar result in the heavy soils of the Palouse area of Washington State, and they showed that nitrate produced during the summer fallow was stored in the subsoil between 60 and 150 cm deep during winter and used by the following wheat crop.

TABLE 24.10   Subsoil storage of nitrate over winter. Broadbalk, 1935–36. Amount of nitrate + ammonium nitrogen in the soil in ppm. Section III fallowed 1934–35, first crop after fallow; section II, fourth crop after fallow.

Depth of sampling in cm	November Section		April Section		June Section	
	III	*II*	*III*	*II*	*III*	*II*
Plot 2B (37t/ha farmyard manure)						
0–25	12	7	7	7	7	6
25–45	16	11	17	8	4	5
45–69	20	10	14	11	3	4
Plot 3 No manure						
0–25	5	5	5	3	4	3
25–45	10	5	6	8	5	4
45–69	9	4	9	6	4	2
Plot 8 145 kg/ha N as sulphate of ammonia $\frac{1}{4}$ in October, $\frac{3}{4}$ in May						
0–25	10	8	6	6	6	4
25–45	12	10	7	6	5	3
45–69	13	7	11	8	4	2

1 *Washington Agric. Expt. Stat. Bull.* 166, 1922.

The availability of the nitrate held in the subsoil in early winter to the following wheat crop depends more on the spring than the winter rainfall. At Rothamsted winter wheat yields are usually depressed by an additional inch of rain above the average at whatever time during the growing season it falls, [1] and correspondingly increased in years when the rainfall is below average; but the effect of the period of the year when the additional rain comes is important. Nitrate is washed out of the soil by winter rain, but a proportion of the nitrate washed out of the surface soil is held in the subsoil. If the spring is dry the wheat roots can take up this nitrate, presumably because a dry spring encourages deeper and more active root growth. The net result is that an autumn applied nitrogen fertiliser,[2] or the nitrate produced during a summer fallow, are more effective if the spring is dry than if it is wet, and the benefit of a dry spring can be more marked than the harm of a wet winter.

This result is probably only valid for fairly well-structured loam and clay soils, and is due to the fact that drainage water does not percolate uniformly through the body of the soil, but moves down cracks and channels. If the crumbs and clods between these cracks contain nitrates, as they will do in the autumn after a summer fallow, the nitrate will only diffuse slowly into the percolating water, and will also diffuse out from this water into the subsoil crumbs when this water ceases to move.[3] Thus nitrate can only be slowly moved out of well-structured soils under conditions of intermittent and not very heavy rainfall, such as often occurs in the United Kingdom. This result does not apply if nitrate is added in autumn when the soil is wet and the weather is persistently wet, for then all the nitrate is washed out of the profile with the percolating water. The importance of this subsoil storage of nitrate during winter in general agricultural practice must not be overemphasised. It is probably only of relevance in well-structured medium to heavy-textured soils, for the general result is that the higher the winter rainfall, the greater the need of the crop for nitrogen in the spring. Thus, over a large series of experiments with sugar-beet in England, mainly on light to medium-textured soils, D. A. Boyd[4] found that an additional inch of rain above the average in winter increased the response of beet to 100 kg/ha N by 310 kg/ha of sugar, and F. van der Paauw[5] in Holland showed that the optimum level of nitrogen to be given in the spring to a variety of crops should be increased, or decreased, by 0·25 to 1·25 kg/ha for every 1 mm of winter rainfall above or below the average, depending on the soil and the crop.

1 R. A. Fisher, *Phil. Trans.* 1925, **213B**, 89.
2 'Alumnus', *J. agric. Sci.*, 1932, **22**, 101.
3 R. K. Cunningham and G. W. Cooke, *J. Sci. Fd. Agric.*, 1958, **9**, 317.
4 With H. V. Garner and W. B. Haines, *J. agric. Sci.*, 1957, **48**, 464. See also D. J. Eagle in *Residual Value of Applied Nutrients*, Min. Agric. Tech. Bull. 20, p. 145, 1971.
5 *Pl. Soil*, 1962, **16**, 361; 1963, **19**, 324.

Nitrogen fertilisers[1]

Most nitrogen fertilisers are water soluble and derived from ammonia, and the more important for general farm use are ammonium sulphate, nitrate and phosphate, ammonia itself either anhydrous or in solution, and urea. If the fertiliser persists in normal soils for any length of time, the ammonium will be oxidised to nitrate, so that fertiliser nitrogen taken up from non-acid soils a few weeks after it has been applied is likely to be in the form of nitrates. In so far as these fertilisers differ in effectiveness, nitrates tend to be some-what more efficient than urea or ammonium sulphate when used as a quick-acting fertiliser. The only two points on which care must be taken in the use of nitrates is that, since they are very soluble in water and not absorbed by the soil, they will raise the osmotic pressure of the soil solution round seedlings to a damaging level if used during dry weather at too high a level, and they should not be used in poorly drained soils and particularly not for padi rice because of loss by denitrification.

Ammonium sulphate added to a soil causes it to lose exchangeable calcium or some calcium from any calcium carbonate present, and in general for every 100 kg of ammonium sulphate that is added to a soil about 45 kg of calcium are removed in the drainage water.[2] This is because the sulphate in the fertiliser, in so far as it is not taken up by the crop, washes out as calcium sulphate, and a part of the nitrate formed from the ammonium also washes out as calcium nitrate. The loss of sulphate accounts for about 30 kg of calcium and of nitrate for about 15 kg. This loss of about 45 kg of calcium per 100 kg of ammonium sulphate has been found over a very wide range of con-ditions, both in temperate and tropical countries, provided the soils are not too acid. The constancy of this figure is a little puzzling because one would expect a considerable variation in the amount of calcium lost as the nitrate. Although all ammonium fertilisers cause a loss of soil calcium as nitrate, ammonium sulphate is the only one in common use whose anion washes out almost completely as the calcium salt, so it causes the greatest drain on the soil's supply of calcium of any of the fertilisers in general use. Because ammonium sulphate causes this large loss of calcium from the soil, it should only be used with caution in field experiments, since many results obtained in the past using this fertiliser have been due as much to the acidity the ferti-liser has caused as to the nitrogen or sulphur it has supplied.

Ammonium sulphate can also be a very inefficient nitrogen fertiliser if broadcast on the surface of calcareous soils, particularly on those containing more than 10 per cent calcium carbonate under English conditions,[3] for the

1 For an excellent review of this subject see G. W. Cooke, *Fertiliser Soc.*, *Proc.* 80, 1964 and his books *The Control of Soil Fertility* (1967) and *Fertilising for Maximum Yield* (1972).
2 See, for example, W. H. Pierre, *J. Amer. Soc. Agron.*, 1928, **20**, 270; E. M. Crowther in *Fifty years of Field Experiments at Woburn*, by E. J. Russell and J. A. Voelcker, 1936, p. 334; F. Leutenegger, *East Afr. Agric. J.*, 1956, **22**, 81.
3 J. R. Devine and M. R. J. Holmes, *J. agric. Sci.*, 1964, **62**, 377. J. K. R. Gasser, *J. Soil Sci.*, 1964, **15**, 258.

ammonium sulphate reacts with the carbonate to give calcium sulphate and free ammonia. Ammonium sulphate appears to suffer a greater loss of ammonia in this way than does ammonium nitrate or phosphate.[1] Ammonium sulphate can also be undesirable for use in some padi rice soils, for the sulphate becomes reduced to sulphide which can be toxic to the rice roots. It is because of this that ammonium chloride is fairly widely used in Japanese rice fields, for it does not suffer this disadvantage. Ammonium nitrate and the ammonium phosphates are widely used nitrogen fertilisers, and in the past the nitrate was usually intimately mixed with calcium carbonate, which reduced its natural hygroscopicity. This carbonate had the incidental effect of neutralising the acidity caused by the loss of calcium nitrate from the soil by leaching.

Urea is the most concentrated solid nitrogen fertiliser on the market, containing about 46 per cent N; in many countries it is the cheapest form of fertiliser nitrogen on the farm.[2] It is soluble in water and is deliquescent, but it is now produced in a form that stores well by coating urea prills or granules with a suitable resin; or it can be bonded with sulphur, so it is both a nitrogen and a sulphur fertiliser.[3] It used to contain biuret, $(NH_2CO)_2.NH$, as an impurity, which is toxic to many crops if present in too large a concentration, but presentday samples of urea contain less than 1 per cent, which is below the danger limit. It is not strongly absorbed by soils, so can be washed out if applied in periods of continuous wet weather, and it is hydrolysed to ammonium carbonate in the course of time. Most of the practical problems of using urea as a fertiliser turn on the fate of this ammonium carbonate, for it hydrolyses easily to release free ammonia. If the urea is spread on the surface of a soil and allowed to get moist, much of the nitrogen is liable to be lost to the atmosphere as ammonia, and even if the urea is worked into the soil surface, ammonia may be lost if the soil is sandy, so only has a low ammonia-absorbing power.[4] J. R. Simpson[5] also found it inefficient on many pastures due to their surface soils having a very high urease activity. Further, the free ammonia formed in the soil is toxic to germinating seeds and young seedlings, so it must not be used in fertilisers which are placed close to the seed; and under some conditions the nitrite concentration in the soil sometimes rises to toxic levels.[6] These harmful consequences can be mitigated if urea

1 S. Larsen and D. Gunary, *J. Sci. Fd. Agric.*, 1962, **13**, 566.
2 For a review of the problems of using urea as a fertiliser see T. E. Tomlinson, *Fertiliser Soc. Proc.* 113, 1970.
3 P. M. Giordano and J. J. Mortvedt, *Agron. J.*, 1970, **62**, 612; D. A. Mays and G. L. Terman, ibid., 1969, **61**, 489. O. R. Lunt, *Trans 9th Int. Cong. Soil Sci.*, 1968, **3**, 377.
4 J. K. R. Gasser, *J. Soil Sci.*, 1964, **15**, 258; M. N. Court, J. C. Dickens, *et al.*, *J. Soil Sci.*, 1963, **14**, 247; and G. M. Volk. *Agron. J.*, 1959, **51**, 746.
5 *Trans. 9th Int. Cong. Soil Sci.*, 1968, **2**, 459.
6 A. J. Low and F. J. Piper. *J. agric. Sci.*, 1961, **57**, 249; M. N. Court, R. C. Stephens and J. S. Waid, *J. Soil Sci.*, 1964, **15**, 42, 49.

phosphate is used.[1] The value of urea as a fertiliser would often be increased if its rate of hydrolysis to ammonium carbonate could be decreased. There are a number of substances which inhibit the activity of urease enzymes in soils under laboratory conditions, such as dihydric phenols and quinones,[2] and hydroxamates but none are yet in use on the farm.

Since, as already mentioned, urea is both concentrated and cheap, it is most important that the conditions should be clearly understood in which it can be used efficiently. General experience has shown that if used in dressings not exceeding 45 to 70 kg/ha N, it is usually, but not always, as effective as ammonium or nitrate fertilisers, but dressings of 100 kg/ha N are usually appreciably less efficient.[4] However, even this generalisation needs to be used with care, for G. C. Mees and T. E. Tomlinson[5] have given an example of 17 kg/ha N as urea harming the very early growth of wheat seedlings on a sandy soil although 70 kg was safe on a clay soil. In general, urea tends to be least suitable as a nitrogen fertiliser for grassland, for it can only be broadcast on the surface of the sward so that ammonia losses to the atmosphere are likely to be large.

A fertiliser of increasing importance in highly developed countries is ammonia itself, either anhydrous or as a saturated or supersaturated solution. This is the cheapest nitrogen fertiliser to manufacture, but it needs special facilities for transport and storage; and it must be injected into the soil, which can be difficult in many cases. It must be injected deep enough so that as little as possible escapes into the air, and the absorptive power of a soil depends on its texture and moisture content.[6] Further, ammonia is toxic to seedlings. It is also toxic to nitrifying organisms if used in sufficient concentration, so may last much longer in the subsoil than ammonium salts.

Water-soluble nitrogen fertilisers given to one crop used to be considered to have no residual effect on the following crop, except in so far as they affected the soil acidity. Nitrogen fertilisers applied to arable crops have little effect on the level of soil organic matter or organic nitrogen in the soil, so that it is not possible to build up the nitrogen reserves in a soil by a generous use of nitrogen fertilisers, as happens with phosphate and potash. But under suitable conditions a heavy dressing of a nitrogen fertiliser to one crop can have a residual effect on the succeeding crop, as one would expect from the ability of some soils to store nitrates in their subsoil over winter. Thus, F. V. Widdowson and A. Penny[7] found at Rothamsted that the average residual

1 J. K. R. Gasser and A. Penny, *J. agric. Sci.*, 1967, **69**, 139. F. V. Widdowson and A. Penny, ibid, 1969, **73**, 125.
2 See J. M. Bremner and L. A. Douglas, *Soil Biol. Biochem.*, 1971, **3**, 297.
3 K. B. Pugh and J. S. Waid, *Soil Biol. Biochem.*, 1969, **1**, 195, 207.
4 See, for example, J. R. Devine and M. R. J. Holmes, *J. agric. Sci.*, 1963, **61**, 391, and P. A. Collier and A. E. M. Hood, ibid., 1970, **74**, 153.
5 *J. agric. Sci.*, 1964, **62**, 199.
6 For reviews on its use in the USA see R. D. Hanck and D. A. Russel, *Outlook Agric.*, 1969, **6**, 3; and in Europe, P. F. J. van Burg, ibid., 53.
7 *J. agric. Sci.*, 1965, **65**, 195. For a review of English results see J. K. R. Gasser, in *Residual Value of Applied Fertiliser*, Min. Agric. Tech. Bull. 20, p. 114, 1971.

effect of 190 kg/ha N as sulphate of ammonia given to potatoes, for the following wheat crop was equivalent to 63 kg/ha N as top dressing in the spring, and that the residual effect of 125 kg/ha N top dressing on winter wheat was equivalent to 25 kg/ha N applied to the following potato crop. As already noted, however, this residual effect depends on the winter rainfall, being larger in years of dry than of wet winters.[1]

The nitrogen present in a soluble nitrogen fertiliser applied to an established ley or permanent pasture during periods of active growth may be taken up almost completely by the grass, if allowance is made for the nitrogen in the root growth as well as in the tops; though if allowance is not made for this, between two-thirds and three-quarters of the applied nitrogen is usually found in the tops;[2] arable crops rarely take up more than between one-third and one-half of that applied.[3]

A part of the difference between arable and established grass is that the superficial roots of the grass permeate the soil intimately, whereas those of the arable crops are much further apart so that the nitrates can wash through the soil without ever coming into contact with a root. Further, the intimate root system of a grass pasture will usually take up the fertiliser from the soil fairly rapidly, whereas, with arable crops, it may last an appreciable time in the surface soil and in consequence has a greater probability of being denitrified. It is difficult to get good quantitative field evidence on the fate of the nitrogen not taken up, but the indications are that the losses of nitrogen are appreciably higher than can be accounted for by leaching.

Considerable interest has been taken in the problem of increasing the efficiency of nitrogen fertilisers, because of the very large amounts of applied nitrogen that are not harvested. First, it is possible to increase the efficiency, particularly when the crop foliage is well developed, by spraying urea on the crop. This will give uptakes of the order of 70 per cent, but it is only possible to apply a relatively low dressing (about 10 kg/ha N per application) if efficiencies as high as this are to be obtained.[4] In practice, the correct timing of the application, after the crop roots have developed and during conditions of active growth, or the placement of the fertiliser near the seed, will increase the proportion of fertiliser taken up. Second, there is also the possibility of reducing the loss of nitrates by leaching through the use of ammonium fertilisers with which a nitrification inhibitor is incorporated. These inhibit the oxidation of ammonium to nitrite, and 2-chloro-6-(trichloromethyl)-pyridine (N-Serve) is an example, but field trials have not yet shown the inhibitor to be sufficiently efficient to affect appreciably the loss of nitrate from the soil.[5] A third method is the use of either granules of fertiliser coated in such

1 F. van der Paauw, *Pl. Soil*, 1963, **19**, 324.
2 T. W. Walker, H. D. Orchiston and A. F. R. Adams, *J. Brit. Grassland Soc.*, 1954, **9**, 249.
3 For a review see F. E. Allison, *Adv. Agron.*, 1966, **18**, 219.
4 G. N. Thorne, *Rothamsted Rept.* 1954, 188.
5 J. K. R. Gasser, *J. agric. Sci.*, 1965, **64**, 299; 1968, **71**, 243; 1970, **74**, 111. G. C. Turner, L. E. Warren *et al.*, *Soil Sci.*, 1962, **94**, 270. C. A. I. Goring, *Soil Sci.*, 1962, **93**, 211, 431.

a way that they dissolve slowly in the soil, or of granules of nitrogen compounds that are only slowly soluble, such as magnesium ammonium phosphate or some substituted ureas, for example, iso-butylidene di-urea which is in commercial production in Japan, or of insoluble substances which hydrolyse to ammonia only slowly, as may happen with some other insoluble

$$CH_3 \diagdown CH{-}CH \diagup NH.CO.NH_2 \qquad NH_2.CO.NH{-}CH \diagup NH \diagdown CO$$

iso-butylidene di-urea

probable structure of crotylidene di-urea

di-ureas, such as the crotylidene which has been manufactured in Germany.[1] The value of some of these slow-release nitrogen fertilisers depends on the soil temperature, for hydrolysis can be very slow in cold but rapid in warm soils, particularly if the level of biological activity in the soil is high.[2]

To conclude this section on nitrogen fertilisers, it is worth emphasising that, to make the best of them, they should be used in amounts that depend on the responsiveness of the crop to nitrogen, on the amount of nitrate in the soil at the beginning of the growing season, and on the amount that will be produced by the mineralisation of the organic matter during the growing season, and on the rainfall. The amount of nitrate in the soil depends on past rainfall and manurial history, as already described. The amount of nitrate that will be produced during the growing season depends on the past cropping history of the land. Crops following a leguminous crop such as a one-year red clover, or a bean crop, or even a trefoil crop undersown in barley and taken as a green manure, need a lower level of nitrogen than those following a ryegrass ley or a cereal crop. Thus at Rothamsted, wheat following a one-year ryegrass ley needed 63 kg/ha more nitrogen than following a red clover ley,[3] and J. C. Holmes[4] found in Scotland that on all-arable farms the optimum nitrogen dressing for barley was about 110 kg/ha, but was only between 30–50 kg where rotations containing a grazing ley were used, and the barley did not need any nitrogen following a long ley.

1 For a review, with particular reference to Japanese work, see M. Hamamoto. *Fertiliser Soc. Proc.* 90, 1966.
2 J. K. R. Gasser, *J. agric. Sci.*, 1970, **74**, 104.
3 F. V. Widdowson, A. Penny and R. J. B. Williams, *J. agric. Sci.*, 1963, **61**, 397. See also D. J. Eagle in *Residual Value of Applied Nutrients*, Min. Agric. Tech. Bull. 20, p. 125.
4 With W. D. Gill and J. A. B. Rodger, *J. agric. Sci.*, 1960, **54**, 291.

Sulphur

The sulphur requirements of crops are very similar to their phosphorus requirements, namely, between 10 and 30 kg/ha of each. On the whole grasses and cereals tend to take up less sulphur than phosphorus, namely, about 10 kg/ha; legumes have about the same demand for each, about 25 to 30 kg/ha, while the brassica crops have a high sulphur demand, about 40 to 45 kg/ha.[1] Sulphur deficiency is most widely found on leguminous crops, both grain legumes, such as groundnuts, and forage legumes, such as clovers and lucerne; but in really sulphur deficient areas, such as parts of western Canada, cereals also have their yields limited. Sulphur deficiency is rarely seen in tree crops, although it was first recognised in Malawi in young tea,[2] before it was fully established.

Sulphur occurs as sulphide in many igneous rocks, particularly basic igneous, and as these weather it is released as sulphate. Since sea water contains sulphates, many coastal alluvial soils will contain sulphates or sulphides, and many sedimentary rocks laid down as coastal marshes are rich in ferrous sulphides or calcium sulphate, as, for example, some of the Lower Lias and Gault clays in Great Britain. Sulphate is also present as an impurity, presumably as a coprecipitate, in some limestones, calcium carbonate concretions and coral,[3] but no general investigation of the sulphate contents of limestones has been made.

The atmosphere contains sulphur compounds, partly as aerosols and partly as gaseous sulphur dioxide. In areas near sea coasts the aerosols contain sulphates derived from sea spray; and in industrial areas, or areas receiving air from such areas, the aerosols contain both sulphates, particularly ammonium sulphate, and free acids derived from the burning of sulphur-containing fuels. Rain carries down some of this sulphur, and soils and plants will absorb at least some of the sulphur present as aerosol or gas.[4] There is little reliable evidence for the annual return of sulphur compounds from the atmosphere, but J. L. Brownscombe[5] has estimated that at Bracknell, in southern England, between 10–15 kg/ha S is returned annually in rainfall[6] and between two to three times this amount from the direct absorption of the aerosol and gas by the soil surface and plant leaves, assuming they are efficient absorbers of sulphates and sulphur dioxide.

Sulphur deficiency in crops is normally only found in regions well away from industrial areas or the sea, where the rainfall returns less than about

1 H. V. Jordan and L. E. Ensminger, *Adv. Agron.*, 1958, **10**, 407, and D. C. Whitehead. *Soils Fert.*, 1964, **27**, 1. These are two useful reviews.
2 H. H. Storey and R. Leach, *Ann. Appl. Biol.*, 1933, **20**, 23.
3 C. H. Williams and A. Steinbergs, *Pl. Soil*, 1962, **17**, 279.
4 D. W. Cowling and L. H. P. Jones, *Soil Sci.*, 1970, **110**, 346.
5 Personal communication. For reviews see R. Coleman, *Soil Sci.* 1966, **101**, 230, and D. C. Whitehead, *Soils Fert.*, 1964, **27**, 1.
6 See C. M. Stevenson, *Quart. J. Roy. Met. Soc.*, 1968, **74**, 56, for other British results.

5 kg/ha S annually.[1] In addition, it normally only occurs on old deeply weathered land surfaces,[2] where the soils have been strongly leached for a long period of time. It is only rarely found in arid areas.

The immediate source of soil sulphur for crops growing on a well-drained soil is the sulphate in the soil solution and the sulphate absorbed on soil colloids. There are no insoluble sulphates present in humid soils which can maintain a moderate concentration of sulphate in the soil solution, for a saturated solution of calcium sulphate is about 10^{-2} M in sulphate, so it only occurs in the root range of crops in semi-arid or arid regions. Thus drainage waters always contain dissolved sulphates, and the rivers in western Europe carry about 15 kg/ha S annually from their catchments to the sea, which is of the same order of magnitude as that which falls in the rain. However, rivers draining sulphur-deficient regions carry away much less, and the upper reaches of the Murray River in Australia removes just under 1 kg/ha.[3]

Some soils, however, can absorb sulphates and so hold them against leaching. The mechanism by which they do this is through the positive charges which develop on iron and aluminium hydroxide films in acid soils; and this absorption is shown by many humid tropical and subtropical soils, particularly by their subsoils; and it may be reduced to a low level by liming to pH 6·5 or over. This is illustrated in Fig. 24.2 for a lateritic soil,[4] and shows that its power to hold sulphate is greatly reduced if the free iron oxides are removed; and it loses this power if it is limed to over pH7. However, as noted on several previous occasions, it is uncertain in what proportion of soils iron oxide films are responsible for this adsorption, for when the effects of iron and aluminium hydroxides are properly separated, it is the aluminium rather than the iron that appears to be the more active.[5]

The principal reservoir of sulphur in soils is that contained in the soil humus, for it contains between 1·0 and 1·5 parts of sulphur to each ten parts of nitrogen by weight (see p. 305), so that as the humus becomes oxidised, nitrates and sulphates are set free in this proportion.[6] As a consequence, if organic matter of too low a sulphur content is mixed with the soil, the process of decomposition will remove some of the inorganic sulphates present in the soil and incorporate them in the humus produced. A content of 0·15 per cent of sulphur in the dry matter of wheat straw is about the minimum re-

1 See, for example, H. V. Jordan, C. E. Bardsley *et al.*, *U.S. Dept. Agric., Tech. Bull.* 1196, 1959. D. P. Drover, *J. Roy. Soc. W. Aust.*, 1960, **43**, 81 and T. W. Walker, A. F. R. Adams and H. D. Orchiston, *Pl. Soils*, 1956, **7**, 290 (for New Zealand).
2 For a description of the areas involved in Central Africa see E. W. Bolle-Jones, *Emp. J. Exp. Agric.*, 1964, **32**, 241.
3 J. R. Freney, N. J. Barrow and K. Spencer, *Pl. Soil*, 1962, **17**, 295.
4 M. E. Harward and H. M. Reisenauer, *Soil Sci.*, 1966, **101**, 326. See also E. J. Kamprath, W. L. Nelson and J. W. Fitts, *Proc. Soil Sci. Soc. Amer.*, 1956, **20**, 463.
5 N. J. Barrow, *Soil Sci.*, 1967, **104**, 342.
6 See, for example, C. H. Williams, *Aust. J. Soil Res.*, 1968, **6**, 131.

quired for its decomposition not to reduce the soil's sulphate status,[1] and contents as low as one-third of this can be found in wheat straw grown on some sulphur-deficient soils.

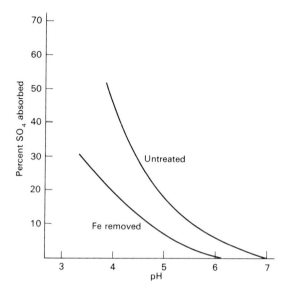

FIG. 24.2. The effect of removing free iron oxides from a Jory soil on its ability to absorb sulphate

There are a number of bacteria in the soil which will reduce sulphates to sulphides whenever conditions of poor aeration set in. Sulphides are, however, very toxic to root growth, so even temporary waterlogging of a soil well supplied with sulphate can have very serious effects on the growth of the crop, unless the soil possesses some mechanism for removing the sulphide from the soil solution. This mechanism will be discussed in detail on p. 688, for it is fundamental to the suitability of flooded soils for rice cultivation.

The ratio of protein nitrogen to organic sulphur in the leaves of a variety of crops is about 15, so if a crop is growing on a low sulphate soil, the amount of leaf protein the crop can synthesise is strictly limited.[2] In general plants need about 0·2 per cent total sulphur in the dry matter of their leaves if growth is not to be limited by its lack.[3] Crops differ in their ability to compete for soil sulphate when it is present at a low concentration, and in particular grasses will take up more sulphate than will associated clovers from low sulphate soils. Thus, T. W. Walker and A. F. R. Adams[4] showed that on a

1 B. A. Stewart, L. K. Porter and F. G. Viets, *Proc. Soil Sci. Soc. Amer.*, 1966, **30**, 355. See also N. J. Barrow, *Aust. J. Agric. Res.*, 1960, **11**, 317, 960; 1961, **12**, 306, who used C:S ratios.
2 B. A. Stewart, *Trans. Int. Soc. Soil Sci.*, *Comm. II* and *IV* (*Aberdeen*), 1966, 131.
3 W. E. Martin and T. W. Walker, *Soil Sci.*, 1966, **101**, 248.
4 *Pl. Soil*, 1958, **10**, 176.

sulphur-deficient soil on the New Zealand Canterbury Plains, grass took up 6·0 kg/ha S, while clover only took up 0·7 kg; but if 12 kg/ha S was added as gypsum, the grass took up 12 kg and the clover 4·5 kg. This example also shows that crops growing on a sulphur-deficient soil may take up a very high proportion of the sulphate added as gypsum, if the dressing used is only light.

The sulphur supplying power of a soil can be estimated from the amount of water-soluble and absorbed sulphate in the root zone, for crops can use this absorbed sulphate quite readily.[1] This total inorganic sulphate is best extracted with a solution that will desorb any absorbed sulphate such as mono-potassium phosphate,[2] though mono-calcium phosphate may be preferable as it is less likely to disperse any organic sulphur.[3] Crops do, in fact, take up more sulphur than corresponds to the drop in this inorganic sulphate[2] due to the mineralisation of organic sulphur in the early stages of the crop's growth.[3,4]

Sulphur is usually added to a soil either as a sulphate or as finely divided sulphur. Sulphur-deficient soils were often not recognised in the past because many were also phosphate deficient and were dressed with single super-phosphate, which contains between 11 and 13 per cent S in addition to its 8 per cent P. Again, the common form of nitrogen fertiliser was sulphate of ammonia, which contains 24 per cent S. Thus, crops requiring both nitrogen and phosphate would receive a substantial dressing of sulphate. If only sulphate is required, calcium sulphate or gypsum is commonly used, which contains between 13 and 23 per cent S, depending on its purity and degree of hydration. But the demand for high analysis fertilisers has encouraged two trends in sulphur fertiliser practice. One, which is only appropriate in areas of highly developed agriculture such as in parts of the USA, is the development of concentrated liquid fertilisers such as ammonium polysulphides and thionates or anhydrous ammonia containing dissolved sulphur,[5] though in this last solution the sulphur was only slowly oxidised.[6] The other trend, which can be used under a much wider variety of conditions, is the use of finely divided elemental sulphur either mixed with triple superphosphates, or mixed with concentrated fertilisers not containing ammonium nitrate.[7] This use of elemental sulphur, however, depends on the presence of sulphur-oxidising micro-organisms in the soil, about which little is known for sulphur-deficient soils. M. I. Virtolins and R. J. Swaby[8] found that about one-third

1 C. H. Williams and A. Steinbergs, *Pl. Soil*, 1964, **21**, 50.
2 K. Spencer and J. R. Freney, *Aust. J. Agric. Res.*, 1960, **11**, 948. M. Cooper, *Trans. 9th Int. Congr. Soil Sci.*, 1968, **2**, 263.
3 R. L. Fox, R. A. Olson and H. F. Rhoades, *Proc. Soil Sci. Soc. Amer.*, 1964, **28**, 243.
4 D. C. Nearpass, M. Fried and V. J. Kilmer, *Proc. Soil Sci. Soc. Amer.*, 1961, **25**, 287. For a review L. E. Ensminger and J. R. Freney. *Soil. Sci.*, 1966, **101**, 283.
5 R. Coleman, *Soil Sci.*, 1966, **101**, 230. J. F. Parr and R. A. Papendick, *Soil Sci.*, 1966, **101**, 336.
6 J. F. Parr and P. M. Giordano, *Soil Sci.*, 1968, **106**, 448.
7 S. L. Tisdale and W. L. Nelson, *Soil Fertility and Fertilisers*, 2nd Edn, Macmillan, 1966.
8 *Aust. J. Soil Res.*, 1969, **7**, 171. For an earlier review see R. L. Starkey, *Soil Sci.*, 1966, **101**, 297.

of the Australian soils they examined did not oxidise sulphur, and of those which did the dominant oxidising organisms depended on the pH of the soil. In general *Thiobacillus thio-oxidans* was dominant if the pH was below 5, *Th. thioparus* if between 5 and 6·5; and if above pH 6·5, as in most of the soils, the organisms were either heterotrophic bacteria or yeasts. The rate of oxidation is sensitive to the phosphate status of the soil[1] and to its temperature, aeration and moisture content, and often has a fairly sharp maximum about field capacity.[2] In some sulphur-deficient soils it may take an appreciable time for sulphur-oxidising bacteria to build up to an appreciable population.

Sulphur-containing fertilisers can have little long-lasting residual effect as there is no mechanism for holding much inorganic sulphate in the soil. Fertilisers containing elemental sulphur are usually somewhat slower-acting and longer-lasting than those containing gypsum, because the sulphur itself is insoluble.[3] M. Greenwood[4] in northern Nigeria found that a dressing of 112 kg/ha of gypsum improved the yield of groundnuts for two years after it was applied with a small effect in the third year, and T. W. Walker and A. F. R. Adams[5] in New Zealand give examples of 45 kg/ha S, given as gypsum to a grass-clover ley repeatedly cut and harvested, increasing the yield and protein content of the grass four years after application, although by this time the clover component, which had earlier been stimulated, had returned to its original poor condition. If the soil is low in sulphur, and the land is under grass, added sulphur will be locked up in the soil organic matter, which will begin to increase.

Selenium

Selenium is often associated with sulphur in rocks and sediments, so that soils high in sulphate are often also high in selenium. The chemistry of selenium in soils has been little studied. Its principal active form in aerated soils is probably selenite, though selenate may also occur; and under reducing conditions both selenides and elemental selenium have been found.[6] Selenite and selenate are both strongly adsorbed on geothite and other ferric oxide surfaces. Many crops growing on such soils will take up some of this selenium, which may substitute for sulphur in amino acids, and some crops are selenium accumulators,[7] but if the content becomes too high, domestic

1 C. Bloomfield, *Soil Sci.*, 1967, **103**, 219.
2 For a review see O. J. Attoe and R. A. Olson, *Soil Sci.*, 1966, **101**, 317.
3 M. B. Jones and J. E. Ruckman, *Agron. J.*, 1966, **58**, 409.
4 *Emp. J. Exp. Agric.*, 1951, **19**, 225.
5 *Pl. Soil*, 1958, **10**, 176.
6 E. E. Cary *et al.*, *Proc. Soil Sci. Soc. Amer.*, 1967, **31**, 21; and for the effect of redox potentials, with H. R. Geering *et al.*, ibid., 1968, **32**, 35.
7 P. J. Peterson and G. W. Butler, *Aust. J. Biol. Sci.*, 1962, **15**, 126. For a review see A. L. Moxon, O. E. Olson and W. V. Searight, *S. Dakota Agric. Exp. Sta. Tech. Bull.*, 2, 1950.

livestock feeding on these crops will suffer from selenium toxicity.[1] At the present time there appears to be no economic method of preventing crops taking up selenium, although S. Ravikovitch and M. Margolin[2] claim that dressings of about 17 t/ha calcium sulphate would reduce the uptake by lucerne considerably, and 80 kg/ha of barium chloride will reduce uptake to a normal level, presumably due to the precipitation of the selenium as barium selenate.

There are a few examples from New Zealand and elsewhere of soils being so low in selenium that lambs feeding on pastures sometimes suffer from a selenium deficiency. The trouble, white muscle disease, occurs on pastures with less than 0·03 ppm Se in their dry matter,[3] but the low selenium appears to be a predisposing agent rather than the direct cause. The trouble can be rectified by using dressings of sodium selenite at levels not exceeding 50 to 250 g/ha Se, depending on the soil.[4] Some rock phosphates, such as Florida pebble phosphate, contain up to 7 ppm Se, and some western USA rock phosphates contain up to 170 ppm, and some of this is present in a plant-available form in both single or triple superphosphates made from them.[5]

Silicon

Silicon is present in soils in the form of silica and aluminosilicates. The most common form of silica in soils is quartz, and this is the dominant mineral in most sandy soils. Quartz is well crystalline and very stable to weathering, particularly when present in sand-sized or coarser particles, but it weathers fairly easily when in clay-sized particles. Quartz in soils is derived from the parent material, and is probably rarely if ever, formed in the weathering profile. Silica is also present in hydrated forms such as chalcedony—a poorly crystalline form, and as opal—an amorphous form normally derived from plant leaves or diatoms. These bodies, or phytoliths, as they are called, may accumulate in soils, and particularly in grassland soils, for grass leaves contain a relatively high amount of silica which, if the grass is burnt, may take centuries to weather away in some soils. There are also amorphous pre-cipitates of silica in some arid soils which receive silica-rich drainage water, or whose surface contains easily weatherable silicate minerals.

Silicon is also present in the soil solution as silicic acid $Si(OH)_4$. Since the first dissociation constant of this acid has a pK of 9·6, the silicate anion is not present in any appreciable concentration in the solution of normal agricul-

1 For an example from Ireland see G. A. Fleming, *Soil Sci.*, 1962, **94**, 28; and with T. Walsh *et al.*, *Nature*, 1951, **168**, 881.
2 *Emp. J. Exp. Agric.*, 1959, **27**, 235.
3 M. R. Gardiner, J. Armstrong *et al.*, *Aust. J. Exp. Agric. Anim. Husb.*, 1962, **2**, 261.
4 E. B. Davies and J. H. Watkinson, *N.Z. J. agric. Res.*, 1966, **9**, 317, 641. F. H. van der Elst and R. Tetley, ibid., 1970, **13**, 945.
5 C. W. Robbins and D. L. Carter, *Proc. Soil Sci. Soc. Amer.*, 1970, **34**, 506.

tural soils. The silicic acid is in equilibrium with a solid phase, for if a soil is extracted with water, and then left moist, the silicic acid concentration in the soil solution will return to its original value.[1]

Silicic acid is being removed constantly from well-drained soils in humid regions, and, as shown on p. 699, annual losses of the order of 15 kg/ha of silicon are common in the temperate regions, but this does not affect the silicic acid concentration in the soil solution, which depends on the soil but is always less than that of a solution of silicic acid in equilibrium with amorphous silica, which is 2 mM. Thus, L. H. P. Jones and K. A. Handreck[2] found the silicic acid concentration in the solution of a number of Australian soils varied from 0·10 mM for a lateritic (krasnozem) soil to 1·1 mM for a black clay vertisol. Further, the solubility of silicic acid in the pure system is independent of pH over the range 2 to 9,[1] whilst its concentration in a soil decreases as the soil's pH rises. Thus, Jones and Handreck found it dropped from 1·0 to 0·4 mM in an Australian soil when its pH was raised from 5·5 to 7·2 by liming, which compares with concentrations of 0·15 to 0·5 mM in the drainage water for different plots on Broadbalk field at Rothamsted, where the soil has a pH of about 7·2.[3]

The reason the silicic acid concentration in the soil solution is less than would be expected from the solubility of amorphous silica is that it is adsorbed on the surface of iron and aluminium hydrous oxide films by ligand exchange, as explained on p. 131. Thus the greater the surface area of the active aluminium and ferric hydroxide films on the soil particles, and the higher the pH up to 9·5, the more strongly will the soil adsorb silicic acid, and the lower will be its concentration in the soil solution.[4]

Several consequences of these processes have been observed. In the first place, waterlogging a soil is likely to increase its water soluble silicic acid, presumably due to its release from hydrated ferric oxide surfaces. In the second, treatments which remove these free sesquioxides from a soil would be expected to release silicic acid, and this is almost universally observed. In the third place, neither quartz nor free silica would be expected to occur in soils having a large area of sesquioxide films, for these films maintain a soluble silicic acid concentration of less than 10^{-4} M, which corresponds to the concentration in equilibrium with quartz, so unless the quartz is coated with sesquioxide which prevents it dissolving, silica will be transferred from the quartz to these sesquioxide surfaces.

The factors that control the solubility of silicic acid in the soil solution will sometimes control the availability of the soil phosphates to crops. Thus,

1 G. B. Alexander, W. M. Heston and R. K. Iler, *J. Phys. Chem.*, 1954, **58**, 453.
2 *Pl. Soil*, 1965, **23**, 79. This result was also found by J. A. McKeague and M. G. Cline, *Canad. J. Soil Sci.*, 1963, **43**, 70.
3 A. Voelcker, quoted by A. D. Hall in *The Book of the Rothamsted Experiments*, 1905 and 1919.
4 J. A. McKeague and M. G. Cline, *Canad. J. Soil Sci.*, 1963, **43**, 83. L. H. P. Jones and K. A. Handreck, *Pl. Soil*, 1965, **23**, 79. R. S. Beckwith and R. Reeve, *Aust. J. Soil Sci.* 1963, **1**, 157; 1964, **2**, 33. For a review see J. A. McKeague and M. G. Cline, *Adv. Agron.*, 1963, **15**, 339.

H. F. Birch[1] found that, for a group of East African soils, the most efficient method of predicting the response of wheat to a phosphatic fertiliser was to determine the amount of water-soluble silica in the soil, or better still, the amount of silica soluble in 1 per cent citric acid; for the response decreased linearly with increases in the amount of silica extracted, up to certain level, after which there was no response to phosphate. This result does not have universal application, and is probably restricted to soils which contain their phosphate associated with sesquioxide films.

Silicon and plant growth

Crops differ considerably in the amount of silicon they take up, but little systematic work has been done on the normal level of silicon in different plants. Grasses and cereals normally have over 1 per cent of SiO_2 in their dry matter, and most dicotyledons under 1 per cent. H. W. Dougal[2] has determined the silicon content of several hundred samples of grass and foliage of bushes and trees belonging to forty-seven families and growing in the drier areas of Kenya, and found that the silica content of grasses varied from 1 to over 10 per cent of the dry matter, with contents of between 2 and 5 per cent being common, while most other plants had silica contents below 1 per cent and many below 0·2 per cent. Coming to British crops, Table 2·1 on p. 24 shows that cereal straw commonly contains between 10 and 15 kg of silicon per ton while clover, beans and root crops remove only a few kilograms per hectare.

Plant roots take up their silica as silicic acid. It is possible that the amount taken up by most gramineous crops is equal to the amount of silicic acid present in the water the roots absorb, so that the greater the amount of water transpired, the greater their uptake of silica.[3] Dicotyledonous crops on the other hand take up much less silica than is present in the water their roots take up, often only about 5 per cent, and L. H. P. Jones and K. A. Handreck[4] consider that crimson clover (*T. incarnatum*), the plant they studied, translocated a constant proportion of the silica that reached the roots to the tops, so again, the greater the transpiration the greater the silica content of their tops.

But the validity of these generalisations is uncertain for uptake both by the gramineous crops and the dicots appears to be a metabolic process since enzyme inhibitors, such as sodium azide and dinitrophenol, will cause a reduction in silica uptake without causing a corresponding reduction in

1 H. F. Birch, *J. agric. Sci.*, 1953, **43**, 229, 329; P. K. Garberg, *E. Afr. Agric. For. J.*, 1970, **35**, 396.
2 With V. M. Drysdale and P. E. Glover, *E. Afr. Wildlife J.*, 1964, **2**, 86.
3 L. H. P. Jones and K. A. Handreck, *Pl. Soil*, 1965, **23**, 79.
4 *Aust. J. biol. Res.*, 1967, **20**, 483.

transpiration.[1] Many varieties of rice[2] and barley will also take up more silica than is carried to the roots by mass flow and dicots, which take up much less silica than comes to the roots by this process, may have a concentration of silica in their xylem sap that is appreciably greater than in the soil solution. Thus, Barber and Shone found that broad beans (*Vicia faba*) had a silica concentration in their sap five times greater than in the solution bathing their roots.

Grasses and cereals have much of their silica content present as a continuous film, either of a hydrated opal or of some silica-organic complex between the walls of contiguous cells, and if the organic matter of the cell wall is oxidised away carefully the opal films show up the fine structure of the fibrils forming the wall very clearly.[3] Grasses and cereals also often possess small cells which are filled with opal. These silica structures break down when dead grass leaves or cereal straw decompose in the soil or are burnt to give the opal phytoliths which are a characteristic feature of grassland soils.

An adequate supply of silica is essential if grasses and cereals are to give a good yield, for it increases the strength and rigidity of their cells. Thus it helps rice leaves to have a more upright habit under conditions of high nitrogen manuring, which may increase the rate of photosynthesis per unit area of land.[4] It increases the oxidising power at the surface of rice roots, probably by increasing the rigidity of the walls of the aerenchyma or gas channels within the plant. It is also essential for a good seed set in some varieties of rice, but its mode of action is still unknown.[5] It has the same type of action in some dicots, for stinging nettles (*Urtica dioica*) lose their power of stinging if grown in a silica-free medium, presumably because silica is necessary for hardening their stinging hairs.[6] Silica also increases the tolerance of some crops to high levels of available soil manganese for reasons that are not fully understood, but it prevents the manganese in the leaf becoming concentrated in a number of spots, which then become necrotic;[7] and in the case of padi rice allows a greater oxygen supply to the root surface, ensuring a more rapid oxidation of manganese within or just outside the root.

An adequate supply of silica to the cereals will thicken those cell walls on which it is deposited, and this may have a number of desirable consequences. An adequate silica supply reduces the tendency of a cereal to wilt during the initial stages of drought, probably because of the reduced permeability to water or water vapour of the walls of the leaf epidermal cells. There is also

1 D. A. Barber and M. G. T. Shone, *J. exp. Bot.*, 1966, **17**, 569.
2 A. Okuda and E. Takahashi, *The Mineral Nutrition of Rice*, International Rice Research Institute, 1965, ch. 10.
3 For some excellent photographs of these films see L. H. P. Jones, A. A. Milne and S. M. Wadham, *Pl. Soil*, 1963, **18**, 358.
4 S. Yoshida *et al.*, *Pl. Soil*, 1969, **31**, 48.
5 A. Okuda and E. Takahashi, *The Mineral Nutrition of Rice* 1965, Ch. 10.
6 D. A. Barber and M. G. T. Shone, *J. exp. Bot.*, 1966, **17**, 569.
7 J. Vlamis and D. E. Williams, *Pl. Soil*, 1967, **27**, 131; *Pl. Physiol.*, 1957, **32**, 404.

evidence that plants adequately supplied with silica have increased resistance to some pests and diseases. Thus, an adequate silica content may increase the resistance of some cereals to powdery mildew (*Erysiphe graminis*)[1] and of rice to blast (*Pyricularica oryzae*),[2] and to some stem borers, such as *Chilo suppressalis*,[3] of sorghums to central shoot fly (*Atherigone indica*)[4] and wheat to hessian fly (*Mayetiola destructor*).[5]

The use of silica fertilisers, in the form of either soluble silicates, or of calcium silicate slags is still very restricted. Slags are used on some padi rice soils low in soluble silica, which in addition to increasing the pH of the soil are also said to increase the silicic acid concentration in the soil solution. Silicates also increase the yield and sugar content in the juice of sugar-cane growing on soils low in soluble silica.[6] Silicate fertilisers can, however, increase crop yields for quite other reasons, for they increase the availability of soil phosphate to the crop,[7] presumably by displacing phosphate absorbed on sesquioxide surfaces. This is illustrated in Table 24.11 for barley on Hoosfield at Rothamsted, which shows that a dressing of 450 kg/ha of sodium silicate annually is still increasing the yield of the no-phosphate plots after a century of use. The effect is unlikely to be due to the sodium in the silicate as the source of nitrogen is sodium nitrate, and the plots receiving potash also

TABLE 24.11   Effect of silicates on the growth of barley. Hoosfield, Rothamsted

	Yield of dressed grain in t/ha					
	1862–1891		*1932–61*		*1964–66*	
	Without silicate	*With silicate*	*Without silicate*	*With silicate*	*Without silicate*	*With silicate*
Nitrogen alone						
	1·85	2·24	1·59	2·07	1·63	2·54
Complete less phosphate						
	1·92	2·45	1·85	2·16	2·66	4·03
Nitrogen plus phosphate						
	2·77	2·89	2·54	2·71	4·06	3·80
Complete						
	2·76	2·98	2·64	2·89	4·85	4·94

Complete: 48 kg/ha N as nitrate of soda. 32 kg/ha P as single superphosphate. 92 kg/ha K as sulphate of potash. 112 kg of magnesium sulphate. 112 kg sodium sulphate. In 1964–66 96 kg/ha N given in place of 48 kg/ha.

1 E. Lowig, *Pflanzenbau*, 1937, **13**, 362. F. Wagner, *Phytopath*, 1940, **12**, 427.
2 R. J. Volk, R. P. Kahn and R. L. Weintraub, *Phytopath.*, 1958, **48**, 179.
3 A. Djamin and M. D. Pathak, *J. econ. Entom.*, 1967, **60**, 347.
4 B. W. X. Ponnaiya, *J. Madras Univ.*, 1951, **21** B, 203.
5 B. S. Miller, R. J. Robinson *et al.*, *J. Econ. Ent.*, 1960, **53**, 995.
6 A. S. Ayres, *Soil Sci.*, 1966, **101**, 216. Y. W. Y. Cheong and P. Halais, *Exp. Agric.*, 1970, **6**, 99.
7 R. A. Fisher, *J. agric. Sci.*, 1929, **19**, 132. O. Lemmermann, *Ztsch. Pflanz. Dung.*, 1929, **13**A, 28.

receive 110 kg/ha of sodium sulphate. It is probably another example showing that the concentration of water-soluble silicic acid and of phosphate are not two independent quantities in soils, but are closely linked since their solubilities are controlled by the sesquioxide and, in particular, the aluminium hydroxide surfaces.

Farmyard manure

Farmyard manure is made by composting the faeces and urine of farm animals, usually cattle, with straw or other bedding material. The plant nutrients it contains depend, therefore, on the nutrient content of the fodder fed to the cattle, the nutrient content of the straw, the degree of composting or rotting of the straw into humus-like compounds, and the losses of ammonia by volatisation and of soluble compounds by drainage or washing out. It is therefore a material of very variable manurial value to crops.

Well-made farmyard manure is a useful source of nitrogen, phosphate and potassium to crops. G. W. Cooke[1] has summarised the results of a large number of investigations on the composition of farmyard manures and of their fertilising value to crops. Twenty-five tons per hectare of farmyard manure, a fairly common dressing, will contain about 160 kg N, 22 kg P and 110 kg K; but if given to a responsive crop will only be equivalent to about 45 kg N, 22 kg P and 90 kg K supplied as normal fertilisers, showing that most of the phosphate and probably all the potassium it contains is as effective as fertiliser phosphate and potassium, but only about one-quarter of the nitrogen is as effective as fertiliser nitrogen. Most of the examples from overseas where dressings of 2 to 5 t/ha of farmyard manure give good crop responses are usually due to the phosphate in the manure.[2] Many of the apparent benefits of FYM compared to fertilisers in the UK are due to inadequate K in the fertiliser, compared with that in FYM.

Farmyard manure differs from a potassium fertiliser in the important respect that if it is used to supply a high level of potassium to a soil it will have a much smaller effect on the osmotic pressure of the soil solution than a corresponding dressing of potassium chloride, and so will not have the harmful effect on seedlings that the fertiliser will. Thus, for certain vegetable seedlings which appear to need a relatively high concentration of potassium in the soil solution, it is preferable to supply a part or even a major part, through the farmyard manure.[3] Farmyard manure can also have an important advantage over phosphate fertilisers, particularly if it is used over a period of years, for the organic matter it adds appears to raise the concentra-

1 *The Control of Soil Fertility*, 1967, p. 147.
2 For an example see K. T. Hartley, *Emp. J. Exp. Agric.*, 1933, **1**, 113; and with M. Greenwood, ibid., 1937, **5**, 254.
3 F. Haworth and T. J. Cleaver, *J. Sci. Fd. Agric.*, 1963, **14**, 264; 1965, **16**, 600; and R. G. Warren and A. E. Johnston, *Rothamsted Rept.* 1960, 43.

tion of phosphate in the soil solution,[1] so seedlings that need a high concentration for fast early growth tend to do better on land that has been regularly treated.

Farmyard manure normally contains all the trace elements needed by plants for growth, and a dressing of 25 t/ha will put back about the amount that has been removed by a crop. This is illustrated in Table 24.12, taken from a summary of the analyses of a large number of samples used in Canadian experiment stations.[2] Assuming the manure has 80 per cent moisture and a 25 ton dressing is used, the number of grams of element added to the soil is obtained by multiplying the figures in the table by two. Naturally, if the soil on the farm is low in any particular trace element, the content of that element in the crops grown on that farm is likely to be low, and so is the manure made. However, adding manure to a soil appears to be a relatively inefficient way of supplying a crop with a trace element, except for some soils low in organic matter and with high pH, where it may be an efficient method of increasing the uptake of the iron and perhaps some other trace elements by the crop.

TABLE 24.12   Content of trace elements in samples of farmyard manure. (Canadian Experiment Stations). Content in parts per million dry matter

	B	Mn	Co	Cu	Zn	Mo
Range	4·5–52	75–549	0·21–4·70	8–41	43–208	0·8–4·2
Average	20	201	1·0	16	96	2·1

It has been very difficult to show under what conditions farmyard manure increases crop yields in any way that a fertiliser will not. An effect on the phosphate and potassium status of soils on the seedling growth of certain vegetable crops which have very high demands if they are to make rapid growth has already been mentioned. There are occasions when the manure appears to give a yield response over and above that given by fertilisers, as, for example, on potato yields in dry summers.[3] Under some conditions, particularly where the soil starts rather poor, the combined use of a high level of fertiliser with farmyard manure gives higher yields than can be obtained by fertilisers alone.[4] This may be due in part to the manure giving many pockets

1 R. G. Warren, A. E. Johnston and J. M. D'Arifat. *Rothamsted Rept.* 1964, 40.

2 H. J. Atkinson, G. R. Giles and J. G. Desjardins, *Canad. J. Agric. Sci.*, 1954, **34**, 76. For comparable figures for Scottish manures see R. G. Hemingway. *Emp. J. Exp. Agric.*, 1961, **29**, 14.

3 For example see R. Holliday, P. M. Harris and M. R. Baba, *J. agric. Sci.*, 1965, **64**, 161, and F. Haworth, T. J. Cleaver and J. M. Bray, *J. Hort. Sci.*, 1966, **41**, 225.

4 A. H. Bunting, *J. agric. Sci.*, 1963, **10**, 113. F. V. Widdowson, A. Penny and G. W. Cooke, *J. agric. Sci.*, 1963, **60**, 347; 1967, **68**, 95.

of soil where the physical conditions are ideal for the roots to take up large amounts of nutrients easily.[1]

Long continued use of high levels of manure can so improve the soil structure that germination and seedling growth is greatly improved in difficult years, but so far no real evidence has been produced that the large number of organic compounds in farmyard manure have any quantitatively measurable effect on yield or composition of the crop when grown in soils, although they may have in some circumstances in water or sand cultures.

Trace elements or minor elements

Crops contain many other chemical elements besides those already discussed, but usually these are present in much smaller quantities, so their contents are given in milligrams of element per kilogram of plant dry matter, or parts per million; although some are present in such small quantities that their contents are given in micrograms per kilo, or parts per billion (10^9). The number of elements recognised is, in fact, largely conditioned by the ease with which low concentrations can be determined spectrochemically. The reason for this is that primary rocks, and most sedimentary rocks, contain most of the stable chemical elements, and as the rocks are transformed to soil so a proportion of each element is usually converted to a form which plant roots can take up. The mere fact of uptake is no evidence that the element plays any role in the development of the plant, but solely that it is present in the soil in a form which the root has no power to reject entirely; though the roots of different species of plants differ very considerably in the proportion of a given element they take up from a soil.

These elements can be divided into those which are known to be needed by crops, those known to be needed by animals feeding on the crops, and those not known to have any appreciable effect on growth, provided they only occur in low concentrations. Many of these elements are known, however, to have undesirable effects on plant and animal growth if present in too high a concentration in their tissues. The principal elements needed by plants are iron, manganese, copper, zinc, boron, molybdenum and, for leguminous plants, cobalt. Chlorine is also needed by plants in very low concentrations,[2] but the author is not aware of any examples of crop yields being limited by lack of chlorine. Mammals need all of these, except possibly boron, though ruminants have a much higher demand for cobalt than plants. But they also require iodine and, for some species at least, small amounts of selenium and chromium. Plants commonly have their growth affected by excess manganese and aluminium in acid soils, and by heavy metals such as nickel, cobalt and chromium on acid soils derived from some ultrabasic igneous rocks. Animals can have their growth and health affected by grazing pastures on soils over-

1 F. Haworth, *J. Hort. Sci.*, 1961, **36**, 202.
2 T. C. Broyer, C. M. Johnson *et al.*, *Pl. Soil*, 1957, **8**, 337.

well supplied with molybdenum and selenium. High molybdenum pastures are called teart, and they cause the grazing animal to scour badly.

The principal reservoir of many of these elements in the soil is as impurities in compounds of other elements.[1] Thus, many of them have isomorphously replaced a very small proportion of the major ions constituting rock and clay silicates, and others are absorbed when precipitates such as iron and manganese oxides[2] are being formed. Thus, much of the soil cobalt occurs in association with manganese precipitates,[3] and much of the molybdenum and probably vanadium, with ferric oxide precipitates;[4] so that soil factors which control the dissolution or formation of these precipitates, or alter their surface properties, affect the pool of active cobalt and molybdenum. Some elements have their availability affected by the presence of soil organic matter, or by microbial or root excretions, for there are many organic compounds present in very low concentration in the soil solution which form soluble coordination or chelation compounds with them. Thus, altering soil conditions by liming or draining the soil will not necessarily have the same effect on the availability of a given element in different soils. However, in general, poor or impeded drainage usually increases the solubility of cobalt, copper, lead, manganese, molybdenum, nickel and zinc,[5] while liming usually reduced the concentration of cobalt, nickel, manganese and sometimes zinc and copper in the crop and increases the molybdenum.[6]

TABLE 24.13   Approximate weights of some trace elements and major elements removed by crops and in the soil. Weights in kg/ha

	Removed by average crop	Extractable by diagnostic reagent	Total content in soil
Cobalt	0·001	0·2–4	2–100
Molybdenum	0·01	0·02–1	0·5–10
Copper	0·1	1–20	2–200
Boron	0·2	1–5	4–100
Zinc	0·2	2–20	20–50
Manganese	0·5	10–200	100–10 000
Iron	0·5	10–200	2 000–100 000
Magnesium	20	100–1 000	2 000–100 000
Phosphorus	20	40–100	1 000–10 000
Potassium	100	40–200	5 000–50 000

1 For a general review see J. F. Hodgson, *Adv. Agron*, 1963, **15**, 119.
2 H. H. Le Riche and A. H. Weir, *J. Soil Sci.*, 1963, **14**, 225.
3 R. M. Taylor and R. M. McKenzie, *Aust. J. Soil Res.*, 1966, **4**, 29.
4 S. Trobisch and G. Schilling, *Chem. d. Erde*, 1963, **23**, 91. H. M. Reisenauer, A. A. Tabikh and P. R. Stout, *Proc. Soil Sci. Soc. Amer.*, 1962, **26**, 23.
5 D. J. Swaine and R. L. Mitchell, *J. Soil Sci.*, 1960, **11**, 347. R. L. Mitchell, J. W. S. Reith and I. M. Johnston, *J. Sci. Fd. Agric.*, 1957, **8**, S 51.
6 R. L. Mitchell, *Scot. Agric.*, 1954, **34**, 139.

The total amounts of these elements taken up by crops varies widely for different elements. Table 24.13, due to R. L. Mitchell,[1] shows the order of magnitude of uptake for typical crops in the United Kingdom, as well as the total content of the element in the top 22·5 cm of soil and the amount extracted by the reagent which gives the best estimate of the availability of that element. In addition, crops commonly contain about the same amount of strontium and barium as zinc, copper and manganese, and more nickel and lead than molybdenum. The table also shows that most soils usually contain far more of these elements than are required for crop growth, and that even the best diagnostic extraction reagents remove far more than the crop will take up. It is not easy to give a wholly satisfactory reason for the great discrepancy between the amount of an element taken up by the crop and the amount extracted by the most suitable reagent, but it is probably partly dependent on the rate of diffusion of the active form of the element through the soil solution. Unfortunately, the concentration of this active form in the soil solution is usually too low to determine directly, and the chemical form is often not known for certain, so quantitative explanations for this divergence are not possible.

It is worth while stressing in this context the importance of a well-developed root system for maintaining an adequate supply of these elements to the crop. A crop whose root system is stunted, or shallow, or restricted in any way is more likely to show a deficiency on a soil whose supply of the element is marginal than one with a well-developed root system. Zinc deficiency has, for example, been noted in fruit orchards where the root system of the trees was limited for some reason, but the deficiency could be cured by allowing deep-rooted indigenous weeds to grow, which were then ploughed into the soil surface. Again boron deficiency has often been noted in young forest plantations growing on low boron soils, but the deficiency was only apparent for a few years, beginning after the young trees had exhausted the available boron in the surface soil and ending a few years later, presumably after the trees had developed a deeper and more extensive root system. On the other hand, it sometimes happens that most of an essential trace element is concentrated in the surface layer of the soil so that when this layer becomes dry, the root system will only be taking up nutrients from the subsoil, and a deficiency of this element will develop in the aerial part of the crop.

A characteristic of many trace elements is that their contents in a range of crops growing on the same soil may vary widely. This can be very marked for manganese and molybdenum. Thus, the manganese content of leaves of forest trees in the East African arboretum at Muguga, Kenya, varied from 90 to 5000 ppm in the dry matter, in different species of lupins from 500 to over 50 000 ppm, while barley had about 100 ppm. Similarly, the leaves of some species of lupin had over 10 ppm of molybdenum compared with about

1 *J. Roy. Agric. Soc. Engl.*, 1963, **124**, 75.

0·04 ppm in barley leaves.[1] It is not known what processes are operative in the tissues of healthy plants which allow them to accumulate these large quantities of trace elements, but these elements must presumably be present in some organic coordination compounds. Further, certain soils can have a very high level of availability of some of these elements, and this will encourage plant species that are tolerant to these elements to accumulate them in their leaves.

The problem of assessing the trace element status of a soil from chemical analysis of the soil is difficult. First of all the uptake of a particular element by a plant from a soil depends not only on the level of the active form of that element in the soil but also on the availability of many other elements, both major and trace, and both essential and non-essential for growth. Second, the level of that element needed for the effective functioning of the plant cells depends to some extent on the level of many other elements. Thus the leaves of a crop may show symptoms that are due to a deficiency of an element, yet if this deficiency is made good, the yield may not be affected, and, conversely, crops may be showing no symptoms of a deficiency yet have their yield appreciably increased if the element is given.

The assessment of trace element deficiencies from field trials can be very difficult, particularly in soils in which several elements are deficient; for supplying only some of the deficient elements may give either little or no improvement in crop growth; and not until all the elements have been given will vigorous healthy growth take place. Diagnosis of multiple deficiencies, when these are suspected, are best made in the glasshouse by growing a test plant in a series of samples of the soil to which either a complete nutrient solution is added, or a series of nutrient solutions each of which is complete except that one element is omitted.

The division of trace elements into essential and non-essential for growth is not absolute. In particular sometimes an element can be beneficial if another element is in short supply. Thus, a certain level of molybdenum is necessary in the soil if leguminous crops are to fix nitrogen actively, but if the level is too low, vanadium or tungsten[2] can take its place, though much less efficiently. Further, some elements may not be essential for growth but may improve crop yields. Aluminium is an example of this for some crops, such as tea,[3] which probably needs a certain level of aluminium for optimum growth, although if the available level becomes too high it will depress growth, as with other crops.

Many elements will be toxic to plants if present in too high a level. Of the non-essential elements, nickel, chromium, cobalt and perhaps lead, can all be present in too high a concentration. The first two are often important con-

1 G. T. Chamberlain and A. J. Searle, *East Afr. agric. and for. J.*, 1963, **29**, 114, and unpublished data.
2 E. B. Davies and S. M. J. Stockdill, *Nature*, 1956, **178**, 866.
3 E. M. Chenery, *Pl. Soil*, 1955, **6**, 174.

stituents of serpentine and other ultrabasic rocks, and are then likely to cause soils derived from these rocks to be very infertile.[1] If the soils are acid, this infertility can usually be reduced by liming.[2]

Iron

It is not known definitely in what form plant roots take up their iron, whether as the ferrous or the ferric ion, nor whether as an ion or as an organic co-ordination compound, nor has the transfer process from the soil to the vascular system of the plant been fully established. It is unlikely that the plant only uses ferric ions present in the soil solution, because of its very low concentration. Moreover, very little is known about the concentration of iron in the soil solution, but it is probably present in organic complexes. In water culture plants need a ferric iron concentration of about 10^{-5} M, which is very much higher than the concentration of ferric ions in the soil solution. It is possible, however, that the roots of the plant itself, or the rhizosphere popu-lation around the root surfaces, may excrete into the soil soluble organic compounds that will form soluble complexes with iron from the surfaces of ferric hydrous oxides. Thus, D. M. Webley and R. B. Duff[3] showed that the rhizosphere population sometimes excretes α-ketogluconic acid, which brings ferric iron into solution, which is then taken up by the plant roots.

Plants differ in their ability to take up iron from soils, particularly from calcareous soils, and it is this differential ability which is probably largely reflected in the division of plants into calcifuges and calcicoles. Calcifuges growing on neutral or calcareous soils typically show symptoms of a trouble known as lime-induced chlorosis, which is due to an upset in the iron nutri-tion of the plant.[4] It is a phenomenon aggravated by poor aeration,[5] by a high concentration of bicarbonate and a high level of available phosphate in the soil,[6] and most crops will show this trouble in alkali soils, though differ-ent varieties of the same crop may vary considerably in their ability to with-stand these conditions.

Considerable interest has been taken in methods of trying to prevent or reduce this chlorosis developing. In the United Kingdom it can seriously affect the productivity of apple and pear orchards on chalk and limestone soils. It can be controlled in apple orchards by grassing the orchard down and

---

1 J. G. Hunter and O. Vergano, *Ann. appl. Biol.*, 1952, **39**, 279 (Aberdeenshire). B. D. Soane and D. H. Saunder, *Soil Sci.*, 1959, **88**, 322 (Rhodesia). G. L. Lyon, R. R. Brooks *et al.*, *Pl. Soil*, 1968, **29**, 225 (New Zealand).
2 W. M. Crooke, *Soil Sci.*, 1956, **81**, 269. A. T. Chang and G. D. Sherman, *Hawaii Agric. Exp. St., Tech. Bull.* 19, 1953.
3 *Pl. Soil.*, 1965, **22**, 307.
4 For a review of this subject see J. C. Brown, *Adv. Agron.*, 1961, **13**, 329.
5 E. F. Wallihan, M. J. Garber and R. G. Sharples, *Pl. Physiol.*, 1961, **36**, 425.
6 G. W. Miller, J. C. Brown and R. S. Holmes, *Pl. Physiol.*, 1960, **35**, 619.

gang mowing the grass between the trees, leaving the mowings on the surface. The reason why this is effective is not known, but it could be due to the production of soluble iron-complexing compounds. In the same way it can often be controlled by adding large quantities of organic manure, such as farmyard manure, to the soil, though care must be taken not to raise the bicarbonate ion concentration to too high a level as can happen if a heavy easily-decomposable green manure is ploughed in, for this will aggravate the chlorosis. It cannot be cured by adding a simple iron salt, such as ferric chloride, to the soil because the iron is immediately converted to an insoluble ferric hydroxide.

Lime-induced chlorosis can now be controlled by the direct application of a suitable iron chelate to the plant itself, and compounds such as ferric citrate and tartrate have been widely used, but they are not suitable for soil application, partly because they are too readily decomposed by the soil organisms. Soil-applied chelates can be used to increase the iron supply provided they fulfil certain definite conditions. They must be stable in calcareous soils, that is, they must remain as iron chelates and not be converted to calcium chelates, they must not be easily decomposable by the soil organisms, they must not be adsorbed by the clay or soil organic matter, and they must not be toxic to the plant, for the whole chelate may be absorbed by the roots. The first of the new synthetic chelates to come on the market was the sodium salt of ferric ethylene diamine tetraacetic acid (EDTA) but it is converted into the calcium chelate in well-aerated calcareous soils. There are now several chelates which are stable in calcareous soils, or soils with a pH above 8, such as those based on ethylene diamine di-(o-hydroxyphenyl acetate) (EDDHA) and cyclohexane diamine tetraacetate (CDTA)[1] though these are still much too expensive for normal commercial use. Mature pear and apple trees need about 100 g per tree if applied to the soil.

The property of a soil that renders it liable to cause lime-induced chlorosis is not the total content of calcium carbonate but the content of a finely divided carbonate that is more soluble than calcite and that reacts rapidly with carbon dioxide dissolved in the soil water. Thus, this liability can be estimated by any test which will measure this active fraction, and the one most widely used was developed by G. Drouineau[2] for the selection of vineyard sites on the calcareous soils around the Mediterranean. It consists of shaking the soil up with an ammonium oxalate solution and determining the amount of oxalate removed by the soil—the larger the amount removed, the more liable is the soil to induce this chlorosis.

1 For a discussion on their stability see W. L. Lindsay, J. F. Hodgson and W. A. Norvell, *Trans. 2nd and 4th Comm. Int. Soc. Soil Sci.*, 1966, 305. *Proc. Soil Sci. Soc. Amer.*, 1969, **33**, 62. For a general discussion J. C. Brown, ibid., 59.
2 *Ann. Agron.*, 1942, **12**, 441; 1943, **13**, 16. P. Boischot and J. Hebert ibid., 1947, **17**, 521. J. and T. Dripuis, *Sci. Soc.*, 1966, No. 1, 31.

Manganese

Manganese is present in the soil in at least two forms: divalent manganous ions held by clay and organic matter and tetravalent manganese present as insoluble oxides usually of poorly defined composition and sometimes associated with ferric oxide concretions. It is also possible that trivalent manganese hydrated oxides may sometimes be present.[1] Unlike iron, divalent manganese can be present in well-aerated soils, and it is a normal constituent of the cations extracted from acid soils by ammonium acetate. It can also be held by soil organic matter, whether in mineral, peat or fens soils, in a form which is not exchangeable with the ammonium ions, but is exchangeable with copper or zinc,[2] for example, and is extractable with chelating agents such as EDTA.[3] A portion of the higher oxides are present in a form that can easily be reduced fairly rapidly to manganous ions by quinols, or by readily oxidisable groups such as thiols, or by ions such as ferrous. This form, therefore, acts as an oxidation-reduction buffer in the soil, and as an oxidiser of any toxic thiols of microbial origin. This fraction, sometimes called the readily-reducible manganese, is not a well-defined fraction, for the amount of reducible manganese in a soil and its rate of production increases as the strength of the reducing agent is increased.[4] A part of this readily-reducible manganese is isotopically exchangeable with added ^{54}Mn, as is a part of the divalent manganese that is strongly held by the organic matter.[5]

Many soils contain visible concretions or thin black surface films on the crumb surfaces that are rich in manganese, but very few analyses of these have been made. R. M. Taylor[6] in Australia have subjected concretions from a wide selection of Australian soils to analysis and have shown that they are all micro-crystalline, are usually based on birnessite $M.Mn^{2+}Mn^{4+}(O.OH)_2$, where M is an alkali or alkaline earth metal; on lithiophorite $Li_2Al_8M_4$, $Mn_{10}^{4+}O_{35}.14H_2O$ where M is divalent manganese, cobalt or nickel; though a few contained hollandite $Ba(Fe^{3+}Mn^{4+})_8O_{16}$ and a very few pyrolusite MnO_2. The first three minerals have a somewhat similar range of composition, with 35 to 60 per cent Mn, 3 to 9 per cent Ba, 0·5 to 1·5 per cent Co with some samples higher, and variable amounts of silicon, iron, aluminium, and usually with a very low content of other elements. Birnessite has an average crystallite size of 0·02 μ and lithiophorite of 0·1 μ. They also found that about 80 per cent of the cobalt in the soil horizons from which the manganese material was extracted is associated with this material, and this must be the case with much of the barium. Manganese is also associated with many

1 H. G. Dion and P. J. G. Mann, *J. agric. Sci.*, 1946, **36**, 239.
2 S. G. Heintze and P. J. G. Mann, *J. agric. Sci.*, 1949, **39**, 80.
3 R. S. Beckwith, *Aust. J. agric. Res.*, 1955, **6**, 299.
4 G. W. Leeper, *Soil Sci.*, 1947, **63**, 79
5 C. C. Weir and M. H. Miller, *Canad. J. Soil Sci.*, 1962, **42**, 105. A. S. J. Reed and M. H. Miller ibid., 1963, **43**, 250.
6 With R. M. McKenzie and K. Norrish, *Aust. J. Soil Res.*, 1964, **2**, 235; 1966, **4**, 29.

iron precipitates in the soil, but no detailed crystallographic or chemical analyses of them has yet been made.

Manganese differs from other trace elements in that both deficiency and toxicity are widespread in agricultural practice. It is present in the soil solution as the divalent manganese cation and as divalent manganese complexed with organic matter. H. R. Geering[1] found concentrations of total divalent manganese in the solutions of the acid and neutral soils examined in the range 10^{-4} to 10^{-6} M, and somewhat lower for calcareous soils, and that between 85 and 99 per cent of this manganese was present in organic complexes. This compares with their results of over 99 per cent of the copper and under 75 per cent of the zinc in the soil solution being in these complexes. Crops probably only take up divalent manganese, and the concentrations and measured diffusion coefficients of the manganese ions in the soil are adequate to supply the crop with all the manganese it takes up.[2]

The concentration of divalent manganese in the soil solution of a given soil is dependent on the soil pH and on its oxidation-reduction potential[3]—low pH and reducing conditions favouring a high concentration of divalent manganese in the solution. Both the oxidation of divalent manganese to insoluble manganese oxides, and the reduction of the higher valency manganese to the divalent form are probably usually carried out by micro-organisms,[4] although ferrous ions and some organic soil thiols will reduce the higher valency manganese under suitable conditions. It is still not established, however, what particular systems determine the manganese ion concentration, or the total divalent manganese content of a given soil under any given acidity and redox conditions.[5]

Liming a soil almost always puts down the amount of manganese a crop

TABLE 24.14   Effect of raising the soil pH on the manganese uptake by winter wheat. Chalk applied 1934. Wheat grown 1945/46. Soil: medium loam on boulder clay. Oaklands, St Albans

Ground chalk applied t/ha	pH of soil	Yield of wheat grain plus straw in t/ha	Uptake of Mn in kg/ha
0	4·3	2·78	0·62
4·8	4·7	4·12	0·61
9·4	5·3	5·25	0·57
14·1	5·8	5·65	0·48
18·8	6·5	5·78	0·38

1 With J. F. Hodgson and C. Sdano, *Proc. Soil Sci. Soc. Amer.*, 1969, **33**, 8.
2 E. H. Halstead and S. A. Barber, *Proc. Soil Sci. Soc. Amer.*, 1968, **32**, 540.
3 For an example see H. L. Bohn, *Proc. Soil Sci. Soc. Amer.*, 1970, **34**, 195.
4 For a review see F. C. Gerretsen, *Adv. Agron.*, 1952, **4**, 222, and for some later work E. G. Mulder and W. L. van Veen, *Trans. 9th Int. Congr. Soil Sci.*, 1968, **4**, 651.
5 See, for example, L. H. P. Jones, *Pl. Soil*, 1957, **8**, 301, 315.

will take up, as is shown in Table 24.14 for winter wheat on a boulder-clay loam in Hertfordshire.[1] In the same way making a soil more acid, by the use of ammonium sulphate as a fertiliser, for example, will increase the amount of manganese taken up by the crop. Similarly, increasing the microbial activity by adding a sugar to the soil, will also increase the manganous ion[2] concentration, as will air-drying or sterilising a soil,[3] but the reasons for this are not fully understood. Manganese toxicity is therefore most likely to be found on acid soils, or soils high in organic matter that have been sterilised.

Manganese deficiency is likely to be induced by overliming, particularly by overliming an acid sand if the pH is brought up to over 6·5 rapidly. However, it is rarely seen on chalk or limestone soils low in organic matter content, though in England it may be seen on chalky soils having a peaty surface whose C/N ratio is appreciably above 10, as in some old water meadows.[4] On these soils it is more commonly seen in spring and early summer when the season is wet, and the symptoms of deficiency typically disappear as soon as a dry spell comes,[5] probably because the crop becomes deeper-rooted and is taking its manganese from the non-peaty subsoil. In the same way, mulching may induce manganese deficiency symptoms in crops growing in a soil low in manganese, for this also encourages shallow-rooting.[6]

Crop roots probably play an active part in bringing about reduction of the manganese in insoluble oxide precipitates through their root excretions. These excretions can be washed out of the soil, but they are readily decomposable by the soil organisms.[7] These exudates may be responsible for the ability of crops to use the manganese of certain oxide precipitates, if they are amorphous, though not if they are well crystalline.[8] It has also been claimed that, in the field, oats may show signs of manganese deficiency when growing on a loose soil but not if it is compacted, the effect of compaction being due to bringing the roots closer to these insoluble oxides, so that they could more easily pick up the manganese their exudates have solubilised.[9]

Bacterial activity in the rhizosphere of a crop's root system may also affect the ability of a crop to take up manganese from a soil. Thus, varieties of oats which are susceptible to grey-speck disease, which is due to manganese deficiency, can have a higher concentration of manganese-oxidising bacteria around their roots than those less susceptible.[10] Manganese deficiency can be

1 W. E. Chambers and H. W. Gardner, *J. Soil Sci.*, 1951, **2**, 246.
2 P. J. G. Mann and J. H. Quastel, *Nature*, 1946, **158**, 154.
3 C. K. Fujimoto and G. D. Sherman, *Soil Sci.*, 1948, **66**, 131, M. I. Timonin and G. R. Giles, *J. Soil Sci.*, 1952, **3**, 145.
4 I am indebted to Dr T. Batey for this observation.
5 For an example see K. Dorph-Petersen, *Tidsskr. Planteaval*, 1950, **53**, 650.
6 E. M. Chenery, *2nd Inter-African Soil Conf.*, 1954, 1157.
7 S. M. Bromfield, *Pl. Soil*, 1958, **9**, 325; 1959, **10**, 147.
8 L. H. P. Jones and G. W. Leeper, *Pl. Soil*, 1951, **3**, 141.
9 J. B. Passioura and G. W. Leeper, *Nature*, 1963, **200**, 29.
10 M. I. Timonin, *Proc. Soil Sci. Soc. Amer.*, 1946, **11**, 284.

cured by spraying manganese sulphate on a crop, when a dressing of 5 to 10 kg/ha is commonly given, or by applying manganese sulphate to the soil. Since this added manganese is usually rapidly converted to a non-available form, it is best given as a side band dressing to the crop and a rate of 35 to 70 kg/ha is common, and it rarely has much residual effect.

The manganese status of a soil is commonly assessed from the divalent manganese that can be extracted with a suitable solvent, but it is unlikely that any one solvent will be the most efficient for all soils of low available manganese status. Thus, ammonium acetate, zinc sulphate and EDTA have all been used to measure the readily-extractable manganese; but the manganese status has also been measured as the sum of the readily-extractable plus the readily-reducible manganese, using hydroquinone as the reducing agent. M. G. Browman[1] compared a number of these methods, using sixty-three soils of contrasting types, and concluded that the simple ammonium acetate extraction, with a correction for the soil pH, appeared to be the most efficient of those used.

Copper, zinc and cobalt

These three metals occur in the divalent form when in an active state in the soil, either as the simple divalent cation or possibly as the monovalent $M(OH)^+$ ion. They can often be found in a form exchangeable with neutral ammonium acetate, but acetic acid or chelating agents such as EDTA will usually extract more, and sometimes very considerably more than the neutral salt. It is probable that soil clays hold small amounts of these cations very tightly, possibly through negatively charged hydroxyls belonging to the second oxygen-hydroxyl sheet of the lattice due to imperfections in the top layer of oxygens.[2] The organic matter in the soil also holds copper very strongly,[3] compared with, say, calcium, sometimes with zinc on some organic soils. Copper and to a lesser extent zinc, iron and aluminium, stabilise microbial and plant polyuronides against decomposition by the soil microorganisms,[4] but it is not known if this is of any relevance for the other copper-humus complexes in the soil.

These ions can also be held in a non-exchangeable form in neutral and, sometimes, in acid soils. If a small amount of one of these cations has been put on a soil, little of the ion will exchange with a neutral salt, such as sodium chloride, when the pH is over 6, though most will when the pH is below 4·5.[5]

1 With G. Chesters and H. B. Pionke, *J. agric. Sci.*, 1969, **72**, 335.
2 L. E. De Mumbrum and M. L. Jackson, *Proc. Soil Sci. Soc. Amer.*, 1956, **20**, 334. J. F. Hodgson, ibid, 1960, **24**, 165, for cobalt.
3 H. Lees, *Biochem. J.*, 1950, **46**, 450.
4 J. P. Martin, J. O. Erwin and R. A. Shepherd, *Proc. Soil Sci. Soc. Amer.*, 1966, **30**, 196.
5 For copper see M. Peech, *Soil Sci.*, 1941, **51**, 473. For zinc A. L. Brown, *Soil Sci.*, 1950, **69**, 349. J. L. Nelson and S. W. Melsted, *Proc. Soil Sci. Soc. Amer.*, 1955, **19**, 439. For cobalt D. K. Banerjee, R. H. Bray and S. W. Melsted, *Soil Sci.*, 1953, **75**, 421.

On the other hand, most of the added cation is extractable with hydrochloric or acetic acid, or with a chelating agent such as EDTA. The charge on the cation also appears to be less than 2, presumably due to a certain proportion being present as the monovalent $M(OH)^+$. But because the ion becomes more difficult to extract with a neutral salt, it does not imply that it becomes less available to the crop. As already discussed, liming a soil does not consistently reduce the uptake of these cations, although it very often does if a sufficiently heavy dressing of lime is given to bring the pH over 6·5–7.

Copper concentrations in the soil solution are usually between 10^{-6} and 10^{-7} M, but a large proportion of this is present as organic complexes, and the cupric ion concentration may be about 10^{-9} M: that is over 99 per cent of the copper in the soil solution may be complexed with organic compounds.[1] Mercer and Richmond have also shown that for a number of soils from southern England the copper associated with compounds having a molecular weight less than 1000 was readily available to crops, but that with compounds whose molecular weights were about 5000 or above was much less available. They found, in fact, that their copper-deficient soils had the same copper concentrations in the soil solution as those adequately supplied with copper, but a much larger proportion of this copper was complexed with the high molecular weight fraction. These results suggest that the principal reservoir of plant-available copper in soils is some fractions of the soil humus.

Copper deficiency is found on many ancient strongly-weathered soils in Australia, in soils derived from granites and rhyolites which are low in copper; and many sandy soils, particularly calcareous sands, are often sufficiently low for worthwhile crop responses to be obtained through the use of a copper-containing fertiliser. The most striking examples of copper deficiency are usually seen on peat or fen soils, low in clay and with a water-table too deep to maintain a sufficiently moist surface soil during dry weather. It is also found on rendzina soils over chalk, particularly when first brought into cultivation, for the humus content of the surface layer is then likely to be high. Copper, like cobalt, zinc, boron and molybdenum, is usually concentrated in clay sediments, so soils derived from such sediments are usually well supplied with these trace elements. A low copper status in the soil may be of importance when the soil is under pasture, for the copper content of the forage may be too low to meet the requirements of the grazing animal.

Copper deficiency can be corrected either by adding a copper salt to the soil or spraying it on to the crop itself. If added to the soil, the recommended dressing can vary a hundredfold depending on the soil. Dressings as high as 70 kg/ha of Cu as copper sulphate have been recommended in the Florida

1 J. F. Hodgson, W. L. Lindsay and J. F. Trierweiler, *Proc. Soil Sci. Soc. Amer.*, 1966, **30**, 723.
E. R. Mercer and J. L. Richmond, *A. R. C. Letcombe Lab.*, *Ann. Rept.* 1968, 46; 1969, 61; 1970, 9; 1971, 20.

Everglades;[1] 9 kg Cu on the peats and fens of eastern England,[2] about 3–15 kg/ha on some calcareous sandy soils in South Australia,[3] and 0·5 kg/ha on some sandy soils in Western Australia on which 2·5 kg is harmful.[4] Copper sulphate is, however, often sprayed on the crop at a rate of 0·5 to 1 kg/ha (0·1 to 0·2 kg Cu); and on some organic soils, particularly in the early years after reclamation, it is often best to use both a soil dressing and a spray. N. H. Pizer[5] found that 9 kg/ha Cu lasted about ten years on the East Anglian fens, although only about 0·5 kg/ha Cu had been removed by the harvested crops, and J. W. S. Reith[6] that about half this dressing was adequate on low-copper soils in Scotland, and this dressing also lasted for about eight years.

Crops differ in their sensitivity to copper deficiency. Wheat, barley, oats and cocksfoot, can all give very strong vegetative growth on low copper soils, but will set no seed, while rye is much more tolerant to low copper. Sugar-beet and carrots are also two crops that often respond well to copper if the level is low. Pasture legumes vary considerably among themselves in their sensitivity to low copper. Thus, C. S. Andrew and P. M. Thorne[7] found that *Desmodium uncinatum* was much more tolerant to low copper status in the soil than was *Stylosanthes bojori* among tropical legumes, and *Trifolium repens* than *T. alexandrinum* among the temperate; and the more sensitive needed a higher dressing of copper for maximum yield than the more tolerant, as they had less ability to take up copper from the soil.

The level of available copper is usually estimated from the amount extracted by a standard EDTA solution because although it extracts from 10 to 50 per cent of the total copper in the soil, which is usually much more than acetic acid does and very much more than do crops, yet the amount it extracts correlates much better with copper uptake or response than does the amount extracted by a weak acid. However, there is often a very marked contrast between the copper content of the surface soil and the subsoil, particularly in uncultivated soils, for the crop residues will have enriched the surface at the expense of the deeper layers. This can result in the dry season growth of pastures having a very much lower copper content than the wet season. At Muguga, in Kenya, for example, the copper content of the star grass, *Cynodon dactylon* pastures was of the order of 12 ppm in the rains, but only 4 ppm towards the end of the dry season.[8] It is also the probable reason why copper deficiency is more noticeable in peat and fen soils when they are deeply

1 O. C. Bryan, *J. Amer. Soc. Agron.*, 1929, **21**, 923. R. V. Allison, O. C. Bryan and J. H. Hunter, *Florida Agric. Exp. Sta. Bull.* 190, 1927.
2 With T. H. Caldwell *et al.*, *J. agric. Sci.*, 1966, **66**, 303.
3 D. S. Riceman and C. M. Donald, *J. Dept. Agric. S. Aust.*, 1939, **42**, 959.
4 A. S. Wild and L. J. H. Teakle, *J. Dept. Agric. W. Aust.*, 1942, **19**, ser 2, 71, 242.
5 With T. H. Caldwell *et al.*, *J. agric. Sci.*, 1966, **66**, 302.
6 *J. agric. Sci.*, 1968, **70**, 39.
7 *Aust. J. agric. Res.*, 1962, **13**, 821.
8 D. A. Howard, M. L. Burdin and G. H. Lampkin, *J. agric. Sci.*, 1962, **59**, 251.

drained, for this will encourage the surface soil to dry out in late spring and summer, and render the copper in the soil unavailable to the crop roots.[1]

The concentration of zinc in the soil solution has not often been determined, but J. F. Hodgson[2] found values between 10^{-6} to 10^{-8} M, of which between 30 and 70 per cent was as inorganic ions. This falls in the concentration range M. D. Carroll and J. F. Loneragan[3] found the roots of plants needed, for they found a concentration of 10^{-7} M was optimal for the crops they used, and 10^{-8} M allowed good growth. Zinc reacts with amorphous silica to give a silicate which maintains zinc concentrations of this order in solution,[4] and it is possible that this is the substance that controls the zinc concentration in the soil solution. This would be consistent with plants taking their zinc from within a few millimetres of their roots over a sixteen-day period, as found by H. F. Wilkinson.[5] Thus the diffusion coefficient for zinc in the soil solution may be about the same as for phosphate. Soluble orthophosphates also form very insoluble compounds with zinc, and the addition of a large dressing of superphosphate can induce a temporary zinc deficiency, which is gradually rectified presumably because the phosphate is being converted to other forms. Pyrophosphates and polyphosphates have a much smaller effect on the zinc solubility.[6]

Zinc deficiency can occur on a very wide range of soils, and is often associated with a low level of total zinc in the soil. It is sometimes due to an unfavourable soil structure which restricts root development, and orchard crops can be susceptible to this. It can then be ameliorated either by growing deep-rooting crops, such as lucerne, or allowing the indigenous weeds to grow and then discing them into the soil. Compacting a soil can also induce zinc deficiency probably through its effect on restricting the root system. Sterilising or partially sterilising a soil in which crops are showing zinc deficiency has been reported to cure the trouble,[7] and if this is a valid observation it also could be due to the better root system the crop would produce on the sterilised soil. Another odd result that has not been explained is that incorporating sugar-beet tops into a soil has sometimes induced zinc deficiency in the following crop of field beans or maize.[8] Zinc deficiency may also sometimes be induced by high phosphate manuring of calcareous soils low in available

1 See, for example, H. H. Nicholson and D. H. Firth, *J. agric. Sci.*, 1953, **43**, 95, and N. H. Pizer, T. H. Caldwell *et al.* ibid., 1966, **66**, 302.
2 With H. R. Geering and W. A. Norvell, *Proc. Soil Sci. Soc. Amer.*, 1965, **29**, 665, and with W. L. Lindsay and J. F. Trierweller, ibid., 1966, **30**, 722.
3 *Aust. J. agric. Res.*, 1968, **19**, 859; 1969, **20**, 457.
4 K. G. Triller, *Trans. 9th Int. Cong. Soil Sci.*, 1968, **2**, 567. W. L. Lindsay and W. A. Norvell, *Proc. Soil Sci. Soc. Amer.*, 1969, **33**, 62.
5 With J. F. Loneragan and J. P. Quirk, *Proc. Soil Sci. Soc. Amer.*, 1968, **32**, 831.
6 L. R. Hossner and R. W. Blanchar, *Proc. Soil Sci. Soc. Amer.*, 1969, **33**, 618.
7 D. R. Hoagland, W. H. Chandler and P. R. Stout, *Proc. Amer. Soc. Hort. Sci.*, 1936, **34**, 210 and P. A. Ark, ibid., 1936, **34**, 216.
8 L. C. Boawn, *Agron. J.*, 1965, **57**, 509. E. D. De Remer and R. L. Smith ibid., 1964, **56**, 67.

zinc.[1] In practice, zinc deficiency in a crop is usually cured by spraying with zinc sulphate, at about 4 kg/ha, and this is more effective than soil dressings of 50 kg/ha.

Cobalt differs from copper and zinc in that the principal reservoir of cobalt in many soils is that adsorbed on the surfaces of manganese oxides;[2] and the availability of cobalt to a crop is decreased by any soil treatment which converts the manganous ion into an insoluble oxide, and is increased by any which converts manganese from the oxide into manganous ions. Thus the cobalt content of a pasture can be reduced by liming, and the cobalt uptake of a crop increased by a temporary waterlogging of the soil.[3] R. L. Mitchell[4] noted that poorly drained soils in Scotland usually have a higher content of extractable cobalt than well-drained soils, although the total cobalt is often lower, and this, in turn, may be sufficiently marked to be reflected in the uptake by a crop.[5]

There are some soils, such as in parts of Australia, in which the cobalt content is too low to allow satisfactory nodulation and nitrogen fixation by subterranean clover and lucerne. This can be corrected by giving 25 to 125 g/ha Co, as cobalt sulphate, to the soil.[6] But the most important aspect of cobalt deficiency in most countries concerns the health of animals grazing pastures which are low in cobalt. In the United Kingdom, for example, sheep grazing pastures containing less than 0·1 ppm Co in their dry matter are subject to a syndrome known locally as pine. This trouble can be rectified either by applying between 100 and 200 g/ha Co as the sulphate or by putting a cobalt bullet in the rumen, which will supply the very small amount needed daily. If too much cobalt is applied to some pastures, the molybdenum content of the herbage may rise so high that the pining pasture is converted into a teart one.[7]

Boron

Boron is probably the trace element which most commonly limits yield and consequently is most widely used in agriculture, horticulture and even in forestry. Many root crops, such as swedes, turnips, sugar-beet and, to a lesser

1 E. J. Langin, R. C. Ward *et al.*, *Proc. Soil Sci. Soc. Amer.*, 1962, **26**, 574; 1963, **27**, 326. See also W. Thorne, *Adv. Agron.*, 1957, **9**, 31.
2 R. M. Taylor, R. M. McKenzie and K. Norrish, *Aust. J. Soil Res.*, 1966, **4**, 29; 1967, **5**, 235; *J. Soil Sci.*, 1968, **19**, 77. K. G. Triller, J. L. Honeysett and E. G. Hallsworth, *Aust. J. Soil Res.*, 1969, **7**, 43.
3 S. N. Adams and J. L. Honeysett, *Aust. J. agric. Res.*, 1964, **15**, 357.
4 In ch. 8, *Chemistry of the Soil*, Ed F. E. Bear, 2nd Edn, Reinhold, 1964.
5 T. Walsh, P. Ryan and G. A. Fleming, *Trans. 6th Int. Congr. Soil Sci.* (Paris) 1956, B 771. L. A. Alban and J. Kubota (*Proc. Soil Sci. Soc. Amer.*, 1960, **24**, 183) give examples based on the uptake by the cobalt-accumulating plant *Nyassa sylvatica* (black gum).
6 P. G. Ozanne, E. A. N. Greenwood and T. C. Shaw, *Aust. J. agric. Res.*, 1963, **14**, 39. K. D. Nicolls and J. L. Honeysett, ibid., 1964, **15**, 609. J. K. Powrie, *Pl. Soil*, 1964, **21**, 81.
7 For an example from Scotland see R. L. Mitchell, R. O. Scott *et al.*, *Nature*, 1941, **148**, 725.

extent, potatoes, many leguminous crops, such as lucerne, red clover, soya-beans, and many fruit and forest trees have a relatively large boron demand; while the cereals and grasses are more tolerant of a low boron supply in the soil, probably because they often contain only about one-tenth the amount of boron in their leaves that boron-sensitive crops do.

Boron is present in soil minerals, partly as tourmaline which is so resistant to weathering that the boron it contains is of no agricultural significance, and partly in more weatherable minerals which have not yet been recognised. These weatherable minerals are usually highest in rocks or deposits laid down in the sea (about 100 ppmB); for seawater contains 4·7 mg/litre B; and are lower in igneous rocks (about 30 ppmB). Rainwater can also be an important source of boron in areas near the sea, and T. Philipson[1] has estimated that at Uppsala in Sweden the rainwater supplies annually about 12 g/ha B, which is comparable to the boron content of a moderate cereal crop.

The chemistry of boron in the soil is still poorly understood. It is probably present in the soil solution as boric acid, $B(OH)_3$, which is a weak acid comparable in strength to silicic acid $Si(OH)_4$; and at high pH is adsorbed by aluminium and ferric hydroxides and oxides through ligand exchange, so has a maximum adsorption on these surfaces at a pH somewhat less than the pK for the first dissociation constant. But there is still no critical evidence on how far it behaves similarly to silicic acid in taking part in ligand exchange at pH values below about pH 7, when the proportion of hydrogen ions dissociated from the free acid in solution would be very low. It is, however, adsorbed in acid conditions, and liming increases its adsorption.[2]

Boric acid differs, however, from silicic acid in that it can be adsorbed by humus, probably through condensation with diol groups, which are probably normally associated with carboxylic acids. This gives the group:

$$\begin{array}{c} -C-O \\ \diagdown \\ \diagup \quad B-OH \\ -C-O \end{array}$$

and not uncommonly the carbons are either bound together giving a five-membered ring, or to another carbon atom giving a six-membered ring.[3] The strength of this bond may be greater than that between boric acid and sesquioxide surfaces under acid or neutral conditions,[4] so the humic colloids are likely to be the principal reservoir of boron in most agricultural soils.

Plants are presumed to take up their boron as the undissociated boric acid, but since it is difficult to determine its concentration in the soil solution, it is not known how the amount taken up compares with the amount present

1 *Acta agric. Scand.*, 1953, **3**, 121.
2 F. J. Hingston, *Aust. J. Soil Res.*, 1964, **2**, 83.
3 J. Boeseken, *Adv. Carbohydrate Chem.*, 1949, **4**, 189. C. A. Zittle, *Adv. Enzym.*, 1951, **12**, 493.
4 F. J. Hingston, *Aust. J. Soil Res.*, 1964, **2**, 83, showed this for mannitol.

in the water taken up by the roots. However, if the concentration is of the order 10^{-6} to 10^{-5} M the amount of water taken up by a crop would contain all the boron in the crop. The amount of available boron in a soil is usually measured by the amount of hot water-soluble boron the soil contains, and there is a good correlation between the amount of hot water-soluble boron in the soil and the uptake of boron by a given crop for soils of similar texture and pH. But for a given level of water-soluble boron, a crop will take up more from a light than a heavy-textured soil, and from an acid than from a limed soil.[1] Since nothing appears to be known on the relationship between hot water-soluble boron and the boric acid concentration in the soil solution, no explanation of these results can yet be given.

Boron deficiency is commonly found on acid light-textured soils low in organic matter, and can often be induced in such soils by liming and particularly by overliming.[2] It is often most pronounced in dry summers, presumably because the subsoil has a much lower boron content than the surface, due to the boron all being held by the organic colloids which are concentrated in the surface. Further, boron does not seem to be mobile in the plant, so that taken up early on in the season is not translocated to the growing points later on.[3]

Boron-demanding crops are likely to respond to boron fertilisers if the hot water-soluble boron falls between 0·3 and 1·0 ppm in the soil, though the actual level depends on the crop, on the soil, for the level is lower for sands than for heavier soils, and on the pH of the soil, since it is lower on acid than on neutral soils. Crop yields can, however, be affected by lack of boron before the crop has used all the hot water-soluble boron in the surface soil. Thus T. Philipson[4] found that sugar-beet growing on a boron-deficient soil in Sweden would remove only about 250 g/ha B although the top 15 cm of soil contained over three times this amount of hot water-soluble boron.

Boron deficiency can be cured either by adding a boron fertiliser to the soil, commonly borax, or by spraying a boric acid preparation on the crop. In boron-deficient areas that are also phosphate deficient, the borax is often added to the sulphuric acid used to make a borated superphosphate. Water-soluble boron fertilisers need, however, to be used with care, as the difference between the optimum level for boron-responsive crops and the toxic level for boron-sensitive crops is not large. For this reason boro-silicate glass frits have been developed, which are sintered glass having a large surface area in contact with the soil solution, and so weather at a controlled rate to release boron. Commonly levels of between 10 and 15 kg/ha of borax (1·2 and

1 J. I. Wear and R. M. Patterson, *Proc. Soil Sci. Soc. Amer.*, 1962, **26**, 344. J. Maurice and S. Trocme, *Ann. Agron.*, 1965, **16**, 579.
2 For an example see T. Walsh and J. D. Golden, *Trans. Comm. 2nd and 4th Int. Soc. Soil Sci.*, 1952, **2**, 167.
3 W. T. Dible and K. C. Berger, *Proc. Soil Sci. Soc. Amer.*, 1952, **16**, 60.
4 *Acta agric. Scand.*, 1953, **3**, 121.

1·9 kg/ha B) are applied to boron-responsive crops such as swedes, sugar-beet and potatoes. In areas where soils are boron-deficient and boron-demanding crops are regularly grown, dressings as low as 100 or 200 g/ha B for each of these crops may be necessary to prevent the boron concentration in the soil building up to too high a level.[1] On sandy soils low in humus, a water-soluble boron fertiliser will only have a small residual effect, as it is rapidly washed out of the soil if the season is wet,[2] but liming the soil, if it is acid, will reduce this rate, and lengthen the period the fertiliser is effective.[3]

Boron toxicity is typically a problem in some areas where irrigation water is bringing in boron, and is discussed further on p. 749; and in semi-arid areas where rocks high in boron are weathering.

Molybdenum

Molybdenum is needed for symbiotic nitrogen fixation, so legumes can only have active nodules in soils adequately supplied with this element. It is also required for the reduction of nitrates in plant tissues, but crops differ very considerably in their ability to take up sufficient to satisfy their requirements. It is taken up as either $HMoO_4^-$ or MoO_4^{2-}. The molybdate concentration in reasonably well-supplied soils is of the order 10^{-7} to 10^{-8} M, and the mass flow of soil solution to roots will bring more molybdate to the root surface than is taken up.[4]

The immediate reservoir of molybdate is that adsorbed on the surface of ferric oxides and hydrated oxides, for these have a very strong affinity for molybdate, which they adsorb by ligand exchange. In many soils almost all the molybdate present is associated with them,[5] and concretionary iron deposits may contain nearly all the soil's supply. Aluminium hydroxide has a much weaker affinity for the molybdate than the ferric,[6] so plays little role except in soils very low in iron. Over the whole range of soil pH the molybdic acid is dissociated, so it has a ligand adsorption curve similar to that for phosphoric acid (Fig. 7.16, p. 132). Thus, raising the pH of a soil by liming causes some ligand exchange between the added hydroxyls and adsorbed molybdate. Molybdenum deficiency is usually found only on acid soils, though J. Delas[7] has found it on some calcareous clays in the Charente district of western France, which are also sulphur deficient. On acid soils

1 For an example from Norway, see M. Odelien, *Soil Sci.*, 1963, **95**, 60.
2 H. W. Winsor, *Soil Sci.*, 1952, **74**, 459. J. I. Wear and C. M. Wilson, *Proc. Soil Sci. Soc. Amer.*, 1954, **18**, 425.
3 A. S. Baker and W. P. Mortensen, *Soil Sci.*, 1966, **102**, 173.
4 L. H. P. Jones, *J. Soil Sci.*, 1957, **8**, 313.
5 S. Trobisch and G. Schilling, *Chem. d. Erde*, 1963, **23**, 91. H. M. Reisenauer, A. A. Tabikh P. R. Stout, *Proc. Soil Sci. Soc. Amer.*, 1962, **26**, 23; and W. O. Robinson and G. Edgington, *Soil Sci.*, 1954, **77**, 237.
6 T. L. Lavy and S. A. Barber, *Proc. Soil Sci. Soc. Amer.*, 1964, **28**, 93.
7 With P. Dutil *et al.*, *Comp. Rend. Acad. Agric.* 1967, 947; 1969, 420.

the deficiency can often be cured by liming to pH 6·5 or 7 on nearly all of them.[1] It is often much cheaper to supply molybdenum in the form of sodium or ammonium molybdate to the crop, either as a spray or to the soil than to lime the soil to neutrality. Dressings of the order of 70 to 250 g/ha Mo are all that are normally required and such a dressing will usually last for five or six years.[2] But dressings as small as 2·5 g/ha Mo sometimes will increase the rate of nitrogen fixation of clovers sufficiently for it to markedly increase the green colour of a grass clover sward.[3] It is naturally important that the molybdenum dressings given to pastures be kept low, because otherwise the molybdenum content of the grass may become so high that the pasture becomes teart.[4] As one would expect manuring with water-soluble phosphate or sulphate fertilisers puts up the molybdenum content of crops,[5] and manganese fertilisers tend to reduce uptake, for reasons that are not known.[6]

High available molybdenum in the soil can occur in practice, and it is pastures on these soils that are called teart (see p. 47). They usually occur on poorly drained neutral clay soils, fairly high in organic matter, such as those derived from the Lower Lias clays in the west of England; but some Irish teart pastures are acid, having pHs between 4·9 and 5·2.[7]

The effect of soil acidity and alkalinity on plant growth

Natural soils differ considerably in their acidity or pH, and these differences are reflected in the vegetation or crops they carry. For a long time it was not clear how far these differences were due to the sensitivity of the plant roots to the hydrogen-ion concentration of the soil or soil solution in which they were growing, and how far to secondary effects brought about by the reaction. Water-culture experiments have now proved conclusively that the harmful effects of acidity are due to secondary causes, except in extreme cases. Thus D. I. Arnon[8] showed that many crops would grow satisfactorily in solution, in which pH ranged from pH 4 to 8, provided precautions were taken to eliminate harmful secondary effects, but that plant roots were definitely injured in solutions as acid as pH 3, and were unable to absorb phosphates at pH 9.

The reason why plants are sensitive to soil pH is that the pH of a soil affects the concentration of different ions in the soil solution, and so their

1 For a review of the effects of soil and fertiliser treatment on the molybdenum uptake by crops see A. J. Anderson, *Adv. Agron.*, 1956, **8**, 164.
2 R. S. Scott, N. A. Cullen and E. B. Davies, *N.Z. J. agric. Res.* 1963, **6**, 538.
3 R. S. Scott, *N.Z. J. agric. Res.*, 1963, **6**, 538, 556, 567.
4 I. J. Cunningham and K. G. Hogan, *N.Z. J. Sci. Tech.*, 1956, **38A**, 248.
5 P. R. Stout, W. R. Meagher *et al.*, *Pl. Soil*, 1951, **3**, 51.
6 A. J. Anderson and D. Spencer, *Aust. J. sci. Res.*, 1950, **3**, 414. E. G. Mulder, *Pl. Soil*, 1954, **5**, 368.
7 T. Walsh, M. Neenan and L. B. O'Moore, *Nature*, 1953, **171**, 1120.
8 *Plant Physiol.*, 1942, **17**, 515, 525.

availability to the plant. Thus, an acid soil is likely to have a higher concentration of aluminium and manganese ions, and a lower concentration of calcium, bicarbonate and molybdate ions than a calcareous. Plants can, for some purposes, be roughly divided into three groups: those that are adapted to and only grow well on calcareous soils, called calcicoles; those adapted to and only grow well on acid soils, called calcifuges; and those that are tolerant of a wide range of soil pH.

I. H. Rorison[1] compared the behaviour of some calcifuges on calcareous soils with some calcicoles on acid soils, and showed that the seedlings of the calcicoles were very sensitive to aluminium, and showed phosphate deficiency symptoms in the acid soil; but they took up potassium, phosphorus and iron easily from the calcareous soil. On the other hand, the calcifuge species on the calcareous soil suffered from potassium and phosphate deficiency as well as from lime-induced chlorosis, presumably due to difficulty in taking up iron properly; on the acid soil they took up phosphate quite easily. The distinction between the calcifuge and calcicole habit need have no specific significance. Thus, R. W. Snaydon[2] found that varieties of white clover (*T. repens*) indigenous to acid soils in Wales were typically calcifuge in habit, while those indigenous to calcareous soils in Hampshire were calcicole.

A high aluminium ion concentration is the most common cause of failure of agricultural crops in acid soils. It probably has two quite distinct effects. A high aluminium ion concentration in the free space in the root surface may prevent the root taking up phosphate, and aluminium inside the living cell may interfere with sugar phosphorylations.[3]

Aluminium tolerant plants differ from non-tolerant plants in their greater ability to take up phosphate from solutions containing aluminium.[4] Some calcifuge plants accumulate aluminium in their foliage, such as tea, for example, in which mature leaves may contain up to 1·7 per cent in their dry matter.[5] But other calcifuges contain no more than typical calcicoles, about 0·01 to 0·003 per cent,[6] and these plants must have some mechanism for preventing aluminium being taken up by the roots. C. D. Foy[7] found in the case of two varieties of wheat which grew equally well on a neutral soil, that the root system of the variety that was tolerant to an acid soil raised the pH in the external solution, and that of the sensitive variety lowered it. A number of crops behave like wheat in showing considerable variations in their tolerance to aluminium and other harmful effects of acidity, but little is yet known about the various mechanisms which control this tolerance.

1 *J. Ecol.*, 1960, **48**, 585, 679.

2 *J. Ecol.*, 1962, **50**, 133, 439.

3 D. T. Clarkson, *Pl. Physiol.*, 1966, **41**, 165; *Pl. Soil*, 1967, **27**, 347.

4 C. D. Foy and J. C. Brown, *Proc. Soil Sci. Soc. Amer.*, 1964, **28**, 27.

5 E. M. Chenery, *Pl. Soil*, 1955, **6**, 174. L. J. Webb, *Aust. J. Bot.*, 1954, **2**, 176.

6 H. Y. Hou and F. G. Merkle, *Soil Sci.*, 1950, **69**, 471.

7 With G. R. Burns *et al.*, *Proc. Soil Sci. Soc. Amer.*, 1965, **29**, 64; with W. H. Armiger *et al.*, *Agron. J.*, 1965, **57**, 413.

The aluminium ion concentration around a root can be very sensitive to pH, for according to the Ratio Law in cation exchange studies (p. 90) the ratio of the activity of the hydrogen ion concentration to the cube-root of the aluminium ion concentration is likely to be a characteristic for the soil if minor changes in pH occur, so that if the hydrogen ion concentration outside the root is doubled, the aluminium ion concentration will be increased eightfold. Root systems of plants differ very considerably in the aluminium ion concentration they can tolerate, ranging from something over 10^{-6} M to something over 10^{-3} M.[1] This should mean that, in so far as the trivalent aluminium ion is the only aluminium ion present, raising the pH of an acid soil on which an acid-tolerant plant can just grow by one unit should make it suitable for an acid-sensitive crop. This inference is not valid in practice, implying either that the root environment is not characterised by the pH of the soil in bulk, or that partially-neutralised aluminium ions, such as $Al(OH)_2^+$ act in the same way as the trivalent ion, or that aluminium is not the sole cause of acidity troubles.

Lowering the pH of a soil nearly always increases the amount of manganese a crop will take up, and crops differ very considerably in their ability to deal with a high manganese supply. Manganese accumulators are probably much more common amongst acid-tolerant crops than are aluminium accumulators, though acid-tolerant crops differ very considerably in the manganese content of their dry matter when growing on the same soil.

Acid soils are typically low in calcium, so crops growing on such soils may suffer from a shortage of calcium. In general the soil supply of calcium is over-abundant for most crops in neutral soils, and many soils must have their pH reduced to a low level before a shortage of calcium becomes important. But a high aluminium ion concentration may interfere with the uptake of calcium,[2] and for many plants a fairly high calcium content will reduce the harmful effects of high uptake of many other ions.[3] There is, as one would expect, a very great variation in the calcium demands of different crops, of different species within the same genus, and of different clones within the same species. Plants adapted to acid soils typically have a lower calcium demand than those adapted to neutral soils.[4] Adding a neutral calcium salt, such as calcium sulphate or chloride, may aggravate the harmful effect of acidity by increasing the uptake of manganese and aluminium.[5]

Since the effects of soil acidity on crop growth is not directly due to the hydrogen ion concentration in the soil solution, there can be no exact relation between the pH of a soil and its suitability for a given crop. However, one

1 F. Adams and Z. F. Lund, *Soil Sci.*, 1966, **101**, 193, cotton roots affected in range 10^{-5} to 10^{-6} M; D. T. Clarkson, *J. Ecol.*, 1966, **167**, some species of *Agrostis* (*A. setacea*) could stand 10^{-3} M, others (*A. Stolonifera*) only 10^{-4} M.

2 R. E. Johnson and W. A. Jackson, *Proc. Soil Sci. Soc. Amer.*, 1964, **28**, 381.

3 A. Wallace, E. Frolich and O. R. Lunt, *Nature*, 1966, **209**, 634.

4 For examples in grasses see A. D. Bradshaw, R. W. Lodge *et al.*, *J. Ecol.* 1960, **48**, 143.

5 J. L. Ragland and N. T. Coleman, *Proc. Soil Sci. Soc. Amer.*, 1959, **23**, 355.

can in practice grade crops in an order of tolerance to soil acidity. Thus, in England, lucerne, barley, beans, sugar-beet and mangolds are only considered suitable for neutral or mildly acid soils; wheat, red clover, peas, and vetches will often succeed in rather more acid soils; white clover, many grasses, oats, rye, lupins and potatoes will often grow on soils too acid for the others. Subtropical crops tolerant of acid soils include lespedeza, soyabeans, some varieties of the *Phaseolus* bean, millets, sorghums, sudan grass, and sweet potatoes.

This grading can only be approximate. E. J. Hewitt[1] has shown that sugar-beet and potatoes, for example, both have an appreciable calcium demand, but that potatoes have a much greater tolerance to aluminium ions than beet, while beet is more tolerant of manganese. In the same way, barley is like beet in being sensitive to aluminium and relatively tolerant of manganese, but it has a much lower calcium demand. Oats is a crop well adapted to acid soils as it has a low calcium demand and is tolerant of both aluminium and manganese.

The effects of soil acidity are usually more noticeable in meadows and pastures than on arable land or short-term leys, for very many species of plants are growing together in competition with each other on these uncultivated soils. Plants which tolerate acidity a little better than their neighbours will spread at their expense and come to dominate the flora of acid soils, even though they do not necessarily thrive best in acid soils in the absence of competition. These effects are very well illustrated in the floristic composition of the long-continued fertiliser plots on Park Grass field at Rothamsted.[2] Soil acidity can have a further important effect, for pastures on soils which are too acid for the larger species of earthworm to flourish will have a surface mat of dead vegetation, since there is no longer any mechanism by which the dead herbage can be mixed with the soil. Liming the soil, to allow earthworms to dominate the soil fauna, will alter the composition of the herbage, will encourage grass species that are more palatable to domestic stock, and will cause the soil to lose its slightly mor-like character and become more mull-like.[3] The use of basic slag on hill pastures in the United Kingdom, by being both a liming material and a phosphatic fertiliser, will cause coarse grasses, such as species of *Nardus* and *Agrostis* to be replaced by more nutritious species, such as ryegrass and the *Poas*.[4]

The pH of a soil can have a further effect on crop production, for it can affect the suitability of the soil as the home for soil-borne disease organisms. Thus, the actiomycete causing scab in potatoes, *Streptomyces scabies*, is much better adapted to life in a neutral than in an acid soil, possibly because

1 *Long Ashton Ann. Rept.* 1947, 82.
2 For an account of this experiment see W. E. Brenchley, *The Park Grass Plots at Rothamsted*, revised by K. Warington, 1958, Rothamsted Experimental Station, Harpenden.
3 E. Crompton, *Agric.*, 1953, **60**, 301; W. H. Pearsall, *J. Soil Sci.*, 1952, **3**, 41.
4 For an example see W. E. J. Milton, *J. Ecol.*, 1940, **28**, 326; 1947, **35**, 65.

it is sensitive to the manganese ion; so that liming a soil tends to encourage scab; this can be rectified to some extent by applying manganese sulphate to the soil.[1] Again, the fungus *Plasmodiophora brassicae*, which causes club-root or finger-and-toe in the cultivated Brassica crops (swedes, turnips, cabbage) tolerates acidity better than its host plant, and is therefore more likely to be injurious on acid soils; so liming the soil will render it less suited to the pathogen, and so increase the resistance of the crop.

Lime requirements of soils

The effect of liming an acid sandy soil on crop yields is illustrated in Tables 24.15 and 24.16. These give the results of two liming experiments on an acid sandy soil of pH 4·6 at Tunstall, Suffolk. Table 24.15 shows the effect of 12 t/ha of soft lump chalk applied in 1925 on crop yields over the seventeen-

TABLE 24.15   Effect of 12·5 tons per hectare lump chalk given to an acid sand on crop yields over the following 17 years Tunstall. Sandy soil: initial pH 4·6, chalked 1925. Yields in t/ha

Crop	Number of Crops	Yield No chalk	Response to Chalk
Lupins grain	4	2·44	−0·23
Oats grain	10	2·09	0·25
Rye grain	8	2·82	0·32
Potatoes roots	10	13·9	13·0
Wheat grain	7	1·11	1·33
Peas grain*	5	4·2	26·8
Sugar beet roots	10	0·5	32·8

* In hectolitres per hectare.

TABLE 24.16   Effect of soil pH on yield of sugar-beet, on an acid sandy soil. Yield of roots in t/ha

pH range of soil	No. of years crop grown	Mean yield	Range of yields
4·5–5·0	5	0·7	0–3·5
5·0–5·5	6	10·4	0–22·6
5·5–6·0	6	31·8	28·4–37·3
6·0–6·5	3	34·8	31·8–36·6
6·5–7·0	5	37·3	32·8–40·8
7·0–7·5	5	38·0	33·9–41·8
over 7·5	2	41·3	39·8–43·0

1 A. J. McGregor and G. C. S. Wilson, *Pl. Soil*, 1964, **20**, 59; 1966, **25**, 3.

year period 1927–43.[1] This dressing probably raised the pH to about 7 for about eight or ten years, and it only gave a small increase in the yield of oats and rye, it doubled the yield of potatoes and wheat, and converted crop failures for peas and sugar-beet into reasonable crops. It does not show how far such a large increase in pH is necessary. Table 24.16 shows this for another experiment on the same station for sugar-beet,[2] and here four levels of calcium carbonate were given, of 2·5, 5, 7·5 and 10 t/ha, in 1932, and sugar-beet was taken at intervals in the period 1933–47. For this experiment there was little benefit in maintaining a pH much above 6·5, and in fact a dressing of 2·5 t/ha allowed a yield of 37t/ha of beet six years after it was given, when the pH was 5·3, while the yield was only 40 t/ha on the plots receiving 5 or 10 t/ha of chalk and the pH was between 6·5 and 7. Even in 1943, eleven years after the lime was given, the yield of beet on the plots which received 0, 2·5, 5 and 10 t/ha of chalk gave yields of 0, 22·7, 30·6 and 33·1 t/ha and the pH of the plots were 4·6, 5·0, 5·4 and 6·4. Barley and beans (*Vicia faba*) are two other crops which in eastern England usually give their maximum yield on plots limed to between pH 6·5 to 7.[3] Table 24.17 illustrates the

TABLE 24.17   Effect of small dressings of burnt lime (CaO) on yields. Macaulay Institute for Soil Research, Aberdeen. Lime applied 1944

Weight of calcium oxide applied t/ha	0	1·1	2·2	4·4	6·6
Barley 1946					
Yield grain t/ha	1·45	2·54	3·03	3·32	3·02
pH of soil	5·3	5·4	5·7	6·0	6·5
Extractable Ca m eq/100 g soil	3·9	4·6	6·4	7·8	12·8
Barley 1951					
Yield grain t/ha	0	0·95	2·36	2·88	3·07
pH of soil	5·0	5·2	5·4	5·8	6·0
Extractable Ca m eq/100 g soil	3·2	4·3	5·7	8·6	10·7
Barley 1957					
Yield grain t/ha	0·24	0·54	1·53	2·09	1·90
pH of soil	4·7	5·0	5·1	5·4	5·8
Extractable Ca m eq/100 g soil	2·2	3·2	3·6	5·0	7·8

1 H. W. Gardner and H. V. Garner, *The Use of Lime*, London, 1953, p. 144. This is a very useful and comprehensive book on the use of lime in British agriculture. Also A. W. Oldershaw and H. V. Garner, *J. Roy. Agric. Soc. Engl.*, 1944, **105**, 98.

2 A. W. Oldershaw and H. V. Garner, *J. Roy. Agric. Soc. Engl.*, 1949, **110**, 89, and see S. G. Heintze, ibid., 98 for the appropriate soil data.

3 For Rothamsted and Woburn data see J. Bolton, *Rothamsted Rept.* 1970, Part II, 98, and for beans, J. R. Moffatt, *Rothamsted Rept.* 1966, 240.

response of barley, another acid sensitive crop, to small applications of calcium oxide at the Macaulay Institute, Aberdeen,[1] and it again shows the small benefit that is derived from heavy dressings of lime compared with light ones; but it also shows that crops tend to be less sensitive to low pH in cool moist climates, for barley appears to give its maximum yield at about pH6.

These results must be interpreted with caution in practice. Thus, the general experience of sugar-beet growers in eastern England and Sweden is that yields tend to be highest on fields with a pH above 7. But this may be simply a reflection of the variability of soil pH over a field, for a field may have an average pH of 6·5 yet contain appreciable areas where the pH is definitely below this, and it could easily be that recommendation of liming to over pH 7 is in fact a way of ensuring that there are no areas with a pH below 6·5. A. Aslander[2] in Sweden has in fact maintained that provided the phosphate and potassium levels of the soil are high enough, liming has little effect on the yields of most farm crops provided the soil is not too acid. This is probably more likely to be true in regions with cool moist than with warm dry summers, where acid tolerant crops are also more likely to be grown, in the west rather than in the east of England for example. In so far as new varieties of crops are bred for regions with acid soils it is likely that breeders will be selecting for, or breeding in, greater tolerance to acid conditions.

These results cannot be applied to many tropical crops, for general experience has shown that they rarely respond to liming if the pH exceeds pH 5·2 to 5·5. Lime is needed on the more acid soils to reduce the exchangeable aluminium to between 4 and 10 per cent of the exchangeable cations, which occurs when the pH has been raised to about 5·2; and raising the pH still further usually has no effect, or slightly reduces yield.[3] Thus the levels of calcium carbonate required on the soils requiring lime rarely exceed 1 t/ha. On some soils low in exchangeable calcium, calcium-demanding crops, such as cotton and groundnuts, may respond to calcium fertilisers, but gypsum is as useful a source of calcium as carbonate for these crops. It is not known why some temperate crops appear to need a much higher pH than this, though the crops requiring a pH nearer neutrality for maximum yield on the farm may all have been developed from wild ancestors adapted to neutral or calcareous soils. The difference could lie in an effect of soil temperature on the aluminium ion activity in the pH range 5·5 to 6·5. It is possible that in cool soils the proportion of aluminium in the form of hydroxylated cations on the surface of the clay particles is higher than in warm soils due to a greater tendency for these to react with silica to give kaolin-type minerals. The temperate crops requiring a neutral soil for

1 J. W. S. Reith and E. G. Williams, *Emp. J. exp. Agric.*, 1949, **17**, 265; 1962, **30**, 27.
2 *Soil Sci.*, 1952, **74**, 181, 436.
3 H. L. Foster, *E. Afr. agr. for. J.*, 1970, **36**, 58 (Uganda); N. G. Reeve and M. E. Sumner, *Proc. Soil Sci. Soc. Amer.*, 1970, **34**, 267, 595 (Natal); E. J. Kamprath, ibid., 1970, **34**, 252 (N. Carolina).

maximum yield are all probably derived from wild ancestors adapted to neutral or calcareous soils.

It is theoretically possible to determine the amount of lime that must be added to a soil in the field to raise its pH to a given level from a determination of the amount that must be added to a sample in the laboratory, when well dispersed and stirred in water. But in general if this dressing is given to the soil in the field, the pH of the soil does not rise to this value; and if a pH of about 6·5 is aimed at, it is commonly necessary to apply two to three times this amount.[1] This is illustrated in Fig. 24.3 for two Danish acid soils,[2] where it was assumed that the added calcium carbonate was distributed uniformly through the top 20 cm of soil, and that this soil weighed 2400 t/ha.

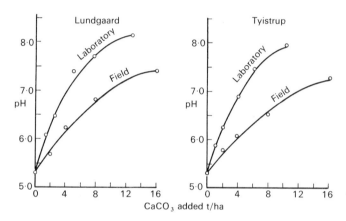

FIG. 24.3. Influence of successive additions of $CaCO_3$ on the pH values of two acid soils in the laboratory and the field respectively

The factor by which the laboratory figure must be multiplied by to get the same soil pH in the field is called the liming factor. No quantitative study has been made of the reasons why it may be as high as two or three for some soils yet be unity for others, but it probably depends on such factors as the evenness with which the limestone is incorporated into the soil, the rate at which it dissolves in the soil solution and reacts with the soil particles, and the rate of leaching of rainwater through the soil. The first two of these factors are responsible for the very large sampling errors found if the pH of the soil is being determined within a year or two of the lime being applied, and the third is rarely important until two to three years after the lime has been applied.

1 See, for example, R. L. Mitchell, *J. agric. Sci.*, 1936, **26**, 664 (Scottish soils); O. de Vries and P. Bruin, *Emp. J. exp. Agric.*, 1947, **15**, 260 (Dutch soils); L. E. Dunn, *Soil Sci.*, 1943, **56**, 341 (Washington, USA soils).
2 H. R. Christensen and S. Tovborg Jensen, *Trans. 2nd Comm. Int. Soc. Soil Sci.*, 1926, A 94.

The rate at which a dressing of lime raises the pH of the soil depends on the composition and fineness of grinding of the lime. Lime is an imprecise word, for it includes calcium oxide or quicklime, calcium hydroxide or slaked lime, and calcium carbonate, which is typically in the form of chalk or limestone although it may be an industrial waste product, such as produced by sugar-beet factories. Some blast furnace slags, high in calcium silicates, are also liming materials, for the calcium silicate hydrolyses in the soil to calcium hydroxide and silica. The calcium oxide and hydroxide are marketed as fine powders, and 56 kg of the oxide or 74 kg of the hydroxide are equivalent to 100 kg of the pure carbonate. Most limestones are calcitic, that is most of the carbonate is present as calcium carbonate, although the actual content of calcium carbonate differs for limestones from different sources. But some limestones are dolomitic, that is a proportion of the carbonate is present as magnesium carbonate, and these are particularly valuable on acid soils low in magnesium.

Natural limestones must be crushed before they can react with the soil. If a chalk is used, crushing to a fine powder is technically simple and the process cheap. But it is more expensive to crush hard limestones, and the finer they are ground, the more expensive the process. On the other hand, the finer they are ground, to within certain limits, the more quickly and completely they will react with the soil when properly incorporated. Limestones crushed to pass a 100 mesh sieve (aperture 0·15 mm) are almost as reactive as the oxide and hydroxide, while a limestone crushed to pass a 20 mesh screen (aperture 0·84 mm), but to be retained by a 40 mesh (aperture 0·42 mm), takes considerably longer to exert its full effect on the soil pH, although in general there is little difference between the effect of the coarse and fine materials after a year.[1] Limestone coarser than 2 mm is, however, relatively inefficient as a liming material. In practice, limestones crushed to pass a 10 (2·0 mm) or 20 mesh screen will normally contain about 25 and 45 per cent of material that will pass a 100 mesh sieve, so the material produced by a crushing mill is agriculturally preferable to material crushed and screened to a uniform size. Natural limestones, however, often contain some silica, and it is possible that the more siliceous, harder limestones may need to be crushed rather finer than the softer, for the silica coatings on the limestone nodules protects their surfaces against dissolution.[2]

Rainwater leaching through a soil containing free calcium carbonate leaches some out as bicarbonate, which will eventually appear in the rivers draining the area. The more finely divided the carbonate and the greater the amount of water leaching through the soil, the greater the loss, and losses of the order of 2·5 kg/ha/mm of water leaching through are to be expected in temperate regions. Thus the Rothamsted soils which contain

1 J. W. S. Reith and E. G. Williams, *Emp. J. exp. Agric.*, 1949, **17**, 265. F. L. Davis, *Agron. J.*, 1951, **43**, 251. T. A. Meyer and G. W. Volk, *Soil Sci.*, 1952, **73**, 37.
2 N. Lahav and G. H. Bolt, *Soil Sci.*, 1964, **97**, 293.

free calcium carbonate lose about 300 to 400 kg/ha annually by this process. Calcium bicarbonate will be lost from soils even if they contain no free carbonate, but the lower the pH of the soil and the less saturated the exchange complex with calcium, the lower the loss. Thus, the rate of loss from a limed soil increases with increasing dressings of lime but it should reach a limit when there is free calcium carbonate uniformly mixed with the soil. Table 24.18, which gives the rate of loss from a liming experiment on an acid sandy soil in Shropshire, illustrates the first part of these conclusions,[1] but there is

TABLE 24.18   The rate of leaching of calcium carbonate out of a limed soil. Harper Adams, Shropshire; Sandy Soil

$CaCO_3$ t/ha	Rate of loss of $CaCO_3$ in 100 kg/year Years after application			Total
	0–5	6–10	11–15	0–15
3·1	3·8	1·9	0·6	31
6·3	6·3	3·1	1·3	53
12·5	10·0	5·0	2·5	88

little experimental evidence for the second part, both because it is not often met with in agricultural practice, and also because it is difficult to mix lime uniformly with the soil, so it is unlikely that all the leaching water is saturated with calcium bicarbonate.

This leaching out of calcium from the soil as calcium bicarbonate is a normal feature of soils in the humid regions if they contain much exchangeable calcium. Most intensive systems of agriculture in temperate regions involve cropping with some acid-sensitive crops, so that the liming of soils is a normal requirement. However, in intensive crop production, based on use of heavy dressings of ammonium or urea-based fertilisers, the loss of calcium, due to the washing out of nitrate, is usually much higher than the loss through bicarbonate.

The neutralising effect of calcium carbonate only appears to be carried down into the subsoil when there is an excess of calcium carbonate in the surface soil. Thus, R. G. Warren and A. E. Johnston[2] showed that on the permanent grass plots on Park Grass at Rothamsted, the neutralising effect was only noticeable in the 22 to 45 cm layer when the pH of the surface soil had risen to 7. This is illustrated in Table 24.19. Nevertheless, some acid sandy soils cultivated to arable crops show a rise in pH and in exchangeable calcium in the layer immediately below the plough layer before the pH of the plough layer has reached neutrality.[3] Heavy dressings of lime can affect

1   T. W. Walker, *J. Soil Sci.*, 1952, **3**, 261.
2   *Rothamsted Ann. Rept.* 1963, 240.
3   For a possible example on the acid sandy soil at Tunstall see S. G. Heintze, *J. Roy. Agric. Soc. Engl.*, 1949, **110**, 98.

the pH of the deeper subsoil, though it takes several years for this to happen. Thus, B. A. Brown[1] found that neither 20 nor 40 t/ha had appreciably affected the pH of fine sandy loam pasture at 12 cm depth after two years, though they raised the pH by 0·4 and 0·7 unit at 15 cm depth after five years, and at 30 cm by about this amount after nine years. By this time the larger dressing had raised the pH of the 45 to 60 cm layer a little, but not the 65 to 75 cm layer.

TABLE 24.19   The effect of surface dressings of calcium oxide on the surface soil and subsoil pH of old grassland. Park Grass Rothamsted. Plots receive 470 kg/ha/year of sulphate of ammonia (which could give a loss of about 1100 kg/ha Ca O in 4 years). Lime applied once in 4 years at 2250 kg/ha Ca O on Plot 9 and 4500 kg/ha on Plot 18. Soil pH in 1:5 water

Soil	Plot 9		Plot 18	
	surface soil *0–22 cm*	*subsoil* *22–45 cm*	*surface soil* *0–22 cm*	*subsoil* *22–45 cm*
Unlimed portion				
1923	4·0	4·8	4·5	5·7
1959	3·8	4·3	4·1	4·4
Limed portion				
1923	4·5	5·2	4·7	5·6
1959	5·3	5·2	7·1	6·6

The effect of soil alkalinity and overliming

The reasons why calcifuge plants have their growth severely affected on neutral or calcareous soils are not yet fully understood. For some species the reason is probably that their ability to take up phosphate from the soil solution decreases as the pH rises, so that they can be grown on calcareous soils only if large dressings of a water-soluble phosphate are used. Some calcifuge grasses come into this category.[2] But the characteristic trouble which develops when a calcifuge grows on a calcareous soil is known as lime-induced chlorosis,[3] and is due to some upset in the iron metabolism of the plant, and is not necessarily due to any inability to take it up.[4] It appears to be related to the bicarbonate ion concentration in the soil solution, and is encouraged by poor drainage and lessened by good drainage, but a high bicarbonate concentration does not necessarily give this chlorosis. It can, however, often be cured by the use of an iron chelate that is stable in a calcareous soil.[5] (see p. 646).

1 With R. I. Munsell *et al.*, *Proc. Soil Sci. Soc. Amer.*, 1956, **20**, 518.
2 See, for example, D. B. James, *J. Ecol.*, 1962, **50**, 521.
3 For a review see J. C. Brown, *Adv. Agron.*, 1961, **13**, 329.
4 For autoradiographs of the distribution of iron in the leaf see P. C. de Kock, *Soil Sci.*, 1955, **79**, 167.
5 V. Q. Hale and A. Wallace (*Soil Sci.*, 1960, **89**, 285) have studied this using Fe^{59}.

For practical purposes the effects of alkalinity on crop production can be divided into two groups: those occurring in calcareous soils containing only small amounts of exchangeable sodium, and those occurring in soils where the sodium ions are the cause of the alkalinity. The pH of calcareous soils in equilibrium with the atmosphere containing 0·03 per cent CO_2 cannot exceed 8·5 (see p. 126), but soils high in exchangeable sodium may contain sodium carbonate in their soil solution and have a pH above 10·0. The management of these sodium alkali soils will be discussed in chapter 27.

Crops adapted to calcareous soils do not usually have their growth or yield reduced if grown on a soil that has been given a dressing of lime appreciably greater than needed to raise its pH to 7, but overliming some acid soils will result in loss of yield due to induced trace-element deficiencies. These troubles are rarely seen on clay or loam soils in temperate regions, but can be serious in acid sands, particularly if an attempt is made to raise the pH by too large an amount in a single step. On many acid sands, boron deficiency is the most likely to be noticed, while on organic soils, manganese is the most likely. But many acid soils in the tropics are particularly liable to suffer if even quite a small excess of lime is used, as already noted. However, little is yet known either about what types of soil are so sensitive or what is the cause of the sensitivity.

The harmful effects of overliming can usually be neutralised by suitable soil dressings of manganese sulphate or borax, but it is always preferable to reduce this hazard by applying the smallest dressing of lime compatible with good yields of the most acid-sensitive crop to be grown, and re-applying light dressings at relatively frequent intervals. This is particularly important in sandy soils low in boron, if boron sensitive crops such as sugar-beet are to be grown, for this deficiency can be a difficult one to correct.

Manganese deficiency can be induced by quite modest dressings of lime, and on some light soils the minimum dressing of lime to allow a good crop of barley, for example, will cause manganese deficiency in acid-tolerant crops. Thus, in western Scotland, liming to allow barley to be taken will increase the incidence of scab on potatoes, due to the actinomycete *Streptomyces scabies*, so badly that the tubers are unsaleable; this can be corrected by applying about 60 kg/ha of manganese sulphate to the soil at planting time.[1]

1 A. J. McGregor and G. C. S. Wilson, *Pl. Soil*, 1964, **20**, 59; 1966, **25**, 3.

25

The chemistry of waterlogged soils

Flooding or waterlogging a soil containing decomposable organic matter causes the onset of anaerobic or partially anaerobic conditions because the soil micro-organisms, in decomposing the organic matter, will use up any free oxygen dissolved in the soil water much faster than atmospheric oxygen can diffuse into the wet soil. This shortage of oxygen will cause some species of bacteria to carry out a number of chemical reductions which may affect plant growth very considerably.

Organisms obtain the energy they need for their vital processes through a series of chemical reactions involving the transfer of electrons from substances which serve as sources of energy to substances which may become products of respiration. If the organisms are respiring aerobically the final electron sink is oxygen, which accepts electrons and combines with hydrogen ions to give water, so aerobic respiration involves the reduction of molecular oxygen to water. This oxygen can also be pictured as a hydrogen acceptor, for the reaction is;

$$O_2 + 4H^+ + 4e^- \rightarrow 2H_2O.$$

In the absence of free oxygen a number of other substances can accept electrons and take part in a reduction reaction. Some oxygen containing compounds such as nitrates and sulphates can accept electrons and lose their oxygens, thus:

$$NO_3^- + 2H^+ + 2e^- \rightarrow NO_2^- + H_2O$$

and:

$$2NO_2^- + 8H^+ + 6e^- \rightarrow N_2 + 4H_2O.$$

Some high valency cations will accept electrons and become reduced to a lower valency state, as, for example, trivalent iron or tetravalent manganese can be reduced to divalent ferrous and manganous ions; and, finally, the hydrogen ion itself can accept an electron to become hydrogen gas:

$$2H^+ + 2e^- \rightarrow H_2.$$

A second consequence of bacterial activity under conditions of oxygen deficiency is that the organic nutrients are no longer fully oxidised to carbon dioxide and water, but instead intermediate products are excreted, such as simple fatty acids, hydroxy-carboxylic and polycarboxylic acids, alcohols and ketones, some of which can reduce ferric oxides bringing the iron into solution as a ferrous chelate. These organic compounds are, in turn, further decomposed with the production of carbon dioxide, methane and other hydrocarbons, and sometimes hydrogen gas. Thus, waterlogged soils may contain both reduced inorganic and organic compounds, that is compounds which will absorb oxygen, or will donate electrons; and the greater the avidity of the compound for oxygen the stronger a reducing agent it is said to be.[1]

This avidity of a solution to donate or to accept electrons to any reducible or oxidisable substance added to it can be measured by its oxidation-reduction or redox potential. The more strongly oxidising a solution the higher its potential, and the more strongly reducing the lower. In the same way as one can carry out an acid-base titration, and define the buffering power of a solution as the amount of hydrogen ions needed to give a unit reduction in pH, so one can define the poise of a solution as the amount of electrons, as a reducing substance, that must be added to give a unit reduction in the redox potential; and in the same way as a solution which can accept a large amount of acid for a small change in pH is said to be well buffered, so a solution that can accept a large amount of a reducing substance for a small change in redox potential is said to be well-poised. Again in the same way that the hydrogen ion concentration of a system can be measured as the electrical potential difference between two suitable electrodes, so can the state of oxidation or reduction; but because the basic reduction process can be considered as the transfer of an electron to a hydrogen ion, the oxidation-reduction or redox potential of a system depends on its hydrogen ion concentration. Thus, the redox potential of a well-aerated solution falls linearly with pH from 630 mV at pH 5·0 to 510 mV at pH 7·0, and this relation is found to hold reasonably well in aerated soils.[2]

The various reduction reactions that can take place in a soil as it becomes more anaerobic depends on the redox potential at which they are most strongly poised—those poised at a high redox potential going to completion before those poised at an appreciably lower potential become important. These potentials are pH dependent, but the effect of a unit change in pH on the potential is not the same for all reduction reactions, and for simple inorganic reductions can vary from -59 mV to -177 mV per unit increase of pH. This means that the relative order in which a series of reductions take place as the soil becomes anaerobic depends on the pH of the soil, if they are poised at about the same potential.

1 For a review of the changes taking place in waterlogged soils, see F. N. Ponnemperuma, *Adv. Agron.*, 1972, **24**, 29.
2 F. H. Redman and W. H. Patrick, *Louisiana Agric. Exp. Sta. Bull.* 592, 1965.

The principal inorganic reductions that poise a soil as it becomes more anaerobic are nitrate to nitrite; manganic salts and manganese dioxide to manganous ions; ferric hydroxide to ferrous ions; hydrogen ions to hydrogen gas and sulphates to sulphites and sulphides. The redox potential of these reductions at maximum poise are given in Table 25.1 both at pH 5·0 and pH 7·0.

TABLE 25.1   Oxidation-reduction potentials of typical soil systems*

System	Oxidation-reduction potential in mV, 25°C	
	At pH 5	*At pH 7*
$O_2 + 4H^+ + 4e^- = 2H_2O$ $E_h = 1·23 + 0·0148 \log P(O_2) - 0·059$ pH	930	820
$NO_3^- + 2H^+ + 2e^- = NO_2^- + H_2O$ $E_h = 0·83 - 0·0295 \log NO_2^-/NO_3^- - 0·059$ pH	530	420
$MnO_2 + 4H^+ + 2e^- = Mn^{2+} + 2H_2O$ $E_h = 1·23 - 0·0295 \log Mn^{2+} - 0·119$ pH	640	410
$Fe(OH)_3 + 3H^+ + e^- = Fe^{2+} + 3H_2O$ $E_h = 1·06 - 0·059 \log Fe^{2+} - 0·177$ pH	170	−180
$SO_4^{2-} + 10H^+ + 8e^- = H_2S + 4H_2O$ $E_h = 0·30 - 0·0074 \log H_2S/SO_4^{2-} - 0·074$ pH	−70	−220
$CO_2 + 8H^+ + 8e^- = CH_4 + 2H_2O$ $E_h = 0·17 - 0·059 \log P(CH_4)/P(CO_2) - 0·059$ pH	−120	−240
$2H^+ + 2e = H_2$ $E_h = 0·00 - 0·059$ pH	−295	−413

* I am indebted to Dr F. N. Ponnemperuma for this table.

Some of these are reversible in the thermodynamic sense in the soil system, such as the iron and manganese oxidation reductions; but nitrite and sulphite are reduced more easily than nitrate and sulphate; one to nitrogen gas or nitrous oxide which is lost from the system as a gas, and the other to sulphide which, in the absence of ferrous iron, may also be lost as gaseous hydrogen sulphide, so that the nitrate reduction is always, and the sulphate sometimes, an irreversible process.

The reduction of nitrates and ferric iron can be carried out by a number of different bacteria, many of which are capable of bringing about both these reductions. Thus J. C. G. Ottow and H. Glathe[1] isolated seventy-one bacteria from three gleyed subsoils, and found all but three of the seventy-one could reduce nitrate to nitrite, but only about half could carry the reduction to

1 *Soil Biol. Biochem.*, 1971, **3**, 43.

nitrous oxide or nitrogen gas. Most of their bacteria were either Pseudo-monads or Bacilli. On the other hand, only few bacteria appear capable of reducing sulphates, and these belong to the genus *Desulphvibrio*. They are most active at neutral reactions and reduction only takes place slowly outside the pH range 6·5 to 8·5.[1]

Some of the oxidation-reduction systems given in Table 25.1 are only partially applicable to field soils, because the compounds present in the soil are not pure. Thus the ferric hydroxide present nearly always contains im-purities, such as manganese; for if this hydroxide is being formed from the oxidation of ferrous iron under acid conditions it will adsorb some divalent manganese under conditions too reducing for the manganous ion to be oxidised. The manganese dioxide type of precipitate often contains fewer impurities because by the time the redox potential has risen sufficiently high to oxidise the manganous ion, most other oxidisable cations have been oxidised.[2] This has the consequence that these precipitates have oxidation-reduction potentials spanning a fairly wide range, so that, in the field, some ferrous ions will be present before all the higher valency manganese com-pounds have been reduced.[3]

The dominant poise in soils containing iron is the ferric-ferrous reaction, and while the ferric ion concentration in soil solutions is very low, due to the insolubility of ferric hydroxide or hydrated oxide, the concentration of ferrous ions can be sufficiently high for plant growth to be affected. The various reac-tions taking place when ferric iron is being reduced have been studied by F. N. Ponnemperuma and his colleagues,[4] who have shown that the normal reversible oxidation-reduction reaction is between the metastable $Fe(OH)_3$ and the metastable ferroso-ferric hydroxide $Fe_3(OH)_8$. The relations con-necting the redox potential, the pH of the solution and the ferrous ion concentration are given by:

$$E_h = 1·06 - 0·059 \log (Fe^{2+}) - 0·18 \, pH \quad (30°C) \qquad (25.1)$$

if the pH is too low for the ferroso-ferric hydroxide to be formed, and

$$E_h = 1·35 - 0·093 \log (Fe^{2+}) - 0·24 \, pH \qquad (25.2)$$

if the pH is sufficiently high for this hydroxide to be precipitated.

Equation 25.1 is found to hold for many soils containing moderate amounts of iron and not too much organic matter, and J. W. O. Jeffery[5] has suggested that the term $E_h + 0·18 \, pH$, which he called $r(E_h)$ could be used as a measure of the severity of the reducing conditions prevailing. He found that, for the

1 W. E. Connell and W. H. Patrick, *Science*, 1968, **159**, 86.
2 For examples, see J. F. Collins and S. W. Buol, *Soil Sci.*, 1970, **110**, 111, 157: N. van Breeman, *Neth. J. agric. Sci.*, 1969, **17**, 246.
3 C. L. Valera and M. Alexander, *Pl. Soil*, 1961, **15**, 268 and J. H. Jordan, W. H. Patrick and W. H. Willis, *Soil Sci.*, 1967, **104**, 129.
4 *Soil Sci.*, 1966, **101**, 421; 1967, **103**, 374 and *Int. Rice Res. Inst., Ann. Rept.* 1965, 126.
5 *J. Soil Sci.*, 1961, **12**, 172.

rice soils he was interested in, $r(E_h)$ was above 1·34 for aerated soils, between 1·34 and 1·20 to 1·15 for soils which were suitable for rice, but if it was below 1·15 very severe reducing conditions prevailed. Since the reduction process removes hydrogen ions from the solution, as reduction proceeds so the pH of the system rises, and once it has risen sufficiently high, the ferroso-ferric hydroxide begins to precipitate out, provided there is sufficient ferrous iron present, and this controls the concentration of ferrous ions in solution. This hydroxide gives the soils its typical greeny-grey colour, and the relations

$$E_h + 0.059 \text{ pH} = 0.429, \quad \text{and} \quad \log(Fe^{2+}) = 10.6 - 2 \text{ pH}$$

should hold. The constant in the first of these equations is found to fall within the range 0·41 to 0·55 for different soils, the higher values being found for soils low in iron but high in manganese, and in the second to fall within the range 10·2 to 10·8. The effect of manganese can also be seen if the validity of eq. 25.2 is tested, for the constant falls within the range 1·34 to 1·39, the lower values being for soils low in iron and high in manganese.

Soils, however, contain carbon dioxide dissolved in the soil solution, and, if the conditions are suitable, either ferrous carbonate will be precipitated, or the carbon dioxide in the solution, through its effect on pH, will alter the concentration of ferrous ions in equilibrium with $Fe_3(OH)_8$, if this is present. Ponnemperuma and his colleagues[1] have investigated these equilibria, both in the presence and absence of soils, and been able to show that, in the presence of sufficient iron, the $Fe_3(OH)_8$ precipitate continues to control the ionic concentrations in the solution. The complete theoretical equations for this system are:

$$E_h = 0.71 + 0.0885 \log P(CO_2) - 0.059 \text{ pH}.$$
$$\text{pH} = 6.19 - 0.33 \log P(CO_2)$$
$$\log(Fe^{2+}) = -1.94 + 0.66 \log P(CO_2),$$

where $P(CO_2)$ is the pressure of carbon dioxide, in bars, in equilibrium with the solution. They tested the relation between pH and the carbon dioxide concentration for three ferruginous soils and found the relation:

$$\text{pH} = 6.45 - 0.33 \log P(CO_2),$$

although if they worked with the $Fe_3(OH)_8$ system in the absence of soil, the relation was

$$\text{pH} = 6.06 - 0.58 \log P(CO_2)$$

These calculations were made on the assumption that no ferrous carbonate was formed, which appears to be valid in the presence of the ferroso-ferric hydroxide; but if some were formed, it would alter the constants somewhat.

It is interesting to note that if a soil contains free calcium carbonate, the

1 *Soil Sci.*, 1966, **101**, 421; 1967, **103**, 90, 374; *Proc. Soil Sci. Soc. Amer.*, 1969, **33**, 239.

relationship between the soil pH and the CO_2 concentration is:

$$pH = 6.03 - 0.67 \log P(CO_2)$$

so the value of pH of a soil containing adequate iron for $Fe_3(OH)_8$ to be formed and of a soil containing free calcium carbonate are almost the same, and decreases from about pH 7·2 to pH 6·7 as the CO_2 pressure rises from 0·01 to 0·1 bar. Thus the pH of both acid ferruginous soils and calcareous soils tends to stabilise at about 6·7–7·0 after flooding.[1]

Ponnemperuma[2] has also examined the behaviour of manganese oxides in the presence of quartz sand under reducing conditions, and compared these results with that of sixteen Philippine rice soils all containing manganese. He interpreted his results as showing that initially MnO_2 is reduced partly to Mn^{2+} and partly to Mn_2O_3, and later Mn_2O_3 and any Mn_3O_4 are reduced to Mn^{2+}, but the redox potentials are all lower, for a given concentration of manganous ions in solution, than expected from the theoretical equations, presumably due to the contamination of the oxides with ferric and other ions. On the other hand, if manganous carbonate is precipitated in the soil, this behaved as the pure salt. However, it is unlikely that the pH of a soil is ever controlled by the manganous carbonate, although the formation of the carbonate rather than the hydroxide is probably the mechanism for removing the manganous ion from the soil solution as the pH rises during reduction.[3]

The sulphate-sulphide reduction which is brought about by species of the bacteria *Desulphvibrio* is of great importance in waterlogged soils, because hydrogen sulphide is extremely toxic to many plants, and a concentration of 10^{-6} M can affect the functioning or growth of many roots. But in the presence of ferrous ions, the very insoluble ferrous sulphide is produced, which will only maintain a hydrogen sulphide concentration of 10^{-8} M in neutral soils when sulphide production is active. If the soil is well supplied with active iron, that is ferric iron that can be easily reduced, it is usually well poised at a redox potential too high for sulphide formation,[4] unless large amounts of easily decomposable organic material have been added.[5]

The reduction products from organic compounds

The reduction of decomposable organic compounds in waterlogged soils differs from inorganic reductions in that a whole sequence of reduction products are likely to be involved. Typically if carbohydrate-rich materials are added to a waterlogged soil, the first products are in part gases and in

1 For an example see A. Mukhopadhyay, T. R. Fisher and G. E. Smith, *Soil Sci.*, 1967, **104**, 107.
2 With T. A. Loy and E. M. Tianco, *Soil Sci.*, 1969, **108**, 48.
3 A. Mukhopadhyay, T. R. Fisher and G. E. Smith, *Soil Sci.*, 1967, **104**, 107.
4 S. Aomine, *Soil Sci.*, 1962, **94**, 6.
5 C. Bloomfield, *J. Soil Sci.*, 1969, **20**, 207.

part a whole range of fatty and hydroxy-acids; and these, in turn, are reduced eventually to other gaseous products.

The principal organic acids produced are acetic, with smaller quantities of proprionic, butyric, lactic, valeric, fumaric and succinic acids;[1] and their total concentration in the soil can exceed 10^{-2} M for several weeks after soil containing much readily decomposable plant residues has been flooded.[2] Since 10^{-2} M acetic acid is toxic to the roots of many plants, and butyric and some of the other acids are toxic at 10^{-4} M, conditions very unfavourable to plant growth will be present during this period.[3] If this soil is to be used for rice, either these acids must be washed out of the soil by flooding and leaching, or the land must be left wet long enough for their concentration to fall to a low level before the crop is planted. It is probable that it is the undissociated acids rather than their anions that are toxic, so these problems are more serious on soils with a pH below about 5·5.[4]

TABLE 25.2   Reduction of hydrocarbons during first week of waterlogging soils. Production in g/kg soil

Soil	% organic matter	CH_4	C_2H_4	C_2H_6	C_3H_6	C_3H_8
Sand	1·4	0·3	0·6	0·1	0·1	1·7
Sandy loam	3·9	3·1	5·5	0·5	1·0	1·6
Gault clay	5·0	12·5	7·6	0·6	1·0	1·6
Loam from basalt	9·8	17·9	13·3	0·5	1·2	1·7

When the soil is first flooded, and decomposition begins a number of gases are given off, but until the advent of gas chromatography it was not possible to identify them. It is now known that apart from nitrogen and nitrous oxide produced from the reduction of nitrates, hydrogen gas[5] and a range of low molecular weight hydrocarbons are also produced. These include methane, ethane, propane and n- and isobutane and ethylene, propylene and butene-1, but not including any acetylenic compounds. The production of these hydrocarbons only occurs in the initial stages of waterlogging, and typically ceases after a few days, and the rate of production roughly parallels that of nitrous oxide, showing it only takes place under rather mild reducing conditions. Table 25.2 gives the amount of the hydrocarbons, up to the propane group for six British soils;[6] and the butane group are not included because

1 F. J. Stevenson in *Soil Biochemistry*, Ed. A. D. McLaren and G. H. Peterson, Arnold, 1967. *Int. Rice Res. Inst., Ann. Rept.* 1969, 135.
2 T. S. C. Wang, S. Cheng and H. Tung, *Soil Sci.*, 1967, **104**, 138. S. Mitsui, S. Aso *et al.*, *5th Int. Congr. Soil Sci.*, 1950, **2**, 364.
3 *Int. Rice Res. Inst., Ann. Rept.* 1969, 135.
4 *Int. Rice Res. Inst., Ann. Rept.* 1965, 62.
5 R. G. Bell, *Soil Biol. Biochem.*, 1969, **1**, 105.
6 K. A. Smith and S. W. F. Restall, *J. Soil Sci.*, 1971, **22**, 430. For an earlier account with R. S. Russell, *Nature*, 1969, **222**, 769.

the measuring technique used was not suited to their quantitative measurement, though they were probably present in quantities comparable with the C_3 hydrocarbons.

The biochemistry of the process is not known. Ethylene production begins within a few hours of the oxygen concentration in the soil falling below about 1 per cent, so is presumably brought about by organisms that are also active in aerated soils, that is, they are presumably facultative anaerobes. The process is in part carried out by enzymes not dependent on cellular activity, for sterilising the soil with gamma radiation only reduced production by 50 per cent compared with a reduction by 90 per cent if sterilised by autoclaving.

Ethylene is the only one of these hydrocarbons to have a marked effect on root development of many crops. The saturated hydrocarbons are almost without effect, and the effect of propylene and butene are less marked than for ethylene, and they occur in lower concentration in the soil. The concentration of ethylene in the soil will be given in terms of the partial pressure of ethylene in the gas phase that is in equilibrium with the solution; and a concentration of 1 ppm refers to the ethylene concentration in the soil solution when its partial pressure in the gas phase is 10^{-6} bar, and this solution contains 1.5×10^{-7} g/litre ethylene at 20°C, so its concentration in the solution is of the order of 5.3×10^{-9} M. Plant roots differ widely in their reaction to ethylene. Thus the seminal roots of tobacco and tomato will have their rate of elongation reduced by 75 per cent if the solution contains 1 ppm of ethylene, and this concentration will reduce the rate of elongation of barley roots by 60 per cent, rye roots by 25 per cent but will not affect the elongation of several varieties of rice; even 0.3 ppm will reduce the rate of growth of barley roots by 50 per cent, though this concentration is without effect on rye.[1] Wheat and oats come intermediate between rye and barley. The sensitivity of these plants to ethylene is in line with field experience, for tomatoes and tobacco are known to be very intolerant of waterlogging, and rye and wheat are known to be more tolerant than barley.

Ethylene will persist in poorly drained soils in the field for periods of weeks, particularly in the subsoil. Figure 25.1 gives the ethylene concentration in the surface and subsoil of a pasture soil in a poorly drained Oxford clay in the Vale of the White Horse in southern England,[2] and it shows quite clearly that the ethylene concentration in the soil water is high enough, over an appreciable period of time, to have a very definite influence on the root development of the crop. It is very probable that the ethylene content of many poorly drained soils is a more common source of damage to crop growth than lack of oxygen or high carbon dioxide content in the soil. It is likely that, with the introduction of gas chromatographic techniques, two of the most useful

1 K. A. Smith and S. W. F. Restall, *A.R.C. Letcombe Lab.*, *Ann. Rept.* 1969, 56; 1970, 5.
2 R. J. Dowdell, K. A. Smith *et al.*, *Soil Biol. Biochem.*, 1972, **4**, 325.

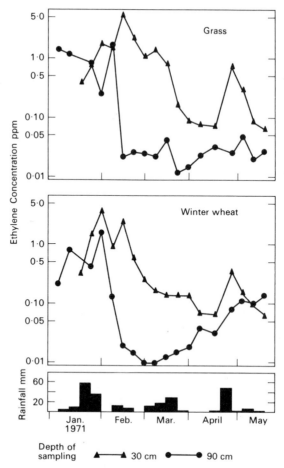

FIG. 25.1. Mean concentration of ethylene in the soil atmosphere of an Oxford clay soil under grass and under winter wheat

parameters to measure the degree of temporary anaerobicity in a soil will be the ethylene and the nitrous oxide content of the soil water.

Three other points about the effect of ethylene should be noted. It affects roots growing in aerated soils, so the damage it does in a poorly drained soil is not confined to the roots present in any anaerobic soil volumes. It causes epinasty and other deformities in plants, probably due to it preventing the root from deactivating indolyl-acetic acid produced in the tops;[1] and epinasty can be mistaken for a wilt if the observer is not looking out for the difference. It can presumably build up very quickly in soils of moderate organic matter content in warm weather when the soil surface is temporarily sealed by heavy

1 D. J. Phillips, *Ann. Bot.*, 1964, **28**, 17, 37.

rain giving a surface crust, which reduces gas interchange between the air and the soil for more than a short period, or if irrigation water stands on the surface of the soil. Field observers have often noted what they considered to be the crop wilting under these conditions, but the collapse of the leaf from the upright position may be due to an epinastic effect rather than to a wilt.

The gases given off after the first few days of active reduction are predominantly carbon dioxide and methane. Methane is therefore produced by two quite different processes—an early-stage process as just described, in which only small amounts of methane are produced, and a late-stage process in which an appreciable proportion of the carbon in the original organic matter is converted into this compound. This methane is produced by a number of strict anaerobes, most of which can only utilise simple organic compounds such as the fatty and hydroxy-acids produced by the earlier group of organisms or by anaerobes capable of decomposing cellulose. Methane is, therefore, liberated over an appreciable period of time in flooded padi soils in amounts sufficiently large to be measurable by the older chemists. W. H. Harrison and P. A. S. Aiyer[1] did pioneer work on the gases liberated from the reducing layer in some Indian padi fields and showed that the gases leaving this layer were predominantly methane, with some hydrogen and carbon dioxide. However, they showed that no methane escaped from the soil if it was flooded, for the methane that was produced in the soil itself was oxidised by bacteria to carbon dioxide in the soil surface, or in the water on the surface, and this was then used by the algal films in photosynthesis, so the principal gases escaping were oxygen and possibly nitrogen.

The changes taking place on waterlogging a dry soil

These will depend on the supply of decomposable organic matter, the various substances present in the soil which can be reduced, and the temperature; for reducing conditions are brought about almost entirely by microbial activity. Thus waterlogging a soil low in organic matter will have little effect on its redox potential. Low soil temperatures only allow sluggish activity, so the redox potentials again will only fall slowly, and the higher the temperature, up to a certain limit, the more rapid its fall. F. N. Ponnamperuma and R. U. Castro[2] have given some excellent examples of the effects of these various factors on the rate of fall of the redox potential. Thus, soils high in nitrate, or soils to which nitrate is added, maintain a redox potential of between $+400$ to $+200$ mV until all the nitrate is reduced, which may only take a few weeks, and then the potential may fall rapidly. The potential in soils well supplied with active manganese and moderately supplied with

1 *India Dept. Agric. Mem., Chem. Ser.*, 1913, **3**, 65; 1914, **4**, 1; 1920, **5**, 181. For later work P. K. De and S. Digar, *J. agric. Sci.*, 1954, **44**, 129; 1955, **45**, 280.
2 *Trans. 8th Int. Congr. Soil Sci.*, 1964, **3**, 379.

organic matter falls fairly rapidly to about $+100$ mV then slowly to about -50 mV, but this may take several months. The potential in soils reasonably supplied with iron and with organic matter falls to about -50 mV and then slowly over a period of months to -200 mV. The potential in soils low in nitrates, manganese and iron but well supplied with organic matter will fall rapidly to about -250 mV and hydrogen sulphide is likely to be present after a short time provided the pH has risen to about neutrality, while in soils low in organic matter the potential will only fall slowly, and may be still positive after several months of waterlogging. C. Bloomfield,[1] however, showed that under very active reducing conditions, brought about by the incorporation of decomposable organic matter, and in the presence of soluble sulphates, a proportion of the sulphide produced is in the form of hydrogen sulphide, even in soils well supplied with ferric iron, due to the mobilisation of ferric iron being the rate-limiting process for the formation of ferrous sulphide.

In some soils the redox potential will show a more complex change, rapidly falling to a low value directly after waterlogging, then rapidly rising to a value well below its initial value, and finally slowly falling. The fast fall is due to high bacterial activity before the ferric-ferrous system is operative, for if the iron is only present in relatively well crystalline forms, the amount of reducible ferric iron is controlled by the rate of solution of ferric iron. This effect would be expected to be most noticeable when a dryland soil is flooded for the first time, because the iron in soils regularly waterlogged and drained usually remains in a form readily accessible for reduction.[2] In some soils nitrates remain much longer than expected, and instead of all the nitrate-nitrogen being lost as gas in a few weeks, as is normal, up to one-half of this nitrogen may become incorporated in the organic matter.[3] The cause of this difference in behaviour is not known.

The pH of an acid soil usually rises after being waterlogged and strongly reducing conditions are often rapidly built up. As already mentioned on p. 675, the pH will stabilise at about pH 6·5 or just above, if sufficient iron is present, but the rise to this pH will be delayed if the temperature, the amount of reducible substances, or the amount of ferric iron are too low to produce sufficient ferrous iron for the $Fe_3(OH)_8$—$FeCO_3$—CO_2 buffer to become operative. It frequently happens in soils being poised by the ferric-ferrous reduction that the theoretical relation between the redox potential of the soil, its pH and the concentration of ferrous iron in the solution does not hold. This is because the redox potential is not uniform throughout the soil, but is much lower on the surface of the bacterial cells, which are the source of the electrons causing the reduction, than in the soil solution a little way away

1 *J. Soil Sci.*, 1969, **20**, 207, and see W. E. Connell and W. H. Patrick, *Proc. Soil Sci. Soc. Amer.*, 1969, **33**, 711.
2 For an example see C. Bloomfield, *J. Soil Sci.*, 1969, **20**, 207.
3 *Int. Rice Res. Inst., Ann. Rept.* 1966, 128.

from these cells. This results in the redox potential of the solution extracted from the soil being higher than the redox potential of the soil itself, and the theoretical relations should be valid for the solution provided the chemical composition of the solid phases controlling the redox potential have been correctly identified, and the solution is free of bacterial cells. It must also be remembered that some of the ferrous iron in solution is likely to be present as a chelate and not as free ions, particularly in the early stages of reduction, so it will not affect the ferric-ferrous ion equilibrium in the solution. This mosaic of actively reducing spots with large redox potential gradients into the solution may involve a range of reductions and oxidations taking place outside the cells involving systems that cannot yet always be predicted; and this has the consequence that the effect of a change of the pH of the soil on the redox potential cannot be predicted. F. N. Ponnamperuma[1] finds that it can vary from -30 to -180 mV per unit increase in pH in different soils, so that it is not justifiable to assume a figure of -60 or -180 mV per pH unit to convert an observed redox potential at an observed pH to its corrected value at a standard pH. However, he also finds that if the solution extracted from the soil is analysed, then the figure of -60 mV per pH unit is found. It is often also found that the concentration of ferrous iron in the soil solution rises rapidly to a maximum and then slowly falls as reduction proceeds, and this is true both for the free and complexed ferrous irons as is shown in Fig. 25.2.[2] The cause of the fall is due to the precipitation of $Fe_3(OH)_8$ brought about by this rise of pH.

Waterlogging a soil, by causing the formation of manganous and ferrous ions which can take part in cation exchange, will increase the amounts of exchangeable calcium, magnesium and potassium that come into solution. The reduction of ferric oxides also typically releases phosphates into solution,[3] for phosphate is often strongly absorbed on ferric hydroxide films formed as oxidising conditions take over from reducing; and it often releases silicate[4] for the same reason. If water slowly percolates through the water-logged soil, as often happens, it will be removing these cations, and, in fact, slow percolation through the soil when waterlogged may impoverish it faster than normal percolation under aerobic conditions. The ferrous and manganous ions present in the percolate may be precipitated lower down the profile. However, land which is only periodically waterlogged may continue to have oxidising conditions present in its subsoil during the whole period of inundation. This is because reduction processes can only occur actively when there is microbial activity, that is, under conditions where there is a supply of decomposable organic matter, and these conditions are typically not found

1 *Mineral Nutrition of the Rice Plant*, Int. Rice Res. Inst., 1965.
2 Taken from F. N. Ponnamperuma, in *The Mineral Nutrition of Rice Soils*, p. 315.
3 See, for example, J. K. R. Gasser and C. Bloomfield, *J. Soil Sci.*, 1955, **6**, 219. W. H. Patrick. *Trans. 8th Int. Congr. Soil Sci.*, 1964, **4**, 605.
4 C. H. Mortimer, *J. Ecol.*, 1941, **29**, 280.

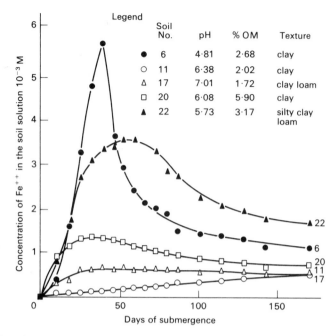

FIG. 25.2. Changes in the ferrous iron concentration in the soil solution during prolonged flooding

in subsoils. Thus, the ferrous iron, which can be oxidised fairly easily, is converted into ferric hydroxide near the interface between the reducing and oxidising regions, and the manganous ions, which need stronger oxidising conditions to be converted into manganese dioxide, are precipitated below the iron hydroxide deeper in the subsoil.

If a soil is subjected to several drying and flooding cycles without the addition of any organic matter, E_h falls increasingly slowly with time after each inundation, because the amount of readily reducible organic matter set free at each inundation decreases rapidly. This is in contrast to the behaviour of well-drained soils that are wetted but not flooded (see p. 237); and this slow build-up of reducing conditions after each inundation means that if nitrates, for example, are added to the soil after it has been through several cycles, they will be reduced increasingly slowly, so will last for an increasingly long time in the flooded soil.[1]

Table 25.3, due to Y. Takai and T. Kamura,[2] summarises the redox potentials usually associated with the formation of the different reduction products in waterlogged soils containing decomposable organic matter. These figures must be treated with some reserve, however, probably owing

1 W. H. Patrick and R. Wyatt, *Proc. Soil Sci. Soc. Amer.*, 1964, **28**, 647.
2 *Folia Microbiol.*, 1966, **11**, 304.

to the number of side-reactions that can take place in soils. Thus R. G. Bell[1] found that, in the soil he was using, denitrification of nitrates took place when $E_h = 200\,mV$ and methane production when $E_h = -250\,mV$, when corrected to pH 7, both of which are lower than the figures in Table 25.3.

TABLE 25.3 The redox potentials associated with the production of different reduced substances in waterlogged soils

Chemical transformation	E_h (at pH 7) in volts	
	From	To
Disappearance of molecular oxygen	0·6	0·5
Disappearance of nitrates	0·6	0·5
Formation of Mn^{2+}	0·6	0·4
Formation of Fe^{2+}	0·5	0·3
Formation of sulphide	0·0	−0·19
Formation of hydrogen gas	−0·15	−0·22
Formation of methane	−0·15	−0·19

Effect of adding decomposable organic matter to waterlogged soils

If organic matter is added to a waterlogged soil, by the puddling in of green weeds, for example, or the ploughing in of a green manure crop before flooding, there is usually a flush of microbial activity, which is greater the more easily decomposable the organic matter. Initially, there is a rapid fall in pH and redox potential, a rapid rise in the partial pressure of carbon dioxide, and a rapid rise in products of reduction such as simple organic acids, which is followed by production of methane, hydrogen and other gases. The higher the soil temperatures the more marked these effects, and in hot climates this flush is over after about two weeks. The severe reducing conditions raise the concentrations of ferrous and manganous ions in the soil solution very rapidly, reaching a maximum in about four weeks, if the temperatures are high, as is shown in Table 25.4 for ferrous iron. It may take ten to twelve weeks for the soil conditions to reach a fairly steady state.

The bacteria carrying out these decompositions release nitrogen surplus to their requirements as ammonium ions, sulphur as sulphate, hydrogen sulphide or mercaptans and phosphorus as phosphate or under some conditions which are not well understood as phosphine. The amount of nitrogen released during decomposition differs from that in aerobic conditions in that the nitrogen factor is lower for anaerobic than aerobic decompositions. Thus,

1 *Soil Biol. Biochem.*, 1969, **1**, 105.

TABLE 25.4   Effect of adding rice straw to a flooded soil on its ferrous iron status

Weeks incubated	*Without rice straw*			*With 5% rice straw*		
	Extractable Fe^{2+} m eq/100 g soil	*Fe^{2+} in solution 10^{-4} M*	*Complexed Fe^{2+} in solution 10^{-4} M*	*Extractable Fe^{2+} m eq/100 g soil*	*Fe^{2+} in solution 10^{-4} M*	*Complexed Fe^{2+} in solution 10^{-4} M*
Acid soil, initial pH 4·81, organic matter content 2·68%						
0	0·68	0·18	—	1·5	0·13	—
4	80	9·0	3·4	100	15·8	4·9
8	78	2·2	—	93	7·7	4·5
12	80	1·6	—	96	5·2	1·1
Neutral soil, initial pH 6·64, organic matter content 2·02%						
0	0·01	0·03		0·36	0·02	—
4	1·35	0·05		37	2·2	1·4
8	0·79	0·09		33	1·6	1·4
12	1·35	0·13		34	0·9	—

Source: *Int. Rice Res. Inst., Ann. Rept.* 1963, 61.

C. N. Acharya[1] found a nitrogen factor (see p. 267) for rice straw decomposing under aerobic conditions of 0·54 after six months, of about 0·3 if decomposing under mildly anaerobic conditions, and of 0·07 if under strict anaerobic conditions. The reason for this difference is that anaerobes obtain much less energy than aerobes per unit of carbohydrate decomposed, for only half the carbon they metabolise is lost as carbon dioxide, the other half is lost as methane. Organic matter of an appreciably higher C/N ratio can therefore be added to a waterlogged than can be added to a well-drained soil before the decomposition begins to remove ammonium ions from the solution; and W. A. Williams,[2] for example, found that incorporating a rice straw containing 0·55 per cent N just did not decrease the grain yield of rice, whereas the critical nitrogen content for dryland crops is between 1·5 and 1·7 per cent. On the other hand, although low-nitrogen organic compounds will release ammonium ions, most such natural products decompose slowly anaerobically, so only substances with a low C/N ratio will release appreciable amounts of ammonium over periods of a few weeks.

Once a waterlogged soil has come to a stable condition, it is not easy to disturb its stability so long as there is no drainage and the soil remains waterlogged. Thus Ponnamperuma[3] has shown that if fresh organic matter is carefully incorporated in the soil, with minimum disturbance, it decomposes very slowly with only small changes in redox potential, and correspondingly if nitrate is added it is only slowly reduced; but if the soil is drained and reflooded, there will be a rapid build-up in the rate of decomposition and of reducing conditions.

1 *Biochem. J.* 1935, **29**, 1116.
2 With D. S. Mikkelsen and K. E. Mueller, *Pl. Soil*, 1968, **28**, 49.
3 *Int. Rice Res. Inst., Ann. Rept.* 1966, 113.

A point of great importance that must constantly be borne in mind when working with either waterlogged soils or very wet soils in the field is that the redox conditions can change very rapidly over short distances. Thus, small pieces of organic matter being actively decomposed by the soil bacteria in an anaerobic volume of soil can be surrounded by a thin zone where very strong reducing conditions have been set up, yet there may be an air pore only a few millimetres away. Analysis of soil air from wet soils with a very restricted air space will not necessarily be of any value in indicating the conditions prevailing in the bulk of the soil. An analysis of the soil water will give a much more useful picture, yet even this will give only an average picture of the conditions prevailing in the volume being sampled. Thus, K. H. Sheikh[1] found that a bulk sample of water extracted from a British wet heath could contain both small amounts of dissolved oxygen (2 ppm) and appreciable amounts of hydrogen sulphide (10 to 20 ppm).

Adaptation of plants to waterlogged soils

The various processes by which some plant species can overcome the effects of poor aeration or of reducing conditions can be studied in plants adapted to moor or marshland conditions. W. H. Pearsall,[2] a pioneer in this field, found that a redox potential of about 320 mV at pH 5 was a good dividing line between oxidising and reducing soils, so that plants adapted to soils with a lower redox potential than this have developed some protective or adaptive mechanism in their root surfaces to allow them to function as organs for the extraction of nutrients and water from the soil. Plants adapted to conditions of poor soil aeration develop a system of interconnected internal air spaces through which oxygen from the atmosphere can diffuse down into the root, and out of the root into the soil immediately outside the root, thus allowing the actual uptake of ions by the root to take place as an aerobic process. This rate of diffusion can be measured electrometrically.[3] It is interesting to note that the seedlings of a number of crops are adapted for this oxygen transport, though the rate of transport becomes much less as they grow.[4] Padi rice varieties are well adapted for this mode of transport, as are many marsh plants, though most varieties of rice have lost this adaptation at the onset of flowering. It is possible that some of the carbon dioxide that accumulates in the soil near the roots may diffuse from the soil through the plant to the atmosphere, for the carbon dioxide content of the air in the plant tissue increases as the carbon dioxide concentration outside the roots increases.[5]

1 *J. Ecol.*, 1969, **57**, 727.
2 *J. Ecol.*, 1938, **26**, 180, 194, 298, and with C. H. Mortimer, 1939, **27**, 483.
3 See, for example, W. Armstrong, *J. Soil Sci.*, 1967, **18**, 27.
4 D. J. Greenwood, *New Phytol.*, 1967, **66**, 337; with D. Goodman, ibid., 1971, **70**, 85, for some vegetable crops; C. R. Jensen *et al.*, *Science*, 1964, **144**, 550, for maize; D. A. Barber *et al.*, *J. exp. Bot.*, 1962, **13**, 397, for barley.
5 J. R. Webster, *J. Ecol.*, 1962, **50**, 619.

A consequence of oxygen diffusing out of plant roots growing in water-logged soils is that they are often surrounded by a sheath of soil having a dark brown colour, due to the conditions being sufficiently oxidising for the formation of ferric hydroxide and manganese dioxide. The thickness of this sheath depends on the rate that oxygen diffuses from the root. In addition to the atmospheric oxygen diffusing out of the root, there may also be a supply of oxygen produced enzymatically outside the root surface from the decomposition of glycollic acid to carbon dioxide through glyoxalate, oxalate and formate, with the liberation of hydrogen peroxide.[1] The sheath may extend for a distance of 4 to 5 mm from the root surface during the active growth of the root, but it breaks up as soon as the oxidising conditions cease.[2] The root surface typically remains white if it is surrounded by an oxidised sheath, but, if the outward rate of oxygen diffusion is too low to give this sheath, it may still be adequate to cause ferric hydroxide to be precipitated on the epidermal cell walls of the root, or within the intercellular spaces around the cortical cells; in either case the root surface is stained brown.[3] Marsh plants differ considerably in their ability to tolerate hydrogen sulphide in the soil outside their roots, probably because of differences in the rate at which oxygen can diffuse out of the roots; though in part it may be due to their ability to precipitate ferric hydroxide in the intercellular spaces in the cortical and epidermal layers which can absorb the hydrogen sulphide before it reaches the inside of the root cells. There is some evidence that the roots of marsh plants have little ability to force their way into soils, and so are restricted to channels large enough for them to enter.[4]

The soil conditions prevailing in a padi field

The principal agricultural use of waterlogged soils is for the production of swamp or padi rice, which is a very important crop of the tropics and sub-tropics. Padi soils are not necessarily soils with a water-table at or near the surface. Many have a fairly deep water-table but their surface soil is made nearly impermeable either by being puddled when wet or by having a compacted layer in the subsurface. Nor are they normally flooded throughout the year, for many are only flooded during the greater part of the growing season. Thus, the typical padi soil is characterised by alternate periods of reducing conditions, when it is cropped to rice, and oxidising conditions during the dry season, when it may be either allowed to dry out with its natural weed population or is planted with a dry-season crop. The depth to which oxidising conditions prevail before the land is next flooded depends on the depth of

1 S. Mitsui, *Mineral nutrition of the rice plant*, 1965, and W. Armstrong, *Physiol. Pl.*, 1967, **20**, 920.
2 O. W. Bidwell, D. A. Gier and J. E. Cipra. *Trans. 9th Int. Cong. Soil Sci.*, 1968, **4**, 683.
3 W. Armstrong and D. J. Boatman, *J. Ecol.*, 1967, **55**, 101.
4 K. H. Sheikh and A. J. Rutter, *J. Ecol.*, 1969, **57**, 713.

the water-table, which may be very shallow for fields on the flood plains of rivers, but which may increase in depth as the fields extend up the valley sides.

A typical soil profile in a fertile padi soil with a deep gley horizon during the middle of the growing season is as follows:[1]

1 There is a film or sheet of water on the soil surface in which there is a growth, and sometimes a copious growth, of algae, which keep this water strongly oxygenated.
2 There is a thin surface film of oxidised soil, brown or yellow-brown in colour, which receives its oxygen from the surface water film, and may be several millimetres thick.
3 Below this layer is a thicker layer of reduced soil, typically grey or blue-grey in colour. The boundary between the oxidised and reduced layer may be fairly sharp, it need not be horizontal, and it may be marked by a thin brown film of ferric hydroxide. As will be described later, this layer can be very heterogeneous.
4 Below this is an oxidised layer, again often with a sharp boundary, and again sometimes with a layer indurated with ferric hydroxide, below which may be a layer enriched with black manganese dioxide, or the layer may be flecked black. This layer probably remains oxidised during the period of flooding because of low microbial activity and of an adequate amount of poising material to prevent serious reducing conditions building up.
5 This lower oxidised layer often overlies a permanently gleyed subsoil.

A healthy young rice plant has its roots entirely in the reducing layer and they are typically either surrounded by a brown sheath of ferric hydroxide or stained brown, and this is brought about by the oxygen that diffuses out from them into the soil and by any oxygen that is produced enzymatically just outside their surfaces. This results in the reducing layer containing pockets of soil that are oxidised. Further, since fresh organic matter has usually been incorporated in this layer when the soil was being prepared for rice, it often contains channels up which the gases escaped that were produced during the fermentation of this organic matter, and down which oxygenated water can diffuse or percolate, resulting in the soil bounding these channels also becoming oxidised. Thus, this reduced layer in a fertile padi rice soil is characterised by being a mosaic of reducing and oxidising conditions.[2] As the rice plant develops, the root system in the reducing layer loses its ability to excrete oxygen, and a new root system develops in the thin oxidised surface

1 See, for example, S. Aomine, *Soil Sci.*, 1962, **94**, 6. F. F. R. Koenigs, *4th Int. Congr. Soil Sci.*, 1950, **1**, 297. J. C. de Gee, ibid., 300. C. J. Grant, *Mineral Nutrition of the Rice Plant*, Int. Rice Res. Inst., 1965.
2 For an experimental demonstration on the magnitude of this, see *Int. Rice Res. Inst., Ann. Rept.* 1966, 110.

layer of the soil; this becomes the active system from heading or flowering time onwards.[1]

The reducing layer in soils less suited to padi rice typically differs from that in a good soil in that it is much more uniform in colour, being grey to the depth of cultivation or disturbance and often bleached below this, and this whole layer often overlies a layer indurated with ferric hydroxide and perhaps silica. Such soils are called degraded by the Japanese workers, and they are characterised by being low in iron and manganese, so are typically found on parent materials derived from siliceous marine deposits, and from granites and rhyolites. The reducing layer in these soils will have a redox potential sufficiently low for hydrogen sulphide to be formed. The rice roots are greyish white, the oxygen diffusion rate from the root to the soil is very low or is zero, due to reducing substances diffusing to the root surface faster than the roots can excrete oxygen, so any ferrous iron that may be close to the surface will remain unoxidised. These undesirable conditions can sometimes be mitigated by draining the surface water off the soil for a period, one or more times during the first part of the growing season, so allowing the crop to dry the soil and oxidising conditions to develop temporarily.[2] These reducing conditions may also encourage the loss of soluble silica and of ferrous iron from the reducing layer, and ferrous iron can also be lost as colloidal particles of ferrous sulphide stabilised by the hydrogen sulphide. Both of these may be precipitated in the indurated layer underlying the reducing layer. The oxidised layer below the reducing layer may be very poorly developed, particularly in poorly drained soils or in some degraded soils. The layer may be grey with flecks of orange or brown, instead of being brown to red with flecks of grey; and if this layer is well structured, the ped faces may be coated with a thin ferric hydroxide film though the interior of the peds remain under reducing conditions.

Rice growing on degraded soils can suffer from a number of physiological diseases,[3] known by local and usually Japanese names, which rarely have very specific symptoms because of the variety of unfavourable soil conditions which can induce them. If the principal trouble is due to hydrogen sulphide or simple fatty acids, the damage can be reduced by ensuring as little fresh organic matter as possible is incorporated in the soil, and that the soil is flooded well before the crop is planted. A shortage of ferric iron can be alleviated by applying a heavy dressing of soil high in iron when this is locally available. These degraded soils are often very acid, and some contain sufficient iron for the ferrous ion concentration in the reducing layer to rise to toxic levels, and this can be ameliorated by liming, for this raises the soil

1 T. Alberda, *Pl. Soil*, 1953, **5**, 1.
2 For an example from northern Australia see R. W. Strickland, *Aust. J. exp. Agric. anim. Husb.*, 1968, **8**, 212.
3 See, for example, I. Baba, K. Inada and K. Tajima, *Mineral Nutrition of the Rice Plant.* S. Mitsui, S. Aso et al., *5th Int. Congr. Soil Sci.*, 1954, **2**, 364. J. P. Hollis, *Louisiana Agric. Exp. Sta. Bull.* 614, 1967. *Int. Rice Res. Inst., Ann. Rept.* 1965, 45.

pH and reduces the ferrous ion concentration. It is possible that the toxic effect of a high ferrous iron concentration can be alleviated, at least to some extent, by the use of a potassium fertiliser, for this increases the potassium and reduces the iron uptake by the plant.[1] Since these soils are often low in available silica, silicate slags are often the preferred liming material. J. W. O. Jeffery[2] suggested a simple method for distinguishing between suitable and degraded soils would be to determine the concentration of ferrous iron in the reducing layer, for he considered it should lie between 10^{-3} and 10^{-4} M; if it was too low sulphides could become important and if too high iron toxicity would be likely to occur.

The management of rice soils

A padi soil before flooding usually carries either a crop of weeds or a green manure crop, and this is incorporated into the soil, either before or after flooding, so that very active microbial decompositions are usually taking place in the soil during the first few weeks after flooding. It is common practice therefore not to transplant or sow rice for two to three weeks after flooding, to allow the first flush of decomposition to pass, so that the rice roots will be in a relatively favourable environment for growth from the beginning. If the soil is well poised, the roots should continue to grow and function effectively; but on degraded or acid soils, the water may be taken off the land once or twice in the early part of the growing season to allow air into the soil to reduce the severity of unfavourable conditions prevailing.

The farmer can control the intensity of the reducing conditions and their effect on crop development in a number of ways. The amount and type of organic matter that is incorporated determines the rate of fall of redox potential, which, in turn, controls the rate of build-up of ferrous iron. Thus, if the soil is rather low in iron, puddling-in a green manure can give a sufficient fall in redox potential to allow the pH to fall and the ferrous ion concentration to build up; and if circumstances are correct, this build-up of ferrous iron will, in turn, cause a rapid rise of pH and allow $Fe_3(OH)_8$ to precipitate with a consequent rapid drop in the ferrous ion concentration. If the green manure was not incorporated the ferrous ion concentration would be maintained at a lower though still at a toxic level during the early stages in the growth of the crop.

The choice of date at which the soil is puddled, and then the crop is planted, can be of great importance, particularly in regions where there is a marked seasonal change of temperature.[3] If the soil temperature is below 15°C decomposition goes relatively slowly, and the rise in soil pH becomes delayed; and if fertilisers are not being used, the rate of release of nitrogen from the

1 A. Tanaka and T. Tadano, *Potash Rev.*, subject 9, suite 21, 1972.
2 *J. Soil Sci.*, 1961, **12**, 172, 317. He used different units from the ones given here.
3 *Int. Rice Res. Inst., Ann. Rept.* 1967, 122.

organic matter and the rate of build-up of available phosphate are low; but these effects can be mitigated by proper fertiliser use. However, if much organic matter was incorporated in cool weather, there may be a flush of decomposition, with harmful strong reducing conditions occurring in mid-summer, when the temperature has risen appreciably and the crop is well established but still susceptible to them.

Rice soils are typically puddled at the beginning of the season, largely to reduce the rate of percolation of water through the soil. But puddling may have another effect, for it causes the softening and break-up of soil crumbs, and ensures the maximum thickness of water films around the soil particles. This ensures the maximum volume of the soil is accessible to bacterial growth and presumably also to the roots of the young rice plant.

Drainage and water control in padi soils

Although padi rice is grown in waterlogged or flooded soils, there is con-siderable experimental evidence that a certain downward movement of water through the soil is desirable. This is probably particularly important early on in the season when there is often a considerable amount of organic matter—weeds and stubbles from the previous season—decomposing. The percolating water carries with it soluble organic substances, which can be oxidised, so that the channels down which the water percolates often have a skin of reduced grey-coloured soil around them. Some of these substances are probably fatty acids and other compounds liable to harm the roots of the young rice plant if their concentration rises to too high a level. In parts of Japan tile drainage is practised, just to maintain an adequate rate of percolation. Green manuring and the use of composts is also valuable in some areas, again probably because they help to maintain a good enough soil structure in the initial stages of flooding to allow adequate percolation.

Ponnemperuma also considers that drainage may decrease the partial pressure of CO_2 in the soil solution, so raising the pH and decreasing the ferrous ion concentration. Drainage rates of about 0·5 to 1·0 cm/day are probably adequate.[1] They will not involve too large a waste of water, nor will they impoverish the soil by keeping the concentration of soluble ions too low. However, over-drainage of calcareous or alkali soils can be un-desirable, for by lowering the CO_2 concentration in the soil water, the pH may rise sufficiently to cause loss of yield through too low an uptake of ferrous iron.

Padi rice is typically grown in soils on which a layer of water is maintained. The full reasons for this water layer may not yet be known, but it is probable its function is to maintain adequate reducing conditions in the soil, so its depth need only be sufficient to ensure it covering the whole field. If this is

1 *Int. Rice Res. Inst., Ann. Rept.* 1967, 134.

correct, the more level the field, the shallower the depth of water over the field need be, but since it is never possible to have an absolutely level field, it is usual to aim for an average depth of 5 to 7·5 cm on well-levelled fields. Flooding is probably more important at certain periods in the crop development, and it may be particularly important in mid season, from panicle initiation to heading, as the reproductive phase of growth may have a much higher ferrous iron demand than the vegetative.[1] Padi rice can, however, be grown without flooding on a number of soils, and some of the high yielding IRRI rices appear to give very satisfactory yields on aerated soils in which the water deficit is always kept small. This may be, in part, because such soils will always contain pockets or volumes, possible inside crumbs, that are anaerobic.

The nutrition and manuring of padi rice

Unmanured rice obtains its nitrogen from five sources: from the ammonium produced in the reduction zone by decomposition of organic matter in that zone; from nitrates produced by decomposition of organic matter in the surface oxidising layer of the soil; from nitrogen fixed by blue green algae and other photosynthetic micro-organisms[2] in the water film above the soil; from nitrogen fixed by heterotrophic nitrogen-fixing organisms in the soil or the water film[3]; and from nitrogen fixed by heterotrophic bacteria in the rhizosphere,[4] which may be different from the previous group which are not rhizosphere organisms. The nitrogen fixed by the algal films may be transferred to the rice as nitrate from mid season onwards after the film has built up strongly and after the rice has developed a surface root system. In good padi soils these sources of nitrogen are adequate to maintain rice yields at a modest level almost indefinitely, in contrast to dry land soils in which crop yields fall to very low levels under continuous food crops. It is not easy to measure the amount of nitrogen fixed by the algae that is taken up by the crop, but about 20 kg/ha per season may be a reasonable estimate for many conditions;[5] but no estimate can yet be given for the amount of nitrogen fixed by heterotrophs that is used by the crop. There is, however, evidence that some of the nitrogen fixed during the growing season may become available to the crop in the following season.

Additional nitrogen must be used if high yields of rice are wanted.[6] Leguminous green manures ploughed in at the beginning of the season have

1 *Int. Rice Res. Inst.*, *Ann. Rept.* 1966, 117.
2 M. Kobayashi, E. Takahashi and K. Kawaguchi, *Soil Sci.*, 1967, **104**, 113.
3 I. C. MacRae and T. F. Castro, *Soil Sci.*, 1967, **103**, 277. F. R. Magdoft and D. R. Bouldin, *Pl. Soil*, 1970, **33**, 49.
4 T. Yoshida and R. R. Ancajas, *Proc. Soil Sci. Soc. Amer.*, 1971, **35**, 156.
5 P. K. De and L. N. Mandal, *Soil Sci.*, 1956, **81**, 453, and A. Watanabe, *Nature*, 1951, **168**, 748.
6 For a review see S. K. De Datta and C. P. Magnaye, *Soils Fert.*, 1969, **32**, 103.

limited value because if too large a weight of readily decomposable green matter is incorporated in the soil, reducing conditions develop which are too severe for the young rice roots. The use of nitrogen fertilisers requires care due to the rapid transitions that can occur between aerobic conditions, when ammonium ions are converted to nitrate, and reducing conditions, when the nitrate is almost quantitatively denitrified. Thus, an ammonium fertiliser broadcast on the surface of a flooded soil is likely to be converted to nitrate in this surface layer. In so far as this nitrate can seep down into the reduced layer through channels of oxidised soil, or the rice plant has developed a root system in the soil surface, it can be taken up by the plant; for rice takes up nitrates at least as easily as ammonium.[1] But in many though not in all padi soils the reducing conditions in the root zone are sufficiently severe that any nitrate entering it, or any nitrate present before the reducing conditions set in, is likely to be denitrified, so that adding an ammonium fertiliser to a damp soil one or two weeks before it is flooded, or broadcasting it on the soil surface, is likely to be an inefficient method. In some soils—possibly of low biological activity—a proportion of the nitrate persists for several weeks, and a proportion becomes incorporated in the soil organic matter.[2] The ideal method for use early in the season is to place an ammonium or urea fertiliser or inject anhydrous ammonia in the zone that will become the reducing zone, that is normally about 10 cm deep, as near the time before the land is flooded as possible.[3] If nitrate fertilisers, and particularly ammonium nitrate must be used, they should only be used as a top-dressing in mid season, because the active rice roots are then in the oxidising layer in the soil surface.

High-yielding varieties of rice require a high level of nitrogen in their early stages of growth, but care must be taken how this is applied, otherwise it may cause such a vigorous growth of green algae that the young plant is smothered. It is common practice not to give the full nitrogen dressing in the seedbed, particularly if a high dressing is to be used, but to give a part later on in the season as a top dressing about the time of panicle initiation, when the rice roots are near the soil surface. High levels of nitrogen can only be used, however, if weed growth can be suppressed, which normally involves both good management and the use of appropriate herbicides. The maximum level of nitrogen that can be used profitably, and the response of the crop to the nitrogen, depends on the level of solar radiation in regions where water and temperature are adequate, particularly during the period from panicle initiation to harvest. Thus at the International Rice Research Institute in the Philippines[4] their short stiff-strawed variety IR8 only gives a response of 15 kg of grain per kg N during the rainy cloudy season, but

1 E. Malavolta, *Pl. Physiol.*, 1954, **29**, 98.
2 I. C. MacRae, R. R. Ancajas and S. Salandanan, *Soil Sci.*, 1968, **105**, 327.
3 For examples see G. V. Simsiman, S. K. De Datta and J. C. Moomaw, *J. agric. Sci.*, 1967, **69**, 189 and Z. Aleksic, H. Broeshart and V. Middelboe, *Pl. Soil*, 1968, **29**, 338.
4 S. K. De Datta, A. C. Tauro, and S. N. Balaoing, *Agron. J.*, 1968, **60**, 643. A. Tanaka, *Proc. 9th Int. Congr. Soil Sci.*, 1968, **4**, 1.

30 kg during the dry sunny season, and the most profitable dressing of nitrogen is about 60 and 120 kg, and the yields about 6·5 and 9·5 t/ha in the two seasons.

The two most commonly used nitrogen fertilisers are ammonium sulphate and urea. The continued use of high levels of ammonium sulphate causes acidity, as in well drained soils, and may give accumulation of sulphides, so that urea is probably the safest fertiliser to use, until anhydrous ammonia is introduced. But there is much interest in the use of slow-acting fertilisers and nitrification inhibitors, particularly in Japan. The Japanese workers have evidence that under their conditions high levels of ammonium nitrogen, as for example 100 to 150 kg/ha of nitrogen, will affect rice growth adversely, so that if a high level of nitrogen is to be given in one dressing rather than in split dressings, slow-acting nitrogen fertilisers are required. These are made either by bonding pellets of urea in a bitumen or plastic base so it can diffuse only slowly into the soil solution, or by using insoluble substituted ureas such as guanyl urea

$$NH_2-C-NH-CO-NH_2$$
$$\|$$
$$NH$$

or crotylidene diurea (see p. 628) which only slowly hydrolyse to urea.[1] If the urea is to be broadcast on the soil surface, after the rice has been planted, mixing a nitrification inhibitor with the urea will help ensure that the ammonium produced in the oxidising layer of the soil will remain in that form, and not be nitrified and so lost if it seeps into the reducing layer. It is possible that this harmful effect of high ammonium is less noticeable on fine than coarse textured soils, for high levels of ammonium nitrogen do not appear to be harmful on the finer-textured soils at the International Rice Research Institute.

Phosphatic fertilisers are not so often required on padi as on dryland soils, probably because the phosphate can be held as a ferric phosphate during dry periods and released as a soluble ferrous phosphate when flooded. Single superphosphate is, like ammonium sulphate, often undesirable because of the calcium sulphate it contains. Rock and bone phosphates are probably useful on a wider range of padi than of dryland soils, and if so this is probably because soluble organic compounds produced in the reducing layer increase the solubility of the phosphate through their ability to chelate calcium. The Japanese have also used a fused magnesium serpentine phosphate made by sintering a serpentine with a phosphate rock. This contains about 8·5 per cent P, and 20 per cent SiO_2 as hydrolysable silicates, so has both a liming action and supplies silicate to the crop as well as phosphate. Potassium is often needed for high yields of rice, and potassium chloride is perfectly suitable.

1 M. Hamamoto, *Fertiliser Soc., Proc.* 90, 1966.

The sulphur nutrition of rice raises interesting problems because ferrous sulphate is so insoluble. Presumably this must be oxidised to ferric sulphate in the immediate vicinity of the root, which hydrolyses to give the free sulphate ion. Ponnamperuma also notes that copper sulphide, particularly in the presence of ferrous sulphide, is extremely insoluble, so rice presumably obtains its copper from soluble organic complexes, as in aerobic soils.

Rice is one of the few agricultural crops that respond to silicate manuring, and some of the reasons for this have already been discussed on pp. 637 and 638. Silicate fertilisers are used fairly widely in Japan, and could probably be profitably used in many other areas of South-east Asia, for many of the soils there contain only low amounts of available silica.[1] The commonest silicate fertilisers used in Japan are some blast furnace (not basic) slags, though the fused magnesium serpentine phosphate is also used fairly widely. Both of these are also liming materials, and since most of the soils low in available silica are also acid, they help to correct this trouble as well.

Salt marsh soils

These typically occur when the sea is retreating and the land surface rising by silt deposition. Most of the soil will have been laid down under anaerobic conditions and in the presence of sulphate derived from the sea water. If the silt being deposited contains ferric iron, this will be reduced and ferrous sulphide, ferrous polysulphide and iron pyrites FeS_2 will be formed and deposited along with the silt.[2] As the sea retreats the surface soil will become aerobic; and if the land is now reclaimed and the soil drained, this ferrous sulphide will begin to be oxidised to ferric hydroxide and sulphuric acid, and the pH can drop to under 3, for some of the sulphur-oxidising bacteria are very tolerant of acidity. But the sulphides are not all equally readily oxidised. Ferrous sulphide itself and the polysulphides probably oxidise easily to ferric hydroxide and elemental sulphur, which is rapidly oxidised to sulphuric acid. But iron pyrites may occur in larger crystals and take several years to oxidise, particularly if the pH is rather high, so that pyrites can exist in the soil some long time after the other sulphides have been oxidised.[3]

Very difficult problems of soil management are raised when these soils are to be reclaimed. In some areas, such as in some of the Dutch sea marshes, calcium carbonate was laid down with the ferrous sulphide so that as the sulphide is oxidised to sulphuric acid, it reacts with the carbonate to give calcium sulphate. But in other areas, such as in many saline mangrove swamps on tropical coastlines, there is no carbonate in the soil. It is technically easy to empolder many of these marshes and use them for rice cultivation, but it can be extremely difficult to prevent acidity building up. In Sierra Leone,

1 K. Kawaguchi and K. Kyuma, *9th Int. Congr. Soil Sci.*, 1968, **4**, 19.
2 G. W. Harmsen, *Pl. Soil*, 1954, **5**, 324.
3 M. G. R. Hart, *Pl. Soil*, 1959, **11**, 215; 1962, **17**, 87.

where much work has been done on this problem, the system being recommended is to drain the land, allow the maximum of oxidation of sulphide to take place during the dry season then wash out the sulphuric acid either by flooding or by the rains, lime the surface soil to put some calcium back, and grow the crop.[1] This method has the weakness that, if the silt contains much iron pyrites, only a portion will oxidise in the first season, so acidity will build up again in subsequent dry seasons.

The behaviour of iron pyrites in marsh soils has recently been studied by C. Bloomfield.[2] He found that ferrous sulphide FeS is rarely present in amounts exceeding 100–200 ppm in the soil, but that iron pyrites may be present in amounts of several percent. Whereas ferrous sulphide oxidises readily in neutral conditions, the pyrites oxidise only under acid conditions, and only rapidly under very acid conditions. If the pH is below 3·5, *Thiobacillus ferro-oxidans* is the dominant organism carrying out the oxidation, and it oxidises ferrous iron to ferric and sulphide to elemental sulphur or sulphate; but it releases more ferrous iron than it oxidises, so that if water is leaching through the soil, ferrous iron will be leached out. When this acid drainage water comes into contact with air, this bacteria will oxidise it to a ferric hydroxide ochre, which will accumulate in the drainage channels. If the conditions are less acid, the pyrites will oxidise slowly in the pH range up to 5, giving elemental sulphur and ferric ions, and the sulphur will be oxidised in this range by *Thiobacillus thio-oxidans*. If reducing conditions subsequently develop, and the drainage water is not strongly acid, any ferrous iron leaching into the drainage ditches or tile drains—if the field is under-drained—will be oxidised by filamentous bacteria, which are intolerant of acid conditions, to give a very voluminous deposit of ferric hydroxide associated with organic matter derived from the bacteria.

1 M. G. R. Hart, A. J. Carpenter and J. W. O. Jeffery, *Afr. Soils*, 1965, **10**, 71. P. R. Hesse, *Pl. Soil*, 1961, **14**, 249.
2 *J. Soil Sci.*, 1972, **23**, 1.

26

The formation and classification of soils

The weathering of rocks

Soils are derived from the decomposition of the mineral particles they contain and from the plant and animal remains added to them. The organic remains are usually being decomposed continuously and relatively rapidly by the soil organisms, and many of the mineral particles are also decomposing, but usually at a very much slower rate. The chemistry of these processes will be discussed in two parts: first, when the influence of biological processes appears to be minimal and, second, when they are a relevant factor.

The words *soil* and *rock* need definition. Rock will be defined as the inorganic mineral covering of the earth's surface, and will therefore commonly include soil. Rock may be hard and solid, such as granite, limestone or other consolidated mineral matter; or it may be loose and unconsolidated, such as a gravel, sand or clay, in which case it is called regolith. A soil is a regolith that shows vertical anisotropy due to biological activity. By this is meant that a soil consists of layers of mineral material whose properties differ from one another in some recognisable respects, and these differences are due to biological activity. This activity includes the effect of the growing plants on the rock in which they are growing and of their associated organisms. Soil formation therefore includes all the processes going on in the rock due to these biological activities. However, chemical changes can take place in the rock below the zone of biological activity, giving vertical anisotropy, or layers in the rock which differ from the rock itself. This material is not usually considered to be soil, although it is produced from the rock by weathering. In so far as the process is entirely a chemical decomposition of the rock, with the complete removal of soluble products of weathering and no accumulation of products from either higher or lower layers, the weathered material is often called saprolite—it is the insoluble residual material left behind during weathering. It is still uncertain if this weathering is entirely inorganic or if it requires small amounts of organic substances being leached from the definite soil layers.

Rocks and mineral particles can suffer physical disintegration without change of chemical constitution. Rock particles may be ground to fine particles under ice sheets, and some of the finer particles in boulder clays may have been produced by this action. They can also have fragments broken off them by the expansive pressure of water in any cracks freezing into ice, or by the pressure of expanding tree roots growing into cracks; and these fragments can be made smaller by abrasion by sand grains in areas where wind erosion is appreciable, or by being knocked together on the bottom of fast flowing streams.

The rock can also be reduced to smaller-sized particles by the dissolving power of water. Thus, water percolating through limestone rock will slowly dissolve the calcium carbonate leaving behind the mineral detritus it contained, and it will dissolve the cements that bind individual sand grains into a sandstone. It will also cause the break-up of igneous rocks into their constituent mineral grains. An example of this comes from a lysimeter installed at Versailles which was filled with granite chips 2 to 4 mm in size to a depth of 60 cm and left exposed to the weather. After thirty years only 9 per cent of the granite remained in chips this size, and the remainder was distributed as 31 per cent in the coarse sand, 62 per cent in the fine sand (0·2 to 0·02 mm), 4·6 per cent in the silt and 2·4 in the clay fraction; and most of this was due to the physical disintegration rather than the chemical decomposition of the rock.[1]

It is commonly said that the solid rock surface can lose some of its constituent mineral grains due to their different coefficients of thermal expansion, for they will expand and contract by different amounts during the diurnal temperature changes of the surface, which can be very large in deserts. But the very slow weathering of granite inselbergs or tors, that is ancient granite rocks left exposed after all the weathered material around them has been removed by erosion which are typical of many old land surfaces in the drier parts of East Africa for example, suggests that this process is too slow to be of any appreciable significance over millenia. These finer particles produced by disintegration may then be carried away by wind, water or ice, and redeposited elsewhere as loess, alluvium or boulder clay. Or those on or near the surface of these deposits may be carried into lower layers with the percolating water.

The mineral particles can also suffer chemical decomposition through the solvent action of the water itself, and through the reactions between the dissolved or suspended substances in the water, as, for example, oxygen, carbon dioxide, alkalis and organic acids, and the surfaces of the particles. Some of these reactions can occur at depths exceeding 10 m below the soil surface, so clearly are of importance in forming the crust of weathering, while the effect of organic acids, and possibly of alkalis also, is only of importance near the surface and hence in the soil forming processes.

The fate of the products of decomposition depends mainly on the rate of water movement through the soil: the greater the amount of downward

1 G. Pedro, *Comp. Rend. Acad. Sci.*, 1961, **253**, 2242.

leaching the higher the proportion of the products of weathering that are removed from the zone in which they were formed and either redeposited in lower layers of the soil or else carried with the percolating water into the ground-water and thus into the rivers. Water movement is not the only agency causing movement of the products of weathering. Plants growing on the soil counteract some of the downward movement brought about by the percolating water, for their roots absorb some of the products of weathering from the subsoil and transfer them to their leaves and stems, which later fall to the ground and decompose, releasing these products of weathering in the surface soil. Burrowing animals, such as earthworms, termites and some rodents, also affect the distribution of the products of weathering consequent upon the mixing of soil from different layers which they bring about, as does the blowing over of forest trees by wind or the filling up of the large spaces once occupied by their larger roots. These processes involving large-scale soil disturbances are known as pedoturbation.[1]

The chemical weathering of rock minerals

Weathering of rocks by water involves two separate processes, the decomposition brought about by the water, and the fate of products of decomposition. In so far as water is percolating through the medium, soluble products of weathering will be removed from the profile and enter the ground-water and so move into rivers. The warmer the soil in contact with water, the stronger is the decomposition of the rock; and the higher the rainfall, provided the soil is well drained, the quicker are the soluble products removed. The rate of decomposition of a given mineral particle depends on its surface area, so the smaller the particles the faster the rate of weathering per unit volume, on its temperature, being very slow at $5°C$, and on the pH of the solution. The relative decomposability of different silicate minerals under comparable conditions differs widely for reasons that are not fully known.

Silicates containing ferrous and manganous ions are usually more weatherable than comparable silicates not containing them, due to these ions being oxidised to a higher valency when they become accessible to the oxidising conditions in the percolating water. The more open silicate structures, that is, the larger the volume occupied by each oxygen or hydroxyl ion in the unit cell, the more easily weatherable is the crystal; and the greater the number of positions in the lattice that could contain an ion but do not, that is, the greater the number of empty sites in the lattice, the more weatherable the mineral is likely to be.[2]

The dissolution of ions from the crystal surface probably depends on the

1 See, for example, F. D. Hole, *Soil Sci.*, 1961, **91**, 375. A. Jongerius, *Geoderma*, 1970, **4**, 311.
2 For reviews on the chemistry of weathering see I. Barshad in *Chemistry of the Soil*, Ed. F. E. Bear, 2nd Edn, Reinhold, 1964; H. Sticker and R. Bach, *Soils Fert.*, 1966, **29**, 321; F. C. Loughman, *Chemical Weathering of the Silicate Minerals*, Elsevier, 1969.

exchange of a hydrogen ion from the solution with a simple cation, such as an alkali or alkaline earth ion neutralising the charge on the lattice. This rate of exchange is, in turn, dependent on the accessibility of the ions to the hydrogen ions, and is increased by them becoming hydrated if the lattice will open up sufficiently for water molecules to enter. Weathering of mica typically begins by the opening up of the sheets along the broken edges by the entry of water molecules and the exchange of potassium for hydrogen. This opening up usually involves a further weathering change which decreases the net charge on the lattice where it has been opened up, possibly due to some of the hydroxyl ions in the lattice taking up a proton to become water.

A second consequence of weathering is that the hydrogen ion neutralising the charge on the lattice decomposes the lattice, setting free silicic acid which goes into solution and aluminium ions which replace the hydrogen ions on the lattice and become complex hydroxylated ions. Normally only aluminium ions do this, for any ferric ions set free forms ferric hydroxide which crystallises readily to goethite Fe O(OH). Weathering by percolating water therefore brings into solution simple metallic ions and silicic acid and leaves behind silicate surfaces coated, at least in part, with hydroxylated aluminium ions and micro-crystals of goethite.

The rate that soluble products of weathering are lost from an area can be estimated from an analysis of the river water draining out of the area, provided the river is not receiving effluents from other sources. Very few good estimates of these rates of loss have been made because they require both the continuous gauging of the river flow and regular chemical analyses of the water, but many years ago F. W. Clarke[1] collected together the data then available. He showed that European and North American rivers carried away about 40 tons of soluble salts per square kilometre of their catchment annually, which corresponds to the solution of about 1 cm of soil per thousand years; while tropical rivers removed only about half this amount, due to the soils in the tropical areas being more strongly weathered than in the temperate. The estimated weights of the various soluble mineral constituents lost per year, in kilograms per hectare are:

	CO_3	SO_4	Cl	NO_3	Ca	Mg	Na	K	$Al_2O_3 + Fe_2O_3$	Si
N. American and European rivers	145	55	19	3·9	82	13	23	9	6[2]	17
Amazon	64	17	11	1·2	37	5	10	4	11	18
Drainage water, Broadbalk, continuous wheat, unmanured plot[3]	—	63	27	170	178	8	11	3	14	13

1 Data of Geochemistry, *U.S. Geol. Survey, Bull.* 70, 1924.
2 Iron and aluminium not separately determined.
3 Calculated from the mean of 5 samples analysed by A. Voelcker, *J. Chem. Soc.*, 1871, **24**, 276.

This shows that the Amazon, draining a tropical region, contains much less calcium, magnesium, potassium and sodium than the rivers of the temperate regions, but the same amount of silicon and probably more aluminium and iron. The drainage water from the top 60 cm of Broadbalk soil is given for comparison. Under humid tropical conditions, basic igneous rocks in the soil zone are weathered very much faster than is indicated in this table.

As weathering proceeds, however, some aluminium also will be precipitated as a hydroxide, and this hydroxide usually contains a small amount of associated ferric and other metal hydroxides in the same way that the ferric hydroxide precipitates contain some associated aluminium and other metal.[1] These hydroxide surfaces will absorb silicic acid by ligand exchange (see p. 129), so could be the precursors of clay minerals.

The mixed aluminium hydroxide-silicic acid films that are formed as a result of weathering rarely constitute any large proportion of the residual products of weathering. The principal exception to this is when amorphous silicates, such as volcanic glass, or poorly crystalline silicates such as finely comminuted volcanic andesitic or rhyolitic ash, weather under moist conditions, for the principal residual product is then a hydrated amorphous mixed gel known as allophane.[2] Now that methods are available for dissolving non-crystalline material from the clay fraction of soils without attacking the crystalline clay, allophane-like hydrated mixed gels have been found to accumulate in many soils derived from crystalline silicate rocks under these climatic conditions.[3] Thus, E. A. C. Follett[4] found they constituted 14 per cent of a soil derived from a gabbro in Scotland.

It is usually assumed that in normal conditions, the mixed aluminium hydroxide-silicic acid films recrystallise to give the clay mineral particles found in crusts of weathering. This need not be the origin of most of the clay particles in a soil for some are formed from the residual ions in the original crystal. This certainly is the source of most of the clay particles formed during the weathering of micas, for during the process the mica crystals will get broken up into smaller fragments, potassium will be lost from between the sheets around their broken edges, and the charge density on the loosened sheets will be reduced by the adsorption of hydrogen ions.[5] If there are many aluminium or magnesium ions present these will give sheets or bits of sheets of aluminium or magnesium vermiculite or chlorite. Biotite micas almost certainly weather to such clays without the oxygen and hydroxyl sheets of the crystal breaking up. Muscovite micas correspondingly tend to weather leaving behind illitic or micaceous clay particles containing a core of mica sheets held together by potassium ions. It is possible that other minerals give clays by a rearrangement of their constituent oxygen tetrahedra and octa-

1 K. Norrish and R. M. Taylor, *J. Soil Sci.*, 1961, **12**, 294.
2 See, for example, M. Fieldes, *N.Z. J. Sci.*, 1966, **9**, 599, 608.
3 For a review see B. D. Mitchell, V. C. Farmer and W. J. McHardy, *Adv. Agron.*, 1964, **16**, 327.
4 With W. J. McHardy *et al.*, *Clay Miner.*, 1965, **6**, 23, 35.
5 A. C. D. Newman and G. Brown, *Clay Miner.*, 1966, **6**, 297.

hedra, as for example felspars may weather to illites and halloysites;[1] and the reason why glasses and poorly crystalline silicates give allophane may be that their structure is too disordered for a layer lattice to be formed.

An important characteristic of deep weathering below or at the very bottom of the root zone under warm moist well-drained conditions is that it proceeds through the loss of soluble material without any change in the volume of the weathering rock, so that the volume and shape of every mineral grain is the same in the fresh as in its fully-weathered state; and one can see from the fully weathered material exactly what the original rock looked like. This is the material known as saprolite, and its bulk density is only one-half to two thirds that of the original rock. It consists largely of kaolinitic clays with some ferric hydroxide or hydrated oxide. The constancy of volume during weathering suggests that clay formation takes place with only minor rearrangement of the oxygen and hydroxyl ions present in the original silicate. Under some conditions of intense weathering and strong leaching bauxitic clays, that is, clays high in free alumina, are formed.

It is very difficult to decide how frequently clay formation involves only minor rearrangement of the oxygen and hydroxyl ions to convert the original silicate lattice to a clay lattice, or how far there is a complete break-up of the silicate lattice and a migration of the elements that will form clay at a new site. Clay formation involving migration can certainly sometimes be proved, but under these conditions no saprolite is found. As an example, when basic rocks are weathering under conditions, where there is little leaching, so where there is a good supply of ferrous iron, magnesium and silicic acid, montmorillonite clays are formed in the soil zone. These clays are also formed in valley bottoms and sumps in tropical and subtropical regions where the water flowing in is well supplied with alkaline earth ions and silicic acid, and where the alluvial material is well supplied with free iron oxides. The soils produced under these two conditions are black, very high in montmorillonite clay, crack very strongly on drying, often have a horizon of secondarily deposited calcium carbonate in the subsoil, and are known by a variety of names, of which vertisol is the presently accepted one and black cotton soil an old one.[2] The cause of the black colour is not fully known, for the clays are low in organic carbon; it is probably due to some humic products produced under anaerobic conditions which are absorbed on the clay possibly after they have complexed divalent cations such as ferrous, manganous or the alkaline earths.[3] A second situation is when drainage conditions cause silicic acid-rich waters to percolate through weathered material containing gibbsite, for this will slowly be converted to kaolinite.[4]

1 For examples from Scotland see W. A. Mitchell, *J. Soil Sci.*, 1955, **6**, 94; from Malawi I. Stephen, *Clay Miner., Bull.*, 1963, **5**, 203, and *J. Soil Sci.*, 1965, **16**, 322.
2 For a description of various types of these clays see R. Dudal, *Soil Sci.*, 1963, **95**, 264.
3 S. Singh, *J. Soil Sci.*, 1954, **5**, 289; 1956, **7**, 43.
4 J. B. Harrison, *The Katamorphism of Igneous rocks under humid tropical conditions*, Commonw. Bur. Soils, 1934.

A comparison of the soils formed from the weathering of basalts and of granites will illustrate some of the points made here. Most of the minerals present in a basalt weather fairly easily, so under conditions of good drainage and strong weathering, they are weathered to kaolinite, ferric hydroxide or hydrous oxide and perhaps some aluminium hydroxide. The soil is low in sand and high in clay but, because of the iron and aluminium hydrous oxides present, is very well structured. Granites, on the other hand, contain a high proportion of minerals that weather much more slowly, so even under conditions of strong weathering give soils high in fine sand, much of which is quartz, and a mixture of micaceous and kaolinitic clays, the latter largely derived from the felspars; and since the parent granite is usually rather low in iron, the clay is less well structured. However, weathering goes much deeper in the granite, with well developed saprolite, while it tends to be more shallow in the basalt, with a much sharper transition between the fully weathered and unweathered material and with much less saprolite. In cool temperate regions, where weathering is slower, these differences in texture are not so marked, but the basalts are more likely to give dark coloured well-structured loam to clay loam soils and the granites sandier soils.

Very rarely is only one type of clay mineral present in the weathered material. This is partly because the material has been developed from a range of silicate minerals, some of which will have a tendency to weather to different clay minerals, and partly because the zone of weathering will contain niches in which different factors are controlling the type of clay being formed.

Clay minerals are not resistant to weathering when the conditions change under which they were formed. Thus as weathering proceeds and the original rock minerals become more completely decomposed, the concentration of soluble cations and silicic acid in the soil solution drops, so any three-layered clay minerals that had been formed will begin to decompose, with the consequence that, after a long period of time, the products of weathering often consist only of kaolinites and iron oxides, for any three-layered clay crystals originally formed will slowly lose silica. Even quartz weathers slowly, as one can often see in the tropics when a quartzite band is present in the rock. The quartz gradually breaks up into smaller particles and finally completely decomposes. Thus the older the land surface, provided there is little erosion and weathering remains active and drainage good, the higher the clay content of the soil.

G. Pedro[1] has given an interesting example of simulated weathering due to water in the laboratory. He refluxed gravels made of crushed basalt, andesite and granite with hot water at about 70°C. for two years, and determined the composition of the weathering crust on the outside of the rock, and of the material that had been dissolved. He showed that the weathering had caused no loss in volume, that the basalt had lost 4·3 and the andesite

---

1 *Clay Min. Bull.*, 1961, **4**, 266, and see *Ann. Agron.*, 1964, **15**, 87.

6·7 per cent of its weight, that no iron had been dissolved, but that an appreciable amount of aluminium had been removed with the silicic acid. Thus the silicon-aluminium ratio in the basalt was 3·2 and in the leachate 5·1, and the corresponding figures for the andesite were 2·6 and 4·6. He did not determine whether the alumina present in the leachate came over in a finely dispersed sol either alone or associated with silica. The residual products of weathering were goethite, gibbsite and boehmite. On the other hand, if the extraction was done at 20°C, the crust contained some montmorillonite-type clay. Table 26.1 gives some of his results, and compares the weathering he found for his basalt and andesite with basalt weathering in the Cameroons and diorite weathering in Guyana.

TABLE 26.1   Laboratory and natural weathering of basalt and andesite. Laboratory weathering at 70°C for 2 years. Composition of the fresh rock and the weathering crust

	Basalt				Andesite			
	Laboratory		Cameroons		Laboratory		Guyana	
	Rock	Crust	Rock	Crust	Rock	Crust	Rock	Crust
SiO$_2$	45·2	23·4	42·0	23·0	57·3	50·4	59·4	54·7
Al$_2$O$_3$	12·2	21·4	19·8	40·0	19·3	22·0	17·2	21·7
Fe$_2$O$_3$	14·7	35·9	11·5	30·3	9·0	12·9	8·1	11·1
MgO	8·3	5·3	8·0	0·3	2·8	2·5	3·2	3·6
CaO	11·0	3·0	7·0	0·3	3·8	3·6	6·3	1·0

Weathering in the soil zone: Soil forming processes

The growing plant and the organisms feeding on it and its litter affect both the type of rock weathering and the fate of the products of weathering. Thus plants take up soluble mineral salts from the soil in their root range, transfer them to their leaves and above ground parts, and these are then returned to the soil surface either in leaf drip or in the fallen leaves, stems and branches of the vegetation. This dead vegetation and the living and dead plant roots are a food source for a whole range of soil organisms, which, in turn, can affect the soil by their physical or biological activities. Further, some of them are capable of inducing severe reducing conditions in the soil if aeration becomes poor, which introduces a new set of processes.

The amount of mineral salts in cycle due to the vegetation can be considerably higher than the amounts leaching out of the profile into rivers. The amounts of mineral elements removed by rain washing the leaves and the amounts returned annually under an oak forest in north Lancashire[1] are

1 A. Carlisle, A. H. F. Brown and E. J. White, *J. Ecol.*, 1967, **55**, 615.

given in Table 26.2, and the figures should be compared with those given on p. 699 for the amount being removed by rivers. Potassium is an example of an element for which the amount in cycle is much higher than the annual loss from the catchment, more silicon is usually in cycle than is lost, but the amounts of calcium and magnesium in cycle may be of the same order. Less is known about the amounts of aluminium and iron in cycle since they are

TABLE 26.2   The quantities of nutrients washed from a *Quercus petraea* woodland by rainfall and reaching the soil in litter in kg/ha/year

	Organic matter	Nitrogen	Phos-phorus	Potas-sium	Calcium	Mag-nesium	Sodium
Incident rainfall	77	8·7	0·3	2·8	6·7	6·1	50·8
Washed from oak canopy	174	0·9	0·5	16·0	12·3	5·6	32·0
Washed from other vegetation and stem flow	27	−0·3	0·1	10·9	2·0	1·5	6·6
Total in rain plus throughfall	277	9·5	0·9	29·7	21·0	13·2	89·4
Total in litter	6670	61·6	3·7	22·2	36·1	6·0	4·0
In throughfall as % in litter	4·1	15·1	25·0	133	58·4	222	2210

Washed from oak canopy: soluble carbohydrates 68·4, polyphenols 11·7

not often measured. The plants themselves can also directly affect the weathering process in the soil, for the rain water not only washes salts off their leaves, but also soluble organic compounds, some of which reduce ferric iron and form stable soluble ferrous-organic complexes. The roots also can hasten weathering by reducing the potassium ion concentration in the soil solution to a sufficiently low level for potassium in the mica interlayer to come into solution, so opening up the mica along its broken edge. Rhizosphere bacteria and fungi can also increase the rate of weathering, and some of the bacteria excrete 2-keto-gluconic acid[1] and some of the fungi citric and oxalic acids[2] which can chelate di- and trivalent cations, so increasing their rate of dissolution from mineral particles.

Plant litter collects on the soil surface, but its method and rate and site of decomposition depends on the soil organisms which are active, and this, in turn, depends on the kind of litter, on the moisture and temperature regime, and on the soil. Thus, if conditions favour the larger species of earthworm which form wormcasts, much of the decomposition will take place in the soil under conditions which favour the production of clay-humus complexes, and the earthworms themselves will cause a mixing of the soil throughout

1 D. M. Webley *et al.*, *J. Soil Sci.*, 1963, **14**, 102; *Soil Sci.*, 1963, **95**, 105.
2 M. E. K. Henderson and R. B. Duff, *J. Soil Sci.*, 1963, **14**, 236.

the depth in which they are active; in England this is about 15 cm. On the other hand, if the soil conditions are unfavourable to all the larger soil fauna, there will be no mechanism for mixing the litter with the soil, and much of the decomposition will take place on the surface. The amount and distribution of the plant roots and their longevity also affect the supply of decomposable organic matter in the soil, and so the distribution and activity of its population.

Finally, many species of soil bacteria continue to decompose plant debris or intermediate products of their decomposition when the oxygen supply is severely restricted, by setting up severe reducing conditions which result in the reduction of ferric to ferrous iron and higher valency manganese to manganous ions. A proportion of these ions are present in the soil solution, often as organic complexes, so move with the soil water. These conditions may also result in the reduction of sulphates to sulphides, which on reoxidation to sulphuric acid will increase the soil acidity very considerably.

Soil classification

A soil must, by definition, exhibit vertical anisotropy due to biological processes operating on the parent rock, so the properties of the soil which are singled out for relevance in most systems of classification must be displayed in the soil profile. A soil profile is the vertical face of a soil pit, and typically contains bands, or layers, or horizons that are visually different from neighbouring horizons, though the visual differences often merge so that there need be no clear divisions between them. Well-drained soils that have suffered relatively little disturbance by man typically possess three or four clearly recognisable horizons: an organic litter layer lying on the soil surface known as the O horizon, a horizon darkened in colour by organic matter but which has lost clay or sesquioxides and is known as the A or eluvial horizon, a horizon that has gained some of this clay and sesquioxide lost from the eluvial horizon and known as the B or illuvial horizon, and the C horizon which has been little affected by the biological activities in the soil and which is often presumed to be similar to the rock originally forming the upper part of the soil profile.

Profile development is not confined to well-drained soils, but the characteristics of the horizons will differ under the influence of impeded drainage or of a fluctuating ground-water level. However, since water moves downwards from the soil surface in most soils for some periods of the year, most soils possess to some degree eluvial and illuvial horizons, though their properties are likely to be strongly affected by the consequences of the water regimes.

Soils are formed under such a variety of conditions that no simple universal specification of horizons can be developed to fit all soils. Under well-drained conditions, rock minerals will weather in the B horizon and the soluble products of weathering removed in the drainage water, so it is both an illuvial and eluvial horizon. Again although typically the A horizon visibly contains more humus than the B, in some conditions humus appears to wash out of the A and be deposited in the B. It is still possible to retain the ABC nomenclature in these circumstances, but a modifier should be added to the horizon designation to show that it differs from the original simple concept. Thus, horizons showing some of the characters of the simple B, but with other characters introduced through impeded drainage or fluctuating groundwater and so showing gley characters, can be designated as a gleyed B or Bg horizon. However, the United States Soil Survey may be abandoning the ABC nomenclature on the grounds that when all the soils in the United States are considered, so many factors can influence the profile that it is preferable to specify the actual characteristics of each soil horizon rather than try to force these characters into a preconceived general scheme.

A further practical problem in classification is that when a number of soil profiles are dug in a fairly small area, say, a typical farm field, it is not unusual for the profiles to differ from each other in a number of readily describable characters. One of the problems in the field classification of soils is therefore to decide how great a variation in specific character is allowable between profiles in soils that are to be grouped together in one class; and this problem can only be answered if the purpose for which the classification is made has been clearly formulated.

This common variability in the appearance of profiles that are dug close to one another is usually a reflection of the variability of the material from which the soil was formed. The land surface of many temperate countries was greatly affected by the ice sheets or by periglacial activity during the last Ice Age, so was left covered with a great mixture of tills, loesses, and outwash material, and so the constitution of the top few metres of the land surface after the retreat of the ice is very heterogeneous, and this is reflected in the heterogeneity of adjacent profiles in these areas. It is because of this great variability of soils at adjacent sites, that the American Soil Survey introduced the concept of the pedon which is thought of as the smallest volume that can represent the soil at a given site. If the variations in the soil profile appear to be random from site to site, it is defined as the soil in an area of one square metre, but if the soil shows any cyclical changes which recur with a linear interval of two to seven metres, the pedon includes one-half of the cycle.

Soil classification is needed as a consistent means of identifying soils, and must therefore be applicable to all soils; and it is essential, for example, to have an adequate system of classification if useful soil maps are to be made. But for this purpose the scale of the map must influence the system of classification used, for one that is well adapted for mapping on a scale of 1:50 000 or larger, which is needed by the farmer or farm adviser, is much too detailed to be of value if a scale of 1:1 million or smaller is used, which is needed to allow pedologists from all over the world to compare their soils. There is still no universally accepted system of nomenclature or classification for this latter purpose, since each country with an active soil survey has tended to develop their own system adopted to their own needs.[1] F. A. O., however, has an office concerned with developing a system which will be universally acceptable.[2]

Types and forms of humus in and on the soil surface

The organic compounds washed off the leaves of the living vegetation by rain, the type of litter falling on the soil, and the processes bringing about its decomposition can have a profound effect on the type of soil being formed. These processes can be seen most clearly in soils that have been uncultivated for long periods of time, such as old forests, old heathland and old grassland. Every undisturbed soil will have a characteristic distribution of undecayed and decaying litter and humus on and in its surface, and these distributions are classified under the unfortunate name of a humus type: unfortunate because the word humus should be restricted to organic colloids. For instance, the humus type under a coniferous forest on an acid sand is different from that under many deciduous forests on loams or clay loams, and these are different from that under grassland on the same site.

Each humus type usually consists of two or more layers or horizons, each layer having a different form of humus. Typically the top layer is fairly fresh leaf litter and the bottom layer fully humified material, and the layers are classified by their morphology, or since it is usually necessary to use a microscope to examine this, by their micro-morphology. There is still no generally accepted classification or nomenclature for the different humus types[3] and humus forms,[4] and the situation is complicated by many workers using the same name for a humus type as for the dominant humus form constituting that type.[5] Further, there is no modern monograph in which the various types and forms found in either the temperate[6] or the tropical regions have been critically described.

Three humus types forming the floor of temperate forests[7] have been widely recognised, two of which have the generally accepted names of mull and mor;

1 The most widely used is probably the American system described in *Soil Classification: A Comprehensive System, 7th Approximation*, Soil Survey Staff, U.S. Dept. Agric. 1960 and subsequent supplements.
2 This classification is being continuously amended, see *World Soil Resources Rept.* 33, 1968; 37, 1969, F. A. O. Rome.
3 For recent proposals see S. A. Wilde, *Soil Sci.*, 1971, **111**, 1.
4 The classical descriptions are given in W. Kubiena, *The Soils of Europe*, Murby, 1953.
5 For a discussion of this see P. J. A. Howard in *The Soil Ecosystem*, Ed. J. G. Sheals, The Systematics Association 1969.
6 See P. Duchaufour, *Précis de Pedologie*, 3rd edn. Masson, (1970) for temperate soils.
7 For a description of humus types in British grassland soils see B. C. Barrett, *J. Soil Sci.*, 1964, **15**, 342; *Geoderma*, 1967, **1**, 209.

the third has no generally accepted name, the names used for it including insect mull, mull-like mor, and moder.[1] There has been much dissatisfaction with the name mor because it is so similar in sound to moor, which is a completely different thing, although a mor can be the humus type on a moor.

Typical mor is a humus type that lies on the surface of a mineral soil. It is composed of three layers or humus forms: a litter layer, known as the L layer, consisting of recently fallen leaves, which have only just started to decompose; a fermentation or F layer, in which the leaves are undergoing active decomposition but in which numerous leaf fragments are still recognisable; and a humus or H layer, composed of dark structureless humus. This humus type forms a compact consolidated blanket, which can be stripped off the soil surface, and it can build up to a thickness of 10 or 15 cm in many forests. Typically, the vegetation giving a mor humus roots strongly in this humus horizon, and most or all of the roots are mycorrhizal.

Mull humus, on the other hand, is a humus type in which most of the humus is mixed with the mineral soil though a part lies on the soil surface. The part lying on the surface is usually recognisable leaves which have suffered little decomposition but which typically disappear within a year, while the part in the soil has lost all trace of the plant remains from which it was derived, though it will often contain many fine dead roots that have not yet been decomposed but are still *in situ*. This mull, or mould to use the old name, is crumbly, brown to dark brown in colour, and the organic matter in the top 5 to 10 cm rarely exceeds 10 per cent. It usually has no clearly defined lower boundary. Both the humus type, and the form of humus in the mineral soil are called mull.

Moder humus is principally above the soil surface; it usually has no clear separation from mineral soil and the bottom of the moder layer contains an appreciable amount of mineral grains. It differs from mor in that it is much looser and more porous, and the litter layer is characterised by the leaves being lacerated by the feeding of soil arthropods. A considerable proportion of the layer below the litter can be seen under the microscope to consist of faecal pellets and exo-skeletons of the fauna. Moder thus involves a considerable mixing of plant fragments brought about by animals smaller than the earthworm. Both this humus type and the humus form in this horizon have been called moder. Moders grade into mulls as the conditions become more suitable for earthworms.[2]

The soil, the climate and the vegetation control which of these humus types will be dominant at any one site through their control on the types of soil organisms that can flourish. Mull humus is confined to soils carrying a vegetation of certain species of broad-leaved trees or grasses, and containing

---

1 This classification, but not the actual names, was first put forward by the Danish forester P. E. Muller in *Tidsskr. f. Skovbrug*, 1878, **3**, 1; 1884, **7**, 1. Also *Studien über die naturlichen Humusformen*, Berlin, 1887, and *Ann. Sci. Agron.*, 1889, **1**, 85.
2 For examples see F. Delecour and G. Manil, *Pedologie*, 1958, **8**, 127.

a high population of earthworms which feed on decaying leaves on the soil surface and at the same time ingest soil, and excrete the mixture of macerated digested leaf tissue and soil as a wormcast. The depth of the mull horizon is the depth to which earthworms are active. Mulls are therefore restricted to soils that are not too acid, usually above pH 5, that do not contain a high proportion of coarse sand and usually of loam to clay texture, and are deep enough to allow the worms to hibernate and aestivate and so usually deeper than 70 cm.

The extreme type of mor is formed on soils where the leaf litter is very unpalatable to most members of the soil fauna and where fungi are the dominant organisms causing decomposition. The reason for the unpalatability is probably that the leaves of the vegetation are high in water-soluble polyphenols which polymerise easily to give tannins, which react with the leaf proteins to give very stable tanned products.[1] The particular type of polyphenol may be leucoanthocyanins which tend to be high in mor-forming vegetation grown under conditions of low soil fertility.[2] It is still uncertain if this is the sole cause of unpalatability, because leaf litter from the same species of tree can give either a mull or a mor depending on soil conditions; and it has not been established if the polyphenol content of their leaves differ sufficiently to cause this difference. The F layer is often characterised by a mass of fine white fungal hyphae. The principal arthropods are mites, which eat out the inside of the leaves. The soils are typically very acid and impoverished, often sandy, and the vegetation coniferous forest. The moder type comes intermediate. The soil or vegetation conditions are unsuitable for earthworms, but the litter is palatable to a number of the larger species of arthropods, such as millipedes, woodlice and beetle larvae, as well as to the mites.

Well-drained soils:

Podsols and podsolisation

This soil type, which is usually, but not necessarily, found on well-drained sites, will be described first because, although it is restricted to a narrow range of conditions, these conditions allow a study of some important processes of translocation with the minimum of complicating factors. The profile is characterised by having a band of light-coloured soil separated, often sharply, from a band of darker soil below it, and with a band of darker soil or humus above it. It is this pale band which has given the podsol its name for the word signifies in Russian ash-coloured soil. The profile can be divided into several clearly defined layers or horizons. Under many types of vegetation, and particularly forests, the upper horizon is a mor humus lying directly on the

1 W. R. C. Handley, *Bull. Forestry Commission*, 23, 1954; *Pl. Soil*, 1961, **15**, 37.
2 R. I. Davies, C. B. Coulson and D. A. Lewis, *J. Soil Sci.*, 1964, **15**, 299, 310.

mineral soil, and typically directly on the light-coloured or bleached eluvial horizon, the A_2, although sometimes some humus washes down or is worked into the top of the mineral soil to give a dark-coloured A_1. The dark-coloured layer below the A_2 is the B or illuvial horizon which usually lightens gradually with depth until it becomes indistinguishable from the parent material or the C horizon. When the surface humus is a mor, the division between the A_0 and A_1, or A_2 if there is no A_1, is sharp, and the divisions between the A_1 and A_2 and between the A_2 and B are often, though not always, sharp, but the divisions in the B horizon itself, and between the B and C horizons are usually gradual.

The relative thickness of these layers is very variable, but the cause of the variability cannot always be given. The A_0 can be very thin and the A_1 well developed, as in heaths and some forests with suitable undergrowth, or the A_0 can be well developed and the A_1 thin or both the A_0 and A_1 can be thin or almost lacking. Thus, in well-developed podsols under coniferous forest on sandy soils there is no A_1 horizon—the mor lies directly on the A_2. Similarly, the thickness of the A_2 and of the B varies very considerably, often in quite small areas; and in particular the boundary between the A_2 and the top of the B is often extremely irregular, varying by over 15 cm in very short distances. The total depth of the profile, that is, from the top of the A_0 to the bottom of the B, can vary from a few centimetres to over a metre. Two typical podsols are illustrated in the Frontispiece and Plate 39. The podsol under the forest may not be very old, for the land may not have been under forest for more than 100 to 150 years; so it is not possible to say if the thin A_1 horizon is a relict of an earlier heathland podsol.

The chemical characteristics of the podsol profile are:

1 The pH is usually lowest in the A_0 horizon and increases down the profile, though in some profiles the A_1 may be the most acid.

2 The A_0 and A_1 horizons have a high cation exchange capacity due to their high content of organic matter, but are very unsaturated. The percentage saturation usually increases down the profile, but because of the low exchange capacity of the A_2 and below, the actual number of milli-equivalents of calcium and magnesium may be highest in the A_0 or A_1. The top of the B horizon may also have a high cation exchange capacity relative to its clay content due to the organic colloids that are held in it.

3 The A_2 horizon is subjected to intense weathering, as can be seen from an examination of the fine sand grains. These consist of either minerals very resistant to weathering, or if weatherable minerals are present, they are obviously being heavily attacked.

4 The B horizon is enriched with material that has been washed out of the A horizons, namely organic matter, clay, and iron and aluminium hydroxides. Typically their zone of accumulation occurs at the top of the B, but under some conditions, the organic matter with perhaps the alumin-

PLATE 39. Podsol developed under *Calluna* heath in a Barton Sand contaminated with Plateau Gravel, in the New Forest, Hampshire.

ium has a maximum at the top, often characterised by having a very dark colour, with the zone of iron accumulation lying below. The former type is sometimes called an iron and the latter an iron-humus podsol. Figure 26.1[1] illustrates this difference between these two types of podsol. The separation was originally made on the difference in colour of the top of the B, a rusty, not very dark colour for the iron, and a much darker colour for the iron-humus. Under some conditions in Great Britain, a thin 'walnut shell' pan,[2] a few millimetres thick, develops either between the bottom of the A_2 and the B, or at the base of the major zone of humus accumulation, in the iron-humus podsol. The organic matter in the B horizon can be almost completely dispersed in solvents such as acetyl acetone, which chelates iron and aluminium ions, a characteristic confined to the humus in podsol B horizons.[3] Although there is a deposition of sesquioxides and clay in the B horizon, there can also be strong weathering of

1 Taken from K. Lundblad, *Soil Sci.*, 1934, **37**, 137. For another example from Germany see H. P. Blume and U. Schwertmann, *Proc. Soil Sci. Soc. Amer.*, 1969, **33**, 438.
2 For a description see E. Crompton, *J. Soil Sci.*, 1952, **3**, 277, and *Trans. 6th Int. Congr. Soil Sci.*, 1956, E. 155.
3 A. E. Martin and R. Reeve, *J. Soil Sci.*, 1957, **8**, 268, 279; M. Schnitzer and J. R. Wright, *Canad. J. Soil Sci.*, 1957, **37**, 89.

mineral grains in this horizon, so there is usually some eluviation from the B, although it is predominantly illuvial.

Probably much, perhaps most, of the iron and aluminium deposited in the B horizon is initially associated with the organic matter. Figure 26.1 shows the amount of iron, aluminium or organic matter extracted by Tamm's acid

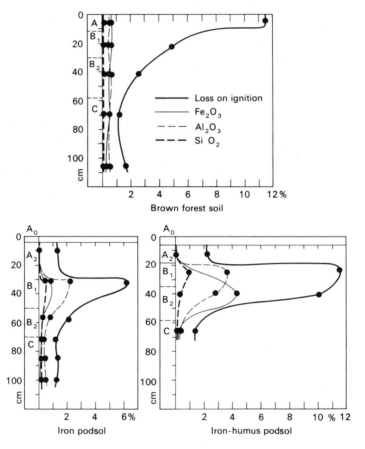

FIG. 26.1. Loss on ignition and amounts of silica and sesquioxides extracted by the acid-oxalate method in a Swedish brown forest soil, iron podsol, and iron-humus podsol

oxalate from two typical podsols as well as from a brown forest soil.[1] This extractant does not distinguish between iron and aluminium in organic complexes and in hydroxides, but J. A. McKeague,[2] using sodium pyrophosphate as the extractant, showed that a considerable proportion of the

1 Taken from K. Lundblad, *Soil Sci.*, 1934, **37**, 137.
2 *Proc. Soil Sci. Soc. Amer.*, 1971, **35**, 33.

TABLE 26.3   Analysis of the various horizons of a podsol profile in Teindland State Forest, Morayshire

(a) Ultimate analysis as percentage of ignited fine earth (<2 mm)

Horizon	A_1 5–7 cm	A_2 7–14 cm	B_1 14–16 cm	B_2 16–26 cm	B_3 26–50 cm	C 60–70 cm
SiO_2	89·43	91·05	82·49	78·86	76·40	83·12
TiO_2	0·14	0·12	0·24	0·21	0·21	0·11
Al_2O_3	6·11	5·81	10·10	10·30	11·08	8·56
Fe_2O_3	0·89	0·55	2·71	3·59	3·32	2·19
CaO	0·63	0 45	0·52	0·68	0·94	0·59
MgO	0·30	0·54	0·40	0·22	0·71	0·46
Undetermined residue*	2·50	1·78	3·54	6·14	7·34	4·87
Total	100·00	100·00	100·00	100·00	100·00	100·00
Loss on ignition	38·73	1·77	7·87	5·66	3·92	1·63
Moisture (105°)	12·02	0·65	5·41	4·72	4·12	1·46
Clay content	—	4·75	12·32	10·12	9·25	4·58

(b) Analysis of clay fraction: as percentage of ignited clay

Horizon	A_2 7–14 cm	B_1 14–16 cm	B_2 16–26 cm	B_3 30–40 cm	C 60–70 cm
SiO_2	59·02	46·28	31·73	33·80	35·20
TiO_2	3·59	1·19	1·82	1·50	1·00
Al_2O_3	24·20	29·92	38·36	31·62	24·56
Fe_2O_3	4·18	14·36	20·42	21·71	18·80
SiO_2/R_2O_3	3·71	2·00	0·95	1·26	1·63
SiO_2/Al_2O_3	4·12	2·61	1·03	1·80	1·92
Al_2O_3/Fe_2O_3	9·09	3·28	4·01	2·30	2·59

(c) Absorbed cations per 100 g of oven-dry soil, and pH value

Horizon	A_0	A_1	A_2	B_1	B_2	B_3	C
m eq. Calcium	7·0	3·0	0·8	1·1	0·5	0·8	0·8
m eq. Magnesium	n.d.	2·6	1·0	1·0	0·9	0·8	0·8
m eq. Hydrogen	n.d.	46·2	1·6	23·1	12·2	6·9	2·9
pH	3·7	3·9	4·3	4·5	4·5	4·8	5·1

* Probably chiefly K_2O and Na_2O.

iron extracted by the acid oxalate from the B horizon of a podsol was in fact complexed with the organic matter; and there are also some free hydroxide films in this horizon because the soil adsorbs acidic dyes, showing it possesses positive charges.[1]

Table 26.3 illustrates the results of these processes on the chemical composition of a podsol developed on flavio-glacial sands and gravels in the Teindland State Forest in Morayshire, Scotland.[2] At the time of sampling the land was under forest, but the podsol was probably developed under heath. Table 26.4 illustrates the point that there can also be weathering in the B horizon of podsols. This is taken from work of A. A. Rode in Russia,[3] and

TABLE 26.4   Weight of soil constituents lost during podsolisation in tons per hectare

Horizon	Si	Al	Fe	Mg	K	Ca	Na
A_1, 0–9 cm	183	109	62	17	25	10	10
A_2, 9–26 cm	342	218	119	35	48	22	17
A_2B, 26–35 cm	107	74	40	12	15	10	8
B_2, 35–50 cm	45	27	10	5	2	10	5
B, 50–63 cm	20	10	5	2	—	—	2
Total, 0–63 cm	697	438	236	71	90	52	42
Average annual loss, assuming 10000 years old, in kg/ha							
A_1 and A_2	54	34	18	6	8	3	2
Total	71	45	24	8	9	6	4
Tamm's estimate for young Swedish podsols							
A	28	11	8	3	4	2	2

is based on the assumption that the profile was originally uniform and that none of the quartz sand had weathered or moved. It shows that in this soil even the B horizon had lost some iron and aluminium. O. Tamm[4] also found a net loss of iron and aluminium from the B horizon.

Rode has calculated the average rate of loss of constituents from the Russian profile, assuming it has taken 10000 years to develop. The figures so found, which are also given in Table 26.4, imply a higher rate of weathering than for Europe as a whole, as deduced from the analyses of river waters given on p. 699, but unfortunately Rode does not say what were the parent mineral particles in the deposit. His results are also rather higher than some

1 K. Lundblad, *Soil Sci.*, 1936, **41**, 383.
2 A. Muir, *Forestry*, 1934, **8**, 25.
3 *The Soils of the USSR*, Ed. L. I. Prasolov, Vol. 1, p. 181, Moscow, 1939.
4 *Medd. Skogsförsöksanst.*, 1932, **26**, 163.

previously obtained by O. Tamm,[1] also given in Table 26.4, and obtained by a similar method, for the rate of removal of constituents from the A horizons of some young Swedish podsols developed on river sand.

The time taken for a podsol to develop its typical profile depends on many conditions, but on sandy soils under coniferous forest between 1000 and 1500 years is adequate,[2] though Tamm also finds that an A_2 horizon 1 cm thick can be formed in 100 years, and A. Muir[3] gives an example of an A_2 horizon 0·5 cm thick being formed in 20 years in Kincardineshire.

The characteristic process in podsolisation is the dissolution of iron and aluminium from the A horizon and their deposition in the B in association with the movement and deposition of organic matter. Initially, it had been assumed that humic acids from the mor dissolved these sesquioxides from the A_2 horizon as they leached down. But podsolisation can occur without any humic acids from the mor dispersing in the water, and if any humic acids are dispersed chemically from the mor they appear to possess no power to solubilise films of iron hydroxide.

Mobilisation of iron and aluminium is brought about by some soluble organic compounds which the rain washes off either living or recently dead leaves of the vegetation growing in the soil. These compounds include polyphenols such as *d*- or *epi*-catechin, which, under acid conditions, will reduce any ferric iron present, either in weatherable minerals or as films of ferric hydroxide on sand grains, and will form water-soluble complexes with both ferrous and aluminium ions.[4] R. L. Malcolm and R. J. McCracken,[5] for example, showed that leaf drip from oak and pine forests in North Carolina contained up to 1 kg/ha of polyphenols which would mobilise up to 1·5 kg/ha of iron and about half this amount of aluminium per year. Plants growing on acid soils low in nitrogen and phosphate, which is typical of these quartz sands, tend to have a higher content of simple polyphenols than those growing on more fertile soils,[6] and their leaves contain their maximum content of these polyphenols in late spring and early summer.[7] Leaves of some plants growing on these soils such as *Pinus sylvestris*[8] and some species of *Vaccinium* and *Calluna*[9] contain hydroxy- and polycarboxylic acids, such as malic, citric and oxalic, and these will also dissolve both iron and aluminium from their hydrated oxides or from the surface of weatherable minerals.

1 *Medd. Skogsförsöksanst.*, 1920, **17**, 49.
2 O. Tamm, *Medd. Skogsförsöksanst.*, 1920, **17**, 49; R. F. Chandler, *Proc. Soil Sci. Soc. Amer.*, 1943, **7**, 454 (for an example from Alaska).
3 *Forestry*, 1934, **8**, 25.
4 C. Bloomfield, *J. Soil Sci.*, 1953, **4**, 5; *J. Sci. Fd. Agric.*, 1957, **8**, 389; C. B. Coulson, R. I. Davies and D. A. Lewis, *J. Soil Sci.*, 1960, **11**, 30; *Proc. Roy. Dublin Soc.*, 1960, **1** A, 183. See also P. Lossaint, *Ann. Agron.*, 1959, **10**, 493; F. J. Hingston, *Aust. J. Soil Res.*, 1963, **1**, 63.
5 *Proc. Soil Sci. Soc. Amer.*, 1968, **32**, 834.
6 R. I. Davies, C. B. Coulson and D. A. Lewis, *J. Soil Sci.*, 1964, **15**, 310.
7 C. B. Coulson, R. I. Davies and D. A. Lewis, *J. Soil Sci.*, 1960, **11**, 20.
8 J. W. Muir, R. I. Morrison *et al.*, *J. Soil Sci.*, 1963, **15**, 220.
9 S. Bruckert and F. Jacquin, *Soil Biol. Biochem.*, 1969, **1**, 275.

There is still some uncertainty about the reasons for the precipitation of the iron and aluminium in the B horizon, though once deposition has started it will build up. It is possible that initially deposition begins by the soil becoming dried out in summer and the ferrous-polyphenol oxidising in these conditions to ferric hydroxide and an oxidised polyphenol polymer similar to some humic acid fractions. Once the B horizon has started to form it will build up partly because the ferrous polyphenol can be absorbed by a ferric hydroxide surface[1] and suffer oxidations, probably helped by a build-up of bacteria capable of hastening this oxidation.[2] These precipitates can be dissolved in 0·1 M alkaline sodium[3] or potassium pyrophosphate,[4] a reagent, which does not easily dissolve iron or aluminium from amorphous or crystalline inorganic hydroxides, so that the organic matter in the B horizon probably remains associated with the iron and aluminium. This may be the reason why the B has a sharp but extremely uneven top, for once deposition has started it will build up on these sites. Since these precipitates possess positive charges, if any humic colloids are dispersed from the mor they would be adsorbed onto these spots. There is no evidence about the fate of iron and aluminium chelates of the carboxylic acids. Those formed in the early summer would presumably be precipitated during any summer drought and decomposed, but the remainder could easily leach out of the profile causing a complete loss of silica, iron and aluminium. At certain times of the year they could also be removing some of the material from the B horizon out of the profile.

The polyphenols washed off the leaves of different species of tree presumably differ appreciably, though this has not been investigated in detail. The polyphenols will reduce ferric to ferrous iron more strongly in acid than neutral conditions, so podsolisation due to these compounds would be expected only in acid soils. Some polyphenols will, however, disperse clay particles,[5] and they will do this even in neutral or calcareous soils,[6] so they may be an important agent in the formation of some luvisols, which are soils from which the clay has been almost completely removed from the A horizon and deposited in the B.

The picture presented here has an important consequence for humus chemistry, for only a part, and possibly none of the organic matter in the B horizon has been derived from the humus in the mor, but most and possibly all has been produced by the polymerisation of simple polyphenols with probably other soluble organic compounds, such as amino acids from the leaves. This means that the humic colloids in this horizon are different in chemical constitution from most soil humic colloids and this could well be

1 C. Bloomfield, *J. Soil Sci.*, 1955, **6**, 284.
2 D. V. Crawford, *Trans. 6th Int. Congr. Soil Sci.*, 1956, C, 197.
3 J. A. McKeague, *Canad. J. Soil Sci.*, 1967, **47**, 95.
4 C. L. Bascomb, *J. Soil Sci.*, 1968, **19**, 251.
5 C. Bloomfield, *J. Soil Sci.*, 1953, **5**, 39. *Nature*, 1953, **172**, 958.
6 See, for example, H. C. Moss and R. J. St Arnaud, *J. Soil Sci.*, 1955, **6**, 293.

the reason for the relatively simple and definite chemical composition M. Schnitzer and J. G. Desjardins[1] have shown that they possess.

The process of lessivage: Brown earths, brown forest soils

Podsols are typically confined to acid quartz sands under mor, yet the same vegetation growing on other soil types will not give a podsol even though it may give a mor humus. Probably the most important reason why well-developed podsols are usually confined to sands is that the ferrous polyphenols are absorbed on clay particles, or can only leach any distance down profiles low in clay. On the other hand, in so far as the free polyphenols increase the tendency of the clay particles to disperse or deflocculate in the soil solution, they will assist in their transfer from the A to the B horizon. Polyphenols are not the only means of moving clay particles down the profile, and in fact it is not proven that they are important in soils containing much clay. A more important process is probably the slaking or dispersion of clay particles in percolating water when the soil is first wet after a dry period, for the surfaces of most dry cracks or channels tend to lose some particles or fine crumbs in this process. The dispersed clay particles move down the wider cracks and channels with the soil water and a proportion at least are adsorbed on their walls, that is, on the walls of the structural units and peds, in so far as these are present. Thus, a textural B horizon is developed, with a higher clay content than the A and the clay particles deposited in the B typically form oriented films. This mechanism need not filter all the clay particles out of the percolating water, but as the channels become narrower due to the clay films, they will become increasingly efficient in filtering them out. The process of the leaching down of clay from the surface soil to give oriented clay films,[2] or argillans,[3] is known by the French word *lessivage*, and soils showing this characteristic *lessivé* soils. This process is usually more active in regions with a continental than with a maritime climate, due to the stronger drying out of the soil in summertime, and is more active in acid than neutral or calcareous soils, for clay particles disperse less readily the higher the calcium ion concentration (see p. 146). However, lessivage can take place in calcareous soils, for the clay covering limestone fragments in the B horizon of such soils is sometimes derived from the A horizon by translocation.[4]

Typical brown earth formation may involve three separate processes; translocation of clay; the weathering of minerals with the setting free of iron and aluminium hydroxides which become adsorbed on clay particles and, if formed in the A horizon will be translocated into the B with the clay

1 See, for example, *Proc. Soil Sci. Soc. Amer.*, 1962, **26**, 362.
2 See, for example, S. B. McCaleb, *Soil Sci.*, 1954, **77**, 319. S. W. Buol and F. D. Hole, *Proc. Soil Sci. Soc. Amer.*, 1961, **25**, 377.
3 R. Brewer, *J. Soil Sci.*, 1960, **11**, 280.
4 For an example see E. C. A. Runge, *Proc. Soil Sci. Soc. Amer.*, 1970, **34**, 534.

particles; and clay formation in the B, a process discussed further on p. 734. A characteristic of many of these brown earths is that the B horizon has a stable crumbly structure and a friable consistency, probably due to the surface coatings of these hydroxides on the clay surfaces. When translocation of iron occurs in these profiles, the B horizon may have a darker colour than the A, and the profile can exhibit some of the characteristics of a podsol; in consequence these soils have been called podsolised, to distinguish them from the true podsol.

These processes become of decreasing importance the more strongly the soil and vegetation can support an active population of the larger earthworms. As already noted, there is little direct evidence that mor humus has any specific effects on podsolisation or lessivage, so earthworms influence the process not because they produce a mull humus as such, but because they cause active mixing of the top 10 to 15 cm of the soil. Their casts also tend to have a more stable structure that the initial soil, so that less clay will disperse or deflocculate in the percolating water. Thus in forest and grassland soils under earthworm mull, the movement of clay particles and sesquioxides out of the mull layer is slow, as earthworm activity tends to counteract their washing down, though this activity does not affect the loss of the soluble products of weathering. Any clay formed from the products of weathering below this layer will only form oriented clay skins, or argillans, if it can be leached from its site of formation and be deposited lower down the profile.

The lower boundary of the mull horizon is usually not well defined, and the darker colour typical of the top 7 to 10 cm lightens gradually. However, if the soil contains some stones, particularly in the smaller gravel sizes, there is often a fairly well defined 'stone line' at about 15 to 20 cm, which is also about the bottom of the A horizon. This represents the bottom of the soil layer in which the earthworms are active, though not the bottom of their burrows. This is commonly seen more clearly in old meadow and pasture soils than forest, possibly because of the greater mixing of forest soils due to wind-throw[1] and to the decay of the large tree roots.

Soils with moder humus come intermediate between those under mor and with mull, for they carry a population of the larger members of the soil fauna which live partly in the soil and partly on it. Further, the soils are typically more fertile than those under mor so the polyphenol content of their leaves is lower, which reduces the intensity of the podsolisation and lessivage processes.

The grassland soils: The prairie soils and chernozems

Grasses differ from forests in that they translocate much of the organic matter they synthesise into the soil as root system. Grasses can add over 2·5 t/ha dry matter to the soil per year as roots, and natural grassland soils

1 For examples see E. P. Stephens, *Proc. Soil Sci. Soc. Amer.*, 1956, **20**, 113.

may contain over 12 t/ha of roots below ground compared with only 2 to 5 t/ha of above-ground material. Thus, J. E. Weaver, V. H. Hougen and M. D. Weldon[1] give the following weights of organic matter and roots in a Nebraska prairie, in tons per hectacre:

Depth in cm	0–15	15–30	30–60	60–90	90–120
Weight of roots	6·5	1·80	1·57	0·75	0·10
Weight of organic matter	77·4*	65·0	65·0	20·2	10·0

* This figure includes the rhizomes of grasses which occurred in the top 15 cm.

This difference between grassland and forest is also illustrated in Plate 40, which shows the profiles, with the distribution of organic carbon down the profile, of a prairie and a deciduous forest soil both on the same parent material, and both from Iowa. Although the prairie soil only contains 50 per cent more organic matter than the forest, yet it is distributed through a considerably greater depth of soil.

The grass roots themselves are extensive and fine, hence, as shown in Table 22.4, on p. 540, they may have a very great length in the soil and many of the finer roots will decompose quickly. These soils also typically have a large population of earthworms, and also of rodents, which may make extensive burrows and nests in the soil well below the soil surface. Hence, little litter accumulates on the surface, most of it being worked into the top layers of the soil, as in the brown forest soils.

Grassland vegetation is found in conditions that are unsuited to forests, though many grassland areas of the world may be a consequence of human activity. They are found above the tree line in many mountainous areas, and very extensively in continental areas where summer drought is sufficiently severe for all the available water in the root zone of the soil to be exhausted before the drought breaks. Grasses can withstand these conditions better than deciduous forests in areas of summer drought and winter frost.

Grasslands occur on many types of soil, but there is one group developed on medium-textured, often calcareous, superficial deposits of Pleistocene age widely distributed throughout the continental areas of North America, central and eastern Europe and central Asia, which have been much studied. The soils under grass in these areas have been divided into prairie, chernozem and chestnut soils.

The typical prairie soils are formed where the summer rainfall is only just too little for deciduous forest, but where there is sufficient rainfall during the rest of the year for considerable leaching to take place. The chernozem soils are developed under rather more arid conditions where little water

1 *Bot. Gaz.*, 1935, **96**, 389. J. E. Weaver and E. Zink (*Ecol.*, 1946, **27**, 115) give further figures for the annual rate of production of roots by three prairie grasses at different depths in the soil.

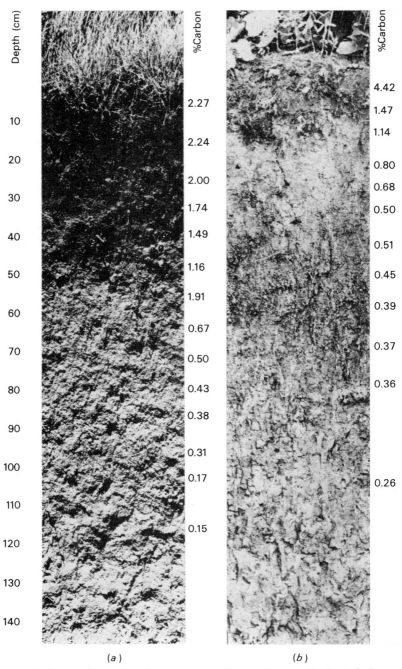

PLATE 40. The profiles and carbon contents of two soils on a silty loam in Iowa:
(a) Prairie (grassland) Soil; Tama Series
(b) Hardwood Forest Soil (Grey-brown podsolic); Fayette Series
Weight of organic carbon in top 1·2 m of soil: (a) 170 t/ha, (b) 105 t/ha.

leaches out of the root zone, and the chestnut under still more arid conditions when the rain water rarely if ever leaches through the profile. These grassland soils have a very characteristic soil structure. The crumbs are small and friable, sometimes breaking down almost to a dust in the surface and become larger, angular and stronger in the deeper layers of the profile, and have considerable water stability. Plate 31 gives a photograph of the characteristic association of soil crumbs and grass roots in the surface soil of a typical chernozem.

The prairie soils are similar in many respects to the brown earths.[1] The surface layers are often somewhat acid, due to leaching, there is usually a washing down of organic matter and clay, and clay formation takes place in the profile. There is a greater annual circulation, and hence a greater movement, of silica in prairie than in forest soils, for grasses have a much higher silicon content than tree leaves.[2] As in the brown forest soils, a pronounced clay pan may be formed both by clay being actively formed in this layer and also by clay formed higher up the profile being washed down into it. Otherwise weathering is not very severe. Thus, J. F. Haseman and C. E. Marshall[3] showed that in the Grundy silt loam, a typical prairie soil developed on loess in Missouri, the sand grains larger than 0·046 mm have barely been weathered at all, the silt particles between 0·025 and 0·002 mm have been washed down from the top 30 cm into the 60 to 120 cm zone, the easily weatherable minerals in the top 60 cm have been attacked, and clay accumulation has taken place in the layer 50 to 150 cm below the surface. These soils are well developed in the corn belt of the Middle Western states of America. They are fertile, or can be made so by adding fertilisers, and the rainfall is adequate for excellent crops of maize and soyabeans if the land is properly farmed. It is to this soil type that the Western European farmer tries to convert his soil by using leys and farmyard manure.

The chernozems, or black earths, to give the English translation of the Russian word, are so called because of their black colour. This black layer may exceed 1 m in depth, though the colour lightens with depth. The climate is moist enough to leach some calcium from the surface soil, so the surface may be slightly acid. Some of the calcium is precipitated lower down as calcium carbonate. If the climate is just sufficiently arid for it to form, the zone of calcium carbonate accumulation begins at about 2 m, and it is present as thin threads or films, which have been called pseudomycelia, because of their resemblance to fungal mycelia; and as the climate becomes more arid, the depth at which these begin to form becomes less, and more of the calcium carbonate is deposited as soft white concretions. Below the layer of calcium

1 For a detailed description of these soils in central USA, see G. D. Smith, W. H. Allaway and F. F. Rieken, *Adv. Agron.*, 1950, **2**, 157.
2 For an example on the accumulation of opal in these soils, see J. E. Witty and E. G. Knox, *Proc. Soil Sci. Soc. Amer.*, 1964, **28**, 685.
3 *Missouri Agric. Exp. Sta., Res. Bull.* 387, 1945.

carbonate accumulation in the more arid chernozems comes a zone in which calcium sulphate is precipitated as gypsum.

The chernozems are well developed on the calcareous loess that extends from central Europe through Russia into central Asia, forming the great belt of the Eurasian steppe. They form the short grass prairies of North America and are also well developed in parts of South America. Their main agricultural use is as the wheat soils of the world. The climate is too dry for crops other than wheat and some sorghums to be grown profitably, and their management will be discussed on p. 786.

The chestnut soils are developed under still more arid conditions than the black earths. Their name describes their colour, the colour of the skin of the edible chestnut, *Castanea sativa*, not the horse chestnut, *Aesculus hippocastanum*. They are formed under conditions where the potential transpiration exceeds the rainfall, so at no time in the year does water leach out of the profile. The effect of the increasing lack of water is to give a grass vegetation with a shorter root system and a smaller annual production of organic matter,[1] so that both the depth to which the organic matter reaches, and its content at each depth, decreases. At the same time the layers in which the calcium carbonate and gypsum are deposited come closer to the surface.

The effect of increasing temperature on these grassland soils is threefold: their organic matter content decreases, their colour reddens,[2] and the calcium carbonate deposits in the subsoil harden; though these last two characteristics are also associated with increasing age of the superficial deposits in which they are formed.

The effect of anaerobic conditions in the profile

Soils with impeded drainage or with a seasonal water-table in the rooting zone of the vegetation will have zones in which reducing conditions arise due to microbial decomposition of organic matter under conditions of oxygen deficiency, whenever the temperature is high enough. The effects of anaerobic conditions on plant growth, and to a large extent on soils are discussed in Chapter 25, so only their effect on the soil profile will be summarised here. Anaerobic conditions involve the reduction of accessible ferric iron to ferrous, and insoluble manganese oxides to manganous ions, and these reduced ions are much more mobile in the soil than the oxidised.

Under temperate conditions soil horizons having a low permeability are likely to be waterlogged, or nearly waterlogged during wet periods, but to be aerated during dry warm periods when the vegetation is transpiring actively. Such horizons have a typical mottled appearance, in which one of the colours of the mottle is an orange brown and the other a grey to blue-grey, and the relative volumes of soil having these colours depends on the relative

1 H. L. Shantz, *Ann. Assoc. Amer. Geog.*, 1923, **8**, 81.
2 J. Thorp and M. Baldwin, *Ann. Assoc. Amer. Geog.*, 1940, **30**, 163.

severity and length of time the reducing conditions last, the amount of drainage that takes place and the content of accessible iron. The bright colour is sometimes due to lepidocrocite,[1] or if dehydrated to maghemite, the γ-forms of FeO(OH) and Fe_2O_3, but it is also sometimes due to goethite; and the blue-grey colour can be due to a number of ferrous compounds such as vivianite (ferrous phosphate), ferrous sulphide or ferroso-ferric hydroxide, Lepidocrocite is certainly sometimes the characteristic form of ferric iron when oxidised from ferrous iron complexes in the presence of organic matter,[2] though the exact conditions under which it, rather than goethite, is formed are not known. The grey layer present in permanently wet soils was originally called gley, but this word is also used for the blue-grey layer, and the mottled layer is often called a gleyed horizon.

Reducing conditions in the gleyed horizon probably start along old root channels in the autumn when the soil becomes waterlogged but still warm, and some of the ferrous iron produced on the walls of the channel will diffuse into the soil ped forming the wall. In the following summer this will be oxidised to ferric hydroxide giving ochreous mottlings inside the ped. Some of the ferrous iron produced during the reducing conditions will be leached out of the profile if there is slow drainage, and it is not uncommon for this slow drainage to be laterally down the slope, so that if the soil is old, it may have lost all its iron and become grey in colour. This process will therefore move ferrous iron into new soil horizons, and when oxidising conditions arise, it will be deposited as a ferric hydroxide. Thus reducing conditions with slow water movement results in some soil volumes losing iron and others gaining iron.

Leaf drip, as already noted, often contains organic compounds capable of reducing ferric to ferrous iron and of forming soluble complexes with the ferrous ions so produced. These complexes move with the soil water, and if the soil is poorly aerated, the organic moiety will be decomposed by the soil bacteria setting free the ferrous iron. This should be found on the surfaces of cracks or channels in the soil, for these are the spaces through which the water moves. There can therefore be three distinct sources of iron in a gleyed horizon: it may have been eluviated from above as a ferrous-organo complex, it may have been brought in by ground water either from below or laterally, or it may have originated in the horizon by weathering of iron minerals and local translocation under reducing conditions.

When the aeration of the surface soil becomes poor for an appreciable proportion of the year, peat will begin to build up; and the longer the periods of poor aeration, the thicker the peat is likely to be. The soil under the peat loses its iron, probably because soluble decomposable organic matter slowly seeps from the peat layer into the mineral soil, reducing any ferric iron and forming soluble ferrous complexes. This ferrous iron will slowly seep out of

1 G. Brown, *J. Soil Sci.*, 1953, **4**, 220.
2 S. M. Bromfield, *J. Soil Sci.*, 1954, **5**, 129.

the soil with the water. In one group of soils, peat formation is due to the surface or subsurface soil becoming impermeable, or very slowly permeable without the permeability of the deeper subsoil being affected, so that while the surface soil is strongly gleyed, there is little or no gleying in the subsoil. The division between the gleyed and ungleyed horizons is often marked by a thin iron pan or a horizon containing ochreous deposits, which may harden to nodules. These soils are often peaty gleyed podsols when they occur in cool humid conditions. Iron pans can also form in valley bottoms where the groundwater level fluctuates, but their morphology depends on local circumstances, such as the texture and permeability of the soil. These pans overlie permanently wet soil, in contrast to the previous type which overlies a soil that appears to be oxidised, and the iron can be encrusted around old root channels and as a coating on the walls of cracks or structured units.

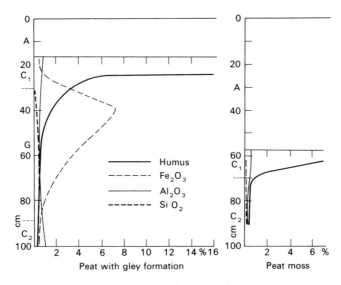

FIG. 26.2. Humus content and amounts of silica and sesquioxides extracted by the acid-oxalate method from two peat profiles. The A layer in each profile represents the peat layer

When the aeration of the surface soil begins to become deficient, peat starts to build up, and the longer the period of the year that these conditions of poor aeration last, the thicker the peat deposits are likely to be. Thin peat often overlies a gley horizon with a pronounced zone of iron accumulation, which may have been formed before peat formation started, or the conditions giving thin peat may also allow appreciable fluctuations in the water-table during the year and the water entering the area carry soluble ferrous compounds. Once the peat becomes deep, and aerated conditions never occur in the mineral soil, this is not seen. Figure 26.2 illustrates this comparison for some Swedish peats. The iron horizon under the thin peat may become de-hydrated giving hardened hydrated ferric oxides either as a fairly continuous and impervious layer or as concretionary nodules. Under peat this is known as bog iron ore, but this horizon is not confined to soils under thin peat. It can also be seen in valley bottoms where the groundwater fluctuates, for a thin layer of hydrated ferric oxides sometimes forms at the top of this fluctuating zone, again probably when the groundwater contains soluble ferrous compounds derived from higher up the catchment.

Leached tropical soils

There are no specific tropical soil weathering processes, for the same pro-cesses are at work in the tropics as in the temperate regions, though there is one important soil-forming process which operates through a much greater depth of soil in the tropics and subtropics than in temperate regions. Tropical soils may contain mound-building termites, and these bring up large quanti-ties of subsoil, mainly from the depths not exceeding 1 m, though some species will bring up a certain proportion from considerably greater depths, which they use to build their mounds above the soil surface. Thus, whereas earthworms in temperate soils only bring about soil mixing in the top 15 to 20 cm of the soil, termite activity extends to four to five times this depth. The intensity of weathering can, however, be much greater under tropical conditions, because the soil temperatures are higher than in high latitudes and they show much less seasonal variation. The intensity of weathering is therefore dependent on the length of time the particular volume of weathering material is moist; and the fate of the products of weathering depends on the intensity of leaching. Thus, on old land surfaces there may be a very deep layer almost devoid of weatherable minerals provided the water-table is deep, and consisting of kaolinite, goethite or haematite, with perhaps some gibb-site. Under suitable conditions of high rainfall, podsols sometimes form on coarse sandy soils under some forests, though the A horizon may be over 1 m thick and be strongly depleted of iron, while the B may be 0·5 m thick and have a high concentration of iron oxides in its upper part.[1]

An important difference between soils in temperate regions and in many tropical regions is that the soils on the tropical land surfaces were not dis-turbed by ice sheets or periglacial action during the last Ice Age, and some of these land surfaces are geologically very ancient. It is possible that some of the existing surfaces in Tanzania are of Jurassic age,[2] but many are certainly Pliocene and some probably Miocene.[3] The result is that weathering

1 For examples from Colombia see H. Jenny, *Soil Sci.*, 1948, **66**, 5, and I. Barshad and L. A. Rojas-Cruz, ibid., 1950, **70**, 221.
2 A. M. M. Spurr, *2nd Int. Afr. Soils Congr.*, 1954, 175.
3 See, for example, R. V. Ruhe, *I.N.E.A.C. Ser. Sci.* 66, 1956, and C. G. Stephens, *Geoderma*, 1971, **5**, 5.

may have been going on for many million years, under a variety of climates, so that not only may the weathered layers be very deep, and there are examples of it being 100 m deep,[1] but the relative land levels may have been subjected to very great changes due partly to tectonic activity and partly to changes in the base level to which the rivers are grading. Thus, the weathered material lying on a hilltop at the present time may have been formed in a valley bottom; and land that was a plain in which deep weathering was taking place may now have a variety of levels due to rifting, faulting and mountain-building on the one hand and river terrace formation on the other.

Consequently one can only interpret the factors forming the existing soil and crust of weathering correctly if the geomorphological history of the area is known; and this, for the time being, must often be based on hypotheses the validity of which can only be approximately tested. Thus, it is easier to study soil-forming processes in the tropics on soils derived from recent volcanic tuffs and lavas, which can often be dated approximately, than on soils derived from the basement rocks. Even here interpretation may involve hypotheses, for the climate, and in particular, the rainfall has almost certainly varied appreciably over the last few millenia. At Muguga in Kenya, for example, with a mean annual rainfall a little under 1000 mm, only in an exceptionally wet season does any water move beyond the root range of vegetation, yet the parent rock is young and the weathering very deep. But if most seasons became as wet as the wettest 10 per cent, considerable amounts of water would percolate well beyond the root range of the vegetation in most seasons, so one need not postulate great changes in climate to explain the deep soil in this semi-arid region. Undisturbed soils on old land surfaces differ from young soils in that typically they have a deep, porous, well-developed oxic horizon, that is, one that has no clearly defined smooth-faced peds and therefore no clay skins, but which is well aggregated with a stable microstructure that does not disperse easily in sodium hydroxide; also the soil has a ratio of silt to clay that is usually below 0·15, and the proportion of weatherable minerals is below 3 per cent in the fine sand fraction.[2] They also tend to have a higher pH in a KCl solution than in water,[3] due to the positive charges on the clay surfaces exceeding the negative at the soil's field pH.

The soil profiles on many tropical land surfaces can only be interpreted if one has a picture of the geomorphological processes that have formed them, and the most important process to be considered is geological erosion which causes the widening of valleys. This is now commonly assumed to take place in semi-arid areas by scarp retreat, that is, the valley wall does not rise uniformly from the river to the interfluve, but the valley floor widens by the valley wall retreating but remaining relatively steep. If the land surface above

1 For an example from Nigeria see M. F. Thomas, *Trans. Inst. Brit. Geogr.*, 1966, **40**, 176.
2 See, for example, A. van Wambeke, *Soil Sci.*, 1967, **104**, 309.
3 J. Bennema, *Soil Sci.*, 1963, **95**, 250.

the valley is capped with a cemented horizon resistant to weathering, the broken cemented material at the edge of the valley wall will slowly work down the slope. If the material is primarily a cemented ironstone, it will be slowly decomposed, so that at the bottom of the slope it may have disappeared,[1] but if it is a quartzite it will remain as a fairly definite stony horizon, known as a stone line, and the soil which covers it may be formed on colluvial material derived from the interfluve rather than from material formed from weathering at its present site; and it is on this colluvial material that the current vegetation grows and the termites may be active. It is often possible to prove that the stone line came from an old land surface if that surface contained an indurated and hardened laterite layer, for the stones will be recognisable bits of the old laterite; but it is rarely possible to prove if the soil above the stone line came from colluvial material or not, because if the stones were originally left on the surface of the valley soil, they would soon be buried under that soil due to termite activity.[2]

Young soils can usually be recognised because they still contain some minerals from the parent rock, so that some of their properties are determined by the properties of the rock from which they have been formed. Thus, the more quartz in the rock, the more sandy the soil is likely to be, because quartz is one of the more resistant minerals. When these soils possess a stone line, it is usually 30 to 60 cm below the surface and it follows the topography fairly closely. It can usually be seen to derive from quartz veins in the underlying rock.[3] If the rock was low in quartz, the soil is likely to be high in clay. If the rock was also high in iron or aluminium, as it is likely to be, the clay will be very well aggregated, friable when dry and not sticky when wet. It is sandy soils derived from fine-grained granites and quartzose basement rocks that are liable to erosion, whereas those derived from basalts and other high-iron lavas are erosion resistant. The rate of weathering of fresh rock under high rainfall conditions in the low altitude tropics is very high. N. Leneuf and G. Aubert[4] estimated that all the silicates in a one metre depth of a granite were decomposed in a period of between 20 and 70000 years in an area of the Ivory Coast, under a rainfall of 1·8 to 2·0 m per year and a mean soil temperature of 26°C.

Young soils on level sites do not necessarily weather to kaolinites. Thus basalts on level sites in semi-arid regions may weather to cracking montmorillonite clays, typically with a horizon of calcium carbonate accumulation. The reason for this is that there is not enough rain to wash out all the silica and cations produced by the weathering of the basic rock, and under these conditions the clay that forms is montmorillonite. J. A. Kittrick[5] has

1 For an example from S. W. Nigeria see R. P. Moss, *J. Soil Sci.*, 1965, **16**, 192.
2 For examples see P. Segalen, *Cah. Pedol. ORSTOM*, 1969, **7**, 113. J. Riquier, ibid., 71. C. D. Ollier, *J. Soil Sci.*, 1959, **10**, 137. R. V. Ruhe, *Soil Sci.*, 1959, **87**, 223.
3 For an example see P. H. Nye, *J. Soil Sci.*, 1955, **6**, 51.
4 *Trans. 7th Int. Congr. Soil Sci.*, 1960, **4**, 225.
5 *Proc. Soil Sci. Soc. Amer.*, 1971, **35**, 450.

given the solubility diagram for the four important components of the products of weathering—gibbsite, kaolinite, montmorillonite and silicic acid —which is reproduced in Fig. 26.3. It shows that when the concentration of silicic acid is very low, gibbsite is the stable compound of aluminium, at moderate levels kaolinite is, at high levels montmorillonite. It also shows that montmorillonite and gibbsite cannot coexist in equilibrium in a soil, nor amorphous silica and kaolinite or amorphous silica and gibbsite.

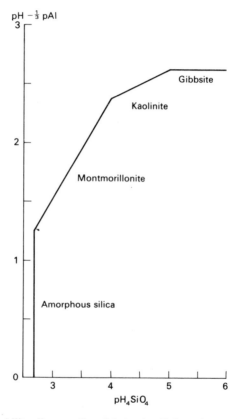

FIG. 26.3. The solubility diagram for gibbsite, kaolinite, a bentonite and amorphous silica. The bentonite (from Belle Fourche) was in equilibrium with a magnesium ion concentration (pMg^{2+} = 3·7) corresponding to the natural water at pH 6·0 and with haematite.

The clays in well-drained tropical soils are predominantly kaolinitic and the soil red, unless the iron content of the rock or soil is low, in which case the colour varies from almost white or grey in the virtual absence of iron, through orange or pink to bright brick red at moderate levels, to dark red at high levels. This is in contrast to the predominantly brown colours of temperate soils, and is due to the greater degree of dehydration of the iron oxides in the tropics. However, there is nearly always some smectite, illite or vermiculite-type clay present, for the exchange capacity of the clay, which is typically over 10 m eq/100 g clay, is usually too high to be due to kaolinite alone.[1] These soils have been given a variety of names depending on their degree of weathering. The more highly weathered friable soils have been called red earths, latosols or oxisols; and the less weathered with a recognisable argillic horizon, i.e., one that contains illuvial clay, red loams or ultisols.

Many tropical soils, particularly in areas with a well-pronounced dry season, have a pH change with depth the opposite of that found in temperate soils. The natural vegetation in these areas is very deep rooting (see p. 533), so the plants take up cations such as calcium, magnesium and potassium from very considerable depths and deposit them on the soil surface either in their litter or as leaf-wash. Leaching is often inadequate to wash these into the subsoil as fast as the roots take these ions up from there, with the consequence that the pH typically falls with depth in the profile,[2] down to 2 to 3 m; whereas in temperate regions the rate of leaching of these cations into the subsoil is faster than the rate of uptake, so the pH rises with depth.

Types of laterite

Some soils on old land surfaces are associated with laterite. This word needs careful definition because different authors have used it for a wide variety of materials.[3] The word was coined by a geologist, F. Buchanan, in 1807, for a particular type of iron-rich material which was quarried and cut up in the shape of bricks; these were then exposed to the weather for a time, which caused them to harden, and then used for buildings. The term he used for this material was laterite from the Latin *later* (a brick).[4]

Buchanan's site, which is in the Kerala district of India, has been visited by other geologists and pedologists, and one of the most accessible descriptions is due to C. G. Stephens.[5] The typical profile has a variable depth of soil, but 50 to 70 cm is common, usually it is dark red with a loam to clay texture, which overlies an indurated horizon, which is again of variable depth but is often between 2 and 10 m, and grades into an earth which is red at the top but which has an increasing proportion of mottles and then flecks of white to grey with increasing depth, until it becomes only a white or grey clay, the so-called pallid zone.[6] The red earth which lies under the indurated

1 See, for example, A. van Wambeke, *Soil Sci.*, 1967, **104**, 309.
2 This was first noted by H. C. Doyne and W. A. Watson, *J. agric. Sci.*, 1933, **23**, 308; 1935, **25**, 192.
3 For a monograph on laterites see R. Maignien, *Review of Research on Laterites*, UNESCO Nat. Res. Ser. IV, 1966.
4 *A Journey from Madras through the countries of Mysore, Canara, and Malabar*, East India Co., 1807.
5 *J. Soil Sci.*, 1961, **12**, 214; *Geoderma*, 1971, **5**, 5.
6 For a description of such soils in W. Australia, see M. J. Mulcahy, *J. Soil Sci.*, 1960, **11**, 206.

horizon is the material that is quarried in the way described by Buchanan, and the word laterite should strictly be applied to this material; but it has been more widely used for the overlying indurated horizon. The lower part of the pallid zone is typically a saprolite which usually grades into the unweathered rock through a layer of rotting rock, though it may have a sharp boundary with the rock.

Laterite, if defined to include both the indurated and the underlying red earth, has a range of morphological structures, varying from vermicular, or worm-like structures to vesicular or slag-like. R. A. Pullan[1] has given a detailed description of these types as they occur in northern Nigeria, and has shown how they grade into each other. The vermicular is usually the softest of the various types of laterite, though it needs an axe to quarry it, and typically it has sinuous worm-like tunnels, 5 to 10 mm in diameter, filled with loose pale-coloured earth which can easily be washed out. The walls of the channels are lined with a harder smooth clay skin, that is probably not a true argillan or well-oriented film, and the colour of the matrix is darkest nearest the channel wall and may contain flecks of yellow or grey colours. This vermicular material is often capped by a harder layer of one of the other two main types of laterite. Vesicular laterite has small bladder-like cavities, up to 5 mm in size, and usually filled with loose pale-coloured earth; though there may be some rather larger cavities containing small ironstone nodules.[2] It has a hard red surface layer, and also contains light-coloured earthy material. The colour of the matrix is redder than for the vermicular, and the layer appears to have a higher iron content. Plate 41a reproduces a photograph of this material from Samaru, N. Nigeria. The laterite itself is harder than the vermicular in its fresh state. The cellular or slag-like laterite has a deeper red to an almost black colour, which is more uniform. The matrix contains pores 1 to 2 mm in diameter that may be interconnected, and contain vesicles and vermicules that have been filled up by iron hydroxides cementing the earthy material originally present. Plate 42 reproduces a photograph of this material also from Samaru. It is probable that these three types represent the same basic material with increasing proportions of iron oxides.

The micro-structure of laterites has been described by L. T. Alexander and J. G. Cady.[3] They noted that if a vermicular or vesicular laterite was to harden the matrix had to contain more than 9 to 11 per cent of iron (Fe) as amorphous ferric hydroxide. The laterite layer is usually below the depth of rooting of the native vegetation and so will always have been permanently moist. It hardens when cut up into bricks and exposed to the weather, because the dryings and wettings it is subjected to cause the amorphous iron hydroxide to crystallise out as an interlocking system of goethite crystals. The significance of the requirement that there must be more than 9 to 11 per

1 *Nigerian J. Sci.*, 1967, **1**, 161.
2 See, for example R. L. Pendleton and S. Sharasuvana, *Soil Sci.*, 1942, **54**, 1; 1946, **62**, 423.
3 *U.S. Dept. Agric. Tech. Bull.* 1282, 1962; and see *Adv. Agron.*, 1962, **14**, 1.

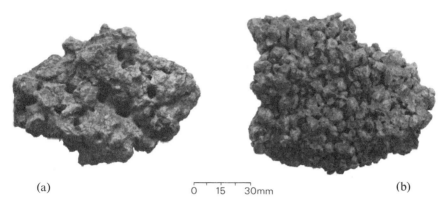

(a)

0 15 30mm

(b)

PLATE 41. (*a*) A lump of vesicular laterite.
(*b*) A lump of re-cemented nodular laterite.
Both from northern Nigeria.

cent of iron present as amorphous ferric hydroxide is that this is the amount needed to cover all the mineral surfaces capable of adsorbing a ferric hydroxide film.[1] It is also possible that only for horizons where the aeration is poor will amorphous ferric hydroxide gels remain stable, for A. K. M. Habibullah and D. J. Greenland[2] found this to hold for poorly drained soils in East Pakistan, while the iron formed small granules in the better aerated soils.

Laterites vary considerably in the additional irreversible hardening they undergo on drying, and there may be laterite layers near to the surface that undergo only a moderate degree of irreversible hardening on drying. Such layers are found under forest in parts of West Africa, and are soft and permeable under forest, but if the forest is cleared and the land goes under cultivation, they will form a hard surface crust, probably after the topsoil has been lost by erosion, making the land quite unusable for agriculture. Fortunately these ironstone or laterite crusts soften again if the land is replanted to forests, though it may take several decades for the whole crust to be softened.[3]

The origin of the iron that has accumulated in laterites is not certain. It could all be residual, that is, what is left after weathering has removed most of the silica and sometimes alumina, from the original rock, or it could have

1 J. J. Fripiat and M. C. Gastuche, *I.N.E.A.C. Ser. Sci. Publ.* 54, 1952; J. D'Hoore, ibid., *Publ.* 62, 1954.
2 *J. Soil Sci.*, 1971, **22**, 179.
3 See, for example, R. D. Rosevear, *Farm Forest*, 1942, **5**, 1, and S. Sivarajasingham, L. T. Alexander *et al.*, *Adv. Agron.*, 1962, **14**, 55.

0   15   30mm

PLATE 42. A lump of cellular laterite from northern Nigeria.

been brought into this horizon by ground water, either from below or, more likely, from higher up the catchment. If it is residual, this would imply the iron has moved over only short distances to give the amorphous ferric hydroxide gel, which crystallises to give the hard laterite; which, in turn, implies that all the iron that has been lost from the pallid zone beneath the laterite has been removed from the profile altogether. This does not mean that the pallid zone has lost all of its iron, for it does not necessarily consist of fully weathered material. P. H. Nye,[1] for example, has shown that the lower part of the pallid zone over a biotite granitic schist near Ibadan, Nigeria, contains some unweathered or partially weathered ferromagnesian minerals, which become scarcer with height above its base. No detailed examination has been made of the characteristic conditions under which vesicular or cellular laterites are formed, and whether the additional iron they contain is due to the parent material being higher in iron or whether they received more iron from above or laterally than the vermicular.

Soils in these regions often contain an ironstone known as pisolithic, nodular, or pea-iron gravel. This usually forms a continuous sheet, which may be on, but is usually below the soil surface, normally with a clearly defined top, but often grading into one of the previously described types of laterite below. The top consists of ironstone gravel or crudely rounded

1 *J. Soil Sci.*, 1955, **6**, 51.

nodules, about 3 to 10 mm thick, which may show a certain amount of cementation between them. Typically, they are dense, very hard, with a dark almost purple smooth surface, and may contain over 35 per cent of iron as oxide or hydrated oxide, including a certain proportion of maghemite. They are composed of soil particles heavily impregnated with iron, without any definite structure, though the outside may show evidence of concentric growth.[1] Plate 41b gives a photograph of a lump of cemented pea-iron gravel also from Samaru. This layer may overlie a laterite layer, or be present on an interfluve if it is an old land surface, but it is usually best seen on gently toe slopes of valley walls when the ground water either seeps out in a spring line, or is drawn on by the vegetation in the dry season. It is usually not seen on the valley floor itself.

The detailed processes reponsible for the formation of nodular ironstone have not been worked out in detail. In some landscapes it is found in all positions from the interfluve to near the valley bottom, being scarce but distributed through a fairly deep horizon near the top of the valley and becoming more frequent, and possibly more concentrated, into a definite horizon as one goes down the slope. It is possible that a proportion of the nodules are simply rounded remnants of old vesicular and cellular laterite, and if some of the nodules are broken, they give this appearance. Iron appears to be fairly mobile within the landscape unit in the tropics. Thus S. A. Radwanski and C. D. Ollier[2] have noted that, in some Uganda soils on steep slopes and almost certain to be subject to erosion, this nodular layer is never found on the surface, and they consider that as the surface soil is removed there is a re-solution of the nodules in the upper part of the soil and its redeposition lower down in the profile. One gets the impression in many tropical catchments, particularly those capped with ironstone, that iron is continuously being transported in the ground water from the upper areas and being deposited in the lower.

Pan formation in soils

A pan is the name given to a soil layer that is appreciably more compact and less permeable than the layers above or below it. It can be produced in agricultural practice by unsuitable methods of cultivation which compact the subsoil when it is moist or wet, or it can be produced pedologically by the entrapment of downward moving particles, such as silt or clay, or by substances dissolved in the water being absorbed as a film on the surface of particles in the subsoil. These two processes are probably aided by the soil being regularly dried to perhaps wilting point by vegetation during the summer.

Clay pans probably develop on level plateau soils that have been subjected

1 See, for example, P. H. Nye, *J. Soil Sci.*, 1955, **6**, 51 (W. Nigeria); G. D. Sherman and Y. Kanehiro, *Soil Sci.*, 1954, **77**, 1 (Hawaii).
2 *J. Soil Sci.*, 1959, **10**, 149.

to weathering for a sufficiently long time. Some of the clay in the pan has probably been washed in from above, for the texture of the soil above is always coarser than that of the pan, and some has probably been synthesised *in situ*, either as a result of local weathering or of a chemical combination between local products and soluble products, such as silicic acid derived from weathering in the A horizon. It is not easy to decide what proportion has come from these two sources, or, indeed, if these are the only two sources of the clay. When clay is washed down from an upper horizon, it is often deposited as an oriented clay skin on the walls of the cracks or channels down which it was being leached, but W. D. Nettleton[1] showed that clay skins are only stable if the peds show relatively little change of volume with moisture content; so that once the clay content in the pan exceeds a certain level, these skins will disappear.

C. E. Marshall and his coworkers[2] studied these pans in some Missouri prairie soils. They found, for example, that in one soil profile a column of soil 1 cm^2 in cross-section and 1·6 m deep contained 68 g of clay, and they estimate that before weathering it would have contained only 57 g, so that 11 g of clay have been formed during the weathering process. In addition, they estimate that 11 g of clay were washed out of the top 45 cm of soil and 22 g were gained in the next 65 cm. There is no means of estimating how much of this extra 22 g is represented by the 11 g lost by the upper layer, how much represents clay formed in the upper layer and translocated into the clay pan, and how much was formed in the clay pan itself. Translocation was probably responsible for most of the 11 g lost by the upper layer because it also lost 11 g of silt, all of which appeared in the clay pan. But they further showed that the clay in the clay pan does not merely fill the interstices between the silt and sand particles, it increases the average distance between neighbouring sand and silt particles, and they found that the soil volume containing a sand particle increased by 50 per cent in the centre of the pan compared with that in the undisturbed soil. In another two profiles, they fractionated the clay in the various horizons, and showed that the principal gain in clay in the pan was in particles smaller than 0·5 μ and were principally composed of a mineral they called beidellite, but which is now considered to be an inter-layer mixture of hydrous mica and montmorillonite. There was some trans-location but little formation of coarse clay particles.

Pans can also be produced in arid regions on valley floors where the ground water periodically rises sufficiently close to the soil surface for appre-ciable amounts of water to be lost by evaporation, for the dissolved salts it contains will be precipitated on or near the surface depending on their solubility. In so far as the salts in the water are soluble, such as chlorides, no pan is formed; but in so far as the water contains calcium bicarbonate or

1 With K. W. Flach and B. R. Brasher, *Proc. Soil. Sci. Soc. Amer.*, 1969, **33**, 121.
2 With E. P. Whiteside, *Missouri Agric. Expt. Sta. Res. Bull.* 386, 1944; with J. F. Haseman, ibid., *Res. Bull.* 387, 1945.

silica, pans of calcium carbonate or silica may be formed. Since these are entirely dependent on strong evaporating conditions for their formation, they are known as evaporite pans.

Extreme examples of calcium carbonate pans, in the form of thick crusts of soil impregnated with calcium carbonate, are found in desert regions, usually when the parent material is a limestone, though this is not always so. L. H. Gile[1] studied these in New Mexico, and described the stages in their formation. In a non-gravelly soil, the first stage is the precipitation of thin filaments or pseudo-mycelia in the soil, or thin coatings on sand grains; and A. Ruellan,[2] who studied their formation in Morocco, considers that this precipitation begins near plant roots. As the amount of precipitated calcium carbonate increases, nodules begin to form which harden with age, carbonate begins to fill the pore spaces, and the upper part of this zone of accumulation —the K horizon—becomes almost filled with carbonate. Finally, further calcium carbonate begins to form an indurated laminar horizon directly on top of the relatively thick plugged layer. This horizon may contain up to 90 per cent of precipitated carbonate. If magnesium carbonates are present in the parent rock, mixed calcium-magnesium carbonates will be precipitated rather than calcium carbonate. Well-developed carbonate pans may be confined to land surfaces older than late Pleistocene. These pans and crusts are sometimes known by the Spanish name, caliche, or their Indian name, kankar, and are petrocalcic horizons in the US system of classification.

Under some conditions thick pans of amorphous silica can develop. Most appear to be fossilised and are found, for example, in the desert areas of central Australia.[3] They are known as silcrete and were first recognised near the Victoria Falls in Zambia by G. W. Lamplugh[4] in 1902, and now frequently cap hill tops giving a mesa-like landscape similar to laterite capped hills. On weathering, they break up into quartzite gravels which have a brownish lustre, giving extremely poverty stricken soils, for some of these are now in regions of fairly high rainfall.

Effect of topography on soil formation: The soil association or catena

Soil formation is very dependent on the way water moves in the soil and on the fate of the compounds dissolved in it or particles dispersed in it. This has the consequence that the soil-forming processes operative at any particular site depend on the position of that site in the morphological landscape. Thus, if in an area of fairly uniform parent material there is a fairly regularly

1 With F. F. Peterson and R. B. Grossman, *Soil Sci.*, 1965, **79**, 74; 1966, **101**, 347.
2 *Cah. ORSTOM Ser Pedol*, 1967, **5**, 421. See also J. H. Bretz and L. Horberg, *J. Geol.*, 1949, **57**, 491 for examples from New Mexico.
3 For a discussion of these see C. G. Stephens, *Geoderma*, 1971, **5**, 5.
4 *Geol. Mag.*, 1902, **9**, 575, and see *Quart. J. Geol. Soc. London*, 1907, **63**, 162.

recurring pattern of landscape morphology, there is likely to be a fairly regularly recurring pattern of soils associated with it. This recurring pattern
was called a catena by G. Milne,[1] though some authors refer to it as an
association. It was introduced as a mapping unit for small-scale soil maps, but
a surveyor can only make full use of it if he can recognise the basic land forms
present in the landscape. The subject of land form analysis and classification
is still a subject of active research. The catenary concept has been generalised
by the introduction of the soil sequence. Thus a toposequence is the sequence
of soils formed on different land form elements in a region of fairly constant
parent material, where the different elements are all of about the same age;
a lithosequence is the soil sequence on different parent materials where the
topography and age of the land form elements are about the same; and a
chronosequence is where age of the elements differ. Actual soil sequences
in the field are nearly always compound, such as a topo-lithosequence where
each land form may have a different parent material, or even a topo-litho-
chronosequence where each land form element may have both a different
parent material and be of a different age.

The basic process producing the catena, or toposequence complex of soils,
is the movement of water from the interfluve and the valley sides into the
ground water or into the river draining the catchment. Rain falling on the
interfluve tends to drain vertically downwards, although if the interfluve is
flat and on an old land surface its drainage may be impeded, in which case
some of the water may seep out laterally. Rain falling on the valley walls
will have a lateral component of flow, unless the profile to the ground water
is very permeable, and this lateral component is likely to cause slow soil
creep downslope, and may also carry some soil particles downhill considerable distances. Thus, the soil profile on the mid- to lower slopes under
conditions of strong or long-continued weathering can often be seen to consist of two parts—an upper part in which the soil mineral soil particles have
been brought in by creep from higher up the slope, and a lower part formed
from the products of weathering of the parent rock *in situ*; though there
need be no sharp boundary separating them. Lower down the slope, as the
valley floor is approached, the land surface may gain considerable quantities
of mineral soil material from higher up the valley wall, and it becomes the
toe slope with a more gentle gradient than the valley wall above it. The
ground water also is likely to come close to the surface, with the consequent
development of a gleyed horizon, first weakly developed at depth, but becoming more strongly developed and closer to the surface towards the valley
floor. The ground water may give rise to a spring line along the lower slopes
of the valley, or may be sufficiently high in the valley bottom for parts of the
year to give temporary waterlogging or peat formation on the valley floor.

The percolating water will remove all soluble products of weathering from

1 *Soil Res.*, 1935, **4**, 183. *Trans. 3rd Int. Congr. Soil Sci.*, 1935, **1**, 345.

the soil profiles on the interfluve and upper slopes, but the laterally moving water will carry some of these into and through the profiles on the mid- and lower slopes. Thus, the profiles in the upper part of the area are losing soluble substances and the soils will tend to become more acid; but lower down the profiles are partly illuvial, so will usually have a higher pH and show less clearly those characteristics produced by eluviation. Thus, in the particular conditions prevailing in some hill lands of northern and western Britain, the soils on the interfluves are well-developed podsols, but on the valley sides are brown earths, although the vegetation and lithology of the soils are the same.[1] If the water moving down slope through the soil is low in oxygen, it may pick up some ferrous iron, which will be deposited as ferric hydroxide if it seeps out of the soil as a spring line near the bottom of the slope, or if it comes sufficiently close to the surface to become oxidising. Thus in the tropics ironstone pans are often found in this position, giving thin poorly drained soils above them.[2]

The soils on the valley floor typically have an appreciably higher pH than those on the upper slopes because they receive the soluble products of weathering from the catchment. Under tropical conditions when there is a pronounced dry season this leads to clay synthesis from these products, perhaps with products of weathering formed in the valley bottom, and the clay is typically a black montmorillonite if the ground water or the soil material contains much ferrous iron and magnesium, giving the black cracking vertisols which are so typical of such regions.[3]

Saline soils

Under hot, arid conditions soluble salts accumulate in the surface of soils whenever the ground water comes within a few feet of the surface, as may happen, under natural conditions, in the flood plains of rivers, the low-lying shores of lakes, and in depressions in which drainage water accumulates—in fact, in any region where marsh, swamp or other ill-drained soil would be found in humid regions. The amount of salts that accumulate depends on the salt content of the ground water and the length of time salts have been entering the region. During dry periods the surface of these soils is covered with an efflorescence, or salt crust, which is dissolved in the soil water each time the soil is wetted.

Saline soils typically have an uneven surface, being covered with small puffed up spots a few inches high that are enriched in salts, for, as explained

1 For British examples see E. Crompton, *J. Soil Sci.*, 1952, **3**, 277.
2 P. H. Nye, *J. Soil Sci.*, 1954, **5**, 7. C. D. Ollier, ibid., 1959, **10**, 137 and J. Makin, *East Afr. agr. for. J.*, 1969, **34**, 485.
3 For descriptions of some African catenas see G. Milne, *J. Ecol.*, 1947, **35**, 192 (Tanzania); P. H. Nye, *J. Soil Sci.*, 1955, **6**, 51 (Nigeria); J. P. Watson, ibid., 1964, **15**, 238, and 1965, **16**, 158 (Rhodesia).

on p. 446 and as illustrated in Plate 43, salts concentrate in the most salty areas because these areas remain moist longest after the onset of drought.

Saline soils normally show no change of structure down the profile, implying that the soil is barely affected by soil weathering and soil-forming processes, except that they may show signs of gleying in the subsoil. Such soils are often called solonchaks, a word used by the Russians for saline soils on recently deposited calcareous river terraces; E. W. Hilgard[1] called them white alkali, but they are now usually simply called saline soils. Usually they are low in humus, because the natural vegetation cannot make much annual growth on them. The salts usually present in the soil are the sulphates and chlorides of sodium and calcium, though nitrates occur in a few places, and magnesium sometimes constitutes an appreciable proportion of the cations. Under these conditions, the pH of the soil is below 8·5 and the colour of the soil surface is light. However, under some conditions an appreciable proportion of the salts present may be sodium carbonate which will raise the pH of the soil to 9 or even up to 10. If other salts are only present in small concentrations, this sodium carbonate may cause the humic matter in the soil to disperse and take on a black colour, giving the black alkali soil of Hilgard, which will be discussed in the next section.

Saline soils may contain over 250 t/ha of salt in the top 120 cm of soil, that is, the salts may constitute over 1 per cent by weight of the soil, though many saline soils contain less than this. The natural vegetation on such soils has a very high ash content, up to one-quarter of the air-dry plant may be ash, and the greater proportion of the ash may be soluble salts, typically sodium chloride. Hence, the vegetation will also bring salts to the soil surface, but its effect is probably small, amounting to under 200 kg/ha annually,[2] owing to the small amount of total growth made per year.

The source of the salts in natural saline soils is usually the ground water, which is enriched with salts from two sources. Part, sometimes all, is derived from the weathering of rocks in the upper reaches of the river, and part is sometimes derived from salt deposits laid down in early geological periods in strata through which the ground water moves. But soils in continental areas can also receive soluble salts in the rain, when it falls, and much of these salts may come from the salt crusts formed elsewhere, which are picked up by the wind after they become dry and loose on the soil surface.[3] Saline soils have also been produced artificially by faulty irrigation, for irrigation always involves putting salts on the land as well as water. Hence, salt control is a fundamental part of irrigation and will be discussed in detail in the next chapter.

1 *Soils*, New York, 1906.
2 V. A. Kovda, *Pedology* 1944, Nos. 4–5, 144.
3 For an example from Australia see J. T. Hutton, *Trans. 9th Int. Soc. Soil Sci.*, 1968, **4**, 313.

(a)

(b)

PLATE 43. (*a*) and (*b*) Salt soils in the Karun delta, Iran. Note the way the salt efflorescences occur in separate patches.

Alkali soils: The solonetz and solod

When the water-table in a natural saline soil falls, so that salts no longer accumulate in its surface, the rainwater washes the salts that were there down the profile, and this process sometimes causes considerable chemical changes to take place in the profile. If the salts are predominantly calcium, or if during the process of washing out over 90 per cent of the exchangeable ions remain calcium, due to the soil being calcareous, then the saline soil is converted into the steppe or semi-desert soil appropriate to its region.

Much more radical changes in the surface soil take place if the calcium reserves in the soil are so low that, during the washing out of the salts, an appreciable proportion of the exchangeable calcium ions are replaced by sodium. Sodium ions only need to constitute 12 to 15 per cent of the exchangeable ions to reduce the water stability of the soil structure sufficiently for the clay and humic particles to disperse. This harmful effect is accentuated by sodium carbonate being formed in the soil solution during the final stages of the washing out of the salts, causing the pH of the soil solution to rise, often above 9, and consequently increasing the ease of dispersion of the fine particles.

For a long time the source of the sodium carbonate produced during this washing out was not understood, although P. de Mondésir[1] in 1888, and K. K. Gedroiz[2] in 1912, gave essentially the correct explanation. Carbonate and bicarbonate anions are being continually produced in the soil by the carbon dioxide given off by the plant roots and soil organisms, and these anions must be neutralised by cations or hydrogen ions. These cations will be obtained from the exchangeable cations in the exchange complex unless there are reserves of calcium carbonate in the soil. Hence, if there is an appreciable proportion of exchangeable sodium ions in the soil, enough will come into the soil solution to give what is in effect a solution of sodium carbonate strong enough to raise the pH of the soil to 9 or over. There is also a second process which produces sodium carbonate in soils, for if the soil contains sulphates, is low in accessible ferric iron, and is subjected to reducing conditions, the sulphate will be reduced to sulphide, which in the absence of ferrous ions or ions of other metals having insoluble sulphides, will be lost as hydrogen sulphide; and electrical neutrality of the solution will be maintained by the dissociation of carbonic acid to give carbonate anions (see p. 763). Reducing conditions will commonly occur in these soils as soon as their permeability drops due to the salt content dropping.

These conditions of high alkalinity and low salt content lead to the clay and organic matter particles becoming deflocculated and the soil structure water-unstable. The soil surface becomes dark-coloured, often black, due to the dispersed humic particles; the surface typically dries into large, very

1 *Comp. Rend.*, 1888, **106**, 459.
2 *J. Expt. Agron.* (*Russian*), 1912, **13**, 363.

hard, prismatic units having well-defined edges and smooth surfaces; and clay particles tend to wash down the profile, giving an incipient clay pan. Soils in this condition were called black alkali by Hilgard—he used this name without regard to the amount of other salts present—and they are extremely difficult soils to handle, for they are very plastic and sticky when wet and form hard compact clods when dry.

The second stage in the washing out of salts, when there is an appreciable proportion of exchangeable sodium in the exchange complex, is for clay and organic matter to move down the profile into the developing clay pan, with the consequence that the profile becomes banded rather like a podsol. The details of the type of soil developed as leaching proceeds depends on local circumstances, particularly on the soil texture and type of clay present. On the Russian steppes, where these were first studied, the surface soil is dark grey, owing to the deflocculated humus, then comes a pale layer, and then another dark, very compact layer having a very sharply defined upper surface and merging gradually into the subsoil with increasing depth. The darker colour of the compact layer compared with the layer above it may be due to its higher clay content, for it does not always have a higher content of organic matter.[1] The top two layers have lost much of their clay, and have a loose, porous, laminar structure, whose upper surfaces may be paler than their lower, possibly because of silica being deposited on them. The clay pan cracks on drying into well-defined vertical columns having a rounded top and smooth, shiny, well-defined sides as shown in Plate 25 on p. 484. These can be broken into units about 10 cm high and 5 cm across with a flat base. Below this the columns break into rather smaller units with a flat top and bottom which on light crushing break up into angular fragments.

As the leaching of these desalinised soils proceeds, the upper two horizons deepen, and often become slightly acid in reaction. Gedroiz, who was one of the first to give a plausible account of the chemical consequences of the process, noticed that the content of amorphous silica in these horizons, and particularly in the top darker horizon, increased. The further development of these soils has not been worked out in detail, though some of B. B. Polynov's work suggested that the very characteristic clay pan becomes less pronounced, possibly because of sandy material from the A horizon washing down in the cracks between the structural units.

These soils are called solonetz in the early stages and solod in the later stages of their development by the Russians, and they have been extensively studied on the river terraces of southern Russia and Central Asia.[2] Gedroiz assumed that they formed under the influence of exchangeable sodium, as

1 A. L. Brown and A. C. Caldwell, *Soil Sci.*, 1947, **63**, 183, give an illustration of this for well-leached Minnesota soils.
2 For some detailed Russian work, see K. K. Gedroiz, *Nosovka Agric. Expt. Sta., Bull.* 38, 1925; 44, 1926; 46, 1928; D. G. Vilensky, *Salinised Soils*, Moscow, 1924; V. A. Kovda, *Solonchaks and Solonetz*, Moscow, 1937. J. S. Joffe, in his book *Pedology*, New Brunswick, 1938, has given a long summary of Gedroiz's and Vilensky's work.

given in the account above, and this assumption has been accepted by many other workers. But there are large areas in western Canada and the United States[1] and in Australia[2] where soils having the morphology of these solonetz and solods are found, yet where sodium forms a minor proportion of the exchangeable ions. It is possible that originally they contained enough exchangeable sodium for the solonetz-solod morphology to develop in the profile, but that most of this sodium has been lost by leaching. However, most of these soils that are now low in exchangeable sodium contain over 40 per cent, or even over 50 per cent, of exchangeable magnesium. It is, therefore, possible that if during the leaching out of soluble salts the exchange complex acquires a high proportion of exchangeable magnesium, the soils will partially deflocculate and undergo the same type of profile development as soils with a high proportion of exchangeable sodium. Hence, much more critical work must be done before one can decide just what changes can only be brought about by exchangeable sodium, and what by either exchangeable magnesium or other causes. The problem is complicated because many claypan and planosol soils have profiles that are similar to some solonetz-solod profiles, and it is not certain what features are found only in soils which had earlier been through a stage of deflocculation due to the washing out of sodium ions.

Under natural conditions, areas of saline and alkali soils usually contain a mixture of solonchak and solonetz, or solonetz and solod, or even all three together, depending on the relief of the area. The low-lying spots may be solonchak, those on moderate relief solonetz, and on higher ground solod; and differences in level of as little as 30 to 60 cm can have a great influence on the stage in the solonchak-solonetz-solod sequence that the soil has reached. Areas containing much solonetz soil may be very misleading to the inexperienced, for in some stages of development the dry soil is black and powdery and looks as if it ought to be fertile. It is only when wet, and particularly after heavy rain or irrigation, when water is held on the soil surface because of the impermeable B horizon, that one realises how intractable they are.

Land use capability classification

One important task the soil surveyor is often given is the production of a land-class or land use capability map of a region. This may be wanted for land-use planning or for agricultural development planning. The suitability of the land for improved or productive agriculture is not the only factor that must be borne in mind when development or improvement schemes or schemes

---

1 See, for example, W. P. Kelley, *Amer. Soil Survey Assoc.*, 1934, **15**, 45 (California); C. E. Kellogg, *Soil Sci.*, 1934, **38**, 483 (N. Dakota); C. O. Rost and K. A. Maehl, ibid., 1943, **55**, 301 (Minnesota); J. M. MacGregor and F. A. Wyatt, ibid., 1945, **59**, 419 (Alberta); C. F. Bentley and C. O. Rost, *Sci. Agric.*, 1947, **27**, 293 (Saskatchewan).
2 *Manual of Australian Soils*, Ed. H. C. T. Stace, G. D. Hubble *et al.*, Rellim, 1968.

involving changes in land use are being considered, but it is an essential factor.

The fundamental principle underlying land use classification is the recognition of the factors that are preventing crops giving high yields or are complicating the problems, or increasing the costs, of crop production compared with productive easy-working soils. Thus, the basic problem of most classifications is to define a class of land units from which a good farmer can expect to get high yields of a wide range of crops regularly, without there being any particularly difficult management problems to be solved. Other soils are then down-graded to a greater or less extent depending on the severity of the limitations on cropping, yield, or management imposed by the local conditions. The usual methods of land classification have been based on the system developed in the United States, in which land is divided into seven classes of decreasing agricultural value. A typical definition of these classes is that at present in use by the Soil Survey of England and Wales and of Scotland, and this is the one that will be summarised here.[1]

Class 1 land has the minimum of physical limitations for agricultural use. The soils should be deep, well-drained, with a texture between sandy to silty loam, though some peats and fen soils are also included, with a good available water-holding capacity; and if their nutrient status is low, crops should give good responses to added fertilisers. The site should be level or only gently sloping, and the climate must be favourable for crop production. These soils allow a wide range of crops to be grown, and the crops should yield well and reliably from year to year.

Class 2 land is not quite so favourable for crop production. The soil may be shallower, the texture may be rather coarser or rather finer, the drainage may be somewhat impeded, the structure rather unstable, the slope may be rather too steep for the easy handling of farm machinery or be sufficiently steep that definite soil erosion control measures must be taken, or the climate may impose some limitation on the cropping. In general, only one or two of these unfavourable factors can be present, otherwise the site becomes still less favourable for productive economic agriculture. In general good yields of a wide range of crops can still be grown, though on some sites harvesting of root crops may become difficult in unfavourable seasons.

Class 3 land has one or more of the above limitations sufficiently strongly developed to limit still further the choice of crop, or the flexibility of the management system. In Great Britain this will mean that grass is likely to become an important crop, and cereal and forage crops will be preferred to root crops that must be harvested in late autumn. But good yields of suitable crops should be obtained in many years if the management is good.

Class 4 land has sufficiently severe limitations that the choice of crops becomes very restricted and management difficult, so yields are likely to be

1 This is described by J. S. Bibby and D. Mackney, *Soil Survey Tech. Monog.* 1, 1969.

very variable. Grass tends to become an important crop if the region has a well-distributed rainfall, so that severe summer droughts are not normal hazards.

Class 5 land has such severe limitations that it can only be used for pasture, forestry or recreation. High rainfall, steep slopes and poor drainage that is difficult or expensive to rectify are typical causes of restricted use and moderate to low yields.

Class 6 land can be used for rough grazing and forestry, but it has such severe limitations that this very unproductive land cannot be improved. Limitations can include a cold wet sunless climate due to altitude, thin peat or humose soils over rock, very poor drainage, boulders or large stones strewn on the surface, or regular liability to severe floods.

Class 7 land can really only be used for recreation. It includes bare rocks, screes, sand-dunes, or land covered with boulders.

It is important in land use capability classification to indicate the factors that limit the use of the land. These factors are grouped into five subclasses, namely, limitations due to wetness (w), soil (s), land gradient (g) or soil pattern limitations, liability to erosion (e) and climatic limitations (c); and the severity of the limitations imposed for each factor are defined to suit local conditions. Wetness can be broken down into subgroups such as the presence of slowly permeable layers, impermeable layers due to pans or indurated layers, ground water, spring lines and flooding. Soil limitations include shallowness, stoniness, unfavourable texture or unstable structure. Gradient can be broken down to gently sloping, moderate, strong, steep, or very steep slope, where the exact definition of each class may depend on local circumstances. Climatic limitations include such factors as liability to frost or drought, length of the growing season, and rainfall amount and distribution. Thus, regions where prolonged rain can usually be expected at the time cereal crops come to harvest are unsuited to these crops, and high altitude hill lands may have such a short growing season due to cool wet sunless weather persisting from early autumn to late spring, often with winter snow, that they can at the best only give low yields of grass.

Any system of land use classification must have an arbitrary element built into it, as there can be no strictly objective quantitative criteria which can be used for deciding into which land class any particular piece of land should be placed. Thus, the limitation due to the slope of the land will depend on the permeability of the soil and the susceptibility to erosion in the particular climate and system of farming being followed, as well as on the uniformity of the slope; for land with uneven slopes is likely to give greater variability in crop yield than with even slopes, provided the average slopes are not too steep. Again, the degree of stoniness that could complicate the mechanical harvesting of potatoes would have little significance for grass-cereal rotations.

Some of these land classes are illustrated in Plates 44–46. Plate 44a illustrates class 1 land—level, well drained, of medium texture and well suited to

(a)

(b)

PLATE 44. Examples of high quality arable land:
(*a*) Class 1 land;
(*b*) Class 2 land, downgraded from class 1 because the soil is shallow. In the background are moderately steep slopes (Class 4g) which restrict cultivations.

(a)

(b)

PLATE 45. Examples of Class 3 quality land.

(a) Coarse sandy soil in an area where summer droughts are common and where stones hinder cultivation.

(b) Improved and unimproved wet fine-textured soils. These are placed in Class 3ws irrespective of differences in present land use.

PLATE 46. Inherently low quality land. Fine-textured soils with poor drainage (Class 4w) receive run-off and seepage water from the adjacent steep slopes (Class 6g). The gently sloping upland surface has shallow stony soils (Class 5c).
West edge of the Longmynd, Salop.

a wide range of arable crops. Plate 44b illustrates otherwise good level well-drained arable land but downgraded because it has a rather shallow soil over chalk, which limits its range of cropping. The slopes in the background are graded 4g because they are too steep for cultivation. Plate 45a illustrates level well-drained arable land downgraded because the soil is a gravelly coarse sand. It is thus very liable to drought during the summer and the stones seriously affect cultivations and cropping. Plate 45b illustrates pasture land on a poorly drained fine textured soil, unsuitable for arable, so graded 3ws, and the photo shows that the grading is the same for the enclosed im-proved pasture as for the unenclosed rough grazing; so it does not depend on the level of farming being practised. Plate 46 is a photograph of an upland area in the English Midlands, where the bottom land is graded 4w because although it is only gently undulating it consists of poorly drained fine-textured soils due to seepage and run-off water from the escarpment slope. The escarpment itself is too steep to justify improvement, and is graded 6g, and the gently-sloping upland surface has shallow stony soils, but it is sufficiently above the level of the lowland to have a cooler, wetter and more sunless climate and hence a shorter growing season, so is graded 5c.

There is a further limitation which concerns the amount of reclamation operations that are necessary to bring the land up to its maximum potential. A low nutrient status or soil acidity are not normally considered to affect the land class, as these can be rectified by fertilisers and liming. A high water-table does not lower the land class if this can be rectified by a field drainage system, but would do so if an arterial scheme is necessary; though if such a scheme is put into operation, then the land class would be raised. This is particularly relevant for valley or low-lying neutral peat soils, which can become very productive if drained. Similarly, land may be downgraded due to the climate giving a serious drought hazard at some critical periods in the growing season, but if supplementary irrigation becomes economically feasible during these periods, the land class would have to be upgraded.

27

The management of irrigated saline and alkali soils

The effect of soluble salts on plant growth

Soluble salts can have two types of effect on the growing plant: specific effects due to particular ions they contain being harmful to the crop, and a general effect due to the raising of the osmotic pressure of the solution around the roots of the crop.

Specific effects fall into two classes: those operative at low and those at high concentrations. Of the former, only two salts are normally of importance—sodium carbonate and soluble borates. The former may be harmful in itself, but its harmful effect is more likely to be due to the consequences of the high pH it brings about. Thus, many nutrients, such as phosphates, iron, zinc and manganese, become unavailable to the plant at these high pHs on the one hand, and the soil structure tends to become water-unstable on the other, thus bringing about conditions of low water permeability, poor aeration and an unkind, almost unworkable tilth.

Irrigation water containing more than 0·3 ppm boron must be treated with caution as a concentration as high as this will affect the yield of many crops; and water containing 2 to 4 ppm restricts cropping to boron-tolerant crops such as some varieties of sugar-beet, lucerne, some brassica crops and the date palm.[1] Some ions have a toxic effect at high concentrations which enhances the harmful effect of the osmotic pressure they cause. Thus many fruit trees are sensitive to high concentrations of chloride or sodium, and flax and some grasses are sensitive to high concentrations of sulphates. Again, at equal fairly high osmotic pressures, magnesium ions are more toxic than calcium, and calcium may be more toxic than sodium, though this is usually in crops that can take up calcium easily but exclude sodium. High concentrations of sodium chloride may also be harmful due to their effect on nutrient uptake. Thus, they may interfere with the uptake of potassium, and some varieties of barley, when young, need appreciably higher concentrations

---

1 For a list of boron-tolerant and boron-sensitive crops see L. V. Wilcox and W. H. Durum in *Irrigation of Arid Lands*, Amer. Soc. Agron. Monogr. 11, 112, 1967.

of potassium in the soil solution if their yield is not to be affected.[1] A number of crops show considerable genetic variability in their reaction to high salt concentrations, so it is sometimes possible to breed new varieties that are more tolerant to them.

The general effect of a high salt content in the soil is to give a dwarfed, stunted plant, but this is often not apparent in the field if there are no patches of low-salt soil to act as controls; and yields can be reduced by over 20 per cent without salt damage being apparent to the farmer. As the salt content becomes higher the stunting becomes more noticeable, the leaves of the crop become dull-coloured and often bluish-green, and they become coated with a waxy deposit. Further, many crops growing in very saline soils do not display the symptoms of wilting very clearly, so considerable loss of yield can occur if irrigation is applied only when the plants are obviously wilted.

The effect of increasing the salt concentration or osmotic pressure in the soil solution on the transpiration, nutrient uptake, and photosynthetic and respiration rates has not been fully worked out, since these are probably dependent, at least in part, on their effect on the production of growth-control substances in the plant. As the osmotic pressure in the soil solution increases, so does the osmotic pressure in the cell sap. The difference between these may remain the same, with the cell sap having a pressure of about 10 bar higher, as R. O. Slatyer[2] found for tomatoes, and L. Bernstein[3] for cotton, or the osmotic pressure of the cell sap may increase more rapidly than the osmotic pressure of the soil solution, as J. S. Boyer[4] found for the variety of cotton he was using. As the osmotic pressure in the external solution increases, the transpiration rate and stomatal resistance in the leaves may remain constant, as Boyer found for cotton up to external pressures of 10 to 12 bar, or it may decrease, as found by F. M. Eaton;[5] but it usually reduces the growth rate of the plant and its rate of photosynthesis, though it sometimes reduces and sometimes increases the dark respiration rate.[6]

C. H. Wadleigh and A. D. Ayers[7] showed that, for some crops at least, crop growth was controlled by the free energy of the soil water, and the decrease in growth rate due to this lowering of free energy was the same whether it was due to a lowering of the matric or the osmotic potential. Figure 27.1 illustrates their results with beans (*Phaseolus vulgaris*). The plants were grown in a loam soil to which no sodium chloride, or 0·1, 0·2

1 H. Greenway, *Aust. J. biol. Sci.*, 1963, **16**, 616. He used water cultures.
2 *Aust. J. biol. Sci.*, 1961, **14**, 519.
3 *Amer. J. Bot.*, 1961, **48**, 909. H. E. Hayward and C. H. Wadleigh (*Adv. Agron.*, 1949, **1**, 1) found this for lucerne also.
4 *Pl. Physiol.*, 1965, **40**, 229.
5 *J. agric. Res.*, 1942, **64**, 357.
6 For reviews see L. Bernstein, *UNESCO Arid Zone Res.*, **18**, 139, 1962.
7 *Pl. Physiol.*, 1945, **20**, 106. For the corresponding result with guayule rubber see *U.S. Dept. Agric., Tech. Bull.* 925, 1946.

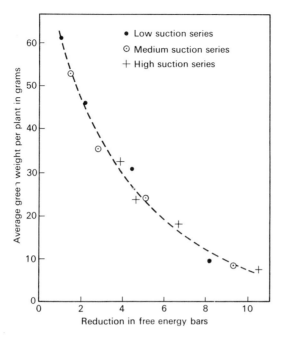

FIG. 27.1. The relation between the green weight of beans and the mean free energy of the water supply

or 0·4 per cent, was added to the soil, and the soil was watered when one-half two-thirds or nine-tenths of the available water, as measured in the salt-free soil has been used. In the field the effect of a moderate concentration of salts on a crop growth depends on the shape of the moisture content–suction curve for the soil. Typically, for light- to medium-textured soils, much of the available water is held at relatively low suctions; so if the soil is non-saline, the crop can use most of the available water before its matric suction has risen above 2 to 3 bar. But if the soil is saline, as the crop uses water the osmotic pressure increases roughly proportionately to the amount of water used, so the effective free energy of the water drops more rapidly in the presence than in the absence of soluble salts. However, in the field, salts are usually unevenly distributed in the soil, so that those crop roots growing in volumes of soil containing less salt than the average will take up relatively more water than those roots growing in volumes containing more than the average.[1]

Plants differ in their ability to withstand the harmful effects of salinity in the field. In the first place, as L. J. Briggs and H. L. Shantz[2] showed in 1912,

1 For an example see W. R. Gardner, *Proc. Symp. FAO/IAEA Istanbul*, 1967, 335.
2 *Bot. Gaz.*, 1912, **53**, 229.

plants have different abilities to extract water from soils in the wilting range, and plants adapted to saline soils tend to have a greater ability to extract water at the drier end of this range. However, salt tolerance and drought tolerance are not necessarily related; the coconut, for example, is salt tolerant but drought sensitive. The greater the salinity of the soil, the less water a crop can remove before it begins to suffer from water shortage, so that irrigated soils with an appreciable salt content need more frequent irrigations than non-saline soils. Crop yields can, in fact, easily be reduced unnecessarily by allowing too long an interval between irrigations; and this is liable to happen because, as already mentioned, the crop may not show signs of wilting as clearly as if it was growing in a low-salt soil. There is also some field evidence that loss of yield due to moderate salinity can be more serious on soils of low than of high fertility, and that a moderate level of salinity sometimes increases the response of a crop to fertilisers, particularly to phosphate and perhaps also to nitrogen.[1] Phosphate fertilisers have the advantage that they do not increase the osmotic pressure of the soil solution as they are usually strongly absorbed by the soil.

Salt tolerance is a complex concept for other reasons. The tolerance of a plant may be low when it is young but high when established—lucerne is an example of this. The plant may be able to keep alive at high salt contents, but will make very little growth under these conditions and only grow slowly under moderate salt contents, and hence be of little commercial value—some of the alkali grasses are examples. Again, though the plant may grow in fairly saline soils, the quality of the part harvested may be affected. Thus, cereals will grow and produce green matter in soils too saline for them to produce any grain; sugar-beet growing in saline soils produces a root low in sugar which is difficult to refine; and forage crops may contain so much salts that they are unpalatable or injurious to livestock. Again, in practice, salt tolerance is often bound up with tolerance to alkali, high pH and low calcium on the one hand, and ability to withstand prolonged waterlogging during irrigation, which is a common consequence of alkali, on the other.

In spite of these limitations, crops can be graded into three categories of salt tolerance when grown under irrigation: tolerant, moderately tolerant and sensitive to salts. Table 27.1 gives such a grading for crops grown under irrigation in the west of the USA.[2] Dates, cotton, sugar-beet and barley fall into the tolerant class, as do bermuda grass (*Cynodon dactylon*) in the warmer and tall wheatgrass (*Agropyron elongatum*) in the cooler regions; while peas, beans (both *Vicia faba* and many *Phaseolus* species) and most clovers are in the sensitive group.

Some general experience is available on the amount of salt a soil can hold before the crops on it begin to suffer appreciably. O. C. Magistad and

---

1 For an example from Israel see S. Ravikovitch and A. Porath, *Pl. Soil*, 1967, **26**, 49.
2 *Diagnosis and Improvement of Saline and Alkaline Soils*, U.S. Dept. Agric. Handb. 60, 1954. See also *U.S. Dept. Agric. Yearbook, 1943–47*, and *Circ. 707*, 1944 for other similar lists.

TABLE 27.1   The relative tolerance of crops to salts in the western states of America

Good tolerance	Moderate tolerance	Sensitive
Date palm	Pomegranate Fig, olive, grape	Pear, apple, orange, grape-fruit, almond, apricot, peach
Barley	Rye, wheat, oats	Field beans
Sugar- and fodder-beet	Rice, sorghum	Green beans
Rape, kale	Maize	
Cotton	Potatoes, peas	
Bermuda and Rhodes grass	Sweet clover	White, alsike and red clover
	Ryegrass	Ladino clover
Bird's-foot trefoil	Strawberry clover	
	Lucerne	
	Cocksfoot	

R. F. Reitemeier[1] showed that if the soil solution held at 15 bar suction has an osmotic pressure less than 2 bar, that is, if it contains less than 0·4 per cent of dissolved salts, no crops suffer from salt trouble; but if the osmotic pressure at this suction is 10 bar, most crops suffer severely; and provided the soil solution does not contain appreciable amounts of sodium carbonate or borate, the composition of the salts is relatively unimportant compared with their effect on the osmotic pressure of the solution. However, osmotic pressures are inconvenient to measure, so salt concentrations are usually measured by the specific electrical conductivity of their solutions, which is a property that is approximately related to their osmotic pressure. In practice the salinity of a soil is specified by the conductivity of its solution at a standard condition, which is usually taken when the soil is just saturated with water at its sticky point. If the specific conductivity of this saturation extract at 25°C is less than 4 millimhos per centimetre (4 mmhos/cm), which corresponds roughly to less than 3000 ppm salts in the solution, no crop is likely to suffer from salinity trouble; if the conductivity exceeds 8 mmhos/cm, corresponding to about 5000 ppm salts, only salt-tolerant crops are grown, and even their yields are likely to be reduced; while if it exceeds 15 mmhos/cm, or 10000 ppm, no agricultural crops are likely to give an economic yield.[2]

The management of irrigated soils

Quality of irrigation water

The water used for irrigation, whether taken from a river or from wells, is never pure water but always contains dissolved salts. Much of the water

1 *Soil Sci.*, 1943, **55**, 351.
2 *Diagnosis and Improvement of Saline and Alkali Soils*, U.S. Dept. Agric. Handb. 60, 1954.

applied to the land will be taken up by the crop and transpired, and if the water contains much dissolved salts, a proportion of the salts will be left behind in the soil. In regions where the rainfall is so light that rain water rarely if ever leaches through the soil profile beyond the root range of the crop—and it is in such regions that the major irrigation systems of the world are situated—continuous irrigation will lead to a build-up of salts in the root range of the crop, unless precautions are taken to leach them out of the profile at regular intervals.

The salt content of irrigation water is commonly specified in practice by its electrical conductivity, for this is a measurement that is easy to make. The electrical conductivity, in the range of concentrations usually met with in irrigation practice, is approximately proportional to the osmotic pressure of the solution and to the concentration of the solution expressed as milli-equivalents per litre. Table 27.2 gives the composition of some river waters in the western part of the United States that are used for irrigation;[1] and it shows that their electrical conductivity varies from 0·1 to 3·2 mmhos/cm,

TABLE 27.2   Chemical composition of some river waters used for irrigation in the western United States. Concentration of ions in m eq/litre

River	*Columbia*	*Rio Grande*		*Gila*	*Pecos*	*Humboldt*
Location	*Wenatchee Wash.*	*Otowi Br. N. Mex.*	*El Paso Texas*	*Florence Ariz.*	*Carlsbad N. Mex.*	*Rye Patch N. Mex.*
Electrical conductivity mmhos/cm at 25°C	0·15	0·34	1·16	1·72	3·21	1·17
Dissolved solids ppm	78	227	754	983	2380	658
Sum of cations	1·5	3·4	11·5	16·8	38·0	11·5
Ca	0·9	1·9	4·2	3·6	17·3	1·7
Mg	0·4	0·7	1·4	2·0	9·2	1·9
Na	0·2	0·8	6·0	11·3	11·5	7·9
HCO$_3$	1·2	1·8	3·6	3·7	3·2	5·2
SO$_4$	0·2	1·5	5·0	3·3	23·1	2·2
Cl	0·1	0·1	3·1	10·0	12·0	4·5
Soluble sodium percentage	13	24	52	67	30	68
Sodium adsorption ratios	0·2	0·7	3·6	6·7	3·2	5·9

corresponding to concentrations between 1·5 and 38 m eq/litre and that sodium constitutes between 13 and 67 per cent of the total cations. The U.S. Salinity Laboratory[2] grades the quality of irrigation water, based on its soluble salt content into four grades, those with an electrical conductivity

1 Taken from *The Diagnosis and Improvement of Saline and Alkali Soils*, U.S. Dept. Agric. Handb. 60, 1954.
2 The results of experimental work of this Laboratory up to 1954 are summarised in *The Diagnosis and Improvement of Saline and Alkali Soils*, U.S. Dept. Agric., Agric. Handb. 60, 1954. This is the standard American text on the subject. For a more recent summary of their results see L. E. Allison, *Adv. Agron.*, 1964, **16**, 139.

below 0·25 mmhos/cm which do not contain enough soluble salts to cause any trouble, those with conductivities between 0·25 and 0·75, between 0·75 and 2·25, and above 2·25, which are so saline that they can only be used under very limited conditions.

Management for salt control

Irrigation water always contains some dissolved salts, and in most irrigation schemes more salts are added to the land in the irrigation water than are taken up by the crop. In regions where there is sufficient rainfall at some period of the year for water to leach out of the soil profile, the salts left behind from the irrigation water will also be leached out. But in regions of light rainfall, these salts will accumulate in the soil. In typical warm arid regions, in which many of the large irrigation schemes in the world are situated, these accumulations can be very large. Thus if the irrigation water contains 500 ppm dissolved salts—a not uncommon figure for many rivers flowing in semi-arid areas—and 50 cm of irrigation water are used for a crop, and most crops will need at least this amount, the water will add 2·5 t/ha of salts during the growing season. Thus, irrigation schemes in semi-arid areas must be so managed that these salts can be removed at regular intervals from the root zone of the crop.

Crop yields suffer if the osmotic pressure of the soil solution rises too high, though the actual value it must attain to affect yield seriously depends on the crop. The U.S. Soil Salinity Laboratory have used the limits of electrical conductivity of 4 and 8 mmhos/cm to separate out the salt-sensitive crops for which the conductivity of the soil solution should remain below 4, the moderately tolerant crops for which it can rise to 8, and the very tolerant crops which will give a yield if it is somewhat above 8, though it must usually be below 16. If crop yields are not to suffer from salinity, the conductivity of the soil solution must be kept below the appropriate value for the crop being grown. Since the soil solution becomes more concentrated as the crop removes water from the soil, the more saline the water, the less water can the crop be allowed to remove before irrigation is reapplied; and sufficient must be reapplied to wash the concentrated soil solution below the root zone of the crop. Naturally this can only be done if the ground-water level is below the root zone and the soil profile is permeable.

The first condition for the success of any irrigation scheme is therefore that the ground-water level be kept deep enough, and in practice very great difficulties arise if it comes much closer than 1·8 m from the surface, and management problems are eased if it is below 3 m. If the profile is reasonably permeable, and the ground water naturally high, pumping of the ground water may have to be used; and if the quality of the river water is good, this pumped ground water can be put into the irrigation ditches to increase the amount of irrigation water available. Since in many schemes land is more plentiful than water, this pumping of ground water not only keeps existing land in cultivation but increases the area of land that can be cultivated. If the permeability of the land is not very good, it may be necessary to install a system of deep drainage ditches, sometimes with an associated tile drainage scheme.

The proportion of applied irrigation water that must be used for leaching out the residual salts from the profile is called the leaching requirement. On simple theory it is the ratio of the salt concentration in the irrigation water to the maximum permitted salt concentration in the soil solution, measured as electrical conductivity and expressed as a percentage. Thus, if two irrigation waters have electrical conductivities of 0·1 and 0·5 mmhos/cm respectively, and it is decided to keep the salt concentration in the soil solution below 4 mmhos/cm, the leaching requirements will be 2·5 and 12·5 per cent respectively, for a fortyfold and an eightfold increase of concentration will be allowed. In actual practice more water than this is commonly needed. In the first place, the leaching water does not move down uniformly through the profile, for the soil usually contains a number of cracks and channels down which some of the water will move rapidly, taking only a little of the salts within the soil crumbs or clods with it. In the second place, the simple picture assumes that the leaching water dissolves no salts from the soil and that no chemical reactions, such as precipitation of calcium carbonate for example, take place between the dissolved salts and the soil. This restriction is usually of little importance if the dissolved salts are predominantly chlorides with small amounts of sulphate, but can be very important if the water contains much sodium carbonate, as will be discussed on p. 763. It is naturally not necessary to leach out residual salts after each irrigation, and the term leaching requirement or leaching factor refers to the proportion of water that must be used for leaching over a period of time, which is shorter the greater the requirement. Some authors restrict the term 'leaching requirement' to the calculated ratio of the amounts of drainage to applied water and leaching factor to the ratio actually used in the field; but this distinction is not generally recognised. The most economic leaching factor to be used on any particular site depends on the amount of salts that can be tolerated, and the distribution of coarse and fine pores in the profile.[1]

Plants are usually most sensitive to salts at the seedling stage, so it is particularly important to keep the soil around the seeds low in salt. This is usually quite easy if the irrigation water is low in that respect, but can be difficult if it is rather saline because the soil surface is bare when the crop is sown, so evaporation of water will be taking place from the soil surface, which will cause salts to concentrate there. If the crop being grown is very high-priced, this can be prevented by covering the surface with polythene sheet; but in general this trouble can be reduced by encouraging the salts

1 For an example of mathematical computation of this factor see W. R. Gardner and R. H. Brooks, *Soil Sci.*, 1957, **83**, 295.

to concentrate in definite bands, and placing the seeds sufficiently far from these bands to allow the rootlets of the young plant to grow in a low salt solution. One method of doing this[1] is to lay the land up in ridges of a suitable shape and run the irrigation water along the furrows between the ridges. If the ridge is symmetrical with flat sides either coming to an apex or having a flat top, salts will concentrate in the apex or in the centre of the flat top, for this is where the wetting fronts from the irrigation water in the furrows will meet. This will leave the flat side or the edge of the flat top with a salt content close to that of the irrigation water, so rootlets from seeds planted in these positions will start growing in a low salt solution.

The control of exchangeable sodium

No irrigation scheme can succeed unless the soil profile remains permeable, and this, in turn, depends both on the proportion of the exchangeable cations held by the soil that are sodium, and on the concentration of soluble salts in the percolating water. A soil can hold a high proportion of exchangeable sodium ions and still remain permeable if the percolating solution with which it is in equilibrium is sufficiently concentrated; but as the solution becomes more dilute, a concentration will be reached when the permeability begins to drop, due to the swelling of the soil crumbs reducing the size of the channels and pores down which the solution is percolating. As the concentration drops still further, the clay particles will begin to deflocculate and completely block these pores, so the soil becomes impermeable (see p. 145). Calcium-saturated soils show this swelling phenomenon to a much smaller extent, but they only deflocculate if subjected to shearing stresses.

The actual proportion of the total exchangeable cations that must be sodium for these swelling and deflocculation processes to become important in a given strength of solution depends on the type of clay, the content and distribution of the sesquioxide films, and the amount and type of organic matter. Ferruginous soils, and particularly ferruginous kaolinitic soils[2] and old pasture soils[3] can contain a high proportion of exchangeable sodium and maintain their permeability in dilute salt solutions, but many alluvial soils high in fine sand and silt, containing micaceous type clays and low in organic matter, will lose permeability in such a solution with quite a low proportion of exchangeable sodium.

It is important to stress that there is no universal value for the minimum proportion of exchangeable sodium a soil must possess for its permeability to be appreciably affected by the irrigation water, for this depends on the

1 L. Bernstein and M. Fireman, *Soil Sci.*, 1957, **83**, 249.
2 S. A. El-Swaify and L. D. Swindale, *Trans. 9th Int. Congr. Soil Sci.*, 1968, **1**, 381; *Soil Sci.*, 1970, **109**, 197. H. M. E. El Rayah and D. L. Rowell, *J. Soil Sci.*, 1973, **24**, 137.
3 W. W. Emerson, *J. Soil Sci.*, 1954, **5**, 233.

soil, on its management and on the concentration of the salts in the percolating water.[1] A figure of 15 per cent exchangeable sodium has however been widely quoted as the critical content above which the soil structure will become unstable, and whilst this figure is a useful guide for many irrigated soils, it must not be taken as being valid for all irrigated soils. The critical figure, or rather the safe content of exchangeable sodium that will not cause sufficient swelling to affect permeability appreciably or will not give a turbid drainage water, should be determined for each major soil group using a solution corresponding to the lowest salt content of the irrigation water. It is likely that for low-salt waters and for alluvial soils containing micaceous and montmorillonitic clay that are high in fine sand and silt, as many such soils are, 15 per cent exchangeable sodium is a useful general guide to the probability of an unstable structure, and a content of 7·5 per cent exchangeable sodium may begin to give trouble due to swelling though not to dispersion. Nevertheless, D. L. Rowell[2] found that a montmorillonite clay containing 12·5 per cent exchangeable sodium with the remainder of the cations calcium behaved almost like the calcium-saturated clay provided it is subjected to no mechanical stress.

The exchangeable cations held by most irrigated soils consist of calcium and magnesium, with a small proportion of exchangeable potassium and a variable proportion of exchangeable sodium, for irrigation water is usually low in potassium but, as shown in Table 27.2, may contain appreciable amounts of sodium salts. As irrigation water percolates through the soil, there will be a redistribution of cations between the solution and the exchange sites, and as the crop dries any given volume of soil, the concentration of the soil solution will increase, which will alter the exchange equilibrium between the mono- and the divalent cations.

The Gapon equation, which was discussed on p. 94, states that the ratio of the exchangeable sodium to exchangeable calcium plus magnesium on a soil is proportional to the ratio of the activity of the sodium ions to the square root of the activities of the calcium plus magnesium ions in the solution; and for many practical applications no great error is made if the activities are replaced by the concentrations. The U.S. Salinity Laboratory in Riverside, California, have used the phrase exchangeable sodium percentage, ESP, for the ratio of exchangeable sodium to the cation exchange capacity, which they determine using sodium acetate at pH 8·2, and the phrase sodium adsorption ratio, SAR, for the ratio of the sodium ion concentration to the square root of the calcium plus magnesium ion concentrations, all expressed in millimoles per litre. This differs from the activity ratios used elsewhere in this book not only because they use concentrations rather than activities, but also because the concentrations are in millimoles per

1 For examples see J. P. Quirk and R. K. Schofield, *J. Soil Sci.*, 1955, **6**, 163; D. L. Rowell, D. Payne and N. Ahmad, ibid., 1969, **20**, 176.
2 *Soil Sci.*, 1963, **96**, 368.

litre. The activity ratio must be multiplied by $\sqrt{1000} = 31\cdot6$ to convert to SAR. It happens that the ESP of many irrigated soils is approximately equal to the SAR of the solution in equilibrium with the soil, when ESP and SAR are measured in the units as defined above.

The SAR of the irrigation water will influence the ESP of the soil, but the relation between the two is not straightforward, for the ESP of the soil is conditioned by the SAR of the soil solution, and this is constantly changing. After irrigation, the solution slowly becomes more concentrated as the crop transpires water, so its SAR rises, and if the effect of concentration is to cause some calcium or magnesium to precipitate out as the carbonate, this will also cause it to rise still further. Again, it takes time for the equilibrium to be established between the solution and the inside of soil crumbs and peds, so that the actual average ESP of the soil is likely to lag behind changes in the SAR of the soil solution. Californian experience, as summarised by the work of the U.S. Salinity Laboratory, has shown that the higher the total salt content of the irrigation water, the lower must be its SAR if the ESP of the

TABLE 27.3   Types of cracks and cracking pattern of Gezira clay soils. Percentage distribution of crack parameters

Depth of cracks			*Width of cracks*			*Spaces between cracks*		
	Site			*Site*			*Site*	
Cm	*1*	*2*	*Cm*	*1*	*2*	*Cm*	*1*	*2*
<16	4	1	<2·6	18	9	<21	22	39
16–35	32	45	2·6–3·5	42	25	21–40	29	53
36–55	58	19	3·6–4·5	27	22	41–60	20	7
56–85	6	21	4·6–5·5	11	28	61–80	20	1
>85	0	14	>5·5	2	16	>80	9	0

Site 1 irrigated. Site 2 not irrigated.

soil is to remain below a given level. Thus using their definitions of low and medium sodium waters, irrigation water with a low conductivity of 0·1 mmhos/cm can have a SAR up to 10 and 18, and with a medium conductivity of 0·75 mmhos/cm up to 6 and 12 if the ESP of the soil is to remain below 7·5 and 15 per cent.

There is a group of alluvial soils for which permeability, in the normal sense of the word, is irrelevant, and these are cracking black clays high in montmorillonite—the so-called vertisols. They cover the Gezira in the Sudan, for example, where they are used successfully for irrigated cotton and other crops. The clays are completely impermeable but, on drying, cracks open up which may become several centimetres wide at the soil surface, as is shown in Table 27.3, and some penetrate to a depth of 0·5 m.[1] When

1 A. Z. El Abedine and G. H. Robinson, *Geoderma*, 1971, **5**, 229.

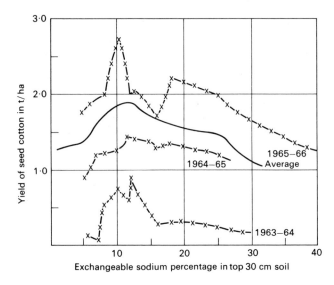

FIG. 27.2. Effect of the exchangeable sodium percentage in the top 30 cm of soil on cotton yields in the Sudan Gezira

such a soil is irrigated, the water runs down the cracks before they have time to seal up—and, in fact, they usually never seal up completely (see p. 760)—and the water becomes distributed along the systems of smaller cracks produced by the drying of the clay by the plant roots. These soils are therefore irrigable even though they have a high ESP and have an unstable structure, and Fig. 27.2 shows that useful yields of cotton can be obtained,[1] even when 40 per cent of the exchangeable ions are sodium. However, although these clays are montmorillonitic, they may contain aluminium hydroxide films (see p. 112) which could help to stabilise the cracks to some extent. On the other hand, these soils can only be irrigated with waters very low in salts because of the small amount of water and salts that drain out of the soil profile. The water of the Blue Nile is very low in salts, and although the soils have a certain salt content, this is kept at a fairly constant level because the crops carried off the land contain sufficient anions and cations to prevent the salts in the soil building up beyond a level which begins to affect crop yield appreciably.

It is sometimes necessary to use a water for irrigation with an SAR which is likely to give an ESP on the soil sufficiently high for the soil permeability to fall to an unacceptably low level. The exchangeable sodium can still be kept low if the SAR of the irrigation water or of the soil water is lowered by adding gypsum (calcium sulphate) either to the irrigation water or to the soil; or sulphur itself can be added to the soil if it contains free calcium carbonate, for it will be oxidised to sulphuric acid which will react with the calcium carbonate to give calcium sulphate. The advantage of adding gypsum to the irrigation water, where this is economically feasible, is that it helps maintain the permeability of the surface soil. Calcium carbonate can also be added to a soil, if it is noncalcareous, but its solubility is much less than that of gypsum, so it has a much slower effect on the SAR of the soil solution.

The maintenance of a high level of decomposable organic matter and the use of crops with a high root respiration rate can be a valuable means of maintaining permeability when high SAR waters must be used, again, provided there are reserves of calcium carbonate in the soil. This will involve either the use of green manures or farmyard manure on the one hand, or of some grass and legume pastures on the other.[1]

The control of alkalinity

Irrigation water usually contains free bicarbonate ions as well as sodium; and if both of these accumulate in the soil, its pH will rise up to values as high as 10. This rise in pH can have two very undesirable consequences—it will increase the salt concentration necessary to keep the soil flocculated and so permeable, and it may cause nutritional troubles in the crop, which show themselves as chloroses,[2] once it has exceeded a certain value, which for many irrigated crops is about pH 8·5. If the irrigation water has a fairly high neutral salt content, the pH of the soil does not normally rise above pH 8·5, so problems of alkalinity are typically problems with waters low in neutral salts.

Waters can contain bicarbonates and give no alkali hazard, for if they also contain about the same concentration of calcium and magnesium ions as bicarbonate and carbonate, sodium bicarbonate cannot accumulate because as the soil solution becomes more concentrated due to transpiration, calcium and magnesium carbonates will precipitate out. Irrigation water in which the total bicarbonate and carbonate concentration is greater than that of the divalents is thus very dangerous, and can only be used in long-term irrigation if precautions are taken to prevent sodium carbonate accumulating in the soil. F. M. Eaton[3] introduced the concept of the residual sodium carbonate in an irrigation water, which he defined as the difference between the concentration of the bicarbonate and carbonate anions and the calcium and magnesium cations. He suggested that if this exceeded 2·5 m eq/litre the water would be unsuitable for irrigation, and if between 1·25 and 2·5 m eq/litre would give a serious danger of alkalinity in the soil building up. The

1 For an example from Nebraska of crops which help to maintain soil permeability to irrigation water see A. P. Mazurak, H. R. Cosper and H. F. Rhoades, *Agron. J.*, 1955, **47**, 490.
2 For an illustration see M. A. Salem and M. A. El Kadi, *Pl. Soil*, 1965, **23**, 377.
3 *Soil Sci.*, 1950, **69**, 123; see also L. V. Wilcox, G. Y. Blair and C. A. Bower, ibid., 1954, **77**, 259.

water of the Humboldt river, whose analysis is given in Table 27.2, p. 754, is an example of water that is dangerous to use for this reason.

Bicarbonate-containing water, however, may increase the level of exchangeable sodium in the soil, even if the concentration of calcium and magnesium ions exceeds that of the bicarbonate and carbonate, because the precipitation of these cations as insoluble carbonates increases the SAR of the soil solution, and hence the level of exchangeable sodium. The reason that it only may and not that it must increase the SAR of the water is that when a bicarbonate-containing water is percolating through a soil containing free calcium carbonate, the water will either dissolve some calcium carbonate or precipitate out calcium carbonate, depending whether the percolating solution is under- or over-saturated with respect to the solubility of the carbonate.

It is possible to predict if a solution is over- or undersaturated, with respect to calcium carbonate, from the solubility product of calcium carbonate, the dissociation constant of bicarbonate to carbonate, and the composition of the irrigation water; and the degree of under- or oversaturation can be expressed as a Precipitation Index PI for that water, which is defined by

$$PI = 8.4 - pH_c$$

where pHc is the pH which the water, with the given amounts of divalent cations and bicarbonate and carbonate anions, would have if it were in equilibrium with solid calcium carbonate. This is given by

$$pH_c = pK_2 - pK_c + p(Ca + Mg) + p(HCO_3^- + CO_3^{2-})$$

where pK_2 is the negative logarithm of the second dissociation constant of carbonic acid, pK_c the negative logarithm of the solubility product for calcium carbonate, both corrected for the ionic strength of the percolating solution, and $p(Ca + Mg)$ and $p(HCO_3^- + CO_3^{2-})$ are the negative logarithms of the concentration of calcium plus magnesium in moles per litre, and of bicarbonate plus carbonate in equivalents per litre in the equilibrium solution.[1] The figure 8.4 is the pH of the system calcium carbonate–carbon dioxide–water when the solution is in equilibrium with air containing 0.03 per cent carbon dioxide (see p. 126). If the precipitation index is positive, percolating the irrigation water through the soil will cause calcium and magnesium carbonates to precipitate out from the solution, thus increasing the SAR of the solution; and if it is negative, it will dissolve some carbonate, so decreasing the SAR.

It should be possible to calculate the effect of this precipitation or dissolution of calcium carbonate on the SAR of the soil solution. If the leaching

---

1 C. A. Bower, L. V. Wilcox *et al.* (*Proc. Soil Sci. Soc. Amer.*, 1965, **29**, 61) give tables to help with the calculation of pH_c, and experimental results showing the accuracy of the Precipitation Index calculation.

factor is LF, that is, if LF is the proportion of applied irrigation water that leaches out of the soil, and if SAR_d is the SAR of the drainage water and SAR_i of the irrigation water, then

$$SAR_d = (1/\sqrt{LF}).SAR_i(1 + 8\cdot4 - pH_c)$$

if this is the complete description of what is happening. The leaching factor enters as the square root since it is assumed to be equal to the ratio of the salt concentrations in the irrigation to the drainage water, and the SAR is proportional to the square root of the salt concentration. The Californian workers have found that this calculation may overestimate the SAR of the drainage water,[1] and J. D. Rhoades[2] suggested this was because some soils contain calcium silicates which slowly dissolve in the soil solution, releasing calcium ions. He finds he can allow for this by multiplying the SAR of the drainage water, as calculated from the above equation, by the factor $y^{(1+2LF)}$ where y is a constant, which for his soils equals $0\cdot7$; and he has published tables which give the calculated value of the leaching factor, LF, for different values of pH_c and the SAR of the irrigation and drainage waters.

P. F. Pratt and F. C. Bair[3] showed that the empirical equality of the ESP of the soil and the SAR of the soil solution with which it is in equilibrium holds for many irrigated alluvial Californian soils even if they contain soluble carbonates provided the pH of the soil saturation extract does not exceed 8·6, though the ESP is appreciably higher than the SAR at higher pH values. Thus provided this soil pH is less than 8·6, the ESP of the soil in equilibrium with the drainage water is approximately given by

$$ESP = \frac{y^{(1+2LF)}}{\sqrt{(LF)}} SAR_i(1 + Pl)$$

and this relation allows the computation of the leaching factor, still un-corrected for non-uniformity of water flow through the soil, that must be used if the ESP of the soil is to be maintained below a given level when the irrigation water contains soluble carbonates.

Soils will sometimes contain appreciable levels of sodium carbonate which is formed in the soil profile. Alkali soils tend to deflocculate and so become anaerobic; and if there is both decomposable organic matter and soluble sulphates in the soil, the reducing conditions set up by the bacteria de-composing the organic matter anaerobically will reduce some of the sulphates to hydrogen sulphide. If sodium ions had been neutralising some of this sulphate, they will now neutralise bicarbonate or carbonate anions, for sodium sulphide decomposes to hydrogen sulphide in the soil. Thus, one

1 C. A. Bower, G. Ogata and J. M. Tucker, *Soil Sci.*, 1968, **106**, 29.
2 *Proc. Soil Sci. Soc. Amer.*, 1968, **32** 643, 648, 652.
3 *Proc. Soil Sci. Soc. Amer.*, 1969, **33**, 880.

equivalent of sodium carbonate can be produced for each equivalent of sulphate reduced.[1]

The problems of managing soils when a high carbonate irrigation water is used are similar to those when a water with a high SAR is used. If the soil does not contain free calcium carbonate, it is usually helpful to add it to help maintain the reserves of calcium, and it may also be necessary to use gypsum or sulphur. Australian experience[2] has shown that there are advantages in adding the gypsum to the irrigation water if the soils are unstable clays high in exchangeable sodium, for this helps to maintain a stable flocculated surface tilth, so a good surface permeability. Again, problems of management are greatly eased if systems of cropping can be used which allow the maintenance of a high level of microbial and root respiration, for the maintenance of a high level of carbon dioxide production will help maintain the concentration of calcium ions in the soil solution.

The reclamation of salt and alkali affected soils

The reclamation of land whose sole defect is that it contains too high a concentration of neutral salts is, in theory, simple; for it is only necessary to supply sufficient water over and above the transpirational needs of the crop to leach the salts out of the soil (see p. 756), and prevent the uncontrolled entry of new salts. But this can only be done if the soil is permeable, and the ground water is well below the root zone or can be brought below the root zone by drainage. This process may, however, take time if the water that leaches out of the profile flows down a restricted number of channels or pores, because only the salts that diffuse out of the body of the soil into this percolating water will be removed. Deep saline clay soils, for example, can only be reclaimed economically if they are naturally well structured and can carry the cost of tile drainage. However the salt content of clay soils can be reduced by growing crops that accumulate sodium, such as some species of *Atriplex*, and removing the crop from the irrigated area. H. Ferguson[3] notes that *Atriplex muelleri* may contain over half a ton of sodium per hectare in the above ground portion of its crop when grown on the Sudan Gezira clay. If the soil is permeable and the ground-water level is deep, well below the root zone of the crop, it is merely necessary to leach the salts within the root zone to below the level from which the roots take water, and to ensure that there is sufficient water available for the regular leaching of any salts that accumulate subsequently.

The usual cause of salinity is, however, a high water-table, and this must

1 L. D. Whittig and P. Janitzky, *J. Soil Sci.*, 1963, **14**, 322; 1964, **15**, 145. G. Ogata and C. A. Bower, *Proc. Soil Sci. Soc. Amer.*, 1965, **29**, 23.
2 D. R. Scotter and J. Loveday, *Aust. J. Soil Res.*, 1966, **4**, 69. J. L. Davidson and J. P. Quirk, *Aust. J. agric. Res.*, 1961, **12**, 100.
3 *Emp. Cotton Gr. Rev.*, 1953, **30**, 241.

be lowered if the land is to be successfully reclaimed. The water-table may be high because there is no effective drainage system, or because water is rapidly leaking out of unlined irrigation channels dug in permeable soil, or because water is seeping in from higher ground. In arid areas the ground water is almost always saline, so a high water-table almost always will cause serious salinity problems. Reclamation may involve the installation of deep drainage ditches and a drainage canal draining into the river well below the irrigation scheme, it may involve lining the main irrigation canals to prevent seepage of water from them into the ground water, or it may involve a properly designed series of cut-off drains to prevent ground water entering the area. However, in many areas water is becoming sufficiently valuable, and power sufficiently cheap, for the ground-water level to be kept low by pumping; the ground water being pumped into the irrigation ditches and diluted with the irrigation water.

Nevertheless, salinity control can have an unfortunate consequence, for if the area being reclaimed is up-stream of other irrigation schemes, and if the salts are to be run into drains discharging lower down into the river, these salts are being exported from the up-stream scheme and imported into the down-stream site. Since salt-affected soils may contain ten tons of salts or more per hectare, it may be necessary to limit the annual export of salts during the reclamation process. Even when the upstream scheme is not badly salinised, irrigation will always lower the quality of the water leaving the scheme, for crop transpiration increases the concentration of salts in the water, and therefore its SAR; and since concentration of the soil solution often leads to precipitation of calcium and magnesium carbonates, this again will raise the SAR.

Table 27.4 illustrates this effect for two adjacent sites on the Rio Grande in New Mexico and Texas, and it also gives the salt balance for these two schemes.[1] In the upper Mesilla area the salts are approximately in balance over the scheme, while the El Paso area is gaining about 150 000 tons of salts a year. In detail the Mesilla area is losing about 40 000 tons of choride and 20 000 tons of sodium annually, while the El Paso area is losing about 50 000 tons of chloride but retaining half of the sodium being lost from the Mesilla scheme. Thus the two areas between them have taken about 75 per cent of the water entering the Mesilla area, have taken about half the calcium and sulphate, but have added 90 000 tons of chloride to the water leaving the El Paso area.

Land affected by alkali is usually black in colour, because the sodium carbonate present deflocculates the soil, and the humic colloids are dispersed from the clay and so colour the soil much more strongly. When wet, the land usually has a fairly characteristic smell, presumably due to the activity of anaerobic bacteria.

1 Taken from C. S. Scofield, *J. agric. Res.*, 1940, **61**, 17.

TABLE 27.4 Annual salt and water balance in two irrigation areas of the Rio Grande (New Mexico and Texas). Water in million tons. Salts and ions in thousand tons

Area under irrigation	Mesilla area about 32 000 hectares			El Paso area about 48 000 hectares		
	Entering	Leaving	% retained	Entering	Leaving	% retained
Water	915	611	33·3	611	213	65·1
Concentration of salts in water, ppm	660	1010		1010	2200	
Total Salts	599·4	608·1	−1·5	608·1	461·8	24·1
Ca	80·6	69·7	14·5	69·7	41·3	40·7
Mg	17·6	15·5	11·9	15·5	10·2	33·9
Na + K	102·0	122·6	−20·3	122·6	111·7	9·0
$HCO_3 + CO_3$*	93·9	80·5	14·3	80·5	28·2	64·9
SO_4	233·4	208·0	10·9	208·0	108·7	47·7
Cl	71·3	111·1	−55·8	111·1	161·4	−45·3
NO_3	1·4	1·5	−3·0	1·5	0·5	65·2

* Computed as carbonate.

The principles underlying the reclamation of land rendered infertile by alkali are: first, to ensure that their drainage is adequate and that saline water is not seeping into them from higher ground; and, second, to replace some of the exchangeable sodium by calcium.[1] If the soil contains gypsum, draining the land and flooding it with water is probably all that is required, although if the soil is heavy, it may initially be sufficiently impermeable for the cropping to have to be chosen very carefully. But if the soil is low in gypsum, the primary trouble, which is either present, or which will develop unless precautions are taken, is the impermeability of the surface soil to water, and one great danger in reclamation that must be guarded against is increasing the impermeability of the soil during the reclamation. Soils containing much exchangeable sodium, or free sodium carbonate, will deflocculate and become quite impermeable to water if wetted with pure water, or with rain, whereas if they contain much soluble salts, or the irrigation water has a high salinity, they may remain flocculated and permeable. Hence, the second great principle in reclaiming alkali soils is to maintain a fairly high salt content in the soil during the process of leaching out the exchangeable sodium, and in some areas high-salt waters are initially used to help

1 For illustrations of successful methods of reclamation, see W. P. Kelley, *J. agric. Sci.*, 1934, **24**, 72; *Hilgardia*, 1934, **8**, 149; *Calif. Agric. Expt. Sta.*, Bull. 619, 1937 (San Joaquin Valley, California); *The Diagnosis and Improvement of Saline and Alkaline Soils*; R. E. McKenzie and J. L. Bolton, *Sci. Agric.*, 1947, **27**, 193 (Val Marie, Saskatchewan); R. Aladzem, *Bull. Un. Agric. Egypte*, 1947, **45**, 37; R. S. Snyder et al., *Idaho Agric. Sta.*, Bull. 233, 1940; W. L. Powers, *Oregon Agric. Expt. Sta.*, Bull. 10, 1946.

get leaching started.[1] Provided the soil remains permeable, drainage, adding gypsum or sulphur, and flushing down the salts will remove exchangeable sodium without difficulty.

Many disused alkali soils are, however, almost impermeable to begin with, and the improvement in permeability is the primary problem.[2] Typically this is done by replacing the exchangeable sodium in the surface layer and so stabilising it, and then deepening this stabilised layer. Adding gypsum to the surface soil, or on some lighter soils working in farmyard manure or compost, and then letting it wet and dry a few times will be enough to give a few inches of stable permeable soil. The soil may then be flooded, provided an adequate

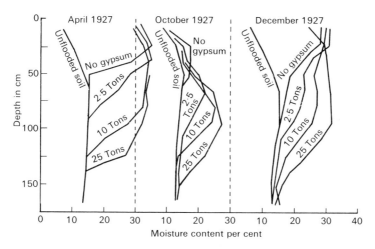

FIG. 27.3. Effect of gypsum on the permeability of soil. Percentage of moisture at different depths of unflooded soil (lefthand line) and of four flooded soils (righthand lines) treated with different quantities of gypsum (tons per hectare) applied

drainage system has been installed, to allow the gypsum to wash down slowly into the subsoil, improving the permeability of every layer into which it penetrates, for rarely does a drained soil have a permeability of less than a few inches of water a month.[3] Figure 27.3, taken from some experiments of H. Greene on an impermeable Gezira clay, in which sodium constitutes 10 per cent of the exchangeable ions, shows this effect of gypsum of increasing the permeability of the subsoil.[4] The land was flooded in April, and shortly

1 For examples see R. C. Reeve and C. A. Bower, *Soil Sci.*, 1960, **90**, 139; *Proc. Soil Sci., Soc. Amer.*, 1966, **30**, 494; S. Muhammed *et al.*, *Soil Sci.*, 1969, **108**, 249.
2 For an American bulletin on methods of doing this, see R. F. Reitemeier *et al.*, *U.S. Dept. Agric., Tech. Bull.* 937, 1948.
3 W. L. Quayle (*Wyoming Agric. Expt. Sta., Bull.* 243, 1941) quotes examples of the value of flooding some Wyoming soils for two months before a crop is taken.
4 *J. agric. Sci.*, 1928, **18**, 531. See also with O. W. Snow, *J. agric. Sci.*, 1939, **29**, 1.

afterwards soil samples were taken for moisture content. The soil was left fallow and then flooded again in December and soil samples again taken. Without gypsum the water remained near the surface: with increasing dressings of gypsum it penetrated deeper and deeper into the soil. The drying out affected the first 60 cm equally whatever the gypsum treatment, but, as would be expected, much of the water that had penetrated below this depth remained in the subsoil. The permeability of the soil can also sometimes be increased by very deep ploughing,[1] particularly if some gypsum is ploughed in at the same time or if gypsum is present in the subsoil. This is probably most efficacious if the subsoil has a compacted layer, for deep ploughing can bring this layer up to the surface, break it up to some extent, and expose it to the weather so its structure can be mellowed by drying and wetting in the hot season.

The second operation is to establish a crop on the land, either sown or of natural weeds, for the plant roots will continue the task of increasing the permeability of the subsoil, both by abstracting water from it, so causing cracks to develop which will let water down quickly, and also by respiring carbon dioxide there which will reduce the alkalinity somewhat. The choice of crop is, however, limited, for it must be able to withstand the prolonged waterlogging necessary for washing down as much sodium as possible into the deeper subsoil. Rice is an ideal crop, if other conditions are suitable, as it can be kept waterlogged throughout much of the season. Sweet clover (*Melilotus alba*) and strawberry clover (*T. fragiferum*) are also suitable, so are many grasses, and under some conditions so is lucerne, though lucerne will not stand waterlogging so well as the others. Once these crops are established, they are encouraged to root deeply by being given heavy irrigations at as long intervals as possible, so the roots take as much water as they can from the subsoil before the next irrigation, though they must not be allowed to dry the soil so much that the soil solution becomes sufficiently concentrated to harm the roots. The first crop to be taken is often ploughed in as green manure, as the plant may contain too much salt for high-feeding quality, and the green manure will not only produce carbon dioxide in the soil during its decomposition, but will also slowly set free plant nutrients such as phosphate, iron, manganese and zinc, which may be very unavailable in alkaline soils.

A good crop rotation is an excellent insurance against alkali trouble, for grass, clover and lucerne leys can all build up the structure of a soil and improve its stability; and if these leys are consumed on the farm by dairy cattle, they will be returned to the land as farmyard manure, which again has a valuable action in maintaining the permeability of the surface soil.

The most usual addition to the soil to help displace exchangeable sodium is gypsum, $CaSO_4 . 2H_2O$, for this is more efficient than calcium carbonate

---

1 V. A. Kovda, *Khim. Sotsial. Zemled.*, 1941, No. 4, 31; I. N. Antipov-Karataev and A. A. Zaitzev, *Dokuchaev Inst. Soils*, 1946, Anth. 14.

because 1 litre of water will dissolve 30 m eq of calcium as calcium sulphate, but at pH 8·4 only 1 m eq of calcium from the carbonate; although at pH 7 it will dissolve 5 m eq. Sulphur is also used when it can be obtained cheaply. If these ameliorating substances are used with complete efficiency 10 tons of gypsum correspond to 5·8 tons of calcium carbonate or 1·9 tons of sulphur, and if applied to a hectare of land would displace 2·2 m eq of sodium per 100 g soil from the top 30 cm layer. Since it is difficult to determine how much exchangeable sodium a soil holds when in the presence of sodium salts, R. K. Schofield and A. W. Taylor[1] suggested it is preferable to determine how much calcium the soil would need to take up to saturate its exchange complex with calcium and they developed a simple method which gives this figure to an adequate accuracy for most practical purposes.

Reclamation of soils damaged by sea water

It happens from time to time that low-lying land by the sea becomes flooded with sea water, usually due to exceptionally high tides overtopping a protective sea wall. This happened in East Anglia in 1897[2] and again over a wider area of eastern England[3] and parts of Holland and Belgium[4] in 1953. This flooding leaves large quantities of salts, principally sodium chloride, in the land which must be got rid of in such a way that the soil does not deflocculate.

After the sea water has been removed, the flooding may kill the crops due to the direct effect of the salt. If the land is ploughed, once it has dried, a good seedbed can usually be made quite easily as the soil works and crumbles well. As rain removes the salt, the soil condition deteriorates, it becomes increasingly difficult to work, water begins to stand on it, and it dries out to hard lumps; the soil has gone from a flocculated to a deflocculated condition.

It is now possible to prevent this deflocculation from occurring by adding the requisite amount of gypsum to the soil. After the 1953 floods in eastern England, gypsum added at the rate of about 6 t/ha annually for two to three years on medium-textured soils kept the soils reasonably permeable and prevented the pH rising much above 8·0; but on heavier soils dressings up to 25 t/ha were used. The gypsum not only kept the soil from deflocculating, it also prevented the formation of the hard surface cap that was characteristic of the soils not receiving gypsum or receiving an inadequate

1 *J. Soil Sci.*, 1961, **12**, 269.
2 See *Report on Injury to Agricultural Land on the Coast of Essex by the Inundation of Sea Water on 29 November 1897*, by T. S. Dymond and F. Hughes, Chelmsford, 1899.
3 See *The East Coast Floods 1953*. Report by the Ministry of Agriculture Fisheries and Food, London, 1962. Also for the Lindsey region, J. W. Blood, *N.A.A.S. Quart. Rev.*, 1955, **27**, 125; T. G. Heafield and G. D. Ashley, ibid., 1956, **32**, 47.
4 For some Dutch experience of reclaiming land flooded with sea water see C. W. C. van Beekom, C. van den Berg *et al.*, *Netherl. J. agric. Sci.*, 1953, **1**, 153, 225.

dressing. In the early stage of reclamation, the land should be cultivated as shallow as possible and only when the surface is dry enough to carry the tractor and implements without the soil sticking together in a compact mass.

The experience in the 1953 floods showed that soils differ considerably in their reaction to flooding.[1] Soils well supplied with organic matter, such as old pastures or soils that had only been out of pasture for a few years, showed much less collapse of structure than soils that had been under intensive arable crops where the soil organic matter had fallen to a low level. Some soils also appeared to be inherently more stable than other, possibly because their structure was still being stabilised by iron oxide cements formed during the process of formation from salt marsh. But many of the finer-textured soils, particularly those high in fine sand, silt or clay, remained in a difficult state for a number of years, and the deflocculation of the subsoil that occurred caused the tile drains to become filled with fine particles, and the sides of the drainage ditches to become unstable.

The soils also differed considerably in the rate that salt was leached out of the profile, being very rapid for the sandy soils but slow to very slow for the fine sandy loams and clays. In fact some of the sandy soils could be brought back into cultivation a few months after the sea water had been drained away.

The reclamation of land from the sea, such as the Dutch have been doing extensively both along their coast and in the old Zuyder Zee, poses much the same problems as reclamation after sea flooding. The first problem is to install a drainage system and encourage the soil to crack, so salts can be washed down. It may then be necessary to add gypsum, though with careful management this is not always necessary if the soil is calcareous. A number of the clays formed in the shallow water under the sea may not only contain calcium carbonate, but ferrous sulphide or iron pyrites as well, and when these soils are drained and air enters, the ferrous sulphide is oxidised and the sulphuric acid set free by hydrolysis is neutralised by the calcium carbonate, so that calcium sulphate is formed in the body of the clay near the cracks down which the air diffused, and this helps to improve the permeability of the profile.

1 N. H. Pizer, *Chem. Ind.*, 1966, 791.

28

The general principles of soil management

A good system of agriculture is required to produce as much food, either human or farm stock, as possible from the land at a reasonable cost without impairing its fertility. A farmer should always aim to leave the land in at least as productive a condition as when he acquired it. A good system of management must therefore ensure that the nutrient status of the soil is maintained; that all factors directly harmful to plant growth, such as high acidity, high alkalinity, or poor drainage are absent; that the land only grows the crops desired and not unwanted ones, that is, that weeds are kept under control; and that the soil particles themselves remain in place and are neither washed nor blown away.

The nutrient status of the soil is now very largely under the direct control of the farmer. If high-priced cash crops are being grown he can, if need be, buy all the plant food needed by the crop. But he should also aim at returning to the land all crop and animal residues that he can so as to minimise the loss of nutrients. Some soils, such as many semi-arid and desert, are naturally well provided with most plant foods except phosphate and nitrogen; others, such as some tropical soils being derived from young, basic igneous rocks, are having their supplies of plant nutrients supplied by the weathering of the rock particles in the soil fast enough to meet the demands of the crops. Still others need to have nutrients added if they are to maintain good yields, and of these nutrients nitrogen and phosphate compounds are far the most important—nitrogen compounds wherever the water supply is sufficient to keep the plant growing all through its season, and phosphates in most parts of the world. These nutrients do not necessarily have to be added as fertilisers. The available nitrogen compounds in the soil can be greatly increased either by growing leguminous crops which contain nodules capable of fixing large quantities of atmospheric nitrogen, or by growing crops which return much nitrogen-poor organic matter to the soils under conditions in which decomposition can proceed rapidly and in which free-living nitrogen-fixing bacteria can flourish. The supply of other nutrients can also sometimes be increased by selecting crops which have a considerable power of extracting the nutrients in short supply from the soil minerals, or which have a particularly deep root system, and which can, therefore, concentrate the available soil supplies. In all these cases, however, if the crop is not ploughed in, it should be fed to animals and their dung and urine put back on the land in a way that gives the smallest possible losses.

The prevention of harmful factors, such as acidity and alkalinity, have already been dealt with, and both require that the calcium status of the soil shall remain satisfactory. Good drainage is also essential to allow the crops to root deeply, hence be able to withstand drought and possibly frost as well, and to prevent the accumulation of harmful substances produced biologically whenever the oxygen tension in the soil falls too low. The soil must also be kept permeable, so water can drain through it at a reasonable rate, and the carbon dioxide content of the air around the roots not rise too high, nor the oxygen content fall too low. To some extent these factors can be controlled by proper cultivations, a subject that will be discussed in a later section.

The principles underlying the control of soil erosion

Soil particles can move by three processes: they may be blown away, they may be washed away, or the whole soil may slide or slump down a hillside.[1] Soil erosion can cause great troubles over large areas: dust storms, once started, may travel great distances; and conditions conducive to water erosion can lead to extensive flooding of valleys after storms, to silting-up of rivers, valleys and reservoirs, and to the great impoverishment of the land above the valley floor. Soil erosion is nearly always caused by an unsuitable method of agriculture being practised, and since among primitive peoples the system of agriculture affects their whole way of life, methods of control involve not merely devising a system of agriculture which is better suited to the area and more productive, which is relatively easy, but also of altering the whole social outlook and sometimes even some of the religious beliefs or practices of the community, which is always an extremely difficult problem in sociology.

Wind erosion and soil drifting

High winds can blow much material out of some bare soils, so that the wind itself becomes a dust storm, and soil material drifts across the land, forming dunes, filling up hollows and drifting against farm buildings and hedges. The physics of this action is now fairly well understood, as the principles have been clearly stated by R. A. Bagnold[2] for sands, and filled in in considerable

---

1 For a recent book on erosion and its control, see N. W. Hudson, *Soil Conservation*, Batsford, 1971.
2 For a summary of his work, see his book *The Physics of Blown Sand and Desert Dunes*, London, 1941.

detail for soils by W. S. Chepil and N. P. Woodruff[1], whose description has been drawn on extensively in the account given here.

Winds move soil and sand particles by three distinct processes: the finer particles are carried in suspension and may be transported as fine dust over very great distances; the coarser particles are rolled along the surface of the soil; and the medium-sized particles move by saltation. Bagnold has shown that saltation is, in fact, the primary process responsible for soil movement. The process of saltation is as follows. A strong eddy of wind at the soil surface picks up a sand grain and carries it up a few centimetres in the air where the wind has a much stronger horizontal component than at the soil surface itself. This wind then gives the sand grain a horizontal acceleration, and as the eddy which picked it up becomes dissipated, the sand grain falls back to the ground after having acquired considerable momentum. On impact it may cause some other sand grains to be shot a little way up in the air, and these, in turn, acquire momentum from the wind, and on hitting the ground may throw up other grains. Thus, the dust storm is due to this stream of sand particles which throw up others as they hit the ground. Soil movement by saltation thus requires a source of sand grains of a suitable size, a wind sufficiently strong to give eddies capable of picking up sand grains, and a clear length of run for the wind to build up a sufficient density of sand grains moving in this way.

The size of particles taking part in saltation movement depends on the wind-speed and the weight, that is the size and bulk density of the particles. If the particles are sand grains most of the particles moving by saltation lie between 0·5 and 0·05 mm, with grains in the size 0·1 to 0·2 mm being particularly active; but if they are soil granules they usually have a lower bulk density than sand and a somewhat larger size. Typically, the bulk of the grains do not rise more than a few centimetres above soil level, but a few rise to a metre or so; they rise steeply to their maximum height and then come down at an angle of between 10° and 16° to the horizontal. Typical paths for these grains are shown in Fig. 28.1.[2] Grains smaller than about 0·1 to 0·05 mm, depending rather on the wind-speed, do not take part in saltation, as they are sufficiently fine to be carried as dust in the turbulent motion of the wind, while grains larger than about 0·5 mm are too heavy to be bounced into the air. Soils containing over 60 per cent of unaggregated sand grains and individual granules in the size range 0·1 to 0·5 mm, are very liable to blow, but those with less than 40 per cent do not usually blow easily. Such blowing soils are either fine sands and loamy fine sands or calcareous clays, for these clays tend to have granules of the appropriate size, or peat and fen soils with over 35 per cent of well-rotted humus.

Movement of soil particles by saltation causes a fractionation of these particles by size. The finest particles are removed from the area altogether

1 *Adv. Agron.*, 1963, **15**, 211.
2 Taken from R. A Bagnold, *The Physics of Blown Sand and Desert Dunes*, London, 1941.

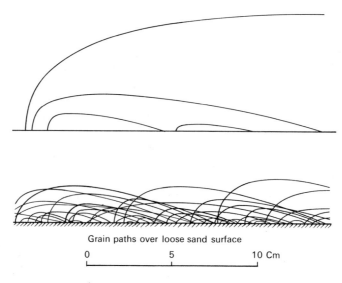

Grain paths over loose sand surface

0   5   10 Cm

FIG. 28.1. *Upper:* Typical paths of sand grains moving by saltation. *Lower:* Paths of sand grains moving over a loose soil surface

in the form of a dust storm, the larger ones between 0·07 and 0·02 mm settling as loess[1] and the finer often being blown over distances of thousands of miles and being brought down in rain.[2] The particles moving in saltation principally in the size range of 0·5 to 0·1 mm, form dunes or drifts up against any obstructions such as hedges or buildings. The particles that are just too large to move by saltation will creep along the soil surface due to the impact of the saltating grains, the larger will remain behind, eventually covering the surface. Thus wind-blowing typically increases the content of coarse sand or gravel in the soil surface, unless the soil itself is a loess or a clay or peat in fine tilth, when the whole of the soil surface will be removed.

The essential condition for wind-blowing to occur is that there should be a supply of particles that can move by saltation. This will not happen if the soil is moist, for the water films between the particles will hold them together too firmly for the wind to dislodge any, or if the soil surface has a compact dry crust, such as may be produced by heavy rain on the bare surface or by a heavy roller, or if the soil has a cloddy structure. On the other hand, a supply of these particles can be created by cultivating the soil surface, particularly when dry or nearly dry, or by wetting and drying, or freezing and thawing a cloddy tilth.

Wind-blowing can be reduced under conditions when the soil surface contains a large number of particles capable of moving by saltation through

1 F. E. Zeuner, *The Pleistocene Period*, Ray Society, London, 1945.
2 For an example of Saharan dust over England see A. F. Pitty, *Nature*, 1968, **220**, 364.

the use of surface mulches.[1] Dead vegetation lying on the soil surface, such as the remains of the previous crop, will reduce the build up of saltation by ensuring that a proportion of the falling sand grains hit a bit of vegetation rather than bare soil. Bitumin emulsions sprayed on the soil surface will also reduce saltation very considerably, and can be of very great help in stabilising blowing sand-dunes so a vegetation cover can become established. Saltation can also be reduced by laying the land up in ridges and furrows running across the direction of the wind, for if the ridge height and spacing is correct, few of the particles moving in saltation will jump the ridge.[2] But such methods can only be of temporary help as the furrows will slowly fill up with sand.

Saltation can be reduced by limiting the area of unprotected soil over which the wind can blow. A fetch of between 50 and 500 metres, depending on conditions, is needed for saltation to build up to its maximum; so provided the wind does not blow over a greater width of soil than this, wind blow can be kept down to an acceptable figure. But this involves being able to trap the soil moving out of this strip, so it does not move any further. A thick stubble, such as a tall wheat stubble from the previous year, or a strip of tall grass, is an excellent trap for sand, while a thin maize or sorgham stubble or short grass is relatively inefficient, the latter because its surface on the windward side will soon be covered with sand. In areas where wheat is grown in alternation with fallow, the land can be laid out in strips 100 to 200 m wide with wheat and fallow on alternate strips. An example of this from Alberta is illustrated in Plate 47b. This method can be fairly effective if the winds causing blowing always come from the same direction, but it is of only limited value if the wind direction is variable.

Saltation will be reduced if the wind velocity is reduced. This can be done by the use of windbreaks, though the effects of windbreaks are strictly limited. A windbreak will reduce wind-speeds at ground level for a distance of about five times the height of the break on the windward and twenty to thirty times the height on the leeward side. The windbreak must, however, have a certain permeability to the wind, which reduces its efficiency; for if the windbreak is solid, it will cause very serious turbulence in the descending air, which will start saltation very strongly. Further, if the windbreak is a belt of trees or a hedge, the plants will use water and be causing shading, so will reduce crop yields for a considerable distance on both sides, particularly if the area is subjected to summer drought. Windbreaks are considered to be of value in areas of eastern England and Sweden where wind-blowing can be serious, though M. T. Spence[3] found they only gave complete protection for a distance of six to ten times their height in the Fens.

1 For a general review of methods for control, see *Soil Erosion by Wind and Methods for its Control on Agricultural Land*, F. A. O. Agric. Devel. Paper, 71, Rome, 1961.
2 D. V. Armbrust, W. S. Chepil and F. H. Siddoway, *Proc. Soil Sci. Soc. Amer.*, 1964, **28**, 557. W. S. Chepil and R. A. Milne, *Soil Sci.*, 1941, **52**, 417.
3 *Met. Mag.*, 1955, **84**, 304.

(a)

(b)

PLATE 47. (*a*) Wheat stubble after being cultivated with a Noble cultivator.
(*b*) Alternate strips of wheat and fallow near Monarch, Alberta.

W. S. Chepil[1] has given an example from Kaifeng in Honan Province, China, where this method is practised to stabilise a sandy soil. Single belts of willow, which grow to about 3·5 m high, are planted in strips 15 to 18 m apart across the direction of the prevailing wind and in strips 30 to 150 m apart at right-angles to this direction, leaving only small rectangular strips of field for cultivation.

A standing crop will also act as a windbreak. Nurserymen on sandy soils in East Anglia and Holland sometimes use strips of autumn sown rye to protect young plants during critical periods in spring, and some American farmers use strips of sorghum to protect autumn sown wheat until it is established. A young crop gives very little protection to the soil, for its leaves cover only a small area of the ground and it cannot reduce wind-speeds appreciably.

Farmers can reduce the liability of land to blow by using appropriate methods of cultivation. One group of these is known as stubble mulch cultivation and will be described in detail on p. 798. It involves using methods of cultivation which leaves the previous year's stubble anchored in the surface, and during the fallow period all cultivations are done below or on the surface in such a way that the previous year's stubble remains anchored in the surface but sticking out above it, as is shown in Plate 47a. Thus, weeds can be kept in control without the soil surface being left bare. Typically, in the wheat-growing areas where these methods were first developed, the wheat is harvested with the combine, a long stubble is left and the straw coming from the combine is also left on the ground. Thus, the wind velocity at ground level is kept very low and the straw and stubble together reduce the chance of a moving sand grain throwing up others by saltation. Chepil[2] found, in fact, that as little as 0·6 t/ha of straw had an appreciable effect in reducing soil drifting. Stubble-mulch farming may introduce some difficult problems in insect control, particularly the wheat stem sawfly and insects of the grasshopper-locust type, because they may over-winter on the stubble, but so far few troubles of this type have arisen.[3]

The use of suitable herbicides allows the minimum of cultivation to be done, and if the land is in suitable condition and the weeds can all be con-trolled economically with herbicides, it may be possible simply to drill the seed into the undisturbed soil, as will be described on p. 798. As a short-term emergency measure, windblow can often be reduced by ploughing the land, for this will bring up clods from the subsoil which will not blow until they have been broken down into finer crumbs by abrasion.

Under English conditions, it may be economic to alter the texture of the soil by adding several hundred tons per hectare of material containing

1 *Agron. J.*, 1949, **41**, 127.
2 *Sci. Agric.*, 1944, **24**, 307.
3 For an example of trouble of this type see D. A. Wilbur *et al.*, *J. Amer. Soc. Agron.*, 1942, **34**, 16.

particles finer than 0·05 mm, such as a clay marl or a power station fly ash.[1] The fly ash is very high in silt-sized particles which are effective in reducing the movement of fine sand particles by saltation.

Two points should be made in conclusion. In the first place increasing the humus content of a blowing fine sand may increase its liability to blow, for the humus will tend to soften the soil clods when they are dry, so causing them to break down by abrasion more easily. However, decomposing plant debris gives compounds of the polysaccharide type (see p. 501) that strengthen soil structure temporarily, but which later on decomposes; so that only if organic matter is added at frequent intervals may it reduce the liability to wind-blow. This is illustrated in Table 28.1 taken from work by W. S. Chepil.[2] The second point is that there are probably no effective methods of control for cultivated land subject to a drought lasting several years, for all surface cover will tend to disappear under these conditions.

TABLE 28.1   Influence of wheat straw at different stages of decomposition on the erodibility of a soil by wind. Chaffed straw added 29 September 1951. Relative erodibility: control soil = 100

	Months after straw incorporated in soil			
Soil type	7	17	29	36
Loamy sand	0·33	7·4	238	157
Sandy loam	0·08	10·1	200	827
Silt loam	0·17	20·2	255	712
Clay	15·4	9·6	156	651

Erosion by running water

Water running over the surface of the soil is a far more serious cause of erosion in most parts of the world than is the wind.[3] Water only runs off the soil surface when the rate of rainfall exceeds the rate of infiltration of the water into the soil, hence every factor that reduces the permeability of the soil increases the likelihood of water run-off. This run-off water has two important consequences: the more the run-off the greater is the 'flash' flood in the rivers draining the area after storms, and the greater is the amount of silt and soil the water is likely to pick up. This type of run-off and erosion gives very difficult problems of flood control on the one hand and silt control on the other. As examples of silt trouble can be mentioned the burying of crops under silt in the flood plains of rivers, the rapid silting up of reservoirs,

1 B. Wilkinson, W. Broughton and J. Parker-Sutton, *Expl. Husb.*, 1969, **18**, 53.
2 *Soil Sci.*, 1955, **80**, 413.
3 For a general account of water erosion and methods for its control in different parts of the world, see *Soil Erosion by Water*, F.A.O. Agric. Devel. Paper 81, 1965.

and the difficulty of keeping irrigation channels clear and irrigated fields level if the river supplying the irrigation scheme is silty.

Run-off can have a third consequence of very great importance, for the rain water, instead of sinking into the soil, runs off the land and hence cannot be used by the vegetation. In areas where crop yields are dependent on an erratic rainfall, run-off can be particularly undesirable as it increases the liability of the crop yield being seriously reduced by lack of water. On the other hand, there are circumstances where run-off is desirable and the problem is to ensure that it does not cause any unnecessary soil erosion. An example is when rain falls on land which is already wet to the bottom of the potential root zone of the crop, and it is preferable to get the surplus water off the soil surface rather than through the soil profile. Thus it is often better to run water off the surface of cropped land, if the soil permeability is low, than to allow it to lie on the surface while it slowly percolates in, since this may kill or seriously set back the crop by waterlogging.

The typical cause of serious erosion is that an unsuitable method of farming is being practised. Well-managed forests and pastures are almost immune from erosion troubles, while any system of agriculture is liable to lead to serious erosion if it involves large areas of bare soil during periods when heavy rainstorms are likely to occur. Thus, in many semi-arid parts of the world liable to long drought and heavy storms, and in which ranching is the natural method of land use, over-grazing in the dry season leaves the soil unprotected against the rain storms when they come, and hence very liable to serious soil erosion.[1] Over-grazed ranges are not only more liable to erosion, but they also suffer more from drought due to this loss of rain water and from the consequent shallow depth of the soil that is wetted by the infrequent rainstorms.[2]

Soil erosion by water involves two distinct processes under normal farming conditions—the break-up of clods into fine granules or particles, and the transport of these fine particles away from the clods.[3] The most important cause of the break-up of clods is the impact of the fast-falling raindrops in a severe storm, for they possess very considerable kinetic energy or momentum. The greater the intensity of the storm, the larger the drops are likely to be and the faster they will fall; and their velocity may even exceed that for free fall because of the air turbulence in the storm. A drop of 1 mm diameter, typical of fairly light rain will have a terminal velocity of almost 3·8 m/sec, and a drop of 4·5 mm in diameter, typical of a heavy storm will have one of 9 m/sec, but its kinetic energy will be 500 times greater than that of the smaller drop. Hence, the more violent the storm,

1 For examples see H. Glover, *Erosion in the Punjab*, Lahore, 1944; K. B. Cumberland, *Soil Erosion in New Zealand*, Wellington, 1944; J. M. Holmes, *Soil Erosion in Australia and New Zealand*, Sydney, 1946.
2 *E. Afr. agr. for. J.*, 1962, **37**, special issue, 69.
3 For a review of American work on this subject see D. D. Smith and W. H. Wischmeier, *Adv. Agron.*, 1962, **14**, 109.

the greater is the shattering effect of the raindrops, and the amount of kinetic energy dissipated in this way is very considerable. Thus, a storm at the rate of 5 cm/hr is dissipating power at the rate of 5·6 kW/ha,[1] but one of 7·5 cm/hr is dissipating power at about 520 kW/ha, nearly 100 times as fast.[2] If these storms last for 1 hour, the energy they dissipate would lift the top 15 cm of soil to a height of 90 cm and 84 m respectively. Raindrops falling in storms of low intensity, less than 2 or 3 cm/hr do not usually have enough energy to shatter clods, though they may cause the dry clod to slake on wetting. Most of the rain in the United Kingdom falls in storms of low intensity, but in areas subject to heavy rainstorms over half of the annual rainfall may fall at higher rates. Hence, liability to water erosion is conditioned by the frequency of serious storms.

Raindrop impact shatters clods and causes splash, and the droplets splashed up carry fine soil granules and sand grains, most of which are smaller than 0·2 mm.[3] Plate 48 is a flash photograph of a raindrop causing splash when hitting a water film, and it shows the production of droplets, each of which can carry fine soil particles. If erosion is not severe, this results in a proportion of these finer particles, which contains much of the soil humus, dispersing in the water and running off the land, leaving the coarser sand particles behind. Thus erosion due to splash tends to coarsen the texture of the soil.[4] In practice the effect of this type of erosion on texture is small because the velocity of run-off of the water in which the fine particles have been dispersed is usually sufficient to carry most of the sand particles with it, and so tends to remove the soil fairly evenly from the soil surface: this is therefore known as sheet erosion. These finer dispersed particles will also clog up all the coarser pores in the soil surface and this, coupled with the levelling of the soil surface due to break up of isolated clods and its compaction by the impact of the raindrops, causes a surface cap to be formed, which then becomes much more resistant to break-up by impact and splash. There is a tendency for clods high in organic matter to be softer and weaker when wet than if low, so they break up more easily in storms, and more of the humus-containing fine fraction is liable to be lost in run-off.[5] Thus increasing the humus content of soil does not increase its resistance to break-up by raindrops because the wet strength of its clods is low and the cohesion between the individual soil granules is weak. On the other hand, soils whose structure is stabilised by iron hydroxide cements, such as those derived from iron-rich basalts or basic igneous rocks, have crumbs which have a very considerable wet strength, so are resistant to break-up by raindrops, while soils derived

1 M. L. Nichols and R. B. Gray, *Agric. Engng.*, 1941, **22**, 341.
2 W. D. Ellison, *Emp. J. exp. Agric.*, 1952, **20**, 81.
3 J. O. Laws, *Agric. Engng.*, 1940, **21**, 431; W. D. Ellison, *ibid*, 1944, **25**, 131, 181; 1947, **28**, 5 papers. P. C. Ekern, *Proc. Soil Sci. Soc. Amer.*, 1950, **15**, 7.
4 N. L. Stoltenberg and J. L. White, *Proc. Soil Sci. Soc. Amer.*, 1953, **17**, 406; L. A. Forrest and J. F. Lutz, ibid., 1945, **9**, 17.
5 D. S. McIntyre, *Soil Sci.*, 1958, **85**, 261.

PLATE 48. Splash created by a water drop falling on a wet surface.

from low-iron rocks, such as granites, which also tend to give soils high in fine sand, have crumbs with a low wet strength, easily broken by raindrops, and hence very susceptible to erosion.

Water, if running over the surface of a soil as a thick sheet, will also bring soil into suspension due to the turbulence in the sheet. The thicker the sheet and the greater its velocity, the greater is its power to disperse and transport soil. Water running off a slope tends to concentrate into runnels, which tend to coalesce, resulting in torrents capable of cutting deep gullies into the land, and carrying or floating large lumps of soil and stones, and even boulders with it. This type of erosion is known as gully erosion, and leaves the land surface uneven, as more soil and subsoil are removed from the runnels and gullies than from the areas between them. It can cause spectacular erosion and can only be controlled by preventing the water concentrating into channels.

The basic principle for reducing erosion or for reducing run-off is to ensure that the water moves slowly over the land. If the policy is to conserve water, it is essential to maintain the permeability of the surface soil for as long as possible, which, in turn, involves keeping the amount of soil dispersed in the water to a minimum; and this latter condition is necessary also if the policy is to shed water, for clear water has less power of picking up soil than water already loaded with soil.

The methods available for the maintenance of permeability are first of all to protect the bare soil crumbs from the direct impact of the raindrops as far as possible. N. W. Hudson[1] did a simple experiment to show the importance of this. He protected some soil with mosquito-netting just above a bare soil surface, for this reduced the run-off from 41 per cent to 1 per cent, and the annual soil loss from 150 t/ha to 5 t/ha. A straw mulch of even 1·2 t/ha can have a marked protective effect,[2] so that stubble-mulch cultivation techniques are very valuable if the mulch itself does not have too many undesirable consequences (see p. 799); so can the leaves of a crop, although if the leaf concentrates the water that falls on it and allows it to run off in a stream, this may itself initiate soil wash. A rough tilth is better than a fine one during the initial stages of a storm,[3] partly because it will slow up the movement of water. The water can be slowed up also by other cultivation techniques, such as drawing contour ridges across the slope at frequent intervals, or by using the technique of tied ridges or basin-listing (see Plate 50 and p. 788).[4] This method, however, is dangerous if the soil has too low a

1 *Rhodesian agric. J.*, 1957, **54**, 297.
2 J. V. Mannering and L. D. Meyer, quoted by D. D. Smith and W. H. Wischmeier, *Adv. Agron.*, 1962, **14**, 109.
3 T. R. Horning and M. M. Oveson, *Agron. J.*, 1962, **54**, 229. W. E. Larson, *Proc. Soil Sci. Soc. Amer.*, 1964, **28**, 118; R. E. Burwell and W. E. Larson, ibid., 1969, **33**, 449.
4 See W. H. Cashmore and J. C. Hawkins, *J. Inst. Brit. Agric. Eng.*, 1957, **13**, 3, for equipment, and H. C. Pereira, P. H. Hosegood and M. Dagg, *Expl. Agric.*, 1967, **3**, 89. M. Dagg and J. C. Macartney, ibid., 1968, **4**, 279, for examples from East Africa.

permeability, for if heavy rainstorms follow each other too closely, the basins may be holding water from a previous storm when the next one arrives, or the soil may still be wet; for this method will allow fine silt and clay to accumulate on the bottom of the basins which reduces their permeability until they have been dried and cracked. Also water standing on the soil surface, whether in furrows or basins, will lead to reducing conditions developing in the soil which will harm or kill the crop.

Methods which maintain permeability or slow up the rate of run-off help to reduce the erosive power of the water. Thus, if a wide-spaced crop is grown in strips of land along the contour which alternate with strips of grass, for example, muddy water running off the relatively unprotected land will be slowed down by the grass, and will drop its soil between the grass blades at the top end of the strip so the clean water will have an opportunity of percolating into the soil before it reaches the lower end. However, the width of the erosion sensitive strips must not be too great otherwise the water will not be flowing uniformly off the strip, and it will also be carrying too large a load of silt for the grass strip to accept.

When surplus water must be removed from the land, this can sometimes be done by growing the crop on ridges with the furrows drawn just off the contour, so the surplus water will run off along them. If the erosion hazard is only small, the furrows can be ten or more metres apart, leaving 'broadlands' between them, which are wide enough to allow mechanical cultivation.[1] But in general the slopes in a field are too uneven to allow this, so terraces are thrown up just off the contour to collect the water running down the slope and make it to run off slowly along the back of the terrace. The terrace length, the spacing between the terraces, and the steepness of fall of the channel behind the terrace must all be designed so that the water running off does not scour the base of the terrace, nor pond up behind the terrace so it overtops or breaches the terrace.

The effect of cropping sequence on reducing soil erosion losses is now well understood in practice, though the full explanation of the results are not always clearly established because of the uncertainty of the effects of humus and organic matter on the factors responsible. As already noted, the present evidence is that increasing the humus content of a soil increases the ease with which the surface clods can be shattered by raindrops, but reduces the amount of slaking due to the wetting of the dry clods. It is likely to give surface crusts that can be broken up more easily into more crumbly clods, and in the subsurface soil it is likely to help maintain the stability of the pores and cracks down which the rain water will percolate, though it will have little protective effect if the subsurface structure is damaged by traffic or unsuitable cultivations when the soil is wet. On the other hand, non-humified organic matter in the soil surface will prolong the life of the surface

1   P. R. Hill, *Emp. J. exp. Agric.*, 1961, **29**, 337 (example from Ghana).

and subsurface channels. Table 28.2 illustrates the effect of cropping sequences on run-off and soil loss for a series of run-off plots at Zanesville, Ohio.[1] The soil is a silt loam on a 12° slope, and the average annual rainfall is 960 mm. The Table shows the magnitude of the reduction of soil and

TABLE 28.2   The effect of crop rotation on the erodability of a soil. Zanesville, Ohio: 9 year mean

Crop rotation	Average run-off of water		Soil loss, t/ha
	Cm	As % rainfall	
Continuous maize	38	40·3	246
Crops in rotation			
Maize	23	23·7	105
Wheat	23	24·8	28·2
1st year ley	18	17·7	1·5
2nd year ley	13	12·8	0·5
Permanent pasture	4	4·3	0·05

water loss which a two-year grass ley can bring about when the land is subsequently cropped to maize; but the experiment cannot answer what particular effect the ley had on the soil to reduce its liability to erosion. The Table also shows a well-known general result that it is often easier to reduce soil loss than water loss to a minor level.

The translation of these general principles to detailed planning of farm systems for erosion control depends very much on local circumstances. Plate 49 gives an example of an area of rather steep slopes that has been carefully planned to allow cultivation but to minimise soil erosion. This area is in central Kenya and can only be cultivated safely because the soils are derived from basic lavas, so have a very strong stable structure (see p. 492). But two points should be made here. In the first place it has been assumed that fields do not receive water from higher up the slope, particularly in the form of flash floods. The second point is that erosion can occur, and be of agricultural significance on very gentle slopes if the soil is impermeable. Thus self-mulching black vertisol clays on valley floors in the tropics can lose their loose surface by erosion, since water falling in heavy rainstorms must run off the surface. Again, some fine sandy soils will lose permeability easily and can give serious run-off and erosion even on slopes as gentle as 1 per cent.

Water conservation in semi-arid regions

The outstanding characteristic of most semi-arid regions of the world where the rainfall is just adequate for arable cropping to be practised is that the

1 *U.S. Dept. Agric.*, *Tech. Bull.* 888, 1945.

PLATE 49. Example of a well-planned soil conservation scheme in an area of small farms on hill slopes in the Central Province of Kenya. The soils are deep tropical red loams derived from basalt.

limited rainfall is usually concentrated in a fairly well-marked rainy season, but the total amount of rain falling in the season, and its length, are extremely variable from year to year; and crops are often liable to almost complete loss through catastrophes, such as plagues of leaf-eating insects— e.g. locusts and chinch-bugs—violent hailstorms and complete failure of the rains. Table 28.3 gives a comparison of the variation in the yield of wheat from year to year in the wheat-growing areas of the Great Plains of the USA with that of the unmanured plot on Broadbalk, Rothamsted. The mean yield of wheat is about the same, but whereas yields of below 5 bushels per acre occur on the average one year in six in the dry-farming area, they only occur in one year in thirty under humid conditions, with a correspondingly far greater probability of famine occurring among populations subsisting under dry than under humid farming conditions.

The crops normally grown between the region of mixed farming and the desert are grain crops, and they are usually only taken during the period of the year when rain is expected. Wheat and barley are the typical grains of the cooler regions and sorghums and millets of the hotter. Wheat needs a longer growing period and rather heavier soils than barley, so barley is the most suitable, and also the typical, crop grown on the edge of deserts having winter rainfall, such as those in North Africa and Arabia. But wheat is the preferred bread corn of the world, so it is much more widely cultivated, and is often taken in places where barley would do better.

TABLE 28.3   Variability of wheat yields in semi-arid and humid regions

Range	Failure	0–5	5–10	10–20	20–30	30–40	Over 40	Mean Yield
Per cent of years in which the yield, in bushels* per acre, within the following ranges								
Great Plain†	4	13	18	29	23	11	2	16
Broadbalk‡ (unmanured)	—	3	24	70	3	—	—	13

*   The American figures are in Winchester, and the Broadbalk in Imperial, bushels. 1 Winchester bushel = 0·97 Imperial bushel = 0·35 hectolitres. 10 bu/ac equals about 670 kg/ha.
†   E. C. Chilcott, *U.S. Dept. Agric., Misc. Circ.* 81, 1927. This is based on 218 crop yields at 16 field stations.
‡   Sixty-year mean, 1852–1911.

Similarly, sorghums and millets can better withstand summer droughts than maize,[1] so are the typical summer cereals of the Arabian, Indian and African semi-deserts, and sorghum is the most reliable grain crop in the southern Great Plains of the United States; but in many areas maize is increasingly being grown in what were considered sorghum areas because it is the preferred human food crop.

Crop yields are not proportional to the rainfall or to the available water; the crop must use a considerable quantity of water before it can give any grain yield at all. Thus, over a run of years in Kansas,[2] wheat only begins to give a yield if the rainfall exceeds 200 to 250 mm, and for every additional 10 mm of rain above this the yield increases by about 50 kg/ha, whereas in the cooler summers of Saskatchewan,[3] 250 mm of rain will give a yield of about 860 kg/ha and every additional 10 mm of rain will increase the yield by about 170 kg/ha. This has the consequence that the grain–straw ratio usually increases with rainfall.

Crops growing in areas of erratic rainfall must be able to withstand drought without their yields being too seriously affected.[4] Sorghums and wheat can recover from a severe wilt occurring at any time up to the onset

1 For examples from East Africa see H. Doggett and D. Jowett, *J. agric. Sci.,* 1966, **67**, 31.
2 J. S. Cole, *U.S. Dept. Agric. Tech. Bull.* 636, 1938. O. R. Mathews and L. A. Brown, *U.S. Dept. Agric., Circ.* 477, 1938.
3 *Soil Res. Lab., Swift Current., Sask., Rept.* 1943.
4 For general reviews on the effect of drought on crop development see *U.N.E.S.C.O. Arid Zones Res.,* **15**, 1960; **16**, 1961.

of heading,[1] but a maize, which is more drought-sensitive, will recover from a wilt occurring up to the three or four leaf stage, but will suffer from a subsequent wilt. This is due in part to the first four maize leaves having a more complete covering of cutin on their outer surface than the later leaves so lose less water when the stomata are closed, while all sorghum leaves are protected; this may also account for the greater resistance to a wilt of the guard cells around the sorghum stomata than around the maize.[2] However, the yield of all cereals is reduced if the crop suffers a wilt during the period from ear emergence to early grain development,[3] as is shown in Table 28.4

TABLE 28.4 Effect of available water at heading time on the yield of milo

Total water available from the end of July to early September	Continuous grain		Alternate grain and fallow	
	Per cent of years	Mean yield kg/ha	Per cent of years	Mean yield kg/ha
Under 10 cm	47	670	28	670
10 cm and over	53	1880	72	2620

for the milo variety of sorghum at Dalhart, Texas.[4] This sensitivity to drought need not be shown by crops with an indeterminate habit of flowering, such as cotton or groundnuts, because if drought kills or upsets the fertilisation of one set of flowers, the crop will produce more flowers after the rains have come again. Thus although cotton and groundnut leaves are less well protected than sorghum to resist a wilt,[5] these crops may give a yield in seasons when sorghums almost fail, if drought hits the sorghum at its critical period.

The grain yield in cereals is dependent on photosynthesis both in the last-formed top, or flag, leaf in wheat and barley or the top one or two leaves in maize and sorghum, and in the green shoot and head. It is therefore independent of the size of the lower leaves, so a crop can still give a good yield even if its growth was seriously set back in the early period, provided the drought did not affect the size of the last-formed leaf or the number of florets that set seed. The yield of dry matter, on the other hand, may have been seriously reduced.

The system of soil management to be adopted must usually ensure that as far as possible all the rain percolates rapidly into the soil where it falls, and that the maximum proportion of this water remains available to the

1 For an example with wheat see J. J. Lehane and W. J. Staple, *Canad. J. Soil Sci.*, 1962, **42**, 180.
2 J. Glover, *J. agr. Sci.*, 1959, **53**, 412.
3 H. A. Nix and E. A. Fitzpatrick, *Agric. Met.*, 1969, **6**, 321 (for wheat and sorghum); A. H. El Nadi, *J. agric. Sci.*, 1969, **73**, 261 (wheat in the Sudan).
4 O. R. Mathews and B. F. Barnes, *U.S. Dept. Agric.*, Circ. 564, 1940.
5 R. O. Slatyer, *Aust. J. agric. Res.*, 1955, **6**, 365.

crop. Stubble-mulch cultivations were designed expressly with this end in view, for they loosen the surface so rain can penetrate, they leave a mulch on the surface to help reduce evaporation, and they will kill weeds to reduce transpiration (see p. 798). If the permeability of the soil surface is inadequate for the soil to accept all the rain that falls in a storm, laying the land up in ridges and furrows drawn on the contour[1] will increase the amount of water percolating in. Still more water can be made to percolate into the land if the

PLATE 50. Maize growing on ridges with the furrows tied to conserve water, on the Cotton Research Station, Namulonge, Uganda.

furrows are tied, that is if small ridges are thrown up across the furrows at regular intervals, so that each furrow becomes a row of basins.[2] This technique, known as tie-ridging by British workers and basin-listing by American, can be a very valuable method of conserving water, as can be seen in Plate 50. J. E. Peat and K. J. Brown[3] showed that land cultivated in this way gave higher yields of cotton at Ukiriguru, western Tanganyika than other methods in the drier years; but because tied ridges will tend to give waterlogging in

1 See for example P. D. Walton, *Emp. Cott. Grow. Rev.*, 1962, **39**, 241, and M. Amemiya, *Agron. J.*, 1968, **60**, 534.
2 See, for example, W. H. Cashmore and J. C. Hawkins, *J. Inst. Brit. Agric. Engrs.*, 1957, **13**, 3; J. D. Lea, R. J. Ofield and R. G. Passmore, *E. Afr. agr. J.*, 1960, **25**, 220.
3 *E. Afr. agr. for. J.*, 1960, **26**, 103. *Emp. Cott. Grow. Rev.*, 1963, **40**, 34. See also M. Dagg and J. C. Macartney, *Exp. Agric.*, 1968, **4**, 279.

wet seasons if the soil has a low permeability, D. A. Lawes suggested tying alternate rows only, allowing the water to run off along the untied furrows.[1]

It is not always desirable to let all the rain penetrate into the soil in regions having a very variable rainfall, for in some years the rainfall during the growing season may greatly exceed the crop transpiration. But even if the potential transpiration exceeds the seasonal rainfall the rain cannot always be used by the crop. In many regions much of the rain falls in the beginning of the season, and the maximum amount of water the soil can store which is available to the crop is that within its root range at the time of maximum root development. The root system of cereals and grain legumes ceases to grow by the time the crop is in flower, and these crops normally do not root much deeper than 120 to 180 cm, depending on the soil. As soon as the soil has been wetted to a depth somewhat greater than this, further rain cannot be stored in the profile for use by the crop later on in the season, so it may be preferable to run any surplus gently off the surface rather than risk it ponding on the surface, so inducing anaerobic conditions in the soil.

The system of manuring used must encourage the crop to root as deeply as possible. If the soil is low in phosphate an adequate dressing, typically of a water-soluble phosphate placed close to the seed, should be used. It may also be necessary to use some nitrogen, if the surface soil is low in nitrates, for this helps the early growth of the root system, and often hastens the time the crop comes to flowering when this is not controlled by day-length or change in day-length. If the dressing is kept fairly low, it does not increase the water demand of the crop too much, and in Nebraska R. A. Olson[2] obtained worthwhile responses with maize at a cost of about 2 to 5 cm extra water transpired; but if too high a dressing of nitrogen is given, it will encourage rapid leaf growth which will not contribute directly to the grain yield but, by increasing the transpiration early on in the season, will increase the risk of water shortage at the critical heading-seedset period of growth.[3] On the other hand, an additional nitrogen supply later on in the season will not increase transpiration appreciably but may increase the size of the flag leaf and of the head, on which the carbohydrate supply to the grain will depend. However, a top dressing can only be effective if good rains occur shortly before heading, for otherwise the nitrogen will remain in the dry soil.

However, the subsoil water typically contains nitrates which were formed either at the end of the previous season, if the soil still contains moisture after the crop has come to maturity, or at the beginning of the current season when the soil is wetted by the early rains. The former source of nitrates is dependent on the length of the growing season of the previous crop: the shorter the season, the greater the amount of subsoil nitrate is

---

1 *Emp. J. exp. Agric.*, 1961, **29**, 307.
2 With C. A. Thompson *et al.*, *Agron. J.*, 1964, **56**, 427.
3 For an example with wheat see R. A. Fischer and G. D. Kohn, *Aust. J. agric. Res.*, 1966, **17**, 269.

likely to be. Thus fallows and short-season crops tend to increase both the subsoil water and nitrates compared with long-season crops; and in the field it may be difficult to decide which of these two effects make the more important contribution to the yield of the following crop.[1]

The water supply to a crop can be increased to some extent by growing the crop in alternate years only and taking a fallow in the other. It must be a weed-free fallow, obtained by stubble-mulch cultivations and herbicides. This practice is common in the wheat belt of the Canadian prairies and the American Great Plains, where the wheat is grown in strips across the direction of the prevailing wind, as shown in Plate 47b, to protect it from wind-blowing. The amount of water that can be stored in a fallow in the Great Plains only amounts, on the average, to about 16 per cent of the rainfall,[2] because the only water that can be stored is that which percolates below about 20 cm depth, for most of the moisture above this depth is lost by evaporation. Thus only when there are a series of storms or rains following in close succession will there be any contribution to the subsoil moisture. The actual figures given by Mathews and Army are that on the average there are 5 cm of water stored in the soil at seeding time with continuous wheat compared with 10 cm with wheat after a year's fallow. This extra water, however, not only increases the yield but reduces the liability to crop failure quite appreciably for, as is shown in Table 28.5,[3] the important factor con-

TABLE 28.5   The effect of soil moisture at sowing time on the yield of spring wheat (Great Plains, Montana to Texas). Yield in kg/ha.

Rotation	Depth to which soil is wet at sowing time						Overall mean yield
	30 cm or less		60 cm		90 cm or more		
	% of years	Mean yield	% of years	Mean yield	% of years	Mean yield	
Continuous wheat	33	440	41	780	26	1050	750
Fallow—wheat	5	463	21	850	74	1340*	1190

* Soil often wet deeper than in continuous wheat.

trolling yield is the depth of moist soil at sowing time. The table shows that if the fallow stored on the average an extra 5 cm of water, each 1 cm has increased the yield by between 80 and 90 kg/ha averaged over the region, which is a little larger than the increases of 50 to 80 kg/ha found by W. C. Johnston[4] for this region. However this table also shows that the total pro-

1 For an example where both factors are operative see R. N. Bennison and D. D. Evans, *J. agric. Sci.*, 1968, **71**, 365.
2 O. R. Mathews and T. J. Army, *Proc. Soil Sci. Soc. Amer.*, 1960, **24**, 414.
3 J. S. Cole and O. R. Mathews, *U.S. Dept. Agric. Circ.* 563, 1940; for the corresponding figures for milo (a grain sorghum) ibid., *Circ.* 564, 1940.
4 *Agron. J.*, 1964, **56**, 29.

duction of wheat from one hectare over the years is higher under continuous wheat than if wheat is alternated with fallow, though the gross cost of keeping the land in cultivation is also higher. W. Baier[1] has given an example from southern Saskatchewan of the calculation of the economic returns from these two systems based on the knowledge of the rainfall variability over the growing season.

The water supply to individual plants can sometimes be increased by increasing the spacing between the plants, that is by reducing the number of plants per hectare, which increases the volume of soil, and so of water available to each plant. The amount of water required to bring a crop to harvest can also be reduced by reducing the length of the growing season and so the number of days it is transpiring water. But while both these methods are likely to be advantageous in seasons when the rainfall is low, they are likely to reduce yields in seasons when it is high. This is illustrated in Table 28.6 for some results by B. D. Dowker[2] with two varieties of maize

TABLE 28.6 The effects of spacing, seasonal rainfall and variety on the yield of maize. Machakos District, Kenya, 2 sites, 2 crops per year, 1958–61. Yield in kg/ha

Yields	Rainfall in cm	No. of seasons	Short season (Taboran) Spacing in m		Long season (Local White) Spacing in m	
			0·9 × 0·6	0·9 × 0·3	0·9 × 0·6	0·9 × 0·3
Failure	Under 38	3	All crops failed			
Low	48–58	4	680	590	300	90
Medium	60–75	2	1230	1580	1100	560
High	65–170	6	1800	2800	2670	3160

in an area with erratic rainfall in the Machakos District of Kenya. The variety Taboran comes to 50 per cent silking about 15 to 20 days before the local white, except in seasons of really poor rainfall when the local white may never reach this stage. It shows that in really dry seasons, the short-season low-plant density crop gives the highest yield, although a poor one; in seasons with an average rainfall, the short-season high-density gives the highest yield; while in wet seasons the long-season high-density gives the highest yield, which is also a really useful one. But his results also show that seasonal rainfall is not the sole factor determining the response to spacing, for in one season 75 cm rainfall was inadequate for the long-season crop, while in three other seasons 65 to 75 cm gave a good yield. This is because there is no necessary relationship between plant number per hectare and transpiration demand, or between seasonal rainfall and water supply at critical periods in the growth of the crop. Table 28.6 also illustrates the

1 *Agric. Met.*, 1972, **9**, 305.
2 *J. agric. Sci.*, 1971, **76**, 523.

important point that the mean rainfall of a district may be adequate for reasonable crops, but in some seasons the crop is growing under great water stress and in others under conditions of a great surplus of water.

The pattern of planting a crop affects crop yields. This is illustrated in Table 28.7 which gives the yield of sorghum at Katherine in northern Australia at two densities of plants, using three different row widths.[1] It shows that when water is short in the early part of the season, it is better to have a close spacing within the rows and a wide spacing between the rows, whereas when it is plentiful it is better to have a more even distribution of plants over the area.

TABLE 28.7   The effect of plant distribution on response to water shortage. Sorghum: Katherine, N.T. Australia. Grain yield in t/ha

Rainfall in first 12 weeks	23 cm		38 cm	
Plants per ha, thousands	50	100	50	100
Row width in cm				
17	1·24	1·51	2·22	2·47
45	1·71	1·91	2·30	2·33
90	2·31	2·26	2·04	2·16

It is becoming possible to compute the amount of evapotranspiration by a crop from seedtime to harvest knowing the climate, the rate of development of the crop and its likely potential evapotranspiration week by week during the season, and the water-holding capacity of the soil down the profile.[2] It is also becoming possible to give quantitative expression to the variability of the more important characteristics of the climate for a region from year to year.[3] This will help the plant breeder specify the growth characteristics needed by a crop if it is to be as well adapted as possible to the region; and given the growth characteristics of an existing crop variety, it will help the agronomist forecast the most appropriate system of management of the crop. In particular crop yields can be very sensitive to planting dates in some years,[4] and in so far as these are caused by climatic factors such as availability of water or temperature, it is possible to forecast the optimum dates for different districts in the region if the variation of climate over the region is known.[5]

1 L. J. Phillips and M. J. T. Norman, *Aust. J. exp. Agric. anim. Husb.*, 1962, **2**, 204. For other examples see D. W. Grimes and J. T. Musick, *Agron. J.*, 1960, **52**, 607; J. J. Bond, T. J. Army and O. R. Lehman, ibid., 1964, **56**, 3.
2 See for an example M. Dagg, *E. Afr. agr. for. J.*, 1965, **30**, 296.
3 For examples see H. L. Manning, *J. agric. Sci.*, 1950, **40**, 171. *Proc. Roy. Soc.*, 1956, **144** B, 460; T. Woodhead, *Exp. Agric.*, 1970, **6**, 81, 87.
4 For examples of this variability for crops in Tanzania see B. C. Akehurst and A. Screedharan, *E. Afr. agr. for. J.*, 1965, **30**, 189.
5 For an example with cotton in Uganda see D. A. Rijks, *Cott. Grow. Rev.*, 1967, **44**, 247.

Removal of excess water by drainage

Artificially draining a soil is a necessary operation under two distinct conditions: if the soil has a high water-table, or if excess surface water cannot penetrate reasonably rapidly to below the root zone of the crop.

Soils having a water-table sufficiently high for the capillary fringe, i.e. for the soil that is kept wet by capillarity above the water-table, to reach the surface always need to have their water-tables reduced. The minimum allowable depth of the water-table below the soil surface depends on the crop to some extent: thus fruit trees need a deeper water-table than grass, for if it is too high their root system is too shallow to anchor them firmly against high winds when the soil is wet. It is probable that agricultural crops can only be cultivated easily in temperate regions if the water-table never rises to within 60 to 90 cm of the surface, but it is usually reckoned that it should be below 90 cm, though rather higher ones are allowable for pastures or meadows.[1]

The optimum depth of the water-table, however, depends on circumstances. On peat and fen soils, which oxidise away rapidly on cultivation, it is usually considered desirable to have as high a water-table as possible, for the higher the water-table the slower the rate of oxidation of the peat.[2] However, English fen farmers, for some reason that has not been established, do not agree with this: they like it to be as deep as possible. Again, if the ground water contains enough dissolved salts to harm the root system of the crop, it should be sufficiently deep for its capillary fringe to be below the bottom of the crop's root system, and depths of less than 3 m are considered undesirable.

The second type of soil needing draining is that in which surface water will not drain away through the soil quickly enough, with the consequence that it either stands in pools on the field, or else it fills all the main soil pores and prevents the plant roots being adequately aerated. The main principle to bear in mind when rectifying defects of drainage under these conditions is that water can only move at an appreciable rate through cracks or other discontinuities in the soil structure: it cannot move through the constitutional pores, i.e. the pores between the constituent soil particles forming the soil structure, of even light loams rapidly enough for practical purposes. Drainage can, therefore, only take place through cracks and spaces between soil crumbs in medium and heavy soils; and soils can be in need of drainage either because there are not sufficient coarse pores to let the water through the superficial layers quickly enough, or because the permeability of the deeper subsoil is too low to let the surface water drain away down to the

1 For experimental results on the effect of depth of water-table on crop yield, see R. Schwarz, *Kulturtech.*, 1932, **35**, 448, for canary grass; H. B. Roe, *Minnesota Agric. Expt. Sta., Bull.* 330, 1937; H. Burgevin and S. Hénin, *Ann. Agron.*, 1943, **13**, 288; H. H. Nicholson, D. H. Firth, *et al.*, *J. Agric. Sci.*, 1951, **41**, 149, 191; 1953, **43**, 95, 265.
2 For an example from the Florida Everglades, see B. S. Clayton and J. R. Neller, *Proc. Soil Sci. Soc. Flor.*, 1943, **5** A, 118; J. R. Neller, *Soil Sci.*, 1944, **58**, 195.

water-table. If the cause of the bad drainage is due to the water being unable to percolate through the surface layers of the soil because of insufficient coarse pores, then proper cultivation or soil management is adequate to remedy this defect. The cause of this trouble may be due either to muddy water running over the land and depositing its load of mud into the coarser pores, or to rain breaking up the surface tilth with the consequent filling up of these pores: preventing muddy water entering the area in the first case, and protecting the surface against the rain in the second, are the proper methods to be employed. If a compact subsoil is the trouble, deep cultivation, or deep ploughing to get underneath the compacted layer, is required. If the cause of the bad drainage is that the deeper subsoil is too impermeable to allow water to percolate below the root range of the crop quickly enough, then the field must be drained artificially. This is done by digging a suitable system of ditches and, if necessary, underdraining the land between them, using either a system of tile or mole drains.[1] Thus, the water falling on the surface of the soil can flow through cracks in the soil into the drains and through the ditches away from the area. If, but only if, the drainage system allows surplus subsoil water to be removed rapidly, its efficiency can often be increased by deep tillage, either ploughing or subsoiling, as this increases the number of channels the water can seep through from the surface into the drains.

The success of a drainage system depends on the stability of the soil structure, i.e. both on the length of life of the various fissures and wide pores in the soil that let the water through into the drains, and also on the clearness of the water entering the drains. Clay soils typically have a stable structure in England and troubles due to silting up of the drains and blocking up of the wide pores are unusual. Their stability may be increased by liming the soil, as the drainage water from unlimed clay soils may be more turbid than from limed,[2] and decreased by heavy dressings of sodium nitrate, using the same criterion of the increased turbidity of the drainage water.[3] The great difficulties in England arise with soils high in fine sand and silt, for these have a weak, unstable structure. So much silt may enter the field drains that devices such as silt traps are essential, and very careful management of the land is needed to keep the wide soil pores open, for they are rapidly filled up with muddy water unless considerable skill is used. The permeability of these soils can be maintained by proper cropping and in particular by the use of potentially deep-rooting crops such as lucerne. Such crops not only open up new

---

1 For a discussion of the principles of field drainage, see M. G. Kendall, *Practical Field Drainage* (a series of articles in *Farm Implement and Machinery Review*, 1944, **70**); H. H. Nicholson, *The Principles of Field Drainage*, Cambridge, 1942; J. L. Russell, *J. agric. Sci.*, 1934, **24**, 544; *The Drainage of Agricultural Lands*, Ed. J. N. Luthin, Agron. Monog. 7, 1957. For a recent review see B. D. Trafford, *J. Roy. Agric. Soc. Engl.*, 1970, **131**, 129.

2 E. C. Childs (*J. agric. Sci.*, 1943, **33**, 136) found 260 ppm of silt and clay in the drainage water from an unlimed clay pasture and only 70 ppm from an adjacent area recently limed.

3 A. D. Hall, *Trans. Chem. Soc.*, 1904, **85**, 964.

channels through the soil but they also tend to stabilise the soil structure. It has not yet been established whether lucerne leys will be helpful on the silty wealden clays of south-east England where they are now being tried, but they can improve the permeability of silt loams in Iowa.[1]

Drainage improves the fertility of a medium to heavy soil for many reasons. It improves the aeration of the soil if the soil contains any appreciable volume of coarse pores that are emptied as a consequence of the draining; but it also improves the aeration of heavy soils having practically no air-space when drained, for it removes all stagnant water from the crevices and cracks in the soil, and allows oxygenated rainwater to penetrate rapidly into the subsoil. In consequence of this improved aeration of the subsoil, and removal of stagnant unoxygenated water, crops can develop a deeper root system, and this is particularly important during the winter when transpiration is low. A young plant of winter wheat can, for example, carry a well-developed root system in these cracks on a heavy soil, so that as soon as the warmer weather of spring comes, it can start growing quickly as its extensive root system can tap a large volume of soil for nutrients. The corresponding plant on the undrained clay will only carry a very restricted root system, of which only a small proportion is capable of absorbing water and nutrients, as explained in chapter 17, so it will make very slow growth in the spring, as it will not be able to tap a sufficient volume of soil for an adequate nutrient supply. The stronger plant on the drained soil will also be able to withstand periods of rapid transpiration during a drought much better than the weaker plant on the undrained, because of its much more extensive root system at the onset of the dry period. This effect of deeper rooting and greater vigour of the crop will also allow it to dry out the soil more during the drought, hence increase the amount of cracking that occurs in the subsoil, and hence again allow water to drain through the soil more easily in the following wet weather. Thus, vigorous crops on heavy land can help create, and stabilise when created, the cracks and fissures which are essential for the downward movement of water and the improvement of the aeration.

Draining surplus water away has other desirable consequences. Drained land may warm up quicker in spring than undrained, particularly if drainage removes an appreciable proportion of the water in the soil, i.e. if the soil has an appreciable specific yield. The surface soil will warm up quicker because its specific heat is reduced by the surplus water being removed, but the same result applies to the subsurface soil, though there appears to be very little information available on the magnitude and importance of this warming-up at depths of, say, 10 cm. The combined effect of improved aeration, and possibly higher temperatures, encourages microbiological decomposition of plant residues, leading eventually to an increased rate of nitrification.

1 G. W. Musgrave and G. R. Free, *J. Amer. Soc. Agron.*, 1936, **28**, 727.

Draining can have one other important effect for autumn-sown crops or for leys and grassland. The surface of a waterlogged soil tends to heave, or be lifted upwards, when it becomes frozen, with the consequence that if any plants are growing in the soil they get lifted up with the surface and may have all their roots broken in the process. Heaving of soil in frosty weather is almost confined to soils containing an appreciable volume of water held in the coarser pores, hence draining a soil always reduces the liability of the surface to heave, and its magnitude if it does.

Principles of soil cultivation

Soil cultivations have a number of functions. They may be used to obtain a seedbed, to kill weeds, to undo the damage done by previous traffic over the land or previous cultivations, or to increase the permeability of the surface soil or subsoil to water, which will allow better drainage and aeration in the soil so better root penetration.

The basic requirements for a good seedbed are that the seed should be placed at a uniform depth, in good contact with the soil so it can take up water easily, yet the soil must be well aerated. The soil above the seed must remain sufficiently loose for the seedling to grow up through the soil, and the pore space around the seed must contain sufficient wide pores to maintain good aeration and to allow the easy growth of the young rootlets. There must also be an adequate supply of essential nutrients close to the seed without the osmotic pressure of the soil solution being raised appreciably. Further there must be an absence of weeds in the seedbed, for the young crop is usually very easily set back, and the final yield reduced, by even quite a small amount of competition at this period.[1] It does not matter what cultivations are used to obtain these conditions, as has been shown by numerous experiments in which different systems of cultivations have been compared;[2] and in so far as these different systems have resulted in different crop yields, the cause has nearly always shown up in the early stages of growth.

The traditional method of obtaining a seedbed in western European agriculture is to plough the land with a mouldboard plough, which turns a furrow slice, then to work this furrow slice down to a suitable tilth for the seedbed with cultivation implements such as cultivators, harrows and rolls, taking advantage of the amount of break-up caused by the weather. The plough, if properly used, is a very efficient implement for levelling land rutted by harvesting equipment, for example, for killing many perennial

---

1 L. Kasasien and J. Secyave, *PANS*, 1969, **15**, 208; J. H. Neto, M. A. Brondo and J. T. Gonzalez, ibid., 1968, **14**, 159.
2 For Rothamsted results see J. R. Moffatt, *J. Inst. Agric. Engnrs.*, 1970, **25**, 161, and E. W. Russell and B. A. Keen, *J. agric. Sci.*, 1941, **31**, 326. For vegetable crops in England J. S. Wolfe, *Expl. Hortic.*, 1965, **12**, 72.

weeds, and for burying surface trash so the subsequent cultivation implements and drills can work properly. It also allows the soil within the depth
of ploughing to be mixed with an efficiency dependent on the type of plough
used and the way it is set, so it allows phosphate fertilisers and lime to be
distributed throughout this depth.

The mouldboard plough has, however, many limitations. If used when the
soil is wet, the mouldboard is likely to compress some or all of the soil in the
furrow slice leaving it in large clods which have subsequently to be broken
and weathered down to a seedbed. An example of this is illustrated in
Plate 28b. This stage can be very difficult to do if the weather is unfavourable. Thus, if the soil lies wet, it cannot carry the tractor without the wheels
compacting the soil badly, and the implements can rarely do a satisfactory
job if the soil is wetter than the lower plastic limit, that is if it is sufficiently
moist for the crumbs to be moulded into a clod under moderate pressure.
The moisture content of a soil at its lower plastic limit is drier, and for many
soils of medium to heavy texture appreciably drier, than that at field capacity,
when it is as dry as it can be under the action of drainage. This limits very
seriously the opportunities for cultivations in rainy periods. On the other
hand if the land was ploughed wet, it can be very difficult to work the furrow
slice down to a seedbed in a period of prolonged dry weather, for it tends to
dry out into hard clods which cannot be broken down into crumbs. Further,
if traditional cultivations are carried out in dry weather, these lead to the
surface layer of soil drying out rapidly, for the cultivations will bring moist
crumbs or clods from below the surface up to the surface where they will
dry out. Ploughing typically has a further limitation if done in moist or wet
conditions, for the plough share will compress the soil below it, and much
more important, the driving wheel of the tractor running on the furrow
bottom will compact it and, if it is slipping, will smear its surface causing all
the coarser pores to collapse. These plough pans can restrict drainage and
root penetration very seriously and, by allowing water to pond up on them,
allow anaerobic conditions to be set up. If the soil structure is weak and
destroyed by these conditions, the soil above the pore will slump down on
to it, thickening it, and so accentuating the anaerobic conditions during wet
weather as noted on p. 505.

There is a great deal of current interest in developing methods of cultivation that do not have the drawbacks of the traditional plough. Ideally, one
would like to go from last season's stubble to next season's seedbed in one
operation. It is, for example, possible to modify the plough by fitting a
hydraulic motor to it to make the mouldboard rotate,[1] which not only helps
to drive the plough forward, so reducing tractor-wheel slip, but breaks up
the furrow slice leaving the land in a fairly uniform tilth without any large
clods. Rotary cultivators can also be used to dig up the soil and break it up

1 *Nat. Inst. Agric. Engng., Ann. Rep.* 1970, 65.

into crumbs and clods. These have the disadvantage that the rotating tines tend to compact and, if the soil is wet, to smear the soil just below the depth of cultivation, and they tend to leave the cultivated soil so loose, that it will slump in wet weather and may not carry the tractor well during the drilling of the seed or the spreading of fertiliser. These disadvantages can be minimised by suitable design of the rotating tines and by suitable adjustments to their speed of rotation.

The introduction of herbicides has increased the opportunities for replacing the mouldboard plough by more rapid and less power-demanding operations and, in particular, it is allowing the wider use of rigid-tine cultivators, or chisel ploughs, as the principal operation for loosening and levelling old stubble land. Under favourable moisture conditions a seedbed can be produced rapidly, but it leaves the land looser than after a plough, with the consequence that the land is less able to carry a tractor if wet weather sets in between the initial loosening and the preparation of the seedbed.

The extreme example of relying on herbicides for weed control, and doing minimum amount of cultivation for a seedbed, is the technique of slit seeding for cereals. Here, either a thin ribbon of soil is loosened into which the seed is placed if the soil is inherently crumbly, otherwise the drill merely cuts a slit in the old stubble or land surface and drops the seed into it. This technique can be quite satisfactory on friable soils,[1] but on stiff soils, such as clays or heavy loams or on compact soils, the seed is not properly covered, and germination tends to be uneven. Further, in these soils there may be only a few pores wide enough for the young rootlets to enter, so the root system is often very restricted in the early part of the season—a time when the crop should be developing a good root system quickly; nor is it possible to get even germination if the soil surface is uneven or the soil wet. In Great Britain an extra 50 kg/ha N is required when cereals are slit-seeded.

There are circumstances when the previous season's stubble and crop remains are wanted on the surface, so they will protect the soil from wind or water erosion and help maintain the permeability of the soil to rain water. The need was particularly felt in the Canadian prairies and the American wheat belt, where water conservation is of vital importance and where cropping alternate years is often practical to allow the soil to become moist to the full root range of the crop under conditions of low erratic rainfall. The technique developed is known as stubble-mulch cultivation, and it is designed to loosen the soil below the surface, and kill weeds, while keeping as much of the stubble and straw on the surface (see Plate 47). This can be accomplished, for example, by cultivating shallow with cultivators fitted with wide sweeps, so they cut and loosen the soil 5 to 10 cm below the surface, or with rotary weeders which have rotating knives that cut and loosen the

1 For some American experience see G. M. Shear and W. W. Moschler, *Agron. J.*, 1969, **61**, 524; J. N. Jones, J. E. Moody and G. M. Shear, ibid., 1968, **60**, 17. For some British experience see A. E. M. Hood, *Outlook on Agriculture*, 1965, **4**, 286.

surface, or with light reciprocating harrows, for example.[1] These implements give less subsoil compaction that does normal disc ploughing,[2] and
the increased biological activity and organic matter in the soil surface
increase the porosity and stability of the surface tilth.[3] Weed control carried
out by these cultivation techniques often results in a smaller loss of water
from the soil than if only herbicides are used on the uncultivated land.[4]

These methods depend on there being an adequate amount of stubble and
residue from the previous crop—and 2 to 4 t/ha of wheat straw gives considerable protection—and on the soil surface being level and the surface soil
moist or dry. They are very helpful in protecting the soil against erosion and
maintaining permeability, but they have two limitations. In regions with
cold winters where spring-sown crops, such as maize, are grown, a stubble
mulch reduces the rates at which the soil warms up and the rate it dries after
the snow melt.[5] Under some conditions, not clearly understood, the processes of decomposition of the straw produce soluble organic substances
which are toxic to young seedlings,[6] particularly if there are anaerobic
pockets in the surface soil.[7] These substances include both phenol carboxylic
acids and certain antibiotics such as patulin, gliotoxin and citrinin; and
F. A. Norstadt and T. M. McCalla[8] consider that their production is
favoured by conditions which allow the patulin-producing fungus *Penicillium
urticae* to flourish. They find that its activity appears to be controlled by
competitive interactions with fungi of the *Trichoderma* group; when conditions favour one of the fungi, the activity of the other is reduced.

Many of the difficulties of maintaining a dead mulch on the soil surface,
while keeping the land free from weeds have been eased by the use of herbicides, for these reduce the need for cultivation and reduce the consequent
loss of water by evaporation, so increase the amount of water stored in the
profile. They also reduce the need for these special machines.[9]

Cultivations are not only needed for preparing a seedbed, and killing
weeds, they can also help to loosen the subsoil, and break any plough pans
or compacted layers within the root range of the young crop. Deep cultivations can be done by deep ploughs, deep cultivators or subsoilers. It is still
difficult to predict when subsoiling will be helpful in British conditions,
probably because the effectiveness of subsoiling in breaking up pans and
loosening compacted layers is very dependent on the moisture content of the

1 See, for example, F. L. Duley and J. C. Russel, *Agric. Engng.*, 1942, **23**, 39; M. B. Cox, ibid.,
  1944, **25**, 175; *U.S. Dept. Agric., Farmers Bull.* 1797, 1938; 1997, 1948.
2 S. Tanchandrphongs and J. M. Davidson, *Proc. Soil Sci. Soc. Amer.*, 1970, **34**, 302.
3 J. W. Turelle and T. M. McCalla, *Proc. Soil Sci. Soc. Amer.*, 1961, **25**, 487.
4 A. L. Black and J. F. Power, *Proc. Soil Sci. Soc. Amer.*, 1965, **29**, 465.
5 For a review see T. M. McCalla and T. J. Army, *Adv. Agron.*, 1961, **13**, 125.
6 See, for example, W. D. Guenzi and T. M. McCalla, *Proc. Soil Sci. Soc. Amer.*, 1962, **26**, 456;
  V. Iswaran and J. R. Harris, *Trans. 9th Int. Congr. Soil Sci.*, 1968, **3**, 501.
7 T. M. McCalla, W. D. Guenzi and F. A. Norstadt, *Z. Allg. Mikrobiol.*, 1963, **3**, 302.
8 *Proc. Soil Sci. Soc. Amer.*, 1968, **32**, 241; *Soil Sci.*, 1969, **107**, 188.
9 D. E. Smika and G. A. Wicks, *Proc. Soil Sci. Soc. Amer.*, 1958, **32**, 591.

PLATE 51. The system of cracks produced by a subsoiler working in compacted soil at a suitable moisture content.

profile. For subsoiling to be successful, the subsoil shoe must lift the soil above it in such a way that the force of lifting produces the maximum number of cracks spreading out as wide as possible, as is illustrated in Plate 51. Further, if the benefit of subsoiling is to last, the cracks must be stable for as long as possible. This means that subsoiling is best carried out when the soil profile is just moist or not too dry, which means that it is best done in years when the autumn is dry.

Table 28.8 gives the results of a large series of experiments made in Great Britain between 1945 and 1951 on the benefit obtained by subsoiling land

TABLE 28.8 Per cent distribution of crop responses to deep tillage compared with 20 cm ploughing by soil textural classes

| | Benefit | | No | Harm | |
	High	Medium	effect	Medium	High
Clay soils (about 30 results):					
Subsoiling	19	34	41	6	0
Deep ploughing	29	24	26	12	9
Both together	27	23	27	15	8
Loams (40 results):					
Subsoiling	17	28	35	17	3
Deep ploughing	22	15	40	18	5
Both together	25	20	35	15	5
Light loams (about 30 results):					
Subsoiling	9	42	37	9	3
Deep ploughing	12	30	37	12	9
Both together	33	17	47	0	3
Sands (about 20 results):					
Subsoiling	16	16	42	5	21
Deep ploughing	5	26	27	21	21
Both together	20	13	40	27	0

High: 3·7 t/ha or over for potatoes and sugar-beet, 250 kg/ha or over for cereals.
Medium: between 3·5 t/ha and 1·2 t/ha for roots, 240 kg/ha and 120 kg/ha for cereals.
No effect: between 1·0 t/ha and −1·0 t/ha for roots, 110 kg/ha and −110 kg/ha for cereals.

to a depth of 15 to 25 cm below the normal depth of ploughing (20 cm), or by ploughing the land to a depth of about 35 cm, or by ploughing it to this depth and subsoiling to an additional 10 or 15 cm.[1] It shows the large variations in response to these treatments, but it was not possible to pick out the reasons why crops responded to deep tillage on some soils and not on others. Deep tillage has, however, been shown to improve drainage conditions and depth of rooting, and in consequence the amount of water in the

1 E. W. Russell, *J. agric. Sci.*, 1956, **48**, 129. See also R. Hull and D. J. Webb, ibid., 1967, **69**, 183.

profile available to the crop, particularly on heavy clay soils; but these improvements do not necessarily have any large effect on yield.[1] A little work has been done on the conditions under which deep incorporation of fertilisers or lime may improve yields, but the general experience is that deep incorporation rarely gives any yield benefit.[2]

Two final points on soil problems arising in cultivations. Untimely cultivations or cultivations done when the soil is too wet often lead to anaerobic conditions in the surface soil, giving a pale nitrogen-starved crop. The loss of yield due to these conditions can be reduced by subsequent top-dressing with a nitrogen fertiliser, which may have to be applied from the air if the soil is too wet to carry a tractor. A seedbed must not be too loose or the individual soil granules too fine, otherwise heavy rain may cause the surface tilth to collapse and the surface to dry out into a crust that is too hard for the seedlings to penetrate it. These crusts may only be a few millimetres thick, as is illustrated in Plate 30, and their strength is reduced if the soil is well supplied with humus, or there is a mulch on the surface. The seed itself must be in good contact with the soil, so it can suck water out of the soil to allow it to germinate without difficulty. This can often be achieved more easily by using a press wheel behind the drill rather than by rolling the land after drilling, for rolling also tends to give a surface that will collapse with heavy rain.

Mulches

The practice of applying a layer of dead vegetable waste material, such as straw, hay or old grass, composts or farmyard manure, to the surface of the soil around trees and bushes has been prevalent for a very long time in many parts of the world. These surface mulches can have very important effects on the conditions in the surface layers of the soil, and in consequence on the crops with shallow root systems. Thus, mulching has been widely used for many fruit trees and bushes and for tropical plantation crops with superficial root systems such as coffee and tea.

A surface mulch has two types of effect on the soil: a characteristic effect, from its being on the surface of the soil, and a general effect, which it would equally well have if it were ploughed into the soil, due to the plant nutrients set free as it decomposes. The primary specific effects of the mulch are confined to the superficial soil layers, which it keeps at a more even temperature, and damper and more permeable to water than the unmulched soil.

1 J. A. Hobbs, R. B. Herring *et al.*, *Agron. J.*, 1961, **53**, 313 (Kansas). E. Burnett and J. L. Tackett, *Trans. 9th Int. Congr. Soil Sci.*, 1968, **3**, 329. W. P. Unger, *Proc. Soil Sci. Soc. Amer.*, 1970, **34**, 492 (Texas). A. J. Thomasson and J. D. Robson, *J. Soil Sci.*, 1967, **18**, 329 (Nottinghamshire).
2 J. Webber, *Expl. Husb.*, 1961, **6**, 8. S. N. Adams, *J. agric. Sci.*, 1961, **56**, 127. T. A. Singh, G. W. Thomas *et al.*, *Agron. J.*, 1966, **58**, 147. G. B. Triplett and D. M. van Doren, *Agron. J.*, 1969, **61**, 637.

A surface mulch affects both the diurnal and seasonal fluctuations in soil temperature. Its principal effect on diurnal temperature is to reduce the midday maximum temperature of a bare soil under hot and dry conditions, without having any large effect on its minimum temperature. Since a bare dry soil under bright sunshine in a continental area can reach temperatures in excess of 40°C, which is too high for the normal functioning of plant roots, this cooling effect can be very valuable. Some results of T. M. McCalla and F. L. Duley[1] in Lincoln, Nebraska, give a less extreme illustration of this effect. The average daily maximum temperature from June to September in the top 2·5 cm of soil under maize was 31·2°C if the soil was clean weeded, and 23·6°C under a mulch of 20 t/ha of straw, while the average daily minimum temperatures were 17·9°C and 19·7°C respectively, giving mean temperatures of about 24·5°C and 21·6°C respectively, with an average daily range of 13·3°C and 3·9°C respectively. The mulched soil is much cooler during the heat of the day and rather warmer during the night.

The seasonal effect of a mulch is that in areas with a cold winter a mulch increases the soil temperature during winter, but reduces the rate at which the temperature rises during spring and early summer. This effect can be very undesirable, for it delays the time during which the soil becomes warm enough for the seed of the next crop to germinate; or if it is an autumn sown crop, the time it starts to make active growth in the spring. This can be particularly serious in regions with cold winters where maize is an important crop, for maize needs a fairly high soil temperature if it is to give a high germination and fast early growth. American experience is that the germination of their varieties of maize is harmed by mulches in spring time if soil temperatures are below about 23°C, but is not affected if above.[2]

The mulch slows down the rate of evaporation from a bare, wet soil very considerably for, as pointed out on p. 443, the rate of evaporation is controlled by the proportion of the energy absorbed by the soil which is used for evaporating water, and by the rate of removal of the water vapour from the region where it is being produced. So long as the wet soil is exposed to the air, the water vapour is rapidly removed by the general turbulence of the air, but if the water vapour must diffuse through a thin dry layer of soil, or through a mulch which keeps the air almost stationary, then the rate of diffusion limits the rate of evaporation. Thus, whereas a wet, bare soil can lose 12 mm of water in three to five days, it takes a mulched soil several weeks to lose this amount, so that in hot climates a mulched soil can retain against evaporation much more of the water falling as rain than an unmulched.

This effect can be important in saline soils, for mulches will reduce the rate at which salts accumulate in the surface by evaporation from the soil

1 *J. Amer. Soc. Agron.*, 1946, **38**, 75.
2 W. C. Burrows and W. E. Larson, *Agron. J.*, 1962, **54**, 19. *Trans. 7th Int. Congr. Soil Sci.*, 1960, **1**, 629. *Proc. Soil Sci. Soc. Amer.*, 1959, **23**, 428.

surface, and this will allow the osmotic pressure of the soil solution around the rootlets of seedlings, or the roots of a young plant, to remain low enough for them to function long enough to let the crop get established. If the soil is alkali, a mulch will either delay the onset of the hard surface crust, which is a typical feature of these soils, or prevent it from hardening.

A vegetable mulch will increase the permeability of a poorly structured surface soil and will help to produce a good stable structure,[1] probably because it serves as a food supply to the larger soil animals, such as earthworms and termites. Its first effect on poorly structured soils may however be to increase the length of time anaerobic conditions persist in the surface soil, for it reduces the rate at which it dries after rain or in the spring.[2] It also helps to keep the soil surface permeable. Further rain water can only reach the soil surface as a gentle stream of clear water, which causes a much smaller drop in permeability than if the soil surface itself was exposed to the beating rain. Thus, the mulch reduces the run-off of the rain and consequently reduces the amount of soil the water can carry away[1] and increases the proportion of the rain water that percolates into the soil. Thus, not only is evaporation from the surface of the mulched soil reduced, but the amount of water infiltrating into it is increased. Hence, the water-supplying power of a soil can be considerably increased by mulching as is well illustrated by some of S. M. Gilbert's results on the mulching of coffee with banana leaves in Moshi, Tanzania, given in Table 28.9.[2] The effect was large enough for the trees to be wilting on the clean weeded plots, but growing well on the mulched plots, at the time the first samples were taken.

A heavy grass or trash mulch can have an important effect on weeds, for weed growth is slow through the mulch. If mulching is done at rather long intervals, the surface soil has usually a sufficiently good structure, when the mulch has nearly all decomposed, to withstand some weed-killing cultivations,[1] when the use of herbicides is not justified or practicable.

The mulch has certain general effects on the soil due to its organic matter decomposing. The moist conditions under the mulch favour biological activity and the improvement and stabilisation of soil structure which this activity brings about. The nutrients in the mulch are returned to the soil as it decomposes. The effect of a mulch on the level of ammonium and nitrate cannot always be predicted. In so far as it keeps the soil moist for a long

1 For an example from Kenya see H. C. Pereira, M. Dagg and P. H. Hosegood, *Emp. J. exp. Agric.*, 1964, **32**, 31.
2 For example, R. B. Tukey and E. L. Schoff, *Proc. Amer. Soc. hort. Sci.*, 1963, **82**, 68, and W. J. Carpenter and D. P. Watson, *Soil Sci.*, 1954, **78**, 225.
3 For an example of the effect of the weight of mulch applied on this control, see H. L. Borst and R. Woodburn, *U.S.* Dept. *Agric.*, *Tech. Bull.* 825, 1942.
4 *E. Afric. agric. J.*, 1945, **11**, 75.
5 For an example in arabica coffee in Kenya, see H. C. Pereira with P. A. Jones, *Emp. J. expt. Agric.*, 1954, **22**, 323, with M. Dagg and P. H. Hosegood, ibid., 1964, **32**, 31.

TABLE 28.9   The effect of mulch on conserving and maintaining the water supply in the soil. Percentage moisture in the soil on the oven-dry basis

Depth of soil in cm	After prolonged drought			After a rainy period		
	0–15	*15–30*	*30–45*	*0–15*	*15–30*	*30–45*
Mulched plots	27·5	35·0	37·5	41·0	39·0	38·0
Clean weeded plots	15·5	22·0	26·0	30·0	33·0	32·5

period of time, it encourages the slow decomposition of the soil organic matter which releases ammonium ions;[1] but in so far as the mulch has a low nitrogen content, due to it consisting of mature plant residues, its decomposition will lock up ammonium and nitrate, so lowering the level of available nitrogen.[2] The effect of the mulch on the potassium supply is usually considerable because the materials used for mulch, such as mature dried grass, grain straw or stover, and dried banana trash, are high in potassium. In fact, mulching a soil with straw or old grass always increases the amount of potassium taken up by the crop, so much so that a magnesium deficiency may be induced. Since grasses take up potassium very strongly from soils, grassing land down and gang mowing the grass to keep it short raises the level of available potassium in the surface soil. This practice is widely used in English apple orchards, for apples do not require a high level of available nitrogen in the soil, and leads to high levels of potassium in the leaves of the trees.[3] Potassium is usually the only cation showing this effect strongly both because the calcium, magnesium and sodium contents of typical mulching material is low compared with the potassium, and because potassium is taken up preferentially by the roots of most crops. It is possible that this is not the full explanation of why mulches increase the potassium uptake by crops, as inorganic mulches, such as gravel or glass fibre, may also increase the uptake.[4]

Mulching also increases the phosphate uptake by crops, partly because it contains phosphates but principally because it encourages the surface rooting of the crop, and keeps the surface soil moist for a longer time. For these reasons, it increases the length of time the roots can take up phosphate from the surface soil,[5] which is where the decomposable organic and fertiliser phosphates are concentrated.

Mulching combined with weed control using herbicides is becoming a

1 For an example for coffee in Tanzania see S. M. Gilbert, *E. Afr. agr. J.*, 1945, **11**, 75.
2 For an example for the Great Plains in the U.S., see T. M. McCalla and F. L. Duley, *Agron. J.*, 1943, **35**, 306.
3 C. Bould, *Trans. 4th Int. Congr. Soil Sci.*, 1950, **1**, 262.
4 R. B. Tukey and E. L. Schoff, *Proc. Amer. Soc. hort. Sci.*, 1963, **82**, 68.
5 For an example with cocoa in Trinidad see T. Wasowicz, *Trop. Agric. Trin.*, 1952, **29**, 163, and arabica coffee in Kenya, J. B. D. Robinson and P. H. Hosegood, *Expl. Agric.*, 1965, **1**, 67.

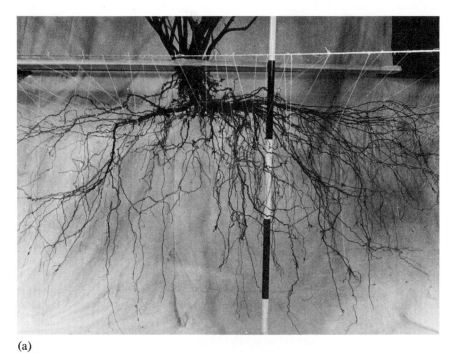

(a)

(b)

PLATE 52. The effect of inter-row cultivation on the root system of a blackcurrant bush: (*a*) weed control by inter-row cultivation; (*b*) weed control by the use of herbicides only.

widely-used management practice in orchards, and for tree and bush crops; and mulching need not involve the addition of dead plant residues to the soil surface. It is usually unnecessary to keep land under these crops weed-free throughout the year, as long as it is free from weeds at critical periods in the growing season, so the weeds that are killed by the herbicides at these periods then form a surface mulch. Further all leaf litter and prunings, if the crop is pruned, can often be left on the surface, and this also forms a valuable mulch. Plate 52 illustrates the way blackcurrant bushes respond to this method of management at Loughgull farm in Co. Armagh, N. Ireland.[1] The extensive surface root system helps to maintain cracks and holes in the actual soil surface, so allowing the surface to remain permeable to rainwater, though if it falls in heavy storms the permeability will be too low to prevent considerable run-off. However the surface itself is sufficiently compact for this run-off to cause little if any soil erosion, and this resistance to soil loss is increased by any crop litter that is lying on the surface and by any dead weeds anchored in the surface. Two other practical advantages of this system are worth stressing. Because the soil is not disturbed by cultivation, it becomes compact though porous due to old root channels, so it can carry traffic much better than can cultivated soil, without the coarser pores being destroyed; and because of the surface rooting habit of the plant, its root system can take up phosphate from a small dressing of fertiliser more effectively than if the surface soil is cultivated.

Synthetic mulching materials

Two types of synthetic mulching material are coming into use: clear polythene and bitumin-in-water emulsions.[2] The polythene sheet is laid on the soil surface between the rows of vegetable crops, or of strawberries, melons or pineapples, and the bitumin emulsion is sprayed over the rows after seeding.[3] Both these materials raise the soil temperature, but the effect of clear polythene is greater than for the bitumin film (see p. 396); and both reduce water losses by evaporation from the soil surface. The mulches may also increase the nitrate content of the soil due to the higher soil temperatures.

1 D. W. Robinson, *Meded. Dir. Tuinb.*, 1965, **28**, 401; and see *Sci. Hort.* 1966, **16**, 52, *Weeds,* 1964, **12**, 4, 245.
2 For a description of some of these see E. W. Lang, *J. Inst. Agric. Engnrs.*, 1968, **23**, 183.
3 F. H. Takatori, L. F. Lippert and F. L. Whiting, *Proc. Amer. Soc. hort. Sci.*, 1964, **85**, 532, 541.

Appendix 1   Description of the Rothamsted and Woburn soils[1]

The predominant soil in most of the experimental fields at Rothamsted has a loamy and flinty surface layer 20 to 60 cm thick, derived partly from loess, over strong brown or yellowish red clay with scattered flints and greyish and reddish mottling (clay-with-flints). The latter material comprises weathered remains of Tertiary sediments and of the underlying Cretaceous chalk with flint, more or less mixed and rearranged during the Pleistocene period. Moderately well-drained soil with these characteristics has been mapped as the Batcombe series, and classed as a gleyed brown earth (*sol lessivé*) in Britain.[2] It would be classified as a Paleudalf in the American system, and as a Chromic or Ferric Luvisol in the F.A.O. system. Deeper loamy soils with few stones (Hook series) and flinty loamy colluvial soils (Nettleden series) occupy smaller areas.

Normal and shallow (eroded) phases of the Batcombe series have been mapped. In the shallow phase, which predominates in Broadbalk and Barnfield,[3] a flinty plough layer of heavy silt loam or silty clay loam texture rests on clay-with-flints at less than 30 cm depth. Mean clay contents of the topsoil, based on 12 systematically sited samples in each field, are 25.3 ± 1.5 per cent in Broadbalk and 30.6 ± 1.4 per cent in Barnfield.

A profile of the shallow phase from Broadbalk has the following particle-size distribution.

Sample no.	1	2	3	4
Depth (cm)	0–20	20–50	50–95	95–170
2 mm–200 μm	6·0	2·0	1·8	4·3
200–50 μm	4·8	2·7	9·5	7·1
50–20 μm	38·9	22·7	7·9	9·6
20–2 μm	26·6	15·9	10·3	9·3
< 2 μm	23·6	56·8	70·6	69·7
$CaCO_3\%$	1·5	tr.	—	—
pH in water	8·1	7·8	7·5	7·4
pH in 0·01 M $CaCl_2$	7·6	7·2	7·0	6·8
Cation-exchange capacity m eq/100 g	15·8	31·5	n.d.	34·1
Free Fe_2O_3 (dithionite-extractable)	2·6	5·2	6·5	5·6

1 I am indebted to Mr B. W. Avery for these descriptions.
2 B. W. Avery, 'Soils and Land Use of the District around Aylesbury and Hemel Hempstead', *Mem. Soil Surv. Gt. Br.*, 1964, HMSO. .
3 B. W. Avery and P. Bullock, 'Morphology and classification of Broadbalk soils' *Rothamsted Rep. 1968, Pt 2*, **63**

The uniformly high pH values presumably result from additions of chalk to the surface over a long period, as uncultivated soils of the Batcombe series are always moderately or strongly acid to depths exceeding 1 m.

The clay (< 2 μm) fractions of Broadbalk and Barnfield soils[1,2] are composed mainly of expanding layer silicates, with subsidiary amounts of mica, kaolinite, and crystalline or amorphous oxides (chiefly $FeO.OH$), and quartz (only in the fraction 0·1 to 2 μm). In Mg-saturated form, the expanding material in the clay-rich subsoil horizons expands from 14 Å to 17 Å when solvated with ethylene glycol, and collapses on heating to give a 10 Å reflection tailing to slightly larger spacings. It therefore resembles smectite, but recent studies show that it is a complex interstratified material containing smectite and other layers. Clay fractions from loamy surface layers derived largely from loess contain a vermiculite component in addition to smectite, more mica, kaolinite and quartz, and also small amounts of chlorite and felspar. X-ray reflections from the expanding minerals in topsoil horizons are broader and less intense, probably because the diffracting units are smaller and the stacking sequences less regular.

The soil at Woburn has not been studied in as much detail as that at Rothamsted. Parts of the farm, including most of Stackyard, Lansome, Butt Close and Great Hill fields, have well-drained, brownish, coarse textured soil, mapped as Cottenham series,[3] in drift of variable thickness over bedded ferruginous sands and sandstones (Lower Greensand) of Lower Cretaceous age. Many of these sandy brown earths have uniformly sandy (loamy sand or sand) texture, with 8 to 12 per cent clay and around 80 per cent sand in the topsoil and decreasing amounts of clay in lower horizons. These would be classed as Udipsamments in the American system, and as Arenosols in the F.A.O. system. Others, usually in thicker drift incorporating material derived from boulder clay, have surface and sub-surface horizons of somewhat finer (sandy loam) texture, and may have a textural B horizon (Hapludalf in the American system and Brown or Chromic Luvisol in the F.A.O. system). The old arable soils have a weak, unstable structure and are subject to accelerated erosion. Clay separates from a Stackyard topsoil[4] are broadly similar in composition to those from Rothamsted topsoils. The main component is on ill-defined mixed-layer aluminosilicate with a high proportion of expanding layers, with kaolinite, goethite and a little mica as subsidiary components.

1 A. H. Weir, J. A. Catt and E. C. Ormerod, 'The mineralogy of Broadbalk soils' *Rothamsted Rep. 1968, Pt 2*, 81.
2 B. W. Avery, P. Bullock, J. A. Catt, A. C. D. Newman, J. H. Rayner and A. H. Weir, 'The soil of Barnfield, *Rothamsted Rep. 1971, Pt 2*, 5.
3 D. W. King, 'Soils of the Luton and Bedford District', *Soil Surv. Special Survey No. 1*, 1969, Harpenden.
4 As studied by A. Islam and J. Bolton, *J. agric. Sci.*, 1970, **75**, 571.

Other parts of the farm have variable and generally finer textured soil with imperfect or poor natural drainage, formed in recent alluvium and older superficial deposits over impervious Jurassic (Oxford) clay or chalky boulder clay.

Appendix 2 Conversion factors for units used in the text

1 metre = 1 m = 39·37 in = 3·280 ft.
1 sq. metre = 1 m^2 = 1·196 yd^2
1 hectare − 1 ha − 2·47 acres
1 kilogram = 1 kg = 2·205 lb
1 ton (metric) = 1 t = 19·7 cwt (of 112 lb)
1 kg/ha = 0·892 lb/acre
1 t/ha = 7·97 cwt/acre
1 t/m^3 = 1 g/cm^3 = 62·4 lb/ft^3

1 joule (J) = 0·239 gram-calories = 10^7 ergs
1 watt = 1 joule/sec
1 bar = 10^6 dynes/cm^2 = 1020 g/cm^2 = 0·987 atm = 750 mm Hg

1 acre foot of water = 1233 m^3 water, and weighs about 1230 tons (metric)

1 cusec = a flow of 1 cubic foot of water per second = 0·0283 cumecs (m^3/sec)
1 cusec per day (24 hours) = 1·98 acre ft = 2440 m^3 water.

To convert K to K$_2$O multiply by 1·20
To convert P to P$_2$O$_5$ multiply by 2·29

Author index

Subject index